Frank Kempken
Editor

Plant Mitochondria

Editor
Prof. Dr. Frank Kempken
Christian-Albrechts-Universität zu Kiel; Botanisches Institut
Olshausenstr. 40
24098 Kiel
Germany
fkempken@bot.uni-kiel.de

ISBN 978-1-4614-2770-4 ISBN 978-0-387-89781-3 (eBook)
DOI 10.1007/978-0-387-89781-3
Springer New York Dordrecht Heidelberg London

Printed on acid-free paper

Springer is part of Springer Science+Business Media (www.springer.com)

Advances in Plant Biology 1

For other titles published in this series, go to
www.springer.com/series/8047

This book is dedicated to Prof. Daryl R. Pring for his lifetime contributions to the field of plant mitochondria

Preface

Mitochondria are the product of a long evolutionary history. It is now a well-established fact that mitochondria did evolve from free living bacteria being the common ancestor of both eukaryotic mitochondria and α-proteobacteria. Advances in genome sequencing, the establishment of in organello and in vitro assays to name only a few, contributed significantly to advances in plant mitochondrial research. Second generation sequencing and the ability to directly sequence and analyze the whole plant transcriptome certainly will help to develop the research on plant mitochondria to another level in the future. In this book the current knowledge about plant mitochondria is presented in a series of detailed chapters, which have been organized in five main parts: (1) dynamics, genes and genomes; (2) transcription and RNA processing; (3) translation and import; (4) biochemistry, regulation and function; and (5) mitochondrial dysfunction and repair. These parts consist of two to five chapters, each written by well-known specialists in the field. The 19 chapters cover the field very well.

In Part I (dynamics, genes and genomes) Volker Knoop (Bonn, Germany) and coworkers provide an insight to the evolution of plant mitochondria which is discussed in the framework of our modern understanding of plant phylogeny. David C. Logan (Saskatchewan, Canada) together with Iain Scott (Bethesda, USA) discuss mitochondrial division and fusion as primary processes controlling mitochondrial form, size, and number. Sally A. Mackenzie together with Maria P. Arrieta-Montiel (Lincoln, USA) reports on the emerging mitochondrial sequence data from early land plants and recent studies of nuclear influence on mitochondrial genome behavior which have provided important insight into the evolutionary trends and possible rationale for the genomic variability that is seen in plant mitochondria.

In Part II (transcription and RNA processing) Thomas Börner and Karsten Liere (Berlin, Germany) discuss the current knowledge about plant mitochondrial transcription, which is sustained by phage-type RNA polymerases that are encoded by a small nuclear encoded gene family. Transcription of most mitochondrial genes is driven by multiple promoters, which may ensure transcription despite possible mitochondrial genome rearrangements. Stefan Binder (Ulm, Germany) and coworkers summarize recent progress made in the understanding of RNA processing and RNA degradation in mitochondria of higher plants. In the complex framework of plant mitochondria posttranscriptional processes play predominant roles. Linda Bonen (Ottawa, Canada) presents recent advances in our understanding of splicing mechanisms, the nature of splicing machinery, and the relationships among splicing and other RNA processing events in plant mitochondria. Anika Bruhs (Kiel, Germany) and myself present an overview of plant mitochondrial (and plastid) RNA editing, its consequences for translation, the current knowledge of its mechanism, and some ideas on its evolution.

In Part III (translation and import) Nicolas L. Taylor and coworkers (Crawley, Australia) compare the proteomes of mitochondria from monocots and dicots plants and highlight the conservation of the mitochondrial electron transfer chain protein complex I. They also provide important insights, directions, and methodology currently utilized in their laboratory. Laurence Maréchal-Drouard (Strasbourg, France) and coworkers address the basic questions on the tRNA mitochondrial import selectivity, regulation, targeting, and translocation in plants. These data are discussed and compared to what has been discovered in tRNA mitochondrial import in evolutionary divergent organisms. Elzbieta Glaser and James Whelan (Stockholm, Sweden) give an overview on mitochondrial protein import in higher plants. Interestingly, many components of the plant protein import apparatus appear to be different to those in yeast and mammalian systems.

In Part IV (biochemistry, regulation and function), Hans-Peter Braun (Hanover, Germany) and coworkers summarize recent insights into the assembly of the OXPHOS system, consisting of five large multisubunit complexes within the inner mitochondrial membrane, the soluble intermembrane space-localized protein cytochrome c and the lipid ubiquinone. Allan G. Rasmusson (Lund, Sweden) together with Ian M. Møller (Aarhus, Denmark) look at the mitochondrial electron transport chain, which can mediate major adjustments in cellular metabolism important for cellular function under a great variety of stress conditions such as low temperature and drought. Keisuke Yoshida together with Ko Noguchi (Tokyo, Japan) describe and discuss the interaction between chloroplasts and mitochondria and review recent advances of understanding about the activity, function, and regulation of the mitochondrial respiratory system during photosynthesis. David M. Rhoads (Tucson, Arizona) describes plant mitochondria as stress sensors that contribute to decisions regarding cell fate during stresses. These are conveyed to the nucleus by mitochondrial retrograde regulation. Paul F. McCabe (Dublin, Ireland) together with Mark Diamond (Piscataway, USA) discuss the plant mitochondrion as a crucial mediator of programmed cell death. While similarities between plant and animal programmed cell death systems have been discovered, current knowledge suggests there are also key differences.

In Part V (mitochondrial dysfunction and repair) Jenny Carlsson together with Kristina Glimelius (Uppsala, Sweden) report on the current knowledge on cytoplamic male sterility (CMS) which is caused by mutations, rearrangements, or recombinations in the plant mitochondrial genome. CMS has important applied aspects in plant breeding and is the prime example of mitochondrial dysfunction in higher plants.

However, this book is intended not only for the specialist in plant mitochondria, but also for colleagues from related fields. As such I have invited a few chapters from authors that work on plastids or non-plant systems. These additional chapters provide very important insights into related areas which are not yet developed in plants or are important to understand the level of differences between the plant mitochondrial machinery and other model systems. It seemed prudent to invite Christian Schmitz-Linneweber (Berlin, Germany) and coworkers to add information on RNA binding proteins in plant chloroplast, as there are clear links between plant plastids and mitochondria, e.g., in the field of plant RNA editing. Wolfgang Voos (Bonn, Germany) and coworkers contribution about animal and fungal protein import is a quite logical extension to plant mitochondrial import. Readers can now directly compare fungal or animal protein import with that of plant mitochondria (written by Elzbieta Glaser and James Whelan). Finally, Susan LeDoux (Mobile, USA) provides important insight in human mitochondrial mutations and repair, a field of research which is somewhat related to cytoplasmic male sterility (written by Jenny Carlsson and Kristina Glimelius).

Graduate student and post-docs who recently joined a new field of research often find it difficult to access the literature. Book chapters and reviews written by specialists may contain a tremendous amount of information which literally may overwhelm the inexperienced reader. For easier access to this book, each of its chapters contains a list of abbreviations and a short glossary with explanations for important keywords. Also each chapter contains at least one text box, where more detailed information or cross-reference is given to specific subjects related to the chapter in question. I am sure these measures will not only aid the newcomers to access a complex field but also make the chapters more comprehensive to all readers.

Finally, I would like to thank all the many authors and coauthors who provided great chapters. It was great fun to get this book together. I also would like to thank Hannah Schorr and the publisher for their help and input. I truly hope that this book will help to attract young scientists to the fascinating and exiting field of plant mitochondria.

July 2010

Contents

Part III Translation & Import

Part IV Biochemistry, Regulation & Function

Part V Mitochondrial Dysfunction & Repair

Part I
Dynamics, Genes & Genomes

Chapter 1
Mitochondrial Genome Evolution in the Plant Lineage

Volker Knoop, Ute Volkmar, Julia Hecht, and Felix Grewe

Abstract Land plants feature particularly complicated mitochondrial genomes. Plant mitochondrial DNAs may be more than 100 times larger than those of animals and are structurally much more complex due to frequent ongoing recombination. The significant increase of plant mitochondrial genome sizes results from a combination of several factors: more genes are encoded, many of these carry introns and, most importantly, the plant mitochondrial genome has a propensity to accept foreign DNA sequences from the chloroplast, the nucleus, or even from other mitochondrial genomes via horizontal gene transfer. Similarly, plant mitochondria are also more complex on the transcriptome level where processes such as frequent RNA editing or *trans*-splicing of disrupted introns contribute to RNA maturation. The evolution of these peculiar features is discussed in the framework of our modern understanding of plant phylogeny to which mitochondrial genome data have contributed significantly.

Keywords Introns • RNA editing • Plant phylogeny • Mitochondrial evolution

Abbreviations

CMS Cytoplasmic male sterility
EGT Endosymbiotic gene transfer
HGT Horizontal gene transfer
LGT Lateral gene transfer

1.1 Introduction

The mitochondrial genomes of land plants (embryophytes) are characterized by a multitude of peculiarities that are counterintuitive to a general understanding of increasingly compacted genomes in endosymbiotic organelles. Some 500 million

V. Knoop (✉)
Abteilung Molekulare Evolution, Institut für Zelluläre und Molekulare Botanik, Universität Bonn, Kirschallee 1, Bonn, 53115, Germany
e-mail: Volker.knoop@uni-bonn.de

F. Kempken (ed.), *Plant Mitochondria*, Advances in Plant Biology 1,
DOI 10.1007/978-0-387-89781-3_1,

years of evolution have created very streamlined, compact and economically organized mitochondrial DNAs (mtDNAs) in the animal (metazoa) lineage only rarely exceeding some 16–17 kbp in size (Lavrov 2007; Gissi et al. 2008). In contrast, it seems that opposite trends rather complicating than simplifying mitochondrial genome structures have taken over during evolution in the plant lineage.

Several features of plant mitochondrial gene arrangements and expression will be dealt with in much detail in other Chapters of this book, e.g., recombinational activity (Chap. 3), intron splicing (Chap. 6), or RNA editing (Chap. 7). We will here mainly give a phylogenetic perspective on several peculiar features in the evolution of plant mtDNAs since Ordovician times when first plants appeared on land and gave rise to one of the most significant evolutionary transitions of multicellular life on this planet.

1.2 Land Plant Mitochondrial DNAs and Their Peculiarities

The first completed sequence of a mitochondrial DNA of a land plant, the one of the liverwort *Marchantia polymorpha* (Oda et al. 1992a, b) at a size of 186 kbp turned out to be more than 10 times larger than the first completely determined mtDNA sequence of the animal lineage, the one of *Homo sapiens* of only 16.6 kbp (Anderson et al. 1981). The *Marchantia* mtDNA revealed several genes not present in the mitochondrial genomes of animals. Moreover, the liverwort mtDNA contains seven introns of group I and 25 of group II type, typical intervening sequences frequently found in the organelle DNAs of fungi, algae, and plants (see Box 1.1 and Chap. 6). Only exceptional occasional discoveries of introns in mtDNAs of animals were made later, beginning with a first group I intron identified in the mtDNA of the sea anemone *Metridium senile* (Beagley et al. 1996). By and large, however, introns are rare in animal mtDNAs, mostly identified in some early branches of the metazoa phylogeny.

By the time the *Marchantia* chondrome was completely determined (note that the term chondriome is alternatively used to describe all mitochondria in a cell, see Chap. 2), it was already clear that yet more surprising features exist in the mitochondrial DNAs of flowering plants (angiosperms) and several others were discovered subsequently:

1. Plant mitochondrial genome sizes of 2 Mbp (mega base pairs = million base pairs) and more, hence exceeding the ones of several free-living bacteria, are present in the angiosperm family of Cucurbitaceae.
2. Endosymbiotic gene transfer (EGT), the functional transfer of genes to the nucleus, is an ongoing process in plant mitochondria in recent times of plant evolution.
3. Simple circular mitochondrial genomes such as the ones found in algae or bryophytes may be more of an exception than a rule in land plants. Active recombination across repeated sequences produces multipartite structure of plant mitochondrial genomes, at least in vascular plants (tracheophytes).
4. The average plant mitochondrial genome is characterized by some 20–30 introns. With a single exception these introns are stable within well-defined monophyletic plant clades, but differ significantly between them and have helped to elucidate land plant phylogeny. Some disrupted genes need to be reassembled on

RNA level via *trans*-splicing group II introns and an example of a *trans*-splicing group I introns has recently been identified in a lycophyte.

5. The peculiar process of RNA editing apparently repairs DNA coding information posttranscriptionally on RNA level by precise pyrimidine exchanges in plant mitochondrial RNAs.
6. Foreign DNA derived from the chloroplast or the nuclear genome is inserted into the mtDNA of tracheophytes as so-called "promiscuous DNA."
7. Several instances of DNA sequences in plant mitochondria acquired by horizontal gene transfer (HGT) across species barriers and over wide phylogenetic distances have recently been reported.

Box 1.1 Group I and group II introns

These two types of introns are defined through their characteristic secondary structures and conserved RNA sequence motifs (see, e.g., Bonen 2008; Cech et al. 1994; Haugen et al. 2005; Pyle et al. 2007 for recent summary updates). Numerous group I and group II introns are typically present in the organelle genomes of plants and fungi. A nomenclature to label them according to their insertion sites has been proposed (Knoop 2004; Dombrovska and Qiu 2004) as detailed in the legend of Fig. 1.2. Group I and group II introns have also been identified in mitochondria of basal metazoan lineages (see, e.g., Fukami et al. 2007), protists, phages, and more frequently also bacteria (Tourasse and Kolstø 2008). Introns of both classes are often referred to as "autocatalytic," "self-splicing," "ribozymes," or "mobile," but this is in fact biochemically demonstrated only for some of them. Some intron species of both classes carry open reading frames (ORFs), which participate in intron mobility or splicing (see Chap. 6): endonucleases in group I introns and the so-called maturases in group II introns. Intron-encoded ORFs are absent in the overwhelming majority of seed plant chloroplast and mitochondrial introns with singular exceptions in both cases: nad1i782g2 in mitochondria carrying *mat*R and trnK37g2 in chloroplasts carrying *mat*K. Absence of other intron-encoded ORFs may explain both the obvious positional stasis of seed plant organelle introns since more than 200 million years of evolution as well as their reliance on nuclear encoded cofactors for splicing (see Chap. 8), which have been identified over the last years (recent examples include Asakura and Barkan 2006; Asakura and Barkan 2007; Keren et al. 2009; Kroeger et al. 2009; Ostheimer et al. 2006; Schmitz-Linneweber et al. 2006; Watkins et al. 2007). Several examples for disrupted group II introns, which require *trans*-splicing of independent primary transcripts for RNA maturation, have been identified in chloroplast and mitochondrial DNAs since quite some time (Bonen 2008; Glanz and Kück 2009). Examples for *trans*-splicing group I introns, however, have only been discovered very recently (Burger et al. 2009; Grewe et al. 2009). For more details see Chap. 6.

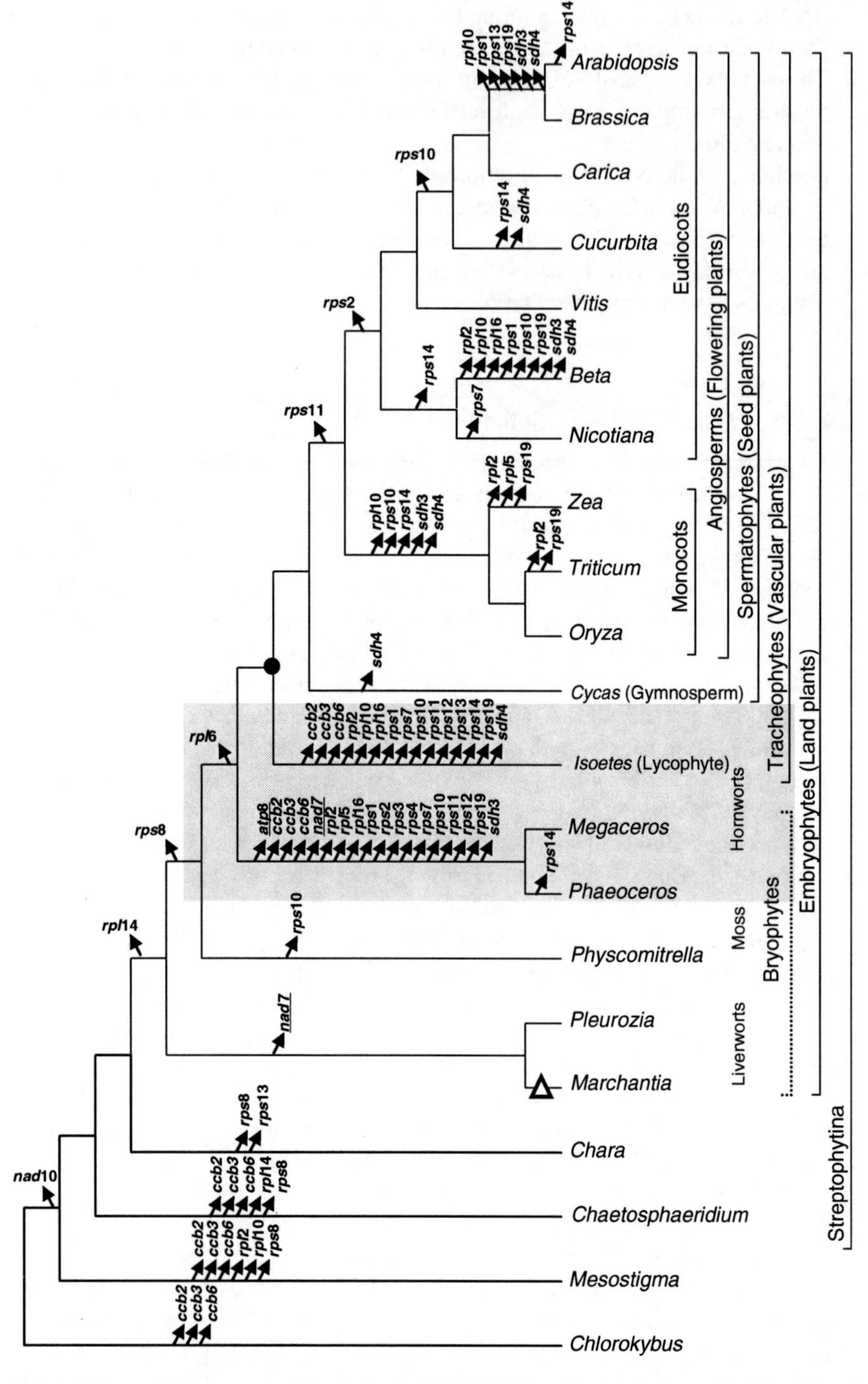

Fig. 1.1 Streptophyte genera with completed mitochondrial genome sequences currently available (references given in the text except for the *Carica papaya* mtDNA, as yet only available as database accession EU431224) are assembled into a cladogram based on modern insights of plant phylogeny (Qiu et al. 2006). The mtDNAs of taxa with gene complements identical to closely related genera included in the figure (e.g., *Citrullus, Tripsacum, Sorghum*) were omitted for clarity. No complete

1.3 Plant Mitochondrial Genomes: Completed MtDNA Sequences

The expanded sizes and genomic complexities of plant mitochondrial DNAs probably explain why only so few mtDNAs of plants have been determined in comparison to the numerous completely sequenced mtDNAs of animals. The first complete mtDNA of a flowering plant, the model angiosperm *Arabidopsis thaliana* (Unseld et al. 1997), was followed by those of more economically important plants such as sugar beet *Beta vulgaris* (Kubo et al. 2000), rice *Oryza sativa* (Notsu et al. 2002), rapeseed *Brassica napus* (Handa 2003), maize *Zea mays* (Clifton et al. 2004), tobacco *Nicotiana tabacum* (Sugiyama et al. 2005), wheat *Triticum aestivum* (Ogihara et al. 2005) and grape *Vitis vinifera* (Goremykin et al. 2009). Finally the mtDNAs from watermelon *Citrullus lanatus* and zucchini (*Cucurbita pepo*) have been determined recently (Alverson et al. 2010), two species of the Cucurbitaceae long known to harbor particularly large chondromes (possibly exceeding 2 Mbp in muskmelon Ward et al. 1981). The availability of complete mtDNAs from different strains or cultivars of the same species (e.g., sugar beet, Satoh et al. 2004; maize, Allen et al. 2007), allows insights on the mtDNA rearrangements accompanying important mitochondrially encoded traits such as cytoplasmic male sterility (CMS, see, e.g., Kubo and Newton 2008 see Chap. 18).

In contrast, only a single gymnosperm mtDNA sequence, the one of the cycad (fern palm) *Cycas taitungensis,* has so far been determined (Chaw et al. 2008). Similarly, it took more than a decade after determination of the *Marchantia polymorpha* mtDNA (Oda et al. 1992a, b) before a second bryophyte mtDNA became available, the one of the model moss *Physcomitrella patens* (Terasawa et al. 2006). The chondrome sequences of two representatives of the third clade of bryophytes, the hornworts, have been determined only very recently with the mtDNAs of *Megaceros aenigmaticus* (Li et al. 2009) and *Phaeoceros laevis* (Xue et al. 2010). Finally, the mtDNA of a second liverwort, *Pleurozia purpurea,* has recently been published (Wang et al. 2009). This currently leaves us with a considerable phylogenetic gap between the bryophytes and the seed plants for mitochondrial genome sequences of early branching (i.e., non-seed) tracheophyte lineages (Fig. 1.1).

Fig. 1.1 (continued) mtDNA sequence for a representative of the monilophytes (ferns *sensu lato,* including whisk ferns and horsetails) is currently available and only singular mitochondrial genome sequences are so far determined for mosses (*Physcomitrella*), lycophytes (*Isoetes*), and gymnosperms (*Cycas*). Taxonomic labeling is indicated to the right. Note that "bryophytes" are paraphyletic (*dotted line*). Losses of functional genes from the respective mtDNAs are indicated by arrows on the branches of the model phylogeny. Losses of ribosomal proteins (*rpl, rps*), subunits of succinate dehydrogenase (*sdh*), and cytochrome c biogenesis (*ccb*) occur independently multiple times. Conversely, independent losses of core respiratory subunits are rare (*atp*8 and *nad7, underlined*). Sequential losses of *rpl*14, *rps*8, and *rpl*6 from the mtDNA deep along the tree's backbone can be taken as support for this model embryophyte phylogeny. RNA editing emerges with embryophytes and is secondarily lost in marchantiid liverworts (*open triangle*). The bryophyte-tracheophyte transition zone (*gray shading*) is characterized by multiple parallel losses of mt genes and an increase of reverse (U to C) RNA editing. A strong increase in DNA recombinational activity in the plant mitochondrial genomes and the integration of promiscuous DNA originating from the chloroplast and nucleus apparently emerges with the tracheophyte lineage (*black dot*)

These lineages, the "pteridophytes" *sensu lato* (ferns and "fern allies"), are now unequivocally distinguished into lycophytes and monilophytes (Pryer et al. 2001). Only the latter is the sister clade to seed plants and comprises the true ferns, the horsetails (*Equisetum*) and the whisk ferns (*Psilotum, Tmesipteris*). Reports on the mtDNAs of ferns and "fern allies" have been very rare (Palmer et al. 1992). As a first mtDNA of this group the one of the quillwort *Isoetes engelmannii* has been reported very recently (Grewe et al. 2009).

A monophyletic group of land plants and related green algal lineages has been recognized as the clade of "streptophytes" (see Fig. 1.1). The available mtDNAs of several streptophyte algae are a very important addition to the collection of embryophyte mtDNAs for insights on phylogeny and molecular evolution in the green lineage: *Chaetosphaeridium globosum* (Turmel et al. 2002a), *Mesostigma viride* (Turmel et al. 2002b), *Chara vulgaris* (Turmel et al. 2003), and *Chlorokybus atmophyticus* (Turmel et al. 2007). Whereas the former three algal mtDNAs are of moderate sizes between 42 and 68 kbp, the mitochondrial DNA of *Chlorokybus* at 202 kbp even exceeds the *Marchantia* mtDNA of 186 kbp in size. As novelties among streptophyte mtDNAs the *Chlorokybus* mtDNA features a *nad*10 and a *trn*L(gag) gene and 13 introns at novel insertion sites with a surprising number of 10 group II introns in tRNA-encoding genes. Somewhat puzzling is the mtDNA of the tiny unicellular chlorophyte green alga *Ostreococcus tauri*: On the one hand the 44 kbp mtDNA is very gene dense, on the other hand it consists of a large segmental duplication of 44% of the chondrome (Robbens et al. 2007).

1.4 Ongoing Gene Transfer to the Nucleus

The functional transfer of genes from an endosymbiont genome to the nuclear genome of the host is frequently referred to as endosymbiotic gene transfer (EGT, see Timmis et al. 2004). An apparent stasis of EGT has been reached in the animal lineage, where largely the same (small) gene complement is conserved across phylogenetic distances dating back some 500 million years. Animal mtDNAs usually encode 22 tRNAs, 2 rRNAs, and 13 genes encoding proteins of the respiratory chain complexes I, III, IV, and V (the NADH dehydrogenase, apocytochrome B, the cytochrome c oxidase, and the ATPase): *nad*1-6 and *nad*4L, *cob*, *cox*1-3, and *atp*6 and 8. Typically, the gene complement of plant mtDNAs is extended by genes for additional subunits of these respiratory chain complexes (*atp*1, 4, and 9, *nad*7 and 9) and for two subunits of complex II, the succinate dehydrogenase (*sdh*3 and 4). Moreover, *ccb* genes encoding proteins of cytochrome c biogenesis, *rps* and *rpl* genes encoding proteins of the small and large ribosomal subunit and *tat*C, a transport protein subunit (twin arginine translocase), are encoded in plant mtDNAs. Plant mitochondrial gene complement varies widely among angiosperms in particular for *rps, rpl* and *sdh* genes (see Fig. 1.1) showing that endosymbiotic gene transfer is an ongoing process also in recent times of plant evolution (Liu et al. 2009). The identification of genes in the mitochondrial DNA of one but not in another plant lineage has generally been taken as a first hint for such recent EGT. The *rps*10 gene was a typical early example along such lines (Knoop et al. 1995), and then found to be transferred to the nucleus independently in many plant lineages (Adams et al. 2000). The same point was made for the *sdh* genes (Adams et al. 2001b) and the *rpl* and *rps* genes similarly turned out to be transferred to the nucleus via EGT many times independently in a systematic survey of angiosperms (Palmer et al. 2000; Adams et al. 2002b; Adams and Palmer 2003).

In contrast, only three examples have as yet been identified for functional EGT of any of the core components of respiratory chain complexes I–V during land plant evolution. The example of *cox*2 transfer among legumes (Fabales) is particularly intriguing given that the necessary steps of nuclear copy establishment, activation of the nuclear copy, and subsequent defunctionalization and disintegration of the mitochondrial copy (Brennicke et al. 1993) can be traced among closely related taxa of this clade (Nugent and Palmer 1991; Covello and Gray 1992; Adams et al. 1999). Secondly, functional *atp*8 genes are apparently lost from *Allium* (Adams et al. 2002b) and are also missing from the two completed hornwort mtDNAs (Li et al. 2009; Xue et al. 2010). It will be interesting to see whether EGT of *atp*8 may turn out as an apomorphy of the entire hornwort clade.

Finally, a third example of gene transfer concerns the interesting issue of the *nad*7 gene in liverworts (Kobayashi et al. 1997), which appears to be a very ancient event in this plant clade. Once a gene is transferred via EGT it normally disintegrates quickly and the identification of a recognizable pseudogene copy in the mitochondrial genome can be taken as evidence for evolutionary recent gene transfer. Strikingly, however, *nad*7 remains present in the liverwort mtDNAs as a pseudogene after functional establishment of the nuclear *nad*7 copy instead of quick pseudogene disintegration as generally observed in the angiosperms (Groth-Malonek et al. 2007b). Whether this long-term pseudogene survival simply reflects particular slow structural mtDNA evolution due to lack of mtDNA recombination in the liverworts (see below) or a remaining functional necessity of the pseudogene in liverworts remains to be seen. The former possibility is supported by the extreme degrees of gene synteny between the two fully sequenced liverwort mtDNAs, in fact also in comparison to moss and hornwort mtDNAs (Wang et al. 2009; Xue et al. 2010). The latter idea of remaining functionality in a pseudogene retained in mitochondria is supported through the second known example for exceptional long-term retention of a pseudogene after EGT, in this case among angiosperms: an *rps*14 pseudogene is conserved for some 80 million years in the mtDNAs of grasses (Ong and Palmer 2006).

Similar to *atp*8, independent EGT of *nad*7 has obviously occurred in the hornworts (see Fig. 1.1). In fact, the recently determined complete mtDNA sequences of two hornworts (Li et al. 2009; Xue et al. 2010) show many pseudogenes or recognizable pseudogene fragments of mitochondrial genes such as the *ccb* genes and most of the *rpl* and *rps* genes, which suggests an increase of EGT activity in the hornwort lineage. Surprisingly, the *ccb* genes and many *rpl* and *rps* genes are also lacking in the lycophyte *Isoetes* mtDNA as well (Grewe et al. 2009). Given our current understanding of embryophyte phylogeny (see Fig. 1.1) a rise in EGT activity may have taken place at the time of bryophyte-tracheophyte transition in land plant evolution. Nevertheless, these gene losses need to be explained as independent events in the hornwort and tracheophyte lineages given that the affected genes are present in seed plants.

Obviously the genes known to be subject to frequent, recent, and independent EGT in embryophytes (*ccb, rpl, rps, sdh*) are the ones that are absent from the modern metazoan mtDNAs altogether, suggesting that similar functional selective pressures exist for retention of genes in the organelles. It is striking to see that the *tat*C gene so far not found subject to recent EGT in the embryophyte lineage has recently also been identified in a very basal metazoan lineage, the sponge *Oscarella* (Wang and Lavrov 2007).

There is good evidence from the study of EGT in plants that functional gene transfer is mediated by RNA (e.g., Covello and Gray 1992; Nugent and Palmer 1991; Wischmann and Schuster 1995). Functional nuclear genes are inserted in the RNA-edited version and certainly lack the organellar introns. The RNA-mediated EGT serving to functionally migrate a gene from mtDNA to the nucleus is, however, accompanied by DNA-based transfer of mitochondrial sequences during evolution, mostly without any discernible function. Small fragments of mitochondrial and plastid DNA (called "numts" and "nupts") have been identified in many nuclear genomes of diverse eukaryotes. In fact, a full mtDNA sequence copy is even present on chromosome 2 in the nucleus of *Arabidopsis* (Lin et al. 1999; Stupar et al. 2001). Exceptionally such insertions of organelle DNA fragments may possibly contribute new functionality in their new nuclear location such as the insertion of a group II intron fragment in one of two lectin genes in *Dolichos biflorus* (Knoop and Brennicke 1991).

Several examples of EGT show that mitochondrial genes must not necessarily be transferred to the nucleus in a 1:1 fashion. In some cases genes are functionally split upon EGT: the transfer of *rpl*2 in flowering plants (Adams et al. 2001a) or of *cox*2 in chlamydomonad algae (Perez-Martinez et al. 2001) are examples. Other particular cases of gene fissions upon EGT are observed in protist groups. The *sdh*2 gene is split in two parts in Euglenozoa upon relocation to the nucleus with the protein subunits independently reimported into the mitochondria (Gawryluk and Gray 2009). In the case of *cox*1 only the C-terminal portion is nuclear encoded in diverse protist groups, whereas the truncated gene remained in the mitochondria (Gawryluk and Gray 2010).

Secondly, a gene may be functionally substituted such as *rps*13 apparently replaced upon EGT by an aminoterminal extension of *rps*19 in *Arabidopsis* (Sanchez et al. 1996). Thirdly, a mitochondrial gene may have been replaced by a homologous gene originating from chloroplast or nuclear DNA as was shown for *rps*13 and *rps*8, respectively (Adams et al. 2002a). Obviously, this substitution process can also work in the opposite direction with mitochondrial genes replacing chloroplast counterparts as was shown for *rps*16 (Ueda et al. 2008).

1.5 Plant Mitochondrial Genomes: Structures

The notion that "plant mitochondrial DNA evolves rapidly in structure, but slowly in sequence" (Palmer and Herbon 1988) still holds generally true. Accelerated primary sequence drift has, however, been observed in angiosperm genera such as *Silene* (Sloan et al. 2009), *Pelargonium,* and *Plantago* (Palmer et al. 2000; Cho et al. 2004; Parkinson et al. 2005), or the fern genus *Lygodium* (Vangerow et al. 1999). Efficient DNA mutation repair mechanisms that are currently being elucidated (Boesch et al. 2009) probably contribute to the generally slow primary sequence drift in plant mitochondrial sequences.

Like in the liverwort *Marchantia,* simple circular mtDNA molecules were deduced as mitochondrial genomes also for the moss, for the hornworts and the streptophyte algae. It has been pointed out though that many simple, circular mtDNAs may actually just be an illusion of mapping studies and the corresponding genomes could in fact rather be populations of coexisting, overlapping, linear, or branched molecules (Oldenburg and Bendich 2001; Bendich 2007).

Recombinationally active mtDNAs were found in angiosperms, resulting in dynamic, multipartite and complex structures of the flowering plant mitochondrial genomes (e.g., Ogihara et al. 2005; Palmer and Shields 1984; Sugiyama et al. 2005). Variations of mitochondrial genome structure may even be observed within different ecotypes, isolates, or strains of single species such as *Arabidopsis* or maize (Ullrich et al. 1997; Allen et al. 2007). Active DNA recombination in flowering plant mitochondria leads to populations of coexisting genomic rearrangements of shifting stoichiometry and possibly also to heteroplasmy (see Woloszynska 2010 for a recent review).

Repeated DNA sequences obviously play a major role in generating plant mitochondrial genome variabilities (Andre et al. 1992; Lilly and Havey 2001; Kubo and Mikami 2007), and it may be assumed that homologous sequences need to have a certain length to serve as substrates for homologous recombination. Small repeated sequences not participating in recombination are particularly striking in the *Cycas taitungensis* chondrome. Some 500 sequence elements of 36 bp in the *Cycas* mtDNA are characterized by the terminal direct repeats of the AAGG tetranucleotide and the internal recognition sites Bpu10I which suggested the label "Bpu sequences" (Chaw et al. 2008). On the other hand, much longer sequence repeats are not per se targets for active recombination. An example is a large part of a *cob* gene intron (cobi783g2)[1] copied into an intergenic region in the mtDNA of *Marchantia* and other liverworts (Groth-Malonek et al. 2007a), without any indication for actively ongoing recombination producing alternative mtDNA gene arrangements (Oda et al. 1992a).

The mitochondrial DNA of the lycophyte *Isoetes engelmannii* in contrast features at least two dozen DNA recombination sites resulting in alternative, coexisting genomic arrangements (Grewe et al. 2009). Similarly, we find strong evidence

[1]Intron nomenclature consists of gene name followed by the letter i, the number of the nucleotide in the mature reading frame preceding the insertion site and a qualifier designating groupI/II introns as g1/2. See Box 1.1 for details.

for mtDNA recombination in the lycophyte sister lineage *Selaginella* (J.H., F.G., and V.K., unpublished observations). Hence, in the light of conserved circular chondromes in bryophytes, mtDNA recombination is obviously a gain in the earliest tracheophyte lineages (see Fig. 1.1). Interestingly, though, in contrast to the dynamically rearranging chondromes of the Brassicaceae (Palmer and Shields 1984; Unseld et al. 1997; Handa 2003), the counter-example of a simple nonrecombining circular mtDNA structure in *Brassica hirta* also exists, even in the same plant family (Palmer and Herbon 1987), suggesting that the recombinational activity may be secondarily lost in evolution.

Nuclear factors involved in plant mtDNA recombination are increasingly being identified (e.g., Manchekar et al. 2006; Shedge et al. 2007; Odahara et al. 2009). Another chapter in this volume will be entirely devoted to the functional analysis of DNA recombinational dynamics of plant mitochondrial genomes (see Chap. 2).

1.6 The Introns in Embryophyte Mitochondrial DNAs

In spite of their many idiosyncrasies, plant mitochondrial genome data have proved useful to help in the phylogenetic analysis of embryophytes. Not only coding sequences but also introns, non-coding intergenic regions, gene rearrangements and a pseudogene have been exploited as phylogenetically informative regions (see Knoop 2010).

The occurrence of mitochondrial group I and group II introns (see Box 1.1 and Chap. 6) differ significantly between land plant clades. So far, some 100 different intron insertion sites have been identified in embryophyte mitochondrial genes (Fig. 1.2). Their appearances have been taken as evidence for a sister group relationship of liverworts to all other land plants (Qiu et al. 1998), as well as for a sister group relationship of hornworts and tracheophytes (Groth-Malonek et al. 2005, see Fig. 1.1). The majority of group I and group II introns appear to be rather stable in presence within a given group, in fact can be typical signatures of a given plant clade (Vangerow et al. 1999; Pruchner et al. 2001; Pruchner et al. 2002; Dombrovska and Qiu 2004). Consequently, such introns can be useful loci for phylogenetic analyses in a given clade and several of them have been used in that way (e.g., Beckert et al. 1999; Beckert et al. 2001; Wikström and Pryer 2005; Volkmar and Knoop 2010; Wahrmund et al. 2010). Rare variability of mitochondrial intron presence among angiosperms (see Fig. 1.2) may in fact suggest that independent intron losses occur even less frequently than total gene losses after EGT, but final conclusions need more systematic and extensive taxon sampling. Most likely, the overall loss of intron-encoded ORFs involved in intron mobility (see Box 1.1) in seed plant mtDNAs plays a major role in determining intron position stabilities. Interestingly, of 20 mitochondrial introns among angiosperms, 19 are also conserved in the mtDNA of the gymnosperm *Cycas taitungensis* (Chaw et al. 2008). Recent studies of the *rps*3 gene in a wide sampling of gymnosperms (featuring intron rps3i257g2[1] absent in angiosperms) suggest that independent

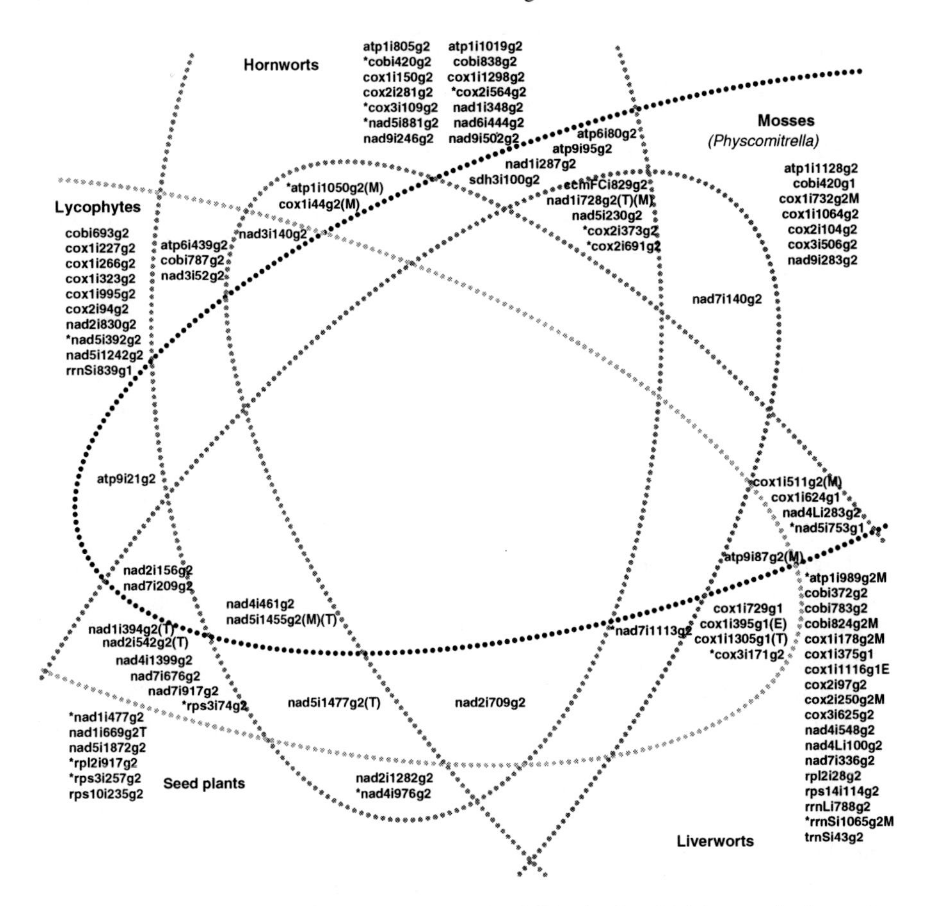

Fig. 1.2 Venn diagram showing the occurrences of 97 mitochondrial introns in the major plant clades, for which complete mitochondrial genome sequences are currently available. The intron nomenclature designates the respective gene, the coding nucleotide number preceding the intron insertion site and the type of intron (g1/g2) as previously proposed (Dombrovska and Qiu 2004; Knoop 2004). The nomenclature is here suggested to be extended by qualifiers "T" for *trans*-splicing and "M" or "E" for intron-borne maturase or endonuclease ORFs, respectively. Known alternative states of *cis*- vs. *trans*-splicing or presence of introns-borne ORFs are indicated with brackets. Variability of intron presence mostly indicating occasional secondary losses within a given group is indicated with an asterisk, e.g., nad1i477g2 absent in Geraniaceae (Bakker et al. 2000), atp1i989g2, and atp1i1050g2 absent in *Treubia lacunosa* (database accession DQ646161), nad5i753g1 absent in *Takakia ceratophylla* (database accession DQ268963), and rps3i257g2 hitherto only identified in some gymnosperms (Ran et al. 2010). Other cases rather suggest rare individual gains: nad5i392g2 in *Huperzia* (Vangerow et al. 1999), nad5i881g2 in *Anthoceros* (Beckert et al. 1999), or cox3i171g2 in *Lycopodium* (Hiesel et al. 1994). Not included in the diagram are introns that are part of promiscuous DNA originating from the chloroplast such as trnVi39g2 in *Cycas* and introns atp1i361g2, cox1i747g2, and nad2i258g2 so far identified only in individual monilophyte taxa (Dombrovska and Qiu 2004; Wikström and Pryer 2005; Bégu and Araya 2009), for which so far no complete mtDNA sequence has been determined

mitochondrial intron losses can occur through retro-processing (Ran et al. 2010) but may be similarly rare as among angiosperms.

As a so far unique exception for the generally conserved presence of plant mitochondrial introns within a given clade, one particular group I intron in the *cox*1 gene, cox1i729g1, originally identified in the angiosperm *Peperomia polybotrya* (Vaughn et al. 1995), has subsequently also been found with a patchy distribution among distant angiosperms. Intron cox1i729g1 has been suggested to originate from a fungal donor source and to invade angiosperm mitochondrial genomes independently (Adams et al. 1998; Cho et al. 1998). A recent reinvestigation with large angiosperm taxon sampling and careful phylogenetic analyses has questioned this idea and instead found that rare gains, if not a single unique gain followed by numerous independent losses are a more satisfying explanation (Cusimano et al. 2008). Yet further taxon sampling, however, reinforced the idea of multiple independent gains of cox1i729g1, now assuming numerous horizontal gene transfers across plant species borders (see below) instead of independent acquisitions from a fungal donor (Sanchez-Puerta et al. 2008).

The mRNAs of three genes in angiosperm mtDNAs – *nad*1, *nad*2, and *nad*5 – are disrupted and need to be assembled via *trans*-splicing group II introns (Chapdelaine and Bonen 1991; Knoop et al. 1991; Pereira de Souza et al. 1991; Wissinger et al. 1991; Binder et al. 1992). Five ancient *trans*-splicing introns in these genes (nad1i394g2, nad1i669g2, nad2i542g2, nad5i1455g2 and nad5i1477g2) are conserved in the gymnosperm *Cycas taitungensis* (Chaw et al. 2008) and their evolutionary histories had earlier been traced back to *cis*-splicing ancestor present in pteridophytes, hornworts, and mosses (Malek et al. 1997; Malek and Knoop 1998; Groth-Malonek et al. 2005). While the disintegrations of these five introns may have been very rare, possibly even singular events, the disruption of intron nad1i728g2 into *trans*-arrangements occurred later in angiosperm evolution multiple times independently (Qiu and Palmer 2004). Obviously the gain of recombinational activity in tracheophyte evolution has played a major role in evolving *trans*-splicing introns. Taken together with the fact that novel group II introns frequently show up during plant evolution (see Fig. 1.2), whereas group I intron history is largely dominated by losses in plant mitochondrial DNA, this may explain why *trans*-splicing group II introns but so far no *trans*-splicing group I introns were discovered. Finally, a first *trans*-splicing group I intron has recently been identified in *Isoetes engelmannii* (Grewe et al. 2009). Preliminary data indicate that yet more *trans*-splicing introns will be discovered in the mtDNA of the sister lycophyte *Selaginella* (J.H, F.G., and V.K., unpublished observations).

1.7 RNA Editing

In plant mitochondria, RNA editing (Box 1.2 and see Chap. 7) comes in the form of pyrimidine nucleotide conversions from cytidine to uridine and in the reverse direction. The main task of RNA editing in plants largely appears to be the restoration of evolutionary conserved codon identities to create functional proteins, which may

Box 1.2 RNA editing

RNA editing defines any changes in the nucleotide sequence of a transcript that results in a nucleotide sequence which could principally be directly encoded in the DNA template. As such it encompasses the insertions, deletions and exchanges of any of the four standard nucleotides of the RNA alphabet (A: adenosine, C: cytidine, G: guanidine, and U: uridine) in a RNA molecule. The definition also includes the deamination of adenosine (A) to inosine (I) nucleotides, which are read as guanidine (G) nucleotides upon translation but excludes chemical modifications leading to nonstandard bases such as for example dihydro-uridine or pseudo-uridine in tRNAs. The term RNA editing was originally coined for the uridine insertion type of RNA editing discovered in trypanosome mitochondria (Benne et al. 1986). Many other types of mitochondrial, cytosolic and chloroplast RNA editing have been discovered subsequently (Knoop and Brennicke 1999). A nomenclature has recently been suggested to label RNA editing sites (Rüdinger et al. 2009) and the software tool PREPACT has been developed to analyze and catalogue the abundant RNA editing sites in plant organelles (Lenz et al. 2009). For more details see Chap. 7.

include the creation of start and stop codons of the respective reading frames. Somewhat more rarely, RNA editing in structural RNAs such as introns or tRNAs may help to stabilize or reestablish base-pairings. Deamination and/or transamination of the pyrimidine bases are most likely the underlying biochemical mechanisms for base conversion. Given that one chapter of this book (see Chap. 7) is exclusively devoted to plant mitochondrial RNA editing, we will here mainly focus on the phylogenetic perspective of RNA editing evolution among embryophytes.

RNA editing in flowering plant chloroplasts (Hoch et al. 1991) was discovered only briefly after its identification in mitochondria (Covello and Gray 1989; Gualberto et al. 1989; Hiesel et al. 1989). Phylogenetic studies subsequently showed that the occurrence of RNA editing in mitochondria and chloroplasts is surprisingly congruent: RNA editing appears to be absent in algae but has been identified in chloroplast and mitochondria of all land plant clades with the unique exception of the marchantiid liverworts (Malek et al. 1996; Freyer et al. 1997; Steinhauser et al. 1999). Similarly the C-to-U type of editing is dominating in both organelles among seed plants, mosses and (nonmarchantiid) liverworts, whereas frequent "reverse" U-to-C editing is only seen in hornworts, lycophytes, and ferns (see Fig. 1.1) in chloroplasts and mitochondria at the same time (Steinhauser et al. 1999; Vangerow et al. 1999; Kugita et al. 2003; Wolf et al. 2004; Grewe et al. 2009). Mitochondrial RNA editing frequencies vary widely across land plant phylogeny, ranging from zero sites in marchantiid liverworts over very few, e.g., only 11 in the moss *Physcomitrella* (Rüdinger et al. 2009) to predictions of over 1,000 RNA editing sites in the gymnosperm *Cycas* or the lycophyte *Isoetes* (Chaw et al. 2008; Grewe et al. 2009).

Biochemical and genetic functional studies suggest very similar mechanisms for recognition of editing sites in transcripts of mitochondria and chloroplasts. In particular, specific members of the vastly extended pentatricopeptide repeat protein (PPR) gene families of plants (Lurin et al. 2004) have been shown to be involved in RNA editing site recognition both in chloroplasts (Kotera et al. 2005) and in mitochondria (Zehrmann et al. 2009). A particular sub-group of PPR proteins with a carboxy-terminal extension ending in the highly conserved DYW tripeptide had previously been connected to RNA editing in plants. The "DYW domain" shows similarity to cytidine deaminases potentially indicating direct involvement in the biochemical process of base conversion through deamination of cytidine to uridine and DYW domain PPR proteins are consistently present in taxa showing RNA editing but could not be identified in those lacking RNA editing (Salone et al. 2007; Rüdinger et al. 2008). Indeed, there seems to be a quantitative correlation between the numbers of organellar editing sites and nuclear DYW-type PPR genes (Rüdinger et al. 2008; Rüdinger et al. 2009). Nevertheless, the existence of plant organellar RNA editing as such remains a mystery given that no convincing evidence for functional gains such as physiological regulation of protein activity or the creation of protein diversity through partial editing (e.g., Mower and Palmer 2006) has been found. RNA editing may largely be a correction mechanism, possibly compensating on RNA level for mutations occurring on DNA level associated with the establishment of embryophytes. The wide variability of editing frequencies including the complete absence in marchantiid liverworts and the rise in reverse U-to-C editing activity in hornworts, lycophytes and monilophytes (see Fig. 1.1) remain, however, mysterious at present. Given that establishment of plant RNA editing obviously comes at the cost of creation and maintenance of large nuclear gene families such as the PPR gene family the phenomenon remains all the more puzzling.

1.8 Gene Transfer Deviations: Promiscuous DNA

Besides EGT other forms of interorganellar gene transfer have been identified as well in plant cells. Flowering plant mitochondrial genomes have been shown to have a surprising disposition to integrate and perpetuate foreign DNA from the chloroplast or nuclear genomes. The initial discovery of a 12 kbp fragment of cpDNA in the maize mitochondrial genome (Stern and Lonsdale 1982) led J. Ellis to suggest the term "promiscuous DNA" at that time (Ellis 1982). Not only chloroplast DNA, but also promiscuous DNA fragments of nuclear origin were subsequently observed in many angiosperm mtDNAs (Schuster and Brennicke 1987), e.g., all different types of retrotransposon fragments in *Arabidopsis* (Knoop et al. 1996). Similarly, sequence inserts originating from the chloroplast and nuclear genome have also been found in the mtDNA of the gymnosperm *Cycas* (Wang et al. 2007). Finally, chloroplast and nuclear promiscuous DNA fragments have been identified in the mtDNA of the lycophyte *Isoetes* (Grewe et al. 2009), showing that the propensity for integrating foreign DNA arose with early tracheophyte evolution.

Notably, an active mechanism of DNA import in potato mitochondria has been described (Koulintchenko et al. 2003), the physiologic relevance of which is unclear at present. In contrast to tracheophytes, no promiscuous DNA has as yet been identified in the mtDNAs of bryophytes. The ability to integrate promiscuous sequences may well depend on the gain of recombinational activity in the tracheophyte lineage (see Fig. 1.1).

1.9 Horizontal Gene Transfer

A series of publications has shown that plant mitochondrial genomes may be the prime examples of donors and acceptors of DNA via frequent horizontal gene transfer (HGT) among eukaryotes (Box 1.3). Evidence for DNA transfer across species borders in plant mitochondria initially came from the surprising identification of genes in certain species that had apparently been regained in the mitochondria after they had previously been transferred to the nucleus via EGT in the respective plant lineage (Bergthorsson et al. 2003). Host–parasite interactions seem to play a major role for mitochondrial HGT in plants (Davis and Wurdack 2004; Mower et al. 2004). Not only angiosperms, but also the gymnosperm *Gnetum* (Won and Renner 2003) and the fern *Botrychium* (Davis et al. 2005) have been identified as acceptor species for mitochondrial DNA transferred from other taxa via HGT. A particularly striking example seems to be the case of the mtDNA in the basal angiosperm *Amborella trichopoda,* which contains many gene sequences of foreign, including bryophyte, origins (Bergthorsson et al. 2004; J. Palmer, personal communication). Intimate physical plant-plant contacts that allow the exchange of DNA into cells

Box 1.3 Horizontal gene transfer

Horizontal gene transfer (HGT) is defined as the migration of DNA sequences across species borders or alternatively as the nonsexual movement of genetic information between two organisms (Keeling and Palmer 2008; Bock 2010). Hence, HGT is an exception to (or extension of) the common vertical transmission of genetic material via cell division or the mating of sexually compatible species. In most cases, HGT is identified as a strong discrepancy between a gene phylogeny to a reliably supported organismal phylogeny. HGT is a common phenomenon in bacterial evolution (Lawrence and Hendrickson 2003), but examples of HGT are increasingly also being identified in eukaryotes, mostly in protists (Andersson 2005). In plants, the majority of identified cases of HGT concern the movement of DNA between mitochondrial genomes in particular (Richardson and Palmer 2006), whereas HGT in plastids seems to occur much more rarely (Rice and Palmer 2006). The term *lateral gene transfer* (LGT) is often used synonymously for HGT.

that develop flowering meristems are an obvious prerequisite. Besides host–parasite interactions epiphytism, illegitimate pollination, or natural grafting may be envisaged for such plant–plant interactions. Once such heterologous cell-to-cell contacts have been established, the obvious question certainly remains why the mitochondrial genome in particular appears to be much more prone to HGT than the chloroplast or nuclear genomes. Most likely the processes of highly active DNA recombination (see Chap. 3) and the readiness for fusion and fission of plant mitochondria, which may rather be seen as a "discontinuous whole" than as continuously separate organelles (Logan 2010; and see Chap. 2), very likely play a major role. A strong gain of DNA recombinational activity during early tracheophyte evolution as reflected with the recently sequenced *Isoetes* mtDNA may at the same time explain restructured genomes as well as the readiness to integrate promiscuous DNA originating from the other two genomic compartments or mtDNA from other plant taxa. If true, one may predict that bryophytes are not prone for acceptance of mtDNA via HGT but only act as donors as was observed for the *Amborella* example (Bergthorsson et al. 2004). Indeed, phylogenetic analyses of several mitochondrial loci at dense sampling of mosses have not given a hint for HGT so far (Beckert et al. 1999; Pruchner et al. 2002; Wahrmund et al. 2009; Volkmar and Knoop 2010; Wahrmund et al. 2010).

1.10 An Extended Perspective: What Else?

What other idiosyncrasies may be identified further down the road in plant mitochondrial genome research? The plethora of peculiarities in plant mitochondrial genomes, including DNA recombination leading to multipartite genome structures, intron gains and losses mainly in early land plant evolution, disrupted *trans*-splicing group II and group I introns, frequent RNA editing, acceptance of promiscuous DNA from the chloroplast and nuclear genomes, and an ongoing endosymbiotic and horizontal DNA transfer is rivaled only by the idiosyncrasies observed in certain protist mtDNAs (Gray et al. 2004). Among the most peculiar mitochondrial genome organizations is the one of the diplonemid protist *Diplonema papillatum,* which consists of more than 100 circular minichromosomes carrying gene modules which give rise to transcripts that have to be matured by an as yet uncharacterized mode of ligation or splicing (Marande and Burger 2007). Similarly idiosyncratic is the organization of the mtDNA in the ichthyosporean *Amoebidium parasiticum,* which features a large population of coexisting linear instead of circular minichromosomes with peculiar terminal sequence motifs (Burger et al. 2003). Extreme degrees of mtDNA recombination as recently observed for the lycophyte *Isoetes* may suggest that yet more complexity of mitochondrial genomes could be expected in other isolated plant lineages (Grewe et al. 2009). The compact and circular mitochondrial genome of the jakobiid protist *Reclinomonas americana* (Lang et al. 1997) reminds of animal mtDNAs in structure but is yet more gene-rich than the ones of plants. Surprisingly though, *rpl*10, a gene that was supposed to be absent from plant mitochondria has only recently been identified in the mitochondrial DNAs of bryophytes, *Chaetosphaeridium* and some angiosperms (Mower and Bonen 2009; Kubo and Arimura 2010). A few more of the small ORFs conserved within or between plant clades may be identified as such missing genes in the future. Although abundant RNA editing of a tRNA species has recently been described in *Isoetes* (Grewe et al. 2009) all edits were of the canonical C-to-U type. Instead, many more types of nucleotide editing converting all types of nucleotides into purines (N-to-R) were observed in mitochondrial tRNA editing in the amoeboid protist *Acanthamoeba* (Lonergan and Gray 1993) and alternative forms of RNA editing may be discovered in plant mitochondria, similar to the adenosine deaminase type editing of a plant chloroplast tRNA recently analyzed (Delannoy et al. 2009; Karcher and Bock 2009). Yet to be identified is also a functional gene within an intron, similar to *rps*3 within a group I intron as in some ascomycetous fungi (Sethuraman et al. 2009). Indeed, a first example for such a peculiar gene arrangement may be verified in our ongoing transcriptome analysis of the lycophyte *Selaginella* (J.H., F.G., and V.K., unpublished observations).

Glossary

Chondrome: Understood as the mitochondrial genome (similar to the term "plastome" for the chloroplast genome). Sometimes mixed up with the term chondriome, now more widely used to circumscribe all mitochondria in a cell (see Chap. 2).

Clade: See monophyletic.

Cytoplasmic male sterility (CMS): The economically most important phenotypic trait connected to plant mitochondrial DNA mutations. The corresponding mitochondrial dysfunctions are revealed as male infertility resulting from nonfunctional pollen grains.

Endosymbiotic gene transfer (EGT): The functional relocation of genes from the organelle endosymbiont genomes into the host cell nucleus as a corollary of the endosymbiont hypothesis. Functional gene transfer may proceed through mature RNA and reverse transcription. Clearly though, such functional gene transfer is accompanied by DNA-based copying mechanisms which lead to insertion of small and large DNA fragments of mitochondrial (numt) and plastid (nupt) origin in the nuclear genome.

Group I and group II introns: Introns that are commonly observed in organelle genomes of plants and fungi, characterized by highly characteristic RNA secondary structures. For details see Box 1.1.

Heteroplasmy: The existence of different organelle genomes within a single individual, a single cell or even a single organelle (given that organelle DNAs are generally present in multiple copies).

Horizontal gene transfer (HGT): The process of DNA sequence migration across species borders. For details see Box 1.3.

Lateral gene transfer (LGT): A term that is frequently used synonymously to >Horizontal gene transfer.

Monophyletic: Literally, of "one stem," a group of organisms (a clade) encompassing all descendants which trace back to a single (usually extinct) ancestor. Indentifying monophyletic clades through shared derived characters (>synapomorphies) is the goal of cladistics as introduced by Willi Hennig (1950) into modern phylogenetics.

Promiscuous DNA: Mostly nonfunctional DNA fragments copied from one genome to another in the eukaryotic cell. Plant mitochondria are particularly prone to accumulation of DNA sequences from the chloroplast or the nuclear genome. Foreign DNA derived from the chloroplast or the nuclear genome is frequently inserted in to the mtDNA of vascular plants.

RNA editing: Modification of nucleotide sequences on transcript level, in plant organelles in the form of cytidine-uridine exchanges. For details see Box 1.2.

Synapomorphy: Any shared, derived (i.e., newly acquired) state of a morphologic, biochemical, developmental, or any other character in a group of organisms helping to identify them as a >monophyletic clade. Examples are the water-conducting tissues and the dominating diploid sporophyte generation for the vascular plant (tracheophyte) clade.

Trans-splicing: The maturation of a RNA molecule through splicing of separate, independent precursor-transcripts encoded by separate, distant genomic loci. In plant organelles, gene arrangements requiring *trans*-splicing have arisen through recombinational activity disrupting ancestral group II or group I intron continuities.

References

Adams, K. L., Palmer, J. D. 2003. Evolution of mitochondrial gene content: gene loss and transfer to the nucleus. Mol. Phylogenet. Evol. 29:380–395.

Adams, K. L., Clements, M. J., Vaughn, J. C. 1998. The *Peperomia* mitochondrial *coxI* group I intron: timing of horizontal transfer and subsequent evolution of the intron. J. Mol. Evol. 46:689–696.

Adams, K. L., Song, K. M., Roessler, P. G., Nugent, J. M., Doyle, J. L., Doyle, J. J., Palmer, J. D. 1999. Intracellular gene transfer in action: Dual transcription and multiple silencings of nuclear and mitochondrial *cox*2 genes in legumes. Proc. Natl. Acad. Sci. U.S.A. 96:13863–13868.

Adams, K. L., Daley, D. O., Qiu, Y. L., Whelan, J., Palmer, J. D. 2000. Repeated, recent and diverse transfers of a mitochondrial gene to the nucleus in flowering plants. Nature 408:354–357.

Adams, K. L., Ong, H. C., Palmer, J. D. 2001a. Mitochondrial gene transfer in pieces: Fission of the ribosomal protein gene rpl2 and partial or complete gene transfer to the nucleus. Mol. Biol. Evol. 18:2289–2297.

Adams, K. L., Rosenblueth, M., Qiu, Y. L., Palmer, J. D. 2001b. Multiple losses and transfers to the nucleus of two mitochondrial succinate dehydrogenase genes during angiosperm evolution. Genetics 158:1289–1300.

Adams, K. L., Daley, D. O., Whelan, J., Palmer, J. D. 2002a. Genes for two mitochondrial ribosomal proteins in flowering plants are derived from their chloroplast or cytosolic counterparts. Plant Cell 14:931–943.

Adams, K. L., Qiu, Y. L., Stoutemyer, M., Palmer, J. D. 2002b. Punctuated evolution of mitochondrial gene content: high and variable rates of mitochondrial gene loss and transfer to the nucleus during angiosperm evolution. Proc. Natl. Acad. Sci. U.S.A. 99:9905–9912.

Allen, J. O., Fauron, C. M., Minx, P., Roark, L., Oddiraju, S., Lin, G. N., Meyer, L., Sun, H., Kim, K., Wang, C., Du, F., Xu, D., Gibson, M., Cifrese, J., Clifton, S. W., Newton, K. J. 2007. Comparisons among two fertile and three male-sterile mitochondrial genomes of maize. Genetics 177:1173–1192.

Alverson, A. J., Wei, X., Rice, D. W., Stern, D. B., Barry, K., Palmer, J. D. 2010. Insights into the evolution of mitochondrial genome size from complete sequences of *Citrullus lanatus* and *Cucurbita pepo* (Cucurbitaceae). Mol. Biol. Evol. 27:1436–1448.

Anderson, S., Bankier, A. T., Barrell, B. G., de Bruijn, M. H., Coulson, A. R., Drouin, J., Eperon, I. C., Nierlich, D. P., Roe, B. A., Sanger, F., Schreier, P. H., Smith, A. J., Staden, R., Young, I. G. 1981. Sequence and organization of the human mitochondrial genome. Nature 290:457–465.

Andersson, J. O. 2005. Lateral gene transfer in eukaryotes. Cell. Mol. Life Sci. 62:1182–1197.

Andre, C., Levy, A., Walbot, V. 1992. Small repeated sequences and the structure of plant mitochondrial genomes. Trends Genet. 8:128–132.

Asakura, Y., Barkan, A. 2006. *Arabidopsis* orthologs of maize chloroplast splicing factors promote splicing of orthologous and species-specific group II introns. Plant Physiol. 142:1656–1663.

Asakura, Y., Barkan, A. 2007. A CRM domain protein functions dually in group I and group II intron splicing in land plant chloroplasts. Plant Cell 19:3864–3875.

Bakker, F. T., Culham, A., Pankhurst, C. E., Gibby, M. 2000. Mitochondrial and chloroplast DNA-based phylogeny of Pelargonium (Geraniaceae). Am. J. Bot. 87:727–734.

Beagley, C. T., Okada, N. A., Wolstenholme, D. R. 1996. Two mitochondrial group I introns in a metazoan, the sea anemone *Metridium senile*: one intron contains genes for subunits 1 and 3 of NADH dehydrogenase. Proc. Natl. Acad. Sci. U.S.A. 93:5619–5623.

Beckert, S., Steinhauser, S., Muhle, H., Knoop, V. 1999. A molecular phylogeny of bryophytes based on nucleotide sequences of the mitochondrial *nad5* gene. Plant Syst. Evol. 218:179–192.

Beckert, S., Muhle, H., Pruchner, D., Knoop, V. 2001. The mitochondrial nad2 gene as a novel marker locus for phylogenetic analysis of early land plants: a comparative analysis in mosses. Mol. Phylogenet. Evol. 18:117–126.

Bégu, D., Araya, A. 2009. The horsetail *Equisetum arvense* mitochondria share two group I introns with the liverwort *Marchantia*, acquired a novel group II intron but lost intron-encoded ORFs. Curr. Genet. 55:69–79.

Bendich, A. J. 2007. The size and form of chromosomes are constant in the nucleus, but highly variable in bacteria, mitochondria and chloroplasts. Bioessays 29:474–483.

Benne, R., Van Den Burg J., Brakenhoff, J. P., Sloof, P., Van Boom, J. H., Tromp, M. C. 1986. Major transcript of the frameshifted *coxII* gene from trypanosome mitochondria contains four nucleotides that are not encoded in the DNA. Cell 46:819–826.

Bergthorsson, U., Adams, K. L., Thomason, B., Palmer, J. D. 2003. Widespread horizontal transfer of mitochondrial genes in flowering plants. Nature 424:197–201.

Bergthorsson, U., Richardson, A. O., Young, G. J., Goertzen, L. R., Palmer, J. D. 2004. Massive horizontal transfer of mitochondrial genes from diverse land plant donors to the basal angiosperm *Amborella*. Proc. Natl. Acad. Sci. U.S.A. 101:17747–17752.

Binder, S., Marchfelder, A., Brennicke, A., Wissinger, B. 1992. RNA editing in trans-splicing intron sequences of nad2 mRNAs in *Oenothera* mitochondria. J. Biol. Chem. 267:7615–7623.

Bock, R. 2010. The give-and-take of DNA: horizontal gene transfer in plants. Trends Plant Sci. 15:11–22.

Boesch, P., Ibrahim, N., Paulus, F., Cosset, A., Tarasenko, V., Dietrich, A. 2009. Plant mitochondria possess a short-patch base excision DNA repair pathway. Nucl. Acids Res. 37:5690–5700.

Bonen, L. 2008. *Cis*- and *trans*-splicing of group II introns in plant mitochondria. Mitochondrion 8:26–34.

Brennicke, A., Grohmann, L., Hiesel, R., Knoop, V., Schuster, W. 1993. The mitochondrial genome on its way to the nucleus: different stages of gene transfer in higher plants. FEBS Lett. 325:140–145.

Burger, G., Forget, L., Zhu, Y., Gray, M. W., Lang, B. F. 2003. Unique mitochondrial genome architecture in unicellular relatives of animals. Proc. Natl. Acad. Sci. U.S.A. 100:892–897.

Burger, G., Yan, Y., Javadi, P., Lang, B. F. 2009. Group I-intron trans-splicing and mRNA editing in the mitochondria of placozoan animals. Trends Genet. 25:381–386.

Cech, T. R., Damberger, S. H., Gutell, R. R. 1994. Representation of the secondary and tertiary structure of group I introns. Nat. Struct. Biol 1:273–280.

Chapdelaine, Y., Bonen, L. 1991. The wheat mitochondrial gene for subunit I of the NADH dehydrogenase complex: a trans-splicing model for this gene-in-pieces. Cell 65:465–472.

Chaw, S. M., Chun-Chieh, S. A., Wang, D., Wu, Y. W., Liu, S. M., Chou, T. Y. 2008. The mitochondrial genome of the gymnosperm *Cycas taitungensis* contains a novel family of short interspersed elements, Bpu sequences, and abundant RNA editing sites. Mol. Biol. Evol. 25:603–615.

Cho, Y., Qiu, Y. L., Kuhlman, P., Palmer, J. D. 1998. Explosive invasion of plant mitochondria by a group I intron. Proc. Natl. Acad. Sci. U.S.A. 95:14244–14249.

Cho, Y., Mower, J. P., Qiu, Y. L., Palmer, J. D. 2004. Mitochondrial substitution rates are extraordinarily elevated and variable in a genus of flowering plants. Proc. Natl. Acad. Sci. U.S.A. 101:17741–17746.

Clifton, S. W., Minx, P., Fauron, C. M., Gibson, M., Allen, J. O., Sun, H., Thompson, M., Barbazuk, W. B., Kanuganti, S., Tayloe, C., Meyer, L., Wilson, R. K., Newton, K. J. 2004. Sequence and comparative analysis of the maize NB mitochondrial genome. Plant Physiol. 136:3486–3503.

Covello, P. S., Gray, M. W. 1989. RNA editing in plant mitochondria. Nature 341:662–666.

Covello, P. S., Gray, M. W. 1992. Silent mitochondrial and active nuclear genes for subunit 2 of cytochrome c oxidase (*cox2*) in soybean: evidence for RNA-mediated gene transfer. EMBO J. 11:3815–3820.

Cusimano, N., Zhang, L. B., Renner, S. S. 2008. Reevaluation of the cox1 group I intron in Araceae and angiosperms indicates a history dominated by loss rather than horizontal transfer. Mol. Biol. Evol. 25:265–276.

Davis, C. C., Wurdack, K. J. 2004. Host-to-parasite gene transfer in flowering plants: phylogenetic evidence from Malpighiales. Science 305:676–678.

Davis, C. C., Anderson, W. R., Wurdack, K. J. 2005. Gene transfer from a parasitic flowering plant to a fern. Proc. Biol. Sci. 272:2237–2242.

Delannoy, E., Le Ret, M., Faivre-Nitschke, E., Estavillo, G. M., Bergdoll, M., Taylor, N. L., Pogson, B. J., Small, I., Imbault, P., Gualberto, J. M. 2009. Arabidopsis tRNA adenosine deaminase arginine edits the wobble nucleotide of chloroplast tRNAArg(ACG) and is essential for efficient chloroplast translation. Plant Cell 21:2058–2071.

Dombrovska, E., Qiu, Y. L. 2004. Distribution of introns in the mitochondrial gene *nad1* in land plants: phylogenetic and molecular evolutionary implications. Mol. Phylogenet. Evol. 32:246–263.

Ellis, J. 1982. Promiscuous DNA–chloroplast genes inside plant mitochondria. Nature 299:678–679.

Freyer, R., Kiefer-Meyer, M.-C., Kössel, H. 1997. Occurrence of plastid RNA editing in all major lineages of land plants. Proc. Natl. Acad. Sci. U.S.A. 94:6285–6290.

Fukami, H., Chen, C. A., Chiou, C. Y., Knowlton, N. 2007. Novel group I introns encoding a putative homing endonuclease in the mitochondrial cox1 gene of Scleractinian corals. J. Mol. Evol. 64:591–600.

Gawryluk, R. M., Gray, M. W. 2009. A split and rearranged nuclear gene encoding the iron-sulfur subunit of mitochondrial succinate dehydrogenase in Euglenozoa. BMC Res. Notes 2:16.

Gawryluk, R. M., Gray, M. W. 2010. An ancient fission of mitochondrial *cox1*. Mol. Biol. Evol. 27:7–10.

Gissi, C., Iannelli, F., Pesole, G. 2008. Evolution of the mitochondrial genome of Metazoa as exemplified by comparison of congeneric species. Heredity 101:301–320.

Glanz, S., Kück, U. 2009. *Trans*-splicing of organelle introns – a detour to continuous RNAs. Bioessays 31:921–934.

Goremykin, V. V., Salamini, F., Velasco, R., Viola, R. 2009. Mitochondrial DNA of *Vitis vinifera* and the issue of rampant horizontal gene transfer. Mol. Biol. Evol. 26:99–110.

Gray, M. W., Lang, B. F., Burger, G. 2004. Mitochondria of protists. Annu. Rev. Genet. 38:477–524.

Grewe, F., Viehoever, P., Weisshaar, B., Knoop, V. 2009. A *trans*-splicing group I intron and tRNA-hyperediting in the mitochondrial genome of the lycophyte *Isoetes engelmannii*. Nucl. Acids Res. 37:5093–5104.

Groth-Malonek, M., Pruchner, D., Grewe, F., Knoop, V. 2005. Ancestors of *trans*-splicing mitochondrial introns support serial sister group relationships of hornworts and mosses with vascular plants. Mol. Biol. Evol. 22:117–125.

Groth-Malonek, M., Rein, T., Wilson, R., Groth, H., Heinrichs, J., Knoop, V. 2007a. Different fates of two mitochondrial gene spacers in early land plant evolution. Int. J. Plant Sci. 168:709–717.

Groth-Malonek, M., Wahrmund, U., Polsakiewicz, M., Knoop, V. 2007b. Evolution of a pseudogene: exclusive survival of a functional mitochondrial *nad7* gene supports *Haplomitrium* as the earliest liverwort lineage and proposes a secondary loss of RNA editing in Marchantiidae. Mol. Biol. Evol. 24:1068–1074.

Gualberto, J. M., Lamattina, L., Bonnard, G., Weil, J. H., Grienenberger, J. M. 1989. RNA editing in wheat mitochondria results in the conservation of protein sequences. Nature 341:660–662.

Handa, H. 2003. The complete nucleotide sequence and RNA editing content of the mitochondrial genome of rapeseed (*Brassica napus* L.): comparative analysis of the mitochondrial genomes of rapeseed and *Arabidopsis thaliana*. Nucl. Acids Res. 31:5907–5916.

Haugen, P., Simon, D. M., Bhattacharya, D. 2005. The natural history of group I introns. Trends Genet. 21:111–119.

Hennig, W. 1950. Grundzüge einer Theorie der Phylogenetischen Systematik. Berlin: Deutscher Zentralverlag.

Hiesel, R., Wissinger, B., Schuster, W., Brennicke, A. 1989. RNA editing in plant mitochondria. Science 246:1632–1634.

Hiesel, R., von Haeseler, A., Brennicke, A. 1994. Plant mitochondrial nucleic acid sequences as a tool for phylogenetic analysis. Proc. Natl. Acad. Sci. U.S.A. 91:634–638.

Hoch, B., Maier, R. M., Appel, K., Igloi, G. L., Kössel, H. 1991. Editing of a chloroplast mRNA by creation of an initiation codon. Nature 353:178–180.

Karcher, D., Bock, R. 2009. Identification of the chloroplast adenosine-to-inosine tRNA editing enzyme. RNA 15:1251–1257.

Keeling, P. J., Palmer, J. D. 2008. Horizontal gene transfer in eukaryotic evolution. Nat. Rev. Genet. 9:605–618.

Keren, I., Bezawork-Geleta, A., Kolton, M., Maayan, I., Belausov, E., Levy, M., Mett, A., Gidoni, D., Shaya, F., Ostersetzer-Biran, O. 2009. AtnMat2, a nuclear-encoded maturase required for splicing of group-II introns in Arabidopsis mitochondria. RNA 15:2299–2311.

Knoop, V. 2004. The mitochondrial DNA of land plants: peculiarities in phylogenetic perspective. Curr. Genet. 46:123–139.

Knoop, V. 2010. Looking for sense in the nonsense: a short review of non-coding organellar DNA elucidating the phylogeny of bryophytes. Trop. Bryol. 31:50–60.

Knoop, V., Brennicke, A. 1991. A mitochondrial intron sequence in the 5′-flanking region of a plant nuclear lectin gene. Curr. Genet. 20:423–425.

Knoop, V., Brennicke, A. 1999. RNA-editing 1999. Biol. i. uns. Zeit 29:336–345.

Knoop, V., Schuster, W., Wissinger, B., Brennicke, A. 1991. *Trans* splicing integrates an exon of 22 nucleotides into the *nad*5 mRNA in higher plant mitochondria. EMBO J. 10:3483–3493.

Knoop, V., Ehrhardt, T., Lättig, K., Brennicke, A. 1995. The gene for ribosomal protein S10 is present in mitochondria of pea and potato but absent from those of *Arabidopsis* and *Oenothera*. Curr. Genet. 27:559–564.

Knoop, V., Unseld, M., Marienfeld, J., Brandt, P., Sünkel, S., Ullrich, H., Brennicke, A. 1996. *copia-*, *gypsy-* and LINE-like retrotransposon fragments in the mitochondrial genome of *Arabidopsis thaliana*. Genetics 142:579–585.

Kobayashi, Y., Knoop, V., Fukuzawa, H., Brennicke, A., Ohyama, K. 1997. Interorganellar gene transfer in bryophytes: the functional *nad7* gene is nuclear encoded in *Marchantia polymorpha*. Mol. Gen. Genet. 256:589–592.

Kotera, E., Tasaka, M., Shikanai, T. 2005. A pentatricopeptide repeat protein is essential for RNA editing in chloroplasts. Nature 433:326–330.

Koulintchenko, M., Konstantinov, Y., Dietrich, A. 2003. Plant mitochondria actively import DNA via the permeability transition pore complex. EMBO J. 22:1245–1254.

Kroeger, T. S., Watkins, K. P., Friso, G., van Wijk, K. J., Barkan, A. 2009. A plant-specific RNA-binding domain revealed through analysis of chloroplast group II intron splicing. Proc. Natl. Acad. Sci. U.S.A. 106:4537–4542.

Kubo, N., Arimura, S. 2010. Discovery of the *rpl10* gene in diverse plant mitochondrial genomes and its probable replacement by the nuclear gene for chloroplast RPL10 in two lineages of angiosperms. DNA Res. 17:1–9.

Kubo, T., Mikami, T. 2007. Organization and variation of angiosperm mitochondrial genome. Physiol. Plant. 129:6–13.

Kubo, T., Newton, K. J. 2008. Angiosperm mitochondrial genomes and mutations. Mitochondrion 8:5–14.

Kubo, T., Nishizawa, S., Sugawara, A., Itchoda, N., Estiati, A., Mikami, T. 2000. The complete nucleotide sequence of the mitochondrial genome of sugar beet (*Beta vulgaris* L.) reveals a novel gene for tRNA(Cys)(GCA). Nucl. Acids Res. 28:2571–2576.

Kugita, M., Yamamoto, Y., Fujikawa, T., Matsumoto, T., Yoshinaga, K. 2003. RNA editing in hornwort chloroplasts makes more than half the genes functional. Nucl. Acids Res. 31:2417–2423.

Lang, B. F., Burger, G., O'Kelly, C. J., Cedergren, R., Golding, G. B., Lemieux, C., Sankoff, D., Turmel, M., Gray, M. W. 1997. An ancestral mitochondrial DNA resembling a eubacterial genome in miniature. Nature 387:493–497.

Lavrov, D. V. 2007. Key transitions in animal evolution: a mitochondrial DNA perspective. Integr. Compar. Biol. 47:734–743.

Lawrence, J. G., Hendrickson, H. 2003. Lateral gene transfer: when will adolescence end? Mol. Microbiol. 50:739–749.

Lenz, H., Rüdinger, M., Volkmar, U., Fischer, S., Herres, S., Grewe, F., Knoop, V. 2009. Introducing the plant RNA editing prediction and analysis computer tool PREPACT and an update on RNA editing site nomenclature. Curr. Genet. 56:189–201.

Li, L., Wang, B., Liu, Y., Qiu, Y. L. 2009. The complete mitochondrial genome sequence of the hornwort *Megaceros aenigmaticus* shows a mixed mode of conservative yet dynamic evolution in early land plant mitochondrial genomes. J. Mol. Evol. 68:665–678.

Lilly, J. W., Havey, M. J. 2001. Small, repetitive DNAs contribute significantly to the expanded mitochondrial genome of cucumber. Genetics 159:317–328.

Lin, X., Kaul, S., Rounsley, S., Shea, T. P., Benito, M. I., Town, C. D., Fujii, C. Y., Mason, T., Bowman, C. L., Barnstead, M., Feldblyum, T. V., Buell, C. R., Ketchum, K. A., Lee, J., Ronning, C. M., Koo, H. L., Moffat, K. S., Cronin, L. A., Shen, M., Pai, G., Van Aken, S., Umayam, L., Tallon, L. J., Gill, J. E., Adams, M. D., Carrera, A. J., Creasy, T. H., Goodman, H. M., Somerville, C. R., Copenhaver, G. P., Preuss, D., Nierman, W. C., White, O., Eisen, J. A., Salzberg, S. L., Fraser, C. M., Venter, J. C. 1999. Sequence and analysis of chromosome 2 of the plant *Arabidopsis thaliana*. Nature 402:761–768.

Liu, S. L., Zhuang, Y., Zhang, P., Adams, K. L. 2009. Comparative analysis of structural diversity and sequence evolution in plant mitochondrial genes transferred to the nucleus. Mol. Biol. Evol. 26:875–891.

Logan, D. C. 2010. The dynamic plant chondriome. Semin. Cell Dev. Biol. 21(6):550–557.

Lonergan, K. M., Gray, M. W. 1993. Editing of transfer RNAs in *Acanthamoeba castellanii* mitochondria. Science 259:812–816.

Lurin, C., Andrés, C., Aubourg, S., Bellaoui, M., Bitton, F., Bruyère, C., Caboche, M., Debast, C., Gualberto, J., Hoffmann, B., Lecharny, A., Le Ret, M., Martin-Magniette, M. L., Mireau, H., Peeters, N., Renou, J. P., Szurek, B., Taconnat, L., Small, I. 2004. Genome-wide analysis of Arabidopsis pentatricopeptide repeat proteins reveals their essential role in organelle biogenesis. Plant Cell 16:2089–2103.

Malek, O., Knoop, V. 1998. Trans-splicing group II introns in plant mitochondria: the complete set of *cis*-arranged homologs in ferns, fern allies, and a hornwort. RNA 4:1599–1609.

Malek, O., Lättig, K., Hiesel, R., Brennicke, A., Knoop, V. 1996. RNA editing in bryophytes and a molecular phylogeny of land plants. EMBO J. 15:1403–1411.

Malek, O., Brennicke, A., Knoop, V. 1997. Evolution of trans-splicing plant mitochondrial introns in pre-Permian times. Proc. Natl. Acad. Sci. U.S.A. 94:553–558.

Manchekar, M., Scissum-Gunn, K., Song, D., Khazi, F., McLean, S. L., Nielsen, B. L. 2006. DNA recombination activity in soybean mitochondria. J. Mol. Biol. 356:288–299.

Marande, W., Burger, G. 2007. Mitochondrial DNA as a genomic jigsaw puzzle. Science 318:415.

Mower, J. P., Bonen, L. 2009. Ribosomal protein L10 is encoded in the mitochondrial genome of many land plants and green algae. BMC Evol. Biol. 9:265.

Mower, J. P., Palmer, J. D. 2006. Patterns of partial RNA editing in mitochondrial genes of *Beta vulgaris*. Mol. Genet. Genomics 276:285–293.

Mower, J. P., Stefanovic, S., Young, G. J., Palmer, J. D. 2004. Plant genetics: gene transfer from parasitic to host plants. Nature 432:165–166.

Notsu, Y., Masood, S., Nishikawa, T., Kubo, N., Akiduki, G., Nakazono, M., Hirai, A., Kadowaki, K. 2002. The complete sequence of the rice (*Oryza sativa* L.) mitochondrial genome: frequent DNA sequence acquisition and loss during the evolution of flowering plants. Mol. Genet. Genomics 268:434–445.

Nugent, J. M., Palmer, J. D. 1991. RNA-mediated transfer of the gene *coxII* from the mitochondrion to the nucleus during flowering plant evolution. Cell 66:473–481.

Oda, K., Kohchi, T., Ohyama, K. 1992. Mitochondrial DNA of Marchantia polymorpha as a single circular form with no incorporation of foreign DNA. Biosci. Biotechnol. Biochem. 56:132–135.

Oda, K., Yamato, K., Ohta, E., Nakamura, Y., Takemura, M., Nozato, N., Akashi, K., Kanegae, T., Ogura, Y., Kohchi, T., Ohyama, K. 1992. Gene organization deduced from the complete sequence of liverwort *Marchantia polymorpha* mitochondrial DNA. A primitive form of plant mitochondrial genome. J. Mol. Biol. 223:1–7.

Odahara, M., Kuroiwa, H., Kuroiwa, T., Sekine, Y. 2009. Suppression of repeat-mediated gross mitochondrial genome rearrangements by RecA in the moss *Physcomitrella patens*. Plant Cell 21:1182–1194.

Ogihara, Y., Yamazaki, Y., Murai, K., Kanno, A., Terachi, T., Shiina, T., Miyashita, N., Nasuda, S., Nakamura, C., Mori, N., Takumi, S., Murata, M., Futo, S., Tsunewaki, K. 2005. Structural dynamics of cereal mitochondrial genomes as revealed by complete nucleotide sequencing of the wheat mitochondrial genome. Nucl. Acids Res. 33:6235–6250.

Oldenburg, D. J., Bendich, A. J. 2001. Mitochondrial DNA from the liverwort *Marchantia polymorpha*: circularly permuted linear molecules, head-to-tail concatemers, and a 5′ protein. J. Mol. Biol. 310:549–562.

Ong, H. C., Palmer, J. D. 2006. Pervasive survival of expressed mitochondrial *rps*14 pseudogenes in grasses and their relatives for 80 million years following three functional transfers to the nucleus. BMC Evol. Biol. 6:55.

Ostheimer, G. J., Rojas, M., Hadjivassiliou, H., Barkan, A. 2006. Formation of the CRS2-CAF2 group II intron splicing complex is mediated by a 22-amino acid motif in the COOH-terminal region of CAF2. J. Biol. Chem. 281:4732–4738.

Palmer, J. D., Herbon, L. A. 1987. Unicircular structure of the *Brassica hirta* mitochondrial genome. Curr. Genet. 11:565–570.

Palmer, J. D., Herbon, L. A. 1988. Plant mitochondrial DNA evolves rapidly in structure, but slowly in sequence. J. Mol. Evol. 28:87–97.

Palmer, J. D., Shields, C. R. 1984. Tripartite structure of the *Brassica campestris* mitochondrial genome. Nature 307:437–440.
Palmer, J. D., Soltis, D., Soltis, P. 1992. Large size and complex structure of mitochondrial DNA in two nonflowering land plants. Curr. Genet. 21:125–129.
Palmer, J. D., Adams, K. L., Cho, Y., Parkinson, C. L., Qiu, Y. L., Song, K. 2000. Dynamic evolution of plant mitochondrial genomes: Mobile genes and introns and highly variable mutation rates. Proc. Natl. Acad. Sci. U.S.A. 97:6960–6966.
Parkinson, C. L., Mower, J. P., Qiu, Y. L., Shirk, A. J., Song, K., Young, N. D., dePamphilis, C. W., Palmer, J. D. 2005. Multiple major increases and decreases in mitochondrial substitution rates in the plant family Geraniaceae. BMC Evol. Biol. 5:73.
Pereira de Souza, A., Jubier, M.-F., Delcher, E., Lancelin, D., Lejeune, B. 1991. A *trans*-splicing model for the expression of the tripartite *nad*5 gene in wheat and maize mitochondria. Plant Cell 3:1363–1378.
Perez-Martinez, X., Antaramian, A., Vazquez-Acevedo, M., Funes, S., Tolkunova, E., d'Alayer, J., Claros, M. G., Davidson, E., King, M. P., Gonzalez-Halphen, D. 2001. Subunit II of cytochrome c oxidase in Chlamydomonad algae is a heterodimer encoded by two independent nuclear genes. J. Biol. Chem. 276:11302–11309.
Pruchner, D., Nassal, B., Schindler, M., Knoop, V. 2001. Mosses share mitochondrial group II introns with flowering plants, not with liverworts. Mol. Genet. Genomics 266:608–613.
Pruchner, D., Beckert, S., Muhle, H., Knoop, V. 2002. Divergent intron conservation in the mitochondrial *nad2* gene: signatures for the three bryophyte classes (mosses, liverworts, and hornworts) and the lycophytes. J. Mol. Evol. 55:265–271.
Pryer, K. M., Schneider, H., Smith, A. R., Cranfill, R., Wolf, P. G., Hunt, J. S., Sipes, S. D. 2001. Horsetails and ferns are a monophyletic group and the closest living relatives to seed plants. Nature 409:618–622.
Pyle, A. M., Fedorova, O., Waldsich, C. 2007. Folding of group II introns: a model system for large, multidomain RNAs? Trends Biochem. Sci. 32:138–145.
Qiu, Y. L., Palmer, J. D. 2004. Many independent origins of trans splicing of a plant mitochondrial group II intron. J. Mol. Evol. 59:80–89.
Qiu, Y. L., Cho, Y. R., Cox, J. C., Palmer, J. D. 1998. The gain of three mitochondrial introns identifies liverworts as the earliest land plants. Nature 394:671–674.
Qiu, Y. L., Li, L., Wang, B., Chen, Z., Knoop, V., Groth-Malonek, M., Dombrovska, O., Lee, J., Kent, L., Rest, J., Estabrook, G. F., Hendry, T. A., Taylor, D. W., Testa, C. M., Ambros, M., Crandall-Stotler, B., Duff, R. J., Stech, M., Frey, W., Quandt, D., Davis, C. C. 2006. The deepest divergences in land plants inferred from phylogenomic evidence. Proc. Natl. Acad. Sci. U.S.A. 103:15511–15516.
Ran, J. H., Gao, H., Wang, X. Q. 2010. Fast evolution of the retroprocessed mitochondrial rps3 gene in Conifer II and further evidence for the phylogeny of gymnosperms. Mol. Phylogenet. Evol. 54:136–149.
Rice, D. W., Palmer, J. D. 2006. An exceptional horizontal gene transfer in plastids: gene replacement by a distant bacterial paralog and evidence that haptophyte and cryptophyte plastids are sisters. BMC Biol. 4:31.
Richardson, A. O., Palmer, J. D. 2006. Horizontal gene transfer in plants. J. Exp. Bot. 58:1–9.
Robbens, S., Derelle, E., Ferraz, C., Wuyts, J., Moreau, H., van de Peer, Y. 2007. The complete chloroplast and mitochondrial DNA Sequence of *Ostreococcus tauri*: organelle genomes of the smallest eukaryote are examples of compaction. Mol. Biol. Evol. 24:956–968.
Rüdinger, M., Polsakiewicz, M., Knoop, V. 2008. Organellar RNA editing and plant-specific extensions of pentatricopeptide repeat (PPR) proteins in jungermanniid but not in marchantiid liverworts. Mol. Biol. Evol. 25:1405–1414.
Rüdinger, M., Funk, H. T., Rensing, S. A., Maier, U. G., Knoop, V. 2009. RNA editing: 11 sites only in the *Physcomitrella patens* mitochondrial transcriptome and a universal nomenclature proposal. Mol. Genet. Genomics 281:473–481.
Salone, V., Rüdinger, M., Polsakiewicz, M., Hoffmann, B., Groth-Malonek, M., Szurek, B., Small, I., Knoop, V., Lurin, C. 2007. A hypothesis on the identification of the editing enzyme in plant organelles. FEBS Lett. 581:4132–4138.

Sanchez, H., Fester, T., Kloska, S., Schroder, W., Schuster, W. 1996. Transfer of rps19 to the nucleus involves the gain of an RNP-binding motif which may functionally replace RPS13 in Arabidopsis mitochondria. EMBO J. 15:2138–2149.

Sanchez-Puerta, M. V., Cho, Y., Mower, J. P., Alverson, A. J., Palmer, J. D. 2008. Frequent, phylogenetically local horizontal transfer of the cox1 group I Intron in flowering plant mitochondria. Mol. Biol. Evol. 25:1762–1777.

Satoh, M., Kubo, T., Nishizawa, S., Estiati, A., Itchoda, N., Mikami, T. 2004. The cytoplasmic male-sterile type and normal type mitochondrial genomes of sugar beet share the same complement of genes of known function but differ in the content of expressed ORFs. Mol. Genet. Genomics 272:247–256.

Schmitz-Linneweber, C., Williams-Carrier, R. E., Williams-Voelker, P. M., Kroeger, T. S., Vichas, A., Barkan, A. 2006. A pentatricopeptide repeat protein facilitates the *trans*-splicing of the maize chloroplast *rps*12 pre-mRNA. Plant Cell 18:2650–2663.

Schuster, W., Brennicke, A. 1987. Plastid, nuclear and reverse transcriptase sequences in the mitochondrial genome of Oenothera: is genetic information transferred between organelles via RNA? EMBO J. 6:2857–2863.

Sethuraman, J., Majer, A., Friedrich, N. C., Edgell, D. R., Hausner, G. 2009. Genes within genes: multiple LAGLIDADG homing endonucleases target the ribosomal protein S3 gene encoded within an rnl group I intron of Ophiostoma and related taxa. Mol. Biol. Evol. 26:2299–2315.

Shedge, V., Arrieta-Montiel, M., Christensen, A. C., Mackenzie, S. A. 2007. Plant mitochondrial recombination surveillance requires unusual RecA and MutS homologs. Plant Cell 19:1251–1264.

Sloan, D. B., Oxelman, B., Rautenberg, A., Taylor, D. R. 2009. Phylogenetic analysis of mitochondrial substitution rate variation in the angiosperm tribe Sileneae. BMC Evol. Biol. 9:260.

Steinhauser, S., Beckert, S., Capesius, I., Malek, O., Knoop, V. 1999. Plant mitochondrial RNA editing: extreme in hornworts and dividing the liverworts? J. Mol. Evol. 48:303–312.

Stern, D. B., Lonsdale, D. M. 1982. Mitochondrial and chloroplast genomes of maize have a 12-kilobase DNA sequence in common. Nature 299:698–702.

Stupar, R. M., Lilly, J. W., Town, C. D., Cheng, Z., Kaul, S., Buell, C. R., Jiang, J. 2001. Complex mtDNA constitutes an approximate 620-kb insertion on *Arabidopsis thaliana* chromosome 2: implication of potential sequencing errors caused by large-unit repeats. Proc. Natl. Acad. Sci. U.S.A. 98:5099–5103.

Sugiyama, Y., Watase, Y., Nagase, M., Makita, N., Yagura, S., Hirai, A., Sugiura, M. 2005. The complete nucleotide sequence and multipartite organization of the tobacco mitochondrial genome: comparative analysis of mitochondrial genomes in higher plants. Mol. Genet. Genomics 272:603–615.

Terasawa, K., Odahara, M., Kabeya, Y., Kikugawa, T., Sekine, Y., Fujiwara, M., Sato, N. 2006. The mitochondrial genome of the moss *Physcomitrella patens* sheds new light on mitochondrial evolution in land plants. Mol. Biol. Evol. 24:699–709.

Timmis, J. N., Ayliffe, M. A., Huang, C. Y., Martin, W. 2004. Endosymbiotic gene transfer: organelle genomes forge eukaryotic chromosomes. Nat. Rev. Genet. 5:123–U16.

Tourasse, N. J., Kolstø, A. B. 2008. Survey of group I and group II introns in 29 sequenced genomes of the *Bacillus cereus* group: insights into their spread and evolution. Nucl. Acids Res. 36:4529–4548.

Turmel, M., Otis, C., Lemieux, C. 2002a. The chloroplast and mitochondrial genome sequences of the charophyte *Chaetosphaeridium globosum*: insights into the timing of the events that restructured organelle DNAs within the green algal lineage that led to land plants. Proc. Natl. Acad. Sci. U.S.A. 99:11275–11280.

Turmel, M., Otis, C., Lemieux, C. 2002b. The complete mitochondrial DNA sequence of *Mesostigma viride* identifies this green alga as the earliest green plant divergence and predicts a highly compact mitochondrial genome in the ancestor of all green plants. Mol. Biol. Evol. 19:24–38.

Turmel, M., Otis, C., Lemieux, C. 2003. The mitochondrial genome of *Chara vulgaris*: insights into the mitochondrial DNA architecture of the last common ancestor of green algae and land plants. Plant Cell 15:1888–1903.

Turmel, M., Otis, C., Lemieux, C. 2007. An unexpectedly large and loosely packed mitochondrial genome in the charophycean green alga *Chlorokybus atmophyticus*. BMC Genomics 8:137.

Ueda, M., Nishikawa, T., Fujimoto, M., Takanashi, H., Arimura, S., Tsutsumi, N., Kadowaki, K. 2008. Substitution of the gene for chloroplast RPS16 was assisted by generation of a dual targeting signal. Mol. Biol. Evol. 25:1566–1575.

Ullrich, H., Lättig, K., Brennicke, A., Knoop, V. 1997. Mitochondrial DNA variations and nuclear RFLPs reflect different genetic similarities among 23 *Arabidopsis thaliana* ecotypes. Plant Mol. Biol. 33:37–45.

Unseld, M., Marienfeld, J. R., Brandt, P., Brennicke, A. 1997. The mitochondrial genome of *Arabidopsis thaliana* contains 57 genes in 366,924 nucleotides. Nat. Genet. 15:57–61.

Vangerow, S., Teerkorn, T., Knoop, V. 1999. Phylogenetic information in the mitochondrial *nad5* gene of pteridophytes: RNA editing and intron sequences. Plant Biol. 1:235–243.

Vaughn, J. C., Mason, M. T., Sper-Whitis, G. L., Kuhlman, P., Palmer, J. D. 1995. Fungal origin by horizontal transfer of a plant mitochondrial group I intron in the chimeric CoxI gene of Peperomia. J. Mol. Evol. 41:563–572.

Volkmar, U., Knoop, V. 2010. Introducing intron locus cox1i624 for phylogenetic analyses in bryophytes: on the issue of *Takakia* as sister genus to all other extant mosses. J. Mol. Evol. 70(5):506–518.

Wahrmund, U., Rein, T., Müller, K. F., Groth-Malonek, M., Knoop, V. 2009. Fifty mosses on five trees: comparing phylogenetic information in three types of non-coding mitochondrial DNA and two chloroplast loci. Plant Syst. Evol. 282:241–255.

Wahrmund, U., Quandt, D., Knoop, V. 2010. The phylogeny of mosses – addressing open issues with a new mitochondrial locus: group I intron cobi420. Mol. Phylogenet. Evol. 54:417–426.

Wang, X., Lavrov, D. V. 2007. Mitochondrial genome of the homoscleromorph *Oscarella carmela* (Porifera, Demospongiae) reveals unexpected complexity in the common ancestor of sponges and other animals. Mol. Biol. Evol. 24:363–373.

Wang, D., Wu, Y. W., Shih, A. C., Wu, C. S., Wang, Y. N., Chaw, S. M. 2007. Transfer of chloroplast genomic DNA to mitochondrial genome occurred at least 300 MYA. Mol. Biol. Evol. 24:2040–2048.

Wang, B., Xue, J., Li, L., Liu, Y., Qiu, Y. L. 2009. The complete mitochondrial genome sequence of the liverwort *Pleurozia purpurea* reveals extremely conservative mitochondrial genome evolution in liverworts. Curr. Genet. 55:601–609.

Ward, B. L., Anderson, R. S., Bendich, A. J. 1981. The mitochondrial genome is large and variable in a family of plants (Cucurbitaceae). Cell 25:793–803.

Watkins, K. P., Kroeger, T. S., Cooke, A. M., Williams-Carrier, R. E., Friso, G., Belcher, S. E., van Wijk, K. J., Barkan, A. 2007. A ribonuclease III domain protein functions in group II intron splicing in maize chloroplasts. Plant Cell 19:2606–2623.

Wikström, N., Pryer, K. M. 2005. Incongruence between primary sequence data and the distribution of a mitochondrial atp1 group II intron among ferns and horsetails. Mol. Phylogenet. Evol. 36:484–493.

Wischmann, C., Schuster, W. 1995. Transfer of rps10 from the mitochondrion to the nucleus in Arabidopsis thaliana: evidence for RNA-mediated transfer and exon shuffling at the integration site. FEBS Lett. 374:152–156.

Wissinger, B., Schuster, W., Brennicke, A. 1991. *Trans* splicing in Oenothera mitochondria: *nad1* mRNAs are edited in exon and *trans*-splicing group II intron sequences. Cell 65:473–482.

Wolf, P. G., Rowe, C. A., Hasebe, M. 2004. High levels of RNA editing in a vascular plant chloroplast genome: analysis of transcripts from the fern Adiantum capillus-veneris. Gene 339:89–97.

Woloszynska, M. 2010. Heteroplasmy and stoichiometric complexity of plant mitochondrial genomes – though this be madness, yet there's method in't. J. Exp. Bot. 61:657–671.

Won, H., Renner, S. S. 2003. Horizontal gene transfer from flowering plants to *Gnetum*. Proc. Natl. Acad. Sci. U.S.A. 100:10824–10829.

Xue, J. Y., Liu, Y., Li, L., Wang, B., Qiu, Y. L. 2010. The complete mitochondrial genome sequence of the hornwort *Phaeoceros laevis*: retention of many ancient pseudogenes and conservative evolution of mitochondrial genomes in hornworts. Curr. Genet. 56:53–61.
Zehrmann, A., Verbitskiy, D., van der Merwe, J. A., Brennicke, A., Takenaka, M. 2009. A DYW domain-containing pentatricopeptide repeat protein is required for RNA editing at multiple sites in mitochondria of *Arabidopsis thaliana*. Plant Cell 21:558–567.

Chapter 2
Mitochondrial Dynamics

Iain Scott and David C. Logan

Abstract Mitochondria cannot be created de novo, meaning that any new mitochondrion must be formed from the division of an existing organelle. In addition to division, mitochondria also undergo fusion where two or more individual organelles join to produce a single mitochondrion. Mitochondrial division and fusion are the primary processes controlling mitochondrial form and together control mitochondrial size and number. Traditionally, the mitochondrion has been portrayed as an immobile, oval-shaped body. In reality, mitochondria are very dynamic organelles, capable of changing size and shape in a matter of seconds. Additionally, they undergo short- and long-distance vectorial transport mediated by association with the cytoskeleton. Advances in bioimaging have allowed scientists to reevaluate the behavior of mitochondria in vivo, stimulating a surge of interest in determining the genes, proteins, and mechanisms that control mitochondrial shape, size, number, and distribution (collectively termed *mitochondrial dynamics*).

Keywords Dynamics • Cytoskeleton • Actin • Microtubules • Fission • Fusion • Motility • Inheritance • Cell death • Morphology

2.1 Introduction

Mitochondrial dynamics is the study of the dynamic behavior of mitochondria in living cells. This behavior includes mitochondrial division and fusion: the two events controlling mitochondrial shape, size, and number, in addition to changes in the morphology of individual organelles, their motility in the cytosol, and their inheritance during cell division. The dynamic nature of mitochondria was evident in some of the earliest published studies of mitochondria in living cells (Cavers 1914) and in

D.C. Logan (✉)
Department of Biology, University of Saskatchewan, WP Thompson Building, 112 Science Place, Saskatoon, S7N 5E2, Canada
e-mail: david.logan@usask.ca

F. Kempken (ed.), *Plant Mitochondria*, Advances in Plant Biology 1,
DOI 10.1007/978-0-387-89781-3_2, © Springer Science+Business Media, LLC 2011

recent years, mainly thanks to the development of green fluorescent protein (GFP) as an in vivo reporter (Chalfie et al. 1994), we have been able to study these dynamic processes in more detail (Rizzuto et al. 1995; Kohler et al. 1997; Sesaki and Jensen 1999; Logan and Leaver 2000).

Visualization of mitochondria in living cells has highlighted some fundamental steady-state differences in chondriome[1] organization and structure in different organisms, and cell types within the same organism. The chondriome of *Saccharomyces cerevisiae* is typically organized as a network of interconnected tubules (Fig. 2.1a), although the structure is partly dependent on growth medium: ramification of the chondriome increases when cells are grown on nonfermentable media, and the volume of the mitochondrial compartment increases (Egner et al. 2002). In animal

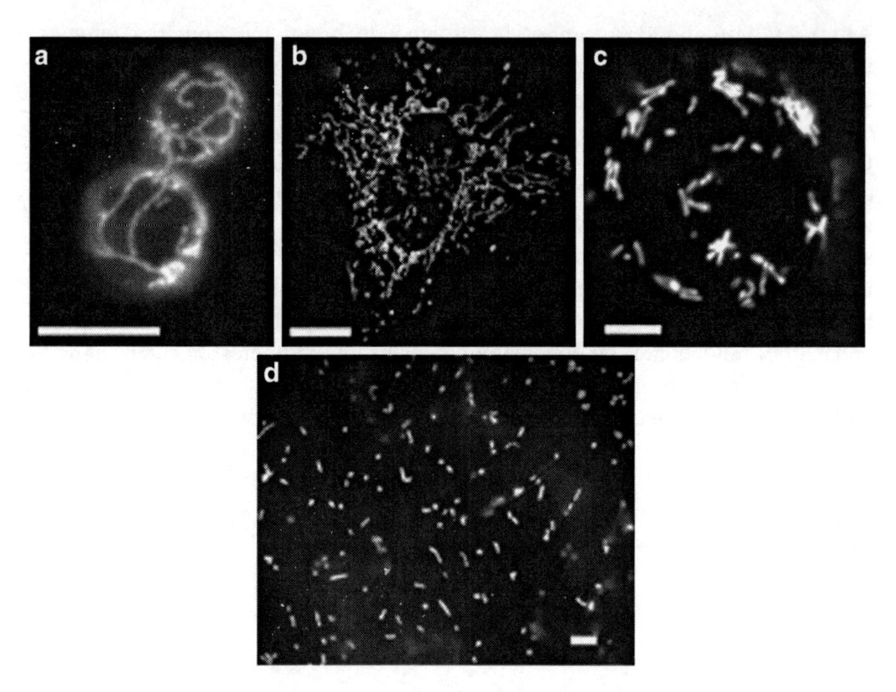

Fig. 2.1 Chondriome structure in yeast, human HeLa cell line, and plants. (**a**) *S. cerevisiae* chondriome labeled with mitochondria-targeted GFP expressed from plasmid pYES-mtGFP (Westermann and Neupert 2000). The cell has been grown in rich medium containing galactose as a carbon source (image by Bastian Seubert and reproduced courtesy of Benedikt Westermann, Universitaet Bayreuth, Germany). (**b**) HeLa cell chondriome labeled with anti-mouse cytochrome c and Invitrogen Alexa 594 goat anti-mouse antibody (image courtesy of Megan Cleland from the laboratory of Richard J. Youle, NINDS, NIH). (**c**) Arabidopsis mesophyll protoplast expressing mitochondria-targeted GFP (Logan and Leaver 2000; image by Sanjaya Ekanayake, Logan Lab., University of Saskatchewan). (**d**) Leaf epidermis of *Arabidopsis* plant expressing mitochondrial-targeted GFP (Logan and Leaver 2000; image by DC Logan). Scale bar in all images = 5 μm

[1] Here, the term chondriome refers to all the mitochondria in a cell, collectively. However, the term is sometimes used to describe the mitochondrial genome. In this book the mitochondrial genome is referred to as the *chondrome*.

cells, although chondriome structure is highly variable depending on animal and cell type, mitochondria are generally more tubular and ramified (Fig. 2.1b) than is typical in higher plants where the chondriome is typically organized as a population of several hundred physically discrete organelles (Logan 2006). Discovery of the genetic and physiological reasons for differences in chondriome structure between the different eukaryotic lineages, and between different cell types of the same organism, are key aims of researchers working on mitochondrial dynamics. Molecular cell biological studies, taking advantage of the tractability of model organisms and their sequenced genomes, have enabled the identification of numerous genes involved in mitochondrial dynamics (Table 2.1). In turn, gene identification has enabled studies aimed at identifying the mechanisms controlling mitochondrial dynamics, and we are now increasing our understanding of how such fundamental cell biological processes are involved in cell and organism physiology, and how mitochondrial dynamics underpins organism health and development. The aim of this chapter is to summarize our knowledge of mitochondrial dynamics in eukaryotes in general, and to relate this to the situation in plants in particular.

2.2 Division

2.2.1 Animal and Yeast Mitochondrial Division

Mitochondrial division is a highly coordinated process that involves the formation of two or more daughter organelles by scission of a single mitochondrion. To an extent the mechanisms that carry out mitochondrial fission reflect the evolutionary history of the organelle. Mitochondria are descended from α-proteobacteria (Zimmer 2009; see also Chap. 1), and studies of extant species, such as *Escherichia coli*, show that cytokinesis is carried out by proteins such as filamentous temperature-sensitive Z (FtsZ), a bacterial tubulin homologue (Bi and Lutkenhaus 1991). FtsZ localizes to the inner surface of the cell membrane at division sites, where it forms a ring structure (Z ring) that enables constriction and scission of the parent into two daughter cells. FtsZ is a GTPase, which can hydrolyze guanosine triphosphate to provide a source of energy. However, it is thought that FtsZ may act as a scaffold for other proteins to effect the final scission, rather than providing the mechanical forces required (Dajkovic et al. 2008). As mitochondria are direct descendents of proteobacteria, it would be logical to assume that they may have retained this system for their own division. Indeed lower eukaryotes, such as the algae *Mallomonas splendens* and *Cyanidioschyzon merolae*, and the slime mould *Dictyostelium discoideum*, use FtsZ for this purpose (Beech et al. 2000; Takahara et al. 2000; Gilson et al. 2003). However, FtsZ has been lost from the mitochondrial division apparatus of most eukaryotes, and in higher plants the protein is primarily involved in plastid division (Osteryoung and Vierling 1995; Stokes et al. 2000). Instead, the eukaryotic cell appears to have evolved a mitochondrial fission system that shares many features with the scission of membrane vesicles during endocytosis. Scission of the plasma

Table 2.1 Genes involved in mitochondrial dynamics

Gene	Organism	Location	Mutant phenotype	Protein properties/role	Reference	Arabidopsis homologue
BIGYIN1 (BGY1)/ FISSION1A	*A. thaliana*	Mitochondrial outer membrane (?); peroxisome (?)	Reduced number of large, spherical mitochondria, peroxisomes	Interacts with dynamin in division (?).	Scott et al. (2006), Lingard et al. (2008), Zhang and Hu (2008), and Fujimoto et al. (2009)	–
BIGYIN2 (BGY2)/ FISSION1B	*A. thaliana*	Mitochondrial outer membrane (?); peroxisome (?)	Reduced number of large, spherical mitochondria, peroxisomes	Interacts with dynamin in division (?).	Scott et al. (2006), Lingard et al. (2008), Zhang and Hu (2008), and Fujimoto et al. (2009)	–
BMT1	*A. thaliana*	?	Fewer but larger mitochondria per cell.	?	Logan et al. (2003)	–
BMT2	*A. thaliana*	?	Fewer but larger mitochondria per cell.	?	Logan (unpublished)	–
cluA/CLU1	*D. discoideum/ S. cerevisiae*	Cytoplasm	Mitochondria aggregates. In yeast they collapse to side of cell	Kinesin-like domain/ outer membrane fission.	Zhu et al. (1997) and Fields et al. (1998, 2002)	*FMT*, see below.
DNM1/Drp1/ DRP-1	*S. cerevisiae/ H. sapiens*	Cytoplasm	Net-like sheet of interconnected tubules	Dynamin-related GTPase/membrane fission	Otsuga et al. (1998) and Smirnova et al. (2001)	*DRP3A & DRP3B* orthologues.

DRP3A (formerly *ADL2a*)	*A. thaliana*	Cytoplasm	Mitochondria form long tubules with many constrictions and protuberances (matrixules)	Dynamin-related GTPase/membrane fission (?)	Arimura et al. (2004) and Logan et al. (2004)	–
DRP3B (formerly *ADL2b*)	*A. thaliana*	Cytoplasm	Mitochondria form long interconnected tubules	Dynamin-related GTPase/membrane fission (?)	Arimura and Tsutsumi (2002)	–
FIS1/MDV2/ hFIS1	*S. cerevisiae/ H. sapiens*	Mitochondrial outer membrane	Net-like sheet of interconnected tubules	Integral membrane protein. Interacts with dynamin & Mdv1p (*S. cerevisiae* only) in fission process.	Mozdy et al. (2000), Tieu and Nunnari (2000), Tieu et al. (2002), Suzuki et al. (2003, 2005), and Yoon et al. (2003)	At3g57090 (*BIGYIN1*), & At5g12390 (*BIGYIN2*) orthologues.
fzo/FZO1/Mfn1/ Mfn2	*Drosophila melanogaster/ S. cerevisiae/ H. sapiens*	Mitochondrial outer membrane	Aberrant mitochondrial fusion; fragmentation of tubules.	GTPase/mitochondrial fusion.	Hales and Fuller (1997), Hermann et al. (1998), Rapaport et al. (1998), Santel and Fuller (2001), and Rojo et al. (2002)	None
KIF5B	*Mus musculus*	Cytoplasm, mitochondrial associated	Mitochondria collapse around nucleus	Kinesin heavy chain protein, attachment to cytoskeleton	Tanaka et al. (1998)	Similarity to At3g63480.
MDM1	*S. cerevisiae*	Cytoplasm	Fragmentation of tubules. Defective transmission to daughter buds.	Intermediate filament-like	McConnell and Yaffe (1992)	Similarity to At2g15900.

(continued)

Table 2.1 (continued)

Gene	Organism	Location	Mutant phenotype	Protein properties/role	Reference	Arabidopsis homologue
MDM10	*S. cerevisiae/ Podospora anserina*	Mitochondrial outer membrane	Large spherical mitochondria. Defective transmission to daughter buds.	Integral membrane protein	Sogo and Yaffe (1994)	None
MDM12	*S. cerevisiae*	Mitochondrial outer membrane	Large spherical mitochondria. Defective transmission to daughter buds.	Integral membrane protein	Berger et al. (1997)	None
MDM14	*S. cerevisiae*	Cytoplasm	Mitochondria aggregate. Defective transmission to daughter buds.	Coiled-coil domain	Shepard and Yaffe (1997)	None
MDM20	*S. cerevisiae*	Cytoplasm	Defective transmission to daughter buds.	Coiled-coil domain/ disrupts actin cables.	Hermann et al. (1997)	None
MDM30	*S. cerevisiae*	Cytosolic?	Fragmented or aggregated; few short tubules.	?	Dimmer et al. (2002)	None
MDM31	*S. cerevisiae*	Predicted mitochondrial inner membrane protein	Compact mitochondrial aggregates	?	Dimmer et al. (2002)	None
MDM32	*S. cerevisiae*	Predicted mitochondrial inner membrane protein	Compact mitochondrial aggregates	?	Dimmer et al. (2002)	None

MDM33	*S. cerevisiae*	Mitochondrial inner membrane protein	Giant ring-like mitochondria	Coiled-coil domains, part of high molecular weight complex; putatively involved in inner membrane fission	Dimmer et al. (2002), Messerschmitt et al. (2003)	Similarity to At3g53350.
MDM34	*S. cerevisiae*	Cytoplasmic?	Spherical mitochondria	?	Dimmer et al. (2002)	None
MDM35	*S. cerevisiae*	Cytoplasmic?	Spherical mitochondria	?	Dimmer et al. (2002)	Similarity to At4g33100.
MDM36	*S. cerevisiae*	Cytoplasmic?	Mitochondrial tubules aggregate/collapse to one side of cell	?	Dimmer et al. (2002)	None
MDM37	*S. cerevisiae*	Predicted mitochondrial inner membrane protein	Fragmented mitochondrial tubules	Rhomboid-like protein	Dimmer et al. (2002)	Similarity to At1g18600.
MDM38	*S. cerevisiae*	Predicted mitochondrial inner membrane protein	Lasso-like mitochondria	Calcium-binding protein	Dimmer et al. (2002)	Similarity to At3g59820.
MDM39	*S. cerevisiae*	Predicted integral membrane protein – no mitochondrial targeting presequence	Fragmented mitochondrial tubules	?	Dimmer et al. (2002)	None
MDV1/FIS2/ GAG3 NET2	*S. cerevisiae*	Cytoplasm/ associates with mitochondrial outer membrane	Netlike sheet of interconnected tubules	Predicted coiled-coil and seven WD-40 repeats/interacts with Dnm1p and Fis1p in fission process.	Fekkes et al. (2000), Mozdy et al. (2000), Tieu and Nunnari (2000), Cerveny et al. (2001), Tieu et al. (2002)	Similarity to WD-40-repeat containing proteins.

(continued)

Table 2.1 (continued)

Gene	Organism	Location	Mutant phenotype	Protein properties/role	Reference	Arabidopsis homologue
Mff	*H. sapiens*	Mitochondrial outer membrane, C-terminal anchor	Elongated mitochondrial tubules	?	Gandre-Babbe and van der Bliek (2008)	None
MGM1/Opa1	*S. cerevisiae/ H. sapiens*	Inner membrane space	Mitochondria fragmentation. Loss of mtDNA. Defective transmission to daughter buds.	Dynamin-related protein/inner membrane modeling, may interact with Ugo1p and Fzo1p.	Jones and Fangman (1992), Guan et al. (1993), Shepard and Yaffe (1999), Alexander et al. (2000), Wong et al. (2000), Delettre et al. (2001), Misaka et al. (2002), and Wong et al. (2003)	Dynamin-like genes in Arabidopsis. Most similar is DRP3A, At4g33650.
MMM1	*S. cerevisiae*	Mitochondrial outer membrane	Large spherical mitochondria. Defective transmission to daughter buds.	Integral membrane protein	Burgess et al. (1994)	None
MMT1	*A. thaliana*	?	Much larger and smaller mitochondria in same cell. Altered thylakoid morphology.	?	Logan et al. (2003)	–
MMT2	*A. thaliana*	?	Much larger and smaller mitochondria in same cell. Giant chloroplasts.	?	Logan et al. (2003)	–

NETWORK1 (ELM1)	*A. thaliana*	Mitochondrial outer membrane	Mitochondria form long tubules and sometimes an interconnected network.	?	Logan et al. (2003) and Arimura et al. (2008)	–
PCP1/RBD1/ hPARL	*S. cerevisiae/ H. sapiens*	Mitochondrial inner membrane	Fragmented mitochondria, aggregated.	Rhomboid-like protein, processing of Mgm1p	Herlan et al. (2003) McQuibban et al. (2003)	Rhomboid domain proteins. Most similar is At1g18600.
UGO1	*S. cerevisiae*	Mitochondrial outer membrane	Fragmentation of mitochondrial tubules.	Involved in mitochondrial fusion.	Sesaki and Jensen (2001)	None
CAF4	*S. cerevisiae*	Cytosolic, associates with mitochondrial outer membrane	No change, partially rescues *mdv1p* mutants	WD-40 repeat protein, mediates Fis1p/ Dnm1p interactions	Griffin et al. (2005)	Similarity to WD-40 repeat proteins
FMT	*A. thaliana*	?	Clusters of mitochondria	TPR-domain; homologue of *cluA* and *CLU1*	Logan et al. (2003)	–
MsftsZ-mt/ CmftsZ1-1/ CmftsZ1-2/ fszA/fszB	*Mallomonas splendens/ Cyanidioschyzon Merolae/D. discoideum*	Mitochondrial division sites and tips.	Elongated tubular mitochondria (*D. discoideum*)	Tubulin-like, ring-forming protein, organelle constriction.	Beech et al. (2000), Takahara et al. (2000), and Gilson et al. (2003)	Chloroplast FtsZ division genes: At5g55280 and At2g36250
YME1L1/YME1	*H. sapiens/ S. cerevisiae*	Mitochondrial inner membrane space	Fragmented mitochondria	AAA-type protease; processing of OPA1	Griparic et al. (2007) and Song et al. (2007)	FtsH proteases, e.g., At2g26140
Paraplegin	*H. sapiens*	Mitochondrial inner membrane	Fragmented mitochondria	AAA-type protease; processing of OPA1	Cipolat et al. (2004, 2006)	FtsH proteases, e.g., At1g07510

membrane during endocytosis is carried out by dynamin, a mechanical enzyme that forms tight spirals around the neck of vesicles to constrict and then cleave them off (Doherty and McMahon 2009). Many of the proteins involved in mitochondrial fission and fusion are members of the dynamin GTPase protein family (see Table 2.1), and they exhibit a similar mode of action to the endocytic form. The dynamin homologues involved in mitochondrial fission, called Dnm1p in yeast (*S. cerevisiae*) and Drp1 in mammals (Fig. 2.2) are predominantly found in the cytosol. During fission events they cycle to the outer mitochondrial membrane at scission site (Labrousse et al. 1999; Smirnova et al. 2001), where they form large homomultimeric complexes that encircle the mitochondrion in spirals, which finally constrict in an energy-dependent manner until membrane fission occurs. Studies of Dnm1p in vitro show that the spirals match the minimal diameter of constricted mitochondria in yeast during fission, indicating that Dnm1p is tailored to the process of mitochondrial division (Ingerman et al. 2005). Mutations in both Dnm1p and Drp1 block mitochondrial fission (Otsuga et al. 1998; Smirnova et al. 2001),

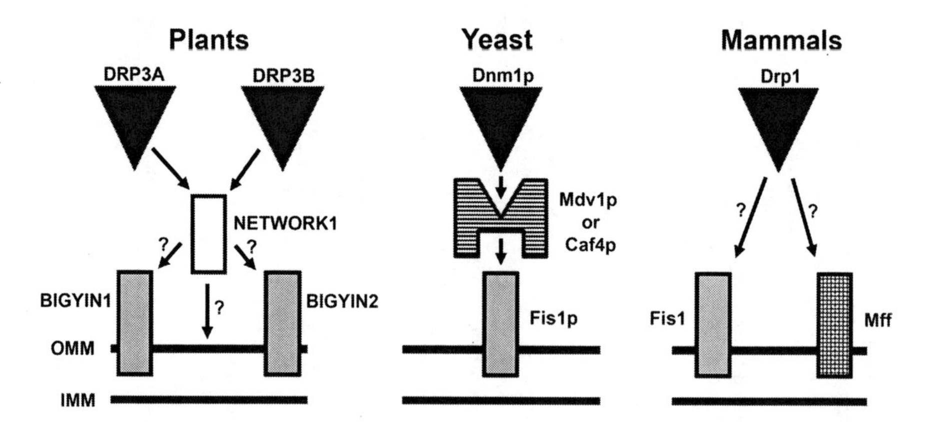

Fig. 2.2 Mitochondrial division machinery. The mitochondrial division machineries of plants, yeast, and animals are composed of multiple proteins, many of which are common to the three groups. The main constriction and scission forces are provided by homologues of dynamin GTPases, called Dnm1p in yeast and Drp1 in mammals, while plants have two homologues, DRP3A and DRP3B. These proteins are directed to the mitochondrial membrane at scission sites by a number of adapters. In yeast, Dnm1p interacts with Fis1p, an outer mitochondrial membrane protein anchored to the organelle by a C-terminal transmembrane domain. This interaction is mediated by a second set of yeast-specific cytosolic adapters, called Mdv1p and Caf4p, which function semi-redundantly. In mammals, Drp1 may interact directly with the homologous hFis1 protein on the outer mitochondrial membrane, or Drp1 may be directed to mitochondria by a mammalian-specific adapter, such as Mff. In plants, the two DRP proteins may, like in mammals, interact directly with the Fis1-type proteins BIGYIN1 (also known as FISSION1A) or BIGYIN2 (FISSION1B), or they may use the plant-specific adapter protein NETWORK1 (ELM1) in a manner analogous to the situation in yeast. Alternatively, DRP3A or DRP3B may be recruited directly to the outer mitochondrial membrane by the mitochondrial-localized NETWORK1. Once localized to the outer mitochondrial membrane by these adapters, the dynamin homologues encircle the organelle, and proceed to constrict and sever the mitochondrion

upsetting the balance between fission and fusion that controls overall mitochondrial morphology. This leads to the formation of networks of highly interconnected mitochondrial tubules, as mitochondrial fusion continues as normal.

Dynamin is recruited to the outer mitochondrial membrane by a series of adapter proteins, which were initially identified by genetic screens in yeast. Fis1p is an outer mitochondrial membrane protein, which is anchored to the organelle by a C-terminal transmembrane domain (Mozdy et al. 2000). The remainder of the protein faces the cytosol, where it interacts with dynamin via a tetratricopeptide repeat (TPR) domain to initiate fission. In yeast, this interaction is mediated by a second set of partially redundant adapters, Mdv1p and Caf4p, which can bind both Fis1p and Dnm1p to allow their association (Tieu and Nunnari 2000; Cerveny et al. 2001; Griffin et al. 2005). Loss of either Fis1p, or Mdv1p, and to a lesser extent Caf4p, prevents Dnm1p from associating with mitochondria, and hence blocks mitochondrial division. In mammals there are no homologues of Mdv1p or Caf4p, therefore the association between Fis1p and Drp1 at the outer membrane may be more direct (Yoon et al. 2003). Knockdown of Fis1 protein levels in mammalian cells leads to an increase in the interconnectivity of mitochondrial tubules, similar to those seen in Drp1 mutants, whereas overexpression of Fis1 has been shown to increase mitochondrial fragmentation (Yoon et al. 2003). Recently a second outer membrane protein involved in division, Mff, has been described (Gandre-Babbe and van der Bliek 2008). Like Fis1, it is anchored by a C-terminal transmembrane domain and faces the cytosol. Knockdown of Mff leads to an increase in elongated mitochondria, indicating that it may function in a similar manner to Fis1. However no direct interaction between Mff and Drp1 has been shown, and Mff and Fis1 appear to reside in different protein complexes, leaving its actual role in mitochondrial division open to speculation (Gandre-Babbe and van der Bliek 2008). In addition to their role in mitochondrial fission, all three mammalian proteins (Drp1, Fis1, and Mff) localize to peroxisomes and regulate their size (Koch et al. 2003a, b; Koch et al. 2005; Gandre-Babbe and van der Bliek 2008). How these proteins are dual-targeted to different organelles, and how this process is regulated, are not currently understood.

2.2.2 *Plant Mitochondrial Division*

The first proteins shown to be involved in plant mitochondrial division were identified through *in silico* homology searches, using mammalian and yeast fission proteins as bait. There are roughly 16 dynamin homologues in the Arabidopsis genome; and two of these, DRP3A and DRP3B, have been implicated in mitochondrial division (Arimura and Tsutsumi 2002; Arimura et al. 2004a, b, c; Logan et al. 2004; Mano et al. 2004). Both DRP3A and DRP3B share 37–41% similarity at the amino acid level with the dynamins involved in animal and yeast mitochondrial division and, like these proteins, localize to constriction sites of dividing mitochondria (Arimura and Tsutsumi 2002; Arimura et al. 2004a, b, c). Disrupting the function of DRP3A or DRP3B by genetic knockout, or overexpression of constructs with a dominant-negative mutation in the GTPase domain, leads to an increase in the number of large, elongated mitochondria, indicating that these proteins are part of the division apparatus (Arimura and Tsutsumi 2002; Arimura et al. 2004a, b, c; Logan et al. 2004). The increased complexity of the plant fission system, in terms of number of homologous proteins involved, is replicated when looking at Fis1-type proteins. In Arabidopsis, there are two orthologues of Fis1p, named BIGYIN1 (also known as FISSION1A) and BIGYIN2 (FISSION1B) (Scott et al. 2006). Both of these proteins localize to the outer mitochondrial membrane (among other locations), and disruption of either leads to a decrease in mitochondrial division, as evidenced by an increase in organelle size and decrease in organelle number per cell (Scott et al. 2006; Fujimoto et al. 2009). As has been found in animals, the proteins involved in mitochondrial division appear to also have a role in peroxisome division. All four proteins (DRP3A, DRP3B, BIGYIN1, and BIGYIN2) have been shown to localize to peroxisomes and, to a greater or lesser extent, affect the size and number of these organelles (Lingard et al. 2008; Fujimoto et al. 2009). Unlike yeast, but in common with animals, there are no sequence homologues of either of the adapter proteins Mdv1p, or Caf4p, in the Arabidopsis genome. As such, it remains unclear whether there is any direct, or indirect, interaction between Drp1- and Fis1-type proteins in the plant division apparatus. However, recent research has identified another plant-specific putative adapter protein involved in mitochondrial fission (Box 2.1).

Box 2.1 The NETWORK protein

Initially identified as NETWORK (now named NETWORK1 – a paralogue was evident following identification of the *NETWORK* gene) in a screen for novel plant mitochondrial morphology proteins (Logan et al. 2003), ELM1 was shown to localize to the outer mitochondrial membrane (Arimura et al. 2008). Plants with mutations in NETWORK1/ELM1 exhibit an elongated mitochondrial phenotype, suggesting a fault in the division process. To further examine the role of NETWORK1/ELM1, Arimura et al. (2008) investigated whether NETWORK1/ELM1 interacts with DRP3A, DRP3B, or BIGYIN1 in yeast two-hybrid assays. These experiments failed to show interaction between BIGYIN1 and NETWORK1, although weak interaction between DRP3A/B and NETWORK1 was reported (interaction between DRP3B and NETWORK1 was greater than between DRP3A and NETWORK1; Arimura et al. 2008). These results suggest that NETWORK1 may act to direct dynamin to mitochondrial fission sites – a hypothesis that was strengthened by the observation that in NETWORK1/ELM1 mutants, DRP3A does not localize to mitochondria (Arimura et al. 2008). Further work is required to see if NETWORK1/ELM1 acts as a functional homologue of Mdv1p/Caf4p, i.e., linking DRP3A and DRP3B to BIGYIN1 (and perhaps BIGYIN2); or whether it acts independently of Fis1-type proteins. Importantly, peroxisome morphology in NETWORK1/ELM1 mutants appears to be unchanged (Arimura et al. 2008), which suggests that this protein may provide selectivity of function for the dual-targeted DRP3A/B division proteins.

2.3 Fusion

2.3.1 Animal and Yeast Mitochondrial Fusion

The counterpoint to mitochondrial division is fusion (Fig. 2.3), where two or more individual organelles join together to create a single, larger mitochondrion. Fusion is a complex process, requiring the tethering of adjacent organelles and the coordinated joining of two independent membranes (the inner and outer mitochondrial membranes), without the significant loss of any mitochondrial proteins (e.g., cytochrome *c*) that could damage or kill the cell. The first genetic component of the mitochondrial fusion machinery was discovered in studies of *Drosophila* spermatogenesis where, following meiosis, the entire mitochondrial population of the cell fuses into two organelles called a Nebenkern. In flies harboring a mutation in an outer mitochondrial membrane GTPase, dubbed *fuzzy onions* (*Fzo*), the Nebenkern structure is altered due to a lack of organelle fusion (Hales and Fuller 1997). Fuzzy onions, so called because the Nebenkern normally resembles an onion slice in cross-section, has homologues in mammals (Mfn1 and Mfn2) and yeast (Fzo1p), which perform a similar function in controlling outer membrane fusion (Hermann et al. 1998; Santel and Fuller 2001). Fzo-like proteins are anchored to the outer mitochondrial membrane by two transmembrane domains, and have a GTPase domain and twin coiled-coil regions facing the cytosol. The coiled-coil domains mediate homotypic binding between Fzo-like proteins on neighboring mitochondria, and the tethering of organelles in this fashion is required to initiate outer membrane fusion (Koshiba et al. 2004). Loss of Fzo-like proteins, or mutations in either the interaction or GTPase domains, leads to an increase in the number of small organelles caused by a lack of mitochondrial fusion (Hermann et al. 1998; Santel and Fuller 2001).

Inner membrane fusion is controlled by a final dynamin-like GTPase, called Mgm1p in yeast, and OPA1 in mammals, which is found in the intermembrane space, or associated with the inner membrane (Wong et al. 2000; Olichon et al. 2002). There are eight variants of OPA1 in human cells, formed as the result of differential transcript splicing from a single gene, and further processing by mitochondrial proteases (Delettre et al. 2001). A similar process occurs in the yeast homologue Mgm1p, and a combination of both the long and short isoforms of either OPA1 or Mgm1p is required to allow inner membrane fusion. A number of proteins have been implicated in mammalian OPA1 processing, including Yme1L1, a AAA-metalloprotease found in the intermembrane space (Griparic et al. 2007; Song et al. 2007); paraplegin, another AAA protease located in the inner mitochondrial membrane (Ishihara et al. 2006); and PARL, a rhomboid protease, whose yeast homologue Rbd1p/Pcp1p performs the same function on Mgm1p (Herlan et al. 2003; McQuibban et al. 2003; Sesaki et al. 2003; Cipolat et al. 2006).

Analysis of the yeast components in vitro have shown that the fusion of both the inner and outer membranes are mechanically distinct (Meeusen et al. 2004).

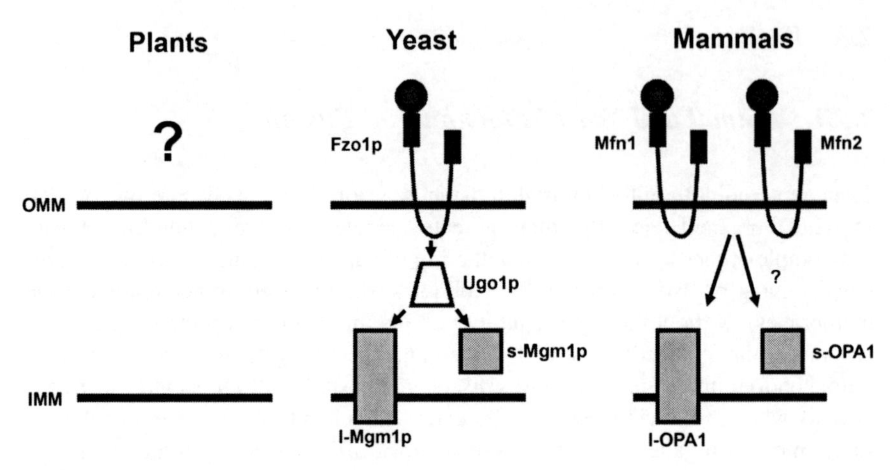

Fig. 2.3 Mitochondrial fusion machinery. The proteins involved in mitochondrial fusion in yeast and mammals are closely related, and appear to act in a similar manner. The main protagonists are dynamin-like GTPases, which provide the energy required for this process. On the outer mitochondrial membrane, homologues of the *Drosophila FUZZY ONIONS* (*FZO*) protein, called Fzo1p in yeast and Mitofusin1 (Mfn1) and Mitofusin2 (Mfn2) in mammals, tether neighboring organelles through homotypic coiled-coil binding domains on opposing proteins (e.g., a Fzo1p molecule on one mitochondrion will bind with a Fzo1p molecule on a second mitochondrion, allowing them to become tethered in the first stage of fusion). On the inner mitochondrial membrane a second GTPase, called Mgm1p in yeast and OPA1 in mammals, provides the energy for inner membrane fusion. These proteins are found in two isoforms (long and short), and the presence of both is required to allow fusion. While the interaction between membranes may be direct in mammals, particularly between Mfn1 and OPA1, this interaction is mediated in yeast by an adapter protein, Ugo1p. While it is known that mitochondrial fusion exists in plants, the proteins involved in this process are currently unknown

However there is an interaction between the outer membrane GTPase Fzo1p and the inner membrane Mgm1p, which is mediated through an adaptor protein, Ugo1p (Sesaki and Jensen 2001). While there may be a direct interaction in human cells with the homologous proteins, particularly Mfn1 and OPA1, there is no human homologue of Ugo1p. The fusion of both membranes requires GTP hydrolysis to provide energy, however, the inner membrane also requires the maintenance of the electrochemical membrane potential (Meeusen et al. 2004). This may be because the homologous human inner membrane GTPase, OPA1, undergoes increased proteolytic processing when membrane potential is lost (Ishihara et al. 2006), leaving only short isoforms that are incapable of mediating fusion alone.

2.3.2 Plant Mitochondrial Fusion

As noted above, homology searches led to the discovery of several plant mitochondrial division proteins. However, the same method has failed to yield any genes involved in plant mitochondrial fusion. There appears to be only weak sequence similarity between Arabidopsis proteins and either OPA1/Mgm1p, or Ugo1p (see Table 2.1), while the closest Fzo/Mitofusin homologue, FZL, is involved in the regulation of chloroplast thylakoid morphology (Gao et al. 2006). Despite a lack of genetic evidence, observations of mitochondrial dynamics in plants have confirmed the existence of this process. Gross observations of plant mitochondria at different stages of the cell cycle show that mitochondria undergo massive fusion events prior to cytokinesis (discussed below). In addition, time-lapse microscopy of plant mitochondria has clearly shown fusion events in a number of cell types (see www.plantcellbiologyondvd.com), indicating that this process also occurs in nondividing cells. To quantify mitochondrial fusion in plant cells (Arimura et al. 2004) transiently expressed, in onion epidermal cells, a matrix-targeted fluorescent protein called Kaede, which can be induced to change its emission spectrum irreversibly from green to red upon by illumination with UV light. By inducing half of the organelles in a cell to fluoresce red, Arimura et al. were able to observe mitochondrial fusion through the appearance of yellow mitochondria as a result of mixing of the matrix-localized fluorescent proteins. All the mitochondria in the cell were fluorescing yellow after 1–2 h (Arimura et al. 2004), indicating that fusion is a relatively rapid and constitutive event. In summary, while plant mitochondria clearly fuse under normal conditions, the mechanisms behind the process remain an enigma.

2.4 Regulation of Chondriome Structure

2.4.1 Temporal Regulation

Changes in mitochondrial ultrastructure during mitochondrial maturation in developing tissues have been recorded in a number of diverse organisms including rat conceptus tissues (Alcolea et al. 2002), rat brown adipose tissue (Justo et al. 2005), and maize embryos (Logan et al. 2001). Such changes in ultrastructure are often accompanied by changes in the gross morphology of mitochondria. Similarly, changes in mitochondrial gross morphology have been reported to occur in various animal tissues during normal differentiation (Chen and Chan 2004). However, most reports of changes in mitochondrial morphology arise from pathological defects in fusion, or division (Zuchner et al. 2005; Yang et al. 2008), or from studies where the normal function of proteins involved in mitochondrial fusion, or division, have been altered in order to determine the role of normal fusion, or division, in the development of a healthy organism (e.g., Chen et al. 2003; Wakabayashi et al. 2009).

Changes observed during the cell cycle provide one of the clearest examples of developmentally related temporal modifications to chondriome structure. During the human cell cycle, mitochondria alternate between two morphologic states (Barni et al. 1996; Karbowski et al. 2001; Margineantu et al. 2002). In the G1 phase mitochondria fuse to form reticula, halving the number of individual organelles relative to the number prior to M phase (Karbowski et al. 2001). As cells proceed from G1 to S phase there is an increase in mitochondrial number as a result of fragmentation (division) of the mitochondrial reticula (Barni et al. 1996; Karbowski et al. 2001; Margineantu et al. 2002). A similar situation has recently been shown to occur in plants. Segui-Simarro and colleagues demonstrated, by means of an analysis of electron micrographs of serial thin-sections prepared from Arabidopsis apical meristems at various developmental stages that, in addition to smaller physically discrete organelles, there existed a large sheet-like mitochondrion that underwent characteristic morphological and architectural changes during the cell cycle (Segui-Simarro et al. 2008). During G1 to S phase this large mitochondrion was described as having a tentaculate morphology and enveloped one pole of the nucleus. In the G2 phase the large and small mitochondria double in volume and the large mitochondrion forms a cage around the nucleus. During M phase ca. 60% of the small mitochondria fuse with the large mitochondrion, thereby increasing its volume to ca. 80% of the total mitochondrial volume. During cytokinesis the cage-like mitochondrion divides to form two physically discrete tentacular mitochondria which then each undergo further division. Segui-Simarro et al. hypothesized that the tentaculate/cage-like mitochondria provide an efficient means to deliver ATP for cell cycle and cytokinesis activities (Segui-Simarro et al. 2008). However, since mitochondria with these morphologies are not seen in root apical meristems, Segui-Simarro et al. proposed that the major role of the tentaculate/cage-like mitochondria is to provide a structure to enable efficient mixing and recombination of mtDNA (Segui-Simarro et al. 2008) which is inherent in the discontinuous whole hypothesis governing plant mitochondrial dynamics (Logan 2006).

2.4.2 Physical Regulation

The proteins involved in mitochondrial fission and fusion are highly regulated, both at the transcriptional and post-transcriptional levels. As changes to mitochondrial morphology can occur over the course of a few seconds, the ability to quickly switch proteins between active and inactive states relies heavily on coordinated post-translational modifications. In mammals, Drp1 has been identified as a substrate for multiple modifications. Phosphorylation of human Drp1 by different kinases has been shown to both promote (CDK/cyclinB; Taguchi et al. 2007) and repress (cAMP-dependent protein kinase A; Chang and Blackstone 2007; Cribbs and Strack 2007) its function, depending on which serine on the protein is modified. Drp1 has also been shown to be a substrate for reversible SUMOylation by SUMO1 (SUMOylation; Harder et al. 2004) and SENP5 (deSUMOylation; Zunino et al. 2007), which promote, or repress Drp1 function, respectively. As such, Drp1 appears to be controlled by a number of rapidly-reversible post-translational processes, indicating that changes in mitochondrial dynamics require a high degree of specific regulation.

Protein turnover is another method used to regulate the state of the mitochondrial population. While addition of SUMO acts as an on-off switch for dynamin, its close homologue ubiquitin is more often used to mark proteins for proteasomal degradation. Steady-state levels of the yeast fusion protein Fzo1p, and possibly mammalian Mfn1 and Mfn2, appear to be regulated by polyubiquitination, as inhibition of the proteosome, either genetically, or chemically, leads to increased expression of these proteins (Neutzner and Youle 2005; Escobar-Henriques et al. 2006; Karbowski et al. 2007). Recently, the first ubiquitin ligase to regulate mitochondrial morphology proteins was identified by three independent groups (Nakamura et al. 2006; Yonashiro et al. 2006; Karbowski et al. 2007). MARCH5, a RING-domain E3 ubiquitin ligase, is localized to the outer mitochondrial membrane and interacts with Drp1. Inhibition of MARCH5 leads to mitochondrial elongation and clustering of an inactive form of Drp1 at the mitochondria (Karbowski et al. 2007), suggesting the MARCH5 either acts as a molecular switch to aid mitochondrial division, or that it ubiquitinates an unknown regulator of Drp1 that inhibits its function.

Presently, there have been no reports of plant mitochondrial dynamics proteins being regulated by post-translational modifications. However, given the ubiquity of this process in animals and yeast, it would appear that this will be a fruitful area for further research.

2.5 Death

The best described link between changes in plant mitochondrial dynamics and cell function occurs during cell death (Logan 2008; see also Chaps. 17–19, and Box 2.2). The documented changes in mitochondrial dynamics are all likely due to changes in the mitochondrial inner membrane permeability (the mitochondrial permeability transition, mPT), although changes in mitochondrial-cytoskeleton interactions are also likely important since cell death induction leads to a cessation of mitochondrial movement, at least in Arabidopsis (Gao et al. 2008; Zhang et al. 2009; Scott and Logan, unpublished). Early evidence for a mPT in plants was obtained from in vitro experiments on Arabidopsis (Tiwari et al. 2002), oat (Curtis and Wolpert 2002), wheat (Virolainen et al. 2002), and potato (Fortes et al. 2001; Arpagaus et al. 2002) mitochondria. Swelling of isolated potato mitochondria was induced by Ca^{2+}, but not Mg^{2+}, while Ca^{2+}-induced swelling could be inhibited by cyclosporin A (CsA), but only in the presence of a reducing agent (Arpagaus et al. 2002). Oxidative stress-induced swelling of isolated Arabidopsis mitochondria was also sensitive to CsA (Tiwari et al. 2002). In other studies swelling was reported to be insensitive to CsA (Curtis and Wolpert 2002; Virolainen et al. 2002), although this may be due to the reported requirement for the addition of CsA under reduced conditions (Arpagaus et al. 2002), a situation that was not tested in the other studies. Swelling led to disruption of the outer, but not inner, mitochondrial membrane and release of cytochrome *c* (Arpagaus et al. 2002). Virolainen et al. (2002) and Arpagaus et al. (2002) both provided evidence for the release of cytochrome *c* as a

Box 2.2 Cell death and mitochondrial morphology

We recently published the results of an investigation into the effects of the induction of cell death on mitochondrial morphology in Arabidopsis protoplasts expressing mitochondrial-targeted GFP (Scott and Logan 2008). Treatment with strong oxidants, or a heat shock, resulted in a rapid and consistent change in mitochondrial morphology, characterized as an increase in the size of individual organelles (termed the *mitochondrial morphology transition*), that preceded by many hours any measureable effect on cell death. Preincubation of protoplasts with the calcium channel-blocker lanthanum chloride, or the permeability transition pore inhibitor cyclosporin A, inhibited both the mitochondrial morphology transition and subsequent cell death (Scott and Logan 2008). In addition, pretreatment of protoplasts with a cell permeable superoxide dismutase analogue, TEMPOL, also blocked the morphology transition and subsequent cell death. The mitochondrial morphology transition was observed much earlier than similar changes in morphology that were detected in previous studies using in vivo assays (Yoshinaga et al. 2005; Zottini et al. 2006). We concluded that the mitochondrial morphology transition that we measured directly in intact protoplasts, or leaf cells, is synonymous with the mPT, as measured in vitro by others, and that perturbation of the permeability of the inner mitochondrial membrane is at least one important mechanism promoting cell death in plants (Scott and Logan 2008).

consequence of mitochondrial swelling. While cytochrome *c* release has a clear role in apoptosis through its role in the formation of the caspase-activating apoptosome, it is not clear if cytochrome *c* release has a specific role to play in the execution of a programmed cell death (PCD) pathway in plant cells, nor is it known if there is a link between mechanisms underlying the mPT and cytochrome-*c* release in plants; no links have been established in vivo between these two events.

The role of a mPT in plant PCD has been demonstrated in a number of studies using a variety of death inducers. Yu and colleagues showed that CsA could inhibit PCD of tracheary elements in *Zinnia elegans*, and that there were subtle changes in mitochondrial ultrastructure, but no report of mitochondrial swelling, prior to loss of tonoplast integrity and cellular autolysis (Yu et al. 2002). Yu et al. commented that the changes in mitochondrial ultrastructure were different from those that occur during apoptosis of animal cells, but in fact an increased density of the matrix and swelling of the intracristal spaces, as reported, are the same as those reported to occur during apoptosis (Zhuang et al. 1998; Martinou et al. 1999). Using a hypersensitive response elicitor, ceramide, or protoporphyrin IX (PPIX), Yao et al. demonstrated that Arabidopsis leaf cell death was accompanied by a loss of mitochondrial membrane potential and nuclear DNA cleavage, and that these effects could be partially prevented by CsA, as could death itself (Yao et al. 2004). The study by Yao et al. is one of the few plant cell death studies in which death was measured rather than inferred. Subsequent work by Yao and Greenberg demonstrated that PPIX caused mitochondria to become swollen, and more round in shape (Yao and Greenberg 2006).

Swelling of mitochondria has also been reported in vivo during the self-incompatibility response in poppy pollen (Geitmann et al. 2004) and during abortion of microspores in CMS sunflower (Laveau et al. 1989). However, although the changes to mitochondrial morphology can be postulated (and indeed are likely) to be linked to the PCD that occurs in these systems, further studies are required to show direct links.

2.6 Motility, Distribution, and Inheritance

One of the most striking phenomena of mitochondria when visualized in living plant cells is their dynamism. Since the plant chondriome is organized as a discontinuous whole (Logan 2006), correct mitochondrial function will require precise control over mitochondrial motility and cellular distribution. In fact it is inherent in the discontinuous whole hypothesis that there is a "need to meet" to enable the exchange and/or complementation of mtDNA, and clearly there is a requirement for movement to drive the meeting of physically discrete organelles. In growing Arabidopsis root hairs, mitochondria have been recorded moving through the cytosol at speeds up to 10 μm s^{-1}, although average speeds range between 0.6 and 3.4 μm s^{-1} depending on location within the hair (Zheng et al. 2009). Variability of the speed of mitochondrial movement has also been reported in maize BY-2 cells in which a single mitochondrion was recorded travelling at speeds between 0.1 and 0.5 μm s^{-1} depending on location within the cell (Watanabe et al. 2007).

2.6.1 Mitochondrial Movement and the Cytoskeleton

There is considerable diversity in the mechanisms underpinning mitochondrial movement in different cell types in different organisms. Movement of mitochondria in mammals and most yeast is microtubule based (for review, see Boldogh and Pon 2007), whereas in *S. cerevisiae*, *Aspergillus*, and plants, mitochondria move predominantly on the actin cytoskeleton (Simon et al. 1995; Olyslaegers and Verbelen 1998; Van Gestel et al. 2002; Koch et al. 2003a, b; Sheahan et al. 2004). There is very little evidence for the involvement of microtubules in the distribution of mitochondria in plants beyond their effects on the geometry of the actin cytoskeleton (Zheng et al. 2009). However, circumstantial evidence from one study suggested that microtubules may play a role in organizing mitochondria into transverse arrays in the cortical cytoplasm in the absence of actin filaments (Van Gestel et al. 2002), but importantly these mitochondria appeared immobilized and so there is no published evidence that microtubules serve as tracts for mitochondrial movement in plants. Recent studies, predominantly using yeast, have shown that there are several mechanisms in operation in addition to the classic motor-protein mediated means of organelle transport on the cytoskeleton (for a review, see Boldogh and Pon (2007)).

2.6.1.1 Mitochondrial Movement and Microtubules

Tubulin is a highly conserved dimeric protein present in all eukaryotes. Tubulin self-assembles into microtubules which are polar dynamic structures regulated by assembly and disassembly mediated by GTP hydrolysis (Kreis and Vale 1998). Two motor proteins are responsible for movement on microtubules: dynein drives mitochondrial movement to the minus end of microtubules (attached to the microtubule organizing center), whereas kinesin drives movement to the plus end (toward the cell periphery). A number of mitochondrial-associated proteins have been implicated in mitochondrial movement on microtubules and the most-studied of these adaptors are Milton and Miro.

Milton was identified in a genetic screen based on defective photoreceptor function. Mutant milton protein failed to deliver mitochondria to *Drosophila* nerve terminals with the result that mutant flies were viable but blind (Stowers et al. 2002). Milton is an essential adaptor protein that recruits kinesin heavy chain to mitochondria and interacts with a second adaptor named Miro (Glater et al. 2006). Although Milton localizes to mitochondria it has neither a mitochondrial targeting signal nor a transmembrane domain and therefore must interact with another protein(s) that is more tightly associated with mitochondria: Miro.

Miro is a Rho-type GTPase that localizes to the mitochondrial outer membrane by means of a C-terminal transmembrane domain and interacts with Milton via two GTPase domains separated by two EF-hands that have the potential to bind calcium (Fransson et al. 2003, 2006). Miro orthologues are present in all eukaryotes for which genome sequences are available, therefore including *S. cerevisiae*, *Aspergillus*, and *Arabidopsis*: organisms in which mitochondrial movement is predominantly actin based. As a result it has been suggested that the function of Miro, at least in *S. cerevisiae*, is different than in *Drosophila* and other animals and that Miro might function as "a more general mitochondrial adaptor that binds to other motile complexes" (Rice and Gelfand 2006). The Arabidopsis genome contains three Miro orthologues, two of which are transcribed ubiquitously (Yamaoka and Leaver 2008). Mutation of *MIRO2* had no apparent effect on plant development but mutation of *MIRO1* led to arrest of embryogenesis at an early stage and an impairment in pollen germination and tube growth (Yamaoka and Leaver 2008). Mitochondria in pollen tubes of *miro1* mutant plants exhibited abnormal morphology, being larger and more tubular than wild type, and they exhibited a disruption in their normal streaming movement within the pollen tube, movement that was dependent on the actin cytoskeleton (Yamaoka and Leaver 2008). Given that mitochondrial movement in pollen tubes is actin dependent (Yamaoka and Leaver 2008), and the apparent absence of Milton homologues from the Arabidopsis genome, MIRO clearly has a different role in plants just as its orthologue, Gem1P, does in *S. cerevisiae* (Frederick et al. 2004). Future research will hopefully uncover the exact role MIRO/Gem1p plays in mitochondrial dynamics.

2.6.1.2 Mitochondrial Movement and Actin

Actin is a highly conserved protein that is a major component of microfilaments in eukaryotes. Actin self-assembles under physiological conditions into long polymers and polymerization, depolymerization, and higher order assembly (e.g., bundling) of filaments results in the production of a variety of transient, or more stable, structures. Actin is important for many types of intracellular motility that are driven by actin polymerization, or by the movement of the motor protein myosin along actin filaments (Kreis and Vale 1998).

Myosin

Myosins are a large superfamily of actin-activated ATPases that act as motors to propel cargo along actin filaments, or propel actin filaments along immobile cell structures. Myosins have been implicated in the movement of other organelles on actin filaments (e.g., peroxisomes, Jedd and Chua 2002); therefore, it seems likely that myosin is involved in some mitochondrial movements. Several recent studies have identified myosin family members that likely serve as motor proteins driving movement along actin filaments. The Arabidopsis myosin gene family comprises 17 members: the majority (13) are in group XI, and the remaining four are in group VIII (Reddy and Day 2001). By means of extensive analyses of myosin-tail fluorescent-protein fusions, four Arabidopsis group XI myosins have been reported to be either directly or indirectly involved in the movement of mitochondria in *Nicotiana benthamia*: XI-C, XI-E, XI-I, and XI-K (Avisar et al. 2008, Avisar et al. 2009; Sparkes et al. 2008). In a colocalization study, again using Arabidopsis proteins but this time transiently expressed in *N. tabacum*, only myosin XI-J (specifically a XI-J tail fusion lacking the motor domain) was found to sometimes colocalize with mitochondria (Reisen and Hanson 2007).

Mechanisms Other Than Myosin-Based

Evidence for mechanisms moving mitochondria on the actin cytoskeleton other than those involving myosin was recently obtained in a study using a pharmaceutical approach (Zheng et al. 2009). Tracking and analysis of mitochondria in Arabidopsis root hairs using a combination of evanescent wave microscopy and spinning disc confocal microscopy following treatment with various drugs that perturb the cytoskeleton has suggested that, as in other organisms, mitochondria move using myosin-motor dependent and myosin-motor independent mechanisms (Zheng et al. 2009). Zheng et al. (2009) concluded that the mechanisms controlling mitochondrial speed, positioning, and direction of movement were the result of the coordinated activity of myosin and the rate of actin turnover. In addition, supporting the conclusions of early work on mitochondrial movement (Van Gestel et al. 2002), microtubule dynamics had a role to play, in this instance by influencing the arrangement of actin filaments (Zheng et al. 2009).

The Arp2/3 complex is a multisubunit ubiquitous regulator of actin nucleation in eukaryotes responsible for the actin-based motility manifest by comet-tail formation (Fehrenbacher et al. 2003). The Arp2/3 complex binds newly polymerized actin, and promotes actin nucleation (Pollard and Beltzner 2002) thereby generating forces to facilitate mitochondrial movement into the developing bud in *S. cerevisiae* (Boldogh et al. 2001, 2005). Disruption of either *ARP2* or *ARP15*, two subunits of the complex, leads to a decrease in mitochondrial motility, and changes to mitochondrial morphology (Boldogh et al. 2001). There is no published evidence of the involvement of the plant Arp2/3 complex in driving mitochondrial motility. Although myosin was initially thought not to be of importance for Arp2/3 mediated mitochondrial movement in *S. cerevisiae*, recent evidence suggests that myosin functions directly and significantly in this process (Altmann et al. 2008). It has been suggested that multiple mechanisms for mitochondrial transport exist to ensure that mitochondria, vital to all mitochondria eukaryotes, are accurately inherited during cytokinesis (Valiathan and Weisman 2008).

2.6.2 Mitochondrial Motility Delivers the Organelle to the Right Places

Whilst it is generally accepted that mitochondrial motility, which enables their close association with other motile structures within the cell, is necessary for mitochondrial or cellular function, there are very few studies linking location to function in plants. Association of mitochondria with energy-consuming structures or organelles has been described in a range of non-plant organisms (see Munn 1974; Tyler 1992; Bereiter-Hahn and Voth 1994). One classic example is the formation of the Nebenkern, a collar around the sperm axoneme, formed during spermatogenesis and comprising two giant mitochondria formed by repeated fusion events (Hales and Fuller 1996, 1997). In living plant tissues containing chloroplasts, visualization of mitochondria has shown the frequent close proximity of these two organelles (Stickens and Verbelen 1996; Logan and Leaver 2000). It is assumed that this facilitates exchange of respiratory gases and possibly metabolites, although direct evidence for this is lacking. In characean internode cells, it has been suggested that the spatiotemporal distribution of mitochondria within the cell promotes their association with chloroplasts (Foissner 2004). However, care needs to be taken in evaluating the experimental evidence required to support the hypothesis that mitochondrial and cellular function are dependent on the juxtaposition of mitochondria with other cell structures when this evidence is simply qualitative, e.g., micrographs showing apparent close association. This is particularly true in highly vacuolated mesophyll cells (Evert and Esau 2006), where the volume of cortical cytoplasm dictates that all cellular organelles are in close association.

A recent paper presented qualitative and semiquantitative evidence that chloroplast movements under different lighting regimes were associated with changes in the location of mitochondria (Islam et al. 2009). In the dark, mitochondria were distributed randomly in palisade mesophyll cells. However, under low-intensity blue light illumination mitochondria moved with chloroplasts to the periclinal walls of the cell, while under high-intensity blue light mitochondria adopted an anticlinal location similar to the light-avoidance response of chloroplasts (Islam et al. 2009). It is not known whether the response of the mitochondria is independent of the chloroplasts, either from a signaling or physical perspective. For example, whether or not a subpopulation of mitochondria becomes physically associated with a chloroplast through cytoskeletal interactions, and then moves with the chloroplast as it reacts to changes in illumination.

2.6.3 Inheritance and Cellular Distribution

Mitochondrial inheritance has been extensively studied during budding in *S. cerevisiae* and is dependent on the cytoskeleton, mitochondrial division, and mitochondrial fusion (Hermann and Shaw 1998; Catlett and Weisman 2000). During the budding process the mitochondrial reticulum moves towards the bud site and a single tubule moves into the newly formed bud. This movement continues until mother and daughter cells have an equal mitochondrial complement, at which point cytokinesis can occur (Catlett and Weisman 2000; Boldogh et al. 2005).

Research into mitochondrial morphology mutants has identified many genes involved in inheritance. The first mutants defective in mitochondrial inheritance were isolated during a screen of temperature-sensitive yeast cell lines for individuals that failed to pass mitochondria to the daughter bud prior to cytokinesis (McConnell et al. 1990). Many of these mutant lines exhibited altered mitochondrial morphology, and were named *mdm*, for mitochondrial distribution and morphology (see Table 2.1). The mutant genes fell into two broad categories: genes encoding integral outer mitochondrial membrane proteins, and genes encoding cytosolic proteins (Yaffe 1999).

Little is known about the mechanisms regulating the cellular distribution of mitochondria in higher plants. However, the actin cytoskeleton has been implicated since disruption of actin polymerization affected the dispersal of mitochondria and resulted in a biased distribution of mitochondria in the daughter cells (Sheahan et al. 2004).

One plant mitochondrial dynamics mutant has been identified with a grossly altered cellular distribution of mitochondria (Logan et al. 2003). Disruption of the *FRIENDLY* gene (*FMT*) results in the formation of large mitochondrial clusters of ten or hundreds of organelles, although some mitochondria remain apparently normally distributed as singletons throughout the cytoplasm (Logan et al. 2003). *FMT* is a conserved eukaryotic gene and disruption of orthologues in *Dictyostelium discoideum* (*cluA*) and *S. cerevisiae* (*CLU1*) also cause aberrant mitochondrial phenotypes (Zhu et al. 1997; Fields et al. 1998). In the *cluA*$^-$ mutant of *D. discoideum* the mitochondria cluster near the cell center (Zhu et al. 1997), while in the *S. cerevisiae clu1* mutant the mitochondrial tubules collapse to one side of the cell (Fields et al. 1998). It is not known how FRIENDLY or its orthologues are involved in the maintenance of normal mitochondrial cellular distribution.

2.7 Conclusions

Advances in light microscopy and genetic techniques over the last two decades have allowed us to gain an unprecedented insight into mitochondrial dynamics. Over this time, we have uncovered many proteins that control mitochondrial shape, size, and position in the cell, allowing us to formulate models of the mechanisms involved. However, as each year passes, another new fission or fusion protein seems to be discovered, indicating that we are far from having the complete picture of what controls mitochondrial dynamics. This is particularly true in plants, where the combination of a more complex wild-type chondriome structure, and a smaller research community, has limited the advances made into understanding mitochondrial dynamics relative to animals and yeast. However, rather being a negative, this should be seen as a future opportunity to discover a uniquely plant-based cell biology system. The discovery of *NETWORK1/ELM1* as a likely mediator of mitochondrial fission, along with the lack of plant fusion protein homologues, suggests that plants have evolved a mitochondrial dynamics system that is distinct from that found in any other kingdom. The quest for the future, therefore, is threefold: (1) to identify the other plant-specific members of the mitochondrial dynamics machinery; (2) to discover how these proteins interact to control mitochondrial form and function; and (3) to uncover why plants evolved such a different dynamics system from the other extant chondriate eukaryotes. As such, it is clear that we still have much to learn about these crucial organelles.

Acknowledgments I.S. is funded by the intramural research program of the National Heart, Lung and Blood Institute, National Institutes of Health, USA. D.C.L. is supported by the University of Saskatchewan, the Natural Sciences and Engineering Research Council of Canada, and the Canada Foundation for Innovation. Thanks to Benedikt Westermann and Richard J. Youle and for images of yeast and human chondriomes, respectively.

Glossary

Chondriome: All the mitochondria in a cell, collectively.

Chondrome: The mitochondrial genome.

Mitochondrial dynamics: Study of the shape, size, number, motility, cellular distribution and cellular inheritance of mitochondria.

Mitochondrial division: Division of parental organelle to form two or more daughter organelles.

Mitochondrial fusion: Fusion of physically discrete mitochondria to form single organelle.

Dynamin: Mechanoenzyme that uses the energy from the hydrolysis of GTP to severe membranes.

Green fluorescent protein: One of two photoproteins (the other being aequorin) that together are responsible for the characteristic luminescence emitted from the light organs in the rim of the bell of the jelly fish *Aequorea victoria.*

Cytoskeleton: Proteinaceous scaffold within the cytoplasm of cells.

References

Alcolea, M., Colom, B., Riera, A., Gianotti, M., Garcia-Palmer, F., Justo, R. 2002. Morphofunctional changes in the mitochondrial subpopulations of conceptus tissues during the placentation process. Cell. Mol. Life Sci. 59:2199–2209.

Alexander, C., Votruba, M., Pesch, U. E., Thiselton, D. L., Mayer, S., Moore, A., Rodriguez, M., Kellner, U., Leo-Kottler, B., Auburger, G., Bhattacharya, S. S., Wissinger, B. 2000. OPA1, encoding a dynamin-related GTPase, is mutated in autosomal dominant optic atrophy linked to chromosome 3q28. Nat. Genet. 26:211–215.

Altmann, K., Frank, M., Neumann,D., Jakobs, S., Westermann, B. 2008. The class V myosin motor protein, Myo2, plays a major role in mitochondrial motility in Saccharomyces cerevisiae. J. Cell Biol. 181:119–130.

Arimura, S., Tsutsumi, N. 2002. A dynamin-like protein (ADL2b), rather than FtsZ, is involved in Arabidopsis mitochondrial division. Proc. Natl. Acad. Sci. U.S.A. 99:5727–5731.

Arimura, S., Aida, G. P., Fujimoto, M., Nakazono, M., Tsutsumi, N. 2004. Arabidopsis dynamin-like protein 2a (ADL2a), like ADL2b, is involved in plant mitochondrial division. Plant Cell Physiol. 45:236–242.

Arimura, S., Yamamoto, J., Aida, G. P., Nakazono, M., Tsutsumi, N. 2004. Frequent fusion and fission of plant mitochondria with unequal nucleoid distribution. Proc. Natl. Acad. Sci. U.S.A. 101:7805–7808.

Arimura, S., Fujimoto, M., Doniwa, Y., Kadoya, N., Nakazono, M., Sakamoto, W., Tsutsumi, N. 2008. Arabidopsis ELONGATED MITOCHONDRIA1 is required for localization of DYNAMIN-RELATED PROTEIN3A to mitochondrial fission sites. Plant Cell 20:1555–1566.

Arpagaus, S., Rawyler, A., Braendle, R. 2002. Occurrence and characteristics of the mitochondrial permeability transition in plants. J. Biol. Chem. 277:1780–1787.

Avisar, D., Prokhnevsky, A. I., Makarova, K. S., Koonin, E. V., Dolja, V. V. 2008. Myosin XI-K is required for rapid trafficking of Golgi stacks, peroxisomes, and mitochondria in leaf cells of *Nicotiana benthamiana*. Plant Physiol. 146:1098–1108.

Avisar, D., Abu-Abied, M., Belausov, E., Sadot, E., Hawes, S. 2009. A comparative study of the involvement of 17 Arabidopsis myosin family members on the motility of golgi and other organelles. Plant Physiol. 150:700–709.

Barni, S., Sciola, L., Spano, A., Pippia, P. 1996. Static cytofluorometry and fluorescence morphology of mitochondria and DNA in proliferating fibroblasts. Biotech. Histochem. 71:66–70.

Beech, P. L., Nheu, T., Schultz, T., Herbert, S., Lithgow, T., Gilson, P. R., McFadden, G. I. 2000. Mitochondrial FtsZ in a chromophyte alga. Science 287:1276–1279.

Bereiter-Hahn, J., Voth, M. 1994. Dynamics of mitochondria in living cells:shape changes, dislocations, fusion, and fission of mitochondria. Microsc. Res. Tech. 27:198–219.

Berger, K. H., Sogo, L. F., Yaffe, M. P. 1997. Mdm12p, a component required for mitochondrial inheritance that is conserved between budding and fission yeast. J. Cell Biol. 136:545–553.

Bi, E. F., Lutkenhaus, J. 1991. FtsZ ring structure associated with division in Escherichia coli. Nature 354:161–164.

Boldogh, I. R., Pon, L. A. 2007. Mitochondria on the move. Trends Cell Biol. 17:502–510.

Boldogh, I. R., Yang, H. C., Nowakowski, W. D., Karmon, S. L., Hays, L. G., Yates, J. R., Pon, L. A. 2001. Arp2/3 complex and actin dynamics are required for actin-based mitochondrial motility in yeast. Proc. Natl. Acad. Sci. U.S.A. 98:3162–3167.

Boldogh, I. R., Fehrenbacher, K. L., Yang, H.-C., Pon, L. A. 2005. Mitochondrial movement and inheritance in budding yeast. Gene 354:28–36.

Burgess, S. M., Delannoy, M., Jensen, R. E. 1994. MMM1 encodes a mitochondrial outer membrane protein essential for establishing and maintaining the structure of yeast mitochondria. J. Cell Biol. 126:1375–1391.

Catlett, N. L., Weisman, L. S. 2000. Divide and multiply:organelle partitioning in yeast. Curr. Opin. Cell Biol.12:509–516.

Cavers, F. 1914. Chondriosomes (mitochondria) and their significance. New Phytol. 13:96–106.

Cerveny, K. L., McCaffery, J. M., Jensen, R. E. 2001. Division of mitochondria requires a novel DNM1-interacting protein, net2p. Mol. Biol. Cell 12:309–321.

Chalfie, M., Tu, Y., Euskirchen, G., Ward, W. W., Prasher, D. C. 1994. Green fluorescent protein as a marker for gene-expression. Science 263:802–805.

Chang, C. R., Blackstone, C. 2007. Cyclic AMP-dependent protein kinase phosphorylation of Drp1 regulates its GTPase activity and mitochondrial morphology. J. Biol. Chem. 282:21583–21587.

Chen, H. C., Chan, D. C. 2004. Mitochondrial Dynamics in Mammals. San Diego: Academic Press.

Chen, H., Detmer, S. A., Ewald, A. J., Griffin, E. E., Fraser, S. E., Chan, D. C. 2003. Mitofusins Mfn1 and Mfn2 coordinately regulate mitochondrial fusion and are essential for embryonic development. J. Cell Biol. 160:189–200.

Cipolat, S., Martins de Brito, O., Dal Zilio, B., Scorrano, L. 2004. OPA1 requires mitofusin 1 to promote mitochondrial fusion. Proc. Natl. Acad. Sci. U.S.A. 101:15927–15932.

Cipolat, S., Rudka, T., Hartmann, D., Costa, V., Serneels, L., Craessaerts, K., Metzger, K., Frezza, C., Annaert, W., D'Adamio, L., Derks, C., Dejaegere, T., Pellegrini, L., D'Hooge, R., Scorrano, L., De Strooper, B. 2006. Mitochondrial rhomboid PARL regulates cytochrome c release during apoptosis via OPA1-dependent cristae remodeling. Cell 126:163–175.

Cribbs, J. T., Strack, S. 2007. Reversible phosphorylation of Drp1 by cyclic AMP-dependent protein kinase and calcineurin regulates mitochondrial fission and cell death. EMBO Rept. 8:939–944.

Curtis, M. J., Wolpert, T. J. 2002. The oat mitochondrial permeability transition and its implication in victorin binding and induced cell death. Plant J. 29:295–312.

Dajkovic, A., Mukherjee, A., Lutkenhaus, J. 2008. Investigation of regulation of FtsZ assembly by SulA and development of a model for FtsZ polymerization. J. Bacteriol. 190:2513–2526.

Delettre, C., Griffoin, J. M., Kaplan, J., Dollfus, H., Lorenz, B., Faivre, L., Lenaers, G., Belenguer, P., Hamel, C. P. 2001. Mutation spectrum and splicing variants in the OPA1 gene. Hum. Genet. 109:584–591.

Dimmer, K. S., Fritz, S., Fuchs, F., Messerschmitt, M., Weinbach, N., Neupert,W., Westermann, B. 2002. Genetic basis of mitochondrial function and morphology in *Saccharomyces cerevisiae*. Mol. Biol. Cell 13:847–853.

Doherty, G. J., McMahon, H. T. 2009. Mechanisms of endocytosis. Annu. Rev. Biochem. 78:857–902.

Egner, A., Jakobs, S., Hell, S. W. 2002. Fast 100-nm resolution three-dimensional microscope reveals structural plasticity of mitochondria in live yeast. Proc. Natl. Acad. Sci. U.S.A. 99:3370–3375.

Escobar-Henriques, M., Westermann, B., Langer, T. 2006. Regulation of mitochondrial fusion by the F-box protein Mdm30 involves proteasome-independent turnover of Fzo1. J. Cell Biol. 173:645–650.

Evert, R. F., Esau, K. P. A. 2006. Esau's Plant Anatomy: Meristems, Cells, and Tissues of the Plant Body: Their Structure, Function, and Development. Hoboken: Wiley-Interscience.

Fehrenbacher, K., Huckaba, T., Yang, H.-C., Boldogh, I., Pon, L. 2003. Actin comet tails, endosomes and endosymbionts. J. Exp. Biol. 206:1977–1984.

Fekkes, P., Shepard, K. A., Yaffe, M. P. 2000. Gag3p, an outer membrane protein required for fission of mitochondrial tubules. J. Cell Biol. 151:333–340.

Fields, S. D., Conrad, M. N., Clarke, M. 1998. The *S. cerevisiae* CLU1 and D. discoideum cluA genes are functional homologues that influence mitochondrial morphology and distribution. J. Cell Sci. 111:1717–1727.

Foissner, I. 2004. Microfilaments and microtubules control the shape, motility, and subcellular distribution of cortical mitochondria in characean internodal cells. Protoplasma 224: 145–157.

Fortes, F., Castilho, R. F., Catisti, R., Carnieri, E. G. S., Vercesi, A. E. 2001. Ca2+ induces a cyclosporin A-insensitive permeability transition pore in isolated potato tuber mitochondria mediated by reactive oxygen species. J. Bioenerget. Biomembr. 33:43–51.

Fransson, A., Ruusala, A., Aspenstrom, P. 2003. Atypical Rho GTPases have roles in mitochondrial homeostasis and apoptosis. J. Biol. Chem. 278:6495–6502.

Fransson, A., Ruusala, A., Aspenström, P. 2006. The atypical Rho GTPases Miro-1 and Miro-2 have essential roles in mitochondrial trafficking. Biochem. Biophys. Res. Commun. 344:500–510.

Frederick, R. L., McCaffery, J. M., Cunningham, K. W., Okamoto, K., Shaw, J. M. 2004. Yeast Miro GTPase, Gem1p, regulates mitochondrial morphology via a novel pathway. J. Cell Biol. 167:87–98.

Fujimoto, M., Arimura, S., Mano, S., Kondo, M., Saito, C., Ueda, T., Nakazono, M., Nakano, A., Nishimura, M., Tsutsumi, N. 2009. Arabidopsis dynamin-related proteins DRP3A and DRP3B are functionally redundant in mitochondrial fission, but have distinct roles in peroxisomal fission. Plant J. 58:388–400.

Gandre-Babbe, S., van der Bliek, A. M. 2008. The novel tail-anchored membrane protein Mff controls mitochondrial and peroxisomal fission in mammalian cells. Mol. Biol. Cell 19:2402–2412.

Gao, C., Xing, D., Li, L., Zhang, L. 2008. Implication of reactive oxygen species and mitochondrial dysfunction in the early stages of plant programmed cell death induced by ultraviolet-C overexposure. Planta 227:755–767.

Gao, H., Sage, T. L., Osteryoung, K. W. 2006. FZL, an FZO-like protein in plants, is a determinant of thylakoid and chloroplast morphology. Proc. Natl. Acad. Sci. 103:6759–6764.

Geitmann, A., Franklin-Tong, V. E., Emons, A. C. 2004. The self-incompatibility response in Papaver rhoeas pollen causes early and striking alterations to organelles. Cell Death Differ. 11:812–822.

Gilson, P. R., Yu, X.-C., Hereld, D., Barth, C., Savage, A., Kiefel, B. R., Lay, S., Fisher, P. R., Margolin, W., Beech, P. L. 2003. Two Dictyostelium orthologs of the prokaryotic cell division

protein FtsZ localize to mitochondria and are required for the maintenance of normal mitochondrial morphology. Eukaryot. Cell 2:1315–1326.
Glater, E. E., Megeath, L. J., Stowers, R. S., Schwarz, T. L. 2006. Axonal transport of mitochondria requires milton to recruit kinesin heavy chain and is light chain independent. J. Cell Biol. 173:545–557.
Griffin, E. E., Graumann, J., Chan, D. C. 2005. The WD40 protein Caf4p is a component of the mitochondrial fission machinery and recruits Dnm1p to mitochondria. J. Cell Biol. 170:237–248.
Griparic, L., Kanazawa, T., van der Bliek, A. M. 2007. Regulation of the mitochondrial dynamin-like protein Opa1 by proteolytic cleavage. J. Cell Biol. 178:757–764.
Guan, K., Farh, L., Marshall, T. K., Deschenes, R. J. 1993. Normal mitochondrial structure and genome maintenance in yeast requires the dynamin-like product of the MGM1 gene. Curr. Genet. 24:141–148.
Hales, K. G., Fuller, M. T. 1996. A novel transmembrane GTPase is required for developmentally regulated mitochondrial fusion during *Drosophila* spermatogenesis. Mol. Biol. Cell 7:3579–3579.
Hales, K. G., Fuller, M. T. 1997. Developmentally regulated mitochondrial fusion mediated by a conserved, novel, predicted GTPase. Cell 90:121–129.
Harder, Z., Zunino, R., McBride, H. 2004. Sumo1 conjugates mitochondrial substrates and participates in mitochondrial fission. Curr. Biol. 14:340–345.
Herlan, M., Vogel, F., Bornhovd, C., Neupert, W., Reichert, A. S. 2003. Processing of Mgm1 by the rhomboid-type protease Pcp1 is required for maintenance of mitochondrial morphology and of mitochondrial DNA. J. Biol. Chem. 278:27781–27788.
Hermann, G. J., King, E. J., Shaw, J. M. 1997. The yeast gene, MDM20, is necessary for mitochondrial inheritance and organization of the actin cytoskeleton. J. Cell Biol. 137:141–153.
Hermann, G. J., Shaw, J. M. 1998. Mitochondrial dynamics in yeast. Annu. Rev. Cell Dev. Biol. 14:265–303.
Hermann, G. J., Thatcher, J. W., Mills, J. P., Hales, K. G., Fuller, M. T., Nunnari, J., Shaw, J. M. 1998. Mitochondrial fusion in yeast requires the transmembrane GTPase Fzo1p. J. Cell Biol. 143:359–373.
Ingerman, E., Perkins, E. M., Marino, M., Mears, J. A., McCaffery, J. M., Hinshaw, J. E., Nunnari, J. 2005. Dnm1 forms spirals that are structurally tailored to fit mitochondria. J. Cell Biol. 170:1021–1027.
Ishihara, N., Fujita, Y., Oka, T., Mihara, K. 2006. Regulation of mitochondrial morphology through proteolytic cleavage of OPA1. EMBO J. 25:2966–2977.
Islam, M. S., Niwa, Y., Takagi, S. 2009. Light-dependent intracellular positioning of mitochondria in *Arabidopsis thaliana* mesophyll cells. Plant Cell Physiol. 50:1032–1040.
Jedd, G., Chua, N. H. 2002. Visualization of peroxisomes in living plant cells reveals acto-myosin-dependent cytoplasmic streaming and peroxisome budding. Plant Cell Physiol. 43:384–392.
Jones, B. A., Fangman, W. L. 1992. Mitochondrial DNA maintenance in yeast requires a protein containing a region related to the GTP-binding domain of dynamin. Gene Dev. 6:380–389.
Justo, R., Oliver, J., Gianotti, M. 2005. Brown adipose tissue mitochondrial subpopulations show different morphological and thermogenic characteristics. Mitochondrion 5:45–53.
Karbowski, M., Neutzner, A., Youle, R. J. 2007. The mitochondrial E3 ubiquitin ligase MARCH5 is required for Drp1 dependent mitochondrial division. J. Cell Biol. 178:71–84.
Karbowski, M., Spodnik, J. H., Teranishi, M., Wozniak, M., Nishizawa, Y., Usukura, J., Wakabayashi, T. 2001. Opposite effects of microtubule-stabilizing and microtubule-destabilizing drugs on biogenesis of mitochondria in mammalian cells. J. Cell Sci. 114:281–291.
Koch, A., Thiemann, M., Grabenbauer, M., Yoon, Y., McNiven, M. A., Schrader, M. 2003. Dynamin-like protein 1 is involved in peroxisomal fission. J. Biol. Chem. 278:8597–8605.
Koch, A., Yoon, Y., Bonekamp, N. A., McNiven, M. A., Schrader, M. 2005. A role for Fis1 in both mitochondrial and peroxisomal fission in mammalian cells. Mol. Biol. Cell 16:5077–5086.

Koch, K. V., Suelmann, R., Fischer, R. 2003. Deletion of mdmB impairs mitochondrial distribution and morphology in *Aspergillus nidulans*. Cell Motil. Cytoskel. 55:114–124.

Kohler, R. H., Zipfel, W. R., Webb, W. W., Hanson, M. R. 1997. The green fluorescent protein as a marker to visualize plant mitochondria in vivo. Plant J. 11:613–621.

Koshiba, T., Detmer, S. A., Kaiser, J. T., Chen, H., McCaffery, J. M., Chan, D. C. 2004. Structural basis of mitochondrial tethering by mitofusin complexes. Science 305:858–862.

Kreis, T., Vale, R. 1998. Guidebook to cytoskeletal and motor proteins. Oxford: Oxford University Press.

Labrousse, A. M., Zappaterra, M. D., Rube, D. A., van der Bliek, A. M. 1999. *C. elegans* dynamin-related protein DRP-1 controls severing of the mitochondrial outer membrane. Mol. Cell. 4:815–826.

Laveau, J. H., Schneider, C., Berville, A. 1989. Microsporogenesis abortion in cytoplasmic male sterile plants from Helianthus-petiolaris or h petiolaris-fallax crossed by sunflower (*Helianthus-annuus*). Ann. Bot. 64:137–148.

Lingard, M. J., Gidda, S. K., Bingham, S., Rothstein, S. J., Mullen, R. T., Trelease, R. N. 2008. Arabidopsis PEROXIN11c-e, FISSION1b, and DYNAMIN-RELATED PROTEIN3A cooperate in cell cycle-associated replication of peroxisomes. Plant Cell 20:1567–1585.

Logan, D. C. 2006. The mitochondrial compartment. J. Exp. Bot. 57:1225–1243.

Logan, D. C. 2008. Having a swell time – mitochondrial morphology and plant cell death programmes. J. Exp. Microsc. 231:215–224.

Logan, D. C., Leaver, C. J. 2000. Mitochondria-targeted GFP highlights the heterogeneity of mitochondrial shape, size and movement within living plant cells. J. Exp. Bot. 51: 865–871.

Logan, D. C., Millar, A. H., Sweetlove, L. J., Hill, S. A., Leaver, C. J. 2001. Mitochondrial biogenesis during germination in maize embryos. Plant Physiol. 125:662–672.

Logan, D. C., Scott, I., Tobin, A. K. 2003. The genetic control of plant mitochondrial morphology and dynamics. Plant J. 36:500–509.

Logan, D. C., Scott, I., Tobin, A. K. 2004. ADL2a, like ADL2b, is involved in the control of higher plant mitochondrial morphology. J. Exp. Bot. 55:783–785.

Mano, S., Nakamori, C., Kondo, M., Hayashi, M., Nishimura, M. 2004. An Arabidopsis dynamin-related protein, DRP3A, controls both peroxisomal and mitochondrial division. Plant J. 38:487–498.

Margineantu, D. H., Cox, W. G., Sundell, L., Sherwood, S. W., Beechem, J. A., Capaldi, R. A. 2002. Cell cycle dependent morphology changes and associated mitochondrial DNA redistribution in mitochondria of human cell lines. Mitochondrion 1:425–435.

Martinou, I., Desagher, S., Eskes, R., Antonsson, B., Andre, E., Fakan, S., Martinou, J. C. 1999. The release of cytochrome c from mitochondria during apoptosis of NGF-deprived sympathetic neurons is a reversible event. J. Cell Biol. 144:883–889.

McConnell, S. J., Stewart, L. C., Talin, A., Yaffe, M. P. 1990. Temperature-sensitive yeast mutants defective in mitochondrial inheritance. J. Cell Biol. 111:967–976.

McConnell, S. J., Yaffe, M. P. 1992. Nuclear and mitochondrial inheritance in yeast depends on novel cytoplasmic structures defined by the MDM1 protein. J. Cell Biol. 118:385–395.

McQuibban, G. A., Saurya, S., Freeman, M. 2003. Mitochondrial membrane remodelling regulated by a conserved rhomboid protease. Nature 423:537–541.

Meeusen, S., McCaffery, J. M., Nunnari, J. 2004. Mitochondrial fusion intermediates revealed in vitro. Science 305:1747–1752.

Messerschmitt, M., Jakobs, S., Vogel, F., Fritz, S., Dimmer, K. S., Neupert, W., Westermann, B. 2003. The inner membrane protein Mdm33 controls mitochondrial morphology in yeast. J. Cell Biol. 160:553–564.

Misaka, T., Miyashita, T., Kubo, Y. 2002. Primary structure of a dynamin-related mouse mitochondrial GTPase and its distribution in brain, subcellular localization, and effect on mitochondrial morphology. J. Biol. Chem. 277:15834–15842.

Mozdy, A. D., McCaffery, J. M., Shaw, J. M. 2000. Dnm1p GTPase-mediated mitochondrial fission is a multi-step process requiring the novel integral membrane component Fis1p. J. Cell Biol. 151:367–380.

Munn, E. A. 1974. The structure of mitochondria. London: Academic Press.

Nakamura, N., Kimura, Y., Tokuda, M., Honda, S., Hirose, S. 2006. MARCH-V is a novel mitofusin 2- and Drp1-binding protein able to change mitochondrial morphology. EMBO Rep. 7:1019–1022.

Neutzner, A., Youle, R. J. 2005. Instability of the mitofusin Fzo1 regulates mitochondrial morphology during the mating response of the yeast *Saccharomyces cerevisiae*. J. Biol. Chem. 280:18598–18603.

Olichon, A., Emorine, L. J., Descoins, E., Pelloquin, L., Brichese, L., Gas, N., Guillou, E., Delettre, C., Valette, A., Hamel, C. P., Ducommun, B., Lenaers, G., Belenguer, P. 2002. The human dynamin-related protein OPA1 is anchored to the mitochondrial inner membrane facing the inter-membrane space. FEBS Lett. 523:171–176.

Olyslaegers, G., Verbelen, J. 1998. Improved staining of F-actin and co-localization of mitochondria in plant cells. J. Exp. Microsc. 192:73–77.

Osteryoung, K. W., Vierling, E. 1995. Conserved cell and organelle division. Nature 376:473–474.

Otsuga, D., Keegan, B. R., Brisch, E., Thatcher, J. W., Hermann, G. J., Bleazard, W., Shaw, J. M. 1998. The dynamin-related GTPase, Dnm1p, controls mitochondrial morphology in yeast. J. Cell Biol. 143:333–349.

Pollard, T. D., Beltzner, C. C. 2002. Structure and function of the Arp2/3 complex. Curr. Opin. Struct. Biol. 12:768–774.

Rapaport, D., Brunner, M., Neupert, W., Westermann, B. 1998. Fzo1p is a mitochondrial outer membrane protein essential for the biogenesis of functional mitochondria in *Saccharomyces cerevisiae*. J. Biol. Chem. 273:20150–20155.

Reddy, A. S., Day, I. S. 2001. Analysis of the myosins encoded in the recently completed *Arabidopsis thaliana* genome sequence. Gen. Biol. 2:0024.1–0024.17

Reisen, D., Hanson, M. 2007. Association of six YFP-myosin XI-tail fusions with mobile plant cell organelles. BMC Plant Biol. 7:6.

Rice, S. E., Gelfand, V. I. 2006. Paradigm lost: milton connects kinesin heavy chain to miro on mitochondria. J. Cell Biol. 173:459–461.

Rizzuto, R., Brini, M., Pizzo, P., Murgia, M., Pozzan, T. 1995. Chimeric green fluorescent protein as a tool for visualizing subcellular organelles in living cells. Curr. Biol. 5:635–642.

Rojo, M., Legros, F., Chateau, D., Lombes, A. 2002. Membrane topology and mitochondrial targeting of mitofusins, ubiquitous mammalian homologs of the transmembrane GTPase Fzo. J. Cell Sci. 115:1663–1674.

Santel, A., Fuller, M. T. 2001. Control of mitochondrial morphology by a human mitofusin. J. Cell Sci. 114:867–874.

Scott, I., Logan, D. C. 2008. Mitochondria and cell death pathways in plants:Actions speak louder than words. Plant Signal. Behav. 3:475–477.

Scott, I., Logan, D. C. 2008. Mitochondrial morphology transition is an early indicator of subsequent cell death in *Arabidopsis*. New Phytol. 177:90–101.

Scott, I., Tobin, A. K., Logan, D. C. 2006. BIGYIN, an orthologue of human and yeast FIS1 genes functions in the control of mitochondrial size and number in *Arabidopsis thaliana*. J. Exp. Bot. 57:1275–1280.

Segui-Simarro, J. M., Jose Coronado, M., Staehelin, L. 2008. The mitochondrial cycle of arabidopsis shoot apical meristem and leaf primordium meristematic cells is defined by a perinuclear tentaculate/cage-like mitochondrion. Plant Physiol. 148:1380–1393.

Sesaki, H., Jensen, R. E. 1999. Division versus fusion: Dnm1p and Fzo1p antagonistically regulate mitochondrial shape. J. Cell Biol. 147:699–706.

Sesaki, H., Jensen, R. E. 2001. UGO1 encodes an outer membrane protein required for mitochondrial fusion. J. Cell Biol. 153:635–635.

Sesaki, H., Southard, S. M., Hobbs, A. E. A., Jensen, R. E. 2003. Cells lacking Pcp1p/Ugo2p, a rhomboid-like protease required for Mgm1p processing, lose mtDNA and mitochondrial structure in a Dnm1p-dependent manner, but remain competent for mitochondrial fusion. Biochem. Biophys. Res. Commun. 308:276–283.

Sheahan, M. B., Rose, R. J., McCurdy, D. W. 2004. Organelle inheritance in plant cell division:the actin cytoskeleton is required for unbiased inheritance of chloroplasts, mitochondria and endoplasmic reticulum in dividing protoplasts. Plant J. 37:379–390.

Shepard, K. A., Yaffe, M. P. 1997. Genetic and molecular analysis of Mdm14p and Mdm17p, proteins involved in organelle inheritence. Mol. Biol. Cell 8(Suppl. S):2585.

Shepard, K. A., Yaffe, M. P. 1999. The yeast dynamin-like protein, Mgm1p, functions on the mitochondrial outer membrane to mediate mitochondrial inheritance. J. Cell Biol. 144:711–720.

Simon, V. R., Swayne, T. C., Pon, L. A. 1995. Actin-dependent mitochondrial motility in mitotic yeast and cell-free systems – identification of a motor-activity on the mitochondrial surface. J. Cell Biol. 130:345–354.

Smirnova, E., Griparic, L., Shurland, D. L., van der Bliek, A. M. 2001. Dynamin-related protein Drp1 is required for mitochondrial division in mammalian cells. Mol. Biol. Cell 12:2245–2256.

Sogo, L. F., Yaffe, M. P. 1994. Regulation of mitochondrial morphology and inheritance by Mdm10p, a protein of the mitochondrial outer membrane. J. Cell Biol. 126:1361–1373.

Song, Z., Chen, H., Fiket, M., Alexander, C., Chan, D. C. 2007. OPA1 processing controls mitochondrial fusion and is regulated by mRNA splicing, membrane potential, and Yme1L. J. Cell Biol. 178:749–755.

Sparkes, I. A., Teanby, N. A., Hawes, C. 2008. Truncated myosin XI tail fusions inhibit peroxisome, golgi, and mitochondrial movement in tobacco leaf epidermal cells: a genetic tool for the next generation. J. Exp. Bot. 59:2499–2512.

Stickens, D., Verbelen, J. P. 1996. Spatial structure of mitochondria and ER denotes changes in cell physiology of cultured tobacco protoplasts. Plant J. 9:85–92.

Stokes, K. D., McAndrew, R. S., Figueroa, R., Vitha, S., Osteryoung, K. W. 2000. Chloroplast division and morphology are differentially affected by overexpression of FtsZ1 and FtsZ2 genes in *Arabidopsis*. Plant Physiol. 124:1668–1677.

Stowers, R. S., Megeath, L. J., Górska-Andrzejak, J., Meinertzhagen, I. A., Schwarz, T. L. 2002. Axonal transport of mitochondria to synapses depends on milton, a novel drosophila protein. Neuron 36:1063–1077.

Suzuki, M., Jeong, S. Y., Karbowski, M., Youle, R. J., Tjandra, N. 2003. The solution structure of human mitochondria fission protein Fis1 reveals a novel TPR-like helix bundle. J. Mol. Biol. 334:445–458.

Taguchi, N., Ishihara, N., Jofuku, A., Oka, T., Mihara, K. 2007. Mitotic phosphorylation of dynamin-related GTPase Drp1 participates in mitochondrial fission. J. Biol. Chem. 282:11521–11529.

Takahara, M., Takahashi, H., Matsunaga, S., Miyagishima, S., Takano, H., Sakai, A., Kawano, S., Kuroiwa, T. 2000. A putative mitochondrial ftsZ gene is present in the unicellular primitive red alga *Cyanidioschyzon merolae*. Mol. Gen. Genet. 264:452–460.

Tanaka, Y., Kanai, Y., Okada, Y., Nonaka, S., Takeda, S., Harada, A., Hirokawa, N. 1998. Targeted disruption of mouse conventional kinesin heavy chain, kif5B, results in abnormal perinuclear clustering of mitochondria. Cell 93:1147–1158.

Tieu, Q., Nunnari, J. 2000. Mdv1p is a WD repeat protein that interacts with the dynamin- related GTPase, Dnm1p, to trigger mitochondrial division. J. Cell Biol. 11:353–365.

Tieu, Q., Okreglak, V., Naylor, K., Nunnari, J. 2002. The WD repeat protein, Mdv1p, functions as a molecular adaptor by interacting with Dnm1p and Fis1p during mitochondrial fission. J. Cell Biol. 158:445–452.

Tiwari, B. S., Belenghi, B., Levine, A. 2002. Oxidative stress increased respiration and generation of reactive oxygen species, resulting in ATP depletion, opening of mitochondrial permeability transition, and programmed cell death. Plant Physiol. 128:1271–1281.

Tyler, D. D. 1992. The Mitochondrion in Health and Disease. New York: VCH.

Valiathan, R. R., Weisman, L. S. 2008. Pushing for answers: is myosin V directly involved in moving mitochondria? J. Cell Biol. 181:15–18.

Van Gestel, K., Kohler, R. H., Verbelen, J. P. 2002. Plant mitochondria move on F-actin, but their positioning in the cortical cytoplasm depends on both F-actin and microtubules. J. Exp. Bot. 53:659–667.
Virolainen, E., Blokhina, O., Fagerstedt, K. 2002. Ca2+-induced high amplitude swelling and cytochrome c release from wheat (*Triticum aestivum* L.) mitochondria under anoxic stress. Ann. Bot. 90:509–516.
Wakabayashi, J., Zhang, Z., Wakabayashi, N., Tamura, Y., Fukaya, M., Kensler, T. W., Iijima, M., Sesaki, H. 2009. The dynamin-related GTPase Drp1 is required for embryonic and brain development in mice. J. Cell Biol. 186:805–816.
Watanabe, W., Shimada, T., Matsunaga, S., Kurihara, D., Fukui, K., Arimura, T., Isobe, K., Itoh, K. 2007. Single-organelle tracking by two-photon conversion. Opt. Exp. 15:2490–2498.
Westermann, B., Neupert, W. 2000. Mitochondria-targeted green fluorescent proteins:convenient tools for the study of organelle biogenesis in *Saccharomyces cerevisiae*. Yeast 16:1421–1427.
Wong, E. D., Wagner, J. A., Gorsich, S. W., McCaffery, J. M., Shaw, J. M., Nunnari, J. 2000. The dynamin-related GTPase, Mgm1p, is an intermembrane space protein required for maintenance of fusion competent mitochondria. J. Cell Biol. 151:341–352.
Wong, E. D., Wagner, J. A., Scott, S. V., Okreglak, V., Holewinske, T. J., Cassidy-Stone, A., Nunnari, J. 2003. The intramitochondrial dynamin-related GTPase, Mgm1p, is a component of a protein complex that mediates mitochondrial fusion. J. Cell Biol. 160:303–311.
Yaffe, M. P. 1999. The machinery of mitochondrial inheritance and behavior. Science 283:1493–1497.
Yamaoka, S., Leaver, C. J. 2008. EMB2473/MIRO1, an *Arabidopsis* Miro GTPase, is required for embryogenesis and influences mitochondrial morphology in pollen. Plant Cell 20:589–601.
Yang, Y., Ouyang, Y., Yang, L., Beal, M. F., McQuibban, A., Vogel, H., Lu, B. 2008. Pink1 regulates mitochondrial dynamics through interaction with the fission/fusion machinery. Proc. Natl. Acad. Sci. 105:7070–7075.
Yao, N., Eisfelder, B. J., Marvin, J., Greenberg, J. T. 2004. The mitochondrion – an organelle commonly involved in programmed cell death in *Arabidopsis thaliana*. Plant J. 40:596–610.
Yao, N., Greenberg, J. T. 2006. Arabidopsis accelerated cell death2 modulates programmed cell death. Plant Cell 18:397–411.
Yonashiro, R., Ishido, S., Kyo, S., Fukuda, T., Goto, E., Matsuki, Y., Ohmura-Hoshino, M., Sada, K., Hotta, H., Yamamura, H., Inatome, R., Yanagi, S. 2006. A novel mitochondrial ubiquitin ligase plays a critical role in mitochondrial dynamics. EMBO J. 25:3618–3626.
Yoon, Y., Krueger, E. W., Oswald, B. J., McNiven, M. A. 2003. The mitochondrial protein hFis1 regulates mitochondrial fission in mammalian cells through an interaction with the dynamin-like protein DLP1. Mol. Cell. Biol. 23:5409–5420.
Yoshinaga, K., Arimura, S. I., Niwa, Y., Tsutsumi, N., Uchimiya, H., Kawai-Yamada, M. 2005. Mitochondrial behaviour in the early stages of ROS stress leading to cell death in *Arabidopsis thaliana*. Ann. Bot. 96:337–342.
Yu, X. H., Perdue, T. D., Heimer, Y. M., Jones, A. M. 2002. Mitochondrial involvement in tracheary element programmed cell death. Cell Death Diff. 9:189–198.
Zhang, X. C., Hu, J. P. 2008. FISSION1A and FISSION1B proteins mediate the fission of peroxisomes and mitochondria in *Arabidopsis*. Mol. Plant 1:1036–1047.
Zhang, L., Li, Y., Xing, D., Gao, C. 2009. Characterization of mitochondrial dynamics and subcellular localization of ROS reveal that HsfA2 alleviates oxidative damage caused by heat stress in *Arabidopsis*. J. Exp. Bot. 60:2073–2091.
Zheng, M., Beck, M., Muller, J., Chen, T., Wang, X., Wang, F., Wang, Q., Wang, Y., Baluska, F., Logan, D. C., Samaj, J., Lin, J. 2009. Actin turnover is required for myosin-dependent mitochondrial movements in *Arabidopsis* root hairs. PLoS One 4:e5961.
Zhu, Q., Hulen, D., Liu, T., Clarke, M. 1997. The cluA-mutant of *Dictyostelium* identifies a novel class of proteins required for dispersion of mitochondria. Proc. Natl. Acad. Sci. U.S.A. 94:7308–7313.
Zhuang, J., Dinsdale, D., Cohen, G. M. 1998. Apoptosis, in human monocytic THP.1 cells, results in the release of cytochrome c from mitochondria prior to their ultracondensation, formation

of outer membrane discontinuities and reduction in inner membrane potential. Cell Death Differ. 5:953–962.

Zimmer, C. 2009. Origins. On the origin of eukaryotes. Science 325:666–668.

Zottini, M., Barizza, E., Bastianelli, F., Carimi, F., Lo Schiavo, F. 2006. Growth and senescence of Medicago truncatula cultured cells are associated with characteristic mitochondrial morphology. New Phytol. 172:239–247.

Zuchner, S., Noureddine, M., Kennerson, M., Verhoeven, K., Claeys, K., De Jonghe, P., Merory, J., Oliveira, S. A., Speer, M. C., Stenger, J. E., Walizada, G., Zhu, D. Q., Pericak-Vance, M. A., Nicholson, G., Timmerman, V., Vance, J. M. 2005. Mutations in the pleckstrin homology domain of dynamin 2 cause dominant intermediate Charcot-Marie-Tooth disease. Nat. Genet. 37:289–294.

Zunino, R., Schauss, A., Rippstein, P., Andrade-Navarro, M., McBride, H. M. 2007. The SUMO protease SENP5 is required to maintain mitochondrial morphology and function. J. Cell Sci. 120:1178–1188.

Chapter 3
Plant Mitochondrial Genomes and Recombination

Maria P. Arrieta-Montiel and Sally A. Mackenzie

Abstract The mitochondrial genome of higher plants provides what is often considered a confusing picture of genome evolution, with extreme variation in its organization, size, and complexity. The genome's recombinogenic nature and chimeric gene content is unusual relative to what is seen in mitochondria of animals and fungi. But emerging mitochondrial sequence data from early land plants and recent studies of nuclear influence on mitochondrial genome behavior have provided insight into the evolutionary trends and possible rationale for the genomic variability that is seen. We review some of these recent findings in the context of plant adaptation. The versatility of the mitochondrial genome structure, in association with mitochondrial cellular signaling capacity, may constitute an important, plant-specific strategy for environmental responsiveness.

Keywords Mitochondrial recombination • Substoichiometric shifting • Nuclear gene control • Plant development

Abbreviations

CMS	Cytoplasmic male sterility
Fr	Fertility restorer
GA	Gibberellic acid
MSH1	MutS Homolog 1
Osb1	Organellar single-stranded DNA binding protein 1
SSS	Substoichiometric shifting

S.A. Mackenzie (✉)
Center for Plant Science Innovation, N305 Beadle Center, University of Nebraska, Lincoln, NE 68588-0660, USA
e-mail: smackenzie2@unl.edu

F. Kempken (ed.), *Plant Mitochondria*, Advances in Plant Biology 1,
DOI 10.1007/978-0-387-89781-3_3,

3.1 Why Study Plant Mitochondrial Genomes?

Studies of mitochondrial sequence data from animals, fungi, plants and protists provide compelling evidence that mitochondria likely arose from a single successful endosymbiotic event (Gray et al. 1999) over one billion years ago (see also Chap. 1). Following this event, much of the original genetic information from the endosymbiont was lost, massively rearranged, or transferred to the host genome, with only a small fraction retained within the organelle. What genome remains in present-day mitochondria is surprisingly similar in the particular gene collection retained across plant, fungal, and animal lineages, yet remarkably divergent in evolution of genome behavior. The mitochondrial genetic complement is generally comprised of about 13 electron transport and oxidative phosphorylation components, together with several tRNAs and ribosomal RNAs and proteins that, in Arabidopsis, comprise approximately 57 identifiable genes (Unseld et al. 1997), in yeast 43 genes (Talla et al. 2004), while in humans, about 37 genes (Anderson et al. 1981).

Animal mitochondrial genomes are characterized by their compactness; small (15–18 kb), often circular structures with dense gene organization, a single replication and transcription origin, and little or no intermolecular genetic variation. These genomes appear to comprise, for the most part, a highly homogenous population of DNA molecules. In contrast, plant and fungal genomes vary much more in size, gene organization, and replication. Perhaps most important to their distinct evolution, these genomes have linearized (Oldenburg and Bendich 2001; Kosa et al. 2006), and display a high propensity to undergo inter- and intramolecular DNA exchange. In plants, this recombination activity appears to be a key source of genetic variation, giving rise to intragenomic rearrangements, stoichiometric variation in copy number, and heteroplasmic organelle populations (reviewed by Woloszynska 2010).

The extensive mitochondrial genomic variation in higher plants has been well documented, but its genetic control and phenotypic consequences are only recently emerging. Such studies have begun to reveal that unusual mitochondrial genome behavior in plants participates directly in the evolution for these highly diverse and environmentally responsive, sessile organisms. Here we review some of these recent investigations.

3.2 The Importance of Double Strand Breaks in Plant Mitochondria

Plant mitochondrial genomes vary widely in size, ranging from 379 to 2,900 kb within the single plant family Cucurbitaceae (Alverson et al. 2010), and markedly in their organization. Even within a species, gene order differs greatly (Kubo et al. 2000; Allen et al. 2007; Kubo and Newton 2008). Replication of the genome is not well understood, although there is some evidence to support strand invasion and rolling circle mechanisms (Backert et al. 1997; Backert and Borner 2000; Oldenburg and Bendich 2001). Early genetic evidence also suggested that at least some portions of the mitochondrial genome might replicate autonomously (Folkerts and Hanson 1991; Levy et al. 1991; Janska and Mackenzie 1993).

The plant mitochondrial genome appears to be comprised of a heterogeneous population of large linear and smaller circular DNA forms, with linear configurations often highly branched (Oldenburg and Bendich 1996; Backert et al. 1997) and circularly permuted, as shown in the lower plant *Marchantia polymorpha* (Oldenburg and Bendich 2001). The distinctive features of plant mitochondrial genomes in size variation, variable gene order, multipartite structure, foreign sequence integrations, and extensive copy number variability all rely on a propensity for double-strand breaks (DSB) within the genome. Recent evidence of a genome organization comprised of a heterogeneous population of largely linear, highly branched molecules is consistent with this model. Reports suggesting that the plant mitochondrial genome replicates by a T4 phage-like mechanism, recombination-dependent via strand invasion and dependent on DSB (Oldenburg and Bendich 2001; Manchekar et al. 2006), would also be consistent with this model.

DSB activity, while vital for genomic processes of recombination, also represents cytotoxic forms of DNA damage. DSB misrepair creates mutations and genome instability. The two major paths of DSB repair involve non-homologous end joining (NHEJ) and homologous recombination (reviewed by Huertas (2010)). NHEJ occurs as ligation of two DNA ends with little or no processing. There is no apparent mechanism to ensure that the two ends being joined were originally contiguous and, thus, NHEJ often yields sequence chimeras. Such end joining can also involve regions of microhomology (McVey and Lee 2008). The second major DSB repair mechanism involves homologous recombination, and comprises a set of pathways that employ homologous stretches of DNA sequence, of genetically definable length, as template to repair the break.

Regions of unusually high DNA exchange activity, characterized by large (>1,000 bp) repeats, provide some indication of likely hotspots for DSB activity. One suggestion for the presence of these repeats is a role as replication origins in strand invasion-mediated replication initiation within large, linear DNA molecules (Zaegel et al. 2006). However, this possibility has not been formally tested.

Data available to date suggest that divergence in mitochondrial genome evolution between plant and animal lineages involved severe reduction of the animal form to minimal size, perhaps relinquishing some components of the original recombination machinery and condensing the genome to a highly homogenous population of circular molecules (Boore 1999). Conversely, mitochondrial genome evolution in the flowering plants apparently elaborated upon original recombination mechanisms to permit accumulation of extensive genomic variation (Bullerwell and Gray 2004). Some of this variation appears to be the consequence of expanded options for management of DSBs. While algal, moss (Terasawa et al. 2006), liverwort (Wang et al. 2009), and hornwort (Li et al. 2009) mitochondrial sequences display a degree of conservation in gene order and genome size, a striking increase in mitochondrial recombination activity in the angiosperms has given rise to highly dynamic genomic properties. Evidence of active recombination is lacking from the Bryophytes, but seems to be present in the lycophyte *Isoetes engelmannii* (Grewe et al. 2009). In *Isoetes*, the mitochondrial genome includes sequence breakpoints similar to asymmetric recombination junctions found in angiosperms. However, this genome does not appear to contain large repeats that participate in reciprocal recombination and subdivision of the genome (described below). The increasingly variable nature of the mitochondrial genome throughout more recent lineages of plants suggests that this versatility may be adaptive. The mitochondrial recombination processes associated with generating this variation are controlled by several recently identified nuclear genes.

3.3 Plant Mitochondrial Recombination Is Under Nuclear Gene Control

Earliest evidence of nuclear influence on plant mitochondrial recombination was surmised from reports of mitochondrial DNA polymorphisms in plants passaged through tissue culture (Li et al. 1988; Hauschner et al. 1998; Hartmann et al. 2000). These data implied that mitochondrial genome stability was reliant on nuclear controls that were relaxed under cell culture conditions. Likewise, particular crosses were shown to produce mitochondrial genome rearrangement (Newton and Coe 1986), implying that nuclear factors might be involved in maintaining nuclear-mitochondrial compatibility. Early reports showed that cytoplasmic male sterility (CMS) (see also Chap. 18) mutants could undergo spontaneous reversion to fertility, linked to mitochondrial DNA rearrangement, and this reversion frequency was influenced by nuclear genetic background (Escote et al. 1985; Mackenzie et al. 1988). Other types of mitochondrial rearrangements were also influenced by nuclear background (Laser et al. 1997). In the case of CMS in common bean, the mitochondrial DNA rearrangement to condition pollen fertility was directed by a single nuclear gene, designated *fertility restorer* (*Fr*) (Mackenzie and Chase 1990). This observation provided evidence that single nuclear factors controlling plant mitochondrial genome behavior would be identifiable.

CMS in the common bean is conditioned by expression of a chimeric mitochondrial sequence designated *pvs-orf239*. Introduction of the nuclear *Fr* allele to a CMS line, by crossing, produces a dramatic depletion of *pvs-orf239* from the genome (Janska et al. 1998). This rapid, dramatic and nuclear-directed change in relative copy number of the *pvs-orf239* sequence occurs by a phenomenon termed substoichiometric shifting (SSS) (Small et al. 1987, 1989; Box 3.1). The CMS-associated *pvs-orf239* sequence is present in all *P. vulgaris* lines tested to date, either substoichiometrically or in high copy number, implying that the identified nuclear-cytoplasmic interaction to control male sterility operates in natural populations (Arrieta-Montiel et al. 2001). This is likewise the case for a CMS determinant in the Brassicaceae, *orf138*, which is also controlled by SSS (Bellaoui et al. 1998).

Box 3.1 The process of substoichiometric shifting (SSS) in higher plants 1/100 cells

The plant mitochondrial genome is comprised of a multi-partite configuration as a consequence of high frequency recombination at large repeats (Fauron et al. 1995). Additional distinct DNA configurations, usually containing novel sequence arrangements, exist in the mitochondrial genome population at substoichiometric concentrations. One report estimated these substoichiometric DNAs at approximately one copy per 100–200 cells (Arrieta-Montiel et al. 2001), providing evidence of mitochondrial heteroplasmy. These novel arrangements can be seen as "ghost" bands, substoichiometric forms detected by DNA gel blot or PCR analysis (see Fig. 3.1). The ratio of predominant mitochondrial forms to substoichiometric can vary, and appears to be controlled by nuclear and/or environmental factors. A dramatic change in concentration of a mitochondrial DNA configuration relative to the predominant form is called substoichiometric shifting (SSS). Recent evidence suggests that SSS occurs as a consequence of changes in recombination activity at particular intermediate size genomic repeats (Arrieta-Montiel et al. 2009). This recombination is generally asymmetric, so that two parental forms of a repeat undergo DNA exchange to produce only one product; the second predicted product does not accumulate (Shedge et al. 2007). What dictates this asymmetric exchange, defining its polarity and frequency, is not yet understood, but it is believed that the novel recombinant product accumulates in direct proportion to SSS recombination activity. In Fig. 3.1, forms A and B represent the two repeat environments present in a WT (Col-0) mitochondrial genome. Upon activation of recombination, the stoichiometry of molecule C, is greatly increased. However, note that molecular C can be visualized as a faint ghost band in Col-0 as well. While asymmetric recombination has been well documented, it is not known whether it comprises the only mechanism underlying SSS activity.

The *orf138* sequence is widely distributed in natural wild radish populations (Yamagishi and Terachi 1996; Murayama et al. 2004), and its influence on male sterility appears to be regulated by SSS, frequency of the fertility restorer gene, and additional cytoplasmic conditions (Giancola et al. 2007). With gynodioecy observed in over 7% of angiosperms, it is reasonable to assume that SSS plays a significant role in its natural regulation (Richards 1997).

In 2003, Abdelnoor et al. reported the cloning of a nuclear gene controlling plant mitochondrial recombination behavior in the Arabidopsis mutant originally designated *chloroplast mutator* (*chm1-1*). The gene, subsequently designated *MSH1*, encodes a homolog of the *E. coli* MutS, a protein that participates in mismatch repair and suppression of ectopic recombination (Schofield and Hsieh 2003). During *MSH1* evolution, the gene fused with a GIY-YIG type homing endonuclease

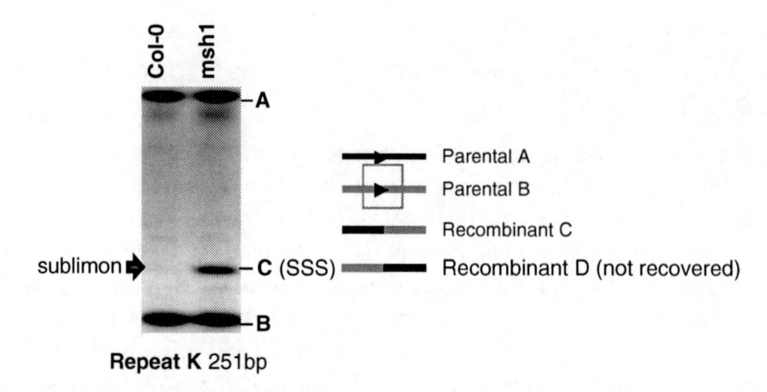

Fig. 3.1 Evidence of SSS in the *msh1* mutant of Arabidopsis. DNA gel blot analysis shows recombination at intermediate repeat K. The two parental configurations are designated A and B. In Col-0, recombination at this repeat occurs at extremely low frequency, producing a ghost band or sublimon (designated by *arrow*). In the *msh1* mutant, repeat K is active in DNA exchange and the recombinant form (C) accumulates

at its 3′ terminus, and it encodes at least three plant-specific domains in addition to the MutS-characteristic DNA binding and ATPase domains; all identified domains appear to be essential to its function (Abdelnoor et al. 2006). *MSH1* is thought to participate in the suppression of illegitimate recombination, and disruption of the gene results in increased DNA exchange activity at more than 36 repeat sites within the Arabidopsis mitochondrial genome (Arrieta-Montiel et al. 2009; Davila et al., unpublished).

Because *MSH1* apparently monitors activity at a large number of mitochondrial recombination sites, and its disruption results in viable mutant plants, suppression of the gene presents a feasible, direct means of genetically perturbing the mitochondrion. RNAi-mediated suppression of *MSH1* in plants produces extensive rearrangement and SSS of the mitochondrial genome, allowing studies of the physiological implications of these events (Sandhu et al. 2007).

Two additional nuclear genes have been cloned and shown to participate in the control of mitochondrial DNA recombination in Arabidopsis. The first, designated *OSB1*, resembles a single-stranded DNA binding protein and possesses a unique PDF domain (Zaegel et al. 2006). Disruption of *OSB1* appears to effect a more gradual accumulation of mitochondrial genomic rearrangements involving recombination activity. Another component of mitochondrial genome maintenance discovered in Arabidopsis shows strong similarity to the bacterial recombination component *RecA*, and is designated *AtRECA3* (Shedge et al. 2007). While the OSB1 protein appears to function exclusively within the mitochondrion, there appear to be at least three expressed *RecA*-homologous gene products, one plastid-targeting (*AtRECA1*), one mitochondrial, and the third dual (*AtRECA2*) (Khazi et al. 2003; Shedge et al. 2007). The mitochondrially-targeted product appears to partially complement bacterial *RecA* function (Khazi et al. 2003), suggesting that the two proteins share some similarity in function. *RecA* proteins have three well-conserved regions, an ATP-binding and hydrolysis domain, a monomer–monomer interaction domain and a carboxy-terminal string of negatively charged amino acids. The higher plant *RECA3* gene displays important modifications in each of these three regions and, similar to the plant *MSH1*, appears to have undergone plant-specific evolution.

A related *RecA* homolog, designated *PpRecA1*, has also been found to function within mitochondria of the bryophyte species *Physcomitrella patens* (Odahara et al. 2007). The *PpRecA1* gene was shown to be induced by treatment with MMS and bleomycin, both of which cause DSBs, suggesting its involvement in DNA repair. No evidence has been found in mosses for the extensive mitochondrial recombination that is characteristic of higher plants (Terasawa et al. 2006), so it appears that *PpRecA1* and *AtRecA3* likely carry out related but distinct functions within the plant lineages where they are found. Likewise, the Arabidopsis *msh1 recA3* double mutant shows evidence of enhanced mitochondrial genomic rearrangement and a pronounced effect on plant development, implying that *MSH1* and *RECA3* function in distinct pathways within higher plants (Shedge et al. 2007).

Recently, it was suggested that members of the *Whirly* gene family might also participate in mitochondrial and plastid genome stability (Maréchal et al. 2008, 2009). The mitochondrially localized, single-stranded DNA binding protein encoded by *AtWhy2*, when overexpressed, leads to reduced mitochondrial DNA content and altered plant morphology (Maréchal et al. 2008). However, no mitochondrial changes are evident when the gene is disrupted. These results were interpreted as indication that the *AtWhy2* gene carries out a redundant function.

3.4 The Genetic Variability of Plant Mitochondria

Several aspects of plant mitochondrial genome structure can be accounted for by the recombination features of the genome. At least 17 plant mitochondrial genomes have been sequenced. These include the two liverworts *Marchantia polymorphia* (Oda et al. 1992) and *Pleurozia purpurea* (Wang et al. 2009), the hornwort *Megaceros aenigmaticus* (Li et al. 2009), the moss *Physcomitrella patens* (Terasawa et al. 2006), a lycophyte (Grewe et al. 2009), the gymnosperm *Cycas taitungensis* (Chaw et al. 2008), and 12 angiosperms (*Arabidopsis thaliana, Beta vulgaris* subs *vulgaris, Brassica napus, Nicotiana tabacum, Oryza sativa, Sorghum bicolor, Tripsacum dactyloides, Triticum aestivum, Zea luxurians, Zea mays*, subsp. *Mays, Zea mays* subsp. *Parviglumis*, and *Zea perennis*) (Unseld et al. 1997; Kubo et al. 2000; Notsu et al. 2002; Handa 2003; Clifton et al. 2004; Ogihara et al. 2005; Sugiyama et al. 2005; Tian et al. 2006). Several additional genomes have been characterized by physical mapping efforts. In all angiosperms investigated, the genomes contain repeated sequences that appear to mediate DNA exchange as well as nonhomologous end joining. These recombination features are not shared in the available bryophyte sequences, with the transition perhaps occurring within the lycophytes (Grewe et al. 2009).

Evidence of recombination activity derived originally from physical mapping efforts, where it is common to observe junction points within overlapping clone contigs. From these original studies emerged an unusually complex picture of recombination activity, including DNA exchange at repeated sequences ranging from as small as 4 bp to as large as 120 kbp in size, often involving sequences of unknown origin or sequences that subsequently disappear, and giving rise to unusual or incomplete recombination products (Kubo and Newton 2008). However, these types of studies, by necessity, often focused on derived or incomplete genome information without benefit of direct analysis of de novo recombination substrate and product. For example, It is now clear that junction points within a genome can represent genomic environments derived by DNA exchange with sequences present substoichiometrically within the genome (Feng et al. 2009; Woloszynska and Trojanowski 2009), or involving multi-step recombination processes (Arrieta-Montiel et al. 2009). Some previous studies have relied on incomplete mitochondrial genome sequence information. In CMS common bean, the SSS process involved in restoration of fertility and spontaneous reversion was originally thought to be effected by differential replication of an autonomous DNA molecule rather than by recombination mechanisms (Janska and Mackenzie 1993). However, recently a 314 bp repeat was identified (Woloszynska and Trojanowski 2009) that appears to mediate the recombination event leading to SSS for induction of the CMS phenotype. Likewise, in Arabidopsis Col-0 ecotype, analysis of recombination within the region surrounding *Atp9* in the *Maternal Distorted Leaf* (*mdl*) mutant suggested DNA exchange at an *Atp9*-associated repeat of only 11 bp in size (Sakamoto et al. 1996). This observation was based on the only mitochondrial sequence available at the time, derived from ecotype C24 (Unseld et al. 1997). In fact, the recombination in question occurred between two much larger repeats (repeats L at 249 bp and K at 251 bp) (Arrieta-Montiel et al. 2009), a detail that could not have been learned until the Col-0 mitochondrial genome was shown to contain a novel *Atp9* environment (Forner et al. 2005) present only substoichiometrically in C24 (Arrieta-Montiel et al. 2009).

Analysis of mitochondrial recombination in *Arabidopsis thaliana*, where it is feasible to assess de novo recombination activity in the *msh1* mutant, has allowed identification of the following three classes of repeated sequences.

3.4.1 Large (>1,000-bp) Repeated Sequences

Large (>1,000-bp) repeated sequences appear to mediate high frequency, reciprocal DNA exchange that can result in subdivision of the genome to a multipartite configuration. In genomes where multiple pairs of repeats are present, this recombination activity produces a complex, inter-recombining population of heterogeneous molecules (Fauron et al. 1995). This type of recombination generally results in nearly equimolar amounts of the parental and recombinant forms. Large repeated sequences have been identified in nearly all of the flowering plants analyzed and, in some cases, they are responsible for most of the observed genome structural complexity. In maize, where five different cytotypes were sequenced to reveal a variable, multi-partite configuration, almost half of the total genome complexity can be accounted for by the presence of large repeats (of 22 repeats identified, 540 bp to 120 kbp in size, 17 were >1 kb) (Allen et al. 2007). An exception is found in the white mustard (*Brassica hirta*) (Palmer and Herbo 1987) where, similar to the liverworts *M. polymorpha* and *P. purpurea*, the mitochondrial genome is comprised of a single species of DNA molecule and no large repeated sequences (Oda et al. 1992; Wang et al. 2009). In fact, *B. hirta* contains only one copy of a large repeat that is recombinationally active in closely related *Brassica* relatives, an observation that will likely prove useful for studying the function of this repeat sequence.

3.4.2 Intermediate (ca. 50 to 500-bp) Repeated Sequences

Intermediate (ca. 50 to 500-bp) repeated sequences have been identified that mediate very low frequency, asymmetric DNA exchange. This rare recombination activity is associated with the emergence of novel DNA polymorphisms, intraspecific genomic variation, and the process of SSS. This SSS genomic rearrangement activity is often accompanied by changes in the plant's phenotype. For example, genomic rearrangements involving SSS are associated with both the induction of cytoplasmic male sterility (Sandhu et al. 2007) and the spontaneous reversion to fertility (Janska et al. 1998). While SSS activity can occur spontaneously at low frequency, similar rearrangements are also observed following passage through tissue culture (Brettell et al. 1980; Li et al. 1988; Vitart et al. 1992).

Intermediate repeats in most plant mitochondrial genomes are much more numerous than large repeats. In Arabidopsis, de novo asymmetric recombination is observed at over 36 repeat pairs (Arrieta-Montiel et al. 2009; Davila et al., unpublished). Similar repeat numbers are predicted in most of the plant mitochondrial genome sequences presently available (Arrieta-Montiel et al. 2009). In beet, 72 repeats range from 50 to 626 bp (Kubo et al. 2000); in wheat, 7 repeated sequences range from 100 to 493 bp (Ogihara et al. 2005); in maize, 32 repeats range from 100 to 830 bp (Clifton et al. 2004); in sorghum, 20 repeats range from 100 to 303 bp, and in tobacco, 26 repeat range from 100 to 405 (Arrieta-Montiel et al. 2009). While spontaneous DNA exchange activity at these repeats occurs only at very low frequency, it is possible to elevate DNA exchange activity in Arabidopsis by disruption of nuclear genes *MSH1*, *RECA3*, or *OSB1* (Zaegel et al. 2006; Shedge et al. 2007). In the case of the *msh1* mutant, all of the intermediate repeats become active simultaneously (Arrieta-Montiel et al. 2009). Less detail is currently available regarding genomic responses to *recA3* or *osb1* mutations. Evidence suggests that genomic rearrangements in *recA3* lines are less extensive than in *msh1*, and recombination can involve reciprocal exchange under some circumstances (Shedge et al. 2007).

An especially peculiar feature of SSS activity by intermediate repeat DNA exchange is that the substrates and/or products of this type of recombination may be present within the plant at substoichiometric levels, relative to the predominant mitochondrial genome. In common bean, substoichiometric forms have been estimated at levels as low as one copy per 100–200 cells of the plant (Arrieta-Montiel et al. 2001). In Arabidopsis it has been shown that sequences amplified by substoichiometric shifting were undetectable even by PCR within vegetative tissues of the plant prior to the rearrangement (Arrieta-Montiel et al. 2009). In pearl millet, recombination at an intermediate repeat pair occurred when one copy of the pair was present at nearly undetectable substoichiometric levels (Feng et al. 2009). Even more stunning is the observation, made in common bean, tobacco and Arabidopsis, that these substoichiometric forms may comprise nearly complete mitochondrial genomic complements (Janska et al. 1998; Arrieta-Montiel et al. 2009). This observation implies that some mitochondrial sequences reside solely within the meristematic cells of the plant, with little or no permeation to vegetative cells of the plant during development. If this is the case, one must assume that the recombination activity that effects substoichiometric shifting is likely confined to a small window of development, presumably during reproduction (Shedge et al. 2007). It has been suggested that the confinement of mitochondrial DNA molecules to extremely low copy number or limited distribution might allow the accumulation or retention of mutations with an insignificant phenotypic consequence, hence, little or no selection pressure for their elimination (Small et al. 1987).

3.4.3 Small (4–25 bp) Repeated Sequences

Small (4–25 bp) repeated sequences also participate in plant mitochondrial genomic changes. However, it appears more likely that these repeats participate in nonhomologous end joining (NHEJ). The small repeated sequences are often found at the junction of unrelated sequences to form chimeric regions within the genome. In a number of cases, these gene chimeras are expressed, and have been associated with the cytoplasmic male sterility trait (Schnable and Wise 1998; Hanson and Bentolila 2004). Likewise, numerous foreign DNA introgressions of plastid, nuclear, viral and unknown origin are observed in plant mitochondrial genomes (Kubo and Newton 2008). These, too, are presumably the consequence of NHEJ activity. Homologous recombination within the plant mitochondrial genome appears to be confined to repeats greater than 1,000 bp in size. Asymmetric recombination at intermediate repeats appears to require a minimum stretch of sequence homology, estimated at ca. 50 bp in Arabidopsis (Arrieta-Montiel et al. 2009; Davila et al., unpublished). Small repeats found at the junctions of sequence chimeras involve even smaller stretches of sequence homology, likely too small to support DNA exchange.

3.5 Other Interpretations of Mitochondrial Genetic Variation

While the majority of studies involving plant mitochondrial genetic variation have implied that mitochondria are strictly maternal in their inheritance, a growing body of literature suggests that particular plant systems may provide evidence of biparental inheritance, or at least of paternal leakage of mitochondria. In natural populations of *Silene vulgaris*, a gynodioecious plant, observed heteroplasmy is thought to reflect, at least in part, a level of biparental mitochondrial contribution (Pearl et al. 2009). Biparental inheritance models dramatically alter interpretation of much of the data presently available regarding mitochondrial genome behavior (McCauley and Olson 2008), and suggest that mitochondrial recombination and heteroplasmy might be more prevalent than originally thought in not only plants but fungal and animal systems (Barr et al. 2005). Paternal leakage, even at low levels, is assumed to sustain a permanent level of cytoplasmic polymorphism within gynodioecious populations (Wade and McCauley 2005).

While *Silene vulgaris* has been a primary focus for much of the studies of paternal leakage, recent studies have also shown biparental inheritance patterns for both plastid and mitochondrial DNA in *Pelargonium* (Weihe et al. 2009). These types of studies may call into question much of the previous data interpretation for plant mitochondrial polymorphism and heteroplasmy studies. However, one complication with these studies is the inherent difficulty of distinguishing mitochondrial genomic heteroplasmy derived from biparental inheritance patterns from the massive genomic rearrangements by SSS that occur in response to suppressed expression of *MSH1* (Arrieta-Montiel et al. 2009). In some cases where biparental inheritance was suspected, the paternal polymorphism was detected substoichiometrically in the maternal parent (Laser et al. 1997). While PCR-based technology is a valuable tool for such analyses, some substoichiometric forms are found to be in such a small proportion of the plant cells as to be undetectable by PCR. For example, in Arabidopsis a sequence is detected within the mitochondrial genome of Col-0 but absent in ecotype C-24 (Forner et al. 2005). While the particular sequence is undetectable in C-24 mitochondria by PCR analysis, it is clearly present in C-24 *msh1* mutants (Arrieta-Montiel et al. 2009). One can envision, within natural plant populations, environmental stress conditions that result in reduced expression of *MSH1* and permit individual flowering plants to undergo SSS, giving rise to a heterogeneous population of mitotypes in the progeny. Not only would such a scenario produce the appearance of biparental inheritance, but it would constitute a remarkable adaptive strategy.

It is possible that plants implement both biparental cytoplasmic inheritance strategies and SSS as mechanisms for enhancing cytoplasmic genetic variation. In fact, it is also possible that some substoichiometric forms that participate in SSS might have originally derived from paternal transmission. To assess these possibilities will require a more thorough investigation of the substoichiometric forms already present within the maternal parent of a cross, as well as the consequences of *MSH1* suppression within plant species where heteroplasmy is well documented.

It will also be important to investigate in more detail mitochondrial behavior within the pollen tube and newly formed zygote of plants suspected of displaying biparental inheritance behavior.

3.6 Mitochondrial Recombination Influences Plant Development

The first, and most prevalent, trait associated with mitochondrial genome dysfunction in higher plants was CMS (see also Chap. 18). CMS mutants are reported in over 150 plant species, displaying a maternally inherited inability to produce or shed viable pollen in an otherwise phenotypically normal plant. Consequently, the trait has been useful commercially for the economical production of hybrid seed. Numerous investigations of the underlying basis of CMS have shown that in each case a novel mitochondrial genomic lesion, comprised of a unique, expressive sequence chimera, appears to condition the phenotype (Schnable and Wise 1998). Some identified chimeras are complex, likely involving multiple steps in their assembly. For example, in the CMS-T cytoplasm maize line, *T-urf13* may have involved as many as seven steps to form the chimeric sequence (Pring et al. 1988). Similarly, some CMS-associated chimeras include sequences of unknown origin (Schnable and Wise 1998). Yet, CMS has been induced via protoplast fusion, tissue culture, and wide hybridization, suggesting that the complex series of events predicted to create the chimeras cannot occur de novo under these conditions. Recently, experiments to transgenically downregulate the *MSH1* nuclear gene in plants by RNAi suppression resulted in cytoplasmic male sterility as a consequence of SSS (Sandhu et al. 2007). Although CMS induction was only reported for tomato and tobacco, similar experiments in sorghum, millet, and soybean have produced identical results (Sandhu et al., unpublished). These observations, together with the detection of CMS-associated sequences substoichiometrically in natural plant populations (Arrieta-Montiel et al. 2001; Murayama et al. 2004), suggest that mitochondrial CMS loci are pervasive in higher plants, but may go undetected in most cases.

Other plant phenotypes have also been associated with SSS activity in plant mitochondria. These include alterations in plastid development to produce leaf variegation and mosaic phenotypes (Jiao et al. 2005; Bartoszewski et al. 2007; Sandhu et al. 2007), altered GA-mediated pathways for growth and flowering (Pellny et al. 2008), changes in stem height and flowering time (Albert et al. 2003), and changes in leaf morphology (Arrieta-Montiel and Mackenzie, unpublished). In *msh1* mutants of Arabidopsis, or with *MSH1* RNAi suppression in other plant species, all of these aberrant phenotypes are observed, sorting into phenotypically distinct plant lineages (Sandhu et al., unpublished). Whether or not these developmental modifications are a consequence of altered redox status or ROS signaling is a question of active investigation (Dutilleul et al. 2003).

In *msh1 recA3* double mutants of Arabidopsis, where mitochondrial genome rearrangement is more extensive than in *msh1* alone, plants display greatly reduced growth rate and altered floral morphologies (Shedge et al. 2007). However, these plants also display a remarkable level of enhanced thermo-tolerance (Shedge et al. 2010). Double mutant plants are altered in both mitochondrial and nuclear transcript profiles, mounting a primed defense response under normal growth conditions. These observations, and others (Adamo et al. 2008; Giraud et al. 2008; Szal et al. 2009), imply mitochondrial participation in plant defense as well.

3.7 Could Mitochondrial Status, Conditioned by Recombination, Influence Plant Adaptation?

Cytoplasmic male sterility is a mitochondrially conferred trait shown to participate in the gynodioecious reproductive patterns of natural plant populations. Yet mitochondrial sequences conditioning the phenotype are distinct in each case, and associated cellular phenotypes differ in sporophytic and gametophytic pollen abortion patterns. This widespread diversity among CMS systems implies that natural selection has occurred for the phenotype throughout flowering plants. Moreover, disruption of *MSH1* expression has given rise to the CMS phenotype in six different plant species tested to date (Sandhu et al. 2007; Sandhu et al., unpublished). These observations, taken together, suggest that angiosperms, in general, retain the capacity for CMS induction through mitochondrial recombination. Interestingly, plants also appear to possess, through mitochondrial SSS, an ability to reverse the CMS phenotype (Bellaoui et al. 1998; Janska et al. 1998). While regulation of this "on-off switch" for pollen sterility is not yet understood, one possibility is that *MSH1* expression is modulated by particular environmental conditions to affect CMS-inducing mitochondrial recombination processes. Similarly, an absence of available pollen producers in the vicinity of a male-sterile plant may result in an altered carbon balance signal within a plant to induce SSS for reversion to fertility. The phenomenon of reversion has been described as a singular pod or revertant sector on an otherwise male sterile plant (Mackenzie et al. 1988; Andersson 1999).

The tobacco CMSII mutant, also derived by SSS, displays evidence of altered carbon and nitrogen sensing. Evidence from this mutant suggests that mitochondria participate both in GA metabolism and in nitrate sensing (Pellny et al. 2008). These observations imply that mitochondria participate directly in programmed developmental pathways.

While further investigation is clearly needed, observations of CMS behavior and the emergence of additional plant phenotypes from mitochondrial DNA recombination are likely evidence of nuclear-mitochondrial coevolution that is distinct to the plant kingdom. Nuclear genes like *MSH1* may participate in an environmental sensing process that allows the plant to capitalize sporadically on the recombinogenic nature of the plant mitochondrial genome. In this way, the plant has the capacity to enhance genetic variation and condition reversible adaptive responses somatically. Whether such a system operates in nature is a subject of future investigation.

Acknowledgments We apologize to any authors whose relevant work was omitted from this review. References were selected to illustrate concepts, but space limitations precluded discussion of all publications relevant to the topic. Work described from the Mackenzie laboratory was supported by funding from the Department of Energy (DE-FG02-07ER15564) and the National Science Foundation (IOS 0820668 and MCB 0744104).

Glossary

Asymmetric recombination: DNA exchange event that produces only one of the two predicted products.

Reciprocal recombination: DNA exchange event that gives rise to two distinct recombinant products.

Heteroplasmy: A heterogeneous mitochondrial population that often undergoes a subsequent process of sorting and may give rise to phenotypically chimeric individuals.

Cytoplasmic male sterility: A variant, maternally inherited plant phenotype characterized by the inability to shed viable pollen.

Gynodioecy: Plant populations in which bisexual (hermaphrodite) flowers are produced on one plant and female flowers are produced on another.

References

Abdelnoor, R. V., Christensen, A. C., Mohammed, S., Munoz-Castillo, B., Moriyama, H., Mackenzie, S. A. 2006. Mitochondrial genome dynamics in plants and animals: convergent gene fusions of a MutS homologue. J. Mol. Evol. 63:165–173.

Adamo, A., Pinney, J. W., Kunova, A., Westhead, D. R., Meyer, P. 2008. Heat stress enhances the accumulation of polyadenylated mitochondrial transcripts in Arabidopsis thaliana. PLoS One 6:e2889.

Albert, B., Lelandais, C., Pla, M., Leuret, C., Vitart, V., Mathieu, C., Sihachakr, D., Godelle, B., De Paepe, R. 2003. Amplification of *Nicotiana sylvestris* mitochondrial subgenomics is under nuclear control and is associated with phenotypic changes. Genetica 117:17–25.

Allen, J. O., Fauron, C. M., Minx, P., Roark, L., Oddiraju, S., Lin, G. N., Meyer, L., Sun, H., Kim, K., Wang, C., Du, F., Xu, D., Gibson, M., Cifrese, J., Clifton, S. W., Newton, K. J. 2007. Comparisons among two fertile and three male-sterile mitochondrial genomes of maize. Genetics 177:1173–1192.

Alverson, A. J., Wei, X., Rice, D. W., Stern, D. B., Barry, K., Palmer, J. D. 2010. Insights into the evolution of mitochondrial genome size from complete sequences of *Citrullus lanatus* and *Cucurbita pepo* (Cucurbitaceae). Mol. Biol. Evol. Epub.

Anderson, S., Bankier, A. T., Barrell, B. G., De Bruijn, M. H. L., Coulson, A. R., et al. 1981. Sequence and organization of the human mitochondrial genome. Nature 290:457–465.

Andersson, H. 1999. Female and hermaphrodite flowers on a chimeric gynomonoecious Silene vulgaris plant produce offspring with different genders: a case of heteroplasmic sex determination? J. Hered. 90:563–565.

Arrieta-Montiel, M., Lyznik, A., Woloszynska, M., Janska, H., Tohme, J., Mackenzie, S. 2001. Tracing evolutionary and developmental implications of mitochondrial stoichiometric shifting in the common bean. Genetics 158:851–864.

Arrieta-Montiel, M. P., Shedge, V., Davila, J., Christensen, A. C., Mackenzie, S. A. 2009. Diversity of the Arabidopsis mitochondrial genome occurs via nuclear-controlled recombination activity. Genetics 183:1261–1268.

Backert, S., Borner, T. 2000. Phage T4-like intermediates of DNA replication and recombination in the mitochondria of the higher plant Chenopodium album (L.). Curr. Genet. 37:304–314.

Backert, S., Nielsen, B. L., Borner, T. 1997.The mystery of the rings: Structure and replication of mitochondrial genomes from higher plants. Trends Plant Sci. 2:477–483.

Barr, C. M., Neiman, M., Taylor, D. R. 2005. Inheritance and recombination of mitochondrial genomes in plants, fungi and animals. New Phytol. 168:39–50.

Bartoszewski, G., Havey, M. J., Ziolkowska, A., Dlugosz, M., Malepszy, S. 2007. The selection of mosaic (MSC) phenotype after passage of cucumber (*Cucumis sativus* L) through cell culture – a method to obtain plant mitochondrial mutants. J. Appl. Genet. 48:1–9.

Bellaoui, M., Martin-Canadell, A., Pelletier, G., Budar, F. 1998. Low-copy-number molecules are produced by recombination, actively maintained and can be amplified in the mitochondrial genome of Brassicaceae: relationship to reversion of the male sterile phenotype in some cybrids. Mol. Gen. Genet. 257:177–185.

Boore, J. L. 1999. Animal mitochondrial genomes. Nucl. Acids Res. 27:1767–1780.

Brettell, R. I. S., Thomas, E., Ingram, D. S. 1980. Reversion of Texas male-sterile cytoplasm maize in culture to give fertile, T-toxin resistant plants. Theor. Appl. Genet. 58:55–58.

Bullerwell, C. E., Gray, M. W. 2004. Evolution of the mitochondrial genome: protist connections to animals, fungi and plants. Curr. Opin. Microbiol. 7:528–534.

Chaw, S. M., Shih, A. C., Wang, D., Wu, Y. W., Liu, S. M., Chou, T. Y. 2008. The mitochondrial genome of the gymnosperm *Cycas taitungensis* contains a novel family of short interspersed elements, Bpu sequences, and abundant RNA editing sites. Mol. Biol. Evol. 25:603–615.

Clifton, S. W., Minx, P., Fauron, C. M., et al. 2004. Sequence and comparative analysis of the maize NB mitochondrial genome. Plant Physiol. 136:3486–3503.

Dutilleul, C., Garmier, M., Noctor, G., Mathieu, C., Chétrit, P., Foyer, C. H., de Paepe, R. 2003. Leaf mitochondria modulate whole cell redox homeostasis, set antioxidant capacity, and determine stress resistance through altered signaling and diurnal regulation. Plant Cell 15:1212–1226.

Escote, L. J., Gabay-Laughnan, S. J., Laughnan, J. R. 1985. Cytoplasmic reversion to fertility in cms-S maize need not involve loss of linear mitochondrial plasmids. Plasmid 14:264–267.

Fauron, C., Casper, M., Gao, Y., Moore, B. 1995 The maize mitochondrial genome: dynamic, yet functional. Trends Genet. 11:228–235.

Feng, X., Kaur, A. P., Mackenzie, S. A., Dweikat, I. M. 2009. Substoichiometric shifting in the fertility reversion of cytoplasmic male sterile pearl millet. Theor. Appl. Genet. 118:1361–1370.

Folkerts, O., Hanson, M. R. 1991. The male sterility-associated pcf gene and the normal atp9-1 gene in Petunia are located on different mitochondrial DNA molecules. Genetics 129:885–895.

Forner, J., Weber, B., Wietholter, C., Meyer, R. C., Binder, S. 2005. Distant sequences determine end formation of *cox3* transcripts in *Arabidopsis thaliana* ecotype C24. Nucl. Acids Res. 33:4673–4682.

Giancola, S., Rao, Y., Chaillou, S., Hiard, S., Martin-Canadell, A., Pelletier, G., Budar, F. 2007. Cytoplasmic suppression of Ogura cytoplasmic male sterility in European natural populations of Raphanus raphanistrum. Theor. Appl. Genet. 114:1333–1343.

Giraud, E., Ho, L. H., Clifton, R., Carroll, A., Estavillo, G., Tan, Y. F., Howell, K. A., Ivanova, A., Pogson, B. J., Millar, A. H., Whelan, J. 2008. The absence of ALTERNATIVE OXIDASE1a in Arabidopsis results in acute sensitivity to combined light and drought stress. Plant Physiol. 147:595–610.

Gray, M. W., Burger, G. and Lang, B. F. 1999. Mitochondrial evolution. Science 283:1476–1481.

Grewe, F., Viehoever, P., Weisshaar, B., Knoop V. 2009. A trans-splicing group I intron and tRNA-hyperediting in the mitochondrial genome of the lycophyte *Isoetes engelmannii*. Nucl. Acids Res. 37:5093–5104.

Handa, H. 2003. The complete nucleotide sequence and RNA editing content of the mitochondrial genome of rapeseed (*Brassica napus* L.): comparative analysis of the mitochondrial genomes of rapeseed and *Arabidopsis thaliana*. Nucl. Acids Res. 31:5907–5916.

Hanson, M., Bentolila, S. 2004. Interactions of mitochondrial and nuclear genes that affect male gametophyte development. Plant Cell 16(suppl):S154–S169.

Hartmann, C., Henry, Y., Tregear, J., Rode, A. 2000. Nuclear control of mitochondrial genome reorganization characterized using cultured cells of ditelosomic and nullisomic-tetrasomic wheat lines. Curr. Genet. 38:156–162.

Hauschner, H., Yesodi, V., Izhar, S., Tabib, Y., Firon, N. 1998. Cytoplasmic diversity caused by mitochondrial (mt) DNA dynamics and mt gene expression in petunia, in Plant Biotechnology and *In Vitro* Biology in the 21st Century, ed. A. Altman. Proceedings of the IX IAPTC International Congress Plant Tissue and Cell Culture, Amsterdam, pp. 147–150.

Huertas, P. 2010. DNA resection in eukaryotes: deciding how to fix the break. Nat. Struct. Mol. Biol. 17(1):11–16.

Janska, H., Mackenzie, S. A. 1993. Unusual mitochondrial genome organization in cytoplasmic male sterile common bean and the nature of cytoplasmic reversion to fertility. Genetics 135:869–879.

Janska, H., Sarria, R.,Woloszynska, M., Arrieta-Montiel, M., Mackenzie, S. A. 1998. Stoichiometric shifts in the common bean mitochondrial genome leading to male sterility and spontaneous reversion to fertility. Plant Cell 10:1163–1180.

Jiao, S., Thornsberry, J. M., Elthon, T. E., Newton, K. J. 2005 Biochemical and molecular characterization of photosystem I deficiency in the NCS6 mitochondrial mutant of maize. Plant Mol. Biol. 57:303–313.

Khazi FR, Edmondson AC, Nielsen BL. 2003. An Arabidopsis homologue of bacterial RecA that complements an E. coli recA deletion is targeted to plant mitochondria. Mol Genet Genomics. 269:454-63.

Kosa, P., Valach, M., Tomaska, L., Wolfe, K. H., Nosek, J. 2006. Complete DNA sequences of the mitochondrial genomes of the pathogenic yeasts *Candida orthopsilosis* and *Candida metapsilosis*: insight into the evolution of linear DNA genomes from mitochondrial telomere mutants. Nucl. Acids Res. 34:2472–2481.

Kubo, T., Newton, K. J. 2008. Angiosperm mitochondrial genomes and mutations. Mitochondrion 8:5–14.

Kubo, T., Nishizawa, S., Sugawara, A., Itchoda, N., Estiati, A., Mikami, T. 2000. The complete nucleotide sequence of the mitochondrial genome of sugar beet (*Beta vulgaris* L.) reveals a novel gene for tRNA ^{Cys}(GCA). Nucl. Acids Res. 28:2571–2576.

Laser, B., Mohr, S., Odenback, W., Oettler, G., Kuck, U. 1997. Parental and novel copies of the mitochondrial orf25 gene in the hybrid crop-plant triticale: predominant transcriptional expression of the maternal gene copy. Curr. Genet. 32:337–347.

Levy, A. A., André, C. P., Walbot, V. 1991. Analysis of a 120-kilobase mitochondrial chromosome in maize. Genetics 128:417–424.

Li, X. Q., Chétrit P., Mathieu C., Vedel F., De Paepe R., Remy, R., Ambard-Bretteville, F. 1988. Regeneration of cytoplasmic male sterile protoclones of *Nicotiana sylvestris* with mitochondrial variations. Curr. Genet. 13:261–266.

Li, Wang. B., Liu, Y., Qiu, Y. L. 2009. The complete mitochondrial genome sequence of the hornwort Megaceros aenigmaticus shows a mixed mode of conservative yet dynamic evolution in early land plant mitochondrial genomes. J. Mol. Evol. 68:665–678.

Mackenzie, S.A., Chase, C.D. 1990. Fertility restoration is associated with loss of a portion of the mitochondrial genome in cytoplasmic male-sterile common bean. Plant Cell 2:905–912.

Mackenzie, S. A., Pring, D. R., Bassett, M. J., Chase, C. D. 1988. Mitochondrial DNA rearrangement associated with fertility restoration and cytoplasmic reversion to fertility in cytoplasmic male sterile *Phaseolus vulgaris* L. Proc. Natl. Acad. Sci. U.S.A. 85:2714–2717.

Manchekar, M., Scissum-Gunn, K., Song, D., Khazi, F., McLean, S. L., Nielsen, B. L. 2006. DNA recombination activity in soybean mitochondria. J. Mol. Biol. 356:288–299.

Maréchal, A., Parent, J. S., Sabar, M., Véronneau-Lafortune, F., Abou-Rached, C., Brisson, N. 2008. Overexpression of mtDNA-associated AtWhy2 compromises mitochondrial function. BMC Plant Biol. 8:42.

Maréchal, A., Parent, J. S., Véronneau-Lafortune, F., Joyeux, A., Lang, B. F., Brisson, N. 2009. Whirly proteins maintain plastid genome stability in Arabidopsis. Proc. Natl. Acad. Sci. U.S.A. 106:14693–14698.

McCauley, D. E., Olson, M. S. 2008. Do recent findings in plant mitochondrial molecular and population genetics have implications for the study of gynodioecy and cytonuclear conflict? Evolution 62:1013–1025.

McVey, M., Lee, S. E. 2008. MMEJ repair of double strand breaks (director's cut): deleted sequences and alternative endings. Trends Genet. 24:529–538.

Murayama, K., Yahara, T., Terachi, T. 2004 Variation of female frequency and cytoplasmic male-sterility gene frequency among natural gynodioecious populations of wild radish (*Raphanus sativus* L.). Mol. Ecol. 13:2459–2464.

Newton, K., Coe, E. H. 1986. Mitochondrial DNA changes in abnormal growth (nonchromosomal stripe) mutants of maize. Proc. Natl. Acad. Sci. U.S.A. 83:7363–7366.

Notsu, Y., Masood, S., Nishikawa, T., Kubo, N., Akiduki, G., Nakazono, M., Hirai, A., Kadowaki, K. 2002. The complete sequence of the rice (*Oryza sativa* L.) mitochondrial genome: frequent DNA sequence acquisition and loss during the evolution of flowering plants. Mol. Genet. Genome 268:434–445.

Oda, K., Yamato, K., Ohta, E., Nakamura, Y., Takemura, M., Nozato, N., Akashi, K., Kanegae, T., Ogura, Y., Kohchi, T., et al. 1992. Gene organization deduced from the complete sequence of liverwort Marchantia polymorpha mitochondrial DNA: a primitive form of plant mitochondrial genome. J. Mol. Biol. 223:1–7.

Odahara, M., Inouye, T., Fujita, T., Hasebe, M., Sekine, Y. 2007. Involvement of mitochondrial-targeted *RecA* in the repair of mitochondrial DNA in the moss, *Physicomitrella patens*. Genes Genet. Syst. 82:43–51.

Ogihara, Y., Yamazaki, Y., Murai, K., et al. 2005. Structural dynamics of cereal mitochondrial genomes as revealed by complete nucleotide sequencing of the wheat mitochondrial genome. Nucl. Acids Res. 33:6235–6250.

Oldenburg, D. J., Bendich, A. J. 1996. Size and structure of replicating mitochondrial DNA in cultured tobacco cells. Plant Cell 8:447–461.

Oldenburg, D. J., Bendich, A. J. 2001. Mitochondrial DNA from the liverwort *Marchantia polymorpha*: circularly permuted linear molecules, head-to-tail concatemers, and a 5′ protein. J. Mol. Biol. 310:549–562.

Palmer, J. D., Herbo, L. A. 1987. Unicircular structure of the *Brassica hirta* mitochondrial genome. Curr. Genet. 11:565–570.

Pearl, S. A., Welch, M. E., McCauley, D. E. 2009. Mitochondrial heteroplasmy and paternal leakage in natural populations of *Silene vulgaris*, a gynodioecious plant. Mol. Biol. Evol. 26:537–5345.

Pellny, T. K., Van Aken, O., Dutilleul, C., Wolff, T., Groten, K., Bor, M., de Paepe, R., Reyss, A., Van Breusegem, F., Noctor, G., Foyer, C. H. 2008. Mitochondrial respiratory pathways modulate nitrate sensing and nitrogen-dependent regulation of plant architecture in *Nicotiana sylvestris*. Plant J. 54:976–992.

Pring, D. R., Gengenbach, B. G., Wise, R. P. 1988 Recombination is associated with polymorphism of the mitochondrial genomes of maize and sorghum. Philos. Trans. R. Soc. Lond. B: Biol. Sci. 319:187–198.

Richards, A. J. 1997. Plant Breeding Systems. London: Chapman and Hall.

Sakamoto, W., Kondo, H., Murata, M., Motoyoshi, F. 1996. Altered mitochondrial gene expression in a maternal distorted leaf mutant of Arabidopsis induced by chloroplast mutator. Plant Cell 8:1377–1390.

Sandhu, A. P., Abdelnoor, R. V., Mackenzie, S. A. 2007. Transgenic induction of mitochondrial rearrangements for cytoplasmic male sterility in crop plants. Proc. Natl. Acad. Sci. U.S.A. 104:1766–1770.

Schnable, P., Wise, R. 1998. The molecular basis of cytoplasmic male sterility and fertility restoration. Trends Plant Sci. 3:175–180.

Schofield, M. J., Hsieh, P. 2003. DNA mismatch repair: molecular mechanisms and biological function. Annu. Rev. Microbiol. 57:579–608.

Shedge, V., Arrieta-Montiel, M., Christensen, A. C., Mackenzie, S. A. 2007. Plant mitochondrial recombination surveillance requires unusual RecA and MutS homologs. Plant Cell 19:1251–1264.

Shedge, V., Davila, J., Arrieta-Montiel, M. P., Mohammed, S., Mackenzie, S. A. 2010. Extensive rearrangement of the Arabidopsis mitochondrial genome elicits cellular conditions for thermotolerance. Plant Physiol. 152:1960–1970.

Small, I. D., Isaac, P. G., Leaver, C. J. 1987. Stoichiometric differences in DNA molecules containing the atpA gene suggest mechanisms for the generation of mitochondrial genome diversity in maize. EMBO J. 6:865–869.

Small, I., Suffolk, R., Leaver, C. J. 1989. Evolution of plant mitochondrial genomes via substoichiometric intermediates. Cell 58:69–76.

Sugiyama, Y., Watase, Y., Nagase, M., Makita, N., Yagura, S., Hirai, A., Sugiura, M. 2005. The complete nucleotide sequence and multipartite organization of the tobacco mitochondrial genome: comparative analysis of mitochondrial genomes in higher plants. Mol. Genet. Genome. 272:603–615.

Szal, B., Lukawska, K., Zdolińska, I., Rychter, A. M. 2009. Chilling stress and mitochondrial genome rearrangement in the MSC16 cucumber mutant affect the alternative oxidase and antioxidant defense system to a similar extent. Physiol. Plant. 137:435–445.

Talla, E., Anthouard, V., Bouchier, C., Frangeul, L., Dujon, B. 2004. The complete mitochondrial genome of the yeast *Kluyveromyces thermotolerans*. FEBS Letts. 579:30–40.

Terasawa, K., Odahara, M., Kabeya, Y., Kikugawa, T., Sekine, Y., Fujiwara, M., Sato, N. 2006. The mitochondrial genome of the moss *Physcomitrella patens* sheds new light on mitochondrial evolution in land plants. Mol. Biol. Evol. 24:699–709.

Tian, X., Zheng, J., Hu, S., Yu, J. 2006. The rice mitochondrial genomes and their variations. Plant Physiol. 140:401–410.

Unseld, M., Marienfeld, J. R., Brandt, P., Brennicke, A. 1997. The Mitochondrial genome of *Arabidopsis thaliana* contains 57genes in 366924 nucleotides. Nat. Genet. 15:57–61.

Vitart, V., De Paepe R., Mathieu C., Chétrit P., Vedel F. 1992. Amplification of substoichiometric recombinant mitochondrial DNA sequences in a nuclear, male sterile mutant regenerated from protoplast culture in *Nicotiana sylvestris*. Mol. Gen. Genet. 233:193–120.

Wade, M. J., McCauley, D. E. 2005. Paternal leakage sustains the cytoplasmic polymorphism underlying gynodioecy but remains invisible by nuclear restorers. Am. Nat. 166:592–602.

Wang, B., Xue, J., Li, L., Liu, Y., Qiu, Y. L. 2009. The complete mitochondrial genome sequence of the liverwort *Pleurozia purpurea* reveals extremely conservative mitochondrial genome evolution in liverworts. Curr. Genet. 55:601–609.

Weihe, A., Apitz, J., Pohlheim, F., Salinas-Hartwig, A., Borner, T. 2009. Biparental inheritance of plastidial and mitochondrial DNA and hybrid variegation in Pelargonium. Mol. Genet. Genome 282:587–593.

Woloszynska, M. 2010. Heteroplasmy ans stoichiometric complexity of plant mitochondrial genomes – though this be madness, yet there's method in't. J. Exp. Bot. 61:657–671.

Woloszynska, M., Trojanowski, D. 2009. Counting mtDNA molecules in *Phaseolus vulgaris*: sublimons are constantly produced by recombination via short repeats and undergo rigorous selection during substoichiometric shifting. Plant Mol. Biol. 70:511–521.

Yamagishi, H., Terachi, T. 1996. Molecular and biological studies on male-sterile cytoplasm in the Cruciferae. III. Distribution of Ogura-type cytoplasm among Japanese wild radishes and Asian radish cultivars. Theor. Appl. Genet. 93:325–332.

Zaegel, V., Guermann, B., Le Ret, M., Andres, C., Meyer, D., Erhardt, M., Canada, Y. J., Gualberto, J. M., Imbault, P. 2006. The plant-specific ssDNA binding protein OSB1 is involved in the stoichiometric transmission of mitochondrial DNA in Arabidopsis. Plant Cell 18:3548–3563.

Part II
Transcription & RNA Processing

Chapter 4
Transcription in Plant Mitochondria

Karsten Liere and Thomas Börner

Abstract Transcription in plant mitochondria is sustained by phage-type RNA polymerases that are encoded by a small nuclear encoded *RpoT* gene family (RNA polymerase of the T-phage type). In angiosperms, it consists of a minimum of two genes, encoding enzymes that are imported into mitochondria (*RpoTm*) and plastids (*RpoTp*). An additional gene for a dually targeted enzyme is found in eudicots (*RpoTmp*). The promoters recognized by mitochondrial RpoT enzymes contain a CRTA-, YYTA-, or DDTA-motif, often accompanied by an upstream A/T-rich region involved in modulating transcription. Transcription of most mitochondrial genes is driven by multiple promoters. Multiple promoters may ensure transcription despite possible mitochondrial genome rearrangements. Transcription factors have not been identified in plant mitochondria. Although proteins related to transcription factors in fungal and animal mitochondria (mtTFA and mtTFB) are encoded in the nuclear genome of *Arabidopsis*, none of these proteins has been shown to function as a factor assisting RNA polymerases in promoter recognition.

Keywords Plant mitochondrial transcription • Phage-type RNA polymerase • Mitochondrial promoters • transcription factors • Evolution

4.1 Introduction to Mitochondrial RNA Polymerases

4.1.1 Evolution

During the course of evolution mitochondrial genomes lost most of their genes by either full loss or transfer to the nucleus (Martin 2003; Timmis et al. 2004; Huang et al. 2005; see also Chap. 1). Consequently, mitochondrial genomes only retained a small set of identified genes encoding, in the case of higher plants, approximately

T. Börner (✉)
Institut für Biologie (Genetik), Humboldt-Universität zu Berlin, Chausseestr. 117, 10115, Berlin, Germany
e-mail: thomasboerner@rz.hu berlin.de

F. Kempken (ed.), *Plant Mitochondria*, Advances in Plant Biology 1,
DOI 10.1007/978-0-387-89781-3_4, © Springer Science+Business Media, LLC 2011

30 proteins, up to 20 tRNAs and 3 rRNAs (Unseld et al. 1997; Kubo et al. 2000; Notsu et al. 2002; Handa 2003; Clifton et al. 2004; Ogihara et al. 2005; Sugiyama et al. 2005; see Chap. 1). Unexpectedly, several mitochondrial replication genes were probably acquired from a bacteriophage early in the evolution of eukaryotes (Shutt and Gray 2006). Similarly, all eukaryotic mitochondria possess a nuclear-encoded phage-type RNA polymerase which replaced the ancestral bacterial-type RNA polymerase (Greenleaf et al. 1986; Masters et al. 1987; Cermakian et al. 1996; Tiranti et al. 1997; Weihe et al. 1997; Gray and Lang 1998; Gaspari et al. 2004; Weihe 2004). The only known exceptions are the Jakobids, primitive protists like *Reclinomonas americana* which encode a bacterial-type RNA polymerase in their mitochondrial genome (Lang et al. 1997; Gray and Lang 1998).

The single-subunit RNA polymerases of bacteriophages are not related to the multi-subunit RNA polymerases used in transcription in bacteria, archea, and the eukaryotic nucleus. Phage RNA polymerases are assumed to descend from a common ancestor, most probably a DNA polymerase or reverse transcriptase (Cermakian et al. 1997). The discovery of cryptic prophages related to bacteriophages like T3 and T7 in several proteobacterial genomes suggested that a prophage containing an RNA polymerase gene might have been already present in the endosymbiotic predecessor of the mitochondrion. During evolution the prophage, together with most mitochondrial genes, was possibly transferred to the host nucleus (Martin 2003; Timmis et al. 2004; Huang et al. 2005). Eventually, the RNA polymerase gene was reactivated and the enzyme retargeted into the mitochondria (Filée and Forterre 2005). Now, the phage RNA polymerase took over transcription from the mitochondrial-encoded bacterial-type RNA polymerase which was subsequently lost during further evolution.

Organellar phage-type RNA polymerases share high sequence homology with each other and the T7 RNA polymerase. The bacteriophage T7 enzyme is a 99 kDa protein

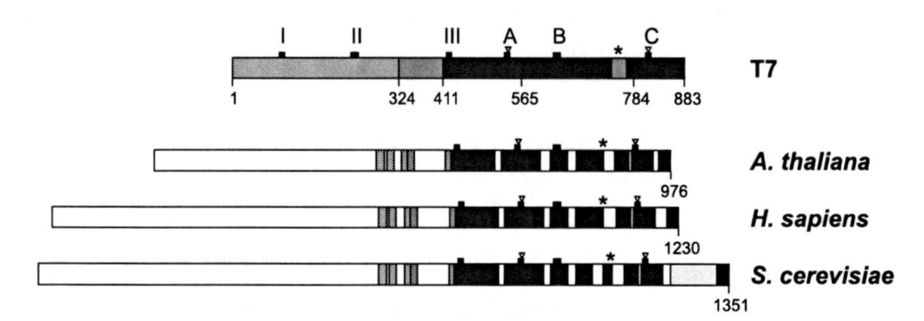

Fig. 4.1 Conserved sequence regions of the T7 phage and eukaryotic phage-type RNA polymerases. *Arabidopsis* RpoTm is shown as one member of the group of mitochondrial and plastidial enzymes of land plants which do not greatly differ in their exon–intron-structure and size of the N-terminus. *Black squares* mark the positions of the motifs T/DxxGR (III), A, B, and C (according to Delarue et al. 1990) that are important for RNA polymerase function and conserved in all enzymes, and of a T7-specific hairpin structure (I) and a subdomain involved in recognition of A/T-rich promoter regions (II). Open triangles denote the invariant residues Asp537 and Asp812 acting as ligands to two catalytic Mg^{2+} ions at the RNA polymerase active site (Woody et al. 1996). The region corresponding to the specificity loop needed for promoter recognition is depicted by an asterisk; a yeast-specific C-terminal insertion is shown in *light gray*

consisting of 883 amino acids (Steitz 2004) while the organellar RNA polymerases are larger, mainly due to N-terminal extensions (Fig. 4.1). Plant phage-type RNA polymerases have a mass of around 110 kDa (Lerbs-Mache 1993; Hess and Börner 1999). The C-terminal amino acid sequences are highly conserved in organellar phage-type RNA polymerases and similar to those of the T7 RNA polymerase (see Fig. 4.1; Chang et al. 1999; Hess and Börner 1999). They form structural domains called "fingers," "thumb," and "palm." This region comprises the catalytically relevant part of the enzyme (McAllister 1993; Sousa et al. 1993). Genes for phage-type RNA polymerases have also been detected on so-called linear plasmids occurring in mitochondria of some fungi and plants. Their function is unknown (Box 4.1).

Box 4.1 RNA polymerases encoded by linear plasmids

In some higher plants and fungi circular and linear extrachromosomal mitochondrial DNA molecules are found in addition to the complex main mitochondrial genome. Sequence analyses of such linear plasmids revealed several genes related to genes of viruses/bacteriophages phages suggesting a viral descent of these DNA elements. Several linear plasmids encode proteins with strong resemblance to the nuclear-encoded phage-type RNA polymerases. Though also the plasmid genes seem to originate from phages, phylogenetic analyses indicate individual evolution of the nuclear genes coding for organellar RNA polymerases and the plasmid genes (reviewed in Kempken et al. 1992; Rohe et al. 1992; Hess and Börner 1999; U. Richter, personal communication; Ikeda and Gray 1999b; Handa 2008). A role of plasmid encoded phage-type RNA polymerases in transcription of mitochondrial genes has yet to be shown and remains questionable (Handa 2008).

4.1.2 Plant Mitochondrial RNA Polymerases

In plants, phage-type RNA polymerases are encoded by the small nuclear encoded *RpoT* gene family (RNA polymerase of the T-phage type). Members of this family have been identified in various angiosperms such as *Chenopodium album* (Weihe et al. 1997), *Arabidopsis thaliana* (Hedtke et al. 1997, 2000), *Nicotiana tabacum* (Hedtke et al. 2002), *Nicotiana sylvestris* (Kobayashi et al. 2001b, 2002), *Zea mays* (Young et al. 1998; Chang et al. 1999), wheat (Ikeda and Gray 1999c), barley (Emanuel et al. 2004), and the moss *Physcomitrella patens* (Kabeya et al. 2002; Richter et al. 2002). Very recently, a single *RpoT* gene was detected and its product characterized as a mitochondrial RNA polymerase in the lycophyte *Selaginella moellendorfii* (Yin et al. 2009). Green algae like *Chlamydomonas reinhardtii* possess also only a single *RpoT* gene which most likely encodes the mitochondrial transcriptase (Maier et al. 2008; A. Weihe, unpublished.).

In eudicotyledonous plant species such as *Nicotiana sylvestris* and *Arabidopsis thaliana,* the *RpoT* gene family consists of three genes, encoding products that are

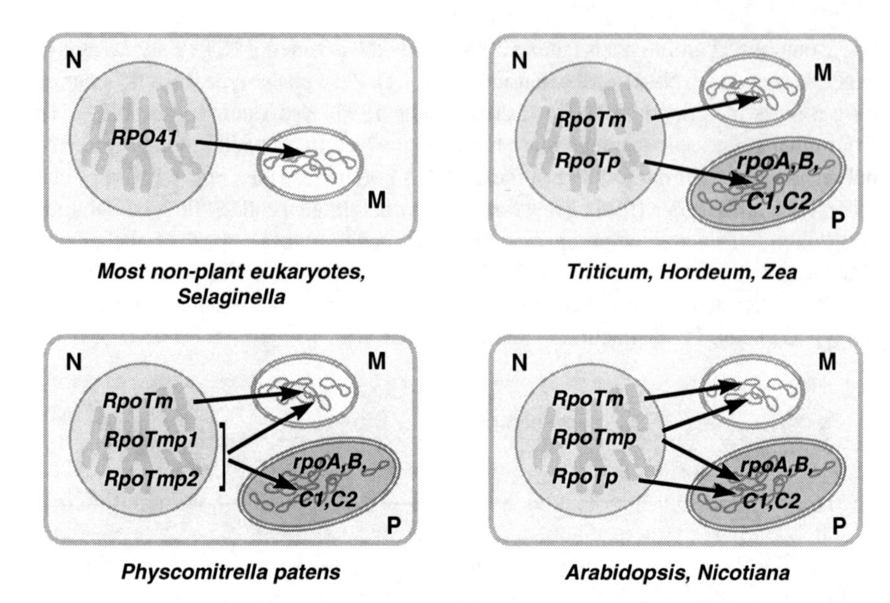

Fig. 4.2 Localization of nuclear genes encoding organellar phage-type RNA polymerases in different organisms. Genes in the nucleus (N) of eukaryotic organisms code for T3/T7 phage-like RNA polymerases which, following their synthesis in the cytoplasm, are imported into mitochondria (M) and plastids (P) as indicated by *arrows*

imported into mitochondria (*RpoTm*), plastids (*RpoTp*) and dually into both organelles (*RpoTmp,* Fig. 4.2; Hedtke et al. 1997 1999, 2000, 2002; Kobayashi et al. 2001a, 2002). The amphidiploid genome of *Nicotiana tabacum* contains six *RpoT* genes with its two sets of three genes from the two diploid parental species (Hedtke et al. 2002). With *RpoTm* and *RpoTp,* however, in monocots so far only *RpoT* genes encoding exclusively targeted enzymes were found (Chang et al. 1999; Ikeda and Gray 1999c; Emanuel et al. 2004). In the moss *Physcomitrella patens,* the nuclear genome harbors three *RpoT* genes. Two of them seem to be capable of dual targeting as a result of translation initiation at two different in-frame AUG start codons (Richter et al. 2002; Box 4.2). Interestingly, the third *RpoT* gene found in the *Physcomitrella* genome project database (http://genomeportal.jgi-psf.org/Phypa1_1/Phypa1_1.home.html) encodes an enzyme which seems to be exclusively targeted to mitochondria (PpRpoTm; U. Richter, unpubl. data). Phylogenetic analyses suggest that the *RpoT* gene families of *Physcomitrella* and higher plants have arisen by independent gene duplication events dating after the separation of bryophytes and tracheophytes (Richter et al. 2002).

For some time distinct functions of *Arabidopsis* RpoTm and RpoTmp in mitochondria remained elusive. The *RpoTm* and *RpoTmp* genes in *Arabidopsis* have been reported to display overlapping expression patterns and the two RNA polymerases were therefore proposed to recognize different types of mitochondrial promoters (Emanuel et al. 2006). Kühn et al. (2007), however, showed RpoTm but not RpoTmp to recognize mitochondrial promoters in vitro. Based on an analysis of *RpoT* transcript levels in an *Arabidopsis* mutant lacking RpoTmp another view

Box 4.2 Dual targeting of Rpot polymerases

Richter et al. (2002) showed that for both PpRpoTmp1 and PpRpoTmp2 translation initiation at the first AUG start codon yields a product targeted to plastids, while initiation at the second start codon leads to a mitochondrially targeted enzyme. Kabeya and Sato (2005) reported promoter: GFP constructs of both *PpRpoTmp1* and *PpRpoTmp2*, as well as *Arabidopsis RpoTmp* in the presence of their native 5′-UTR to be exclusively translated from the second AUG start codon and targeted to mitochondria in vivo. However, an exclusive targeting to mitochondria is not in agreement with the observation that the RpoTmp homolog of spinach was immunochemically detected in chloroplasts but not in mitochondria (Azevedo et al. 2006). Moreover, studies on *Arabidopsis* mutants lacking RpoTmp not only revealed a function of this enzyme in mitochondrial (Kühn et al. 2009) but also in plastid transcription (Baba et al. 2004; Courtois et al. 2007; Swiatecka-Hagenbruch et al. 2008). Therefore, one might speculate the 5′-UTR to be involved in regulating the localization of RpoTmp on translational level in a tissue- and/or developmental-specific manner (Christensen et al. 2005). Notwithstanding the mentioned conflicting results with regard to the localization of RpoTmp polymerases to plastids, all available data indicate that three phage-type RNA polymerases operate in the mitochondria of *Physcomitrella.*

on RpoTm and RpoTmp functions has been presented by Baba et al. (2004). In mitochondria the lack of RpoTmp only affected *atp1* transcript levels. However, the induction of several plastid genes in dark-grown seedlings upon illumination was significantly delayed in the mutant. Therefore, RpoTmp was suggested to be the key RNA polymerase transcribing plastid (and possibly mitochondrial) genes during early seedling development, while both RpoTm and RpoTp take over transcription in their respective organelle later in development.

In plastid transcription, recent studies in transgenic lines lacking RpoTmp or RpoTp provided evidence for two distinct promoters being utilized by RpoTmp (*Prrn16*-139, *PclpP*-58; Courtois et al. 2007; Swiatecka-Hagenbruch et al. 2008). While P*rrn16*-139 is active only during germination and early seedling development (Courtois et al. 2007), which coincides with the idea of RpoTmp becoming inactivated later in plastid development (Azevedo et al. 2008), P*clpP*-58 is active throughout plant development (Zoschke et al. 2007). Similarly, distinct roles of RpoTm and RpoTmp in mitochondrial transcription were very recently demonstrated by re-examination of *Arabidopsis* mutants lacking RpoTmp (Kühn et al. 2009). The study by Kühn et al. (2009) confirmed RpoTm to function as the basic RNA polymerase in mitochondria of dicots required for the transcription of most, if not all, mitochondrial genes. The lack of RpoTmp, however, led to decreased amounts of *nad6* and *cox1* transcripts, which correlated with reduced abundances of the respiratory chain complexes I and IV. Interestingly, decreased transcription in the *rpoTmp* mutant was not associated with changes in promoter utilization for these genes, therefore suggesting an importance of additional gene-specific *cis*-elements in recruiting RpoTmp and not RpoTm. By enhancing the levels of specific mitochondrial RNAs, RpoTmp may represent a transcriptional mechanism allowing mitochondria to independently control the abundances of complexes I and IV to fine-tune the electron flow through these complexes in response to developmental or metabolic requirements of the organelle and the cell (Kühn et al. 2009).

4.2 Plant Mitochondrial Promoters

Vertebrate mitochondrial genomes contain only three promoters. A single promoter (LSP) directs transcription of the complete L-strand into a polygenic primary transcript that is processed to one mRNA and eight tRNAs. Transcription of the complete H-strand is initiated from promoter HSP2. Processing of the long polygenic transcripts results in twelve mRNAs, fourteen tRNA and two rRNAs. Additional rRNA transcripts are achieved by transcription from HSP1 (located in close neighborhood to HSP2) and specific termination at the end of gene encoding the 16S rRNA under involvement of a protein factor, mTERF1 (see below and Box 4.3; Asin-Cayuela and Gustafsson 2007; Scarpulla 2008). In contrast, transcription in mitochondria of higher plants starts from many promoters. Therefore, a high percentage of monogenic/monocistronic primary transcripts should be generated in addition to the proposed polygenic/polycistronic RNAs (Dombrowski et al. 1998).

Historically, mapping primary transcript initiation sites and subsequent comparison of upstream flanking regions identified several mitochondrial promoter sequences in the plant mitochondrial genome, thereby revealing a CRTA-motif similar to the plastid YRTA-motif of plastid NEP promoters (see reviews in Fey and Maréchal-Drouard 1999; Binder and Brennicke 2003; Liere and Börner 2007). In dicots the CRTA-motif is part of a nona-nucleotid sequence overlapping the initiation site (see Fig. 4.3c; Binder et al. 1996). In monocots, however, the mitochondrial promoter sequences exhibit considerably less conservation (see Fig. 4.3a; Fey and Maréchal-Drouard 1999; Hess and Börner 1999).

In vitro analyses with transcriptionally competent mitochondrial extracts not only confirmed the relevance of the YRTA-motif, but also identified promoter sequences of up to 25 nucleotides comprising the transcription start site to be required for correct and efficient initiation of transcription (Hanic-Joyce and Gray 1991; Rapp and Stern 1992; Rapp et al. 1993; Caoile and Stern 1997; Dombrowski et al. 1999; Hoffmann and Binder 2002). To achieve full transcriptional activity in vitro the pea *atp9* promoter required additional nucleotides from the position –14 to +4, and the maize *atp1* promoter nucleotides –12 to +5. Mutagenesis of nucleotides of both the pea *atp9* and the maize *atp1* promoter at positions –3, –2, and +1, although not identical, led to a drastic reduction of the transcriptional activity in vitro in both cases (Rapp and Stern 1992; Rapp et al. 1993; Caoile and Stern 1997; Dombrowski et al. 1999). However, exchanging the highly conserved

Box 4.3 The mTERF protein family

The mTERF-related proteins may form another family of plant proteins that might contain transcription factors. Initially identified in human mitochondria, the mTERF1 protein was shown to bind to a strong bidirectional terminator sequence of 28 nucleotides downstream of the 16S rRNA gene and to specify site-specific transcription termination in mitochondrial lysates (Kruse et al. 1989; Daga et al. 1993; Roberti et al. 2003). More recently, the human mTERF1 protein was found to bind simultaneously to the terminator and the HSP1 initiation site likely accounting for observed stimulatory effects of mTERF on rRNA transcription (Martin et al. 2005). Furthermore, studies on the function of the *Drosophila* mTERF3 showed that this protein may also act as a negative regulator of mitochondrial transcription (reviewed in Roberti et al. 2009). In plants, the complexity of mTERF function is reflected by the in silico identification of 33 and 28 genes for mTERF-related proteins in *Arabidopsis* and rice, respectively, exhibiting putative mitochondrial or chloroplast targeting sequences (Linder et al. 2005). In chloroplasts, the mTERF-related protein SOLDAT10 seems to be involved in retrograde signaling (Meskauskiene et al. 2009). A mitochondrial localized mTERF homologue, MOC1, has been identified in the green alga *Chlamydomonas reinhardtii* (Schönfeld et al. 2004). Loss of the MOC1 gene causes a high light-sensitive phenotype and, interestingly, an aberrant pattern of mitochondrial transcription of genes encoding cytochrome c oxidase subunits and ubiquinone-NAD subunits also affected in rpoTmp mutants (Kühn et al. 2009). Yet, a function of these proteins in plant mitochondria is so far unknown.

nucleotides at positions –7, –1, and +2 had no influence on the transcriptional activity from the pea *atp9* promoter. Therefore, a strict correlation between the evolutionary conservation of the nucleotides and their functional relevance in determining the promoter activity seems unlikely. Furthermore, most mitochondrial promoters in monocots comprise an A/T-rich sequence element upstream of the promoter core motif, which was shown to be involved in positively modulating in vitro transcription (see Fig. 4.3; Rapp et al. 1993; Binder et al. 1995). Substituting the nucleotides within this A/T-rich sequence of the pea *atp9* promoter region with the complementary bases, i.e., not changing the A/T content, still led to a drastic reduction in promoter activity. Thus, rather the identity of the nucleotides in this region seems to be important for its function then the A/T content itself (Dombrowski et al. 1999; Hoffmann and Binder 2002).

Recently, an extensive study on identifying mitochondrial promoters in *Arabidopsis* showed further alternative promoter motifs (Kühn et al. 2005). Apart from CRTA-type consensus motifs, loose or unusual sequences for the tetra-nucleotide core-motif such as ATTA and RGTA were also found (see Fig. 4.3c, DDTA-type), as well as promoters

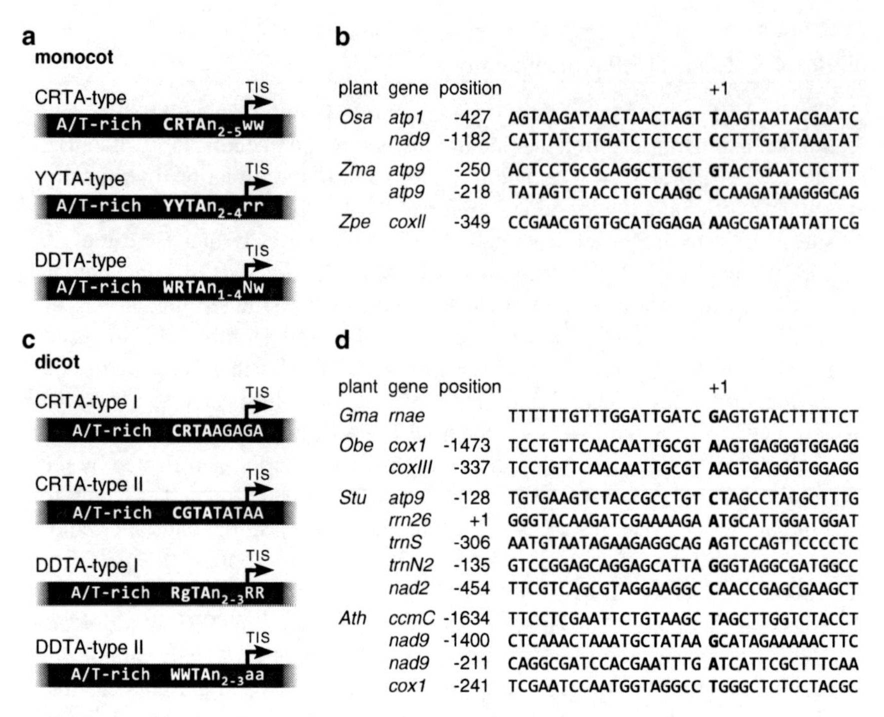

Fig. 4.3 Plant mitochondrial promoters. Conserved core promoter elements in plant mitochondria of (**a**) monocots and (**c**) dicots. Typical motifs frequently found are printed bold and bent arrows depict transcription initiation sites (TIS). Examples of promoter regions without conserved elements are shown for (**b**) monocots and (**d**) dicots with transcription initiation sites marked in bold type (Fey and Maréchal-Drouard 1999; Hess and Börner 1999; Kühn et al. 2005; Forner et al. 2007, 2008; Kühn et al. 2009). R=A or G; W=A or T; K=T or G; Y=C or T; D=A, G, or T; N=any nucleotide; lower case letters depict less strong conservation. *Osa*, *Oryza sativa*; *Zma*, *Zea mays*; *Zpe*, *Zea perennis*; *Gma*, *Glycine max*; *Obe*, *Oenothera berteriana*; *Stu*, *Solanum tuberosum*; *Ath*, *Arabidopsis thaliana*

without consensus sequence (Kühn et al. 2005). Functional relevance of one of these putative promoter motifs was shown in vitro by mutating the CATA-motif of the *atp8* promoter P-228/226 to CtgA, which drastically reduced levels of correctly initiated transcripts (Kühn et al. 2007). The exact parameters determining individual promoter strength in vivo are unclear. Since the exchange of certain nucleotide identities in the promoter regions strongly influences the strength of the promoter in vitro, it seems reasonable that certain sequences or sequence structures determine the promoter strength in vivo. However, the growing availability of mapped primary transcription start sites in mitochondria shows an increasing amount of promoter regions without obvious consensus sequences (Fig. 4.3b,d; Kühn et al. 2009; K. Kühn, U. Richter, personal communication). Together with the observation of RpoTmp promoter recognition in mitochondria being rather gene than promoter specific (Kühn et al. 2009), it is tempting to speculate that the specificity of transcription initiation by phage-type RNA polymerases may be due to different mechanisms: one being recognition of promoter consensus sequences, another one may involve gene-specific *cis*-elements.

A common feature of mitochondrial transcription in both mono- and dicot plants is the existence of multiple promoters in the upstream regions of many genes (Mulligan et al. 1988; Lupold et al. 1999a; Tracy and Stern 1995; Kühn et al. 2005). One might speculate that the principle of multiple promoters is a way to ensure transcription, and therefore gene expression, despite possible mitochondrial genome rearrangements (Kühn et al. 2005). The activity of maize *cox2* promoters was shown to respond to their genomic context thereby indicating consequences of intragenomic and intergenomic recombination for plant mitochondrial gene expression (Lupold et al. 1999b).

4.3 *Trans*-Acting Factors Involved in Plant Mitochondrial Transcription

4.3.1 General Transcription Factor(s)

The T7 RNA polymerase, the best studied example of single-subunit-RNA polymerases, recognizes and melts the promoter and correctly initiates transcription without the assistance of auxiliary factors. The T7 enzyme uses a so-called specificity loop that emerges from its C-terminal domain to recognize the sequence from –3 to –11 of the promoter. The N-terminal part of the protein (see Fig. 4.1) contributes to promoter recognition in phage RNA polymerases by interacting with nucleotides –12 to –17 (Sousa 1996; Cheetham et al. 1999). The N-terminal domains are poorly conserved between organellar and phage polymerases and are unlikely to play a role in promoter recognition by the organellar polymerases. However, the mitochondrial RNA polymerase of *S. cerevisiae* (Rpo41) has recently been shown to utilize, like the T7 protein, a C-terminal loop for promoter recognition (Nayak et al. 2009). Moreover, the yeast enzyme may act as a single-subunit-polymerase and initiate transcription without accessory factors provided the promoter is supercoiled or modified to form a bubble around the initiation site to facilitate melting (Matsunaga and Jaehning 2004). Similar to the phage and yeast enzymes, plant *Arabidopsis* RpoTp and RpoTm are able to specifically recognize several promoters, initiate transcription and catalyze transcript elongation without additional cofactors in vitro on supercoiled but not linear DNA templates (Kühn et al. 2007). Thus, the ability to recognize promoter sequences is an inherent feature of the phage-type enzymes (Gaspari et al. 2004; Matsunaga and Jaehning 2004; Kühn et al. 2007), a property they might have been preserved during evolution from the ancestral phage polymerases to organellar RNA polymerases. In contrast to their ancestor, however, the organellar polymerases require auxiliary factors for correct and efficient transcription initiation in vivo (Fig. 4.4).

In the light of the ability of the RNA polymerases itself to recognize promoters on supercoiled DNA, such general transcription factor(s) may in first line support melting the DNA in the promoter region (Shadel and Clayton 1995; Matsunaga and Jaehning 2004; Kühn et al. 2007). However, since *Arabidopsis* RpoTm recognizes only part of the promoters and *Arabidopsis* RpoTmp virtually none promoter in vitro transcription assays, mitochondrial RNA polymerases of plants may require

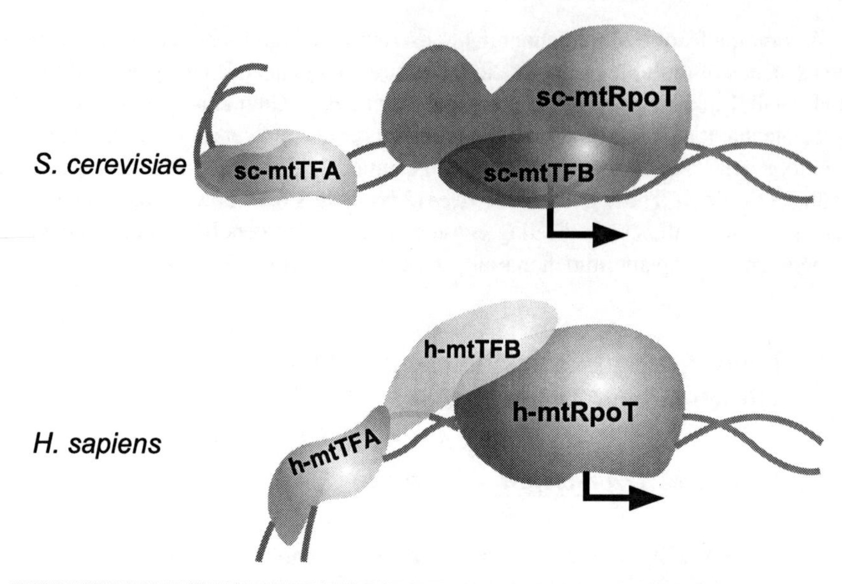

	TFA	TFB
S. cerevisiae	transcriptional enhancer (HMG box)	specificity factor (methyltransferase-like)
X. laevis	transcriptional enhancer (HMG box)	specificity factor (methyltransferase-like)
H. sapiens	specificity factor (HMG box)	specificity factor (methyltransferase-like)

Fig. 4.4 Components of the mitochondrial transcription machineries in *S. cerevisiae* and humans. Yeast transcription initiation model is based on Shadel and Clayton (1993); transcription initiation in human mitochondria as proposed by McCulloch et al. (2002). mtTFB and mtTFA functions are specified in the table below. Details on mtTFB and mtTFA functions and cofactor interactions with the mitochondrial phage-type RNA polymerases Rpo41 and h-mtRNApol are outlined in the text. DNA bending induced by mtTFA is indicated and transcriptional start sites are marked by *bent arrows*

transcription factors also for promoter recognition. Transcription initiation at mitochondrial promoters in *S. cerevisiae* requires a single accessory protein of 43 kDa first described as Mtf1 (Schinkel et al. 1987; Lisowsky and Michaelis 1988) and also referred to as sc-mtTFB (Shadel and Clayton 1993). The nonspecifically transcribing core RNA polymerase Rpo41 initiated correctly on a linear DNA template when complemented with sc-mtTFB in in vitro transcription experiments (Schinkel et al. 1987; Matsunaga and Jaehning 2004). Rpo41 core and sc-mtTFB form a holoenzyme in solution prior to DNA binding and promoter recognition (Mangus et al. 1994). Upon binding to the promoter, the RNA polymerase holoenzyme bends the DNA thereby enhancing promoter activity in vitro (Schinkel et al. 1988). Animals have two mitochondrial homologs of sc-mtTFB, mtTFB1 and mtTFB2 (Falkenberg et al. 2002; McCulloch et al. 2002; Rantanen et al. 2003; Matsunaga and Jaehning 2004). mtTFBs are related to a family of rRNA methyltransferases which dimethylate two adenosines near the 3′ end of the rRNA in the small ribosomal subunit (Schubot et al. 2001; Park et al. 2009; Richter et al. 2009). Although both human mtTFB1 and human mtTFB2 were shown to exhibit rRNA dimethyltransferase activity and to support mitochondrial transcription in vitro, recent data suggest that mtTFB1 functions *in organello* as an methyltransferase while mtTFB2 has evolved to act as an transcription factor (reviewed in Asin-Cayuela and Gustafsson 2007; Scarpulla 2008).

In order to accurately and efficiently initiate transcription at the LSP and HSP promoters of human mitochondrial DNA, the holoenyzme (mtTFB2 and core RNA polymerase, h-mtRNApol) needs in addition the accessory factor mtTFA, a protein of 25 kDa that comprises two HMG boxes (Fisher and Clayton 1985; Parisi and Clayton 1991a, b; Falkenberg et al. 2002). Specific binding of human mtTFA to distal promoter elements upstream of the transcription initiation sites as well as exact spacing between distal promoter elements and transcription start sites are crucial for transcription activation in human and mouse mitochondria (Dairaghi et al. 1995a, b; Gaspari et al. 2004). A protein homologous to human mtTFA is abundantly found in mitochondria of *Saccharomyces cerevisiae* (Diffley and Stillman 1991). In yeast, mtTFA appears to play a major role in structural organization and maintenance of the mtDNA. It stimulates, but is not necessary for transcription initiation (Diffley and Stillman 1992; Xu and Clayton 1992; Parisi et al. 1993). Also the homologous protein in *X. laevis* is dispensable for mitochondrial promoter recognition in vitro, yet the protein specifically binds to a promoter-proximal site and is more efficient in activating transcription than the yeast mtTFA (Antoshechkin and Bogenhagen 1995; Antoshechkin et al. 1997).

The existence of homologous general transcription factors in fungal and animal mitochondria led to the assumption that mtTFA- and/or mtTFB-related proteins may also operate in transcription initiation by phage-type RNA polymerases in plant organelles. However, computational predictions of subcellular localization of *Arabidopsis* HMG-box proteins did not reveal an mtTFA-homolog potentially targeted to mitochondria (Elo et al. 2003; K. Kühn unpubl. data). The nuclear genomes of *Arabidopsis* and other higher plants encode also homologues of fungal and animal mtTFB. The *Arabidopsis* mtTFB homolog was shown to methylate the conserved adenosines in the mitochondrial rRNA, whereas neither in vitro transcription assays nor analysis of a respective mutant line did support these proteins having a function in mitochondrial transcription (Richter et al. 2009).

4.3.2 *Specific Transcription Factors*

The knowledge that in plant mitochondria two or even more types of promoters exist raised the question about their activation by different sets of proteins. Transcription and transcript accumulation depend on ATP levels and are therefore coupled to respiration in *S. cerevisiae* mitochondria. The RNA polymerase Rpo41 itself might sense the ATP concentration (Amiott and Jaehning 2006). Recently, two proteins, Mss116p and Pet127p were found associated with the yeast mitochondrial RNA polymerase and suggested to modulate the activity of the enzyme (Markov et al. 2009). Nuclear encoded transcription factors that might bind to specific *cis*-elements were identified in human mitochondria, including the tumor repressor p53, the retinoid-X receptor and the thyroid hormone receptor (Asin-Cayuela and Gustafsson 2007).

The present knowledge about proteins involved in transcription initiation and regulation in plants is still scarce. One of the discussed candidates is MCT, a maize nuclear gene. Its target is the mitochondrial *cox2* promoter, which is active only when the dominant MCT allele is present (Newton et al. 1995). Furthermore, two proteins of 43- and 32-kDa in size binding to the *atp9* promoter have been purified in pea (Hatzack et al. 1998). While the 43-kDa protein is an isovaleryl-CoA dehydrogenase involved in leucine catabolism (Däschner et al. 1999, 2001), identity and function of the 32-kDa protein remained unclear. In wheat, a cDNA for a factor stimulating transcription initiation in vitro has been cloned (Ikeda and Gray 1999a). The encoded 69-kDa protein shows not only similarities to regions 2 and 3 of bacterial σ factors and the *S. cerevisiae* mtTFB (MTF1), but is also a member of the large family of PPR proteins which function in RNA metabolism (Schmitz-Linneweber and Small 2008; see Chaps. 7 and 8). Although three homologous PPR proteins from *Arabidopsis* were shown to be mitochondrially targeted, the recombinant expressed proteins bound to mitochondrial promoter fragments in an unspecific manner and neither interacted with RpoTm nor with RpoTmp to allow for correct initiation of transcription in an in vitro assay (K. Kühn unpubl. data; Kühn et al. 2007). Therefore, the function of the wheat p69 and its *Arabidopsis* homolog as a transcription factor in vivo still remains uncertain.

The mTERF1 protein is involved in initiation and termination of transcription in human mitochondria. There exists also a family of mTERF proteins in higher plants (see Box 4.3). However, terminators of transcription have not been identified in plant mitochondrial genomes and formation of 3′ ends of mature transcripts are generated by processing events (Forner et al. 2007; see Chap. 5). Apart from termination of RNA synthesis, however, it is also tempting to speculate that plant mTERFs may also have a function in initiating mitochondrial transcription . Further analysis of this group of proteins will inevitably reveal their function in plant organelles.

4.4 Transcriptional Regulation of Mitochondrial Gene Expression

Larger protein complexes in mitochondria comprise mitochondrial- and nuclear-encoded subunits. Thus, the expression of the genes encoded in the different compartments needs to be coordinated to warrant the synthesis of stoichiometric amounts of the different subunits. Generally, the expression of genes is regulated at various levels, i.e., transcriptional, posttranscriptional, translational, and posttranslational. The regulation of mitochondrial-encoded genes is not fully understood yet. Different studies investigating the rate of transcription *in organello* (run-on analyses) showed that rRNA genes are transcribed at rates 2- to 14-fold higher than protein coding genes (Finnegan and Brown 1990; Mulligan et al. 1991). Muise and Hauswirth (1992) found the *rps12* gene to be most strongly transcribed. However, transcription for several other genes was found to be considerably different from the previous reports. An extensive study comprising all mitochondrial-encoded genes in *Arabidopsis* found no enhanced transcription of rRNA genes, but detected even distinct transcription rates of genes encoding components of the same multisubunit complex (Giegé et al. 2000). These differences are, at least partially, counterbalanced in the steady-state RNA pool, most likely by posttranscriptional processes and different RNA stabilities. This is reminiscent to the finding that transcript synthesis in *Arabidopsis* mitochondria cycles in a diurnal rhythm, while steady-state transcript levels do not vary between light and dark phases (Okada and Brennicke 2006). Furthermore, a role of multiple promoters in a developmental or tissue-specific regulation of mitochondrial gene expression seems rather unlikely (Kühn et al. 2005). The relaxed transcription is counterbalanced by posttranscriptional processes representing thereby a major determinant of mitochondrial RNA levels (Mulligan et al. 1988; Tracy and Stern 1995; Lupold et al. 1999a; Giegé et al. 2000; Leino et al. 2005; Holec et al. 2008). Interestingly, the comparison of mitochondrial transcriptional rates in *Arabidopsis, Brassica napus,* a cytoplasmic male sterile line as well as its fertility restored line by Leino et al. (2005) identified species-specific rates for genes like *cox1, nad4L, nad9, ccmB, rps7, and rrn5,* which are most likely determined by different promoter strength in the mitochondrial DNA. However, different transcription rates were found for some genes in the alloplasmic CMS (see Chap. 18) line with the same mitochondrial promoter sequences but different nuclear background. This held also true for posttranscriptional processes which were less effective in the CMS line. Thus, the influence of the nuclear background on both transcriptional rates and posttranscriptional mechanisms shows that both processes are not only depending on mitochondrial *cis*-elements (see above) but also on nuclear *trans*-factors (see below; Edqvist and Bergman 2002).

A few reports suggested tissue- or cell-specific differences in mitochondrial gene expression to be due to transcriptional levels (Topping and Leaver 1990; Li et al. 1996). However, a comprehensive analysis of mitochondrial transcription initiation sites in *Arabidopsis* revealed no qualitative differences in promoter usage between leaves and flowers (Kühn et al. 2005). Therefore, it has been suggested that posttranscriptional processes are the dominant mechanism for tissue-specific differences in steady-state levels of mitochondrial transcripts (Monéger et al. 1994; Smart et al. 1994; Gagliardi and Leaver 1999). Correspondingly, the mitochondrial gene expression remained in response to sugar starvation more or less unaffected at the transcriptional, posttranscriptional, and translational levels (Giegé et al. 2005). The observed reduction of ATPase complexes, however, was a result of nuclear-encoded components of the ATPase being downregulated. Becoming the rate-liming factor in the assembly of new complexes, correct stoichiometric proportions seemed to be achieved posttranslationally. The recent finding that the activity of one of the mitochondrial RNA polymerases in *Arabidopsis,* RpoTmp, is gene- rather than promoter-specific, on the other hand, suggests that RpoTmp-dependent

RNA synthesis may present a transcriptional mechanism allowing mitochondria to control the expression of certain mitochondrial genes (Kühn et al. 2009). Taken together, gene-specific or global regulation of mitochondrial gene expression at the level of transcription initiation in response to certain stimuli, i.e., developmental and environmental cues, it is still unclear.

Acknowledgments The work of the authors is supported by Deutsche Forschungsgemeinschaft (SFB 429). We thank Uwe Richter for helpful discussions and Kristina Kühn for providing initial artwork.

Glossary

Alloplasmic: Describes an individual having a common nucleus, but an "alien" cytoplasm (e.g., alloplasmic rye containing a wheat cytoplasm); may lead to meiotic disturbances and male sterility (CMS).

Bacterial-type RNA polymerase: RNA polymerase found in bacteria. The core enzyme has 4 subunits. α: two α-subunits assemble the enzyme and bind regulatory factors. β: catalyzes the RNA synthesis. β': binds to DNA (nonspecifically). To bind promoter-specific regions, the core enzyme requires the sigma factor (σ), which together form the holoenzyme. Related RNA polymerases are found in plastids (β' is split into two subunits in the cyanobacterial and plastid enzymes) and mitochondria of Jakobids.

Bryophyte: Nonvascular, seedless plants. Bryophytes are among the simplest of the terrestrial plants and comprise mosses, hornworts, or liverworts; e.g., *Physcomitrella*, *Marchantia.*

Cis-element: Defines a region of DNA or RNA that regulates the expression of genes located on the same strand. *Cis*-elements are often binding sites of one or more *trans*-factors.

CMS: Cytoplasmic male sterility (see Sect. 18.1).

GFP: Green fluorescent protein; exhibits bright green fluorescence when exposed to blue light and is frequently used as a reporter of expression.

HSP: Heavy-strand promoter; part of the bidirectional promoter in mammalian mitochondria initiating transcription of the heavy-strand (in opposite the light-strand, defined by their differing G+T content).

LSP: Light-strand promoter; promoter in mammalian mitochondria initiating transcription of the light-strand.

Lycophyte: Plantae subdivision comprising the oldest living vascular plants at around 410 million years old, e.g., *Selaginella.*

NEP: Nuclear-encoded plastid RNA polymerase; a phage-type RNA polymerase that originated by gene duplication from the gene encoding the mitochondrial RNA polymerase.

PPR protein: Pentatricopeptide repeat protein (see Chap. 4).

Promoter: A region of DNA that facilitates the transcription of a particular gene. In plant mitochondria, most promoters were defined by mapping transcript initiation sites and identification of consensus motifs. Therefore, promoters are often specified by the gene's name and position of the initiating nucleotide in respect to the start of the coding sequence or mature rRNA.

Prophage: Latent form of a bacteriophage inserted as part of the chromosome of a bacterium.

Tracheophyte Vascular plants: those plants that have lignified tissues for conducting water, minerals, and photosynthetic products. Tracheophytes include lycophytes, ferns, and seed plants.

Trans-factors: Mostly a protein factor that binds to specific DNA or RNA sequences (*cis*-elements) thereby controlling the regulation/expression of a certain gene.

References

Amiott, E. A., Jaehning, J. A. 2006. Mitochondrial transcription is regulated via an ATP "sensing" mechanism that couples RNA abundance to respiration. Mol. Cell 22:329–338.

Antoshechkin, I., Bogenhagen, D. F. 1995. Distinct roles for two purified factors in transcription of *Xenopus* mitochondrial DNA. Mol. Cell. Biol. 15:7032–7042.

Antoshechkin, I., Bogenhagen, D. F., Mastrangelo, I. A. 1997. The HMG-box mitochondrial transcription factor xl-mtTFA binds DNA as a tetramer to activate bidirectional transcription. EMBO J. 16:3198–3206.

Asin-Cayuela, J., Gustafsson, C. M. 2007. Mitochondrial transcription and its regulation in mammalian cells. Trends Biochem. Sci. 32:111–117.

Azevedo, J., Courtois, F., Hakimi, M.-A., Demarsy, E., Lagrange, T., Alcaraz, J.-P., Jaiswal, P., Maréchal-Drouard, L., Lerbs-Mache, L. 2008. Intraplastidial trafficking of a phage-type RNA polymerase is mediated by a thylakoid RING-H2 protein. Proc. Natl. Acad. Sci. USA 105:9123–9128.

Azevedo, J., Courtois, F., Lerbs-Mache, S. 2006. Sub-plastidial localization of two different phage-type RNA polymerases in spinach chloroplasts. Nucleic Acids Res. 34:436–444.

Baba, K., Schmidt, J., Espinosa-Ruiz, A., Villarejo, A., Shiina, T., Gardestrom, P., Sane, A. P., Bhalerao, R. P. 2004. Organellar gene transcription and early seedling development are affected in the *RpoT;2* mutant of *Arabidopsis*. Plant J. 38:38–48.

Binder, S., Brennicke, A. 2003. Gene expression in plant mitochondria: transcriptional and post-transcriptional control. Philos. Trans. R. Soc. Lond., B, Biol. Sci. 358:181–188; discussion 188–189.

Binder, S., Hatzack, F., Brennicke, A. 1995. A novel pea mitochondrial in vitro transcription system recognizes homologous and heterologous mRNA and tRNA promoters. J. Biol. Chem. 270:22182–22189.

Binder, S., Marchfelder, A., Brennicke, A. 1996. Regulation of gene expression in plant mitochondria. Plant Mol. Biol. 32:303–314.

Caoile, A. G. F. S., Stern, D. B. 1997. A conserved core element is functionally important for maize mitochondrial promoter activity *in vitro*. Nucleic Acids Res. 25:4055–4060.

Cermakian, N., Ikeda, T. M., Cedergren, R., Gray, M. W. 1996. Sequences homologous to yeast mitochondrial and bacteriophage T3 and T7 RNA polymerases are widespread throughout the eukaryotic lineage. Nucleic Acids Res. 24:648–654.

Cermakian, N., Ikeda, T. M., Miramontes, P., Lang, B. F., Gray, M. W., Cedergren, R. 1997. On the evolution of the single-subunit RNA polymerases. J. Mol. Evol. 45:671–681.

Chang, C.-C., Sheen, J., Bligny, M., Niwa, Y., Lerbs-Mache, S., Stern, D. B. 1999. Functional analysis of two maize cDNAs encoding T7-like RNA polymerases. Plant Cell 11:911–926.

Cheetham, G. M., Jeruzalmi, D., Steitz, T. A. 1999. Structural basis for initiation of transcription from an RNA polymerase-promoter complex. Nature 399:80–83.

Christensen, A. C., Lyznik, A., Mohammed, S., Elowsky, C. G., Elo, A., Yule, R., Mackenzie, S. A. 2005. Dual-domain, dual-targeting organellar protein presequences in *Arabidopsis* can use non-AUG start codons. Plant Cell 17:2805–2816.

Clifton, S. W., Minx, P., Fauron, C. M.-R., Gibson, M., Allen, J. O., Sun, H., Thompson, M., Barbazuk, W. B., Kanuganti, S., Tayloe, C., Meyer, L., Wilson, R. K., Newton, K. J. 2004. Sequence and comparative analysis of the maize NB mitochondrial genome. Plant Physiol. 136:3486–3503.

Courtois, F., Merendino, L., Demarsy, E., Mache, R., Lerbs-Mache, S. 2007. Phage-type RNA polymerase RPOTmp transcribes the *rrn* operon from the PC promoter at early developmental stages in *Arabidopsis*. Plant Physiol. 145:712–721.

Daga, A., Micol, V., Hess, D., Aebersold, R., Attardi, G. 1993. Molecular characterization of the transcription termination factor from human mitochondria. J. Biol. Chem. 268:8123–8130.

Dairaghi, D. J., Shadel, G. S., Clayton, D. A. 1995a. Human mitochondrial transcription factor A and promoter spacing integrity are required for transcription initiation. Biochim. Biophys. Acta 1271:127–134.

Dairaghi, D. J., Shadel, G. S. and Clayton, D. A. 1995b. Human mitochondrial transcription factor A and promoter spacing integrity are required for transcription initiation. Biochim. Biophys. Acta 1271:127–134.

Däschner, K., Couée, I., Binder, S. 2001. The mitochondrial isovaleryl-coenzyme a dehydrogenase of *Arabidopsis* oxidizes intermediates of leucine and valine catabolism. Plant Physiol. 126:601–612.

Däschner, K., Thalheim, C., Guha, C., Brennicke, A., Binder, S. 1999. In plants a putative isovaleryl-CoA-dehydrogenase is located in mitochondria. Plant Mol. Biol. 39:1275–1282.

Delarue, M., Poch, O., Tordo, N., Moras, D., Argos, P. 1990. An attempt to unify the structure of polymerases. Protein Eng. 3:461–467.

Diffley, J. F., Stillman, B. 1991. A close relative of the nuclear, chromosomal high-mobility group protein HMG1 in yeast mitochondria. Proc. Natl. Acad. Sci. USA 88:7864–7868.

Diffley, J. F., Stillman, B. 1992. DNA binding properties of an HMG1-related protein from yeast mitochondria. J. Biol. Chem. 267:3368–3374.

Dombrowski, S., Hoffmann, M., Guha, C., Binder, S. 1999. Continuous primary sequence requirements in the 18-nucleotide promoter of dicot plant mitochondria. J. Biol. Chem. 274:10094–10099.

Dombrowski, S., Hoffmann, M., Kuhn, J., Brennicke, A., Binder, S. 1998. On mitochondrial promoters in *Arabidopsis thaliana* and other flowering plants. In Plant Mitochondria: From Gene to Function, I. M. Möller, P. Gardeström, K. Glimelius, E. Glaser eds., pp. 165–170. Leiden, Netherlands: Backhuys Publishers.

Edqvist, J., Bergman, P. 2002. Nuclear identity specifies transcriptional initiation in plant mitochondria. Plant Mol. Biol. 49:59–68.

Elo, A., Lyznik, A., Gonzalez, D. O., Kachman, S. D., Mackenzie, S. A. 2003. Nuclear genes that encode mitochondrial proteins for DNA and RNA metabolism are clustered in the *Arabidopsis* genome. Plant Cell 15:1619–1631.

Emanuel, C., von Groll, U., Müller, M., Börner, T., Weihe, A. 2006. Development- and tissue-specific expression of the *RpoT* gene family of *Arabidopsis* encoding mitochondrial and plastid RNA polymerases. Planta 223:998–1009.

Emanuel, C., Weihe, A., Graner, A., Hess, W. R., Börner, T. 2004. Chloroplast development affects expression of phage-type RNA polymerases in barley leaves. Plant J. 38:460–472.

Falkenberg, M., Gaspari, M., Rantanen, A., Trifunovic, A., Larsson, N.-G., Gustafsson, C. M. 2002. Mitochondrial transcription factors B1 and B2 activate transcription of human mtDNA. Nat. Genet. 31:289–294.

Fey, J., Maréchal-Drouard, L. 1999. Compilation and analysis of plant mitochondrial promoter sequences: an illustration of a divergent evolution between monocot and dicot mitochondria. Biochem. Biophys. Res. Commun. 256:409–414.

Filée, J., Forterre, P. 2005. Viral proteins functioning in organelles: a cryptic origin? Trends Microbiol. 13: 510–513.

Finnegan, P. M., Brown, G. G. 1990. Transcriptional and post-transcriptional regulation of RNA levels in maize mitochondria. Plant Cell 2:71–83.

Fisher, R. P., Clayton, D. A. 1985. A transcription factor required for promoter recognition by human mitochondrial RNA polymerase. Accurate initiation at the heavy- and light-strand promoters dissected and reconstituted in vitro. J. Biol. Chem. 260:11330–11338.

Forner, J., Hölzle, A., Jonietz, C., Thuss, S., Schwarzländer, M., Weber, B., Meyer, R. C., Binder, S. 2008. Mitochondrial mRNA polymorphisms in different *Arabidopsis* accessions. Plant Physiol. 148:1106–1116.

Forner, J., Weber, B., Thuss, S., Wildum, S., Binder, S. 2007. Mapping of mitochondrial mRNA termini in *Arabidopsis thaliana*: t-elements contribute to 5′ and 3′ end formation. Nucleic Acids Res. 35:3676–3692.

Gagliardi, D., Leaver, C. J. 1999. Polyadenylation accelerates the degradation of the mitochondrial mRNA associated with cytoplasmic male sterility in sunflower. EMBO J. 18:3757–3766.

Gaspari, M., Falkenberg, M., Larsson, N. G., Gustafsson, C. M. 2004. The mitochondrial RNA polymerase contributes critically to promoter specificity in mammalian cells. EMBO J. 23:4606–4614.

Giegé, P., Hoffmann, M., Binder, S., Brennicke, A. 2000. RNA degradation buffers asymmetries of transcription in Arabidopsis mitochondria. EMBO Rep. 1:164-170.

Giegé, P., Sweetlove, L. J., Cognat, V., Leaver, C. J. 2005. Coordination of nuclear and mitochondrial genome expression during mitochondrial biogenesis in *Arabidopsis*. Plant Cell 17:1497–1512.

Gray, M. W., Lang, B. F. 1998. Transcription in chloroplasts and mitochondria: a tale of two polymerases. Trends Microbiol. 6:1–3.

Greenleaf, A. L., Kelly, J. L., Lehman, I. R. 1986. Yeast RPO41 gene product is required for transcription and maintenance of the mitochondrial genome. Proc. Natl. Acad. Sci. USA 83:3391–3394.

Handa, H. 2003. The complete nucleotide sequence and RNA editing content of the mitochondrial genome of rapeseed (*Brassica napus* L.): comparative analysis of the mitochondrial genomes of rapeseed and *Arabidopsis thaliana*. Nucleic Acids. Res. 31:5907–5916.

Handa, H. 2008. Linear plasmids in plant mitochondria: peaceful coexistences or malicious invasions? Mitochondrion 8:15–25.

Hanic-Joyce, P. J., Gray, M. W. 1991. Accurate transcription of a plant mitochondrial gene *in vitro*. Mol. Cell. Biol. 11:2035–2039.

Hatzack, F., Dombrowski, S., Brennicke, A., Binder, S. 1998. Characterization of DNA-binding proteins from pea mitochondria. Plant Physiol. 116:519–527.

Hedtke, B., Börner, T., Weihe, A. 1997. Mitochondrial and chloroplast phage-type RNA polymerases in *Arabidopsis*. Science 277:809–811.

Hedtke, B., Börner, T., Weihe, A. 2000. One RNA polymerase serving two genomes. EMBO Rep. 1:435–440.

Hedtke, B., Legen, J., Weihe, A., Herrmann, R. G., Börner, T. 2002. Six active phage-type RNA polymerase genes in *Nicotiana tabacum*. Plant J. 30:625–637.

Hedtke, B., Meixner, M., Gillandt, S., Richter, E., Börner, T., Weihe, A. 1999. Green fluorescent protein as a marker to investigate targeting of organellar RNA polymerases of higher plants *in vivo*. Plant J. 17:557–561.

Hess, W. R., Börner, T. 1999. Organellar RNA polymerases of higher plants. Int. Rev. Cytol. 190:1–59.

Hoffmann, M., Binder, S. 2002. Functional importance of nucleotide identities within the pea *atp9* mitochondrial promoter sequence. J. Mol. Biol. 320:943–950.

Holec, S., Lange, H., Canaday, J., Gagliardi, D. 2008. Coping with cryptic and defective transcripts in plant mitochondria. Biochim. Biophys. Acta 1779:566–573.

Huang, C. Y., Grunheit, N., Ahmadinejad, N., Timmis, J. N., Martin, W. 2005. Mutational decay and age of chloroplast and mitochondrial genomes transferred recently to angiosperm nuclear chromosomes. Plant Physiol. 138:1723–1733.

Ikeda, T., Gray, M. 1999a. Characterization of a DNA-binding protein implicated in transcription in wheat mitochondria. Mol. Cell. Biol. 19:8113–8122.

Ikeda, T., Gray, M. 1999b. Genes and proteins of the transcriptional apparatus in mitochondria. J. Hered. 90:374–379.

Ikeda, T. M., Gray, M. W. 1999c. Identification and characterization of T7/T3 bacteriophage-like RNA polymerase sequences in wheat. Plant Mol. Biol. 40:567–578.

Kabeya, Y., Hashimoto, K., Sato, N. 2002. Identification and characterization of two phage-type RNA polymerase cDNAs in the moss *Physcomitrella patens*: implication of recent evolution of nuclear-encoded RNA polymerase of plastids in plants. Plant Cell Physiol. 43:245–255.

Kabeya, Y., Sato, N. 2005. Unique translation initiation at the second AUG codon determines mitochondrial localization of the phage-type RNA polymerases in the moss *Physcomitrella patens*. Plant Physiol. 138:369–382.

Kempken, F., Hermanns, J., Osiewacz, H. D. 1992. Evolution of linear plasmids. J. Mol. Evol. 35:502–513.

Kobayashi, Y., Dokiya, Y., Kumazawa, Y., Sugita, M. 2002. Non-AUG translation initiation of mRNA encoding plastid-targeted phage-type RNA polymerase in *Nicotiana sylvestris*. Biochem. Biophys. Res. Commun. 299:57–61.

Kobayashi, Y., Dokiya, Y., Sugita, M. 2001a. Dual targeting of phage-type RNA polymerase to both mitochondria and plastids is due to alternative translation initiation in single transcripts. Biochem. Biophys. Res. Commun. 289:1106–1113.

Kobayashi, Y., Dokiya, Y., Sugiura, M., Niwa, Y., Sugita, M. 2001b. Genomic organization and organ-specific expression of a nuclear gene encoding phage-type RNA polymerase in *Nicotiana sylvestris*. Gene 279:33–40.

Kruse, B., Narasimhan, N., Attardi, G. 1989. Termination of transcription in human mitochondria: identification and purification of a DNA binding protein factor that promotes termination. Cell 58:391–397.

Kubo, T., Nishizawa, S., Sugawara, A., Itchoda, N., Estiati, A., Mikami, T. 2000. The complete nucleotide sequence of the mitochondrial genome of sugar beet (*Beta vulgaris* L.) reveals a novel gene for $tRNA^{Cys}$(GCA). Nucleic Acids Res. 28:2571–2576.

Kühn, K., Bohne, A.-V., Liere, K., Weihe, A., Börner, T. 2007. *Arabidopsis* phage-type RNA polymerases: accurate *in vitro* transcription of organellar genes. Plant Cell 19:959–971.

Kühn, K., Richter, U., Meyer, E., Delannoy, E., Falcon de Longevialle, A., O'Toole, N., Börner, T., Millar, A., Small, I., Whelan, J. 2009. Phage-type RNA polymerase RPOTmp performs gene-specific transcription in mitochondria of *Arabidopsis thaliana*. Plant Cell 21:2762–2779.

Kühn, K., Weihe, A., Börner, T. 2005. Multiple promoters are a common feature of mitochondrial genes in *Arabidopsis*. Nucleic Acids Res. 33:337–346.

Lang, B. F., Burger, G., O'Kelly, C. J., Cedergren, R., Golding, G. B., Lemieux, C., Sankoff, D., Turmel, M., Gray, M. W. 1997. An ancestral mitochondrial DNA resembling a eubacterial genome in miniature. Nature 387:493–497.

Leino, M., Landgren, M., Glimelius, K. 2005. Alloplasmic effects on mitochondrial transcriptional activity and RNA turnover result in accumulated transcripts of *Arabidopsis* ORFs in cytoplasmic male-sterile *Brassica napus*. Plant J. 42:469–480.

Lerbs-Mache, S. 1993. The 110-kDa polypeptide of spinach plastid DNA-dependent RNA polymerase: single-subunit enzyme or catalytic core of multimeric enzyme complexes? Proc. Natl. Acad. Sci. USA 90:5509–5513.

Li, X. Q., Zhang, M., Brown, G. G. 1996. Cell-specific expression of mitochondrial transcripts in maize seedlings. Plant Cell 8:1961–1975.

Liere, K., Börner, T. 2007. Transcription and transcriptional regulation in plastids. In Topics in Current Genetics: Cell and Molecular Biology of Plastids, R. Bock ed., pp. 121–174. Berlin: Springer

Linder, T., Park, C. B., Asin-Cayuela, J., Pellegrini, M., Larsson, N.-G., Falkenberg, M., Samuelsson, T., Gustafsson, C. M. 2005. A family of putative transcription termination factors shared amongst metazoans and plants. Curr. Genet. 48:265–269.

Lisowsky, T., Michaelis, G. 1988. A nuclear gene essential for mitochondrial replication suppresses a defect of mitochondrial transcription in *Saccharomyces cerevisiae*. Mol. Gen. Genet. 214:218–223.

Lupold, D. S., Caoile, A. G., Stern, D. B. 1999a. The maize mitochondrial *cox2* gene has five promoters in two genomic regions, including a complex promoter consisting of seven overlapping units. J. Biol. Chem. 274:3897–3903.

Lupold, D. S., Caoile, A. G. F. S., Stern, D. B. 1999b. Genomic context influences the activity of maize mitochondrial *cox2* promoters. Proc. Natl. Acad. Sci. USA 96:11670–11675.

Maier, U. G., Bozarth, A., Funk, H. T., Zauner, S., Rensing, S. A., Schmitz-Linneweber, C., Börner, T., Tillich, M. 2008. Complex chloroplast RNA metabolism: just debugging the genetic programme? BMC Biol. 6:36.

Mangus, D. A., Jang, S. H., Jaehning, J. A. 1994. Release of the yeast mitochondrial RNA polymerase specificity factor from transcription complexes. J. Biol. Chem. 269:26568–26574.

Markov, D. A., Savkina, M., Anikin, M., Del Campo, M., Ecker, K., Lambowitz, A. M., De Gnore, J. P., McAllister, W. T. 2009. Identification of proteins associated with the yeast mitochondrial RNA polymerase by tandem affinity purification. Yeast 26:423–440.

Martin, W. 2003. Gene transfer from organelles to the nucleus: frequent and in big chunks. Proc. Natl. Acad. Sci. U.S.A 100:8612–8614.

Martin, M., Cho, J., Cesare, A. J., Griffith, J. D., Attardi, G. 2005. Termination factor-mediated DNA loop between termination and initiation sites drives mitochondrial rRNA synthesis. Cell 123:1227–1240.

Masters, B. S., Stohl, L. L., Clayton, D. A. 1987. Yeast mitochondrial RNA polymerase is homologous to those encoded by bacteriophages T3 and T7. Cell 51:89–99.

Matsunaga, M., Jaehning, J. A. 2004. Intrinsic promoter recognition by a "core" RNA polymerase. J. Biol. Chem. 279:44239–44242.

McAllister, W. 1993. Structure and function of the bacteriophage T7 RNA polymerase (or, the virtues of simplicity). Cell. Mol. Biol. Res. 39:385–391.

McCulloch, V., Seidel-Rogol, B. L., Shadel, G. S. 2002. A human mitochondrial transcription factor is related to RNA adenine methyltransferases and binds S-adenosylmethionine. Mol. Cell. Biol. 22:1116–1125.

Meskauskiene, R., Würsch, M., Laloi, C., Vidi, P., Coll, N., Kessler, F., Baruah, A., Kim, C., Apel, K. 2009. A mutation in the *Arabidopsis* mTERF-related plastid protein SOLDAT10 activates retrograde signaling and suppresses O-induced cell death. Plant J. 60:399–410.

Monéger, F., Smart, C. J., Leaver, C. J. 1994. Nuclear restoration of cytoplasmic male sterility in sunflower is associated with the tissue-specific regulation of a novel mitochondrial gene. EMBO J. 13:8–17.

Muise, R. C., Hauswirth, W. W. 1992. Transcription in maize mitochondria: effects of tissue and mitochondrial genotype. Curr. Genet. 22:235–242.

Mulligan, R. M., Lau, G. T., Walbot, V. 1988. Numerous transcription initiation sites exist for the maize mitochondrial genes for subunit 9 of the ATP synthase and subunit 3 of cytochrome oxidase. Proc. Natl. Acad. Sci. U.S.A 85:7998–8002.

Mulligan, R. M., Leon, P., Walbot, V. 1991. Transcription and posttranscriptional regulation of maize mitochondrial gene expression. Mol. Cell. Biol. 11:533–543.

Nayak, D., Guo, Q., Sousa, R. 2009. A promoter recognition mechanism common to yeast mitochondrial and phage T7 RNA polymerases. J. Biol. Chem. 284:13641–13647.

Newton, K. J., Winberg, B., Yamato, K., Lupold, S., Stern, D. B. 1995. Evidence for a novel mitochondrial promoter preceding the *cox2* gene of perennial teosintes. EMBO J. 14:585–593.

Notsu, Y., Masood, S., Nishikawa, T., Kubo, N., Akiduki, G., Nakazono, M., Hirai, A., Kadowaki, K. 2002. The complete sequence of the rice (*Oryza sativa* L.) mitochondrial genome: frequent DNA sequence acquisition and loss during the evolution of flowering plants. Mol. Genet. Genomics 268:434–445.

Ogihara, Y., Yamazaki, Y., Murai, K., Kanno, A., Terachi, T., Shiina, T., Miyashita, N., Nasuda, S., Nakamura, C., Mori, N., Takumi, S., Murata, M., Futo, S., Tsunewaki, K. 2005. Structural dynamics of cereal mitochondrial genomes as revealed by complete nucleotide sequencing of the wheat mitochondrial genome. Nucleic Acids Res. 33:6235–6250.

Okada, S., Brennicke, A. 2006. Transcript levels in plant mitochondria show a tight homeostasis during day and night. Mol. Genet. Genomics 276:71–78.

Parisi, M. A., Clayton, D. A. 1991a. Similarity of human mitochondrial transcription factor 1 to high mobility group proteins. Science 252:965–969.

Parisi, M. A., Xu, B., Clayton, D. A. 1993. A human mitochondrial transcriptional activator can functionally replace a yeast mitochondrial HMG-box protein both *in vivo* and *in vitro*. Mol. Cell. Biol. 13:1951–1961.

Park, A. K., Kim, H., Jin, H. J. 2009. Comprehensive phylogenetic analysis of evolutionarily conserved rRNA adenine dimethyltransferase suggests diverse bacterial contributions to the nucleus-encoded plastid proteome. Mol. Phylogenet. Evol. 50:282–289.

Rantanen, A., Gaspari, M., Falkenberg, M., Gustafsson, C., M., Larsson, N.-G. 2003. Characterization of the mouse genes for mitochondrial transcription factors B1 and B2. Mamm. Genome 14:1–6.

Rapp, W. D., Stern, D. B. 1992. A conserved 11 nucleotide sequence contains an essential promoter element of the maize mitochondrial *atp1* gene. EMBO J. 11:1065–1073.

Rapp, W. D., Lupold, D. S., Mack, S., Stern, D. B. 1993. Architecture of the maize mitochondrial *atp1* promoter as determined by linker-scanning and point mutagenesis. Mol. Cell. Biol. 13:7232–7238.

Richter, U., Kiessling, J., Hedtke, B., Decker, E., Reski, R., Börner, T., Weihe, A. 2002. Two *RpoT* genes of *Physcomitrella patens* encode phage-type RNA polymerases with dual targeting to mitochondria and plastids. Gene 290:95–105.

Richter, U., Kühn, K., Okada, S., Brennicke, A., Weihe, A., Börner, T. 2009. A mitochondrial rRNA dimethyltransferase in *Arabidopsis*. Plant J. Published Online DOI: 10.1111/j.1365-313X.2009.04079.x.

Roberti, M., Polosa, P. L., Bruni, F., Manzari, C., Deceglie, S., Gadaleta, M. N., Cantatore, P. 2009. The MTERF family proteins: mitochondrial transcription regulators and beyond. Biochim. Biophys. Acta 1787:303–311. Roberti, M., Polosa, P. L., Bruni, F., Musicco, C., Gadaleta, M. N., Cantatore, P. 2003. DmTTF, a novel mitochondrial transcription termination factor that recognises two sequences of *Drosophila melanogaster* mitochondrial DNA. Nucleic Acids Res. 31:1597–1604.

Rohe, M., Schründer, J., Tudzynski, P., Meinhardt, F. 1992. Phylogenetic relationships of linear, protein-primed replicating genomes. Curr. Genet. 21:173–176.

Scarpulla, R. C. 2008. Transcriptional paradigms in mammalian mitochondrial biogenesis and function. Physiological Reviews 88:611–638.

Schinkel, A. H., Koerkamp, M. J., Touw, E. P., Tabak, H. F. 1987. Specificity factor of yeast mitochondrial RNA polymerase. Purification and interaction with core RNA polymerase. J. Biol. Chem. 262:12785–12791.

Schinkel, A. H., Groot Koerkamp, M. J., Teunissen, A. W., Tabak, H. F. 1988. RNA polymerase induces DNA bending at yeast mitochondrial promoters. Nucleic Acids Res. 16:9147–9163.

Schmitz-Linneweber, C., Small, I. 2008. Pentatricopeptide repeat proteins: a socket set for organelle gene expression. Trends Plant Sci. 13:663–670.

Schönfeld, C., Wobbe, L., Borgstadt, R., Kienast, A., Nixon, P. J., Kruse, O. 2004. The nucleus-encoded protein MOC1 is essential for mitochondrial light acclimation in *Chlamydomonas reinhardtii*. J. Biol. Chem. 279:50366–50374.

Schubot, F. D., Chen, C. J., Rose, J. P., Dailey, T. A., Dailey, H. A., Wang, B. C. 2001. Crystal structure of the transcription factor sc-mtTFB offers insights into mitochondrial transcription. Protein Sci. 10:1980–1988.

Shadel, G. S., Clayton, D. A. 1993. Mitochondrial Transcription. J. Biol. Chem. 268:16083–16086.

Shadel, G. S., Clayton, D. A. 1995. A *Saccharomyces cerevisiae* mitochondrial transcription factor, sc-mtTFB, shares features with sigma factors but is functionally distinct. Mol. Cell. Biol. 15:2101–2108.

Shutt, T., Gray, M. 2006. Homologs of mitochondrial transcription factor B, sparsely distributed within the eukaryotic radiation, are likely derived from the dimethyladenosine methyltransferase of the mitochondrial endosymbiont. Mol. Biol. Evol. 23:1169–1179.

Smart, C. J., Moneger, F., Leaver, C. J. 1994. Cell-specific regulation of gene expression in mitochondria during anther development in sunflower. Plant Cell 6:811–825.

Sousa, R. 1996. Structural and mechanistic relationships between nucleic acid polymerases. Trends Biochem. Sci. 21:186–190.

Sousa, R., Chung, Y. J., Rose, J. P., Wang, B. C. 1993. Crystal structure of bacteriophage T7 RNA polymerase at 3.3 A resolution. Nature 364:593–599.

Steitz, T. A. 2004. The structural basis of the transition from initiation to elongation phases of transcription, as well as translocation and strand separation, by T7 RNA polymerase. Curr. Opin. Struct. Biol. 14:4–9.

Sugiyama, Y., Watase, Y., Nagase, M., Makita, N., Yagura, S., Hirai, A., Sugiura, M. 2005. The complete nucleotide sequence and multipartite organization of the tobacco mitochondrial genome: comparative analysis of mitochondrial genomes in higher plants. Mol. Genet. Genomics 272:603.

Swiatecka-Hagenbruch, M., Emanuel, C., Hedtke, B., Liere, K., Börner, T. 2008. Impaired function of the phage-type RNA polymerase RpoTp in transcription of chloroplast genes is compensated by a second phage-type RNA polymerase. Nucleic Acids Res. 36:785–792.

Timmis, J. N., Ayliffe, M. A., Huang, C. Y., Martin, W. 2004. Endosymbiotic gene transfer: organelle genomes forge eukaryotic chromosomes. Nat. Rev. Genet. 5:123–135.

Tiranti, V., Savoia, A., Forti, F., D'Apolito, M. F., Centra, M., Rocchi, M., Zeviani, M. 1997. Identification of the gene encoding the human mitochondrial RNA polymerase (h-mtRPOL) by cyberscreening of the Expressed Sequence Tags database. Hum. Mol. Genet. 6:615–625.

Topping, J., Leaver, C. 1990. Mitochondrial gene expression during wheat leaf development. Planta 182:399–407.

Tracy, R. L., Stern, D. B. 1995. Mitochondrial transcription initiation: promoter structures and RNA polymerases. Curr. Genet. 28:205–216.

Unseld, M., Marienfeld, J. R., Brandt, P., Brennicke, A. 1997. The mitochondrial genome of *Arabidopsis thaliana* contains 57 genes in 366,924 nucleotides. Nat. Genet. 15:57–61.

Weihe, A. 2004. The transcription of plant organelle genomes. In Molecular Biology and Biotechnology of Plant Organelles, H. Daniell and C. D. Chase eds., pp. 213–237. Dordrecht: Kluwer.

Weihe, A., Hedtke, B., Börner, T. 1997. Cloning and characterization of a cDNA encoding a bacteriophage-type RNA polymerase from the higher plant *Chenopodium album*. Nucleic Acids Res. 25:2319–2325.

Woody, A. Y. M., Eaton, S. S., Osumi-Davis, P. A., Woody, R. W. 1996. Asp537 and Asp812 in bacteriophage T7 RNA polymerase as metal ion-binding sites studied by EPR, flow-dialysis, and transcription. Biochemistry 35:144–152.

Xu, B., Clayton, D. A. 1992. Assignment of a yeast protein necessary for mitochondrial transcription initiation. Nucleic Acids Res. 20:1053–1059.

Yin, C., Richter, U., Börner, T., Weihe, A. 2009. Evolution of phage-type RNA polymerases in higher plants: characterization of the single phage-type RNA polymerase gene from *Selaginella moellendorffii*. J. Mol. Evol. 68:528–538.

Young, D. A., Allen, R. L., Harvey, A. J., Lonsdale, D. M. 1998. Characterization of a gene encoding a single-subunit bacteriophage-type RNA polymerase from maize which is alternatively spliced. Mol. Gen. Genet. 260:30–37.

Zoschke, R., Liere, K., Börner, T. 2007. From seedling to mature plant: *Arabidopsis* plastidial genome copy number, RNA accumulation and transcription are differentially regulated during leaf development. Plant J. 50:710–722.

Chapter 5
RNA Processing and RNA Stability in Plant Mitochondria

Stefan Binder, Angela Hölzle, and Christian Jonietz

Abstract In angiosperms, mitochondrial genomes encode about 60 genes. The corresponding gene products are all directly or indirectly involved in oxidative phosphorylation indicating their vital function. To realize the mitochondrial genetic information, mitochondria maintain comprehensive genetic systems. In these complex frameworks posttranscriptional processes play predominant roles. They decisively contribute to the formation of functional RNAs and might be important for regulation of mitochondrial gene expression. In plant mitochondria RNA maturation includes the generation of mature extremities not only of tRNAs and rRNAs but also of most mRNAs. While the function of 5′ and 3′ end processing of the structural RNAs is quite evident, still little is known about the importance of mRNA processing and its impact on RNA degradation. This chapter summarizes recent progress made in the understanding of RNA processing and RNA degradation in mitochondria of higher plants.

Keywords Posttranscriptional processing • Maturation of 5' and 3' ends • Mitochondrial promoters • RNA stability • RNA degradation

Abbreviations

CMS	Cytoplasmic male sterility
CR-RT-PCR	Circularized RNA reverse transcriptase PCR
L*er*	Landsberg *erecta*
mTERF	Mitochondrial transcription termination factor
PNPase	Polynucleotide phosphorylase
PPR	Pentatricopeptide repeat
RF	Restorer of Fertility
RNR1	RNase R homolog 1
RPF2	RNA processing factor 2

S. Binder (✉)
Institut Molekulare Botanik, Universität Ulm, D-89069 Ulm, Germany
e-mail: stefan.binder@uni-unl.de

F. Kempken (ed.), *Plant Mitochondria*, Advances in Plant Biology 1,
DOI 10.1007/978-0-387-89781-3_5,

5.1 Introduction

After several decades of intensive molecular analysis our knowledge about the genetic system of plant mitochondria has substantially increased. These investigations showed that numerous posttranscriptional processes are required to form mature mRNAs, rRNAs, and tRNAs. Apart from splicing of introns and RNA editing, two posttranscriptional processes addressed in other chapters of this book, almost all mitochondrial RNAs undergo maturation processes at the 5′ and/or 3′ ends. While the role of 5′ and 3′ end maturation of tRNA and rRNAs is quite evident, the biological role of posttranscriptional 5′ and 3′ end processing of mRNAs remains mysterious. Both processes might influence the steady-state level of mitochondrial RNA species and might therefore have an influence on the expression of plant mitochondrial genes. Whether these processes play a role in regulation of gene expression is unclear since it is still uncertain whether there is regulation of gene expression in plant mitochondria at all.

In this chapter we summarize recent progress made in the characterization of the plant mitochondrial transcriptome, in the analysis of 5′ and 3′ processing of mRNAs, rRNA and tRNAs as well as in RNA degradation.

5.2 The Mitochondrial Transcriptome in *Arabidopsis thaliana*

In recent years the mitochondrial genomes of several angiosperms have been completely sequenced (Kubo and Mikami 2007; Kubo and Newton 2008). The gene content is fairly similar in all of these genomes and sequence conservation of the genes and introns is high. To the contrary, the sizes of these mitochondrial genomes vary considerably. This variation is mainly due to distinct, species-specific intergenic regions, which are composed of repeated mitochondrial DNA segments, of DNA fragments originating from plastids and the nucleus and for the most part of sequences of unknown origin. Large parts of these intergenic regions seem to have no function and are considered to be "junk DNA."

In contrast to the wealth of information about angiosperm mitochondrial genome sequences much less systematic data are available about the transcriptomes of these organelles. For several of the completely sequenced mitochondrial genomes RNA editing sites (see also Chap. 7) have been identified confirming transcription of the annotated mitochondrial genes. To further characterize these transcripts, extremities of the mature mitochondrial steady-state transcripts have been systematically mapped in the model species *Arabidopsis thaliana*, from now on referred to as Arabidopsis (Forner et al. 2007). Using CR-RT-PCR analysis major 5′ and 3′ transcript ends of all protein coding genes have been mapped first in the accession Columbia (Col) and later on also in C24 and Landsberg *erecta* (L*er*) (Forner et al. 2005, 2007) (Box 5.1). Clear mRNA ends have been found for each of the reading frames except the intron-internal *matR* (Table 5.1) (Kuhn and Binder 2002; Forner et al. 2007).

Box 5.1 CR-RT-PCR analysis

For a CR-RT-PCR approach, mitochondrial RNA (mtRNA) or total RNA is self-ligated by RNA ligase in a bulk reaction. RNA ligase covalently links 3′ and 5′ ends of single or different RNA molecules. The ligated RNA is then used as template for RT-PCR across the ligation site (Kuhn and Binder 2002; Forner et al. 2007). This is achieved by the PCR strategy and primer design, which selects molecules with linked 5′ and 3′ ends of a given RNA. Direct sequencing of the resulting PCR product(s) or sequence analyses of individual cloned products then allows the identification of RNA extremities (both 5′ and 3′ ends) on the nucleotide level in respect to mitochondrial genome sequences (Kuhn and Binder 2002; Forner et al. 2007); Unseld et al., 1997.

Table 5.1 Mitochondrial mRNA in Arabidopsis

Gene	5′ ends	3′ ends	Transcript lengths	Comments
atp1	**–124**, –267, –361	+160	1,808, 1,951, 2,045	
atp6-1	–198	**+45**, +46, +47, +48	1,401	
atp6-2	–44, –63, **–268**	+45	1,139, 1,158, 1,363	
atp8	**–154**, –224, **–228**	+121	755, 826	
atp9	**–84**, –83, –239	+179, **+180**	522	
ccmB	–140, –346, **–347**	+80	841, 1,048	Additional 5′ end in L*er* at –202
ccmC	from –482 to 487	**–46**, –47, +33, **+37**	between 1,207 and 1,213	5′ end in Col not clear, main 3′ end located within ORF, in C24 5′ end at –390/–391
ccmF$_C$	–124	+115	1,598	
ccmF$_{N1}$	–66, –149	+468	1,683, 1,766	
ccmF$_{N2}$	**–57**, –58	+170	839	
cox1	–241	+47	1,872	
cox2	–151, –205	+315, **+316**	1,250, 1,304	
cox3	**–378**, –379	+313, +314	1,490	In C24 auch 5′ Enden bei -447, -540, -560 and -640
nad1	–149, –355, –645	**+88**, +89	1,711, 1,917, 2n207	
nad2	–121, **–122** –229	**+35,** +36	1,657	
nad3- rps12		+15	1,027	
nad4	–226, **–228**	+30	1,746	Main 5′ end in L*er* at –390
nad4L-orf25	–294, –316, **–318**	+72	1,514, 1,538	
nad5	–21, –72	**+86,** +87	2,117, 2,168	
nad6	–179	–17	781	3′ end within ORF
nad7	–68, –109, –375	+52	1,305, 1,346, 1,612	

(continued)

Table 5.1 (continued)

Gene	5′ ends	3′ ends	Transcript lengths	Comments
nad9	–202, –243	+55	830, 871	Main 5′ end in C24 at –243 and –133
rpl2-orfx	–198	–10	2,244	Additional low abundant 5′ end between –115 and –119
rps3/rpl16	**–163**, –164	+67	2,307	
rps4	+1, **+2**	+498	1,586	
rpl5/cob	–406, –459	+58	4,107, 4,160, see comments	Also co-transcribed with tRNA Ser: transcript length 5,034
rps7	–448	+351	1,246	

Bold numbers indicate the predominant end when several termini were identified

For almost all mitochondrial genes a single 3′ mRNA ends (scattering over a few nucleotides) are found. The only exceptions are *ccmF*$_c$ transcripts with two different 3′ ends at positions +10 and +115. The 3′ termini of almost all transcripts are located downstream of the annotated reading frames with most 3′ untranslated regions (UTRs) being smaller than 180 nucleotides. In some instances 3′ UTRs can reach up to about 500 nucleotides, while no 3′ UTRs at all are present in the *ccmC* and *nad6* mRNAs. Here the major 3′ ends of are located within the predicted reading frames (*ccmC*: –46 and *nad6*: –17 in respect to the translation stop codon). Nevertheless proteins were translated from these mRNAs lacking translation stop codons (Raczynska et al. 2006). Almost no conserved sequences are present at the 3′ ends of mRNAs and only very few 3′ termini map to obvious stem-loop structures. This observation indicates that conserved primary sequence elements as well as RNA hairpins have no general function in posttranscriptional formation of mature 3′ ends of mitochondrial mRNAs (see below).

In contrast to the clear and relatively simple patterns of 3′ termini, a far more complex situation is observed at the 5′ extremities (Forner et al. 2007). Here, often more than one major end was identified. In addition, several 5′ termini of low abundant mRNAs were detected (see Table 5.1). Only a few of the major 5′ ends directly originate from transcription initiation. The vast majority of 5′ termini of main mature ends are generated by posttranscriptional events. Likewise many 5′ termini of low abundant transcripts are derived from processing; however, there are also a number of low abundant 5′ ends that are directly generated by transcription initiation (Kühn et al. 2005, 2009; Forner et al. 2007). The 5′ UTRs of the mature transcripts range up to 646 nucleotides. No 5′ UTR at all is present in the *rps4* transcript whose main 5′ terminus maps to position +2 within the translation start codon.

Apart from conserved promoter elements surrounding some of the 5′ termini of primary transcripts, sequence conservation at the posttranscriptionally generated ends is only given at the ends of 26 S RNA (mature 5′ end), the –84 5′ end of *atp9* and the −179 5′ terminus of *nad6*. Here 24 bp ranging from −15 to +9 (first nucleotide of the RNA = +1) are conserved. Whether this conserved sequence element has any biologic role is presently unclear.

As mentioned, only very few stem-loop structures are present at mRNA 3′ termini in Arabidopsis mitochondria. The same holds true for the 5′ ends, which coincide only with RNA hairpins upstream of the *rpl5* (−459 and −406), the *atp6–2* (−268) and *nad7* gene (−375). Whether these higher order structures are of functional importance is unknown.

Besides stem-loops several tRNA-like structures, so-called t-elements have been identified at both mitochondrial 5′ and 3′ transcript ends. Such elements had originally been described in wheat mitochondria, where their function remained unclear (Hanic-Joyce et al. 1990). Now the identification of t-elements upstream and downstream of mRNA extremities shows that the tRNA processing apparatus is involved in mRNA processing, where it might play an important role (see below).

In addition to the transcriptome data from Arabidopsis, a comprehensive northern analysis had been performed with mitochondrial transcripts from *Brassica campestris* (turnip). Mainly simple transcript patterns had been observed in this plant species, which is in contrast to many more complex patterns found in other plants (Makaroff and Palmer 1987). Apart from that, a huge amount of 5′ and 3′ ends have been mapped for many mitochondrial genes in many plant species. Unfortunately, it is not possible to include this tremendous amount of data in this chapter.

5.3 Formation of Mature Mitochondrial mRNAs in Higher Plants

Already in the early 1990s it was realized that noncoding sequences are transcribed in maize mitochondria (Finnegan and Brown 1990). Pulse labeling experiments revealed transcription of regions in a distance of more than 20 kb from rRNA genes, while no corresponding stable transcripts had been detected. These findings already suggested that posttranscriptional processes might determine the fate of mitochondrial RNA. Now almost 20 years later, other reports confirmed transcription of large noncoding intergenic region resulting in a considerable amount of cryptic transcripts (Perrin et al. 2004b; Holec et al. 2008). For instance, an RNA of approximately 500 nucleotides corresponding to a "noncoding" region in the Arabidopsis mitochondrial genome accumulates in plants, in which the function of mitochondrial polynucleotide phosphorylase (PNPase) is impaired.

Several features contribute to the generation of cryptic transcripts. First, there seems to be an insufficient transcriptional control. Promoter sequences contain only short conserved tetranucleotide- or nonnucleotide motifs or have more or less unique structures, suggesting a relaxed control of transcription initiation (Mulligan et al. 1988a, b; Brown et al. 1991; Covello and Gray 1991; Hanic-Joyce and Gray 1991; Rapp and Stern 1992; Binder and Brennicke 1993a,b; Rapp et al. 1993; Giese et al. 1996; Caoile and Stern 1997; Dombrowski et al. 1999; Hoffmann and Binder 2002). Most genes are transcribed from multiple promoters, which can be located far upstream of a given gene. Moreover, plant mitochondrial promoters are located even in large noncoding regions (Dombrowski et al. 1998; Kühn et al. 2005, 2009). Second, transcription termination seems to be hardly controlled as well. Therefore, once started, transcription can generate huge transcripts. For instance, nascent *atp9* or *atp8* RNAs can extend far beyond the mature 3′ ends with sizes exceeding several times the mature transcripts (Perrin et al. 2004b; Holec et al. 2008). Thus it seems likely that a great portion if not all of a plant mitochondrial genome is transcribed. But finally only a minor part of these transcripts form the steady-state pool, which is dominated by intact mature RNAs "contaminated" by minor amounts of cryptic or defective transcripts. Yet, what determines whether a given RNA accumulates or whether it is degraded?

The analysis of numerous genes and their flanking sequences as well as the complete mitochondrial genome sequences from several angiosperm species clearly show that sequences within the reading frames are extremely conserved while flanking or intergenic regions are highly variable. Even conservation of 5′ and 3′ untranslated regions is rare. Only very few sequence stretches upstream of different mitochondrial genes are duplicated within a species and some are even conserved between species (Hazle and Bonen 2007; Kubo and Mikami 2007; Kubo and Newton 2008).

The observation that flanking regions including 5′ and 3′ UTRs are extremely variable leads to two possible conclusions about *cis*-elements required for the formation and accumulation of mature steady-state transcripts: First, the *cis*-elements reside in the conserved reading frames and are thus also conserved. This seems to be rather unlikely since one would expect 5′ and 3′ ends of a given gene to be identical in different species, which is not the case. Second, these *cis*-elements are present in the highly variable flanking regions and are thus also highly variable. This scenario is consistent with the many different transcript termini found for a given gene in different plant species. This variability hampers the straight forward identification and characterization of such *cis*-elements.

Apart from that it is still unclear which processes influence or determine the stability of a plant mitochondrial transcript. Many posttranscriptional processes including translation might play a role, but it seems plausible that processing of 5′ and/or 3′ termini has a particularly important impact on transcript stability. In addition, the 3′ adenylation state of a mitochondrial transcript is a key determinant of RNA stability in plant mitochondria.

5.4 5′ End Processing of Mitochondrial mRNAs

The 5′ RNA termini can be generated either by de novo transcription initiation or by posttranscriptional processing. To identify and map true 5′ processing sites it is important to discriminate between these two possibilities. This is usually done by means of the structure of the 5′ terminal phosphate groups. Triphosphates are normally present at the 5′ ends of primary transcripts derived from transcription initiation, while monophosphates are found at processed ends. Distinction between these ends is achieved by in vitro capping experiments using guanylyltransferase or by 5′ RACE experiments using RNA ligase. While guanylyltransferase selects for RNAs species with di- or triphosphates, the reaction mechanism of RNA ligation excludes RNA species with more than one 5′ phosphate group. Unfortunately, both experimental approaches are susceptible to inconclusive or false results since triphosphate groups undergo spontaneous or even enzymatic conversion to monophosphate groups. Thus the origin of the 5′ ends of some mitochondrial RNAs cannot be unambiguously be determined (Kühn et al. 2005; Forner et al. 2007).

Processed 5′ ends of RNAs can be generated exonucleolytically or endonucleolytically. Up to date, no experimental data support the presence of a 5′ to 3′ exonuclease in plant mitochondria. Therefore it is assumed that 5′ processing occurs endonucleolytically. This hypothesis is supported by a number of experimental results: First, RNase P (5′ end processing) and RNase Z (3′ end processing) activities have been detected for the generation of both 5′ and 3′ ends of tRNAs. In Arabidopsis, four RNase Z genes have been analyzed. Two of these genes (AtTRZ^{L1} and AtTRZ^{L2}) encode proteins, which are imported into mitochondria (Canino et al. 2009). Plant mitochondrial RNase P is similar to one of the three RNase P subunits described in animal mitochondria (Philippe Giegé, personal communication). Interestingly, both mitochondrial RNase P proteins do not require a *trans*-acting RNA component for catalysis (Holzmann et al. 2008). These data show that plant mitochondria contain at least three different endonucleases, which could also be involved in processing of mRNAs and rRNAs. Second, in agreement with this assumption, several 5′ and 3′ ends of mRNAs are generated by processing of co-transcribed tRNAs. This is indicated by the identification of mRNA extremities exactly upstream or downstream of the tRNA ends. In Arabidopsis mitochondria, this kind of processing has been observed for the 5′ ends of $ccmF_C$ (tRNA^{Gly}), *rps3-rpl16* (tRNA^{Lys}), and the 3′ ends of *atp6-1* and *atp6-2* (tRNA^{Ser}) (Forner et al. 2007). Third, beside true tRNAs also t-elements contribute to 5′ (*cox1*, $ccmF_{N1}$, and *rps4*) and 3′ (*ccmC* and *nad6*) end processing of Arabidopsis mitochondrial mRNAs (Fig. 5.1). Similarly, a t-element promotes 3′ end processing of a CMS-related transcript in rape seed (Bellaoui et al. 1997). Like the true tRNAs these t-elements are located adjacent to 5′ and 3′ termini of mRNA ends. Accordingly 5′ leader and 3′ trailer molecules have been detected, whose extremities map directly upstream or downstream of the mRNA end. These mapping data unambiguously demonstrate endonucleolytic processing of the respective mRNA ends (Forner et al. 2007). Fourth, in agreement with this in vivo observation, in vitro experiments with recombinant proteins confirmed the participation of RNase Z in mRNA 5′ or 3′ processing. Both mitochondrial RNase Z enzymes (*AtTRZ^{L1}* and *AtTRZ^{L2}*) faithfully cleave respective *cox1* and *ccmC* precursor molecules at sites determined by the tRNA-like structures (Canino et al. 2009). Fifth, while aforementioned endonucleolytic processing events involve true tRNAs or t-elements, the main 5′ end of the Arabidopsis *atp9* mRNA (at position –84 relative to the ATG) is also generated by

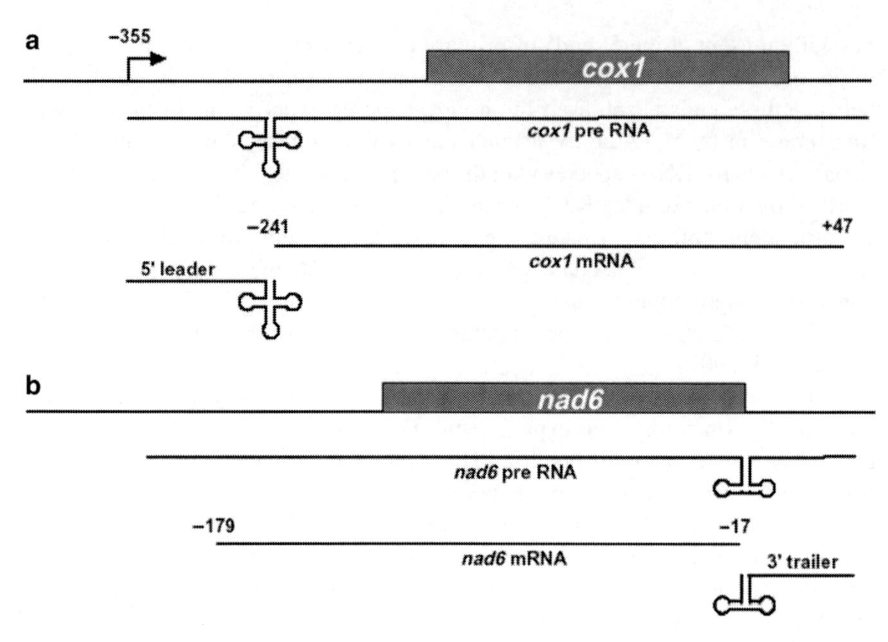

Fig. 5.1 T-elements contribute to endonucleolytic formation of both 5′ and 3′ ends of Arabidopsis mitochondrial mRNAs. (**a**) Transcription of *cox1* precursor RNAs is initiated 355 nucleotides upstream of the reading frame. The pre-mRNA includes a t-element, which is endonucleolytically cleaved at least at its 3′ end. The 3′ processing of this t-element by RNase Z produces the −241 5′ end of the mature *cox1* mRNA (Forner et al. 2007; Canino et al. 2009). The 3′ terminus of the *cox1* transcript is located 47 nucleotides downstream of the translation stop codon. (**b**) A t-element is also present in the 3′ region of the *nad6* precursor RNAs. Cleavage of this t-element most likely by RNase P generates the mature 3′ end of the *nad6* mRNA. In Arabidopsis mitochondria, this transcript is one of two mRNAs having a 3′ end upstream of the translation start codon (Forner et al. 2007)

endonucleolytic cleavage, although no such elements are seen at or in the vicinity of this end. Here 5′ leader molecules with 3′ ends immediately upstream of the 5′ terminus of *atp9* mRNA are detectable, providing evidence for the endonucleolytic generation of this end (Forner et al. 2007).

Apart from t-elements the sequence requirements for 5′ processing mitochondrial transcripts are unclear. In Arabidopsis, investigations of *cox3* mRNAs showed that potential *cis*-elements might be present more than 140 bp upstream of the processing site (Fig. 5.2) (Forner et al. 2005). Unfortunately, no in vitro systems are available that would allow the functional characterization of *cis*-elements for mRNA processing in plant mitochondria.

Taken together, the observations described above clearly demonstrate that several 5′ ends of mRNAs are generated by endonucleolytic cleavage suggesting a predominant role of this type of 5′ maturation. However, an involvement of a 5′ exonuclease cannot ultimately be excluded. In addition, several 3′ termini are generated by endonucleolytic processing, but it seems that this mode of 3′ processing is rather an exception than the rule (see below).

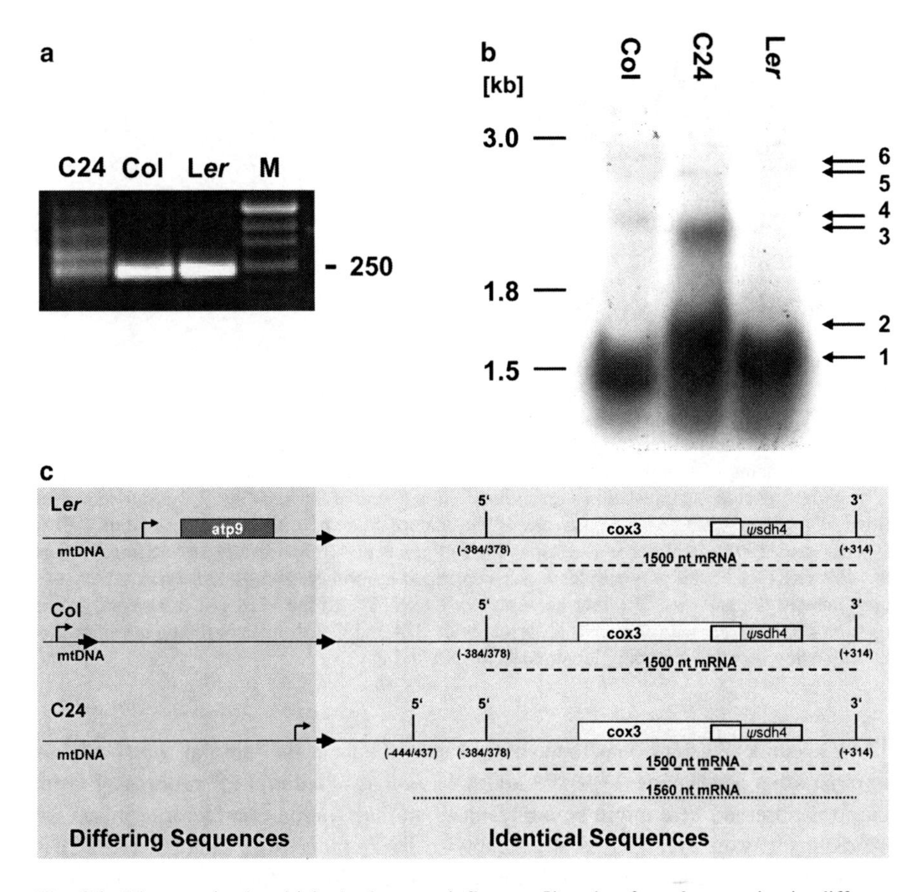

Fig. 5.2 Distant mitochondrial *cis*-elements influence 5′ ends of *cox3* transcript in different *Arabidopsis* accessions. (**a**) CR-RT-PCR analysis of *cox3* mRNAs in accessions C24, Col and L*er*. The major end at –384 is detected in all accessions, however, additional further upstream located 5′ termini are identified in C24. The generation of these larger RNA is due to an accession-specific mitochondrially encoded *cis*-element (Forner et al. 2005). (**b**) Larger transcript species are also seen in a northern blot analysis. (**c**) Different genomic arrangements around the *cox3* genes in C24, Col, and L*er*. At least some of the additional 5′ ends (–444, –437) are found in a sequence region identical between these accessions (*light gray*). This indicates that the *cis*-element(s) required for *cox3* mRNA processing are located more than 140 upstream of the processed ends, where sequences differ between the three accessions (*dark gray*)

As indicated, 5′ ends of mature mRNAs originate predominantly from processing. Probably many more 5′ ends of maturation intermediates are generated also by processing (A. Hölzle et al. 2008). Thus, the posttranscriptional generation of 5′ termini is a frequent process. Nevertheless the functional importance of 5′ processing is still unclear. A loss of processing of a particular 5′ end, as for instance observed for the *nad4* mRNA in the accession Landsberg *erecta* (L*er*), results in the accumulation of another mRNA species, which is normally a maturation intermediate (Fig. 5.3) (Forner et al. 2008). Accordingly, in this accession an mRNA with a

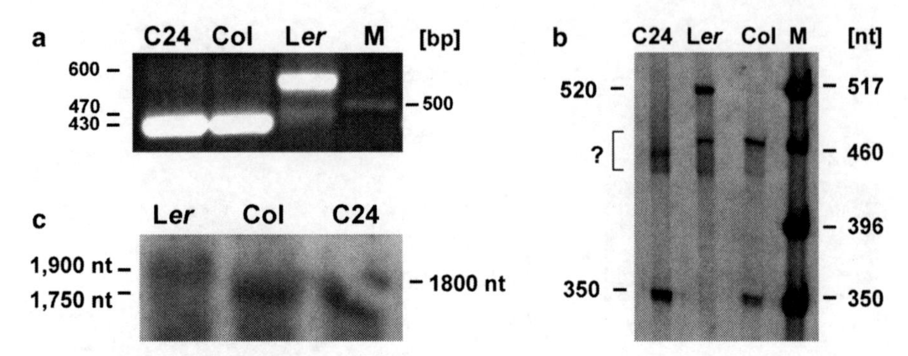

Fig. 5.3 Identification and characterization of the *nad4* 5′ transcript end polymorphism. (**a**) CR-RT-PCR analysis of *nad4* transcripts generated products of about 435 bp in C24 and Columbia (Col) while a product of 597 bp in Landsberg *erecta* (L*er*) indicated the presence of other ends. Direct sequencing of this product revealed identical 3′ termini in the three accessions (position +30, with respect to the translation stop codon), while 5′ end of the *nad4* mRNA in L*er* was identified at position –390 162 bp upstream of the mature 5′ ends in the other accessions (–228) (Forner et al. 2008). In all three accessions PCR products of 476 bp identify an additional 5′ end at –269. (**b**) This 5′ end polymorphism was confirmed by primer extension analysis, which also corroborated the presence of minor amounts *nad4* mRNA with the –228 end in L*er* and of the *nad4* mRNA species with the –390 5′ terminus in C24 and Col. (**c**) Accordingly, northern blot hybridization detected a slightly larger *nad4* mRNA in L*er*

170-nucleotide 5′ extension is found at the same amount as the "normal" *nad4* mRNA in most other accessions (Figs. 5.3 and 5.4). A loss of a certain 5′ processing event like it is observed here might be the result of an inactivation of a factor required for efficient cleavage (see Fig. 5.5b). Likewise RNA processing factor 2 (RPF2) is required for the generation of a 5′ end 202 nucleotides upstream of the *nad9* gene (Jonietz et al. 2010). In addition, this RESTORER OF FERTILITY (RF)-like PPR protein enhances 5′ processing of *cox3* transcripts.

The shift in the abundance of the different RNA species has no detrimental effect on L*er* plants, since they exhibit normal growth, development, and fertility. This strongly suggests that accumulation of the corresponding proteins remains unaffected. Therefore it can be assumed that for a given gene the mitochondrial translation machinery accepts mRNAs with different 5′ ends. However, this observation does not exclude a general importance of 5′ processing for translation. As seen for *nad4* mRNA maturation several parallel and/or consecutive 5′ processing steps occur during the generation of the mature mRNA (see Fig. 5.4). The loss of one (or even a few) of the processing steps is apparently accepted. However, the loss of other particular steps or all steps might be not, as this might have severe consequences for plant survival. Thus mutants with defects in essential processing events cannot be established, which prevents unambiguous conclusions about the importance of 5′ end processing and its potential impact on translation.

Beside a potential role for translation initiation, 5′ processing might also be important for RNA stability. However, there are presently no experimental data supporting this assumption.

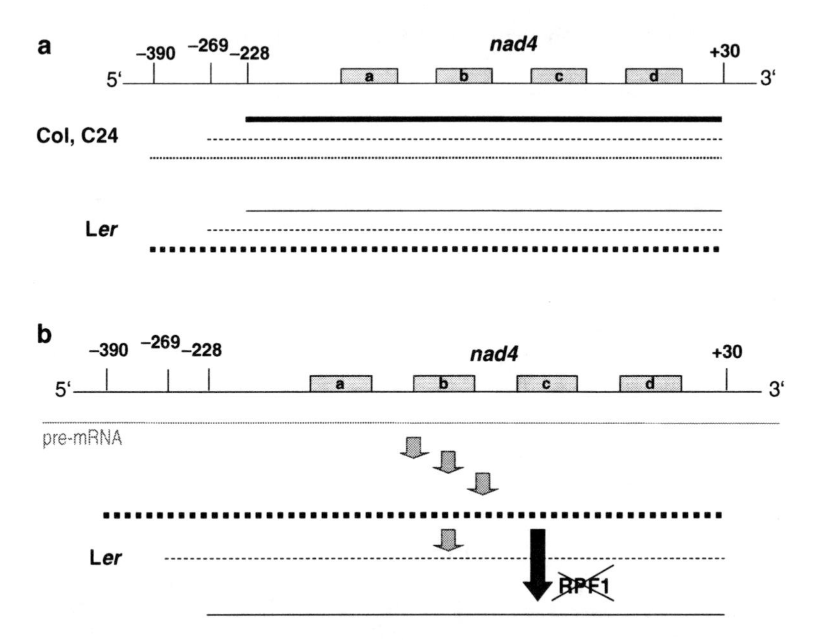

Fig. 5.4 Schematic presentation of *nad4* mRNA maturation. (**a**) Transcript species present in the different accessions given as *black*, *dotted*, or *dashed* lines. Highly abundant transcripts are indicated as *bold*, low abundant RNA species as *thin lines*. Position of the ends are given in respect to the translation start codon (NATG, N=–1) or to the translation stop codon (TAAN, N=+1). (**b**) Maturation of *nad4* mRNA in Landsberg *erecta*. A *nad4* mRNA with a –390 5′ end is formed via multiple processing steps. In this accession, this RNA species represents the mature transcripts, which is to a very little extend converted into an RNA species with the –269 5′ end. In Landsberg *erecta* the inefficient generation of a *nad4* mRNA species is most likely due to an inactivated version of the RNA PROCESSING FACTOR1 (RPF1) gene

5.5 Posttranscriptional Generation of 3′ Ends

The 3′ termini of mRNAs can be generated directly by transcription termination or by posttranscriptional processing. Recently, genes with similarities to animal mitochondrial transcription termination factors (mTERFs) have been identified in Arabidopsis, most of them being predicted to be localized to mitochondria or chloroplasts (Linder et al. 2005). In animal mitochondria, mTERF proteins promote transcription termination, but they are also important for transcription initiation and DNA replication (Roberti et al. 2006, 2009). It is possible that the mTERF proteins from Arabidopsis or plants in general have similar roles as in animal mitochondria including an active or governed termination of transcription. Thus 3′ termini of plant mitochondrial mRNAs could theoretically be generated directly by transcription termination.

On the other hand, several reports suggest only a minor importance of transcription termination in plant mitochondria. As mentioned, transcription proceeds into regions far away from coding sequences and indirect evidence suggest that this might

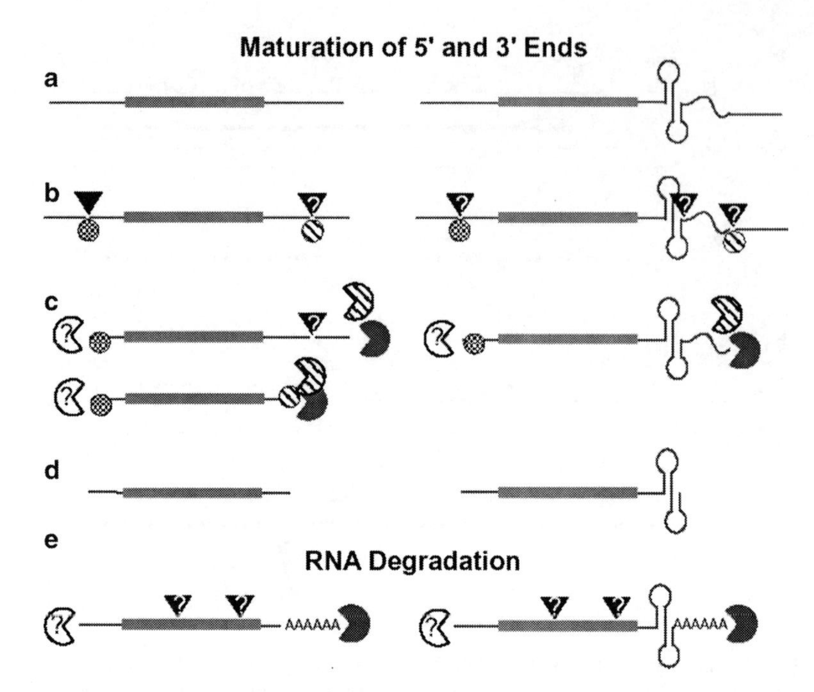

Fig. 5.5 Processing and degradation of mRNAs in plant mitochondria. (**a**) Precursor RNAs with (*right panel*) and without 3′ stem-loop structure (*left panel*). The reading frame encoded in a given RNA is indicated by a flat box. (**b**) A mature 5′ end might be generated by an endonuclease (*black arrowhead*) probably guided by one or several specificity factors (*checked circle*). An endonuclease and other proteins might be also involved in 3′ processing. This step might either directly generate the mature 3′ end or an intermediate end. (**c**) Alternatively, the 5′ terminus of an mRNA could be also generated by an exonuclease (*open mouth with question mark*); however, no experimental evidence for the existence of such an enzyme has been found in plant mitochondria. In contrast, there is good evidence that PNPase (*gray open mouth*) and RNR1 (hatched open mouth), two 3′ exonucleases are involved in 3′ processing. These enzymes might also be assisted by other factors (*hatched circles*). (**d**) Mitochondrial mRNAs with mature 5′ and 3′ ends. (**e**) Degradation of mRNAs requires 3′ polyadenylation and probably endonucleolytic as well as exonucleolytic activities. The role of PNPase in RNA degradation has been experimentally shown (Perrin et al. 2004a, b)

indeed be linked to the lack of a true transcription termination process (Finnegan and Brown 1990; Gagliardi and Binder 2007). This has also been observed in a PNPase knockdown mutant, in which *atp9* and *atp8* precursor RNAs extend far beyond the mature 3′ ends of mRNAs (Perrin et al. 2004b). In addition, in vitro assays have shown that transcription passes a double stem-loop structure located at the 3′ end of the *atp9* mRNA in pea (Dombrowski et al. 1997). Thus up to now, no data indicate a direct generation of mature 3′ termini by transcription termination.

Consequently, the 3′ ends of plant mitochondrial mRNAs have to be generated posttranscriptionally most likely from large precursor RNAs. Posttranscriptional 3′ processing has recently been reviewed and is thus only briefly summarized here (Gagliardi and Binder 2007).

Mapping of mitochondrial transcript extremities identified several 3′ ends immediately downstream of single or double stem-loop structures (Saalaoui et al. 1990; Kaleikau et al. 1992; Morikami and Nakamura 1993). Later functional studies have shown that such secondary structures could indeed function as *cis*-elements in 3′ processing by protecting the transcripts from exonucleolytic degradation resulting in an RNA with a mature stable 3′ end (see Fig. 5.5c) (Bellaoui et al. 1997; Dombrowski et al. 1997; Kuhn et al. 2001; Perrin et al. 2004b). Thus these structures are not only important for 3′ processing they can also influence the stability of a given transcript.

The functional importance of exonucleases in the plant mitochondrial RNA metabolism has been established by excellent in vivo studies. As exemplarily shown for the *atp9* transcripts in Arabidopsis mitochondria, two 3′ to 5′ exonucleases, PNPase and RNR1, an RNase II-like enzyme, participate in the 3′ maturation process (Perrin et al. 2004b). Studies of PNPase knockdown plants revealed an accumulation of *atp9* transcript species, which extend far beyond the mature 3′ ends. Yet there is no direct evidence that these RNA species are indeed precursor molecules and that sequences downstream of the mature 3′ ends are indeed removed by PNPase, but such a function of this enzyme is highly likely (see Fig. 5.5c).

RNR1 is involved in the generation of mature 3′ ends in mitochondria and chloroplasts (Perrin et al. 2004b; Bollenbach et al. 2005). This enzyme removes short 3′ extension, for instance downstream of a double stem-loop structure in the Arabidopsis *atp9* transcript.

While the general function of the two 3′ to 5′ exonucleases in 3′ processing is more or less established, it is unclear whether endonucleolytic activities are involved in this process (see Fig. 5.5c). As mentioned earlier endonucleolytic processing of tRNAs and t-elements in mRNA precursor molecules indeed generates 3′ termini of some mRNAs (see Fig. 5.1) (Bellaoui et al. 1997; Forner et al. 2007). However, these tRNA-like structures are present only in a few precursor mRNAs demonstrating that this mode of processing is the exception rather than the rule. In addition, a comprehensive study of mRNA extremities in Arabidopsis mitochondria did not reveal any 3′ processing intermediate, as it would be expected from endonucleolytic processing (J. Forner et al. 2008). Thus there are presently no data that would support a general function of endonucleolytic activities in mRNA 3′ processing in plant mitochondria.

As tRNAs or t-elements, stem-loop structures are only found in a few plant mitochondrial mRNAs. When present they clearly have a function in 3′ end formation (see above). However, the vast majority of plant mitochondrial mRNAs does not contain obvious RNA hairpin structures excluding a general role of these secondary structures in maturation of 3′ ends. Whether other *cis*-elements for 3′ processing exists and how they look like is presently not known. Alternatively, specific *trans*-acting factors might be involved in generation of mature 3′ termini (see Fig. 5.5c). However, the lack of any polymorphisms at plant mitochondrial 3′ ends suggests that probably only very few, but essential proteins are involved in maturation of 3′ ends of plant mitochondrial mRNAs (Forner and Binder 2007; Forner et al. 2008).

5.6 mRNA Stability

Like in all other genetic systems, steady-state levels of functional plant mitochondrial RNAs are balanced between the rates of RNA synthesis and RNA degradation. Transcriptional rates vary substantially between the different mitochondrial genes, mainly depending on the individual promoter strength. Probably also degradation rates vary considerably (Mulligan et al. 1991; Giegé et al. 2000; Kühn et al. 2009). The rate of degradation depends on various parameters, which determine the stability of a given RNA. Apart from polyadenylation, most of these parameters and their individual impact on RNA stability are unknown. They might include the presence or absence of certain secondary structure elements but may also be influenced by translation, distinct processing steps, or RNA binding factors. Different half-lives of plant mitochondrial RNA have been concluded from comparative studies of *in organello* RNA synthesis and the steady-state levels of mitochondrial transcripts. Here, clear discrepancies have been found indicating the great impact of RNA stability on steady-state levels of most mitochondrial transcripts (Mulligan et al. 1991; Giegé et al. 2000; Leino et al. 2005).

Higher order structure contributes substantially to the high stability of tRNAs and rRNAs. Likewise secondary or tertiary structure might also influence the half-lives of mRNAs. Indeed, in several genetic systems relatively simple secondary structures like the RNA hairpins at the 3′ ends of mRNA from chloroplasts are important determinants of mRNA stability (Gruissem 1989). As mentioned in the previous section similar structures can fold in the 3′ UTRs of some mitochondrial mRNAs as for instance *cob* transcripts in rice and in certain wheat lines, where they seem to increase the stability of these RNAs (Saalaoui et al. 1990; Kaleikau et al. 1992). Similarly the stability of a CMS-specific *orf138* transcript described in rape seed is increased by a stem-loop structure (Bellaoui et al. 1997). Also functional studies showed that a double stem-loop structure in the 3′ UTR of the pea *atp9* transcript protects synthetic RNAs from degradation in vitro (Dombrowski et al. 1997; Kuhn et al. 2001). As the comprehensive mapping studies of mitochondrial transcripts from Arabidopsis and many other plant species showed only very few mRNAs contain such structures. This suggests that either less obvious higher order structures stabilize mitochondrial transcripts or that mechanisms independent from secondary structure influence half-lives of plant mitochondrial mRNAs. The latter has been indicated by the dependence of mitochondrial transcript stability on the nuclear background, which was observed in plant material obtained from protoplast fusion between Arabidopsis and *Brassica napus* (Leino et al. 2005). In line with this, a PPR protein (Rf1b) affects the stability of the *atp6-orf79* mRNA, which is associated with the Boro II CMS system in rice (Wang et al. 2006) (see also Chaps. 8 and 18). But the exact molecular function of this protein is still unclear. A relatively simple mechanism to control the stability of an mRNA was very recently found in chloroplasts from maize. Here PPR10 binds to the 5′ and 3′ UTRs of several mRNAs. These protein "caps" protect these RNAs from both 5′ to 3′ and 3′ to 5′ exonucleolytic degradation. This is accomplished by a sequence-specific interaction of this protein with an approximately 25-nucleotide conserved sequence element present at the extremities of these RNAs (Pfalz et al. 2009). Whether such a mechanism also exists in plant mitochondria is unclear but rather unlikely. Mapping of major 5′ and 3′ ends of all mitochondrial transcripts in Arabidopsis revealed sequence conservation only in a very few instances. A 80-nucleotide repeat is present at the 3′ termini of *atp9* and *nad1,* while sequence element of approximately 25 nucleotides is conserved at the 5′ transcript ends of 26 S rRNA, *nad6* and *atp9* (Forner et al. 2007). Interestingly, the latter sequence element has an almost identical size as the conserved binding sites identified at the mRNA termini in maize chloroplasts.

Taken together, the general means to positively influence RNA stability in plant mitochondria are still unclear. To the contrary, a lot more is known about RNA decay in these organelles. Two major players in this process have been identified: The mitochondrial PNPase, one of the two 3′ to 5′ exonucleases, which is probably also involved in 3′ processing and polyadenylation of mitochondrial transcripts, which tags RNA for degradation (see Fig. 5.5e).

Degradation of plant mitochondrial RNA as well as polyadenylation has been reviewed recently and will thus only shortly be summarized here (Gagliardi and Binder 2007; Schuster and Stern 2009). The presence of non-encoded poly(A) tails has been reported for a number of mitochondrial mRNAs in various plant species (Gagliardi and Leaver 1999; Lupold et al. 1999; Gagliardi et al. 2001; Kuhn et al. 2001; Perrin et al. 2004a, b). Studies in vivo and in vitro demonstrated that poly(A) tails enhance degradation of RNAs. In plant mitochondria these nonencoded extensions can be present at different sites of the RNA but are predominantly found at mature 3′ ends, including those ends with stem-loops. Since poly(A) tails promote degradation of RNA they probably target an exonuclease, most likely the mitochondrial PNPase, to RNA. This is supported by similar degradation systems in *E. coli* and chloroplasts, where homologous PNPases are active in degradation of polyadenylated RNA. In addition, three different endonucleases (RNase E, RNase J, and CSP41) are involved in degradation of RNA in chloroplasts (Schuster and Stern 2009). This suggests that endonucleolytic degradation of RNA might also be feasible in mitochondria, although experimental data supporting this assumption are not available at present (Marchfelder 1995; Canino et al. 2009).

The enzymes required for polyadenylation of plant mitochondrial RNA are still unknown. Some candidate genes coding for poly(A) polymerases (PAP) have been identified (Martin and Keller 2004), but not yet functionally analyzed. In addition, PNPase itself is a polyadenylating enzyme. In chloroplasts, both PAP and PNPase contribute to the tail population and it is suggested that strict homopolymeric tails are synthesized by PAP, while poly(A) extension containing some other nucleotide identities are supposed to be generated by PNPase (Yehudai-Resheff et al. 2001). In mitochondria, the detection of homopolymeric tails suggests an exclusive function of a PAP. In contrast to PNPase, there is no evidence that RNR1 plays a substantial role in degradation of poly(A) RNA. However, this enzyme efficiently removes the poly(A) extensions suggesting that the polyadenylation state of a given RNA is balanced between RNR1 and the tailing enzyme.

Apart from true poly(A) tails short "CCA"-like extensions have been detected at the 3′ ends of several plant mitochondrial RNAs (Williams et al. 2000; Kuhn et al. 2001; Kuhn and Binder 2002). These nucleotides might represent intermediates of poly(A) addition or removal. But it is also possible that they are synthesized independently from true poly(A) tails by a nucleotidyl transferase normally involved in maturation of tRNAs. Whether these short mRNA extensions have any biological function is unclear.

5.7 CMS, Posttranscriptional Processes, and PPR Proteins

Pentatricopeptide repeat (PPR) proteins are defined by the presence of a canonical 35-amino acid degenerate motifs, which are found in tandem arrays of up to 30 repeats (Small and Peeters 2000; Lurin et al. 2004; Schmitz-Linneweber and Small 2008). In Arabidopsis about 450 PPR proteins have been identified, most of them are predicted to be targeted to mitochondria and/or chloroplasts. A number of functional studies revealed that PPR proteins exhibit sequence-specific RNA binding activity, supporting their general importance for the organellar RNA metabolism (Delannoy et al. 2007).

In recent years a number of PPR proteins have been identified as RF in cytoplasmic male sterility (CMS) systems from various plant species (Bentolila et al. 2002; Brown et al. 2003; Desloire et al. 2003; Koizuka et al. 2003; Wang et al. 2006). CMS is a maternally inherited deficiency to produce or release viable pollen (Budar and Pelletier 2001; Chase 2007). This trait, which is linked to the expression of CMS-specific mitochondrial genes, can be used to generate F1 hybrids in many crops. Here crosses between male sterile lines with fertile lines containing nuclear encoded *RF* genes produce hybrid seeds (see Box 5.2 and Chap. 18).

Most PPR proteins encoded by *RF* genes are highly similar to each other and to a group of PPR proteins from the autogamous plant species *Arabidopsis,* where CMS has not been found. In this model species the function of most of these RF-like PPR genes is unknown (Desloire et al. 2003; Schmitz-Linneweber and Small 2008). However, very recently one of these genes was found to encode RPF2, a protein required for 5′ processing of *nad9* and *cox3* mRNAs (Jonietz et al. 2010).

5.8 Mitochondrial tRNA Processing

Plant mitochondrial DNAs encode about 20 tRNA genes. These genes are distributed over the genomes and are found as single units flanked by large noncoding regions or in clusters, sometimes linked to rRNAs or protein coding genes. Plant mitochondrial tRNA genes are generally transcribed as large precursors that undergo a number of processing steps to obtain their functional structure. Maturation of tRNA requires

Box 5.2 Cytoplasmic male sterility

Cytoplasmic male sterility (CMS) is a maternally inherited deficiency to produce or release viable pollen (Budar and Pelletier 2001; Chase 2007). This trait is linked to the expression of CMS-specific mitochondrial genes, which are chimeras composed of fragments of canonical mitochondrial genes, of intergenic regions and of DNA sequences of unknown origin. Expression of these genes has detrimental effects, which interfere with production of release of fertile pollen (Budar and Pelletier 2001; Chase 2007). The CMS genes are often expressed in dicistronic transcripts located upstream or downstream of normal mitochondrial genes. Restoration of fertility is achieved by reduced accumulation of the deleterious CMS-associated protein, which is almost exclusively accomplished by changes of gene expression on the post-transcriptional level. In this context, the restoring PPR proteins play some role in RNA cleavage, RNA stability, RNA editing, and translation, though the exact molecular mode of operation of most of these PPR proteins is often unclear. One of the best characterized restoring PPR proteins is RF1A of the rice Boro II CMS system. In the sterile line the *orf79* protein is translated from the dicistronic *atp6-orf79* transcript. In the presence of RF1A this RNA is endonucleolytically cleaved into an *atp6* and an *orf79* mRNA. The CMS-specific transcript, which partially degraded in the restored line, contains the complete *orf79* reading frame, but it seems to lack *cis*-elements required for efficient translation. In addition, RF1A is also involved for RNA editing (Wang et al. 2006; Kazama et al. 2008). A second restorer gene of the rice Boro II CMS system, RF1B, promotes degradation of the CMS-related transcripts by an as-yet unknown mechanism. For more details see Chap. 18.

an endonucleolytic cleavage, which directly generates the mature 5′ end. This cut is catalyzed by an RNase P, which in most genetic systems is a ribonucleoprotein (Frank and Pace 1998). This seems to be different in plant mitochondria, where RNase P is similar to one of the three subunits of the animal mitochondrial RNase P, which does not require an RNA moiety for catalytic activity (Holzmann et al. 2008).

The 3′ ends of tRNAs are generated in two consecutive steps. First, RNase Z cleaves downstream of the discriminator nucleotide, either directly or one or two nucleotides further downstream. The genes encoding four different RNase Z proteins have recently been characterized in Arabidopsis (Canino et al. 2009). Two of the genes encode long versions of this endonuclease (about 900 amino acids), both being transported to mitochondria. Second, 3′ end maturation is completed by the addition of the 5′ -CCA-3′, most likely by a terminal nucleotidyl transferase, for which several candidate genes have been suggested in plants (Martin and Keller 2004). In addition, base modifications and sometimes true substitutional C to U RNA editing are necessary to generate functional tRNA molecules (Maréchal-Drouard et al. 1993, 1996a, b; Binder et al. 1994b; Marchfelder et al. 1996).

As in vitro analyses showed, the mitochondrial RNase Z enzymes are capable of cleaving tRNA precursors without any auxiliary factors (Canino et al. 2009), which might also be true for plant mitochondrial RNase P. Also t-elements are cut by these enzymes without the necessity of additional protein factors. One can speculate that an interaction of RNase Z (and maybe also RNase P) with other factors might be required to cleave other RNA substrates with less or no similarity to tRNAs; however, no experimental data supporting this hypothesis have been obtained. As mentioned above, a function of the tRNA processing enzymes seems plausible since no other endonucleases have been described in plant mitochondria.

A few tRNAs also undergo RNA editing, which is frequently but not always observed in double-stranded regions of the tRNA molecules. Here it can correct mismatches to canonical Watson and Crick base-pairs. RNA editing of tRNA can be a prerequisite to efficient processing of its extremities. (Marchfelder et al. 1996; Maréchal-Drouard et al. 1996a, b). Details on tRNA editing are summarized in Chap. 7.

5.9 Generation of Mature rRNAs in Plant Mitochondria

Plant mitochondria encode three rRNA genes, *rrn26* for 26 S, *rrn18* for 18 S and *rrn5* for 5 S rRNA. In all plant species investigated the *rrn18* and *rrn5* are cotranscribed, including a spacer of about 110–150 nucleotides. The 26S rRNA is transcribed independently from a separate locus. Promoter studies revealed that transcription of 18 S/5 S and 26 S rRNAs usually starts upstream of the mature 5′ ends resulting in 5′ extended precursors (Giese et al. 1996; Kühn et al. 2005). Only the 5′ terminus of 26 S rRNA might originate directly from transcription initiation (Binder et al. 1994a). Up to now, very little is known about maturation of rRNAs; however, the analysis of 18 S/5 S processing intermediates and maturation byproducts in an Arabidopsis PNPase knockdown line revealed interesting insights into the generation of mature 5′ termini of these rRNA species (Perrin et al. 2004a). In these mutant plants, leader molecules removed from the 5′ end of the 18S rRNA accumulate to substantial amounts, which allowed the precise mapping of their 3′ ends. These ends were found almost exclusively directly upstream of the mature 5′ end, providing clear evidence for the endonucleolytic generation of the 18S rRNA 5′ extremity. Likewise, evidence was obtained for an endonucleolytic processing of the 5′ end of 5S rRNA. These observations strongly suggest a similar scenario for the generation the 5′ end of 26S rRNA. The studies of the PNPase knockdown plants further indicated that mitochondrial PNPase might play a role in the exonucleolytic formation of the 3′ ends of 18S rRNA. Thus it seems that maturation of rRNAs follows a pattern similar to what is suggested for mRNAs, with endonucleolytic generation of the 5′ ends and exonucleolytic formation of 3′ termini.

Presently nothing is known about the proteins involved in 5′ processing of rRNA. In analogy to mRNA maturation it seems possible that one or both of the tRNA processing enzymes participate in this step. In Triticum a tRNA gene (fMet) is located only one nucleotide upstream the 18 S rRNA gene (Coulthart et al. 1993), indicating that RNase Z is indeed involved in both processing of the tRNA as well as in the generation of the mature 5′ end of 18 S rRNA. Apart from endonucleases, which might also be completely unrelated to RNase Z and RNase P other factors like PPR protein or ribosomal proteins might be required for the generation of mature 5′ and maybe also 3′ ends of plant mitochondrial ribosomal rRNA (Gagliardi and Binder 2007).

5.10 Conclusions

Over the last decade substantial progress in the understanding of plant mitochondrial RNA processing and degradation has been made. But still numerous processes have to be resolved in detail and most of the components involved in these processes have to be identified. Nevertheless a number of features of the mitochondrial transcriptome, some posttranscriptional processes required for the formation of mature RNAs and last but not least some proteins involved in these processes have been identified and/or characterized (Box 5.3).

Box 5.3 Main characteristics of mitochondrial mRNA processing and degradation

mRNA processing:

- Many mitochondrial protein coding genes are represented by several mRNA species.
- These mRNAs species have a single 3′ end, but differ at their 5′ termini.
- Most major 5′ and 3′ ends of mRNAs are generated posttranscriptionally. Thus posttranscriptional generation of mRNA extremities is a frequent and important process.
- Sequence conservation at 5′ and 3′ ends is observed only in very rare cases. Also only very few mRNAs contain 5′ or 3′ stem-loop structures.
- The 5′ ends of mRNAs are generated by endonucleolytic processing. This has been shown for mRNAs co-transcribed with tRNAs or t-elements and has also been indicated for the *atp9* mRNAs lacking tRNAs or t-elements. In plant mitochondria no evidence for the presence of a 5′ exonuclease has been found.
- tRNAs and t-elements can also contribute to endonucleolytic formation of 3′ ends. However, there is compelling evidence that these termini a mainly generated by 3′ exonucleases. These enzymes might in some cases be supported by other proteins.
- There is evidence that PNPase and RNR1, two 3′ exonucleases are involved in 3′ processing.

(continued)

Box 5.3 (continued)

tRNA processing:

- Maturation of 5′ and 3′ ends of tRNAs requires RNase P and RNase Z, two endonucleases. These enzymes also contribute to mRNA processing.
- tRNA processing is completed by the addition of 5′-CCA-3′ by a nucleotidyl transferase, by substitutional C to U RNA editing and by normal base modifications.

rRNA processing:

- The mature 5′ ends of 18S and 5S rRNAs are generated by endonucleolytic cleavage.

RNA degradation:

- Poly(A) tails tag RNA for degradation
- Plant mitochondrial PNPase is a major component of the degradation machinery

Glossary

Cytoplasmic male sterility (CMS): CMS is a maternally inherited deficiency to produce or release viable pollen. This economically important trait is linked to the expression of CMS-specific mitochondrial genes, which are chimeras composed of fragments of canonical mitochondrial genes, intergenic regions, and DNA sequences of unknown origin. Expression of these genes has detrimental effects, which interfere with production of release of fertile pollen.

Guanylyltransferase: Is an enzyme that catalyzes the addition of GTP to 5′ ends of RNA molecules, which carry di- or triphosphate groups. This eukaryotic protein is required for the addition of 5′ caps to nuclear encoded mRNAs.

5' RACE: Rapid amplification of cDNA ends: Anchor oligonucleotides are ligated to the 5′ ends of RNAs. Then an RT-PCR is performed with an mRNA-specific primer (which selects for the RNA of interest) and the anchor primer. Sequencing of the PCR product reveals the unknown 5′ end

Restorer of fertility (RF): A RF is a nuclear gene, that restores the detrimental effect linked to expression the CMS-related mitochondrial genes. Restoration of CMS is most often achieved by altering the expression of the CMS-specific mitochondrial gene at the posttranscriptional level (see also Box 5.2).

RNase P: Is an endoribonuclease that generates the mature 5′ ends of transfer RNAs (tRNAs). In the majority of cases RNase P is a ribonucleoprotein composed of a protein and a RNA subunit.

RNase Z: Is an endoribonuclease which is required for the generation of the mature 3′ ends of transfer RNAs (tRNAs).

Steady-state level: The steady-state level of an RNA corresponds to the amount of RNA that is determined by the rates of transcription and degradation.

Stem-loop structure (also RNA hairpin): A stem-loop structure is formed when a single-stranded nucleic acid (RNA) molecule folds back on itself to form a complementary double helix (stem) topped by a loop. Stem-loop structures are encoded by short inverted repeats (which fold back into the double stranded stem) separated by a few nucleotides (which represent the unpaired loop).

T-element: Is a tRNA-like structure that does not function as true tRNA, but which exhibits some conserved features of a tRNA most often the pseudo uridine arm the acceptor stem.

References

Bellaoui, M., Pelletier, G., Budar, F. 1997. The steady-state level of mRNA from the *Ogura* cytoplasmic male sterility locus in *Brassica* cybrids is determined post-transcriptionally by its 3′ region. EMBO J. 16:5057–5068.

Bentolila, S., Alfonso, A. A., Hanson, M. R. 2002. A pentatricopeptide repeat-containing gene restores fertility to cytoplasmic male-sterile plants. Proc. Natl. Acad. Sci. USA 99:10887–10892.

Binder, S., Brennicke, A. 1993a. A tRNA gene transcription initiation site is similar to mRNA and rRNA promoters in plant mitochondria. Nucleic Acids Res. 21:5012–5019.

Binder, S., Brennicke, A. 1993b. Transcription initiation sites in mitochondria of *Oenothera berteriana*. J. Biol. Chem. 268:7849–7855.

Binder, S., Thalheim, C., Brennicke, A. 1994a. Transcription of potato mitochondrial 26 S rRNA is initiated at its mature 5′ end. Curr. Genet. 26:519–523.

Binder, S., Marchfelder, A., Brennicke, A. 1994b. RNA editing of tRNA(Phe) and tRNA(Cys) in mitochondria of *Oenothera berteriana* is initiated in precursor molecules. Mol. Gen. Genet. 244:67–74.

Bollenbach, T. J., Lange, H., Gutierrez, R., Erhardt, M., Stern, D. B., Gagliardi, D. 2005. RNR1, a 3′ –5′ exoribonuclease belonging to the RNR superfamily, catalyzes 3′ maturation of chloroplast ribosomal RNAs in *Arabidopsis thaliana*. Nucleic Acids Res. 33:2751–2763.

Brown, G. G., Auchincloss, A. H., Covello, P. S., Gray, M. W., Menassa, R., Singh, M. 1991. Characterization of transcription initiation sites on the soybean mitochondrial genome allows identification of a transcription-associated sequence motif. Mol. Gen. Genet. 228:345–355.

Brown, G. G., Formanova, N., Jin, H., Wargachuk, R., Dendy, C., Patil, P., Laforest, M., Zhang, J., Cheung, W. Y., Landry, B. S. 2003. The radish Rfo restorer gene of Ogura cytoplasmic male sterility encodes a protein with multiple pentatricopeptide repeats. Plant J. 35:262–272.

Budar, F., Pelletier, G. 2001. Male sterility in plants: occurrence, determinism, significance and use. C. R. Acad. Sci. III 324:543–550.

Canino, G., Bocian, E., Barbezier, N., Echeverria, M., Forner, J., Binder, S., Marchfelder, A. 2009. Arabidopsis encodes four tRNase Z enzymes. Plant Physiol. 150:1494–1502.

Caoile, A. G., Stern, D. B. 1997. A conserved core element is functionally important for maize mitochondrial promoter activity *in vitro*. Nucleic Acids Res. 25:4055–4060.

Chase, C. D. 2007. Cytoplasmic male sterility: a window to the world of plant mitochondrial-nuclear interactions. Trends Genet. 23:81–90.

Coulthart, M. B., Spencer, D. F., Gray, M. W. 1993. Comparative analysis of a recombining-repeat-sequence family in the mitochondrial genomes of wheat (*Triticum aestivum* L.) and rye (*Secale cereale* L.). Curr. Genet. 23:255–264.

Covello, P. S., and Gray, M. W. 1991. Sequence analysis of wheat mitochondrial transcripts capped in vitro: definitive identification of transcription initiation sites. Curr. Genet. 20:245–251.

Delannoy, E., Stanley, W. A., Bond, C. S., Small, I. D. 2007. Pentatricopeptide repeat (PPR) proteins as sequence-specificity factors in post-transcriptional processes in organelles. Biochem. Soc. Trans. 35:1643–1647.

Desloire, S., Gherbi, H., Laloui, W., Marhadour, S., Clouet, V., Cattolico, L., Falentin, C., Giancola, S., Renard, M., Budar, F., Small, I., Caboche, M., Delourme, R., Bendahmane, A. 2003. Identification of the fertility restoration locus, Rfo, in radish, as a member of the pentatricopeptide-repeat protein family. EMBO Rep. 4:588–594.

Dombrowski, S., Brennicke, A., Binder, S. 1997. 3′-Inverted repeats in plant mitochondrial mRNAs are processing signals rather than transcription terminators. EMBO J. 16:5069–5076.

Dombrowski, S., Hoffmann, M., Guha, C., Binder, S. 1999. Continuous primary sequence requirements in the 18-nucleotide promoter of dicot plant mitochondria. J. Biol. Chem. 274: 10094–10099.

Dombrowski, S., Hoffmann, M., Kuhn, J., Brennicke, A., Binder, S. 1998. On mitochondrial promoters in *Arabidopsis thaliana* and other flowering plants. In Plant Mitochondria: From Gene to Function, I. M. Møller, P. Gardeström, K. Glimelius, E. Glaser, eds. Leiden, Netherlands: Backhuys Publications, pp. 165–170.

Finnegan, P. M., Brown, G. G. 1990. Transcriptional and post-transcriptional regulation of RNA levels in maize mitochondria. Plant Cell 2:71–83.

Forner, J., Binder, S. 2007. The red fluorescent protein eqFP611: application in subcellular localization studies in higher plants. BMC Plant Biol. 7:28.

Forner, J., Weber, B., Wietholter, C., Meyer, R. C., Binder, S. 2005. Distant sequences determine 5′ end formation of *cox3* transcripts in *Arabidopsis thaliana* ecotype C24. Nucleic Acids Res. 33:4673–4682.

Forner, J., Weber, B., Thuss, S., Wildum, S., Binder, S. 2007. Mapping of mitochondrial mRNA termini in *Arabidopsis thaliana*: t-elements contribute to 5′ and 3′ end formation. Nucleic Acids Res. 35:3676–3692.

Forner, J., Hölzle, A., Jonietz, C., Thuss, S., Weber, B., Schwarzländer, M., Meyer, R. C., Binder, S. 2008. Mitochondrial mRNA polymorphisms in different *Arabidopsis thaliana* accessions. Plant Physiol. 148:1106–1116.

Frank, D. N., Pace, N. R. 1998. Ribonuclease P: unity and diversity in a tRNA processing ribozyme. Annu. Rev. Biochem. 67:153–180.

Gagliardi, D., Binder, S. 2007. Expression of the plant mitochondrial genome. In Plant Mitochondria, D. Logan, ed. Ames, IA: Blackwell, pp. 50–95.

Gagliardi, D., Leaver, C. J. 1999. Polyadenylation accelerates the degradation of the mitochondrial mRNA associated with cytoplasmic male sterility in sunflower. EMBO J. 18:3757–3766.

Gagliardi, D., Perrin, R., Marechal-Drouard, L., Grienenberger, J. M., Leaver, C. J. 2001. Plant mitochondrial polyadenylated mRNAs are degraded by a 3′-5′ exoribonuclease activity, which proceeds unimpeded by stable secondary structures. J. Biol. Chem. 276:43541–43547.

Giegé, P., Hoffmann, M., Binder, S., Brennicke, A. 2000. RNA degradation buffers asymmetries of transcription in *Arabidopsis* mitochondria. EMBO Rep. 1:164–170.

Giese, A., Thalheim, C., Brennicke, A., Binder, S. 1996. Correlation of nonanucleotide motifs with transcript initiation of 18 S rRNA genes in mitochondria of pea, potato and *Arabidopsis*. Mol. Gen. Genet. 252:429–436.

Gruissem, W. 1989. Chloroplast gene expression: how plants turn their plastids on. Cell 56:161–170.

Hanic-Joyce, P. J., Gray, M. W. 1991. Accurate transcription of a plant mitochondrial gene *in vitro*. Mol. Cell. Biol. 11:2035–2039.

Hanic-Joyce, P. J., Spencer, D. F., Gray, M. W. 1990. *In vitro* processing of transcripts containing novel tRNA-like sequences ('t-elements') encoded by wheat mitochondrial DNA. Plant Mol. Biol. 15:551–559.

Hazle, T., Bonen, L. 2007. Comparative analysis of sequences preceding protein-coding mitochondrial genes in flowering plants. Mol. Biol. Evol. 24:1101–1112.

Hoffmann, M., Binder, S. 2002. Functional importance of nucleotide identities within the pea *atp9* mitochondrial promoter sequence. J. Mol. Biol. 320:943–950.

Holec, S., Lange, H., Canaday, J., Gagliardi, D. 2008. Coping with cryptic and defective transcripts in plant mitochondria. Biochim. Biophys. Acta 1779:566–573.

Hölzle A, Fischer S, Heyer R, Schütz S, Zacharias M, Walther P, Allers T, Marchfelder A. 2008. Maturation of the 5S rRNA 5′ end is catalyzed in vitro by the endonuclease tRNase Z in the archaeon H. volcanii. RNA. 14:928–937.

Holzmann, J., Frank, P., Loffler, E., Bennett, K. L., Gerner, C., Rossmanith, W. 2008. RNase P without RNA: identification and functional reconstitution of the human mitochondrial tRNA processing enzyme. Cell 135:462–474.

Jonietz, C., Forner, J., Hölzle, A., Thuss, S., Binder, S. 2010. RNA PROCESSING FACTOR 2 is required for 5′ end processing of *nad9* and *cox3* mRNAs in mitochondria of *Arabidopsis thaliana*. Plant Cell 22:443–453.

Kaleikau, E. K., Andre, C. P., Walbot, V. 1992. Structure and expression of the rice mitochondrial apocytochrome b gene (*cob-1*) and pseudogene (*cob-2*). Curr. Genet. 22:463–470.

Kazama, T., Nakamura, T., Watanabe, M., Sugita, M., Toriyama, K. 2008. Suppression mechanism of mitochondrial ORF79 accumulation by Rf1 protein in BT-type cytoplasmic male sterile rice. Plant J. 55:619–628.

Koizuka, N., Imai, R., Fujimoto, H., Hayakawa, T., Kimura, Y., Kohno-Murase, J., Sakai, T., Kawasaki, S., Imamura, J. 2003. Genetic characterization of a pentatricopeptide repeat protein gene, orf687, that restores fertility in the cytoplasmic male-sterile Kosena radish. Plant J. 34:407–415.

Kubo, T., Mikami, T. 2007. Organization and variation of angiosperm mitochondrial genome. Physiol. Plant. 129:6–13.

Kubo, T., Newton, K. J. 2008. Angiosperm mitochondrial genomes and mutations. Mitochondrion 8:5–14.

Kuhn, J., Binder, S. 2002. RT-PCR analysis of 5′ to 3′ -end-ligated mRNAs identifies the extremities of *cox2* transcripts in pea mitochondria. Nucleic Acids Res. 30:439–446.

Kuhn, J., Tengler, U., Binder, S. 2001. Transcript lifetime is balanced between stabilizing stem-loop structures and degradation-promoting polyadenylation in plant mitochondria. Mol. Cell. Biol. 21:731–742.

Kühn, K., Weihe, A., Börner, T. 2005. Multiple promoters are a common feature of mitochondrial genes in *Arabidopsis*. Nucleic Acids Res. 33:337–346.

Kühn, K., Richter, U., Meyer, E. H., Delannoy, E., Falcon de Longevialle, A., O'Toole, N., Börner, T., Millar, A. H., Small, I. D., Whelan, J. 2009. Phage-type RNA polymerase RPOTmp performs gene-specific transcription in mitochondria of *Arabidopsis thaliana*. Plant Cell 21:2762–2779.

Leino, M., Landgren, M., Glimelius, K. 2005. Alloplasmic effects on mitochondrial transcriptional activity and RNA turnover result in accumulated transcripts of *Arabidopsis* orfs in cytoplasmic male-sterile *Brassica napus*. Plant J. 42:469–480.

Linder, T., Park, C. B., Asin-Cayuela, J., Pellegrini, M., Larsson, N. G., Falkenberg, M., Samuelsson, T., Gustafsson, C. M. 2005. A family of putative transcription termination factors shared amongst metazoans and plants. Curr. Genet. 48:265–269.

Lupold, D. S., Caoile, A. G., Stern, D. B. 1999. Polyadenylation occurs at multiple sites in maize mitochondrial *cox2* mRNA and is independent of editing status. Plant Cell 11:1565–1578.

Lurin, C., Andres, C., Aubourg, S., Bellaoui, M., Bitton, F., Bruyere, C., Caboche, M., Debast, C., Gualberto, J., Hoffmann, B., Lecharny, A., Le Ret, M., Martin-Magniette, M. L., Mireau, H., Peeters, N., Renou, J. P., Szurek, B., Taconnat, L., Small, I. 2004. Genome-wide analysis of Arabidopsis pentatricopeptide repeat proteins reveals their essential role in organelle biogenesis. Plant Cell 16:2089–2103.

Makaroff, C. A., Palmer, J. D. 1987. Extensive mitochondrial specific transcription of the *Brassica campestris* mitochondrial genome. Nucleic Acids Res. 15:5141–5156.

Marchfelder, A. 1995. Plant mitochondrial RNase P. Mol. Biol. Rep. 22:151–156.

Marchfelder, A., Brennicke, A., Binder, S. 1996. RNA editing is required for efficient excision of $tRNA^{Phe}$ from precursors in plant mitochondria. J. Biol. Chem. 271:1898–1903.

Maréchal-Drouard, L., Kumar, R., Remacle, C., Small, I. 1996a. RNA editing of larch mitochondrial tRNA(His) precursors is a prerequisite for processing. Nucleic Acids Res. 24:3229–3234.

Maréchal-Drouard, L., Ramamonjisoa, D., Cosset, A., Weil, J. H., Dietrich, A. 1993. Editing corrects mispairing in the acceptor stem of bean and potato mitochondrial phenylalanine transfer RNAs. Nucleic Acids Res. 21:4909–4914.

Maréchal-Drouard, L., Cosset, A., Remacle, C., Ramamonjisoa, D., Dietrich, A. 1996b. A single editing event is a prerequisite for efficient processing of potato mitochondrial phenylalanine tRNA. Mol. Cell. Biol. 16:3504–3510.

Martin, G., Keller, W. 2004. Sequence motifs that distinguish ATP(CTP):tRNA nucleotidyl transferases from eubacterial poly(A) polymerases. RNA 10:899–906.

Morikami, A., Nakamura, K. 1993. Transcript map of oppositely oriented pea mitochondrial genes encoding the alpha-subunit and the subunit 9 of F_1F_0-ATPase complex. Biosci. Biotechnol. Biochem. 57:1530–1535.

Mulligan, R. M., Lau, G. T., Walbot, V. 1988a. Numerous transcription initiation sites exist for the maize mitochondrial genes for subunit 9 of the ATP synthase and subunit 3 of cytochrome oxidase. Proc. Natl. Acad. Sci. USA 85:7998–8002.

Mulligan, R. M., Maloney, A. P., Walbot, V. 1988b. RNA processing and multiple transcription initiation sites result in transcript size heterogeneity in maize mitochondria. Mol. Gen. Genet. 211:373–380.

Mulligan, R. M., Leon, P., Walbot, V. 1991. Transcriptional and posttranscriptional regulation of maize mitochondrial gene expression. Mol. Cell. Biol. 11:533–543.

Perrin, R., Lange, H., Grienenberger, J. M., Gagliardi, D. 2004a. AtmtPNPase is required for multiple aspects of the 18 S rRNA metabolism in *Arabidopsis thaliana* mitochondria. Nucleic Acids Res. 32:5174–5182.

Perrin, R., Meyer, E. H., Zaepfel, M., Kim, Y. J., Mache, R., Grienenberger, J. M., Gualberto, J. M., Gagliardi, D. 2004b. Two exoribonucleases act sequentially to process mature 3′ -ends of atp9 mRNAs in *Arabidopsis* mitochondria. J. Biol. Chem. 279:25440–25446.

Pfalz, J., Bayraktar, O. A., Prikryl, J., Barkan, A. 2009. Site-specific binding of a PPR protein defines and stabilizes 5′ and 3′ mRNA termini in chloroplasts. EMBO J. 28:2042–2052.

Raczynska, K. D., Le Ret, M., Rurek, M., Bonnard, G., Augustyniak, H., Gualberto, J. M. 2006. Plant mitochondrial genes can be expressed from mRNAs lacking stop codons. FEBS Lett. 580:5641–5646.

Rapp, W. D., Stern, D. B. 1992. A conserved 11 nucleotide sequence contains an essential promoter element of the maize mitochondrial atp1 gene. EMBO J. 11:1065–1073.

Rapp, W. D., Lupold, D. S., Mack, S., Stern, D. B. 1993. Architecture of the maize mitochondrial *atp1* promoter as determined by linker-scanning and point mutagenesis. Mol. Cell. Biol. 13:7232–7238.

Roberti, M., Bruni, F., Polosa, P. L., Gadaleta, M. N., Cantatore, P. 2006. The Drosophila termination factor DmTTF regulates in vivo mitochondrial transcription. Nucleic Acids Res. 34:2109–2116.

Roberti, M., Polosa, P. L., Bruni, F., Manzari, C., Deceglie, S., Gadaleta, M. N., Cantatore, P. 2009. The MTERF family proteins: mitochondrial transcription regulators and beyond. Biochim. Biophys. Acta 1787:303–311.

Saalaoui, E., Litvak, S., Araya, A. 1990. The apocytochrome from alloplasmic line wheat (*T. aestivum*, cytoplasm-*T. timopheevi*) exists in two differently expressed forms. Plant Sci. 66:237–246.

Schmitz-Linneweber, C., Small, I. 2008. Pentatricopeptide repeat proteins: a socket set for organelle gene expression. Trends Plant Sci. 13:663–670.

Schuster, G., Stern, D. 2009. RNA polyadenylation and decay in mitochondria and chloroplasts. Prog. Mol. Biol. Transl. Sci. 85:393–422.

Small, I. D., Peeters, N. 2000. The PPR motif - a TPR-related motif prevalent in plant organellar proteins. Trends Biochem. Sci. 25:46–47.

Unseld, M., Marienfeld, J. R., Brandt, P., Brennicke, A. 1997. The mitochondrial genome of *Arabidopsis thaliana* contains 57 genes in 366,924 nucleotides. Nat. Genet. 15:57–61.

Wang, Z., Zou, Y., Li, X., Zhang, Q., Chen, L., Wu, H., Su, D., Chen, Y., Guo, J., Luo, D., Long, Y., Zhong, Y., Liu, Y. G. 2006. Cytoplasmic male sterility of rice with boro II cytoplasm is caused by a cytotoxic peptide and is restored by two related PPR motif genes via distinct modes of mRNA silencing. Plant Cell 18:676–687.

Williams, M.A., Johzuka, Y., Mulligan, R. M. 2000. Addition of non-genomically encoded nucleotides to the 3′-terminus of maize mitochondrial mRNAs: truncated *rps12* mRNAs frequently terminate with CCA. Nucleic Acids Res. 28:4444–4451.

Yehudai-Resheff, S., Hirsh, M., Schuster, G. 2001. Polynucleotide phosphorylase functions as both an exonuclease and a poly(A) polymerase in spinach chloroplasts. Mol. Cell. Biol. 21:5408–5416.

Chapter 6
RNA Splicing in Plant Mitochondria

Linda Bonen

Abstract Introns within the mitochondrial genes of land plants belong mostly to the group II category of ribozyme mobile retroelements, although a minor number (particularly in nonvascular plants) are members of the group I intron family of "homing" ribozymes. In vascular plants, the mitochondrial introns often lack conventional structural features, and their degenerate nature is correlated with unusual splicing pathways. In addition, a subset of introns has undergone mitochondrial DNA rearrangements so that coding segments are scattered around the genome, and the mRNAs for such genes are generated through *trans*-splicing. This chapter summarizes recent advances in our understanding of splicing mechanisms, the nature of splicing machinery, and the relationships among splicing and other RNA processing events (such as editing) in plant mitochondria.

Keywords Intron splicing ribozyme editing • Intron evolution • Intron plant mitochondria

6.1 Introduction

Introns have intrigued scientists ever since their discovery in 1977, when noncoding sequences were found to interrupt protein-coding information in viral and nuclear genes (reviewed in Witkowski 1988). It was puzzling why genes would have sequences that must be precisely spliced from precursor RNAs to generate functional messenger RNAs. Where did introns come from? How are they removed? Do they have any biologic function? For their work on introns and splicing, Phillip Sharp and Richard Roberts were jointly awarded the Nobel Prize in Medicine in

L. Bonen (✉)
Biology Department, University of Ottawa, 30 Marie Curie, Ottawa, Canada K1N 6N5
e-mail: lbonen@uottawa.ca

F. Kempken (ed.), *Plant Mitochondria*, Advances in Plant Biology 1,
DOI 10.1007/978-0-387-89781-3_6,

1993. The initial discovery of introns was soon followed by their detection in organellar genes, such as yeast mitochondrial (Bos et al. 1978) and *Chlamydomonas* chloroplast ribosomal RNA genes (Rochaix and Malnoe 1978).

In 1981, the very first angiosperm mitochondrial protein-coding gene to be sequenced, maize *cox2*, was found to contain an intron (Fox and Leaver 1981), and as sequencing data accumulated over subsequent years it became evident that a substantial subset of mitochondrial genes contain introns in both vascular and nonvascular plants (reviewed in Knoop 2004; Bonen 2008). Organellar introns turn out to have elaborate and distinctive secondary structures, and this led to their classification as group I and group II, as well as further subgroups (reviewed in Michel and Ferat 1995; Bonen and Vogel 2001; Lambowitz and Zimmerly 2004; Haugen et al. 2005; Lang et al. 2007). Early on, it was also appreciated from the analysis of respiratory-deficient yeast mutants that mitochondrial splicing in vivo depends on both intron structural features and intronic-encoded protein information.

Group I and group II introns have a number of special features that distinguish them from other classes of introns. (1) They fold into conserved higher-order structures which are crucial for splicing. (2) Some act as ribozymes in vitro, that is, they are self-splicing in the absence of any proteins in the test tube (albeit under non-physiologic conditions). Incidentally, for his work on group I intron ribozymes, Tom Cech received the Nobel Prize in Chemistry in 1989, an honor he shared with Sidney Altman for his contributions to our understanding of RNase P, another type of catalytic RNA. (3) Some group I and group II introns behave as mobile genetic elements capable of migrating to new sites within a genome or invading new genomes. (4) Such introns encode their own protein machinery which assists in splicing and mobility. For the latter function, group I introns have homing DNA endonucleases and group II introns encode reverse transcriptase activity. Both group I and group II intronic reading frames (ORFs) also encode RNA maturase/splicing activity, which ensures that their integration into a preexisting host gene is not detrimental.

These families of mobile intron ribozymes are quite widespread in nature. Both categories have been found in the mitochondrial genomes of plants, fungi, and protists, in plastid genomes and at low frequency in a wide range of prokaryotes (both eubacteria and archaea) where they are often located within intergenic regions (reviewed in Haugen et al. 2005; Simon et al. 2008; Nielsen and Johansen 2009). Animal mitochondrial genomes are generally devoid of introns, although group I introns have been detected in certain members of basal groups of metazoans, such as corals, sponges and sea anemones (Nielsen and Johansen 2009), and several group II introns have been observed in placozoan and annelid worm mitochondria (cf. Signorovitch et al. 2007; Valles et al. 2008). Group I introns are common in nuclear ribosomal RNA genes of unicellular eukaryotes, whereas group II introns have not been found in the nuclear genes of any eukaryotes. Certain group I and group II introns exhibit unusual features and several examples are given in Box 6.1. In addition, the phenomenon of discontinuous trans-splicing introns is discussed in Sect. 6.3.

Box 6.1 Examples of unusual features of Group I and Group II introns

Certain group I introns have atypically complex structures. For example, "twin-ribozyme" introns found within nuclear ribosomal RNA genes in the protist *Didymium iridis* possess two ribozymes (Decatur et al. 1995), a self-cleaving one which is believed to be important in generating the mRNA for the intron-encoded endonuclease, and the conventional self-splicing ribozyme. There is also a very curious group I intron within the *nad5* gene in the mitochondria of scleractinian corals (Medina et al. 2006). This intron is over 18 kb long in the mushroom coral *Discosoma* and it contains 15 other mitochondrial genes, which were relocated there through mitochondrial DNA rearrangements (reviewed in Nielsen and Johansen 2009).

Group II introns can also deviate from the conventional structure. For example, group II introns have been found to be nested inside other introns in bacteria as well as in euglenoid plastid genomes (reviewed in Lambowitz and Zimmerly 2004). The latter have extremely high numbers of introns (about 155 in *Euglena gracilis*, cf. Thompson et al. 1995), some of which have very degenerate group II intron features and were separately categorized as group III. Such introns-within-introns (sometimes called twintrons) show a strict order in their splicing, starting with the innermost one. There are also examples of group II introns which encode group I-type homing endonucleases (with LAGLIDADG motifs) in fungal mitochondria (reviewed in Lambowitz and Zimmerly 2004) and certain plant mitochondrial group II introns exhibit atypical folding or weakened helices (see Sect. 6.4) consistent with reduced functional constraint.

An appealing hypothesis which is gaining widespread acceptance is that spliceosomal introns originated from group II introns that spread to the host (nuclear) genome from the respiring α-proteobacterial endosymbiont which evolved into the mitochondrion. In this scenario, the structural group II intronic information was fractured into small RNAs (encoded as separate genes) which form essential components of the present-day splicing machinery (cf. Cavalier-Smith 1991; Sharp 1991). This proposal is attractive because spliceosomal and group II introns have the same biochemical pathway of splicing and release excised introns as lariats. The role that group II introns might have played in the evolution of the early eukaryotic cell has been taken even further with the suggestion that their presence was an impetus for the creation of the nuclear membrane in order to solve the predicament of translation occurring before splicing has been completed (Martin and Koonin 2006). Interestingly, present-day plant mitochondria face the same dilemma in that RNA processing events such as splicing and editing occur in the same compartment as protein synthesis (see also Chap. 7), and it is presently unknown how (or if) mechanisms exist to spatially or temporally prevent immature transcripts from undergoing nonproductive translation.

6.2 Distribution of Introns in Mitochondrial Genes of Land Plants

A wealth of organellar sequencing information has accumulated during the 30 years since introns were first discovered and these data are providing valuable insights into the distribution and evolution of mitochondrial introns in land plants (Fig. 6.1a). Although introns have had a long history in plant mitochondrial genomes (and indeed in their green algal-type predecessors), their locations vary considerably among plant lineages (see Fig. 6.1b). In fact for the plants shown in Fig. 6.1, even though they all have a rather similar number of introns (ranging from 23 to 30), there is not even one intron which is at exactly the same position in all of them. The single group II intron which is shared between the liverwort *Marchantia polymorpha* and seed plants (namely *nad*2i709) is absent from the moss *Physcomitrella patens nad*2 gene. Surveys of intron distribution in diverse lineages of land plants point to a complex history with multiple invasions and the intragenomic spread of intron copies, as well as sporadic intron losses (reviewed in Knoop 2004). Certain mitochondrial introns within a genome show closer sequence similarity than do others (cf. Haouazine et al. 1993; Ohyama and Takemura 2008), consistent with their more recent origin. However events such as gene conversion between paralogues or even horizontal movement of intron orthologues between diverse plant species (cf. *nad*1i477; Won and Renner 2003) might complicate interpretation. To more easily distinguish the positions of mitochondrial introns among land plants, a nomenclature has been adopted in which numbering is based relative to the initiation codon in *Marchantia polymorpha* mitochondrial homologues (Dombrovska and Qiu 2004).

Among angiosperms and gymnosperms (which shared an ancestor about 300 million years ago), the positions of mitochondrial introns are in fact quite conservative (see Fig. 6.1b), although not all plants examined have exactly the same complement. To date, a total of 25 group II introns have been identified within mitochondrial protein-coding genes in seed plants. *Arabidopsis* and rice each have 23 group II mitochondrial introns, but their *cox*2 introns are at different positions. Incidentally, the *Physcomitrella cox*2 gene has introns at both these sites, whereas some flowering plants, such as *Oenothera* and pea have none. The mode of intron loss in such cases is believed to be through integration of a retro processed cDNA copy derived from spliced and edited mRNA. This has convincing support from RNA editing patterns (especially near intron borders), as demonstrated by examination of *nad4* in angiosperms (Geiss et al. 1994) and *rps3* in gymnosperms (Ran et al. 2009). Moreover, the direct involvement of the group II intron-encoded reverse

Fig. 6.1 (continued) as cited in Terasawa et al. 2007). (**b**) Pairwise comparisons showing the number of introns at homologous sites for the plants shown in panel (**a**). Group II introns are shown *above* the diagonal and group I introns *below* the diagonal. Note that identical numbers do not necessarily reflect shared introns. For example, only two of the four group II introns held in common between *Marchantia* and *Isoetes* (namely, *nad*2i709 and *nad3*i140) are also shared between *Marchantia* and *Megaceros*

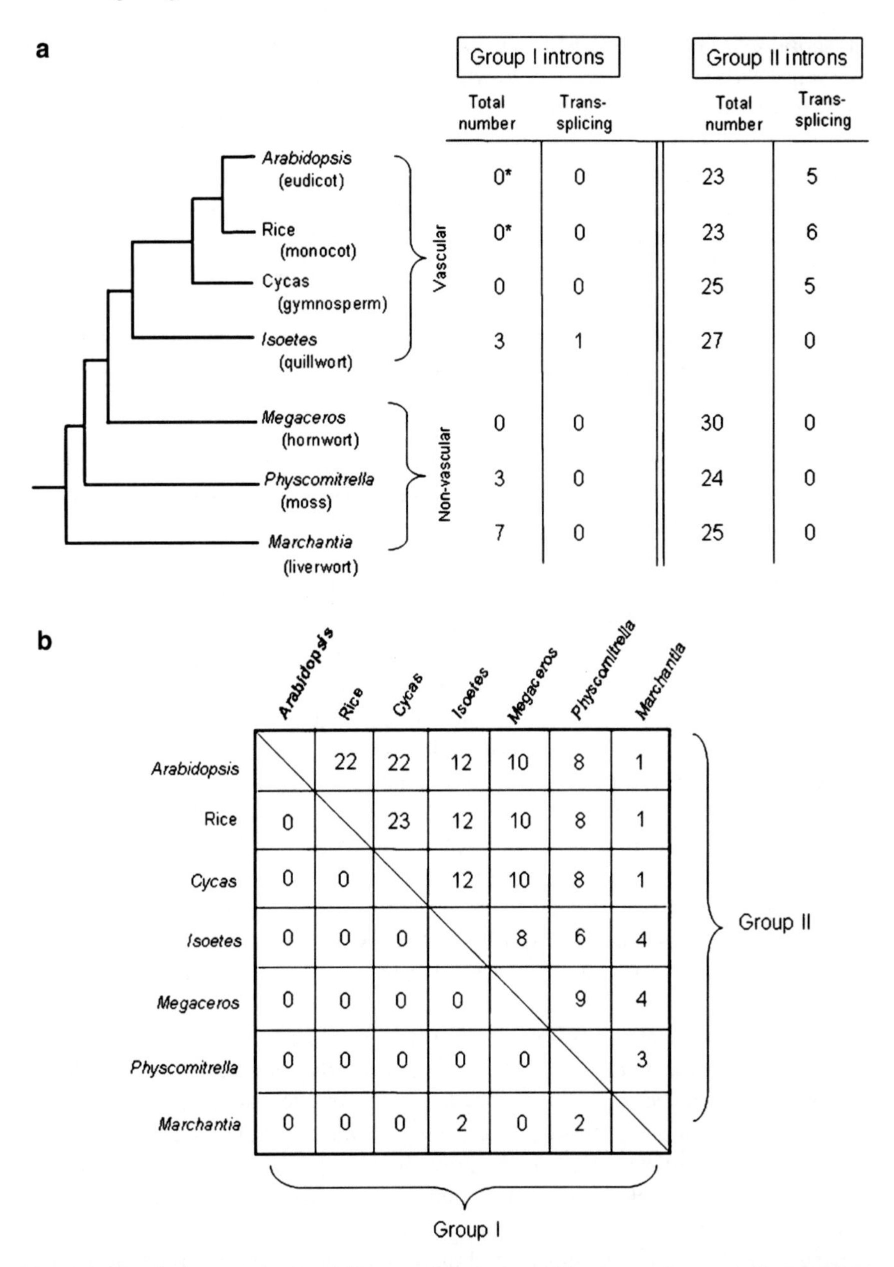

Fig. 6.1 Distribution of mitochondrial group I and group II introns in diverse land plants. Data are from *Arabidopsis thaliana* (Unseld et al. 1997), rice (Notsu et al. 2002), *Cycas taitungensis* (Chaw et al. 2008), *Isoetes engelmannii* (Grewe et al. 2009), *Megaceros aenigmaticus* (Li et al. 2009), *Physcomitrella patens* (Terasawa et al. 2007), and *Marchantia polymorpha* (Oda et al. 1992). (**a**) Cladogram showing plants which diverged from a common ancestor about 400 million years ago (not to scale). Total number of introns is given, as well as number of trans-splicing introns. The asterisk indicates the presence of a group I intron in certain angiosperms (cf. Cho et al. 1998). The *Physcomitrella cob* intron (*cob*i420) is scored as group I (rather than group II

transcriptase in such retro-deletion events has been established in yeast mitochondria (Gargouri 2005). Another reason for differences in the number of introns among flowering plants, is that some genes are simply absent from the mitochondrion in certain plant lineages because they have undergone successful transfer to the nucleus. For example, *rps10* is absent from *Arabidopsis* and rice mitochondria, whereas the functional gene has a group II intron in potato mitochondria (Knoop et al. 1995). Incidentally, genes which have been transferred to the nucleus are seen to lack group II introns, consistent with an RNA intermediate in gene migration (reviewed in Adams and Palmer 2003). It is worth noting that mitochondrial DNA fragments containing introns have also been identified in plant nuclear genomes (virtually the whole mitochondrial genome in the case of *Arabidopsis*, cf. Stupar et al. 2001), but these reflect the recent integration of nonfunctional copies which will presumably decay away over time.

Introns have been found in a wide variety of mitochondrial protein-coding genes in land plants (Table 6.1), and for seed plants most of them are located within *nad* genes which encode NADH dehydrogenase subunits in complex I of the respiratory chain. In contrast, there is an underrepresentation of introns within ribosomal protein genes, in that only 4 out of 16 such genes contain introns among the plants shown in Table 6.1. Although most are within mitochondrial protein-coding genes, a few group I or group II introns have been identified within ribosomal RNA or transfer RNA genes in certain nonseed plants (see Table 6.1). In addition, some plant mitochondrial genomes have acquired segments of chloroplast DNA which harbor intron-containing tRNA genes, for example, a split chloroplast tRNA-Ala gene in wheat mitochondria (Ogihara et al. 2005). However, based on direct tRNA analysis, it appears that the active Ala-tRNAs are nucleus-encoded and imported into the mitochondrion in wheat (Glover et al. 2001). Hence the chloroplast copy is presumably nonfunctional, although nonsplit chloroplast-origin tRNA genes have indeed been shown to be expressed and used in mitochondrial translation (reviewed in Gray 1992). Notably, in the chloroplasts of flowering plants, there is a bias for group II introns in transfer RNA genes rather than protein-coding genes.

Although almost all mitochondrial introns in seed plants fall in the group II category, there is one notable exception, namely a group I intron within *cox1* found in a subset of angiosperms (Cho et al. 1998). The sporadic distribution of this intron among plant species is consistent with multiple independent acquisitions. It contains an intron-encoded ORF which shows high sequence similarity with fungal mitochondrial endonuclease-maturase counterparts, suggestive of a fungal-to-plant transfer of this mobile genetic element. Moreover its position is at a particularly active group I "homing" target site, and there is a group I intron at the homologous position (*cox*1i730) in the *Marchantia cox*1 gene as well. That said, it should be noted that the overall number of group I introns is still minor compared to the group II type. In Fig. 6.1b, it can be seen that *Isoetes* and *Physcomitrella* do not have any group I introns in common, even though each shares two with *Marchantia,* again reflecting their mobile nature. Incidentally, because the homing DNA endonucleases encoded within group I introns have very long recognition sites (~30 bp),

Table 6.1 Summary of group I and group II intron distribution in mitochondrial genes from diverse land plants

	Arabidopsis	Rice	Cycas	Isoetes	Megaceros	Physcomitrella	Marchantia
Group I							
Cob						1	
cox1				2*		1	6 (3 ORFs)
nad5						1	1
rrnS				1			
Group II							
atp1					3	1	2 (2 ORFs)
atp6				1	2	1	
atp9				2	1	3	1 (1 ORF)
Cob				2	2		3 (1 ORF)
cox1				4	3	3 (1 ORF)	3 (3 ORFs)
cox2	1	1	1	1	2	3	2 (1 ORF)
cox3						1	2
nad1	4* (1 ORF)	4* (1 ORF)	4* (1 ORF)	1	3	2	
nad2	4*	4*	4*	4	2	1	1
nad3				2	2		1
nad4	3	3	3	2	2	1	1
nad4L						1	2
nad5	4*	4*	4*	3	3	2 (1 ORF)	
nad6					1		
nad7	4	4	4	4		2	2 ψ
nad9					2	1	
rpl2	1	1	1		–	–	1
rps3	1	1	2 (1 ORF)	1	–		
rps10	–	–	1	–	–	–	
rps14	ψ	ψ		–			1

(continued)

Table 6.1 (continued)

	Arabidopsis	Rice	Cycas	Isoetes	Megaceros	Physcomitrella	Marchantia
sdh3	–	–			1 ψ	1	
ccmFc	1	1	1	–	1 ψ	1	
rrnS							1 (1 ORF)
rrnL							1
trnS							1

Data sources are as in Fig. 6.1. The number of introns which possess ORFs is shown in parentheses, however, because *Megaceros aenigmaticus* intronic ORF homologous sequences (Li et al. 2009) appear interrupted by frameshift mutations they have not been included. *Dash* indicates that the gene is absent from the mitochondrial genome, whereas no symbol indicates that the gene is present but lacks introns. *Asterisks* indicate that at least one intron in this gene is *trans*-splicing. Intron-containing pseudogenes are shown by ψ. Abbreviations for these mitochondrial genes are: *atp* (ATP synthase subunits), *cob* (cytochrome b), *cox* (cytochrome oxidase subunits), *nad* (NADH dehydrogenase subunits of complex I), *rpl* and *rps* (large and small subunit ribosomal proteins), *sdh* (succinate dehydrogenase subunit), *ccm* (cytochrome c maturation subunit), *rrnL* and *rrnS* (large and small subunit ribosomal RNAs), *trn* (transfer RNA) they can be valuable restriction enzyme tools in genomic analysis and this has led to the commercialization of ones such as *Sce*-I (from yeast mitochondria) and *Ceu*-I (from *Chlamydomonas* chloroplast).

6.3 *Trans*-Splicing Introns

The recombinogenic nature of seed plant mitochondrial genomes has had a dramatic impact on certain intron-containing genes. DNA rearrangements have resulted in protein-coding segments (and their flanking intron pieces) being widely dispersed in the genome. Remarkably, such fractured genes are still functional because the separately transcribed RNA molecules can reconstitute proper intron structure so that splicing occurs *in trans* (reviewed in Bonen 2008; Glanz and Kuck 2009). There are seven examples of *trans*-splicing group II introns in seed plant mitochondria, located within the *nad*1, *nad*2, *nad*5, and *cox*2 genes. The latter is found only in *Allium cepa* (Kim and Yoon 2010), and its disruption is presumed to have occurred very recently because in other *Allium* species which were examined, the *cox*2 intron is *cis*-splicing. The approximate timing of the conversion of *nad* introns from *cis*-splicing to *trans*-splicing during evolution has also been tracked by comparative analysis with orthologues from early-diverging plant lineages such as ferns, hornworts, and mosses (Malek and Knoop 1998). For example, in all seed plants and ferns examined the first *nad*1 intron (*nad*1i394) is *trans*-splicing, whereas in the quillwort *Isoetes engelmannii*, this intron is *cis*-splicing. Incidentally, this intron also has a very degenerate core structure in flowering plants (Carrillo et al. 2001) but not in *Isoetes* (Malek and Knoop 1998). In addition, three other *trans*-splicing introns in seed plants (namely, *nad*2i542, *nad*5i1455, and *nad*5i1477) have *cis*-splicing counterparts in *Isoetes*. In contrast, the fourth intron in the angiosperm *nad*1 gene (*nad*1i728) has undergone DNA rearrangements much more recently. It is *cis*-splicing in some flowering plants, such as *Arabidopsis*, but *trans*-split in others. Notably, this intron has been disrupted a number of independent times during seed plant evolution, sometimes upstream of the maturase-type ORF (*mat*R) and in other cases downstream of it (Qiu and Palmer 2004). A few group II introns are further fractured, and this is the case for the third intron (*nad*5i1477) in the *Oenothera nad*5 gene which has a tripartite structure (Knoop et al. 1997), as does a *psa*A intron in *Chlamydomonas* chloroplast (reviewed in Glanz and Kuck 2009). Comparative analysis suggests that plant mitochondrial intron disruption is through intragenomic homologous recombination across short sequences of about 10–12 nucleotides (Chapdelaine and Bonen 1991; Knoop et al. 1997).

An eighth case of *trans*-splicing in land plant mitochondria involves a group I intron in the *cox*1 gene in *Isoetes engelmannii* (Grewe et al. 2009). This was the first naturally occurring *trans*-splicing group I intron to be reported and it is located at the same position as the ninth (and final) intron (*cox*1i1305) in the *Marchantia* mitochondrial *cox*1 gene. Two *trans*-splicing group I introns have also recently been observed within the *cox*1 gene of a primitive placozoan animal (Burger et al.

2009), but they are located at positions different from the *cis*-splicing *cox*1 introns in the plants shown in Fig. 6.1.

Although hundreds of group I and group II introns have now been identified in nature, remarkably few of them are *trans*-split (reviewed in Glanz and Kuck 2009). There are several examples of *trans*-splicing in the chloroplasts of algae (e.g., *Chlamydomonas psa*A), one in land plant chloroplast (*rps*12), and two *trans*-splicing group II introns were identified in the mitochondrial *nad*3 gene of the green alga *Mesostigma viride*. Aside from those, there are several reports of disrupted *cox*1 and *cox*3 genes in the mitochondria of dinoflagellates and diplonemids; however, they are not flanked by group I or group II-type sequences and their mode of splicing is as yet unknown. Given the rearranging nature of angiosperm mitochondrial genomes (especially compared to plant chloroplast genomes), it is perhaps surprising that more cases of *trans*-splicing have not arisen. Indeed rearrangements within plant mitochondrial introns can lead to mutant phenotypes, as is the case for certain nonchromosomal stripe maize plants (cf. *nad*4-*nad*7 fusion, Marienfeld and Newton 1994) and cytoplasmic male sterile tobacco plants (cf. *nad*7 truncation, Pla et al. 1995).

The phenomenon of *trans*-splicing also raises logistical problems for successful expression of the internal sections of such genes. Sometimes they are positioned physically close to other genes and become part of pre-existing transcriptional units, whereas in other cases they apparently recruit new promoters from spacer regions. There are cases in which a single precursor transcript contains coding information for two different *trans*-spliced genes (and their intron pieces), for example, *nad*1 and *nad*5 exons in wheat (Farré and Araya 1999). This raises interesting questions about obligatory (or preferred) orders of splicing to generate mature mRNAs. Notably, the *trans*-splicing second intron of *nad*5 (*nad*5i1455) was observed to undergo extensive mis-splicing (apparently due to incorrect folding) using cryptic splice sites in the upstream exon in plants such as *Brassica* and wheat, whereas the tripartite structure of this intron in *Oenothera* alleviates such mis-splicing (Elina and Brown 2009). This apparent selective pressure for further fragmentation of disrupted introns fits well with evolutionary models in which group II intron pieces gave rise to the small RNAs (snRNAs) which make up the catalytic core of present-day nuclear spliceosomes (cf. Sharp 1991). Interestingly, when the *Lactococcus lactis* group II intron was fragmented at numerous different sites within the core structure, it could still undergo *trans*-splicing, as long as expression of the intronic maturase ORF was not inactivated (Belhocine et al. 2008).

6.4 Mechanism of Splicing in Plant Mitochondria

Conventional group I and group II introns are ribozymes and can be excised from precursor RNA in the absence of any proteins in vitro. Their self-splicing properties have been extensively studied in model systems such as the *Tetrahymena* nuclear

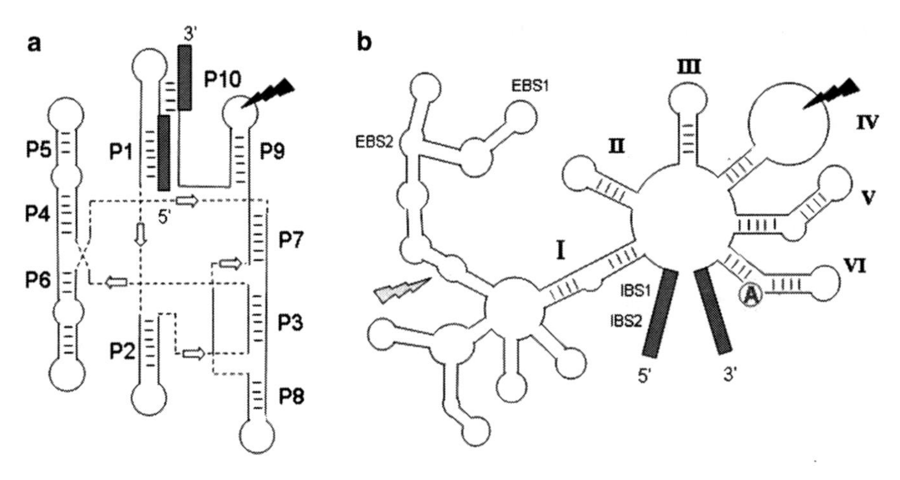

Fig. 6.2 Secondary structure models of classical group I and group II introns. (**a**) Group I intron shown with 10 helical regions (P1–P10), as adapted from Vicens and Cech 2006; Nielsen and Johansen 2009. Flanking exons are shown as *grey boxes*. *Black lightning rod* shows position of disruption in *Isoetes engelmannii cox*1 *trans*-splicing intron (Grewe et al. 2009). (**b**) Group II intron shown with six helical domains (I–VI), as adapted from Michel and Ferat 1995; Lambowitz and Zimmerly 2004. Flanking exons are shown as *grey boxes*. The bulged adenosine in domain VI is *circled*. *EBS* and *IBS* denote exon and intron binding sites, respectively, which base-pair with each other. Domain IV is variable in length among plants and *black lightning rod* shows predicted position of disruptions in bipartite *trans*-splicing mitochondrial introns in plants. *Grey lightning rod* denotes position of additional break in tripartite *nad*5 intron in *Oenothera* (Knoop et al. 1997)

ribosomal RNA intron (for group I) or *Lactococcus lactis* and yeast mitochondrial introns (for group II). Notably, no introns from land plant mitochondria or chloroplasts have been demonstrated to be autocatalytic in vitro, although several algal organellar ones do show self-splicing in the test tube, for example the green algal *Chlamydomonas* chloroplast rRNA group I intron (Herrin et al. 1990) and brown algal *Pylaiella littoralis* mitochondrial rRNA group II introns (Costa et al. 1997). Both classes of intron ribozymes catalyze two sequential transesterification reactions and they require Mg^{2+} for this splicing activity. However, their structures (Fig. 6.2) and biochemical pathways (Fig. 6.3) are distinctly different from each other. The ribozyme core of group I introns is typically much smaller (~250 nt) than for group II introns (>500 nt), and its x-ray crystal structure was solved prior to that of any group II intron (reviewed in Vicens and Cech 2006; Toor et al. 2009). Group I and group II ribozymes have also attracted considerable attention for their potential use as molecular therapeutic tools, for example in targeted gene disruption or repairing defective RNA (reviewed in Long et al. 2003; Lambowitz and Zimmerly 2004; Toro et al. 2007).

Group I intron ribozymes fold into ten helices (P1–P10), which are organized into three helical stacks (see Fig. 6.2), and these have been called the scaffolding domain (P4-P5-P6), catalytic domain (P3-P7-P8-P9 with its pseudoknot) and substrate domain (P1-P2). The P1 helix is composed of the 5′ end of the intron which

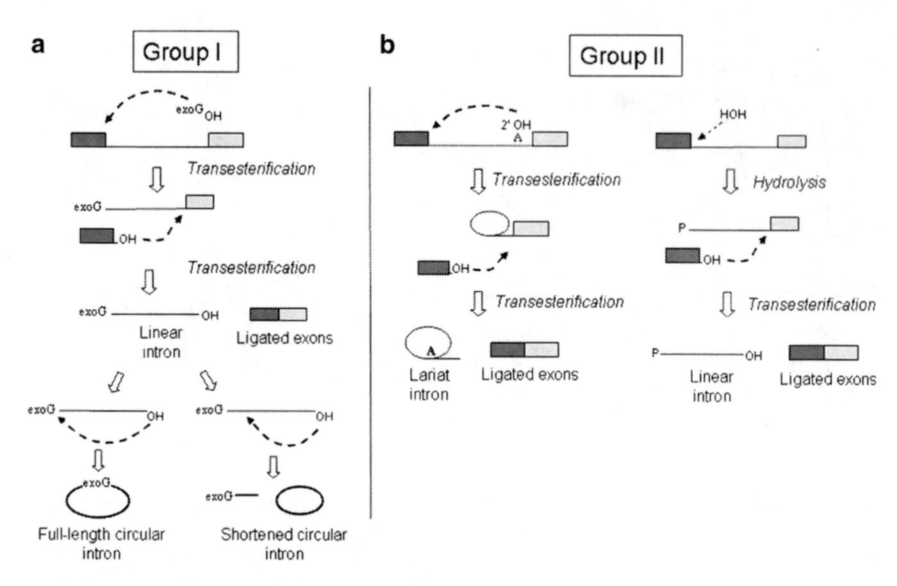

Fig. 6.3 Splicing pathways of conventional group I and group II introns. (**a**) Group I intron splicing via two transesterification steps to yield ligated exons and linear excised intron. An exogenous guanosine (*exoG*) acts as the first attacking nucleophile. The excised intron can undergo subsequent transesterification events to generate either a full-length circular intron (plus *exoG*) or a shortened circular form. (**b**) Group II intron splicing via two-step transesterification (with the adenosine near 3′ end of intron acting as first attacking nucleophile) to yield ligated exons and lariat intron [*on left*] or via a minor pathway of first-step hydrolysis (with water acting as the attacking nucleophile) and second-step transesterification to give ligated exons and linear intron (*on right*). Note that plant mitochondrial splicing yields other additional physical forms of excised introns (cf. Li-Pook-Than and Bonen 2006)

pairs with the upstream exon. The first splicing step is initiated by an external guanosine (exoG) cofactor which attacks the 5′ splice site (Fig. 6.3a) and after this reaction there is a conformational shift to form P10 (with base-pairing between the intron and downstream exon). In the second transesterification step, the 3′OH of the upstream exon attacks the 3′ splice site, so that the exons are ligated together and the intron is released in a linear form. This can be followed by the 3′OH of the linear intron attacking a phosphodiester bond within the intron (near the 5′ end) to yield a circularized, but shortened, form of the intron. Alternatively, if the nucleophilic attack is on the exoG at the 5′end of the intron, then the full-length intron (plus exoG) becomes circularized. The splicing properties of the few group I introns found in plant mitochondria have not been well studied, but they typically contain conventional structural features. Notably the *trans*-splicing group I intron in the *Isoetes cox*1 gene is disrupted within the P9 loop (Grewe et al. 2009; see Fig. 6.2a), which incidentally differs from the placozoan animal *cox*1 introns which appear "broken" within the loop extending from the P8 helix (Burger et al. 2009). The homing endonuclease-maturase ORFs which are present in a subset of the plant mitochondrial group I introns (see Table 6.1) are positioned within peripheral loop regions outside the core.

Classical group II introns are depicted as six helical domains (I–VI) extending from a central hub (see Fig. 6.2) and the catalytic core is comprised primarily of the 34-nt highly conserved domain V and sequences embedded within the elaborate domain I. Unlike group I introns where an exogenous guanosine initiates splicing, for group II introns the 2′ OH of an internal bulged adenosine near the 3′ end of the intron (see Fig. 6.2, circled A in domain VI) attacks the 5′ splice site and forms a 2′–5′ phosphodiester bond at the branchpoint (see Fig. 6.3b). The second transesterification step, in which the 3′OH of the liberated upstream exon attacks the 3′ splice site, results in exon ligation and release of the intron as a lariat with a tail of either 6 nt (group IIB) or 7 nt (group IIA), depending on the position of the bulging adenosine within domain VI. Many aspects of the higher-order interactions have been elucidated, one notable example being base-pairing between domain I sequences (EBS1 and EBS2) and the 3′ end of the upstream exon (IBS1 and IBS2). In plant mitochondria, there is experimental support for this interaction from transient expression studies where wheat mitochondrial *cox*2 intron constructs were examined using an *in organello* electroporation system (Farré and Araya 2002). A few group II introns (particularly in bacteria) have been shown to exhibit a second in vivo splicing mechanism whereby water acts as the first attacking nucleophile and the intron is excised as a linear form (reviewed in Lambowitz and Zimmerly 2004). This hydrolytic pathway has also been observed for a barley chloroplast tRNA-Val intron which possesses an atypical domain VI unlike other chloroplast introns which are released as conventional lariats (Vogel and Borner 2002) and it has been proposed as one of several splicing pathways for the wheat mitochondrial *cox*2 (*cox*2i691) intron (Li-Pook-Than and Bonen 2006). Moreover, the presence of extra nonencoded nucleotides (primarily adenosines) suggests that linear intron forms in organelles are being tagged for RNA turnover (reviewed in Lange et al. 2009).

Although some group II introns in angiosperm mitochondria exhibit conventional secondary structure and indeed several of these have been documented to be excised as lariats (cf. Carrillo et al. 2001), many others show atypical folding or they lack expected features (reviewed in Bonen 2008). Such alterations appear to have originated in the seed plant lineage, based on comparisons with orthologues in plants such as *Isoetes* (quillwort) or *Megaceros* (hornwort), and this suggests that additional compensatory machinery may well be required for splicing. In some cases, helices weakened by A-C mispairs are restored to conventional structure by C-to-U type RNA editing (cf. Wissinger et al. 1991; Binder et al. 1992; Lippok et al. 1994), and the importance of such intron editing is supported by experimental analysis (cf. Borner et al. 1995; Farré and Araya 2002). Curiously though, such predicted corrections are not always observed and intron editing status can vary among plant species (cf. Carrillo and Bonen 1997; Carrillo et al. 2001). Sometimes core intron features are simply missing, an example being within dVI of the second *nad*1 intron (*nad*1i477) where the short tight helix lacks a bulged adenosine. These excised introns comprise a mixed population of circular forms, some of which include a block of nonencoded nucleotides or several nucleotides from the downstream exon (Li-Pook-Than and Bonen 2006). The wheat *cox*2 excised intron shows yet other physical forms, including apparent lariats with unusual tail lengths. In addition to employing novel biochemical pathways of splicing, plant mitochondria also have the distinction of exhibiting the greatest range in size of group II introns seen in nature to date. The introns in *Isoetes engelmannii* mitochondria are exceptionally small (cf. only 327 nt for *cox*1i266; Grewe et al. 2009), whereas the ORF-less group II introns in *Cycas taitungensis* mitochondria tend to be very large (cf. 5.7 kb for *nad*4i976; Chaw et al. 2008), their size inflated in part by Bpu-type tandem repeat arrays. Length variation is typically within the domain IV loop, which is the location of the RT-maturase ORF, when present. This is also the region in which *trans*-splicing group II introns are typically disrupted and the tripartite *nad*5 intron in *Oenothera* has a second break within domain I (Knoop et al. 1997).

Alternative splicing, although an important aspect of nuclear gene expression in plants and animals, has rarely been observed for any group I or group II introns. Perhaps this is not surprising given the expected pressure on these mobile genetic elements to preserve accurate splicing from their "host" transcripts. However a group II intron in *Bacillus anthracis* which has a deviant dV structure was observed to show an atypical choice of 3′ splice site (Robart et al. 2004) and alternative splicing of unusual *Euglena* chloroplast group III introns generates two forms of *roa*A mRNAs (Jenkins et al. 1995). The few documented cases of cryptic splice sites being utilized in plant mitochondria (cf. Li-Pook-Than and Bonen 2006; Elina and Brown 2009) would not be expected to generate productive mRNAs.

6.5 Splicing Machinery for Plant Mitochondrial Introns

Classical group I and group II introns, when fully competent, encode their own machinery for mobility and splicing. In plant mitochondria, relatively few introns possess such ORFs, although there is support (albeit indirect as yet) that the encoded maturases are involved in splicing and assist intron folding or help stabilize the active structure. Among the plants shown in Fig. 6.1, *Marchantia polymorpha* has the largest proportion of mitochondrial introns with ORFs (see Table 6.1), and comparative analysis suggests a recent origin of some via horizontal transfer from fungal mitochondrial sources, as well as others having arisen through intragenomic duplication events (Ohyama and Takemura 2008). Likewise, the recently acquired group I intron observed in certain disparate lineages of flowering plants (Cho et al. 1998) contains a homing endonuclease-maturase ORF with the classic LAGLIDADG motifs (Fig. 6.4). In contrast, none of the tiny group I or group II introns in *Isoetes* mitochondria contain ORFs, nor do there appear to be freestanding ORF homologues elsewhere in that mitochondrial genome (Grewe et al. 2009). In the hornwort, *Megaceros aenigmaticus*, there are 30 mitochondrial group II introns (Li et al. 2009), some of which show stretches of detectable homology to RT-maturase ORFs; however, they appear to contain multiple frame-shift mutations.

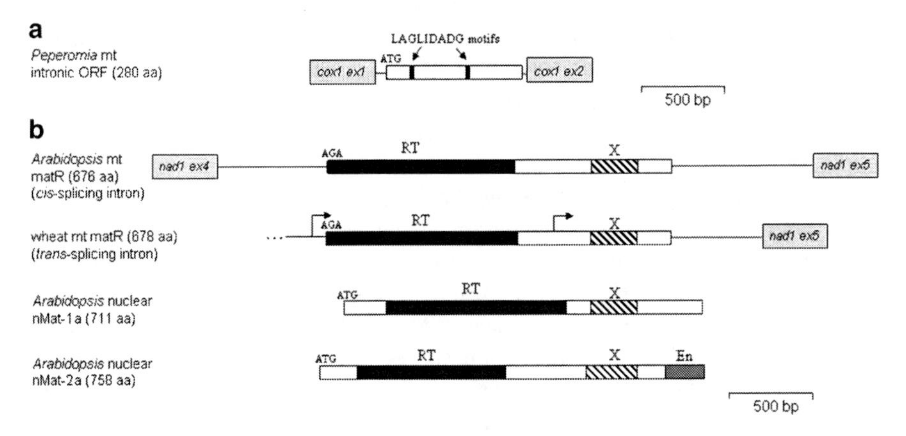

Fig. 6.4 Intron-encoded proteins in group I and group II plant mitochondrial introns and nuclear homologues. Coding sequences are shown by *boxes* and noncoding regions of introns are shown by *lines*. *Light grey boxes* denote flanking exons (not to scale). (**a**) Group I homing endonuclease-maturase intronic ORF (*white box*) with LAGLDADG motifs (*black*) in the *cox*1 intron (*cox*1i730) first identified in the flowering plant *Peperomia polybotrya* (Adams et al. 1998). (**b**) Group II *mat*R-type ORF encoded within the fourth *nad*1 intron (*nad*1i728) which is a *cis*-splicing intron in *Arabidopsis*, but *trans*-splicing in wheat, as well as autonomous *mat*R-homologous genes encoded in the nucleus in *Arabidopsis* (Mohr and Lambowitz 2003). Domains within the ORF are shown in *black* (*RT*, reverse transcriptase), *hatched* (X domain, with RNA maturase function), and *dark gray* (En, Zn-finger type endonuclease distinctive from the group I "homing" type), as adapted from Bonen and Vogel 2001; Lambowitz and Zimmerly 2004. *Bent arrows* indicate promoter positions in wheat mitochondria (Farré and Araya 1999). Lengths of intronic ORFs are given in amino acids, although the mitochondrial *mat*-R intronic ORFs lack a conventional initiation codon and the AGA triplet is preceded by an in-frame stop codon

In other plants, remnant traces of ORFs can sometimes be detected within mitochondrial introns (cf. *Ginkgo bilboa cox*2 intron, Ahlert et al. 2006 and *Equisetum arvense* (horsetail) *cox*1 intron, Begu and Araya 2009), and this is consistent with the view that as introns "age," their ORFs can decay and be lost. Implicit in this model is a greater dependence on splicing machinery either encoded by other mitochondrial introns or provided by host (nuclear) factors.

In seed plant mitochondria, only one potentially functional group II intronic ORF (named *mat*R) has been identified (see Fig. 6.4). It is located within the fourth (and final) intron of *nad*1 (*nad*1i728) and possesses homology to reverse transcriptase and RNA maturase (domain X), which together bind precursor RNA in classical group II intron splicing. However *mat*R lacks other domains involved in mobility that are found in full-length group II-type retroelements. The mode of expression of this intronic ORF is rather puzzling in that it lacks a conventional AUG start codon. That said, comparative sequence analysis and editing status provide compelling support for it being functional. Promoters have been identified both within as well as upstream of the *mat*R reading frame in wheat (Farré and Araya 1999), and this raises the possibility that the maturase region (domain X) may undergo expression as an independent protein. Interestingly, in *Arabidopsis* mutants lacking the activity of a dual-organelle-targeted RNA polymerase

(RpoTmp), the relative steady-state levels of RNA from the *mat*R locus appear markedly lower than for many other protein-coding genes (Kuhn et al. 2009).

The absence of maturase-type ORFs within the vast majority of plant mitochondrial introns was somewhat perplexing (considering that classical group I and group II intron ribozymes encode their own customized maturases required for splicing themselves out in vivo), until it was discovered that there are freestanding *mat*R-homologous genes (see Fig. 6.4) in the nuclear genomes of plants (Mohr and Lambowitz 2003), and that their encoded proteins are indeed targeted to the mitochondria (Keren et al. 2009). They appear to have arisen through gene transfer from the mitochondrion to the nucleus at least 200 million years ago, that is prior to the monocot/dicot split. One of four such copies in *Arabidopsis,* namely nMat-1a (At1g30010), has been shown by mutational analysis to be important for accumulation of mature *nad*4 transcripts (Nakagawa and Sakurai 2006). Incidentally, in tobacco *nms*1 nuclear mutants the first intron in *nad*4 was found to be the only intron defective in splicing (Brangeon et al. 2000). Another member of this maturase family in *Arabidopsis*, namely AtnMat2 (At5g46920) has been shown to be involved in the splicing of a subset of mitochondrial introns, including the second *nad*1 intron (*nad*1i477), the second *nad*7 intron (*nad*7i209), and the single *cox*2 intron (*cox*2i691) (Keren et al. 2009). This AtnMat2 protein was estimated to be part of a very large ribonucleoprotein complex of about 900 kDa.

Aside from group II-type RNA maturases, encoded either in the mitochondrial intron itself or as autonomous genes in the nucleus, it is predicted that a number of other nucleus-encoded proteins are involved in mitochondrial splicing. The most notable candidates are the PPR proteins (possessing pentatricopeptide repeat elements) which comprise a huge multi-gene family with more than 450 members in angiosperms (reviewed in Schmitz-Linneweber and Small 2008; see Chaps. 7 and 8). These RNA binding proteins are known to be involved in many posttranscriptional events, including editing (see Chap. 7) and RNA cleavage (see Chap. 5), in plant organelles. Indeed, analysis of *Arabidopsis* mutants defective in splicing of the first *trans*-splicing intron of *nad*1 (*nad*1i394) demonstrated that a PPR protein named OTP43 is required for its splicing, perhaps by stabilizing correct intron folding (de Longevialle et al. 2007). It is anticipated that numerous other members of the PPR family will also be involved in mitochondrial splicing in plants. Some have been implicated in the splicing of various chloroplast introns (both group I and group II) in flowering plants such as maize and *Arabidopsis,* as well as in the moss *Physcomitrella* (reviewed in Schmitz-Linneweber and Small 2008). One intriguing example is the *Arabidopsis* OTP51 gene, which is required for the splicing of a chloroplast *ycf*3 group II intron (de Longevialle et al. 2008). This PPR protein also contains LAGLIDADG motifs which are diagnostic of group I (but not group II) intronic maturase ORFs. An orthologous gene is present in the moss *Physcomitrella* so that this fusion PPR protein gene has had a long evolutionary history. Interestingly, in yeast mitochondria, a PPR-type protein (from one of three such genes identified in the yeast nuclear genome) was shown to be required for removal of specific group I introns in *cob* and *cox*1 transcripts (Moreno et al. 2009).

It is predicted that plant mitochondrial splicing machinery will also include accessory factors which have been recruited to perform extra "moonlighting" functions, as is turning out to be the case for plant chloroplast and fungal mitochondrial splicing (reviewed in Lehmann and Schmidt 2003; Barkan 2004). For example, aminoacyl tRNA synthetases (tyrosyl tRNA synthetase in Neurospora and leucyl tRNA synthetase in yeast) are involved in fungal mitochondrial group I splicing and peptidyl-tRNA hydrolase homologues in maize chloroplast group II splicing. There are also DEAD-box protein RNA helicases required for fungal mitochondrial group II splicing (such as CYT-19 in Neurospora and Mss116p in yeast) and they act as RNA chaperones to promote proper folding or to correct mis-folding (cf. del Campo et al. 2009). Likewise, candidate DEAD-box RNA helicases for splicing have been shown to be imported into mitochondria in *Arabidopsis* (Matthes et al. 2007) and one of these (designated PMH2) was shown by mutational analysis to be important for efficient splicing of 15 of the 23 mitochondrial introns in *Arabidopsis* (Kohler et al. 2010). Ion transporters that influence Mg^{2+} uptake in *Arabidopsis* can complement yeast mutants defective in mitochondrial splicing in keeping with the importance of Mg^{2+} for catalytic activity (Schock et al. 2000). Certain mitochondrial splicing components appear to be conserved during plant evolution, as supported by *in organello* expression studies in which potato or wheat *cox2* intron-containing (*cox*2i373) constructs were properly spliced and edited in heterologous systems (Choury and Araya 2006). In fact, a monocot mitochondrial system can support splicing of a nonorthologous intron from a eudicot, as evidenced by experiments in which *Arabidopsis cox*2 (*cox*2i691) constructs were introduced into maize mitochondria (Staudinger and Kempken 2003). The full complexity of the splicing machinery for any group I or group II intron in vivo is as yet unknown but it is notable that in the case of the *Chlamydomonas trans*-splicing *psa*A gene, 14 complementation groups have been identified and at least some of them are multifunctional in origin (cf. Merendino et al. 2006). Similarly, in yeast mitochondria the Pet54 protein plays several roles as it is involved in translational activation of *cox*3 mRNA as well as the splicing of a *cox*1 group I intron (Kaspar et al. 2008). In this regard, it is perhaps also worth noting that a number of plant mitochondrial ribosomal proteins possess extra domains compared to their eubacterial counterparts, and such modules may perform additional functions (Bonen and Calixte 2006).

The presence of group I and group II introns in the chloroplast genomes of land plants raises the possibility that some nucleus-encoded splicing factors might be shared in common; that is, either dual-targeted to both organelles or arising from duplicated gene products having specific targeting. For example, members of the DUF860 domain family of RNA binding proteins have been demonstrated to be important for chloroplast group II splicing and paralogues are predicted to be targeted to the mitochondrion (Kroeger et al. 2009). That said, it is worth noting that *in organello* studies in which maize plastid *ycf*3 and *ndh*B precursor constructs were introduced into the mitochondria of maize or cauliflower, no splicing or editing was observed (Bolle et al. 2007).

6.6 Relationship Between Splicing and Other RNA Processing Events in Plant Mitochondria

Because the expression of intron-containing plant mitochondrial genes requires multiple RNA processing steps such as editing and transcript end-cleavage in addition to splicing, this requires coordination to generate mature mRNAs for translation. Relatively little is known about the temporal or spatial relationships among these events as yet, although experiments in which *cox*2 precursor RNAs were directly electroporated into isolated mitochondria have shown that splicing and editing are not obligatorily coupled with transcription (Hinrichsen et al. 2009). In addition, unspliced precursor RNAs exhibit editing, although perhaps not to the full extent (Sutton et al. 1991; Yang and Mulligan 1991) and this supports the view that RNA editing is an early event. However certain exon sites which are located very close to intron borders (either upstream or downstream) were seen to remain completely unedited in intron-containing precursors (Li-Pook-Than et al. 2007), suggesting that the adjacent intron may sterically hinder access of the editing machinery to such sites (for example, EBS-IBS interactions, Fig. 6.2) or alternatively that editing recognition motifs may be created by splicing. Interestingly, during cold stress in wheat seedlings, splicing of the *cox*2 intron (*cox*2i373) as well as exon editing (especially at sites very close to intron borders) was seen to be reduced (Kurihara-Yonemoto and Handa 2001). A connection between exon editing and splicing has also been observed in the processing of the moss *Takakia lepidozioides* chloroplast tRNA precursors in that editing within the anticodon (adjacent to the intron) was found to be necessary for group I splicing and it improves base-pairing within the P1 helix (Miyata et al. 2008). As mentioned, editing has also been observed within intron sequences at positions where it can strengthen group II helical structure. However, it occurs at much lower frequency within introns than in coding sequences, based on the rather limited amount of intron RNA sequence data available as yet.

It will be of interest to learn more about the relationships among splicing and other expression events during plant development. The coupling between transcription and splicing appears to be less efficient in germinating wheat embryos than in seedlings, based on the relative abundances of intron-containing precursors observed by northern analysis (Li-Pook-Than et al. 2004). Steady state levels of excised group II introns also appear higher in embryos than in seedlings (Li-Pook-Than et al. 2004), and this may reflect differences in efficiency of intron turnover or use of different splicing pathways. Interestingly, a screen for nucleus-encoded proteins involved in group II splicing in yeast mitochondria revealed mutants which showed reduced stability of excised introns (Luban et al. 2005). The kinetics of splicing in plant organelles also appears to differ from classical group II introns in that uncoupled splicing intermediates (in which only the first splicing step has occurred) are detectable in plant mitochondria (Carrillo et al. 2001) and chloroplasts (Kim and Hollingsworth 1993).

6.7 Concluding Remarks

Our knowledge about splicing in plant mitochondria is still at an early stage and little is known about the nature of the splicing machinery or how this important aspect of RNA maturation is regulated. Are core mitochondrial "spliceosomes" supplemented with a variety of customized components to accommodate the idiosyncrasies of individual introns? How do the nonconventional structures of group II introns in flowering plant mitochondria impact on splicing mechanisms? Do *trans*-splicing introns require extra chaperones in order for RNAs to correctly recognize (and fold with) their partners? How is cross talk between splicing and editing (as well as other expression events) managed? These issues have been particularly challenging to address because of the lack of plant mitochondrial transformation systems or in vitro splicing assays, although exploitation of *in organello* electroporation systems and the availability of splicing-defective mutants is providing exciting new information. The emerging picture is that the plant mitochondrial splicing machinery will be comprised of large RNP complexes in which a cohort of specialized and common RNA binding proteins (particularly PPR proteins) play prominent roles. The presence of small regulatory RNAs in mitochondrial splicing is also not ruled out. Indeed, the mitochondrial "spliceosome" may well turn out to be part of a super-complex containing other RNP-based (and membrane-associated) expression machines such as "editosomes," "cleavosomes," as well as ribosomes for the translation of mature mRNAs.

Another question of interest has been whether mitochondrial splicing might serve as a control point in regulating respiratory function during plant development or under environmental stress. This idea has been fueled by the disproportionately high number of introns within *nad* genes (encoding NADH dehydrogenase subunits of complex I in the respiratory chain), as well as the ability of complex I to be bypassed in plants because of the presence of alternative respiratory pathways. Indeed for all the plants shown in Table 6.1, the majority of mitochondrial *nad* genes have at least one intron, and a subset of them are shared in common among many of these plants.

The presence of introns in plant mitochondrial genes confers an additional layer of complexity to gene expression, and because of their unusual features and novel splicing behaviors, these ribozyme-family introns have attracted the interest not only of plant molecular biologists, but also RNA biochemists and evolutionary biologists. No doubt they will continue to provoke discussion and stimulate investigation in the future.

Glossary

Intron-encoded proteins: Group I introns encode RNA maturase and "homing" DNA endonuclease activities, whereas classical group II introns encode reverse transcriptase (and other mobility functions) as well as RNA maturase. In seed plant mitochondria, only one group II intron (*nad*1i728) contains a maturase-type *mat*R ORF (open reading frame).

Mobile genetic element: DNA which can move to a new location (within or between genomes/organisms) due to the genetic information it possesses. Group I introns move in a site-specific manner by means of "homing" endonucleases. Group II intron retroelements move via reverse splicing into target DNA followed by reverse transcription into DNA. Mobility is mediated by protein machinery encoded within the introns themselves.

Plant groups: Nonvascular land plants (bryophytes) include hornworts (such as *Megaceros*), mosses (such as *Physcomitrella*) and liverworts (such as *Marchantia*). Vascular plants include non-seed plants (such as the lycophyte *Isoetes*) and seed plants, which can be subdivided into gymnosperms (such as *Cycas*) and angiosperms or flowering plants, including monocots (such as rice) and dicots (such as *Arabidopsis*).

Ribozyme: Catalytic RNA which can perform in vitro in the absence of proteins. Classical group I and group II introns have autocatalytic activity, although none in plant mitochondria have been demonstrated to be self-splicing in the test tube.

RNA editing: Conversion of specific cytidine residues to uridine (C-to-U) in mitochondrial precursor RNAs of vascular plants. Editing occurs primarily within coding sequences, although it is also observed at certain sites within introns and can improve intron helical strength by converting A-C mispairs to A-U. In nonvascular plants, mitochondrial editing also includes U-to-C conversion.

Spliceosomal introns: Introns in nuclear protein-coding genes that are removed from precursor RNAs by spliceosomal machinery (composed of five small nuclear RNAs and numerous proteins) using the same biochemical mechanism of splicing as group II introns.

Trans-splicing: RNA processing whereby exons from separate (independently transcribed) RNA precursor molecules are ligated together. In contrast, *cis*-splicing involves intron excision from a single contiguous RNA precursor molecule.

References

Adams, K. L., Palmer, J. D. 2003. Evolution of mitochondrial gene content: gene loss and transfer to the nucleus. Mol. Phylogen. Evol. 29:380–395.

Adams, K. L., Clements, M. J., Vaughn, J. C. 1998. The *Peperomia* mitochondrial *cox*1 group I intron: timing of horizontal transfer and subsequent evolution of the intron. J. Mol. Evol. 46:689–696.

Ahlert, D., Piepenburg, K., Kudla, J., Bock, R. 2006. Evolutionary origin of a plant mitochondrial group II intron from a reverse transcriptase/maturase-encoding ancestor. J. Plant Res. 119:363–371.

Barkan, A. 2004. Intron splicing in plant organelles. In Molecular Biology and Biotechnology of Plant Organelles, H. Daniell, C. Chase, eds. pp. 295–322. Dordrecht: Springer.

Begu, D., Araya, A. 2009. The horsetail *Equisetum arvense* mitochondria share two group I introns with the liverwort *Marchantia*, acquired a novel group II intron but lost intron-encoded ORFs. Curr. Genet 55:69–79.

Belhocine, K., Mak, A.B., Cousineau, B. 2008. *Trans*-splicing versatility of the Ll.LtrB group II intron. RNA 14:1782–1790.

Binder, S., Marchfelder, A., Brennicke, A., Wissinger, B. 1992. RNA editing in *trans*-splicing intron sequences of *nad*2 mRNAs in *Oenothera* mitochondria. J. Biol. Chem. 267:7615–7623.

Bolle, N., Hinrichsen, I., Kempken, F. 2007. Plastid mRNAs are neither spliced nor edited in maize and cauliflower mitochondrial *in organello* systems. RNA 13:1–5.

Bonen, L. 2008. *Cis*- and *trans*-splicing of group II introns in plant mitochondria. Mitochondrion 8:26–34.

Bonen, L, Calixte, S. 2006. Comparative analysis of bacterial-origin genes for plant mitochondrial ribosomal proteins. Mol. Biol. Evol. 23:701–712.

Bonen, L., Vogel, J. 2001. The ins and outs of group II introns. Trends Genet. 17:322–331.

Borner, G.V., Morl, M., Wissinger, B., Brennicke, A., Schmelzer, C 1995. RNA editing of a group II intron in Oenothera as a prerequisite for splicing. Mol. Gen. Genet. 246:739–744.

Bos, J. L., Heyting, C., Borst, P., Arnberg, A. C., van Bruggen, E. F. J. 1978. An insert in the single gene for the large ribosomal RNA in yeast mitochondrial DNA. Nature 275:336–338.

Brangeon, J., Sabar, M., Gutierres, S., Combettes, B., Bove, J., Gendy, C., Chetrit, P., Colas des Francs-Small, C., Pla, M., Vedel, F., de Paepe, R. 2000. Defective splicing of the first *nad*4 intron associated with lack of several complex I subunits in the *Nicotiana sylvestris* NMS1 nuclear mutant. Plant J. 21:269–280.

Burger, G., Yan,Y., Javadi, P., Lang, B. F. 2009. Group I intron *trans*-splicing and mRNA editing in the mitochondria of placozoan animals. Trends Genet. 25:381–386.

Carrillo, C., Bonen, L. 1997. RNA editing status of *nad*7 intron domains in wheat mitochondria. Nucl. Acid Res. 25:403–409.

Carrillo, C., Chapdelaine, Y., Bonen, L. 2001. Variation in sequence and RNA editing within core domains of mitochondrial group II introns among plants. Mol. Gen. Genet. 264:595–603.

Cavalier-Smith, T. 1991. Intron phylogeny: a new hypothesis. Trends Genet. 7:145–148.

Chapdelaine, Y., Bonen, L. 1991. The wheat mitochondrial gene for subunit I of the NADH dehydrogenase complex: a *trans*-splicing model for this gene-in-pieces. Cell 65:465–472.

Chaw, S. M., Shih, A. C., Wang, D., Wu, Y. W., Liu, S. M., Chou, T. Y. 2008. The mitochondrial genome of the gymnosperm *Cycas taitungensis* contains a novel family of short interspersed elements, Bpu sequences, and abundant RNA editing sites. Mol. Biol. Evol. 25:603–615.

Cho, Y., Qiu. Y. L., Kuhlman, P., Palmer, J. D. 1998. Explosive invasion of plant mitochondria by a group I intron. Proc. Natl. Acad. Sci. U.S.A. 95:14244–14249.

Choury, D., Araya, A. 2006. RNA editing site recognition in heterologous plant mitochondria. Curr. Genet. 50:405–416.

Costa, M., Fontaine, J. M., Loiseaux-de Goer, S., Michel, F. 1997. A group II self-splicing intron from the brown alga *Pylaiella littoralis* is active at unusually low magnesium concentrations and forms populations of molecules with a uniform conformation. J. Mol. Biol. 274:353–364.

de Longevialle, A. F., Meyer, E. H., Andres, C., Taylor, N. L., Lurin, C., Millar, A. H., Small, I. D. 2007. The pentatricopeptide repeat gene OTP43 is required for *trans*-splicing of the mitochondrial *nad1* intron 1 in *Arabidopsis thaliana*. Plant Cell 19:3256–3265.

de Longevialle, A. F., Hendrickson, L., Taylor, N. L., Delannoy, E., Lurin, C., Badger, M., Millar, A. H., Small, I. 2008. The pentatricopeptide repeat gene OTP51 with two LAGLIDADG motifs is required for the *cis*-splicing of plastid *ycf*3 intron 2 in *Arabidopsis thaliana*. Plant J. 56:157–168.

Decatur, W. A., Einvik, C., Johansen, S., Vogt, V. M. 1995. Two group I ribozymes with different functions in a nuclear rDNA intron. EMBO J. 14:4558–4568.

del Campo, M., Mohr, S. Jiang, Y., Jia, H., Jankowsky, E., Lambowitz, A. M. 2009. Unwinding by local strand separation is critical for the function of DEAD-box proteins as RNA chaperones. J. Mol. Biol. 389:674–693.

Dombrovska, O., Qiu, Y. L. 2004. Distribution of introns in the mitochondrial gene *nad1* in land plants: phylogenetic and molecular evolutionary implications. Mol. Phylogen. Evol. 32:246–263.

Elina, H., Brown, G. G. 2009. Extensive mis-splicing of a bi-partite plant mitochondrial group II intron. Nucl. Acids Res. 38(3):996–1008.

Farré, J. C., Araya, A. 1999. The *mat-r* open reading frame is transcribed from a non-canonical promoter and contains an internal promoter to co-transcribe exons *nad1e* and *nad5III* in wheat mitochondria. Plant Mol. Biol. 40:959–967.

Farré, J. C., Araya, A. 2002. RNA splicing in higher plant mitochondria: determination of functional elements in group II intron from a chimeric *cox II* gene in electroporated wheat mitochondria. Plant J. 29:203–213.

Fox, T. D., Leaver, C. J. 1981. The *Zea mays* mitochondrial gene coding cytochrome oxidase subunit II has an intervening sequence and does not contain TGA codons. Cell 26:315–323.

Gargouri, A. 2005. The reverse transcriptase encoded by ai1 intron is active *in trans* in the retro-deletion of yeast mitochondrial introns. FEMS Yeast Res. 5:813–822.

Geiss, K. T., Abbas, G. M., Makaroff, C. A. 1994. Intron loss from the NADH dehydrogenase subunit 4 gene of lettuce mitochondrial DNA: evidence for homologous recombination of a cDNA intermediate. Mol. Gen. Genet. 243:97–105.

Glanz, S., Kuck, U. 2009. *Trans*-splicing of organelle introns – a detour to continous RNAs. BioEssays 31:921–934.

Glover, K. E., Spencer, D. F., Gray, M. W. 2001. Identification and structural characterization of nucleus-encoded transfer RNAs imported into wheat mitochondria. J. Biol. Chem. 276:639–648.

Gray, M. W. 1992. The endosymbiont hypothesis revisited. Int. Rev. Cytol. 141:233–357.

Grewe, F., Viehoever, P., Weisshaar, B., Knoop, V. 2009. A *trans*-splicing group I intron and tRNA-hyperediting in the mitochondrial genome of the lycophyte *Isoetes engelmannii*. Nucl. Acids Res. 37(15):5093–5104.

Haouazine, N., Takvorian, A., Jubier, M.F., Michel, F., Lejeune, B. 1993. The *nad6* gene and the exon d of *nad1* are co-transcribed in wheat mitochondria. Curr. Genet. 24:533-538.

Haugen, P., Simon, D. M., Bhattacharya, D. 2005. The natural history of group I introns. Trends Genet. 21:111–119.

Herrin, D. L., Chen, Y. F., Schmidt, G. W. 1990. RNA splicing in *Chlamydomonas* chloroplast: self-splicing of 23 S preRNA. J. Biol. Chem. 265:21134–21140.

Hinrichsen, I., Bolle, N., Paun, L., Kempken, F. 2009. RNA processing in plant mitochondria is independent of transcription. Plant Mol. Biol. 70:663–668.

Jenkins, K. P., Hong, L., Hallick, R. B. 1995. Alternative splicing of the *Euglena gracilis* chloroplast *roa*A transcript. RNA 1:624–633.

Kaspar, B. J., Bifano, A. L., Caprara, M. G. 2008. A shared RNA-binding site in the Pet54 protein is required for translational activation and group I intron splicing in yeast mitochondria. Nucl. Acids Res. 36:2958–2968.

Keren, I., Bezawork-Geleta, A., Kolton, M., Maayan, I., Belausov, E., Levy, M., Mett, A., Gidoni, D., Shaya, F., Ostersetzer-Biran, O. 2009. AtnMat2, a nuclear-enocded maturase required for splicing of group-II introns in *Arabidopsis* mitochondria. RNA 15:2299–2311.

Kim, J. K., Hollingsworth, M. J. 1993. Splicing of group II introns in spinach chloroplasts (*in vivo*): analysis of lariat formation. Curr. Genet. 23:175–180.

Kim, S., Yoon, M. K. 2010. Comparison of mitochondrial and chloroplast genome segments from three onion (*Allium cepa* L.) cytoplasm types and identification of a *trans*-splicing intron of *cox*2. Curr. Genet. 56:177–188.

Knoop, V. 2004. The mitochondrial DNA of land plants: peculiarities in phylogenetic perspective. Curr. Genet. 46:123–139.

Knoop, V., Erhardt, T., Lattig, K., Brennicke, A. 1995. The gene for ribosomal protein S10 is present in mitochondria of pea and potato but absent from those of *Arabidopsis* and *Oenothera*. Curr. Genet. 27:559–564.

Knoop, V., Altwasser, M., Brennicke, A. 1997. A tripartite group II intron in mitochondria of an angiosperm plant. Mol. Gen. Genet. 255:269–276.

Kohler, D., Schmidt-Gattung, S., Binder, S. 2010. The DEAD-box protein PMH2 is required for efficient group II intron splicing in mitochondria of *Arabidopsis thaliana*. Plant Mol. Biol. 72:459–467.

Kroeger, T. S., Watkins, K. P., Friso, G., van Wijk K. J., Barkan, A. 2009. A plant-specific RNA-binding domain revealed through analysis of chloroplast group II intron splicing. Proc. Natl. Acad. Sci. U.S.A. 106:4537–4542.

Kuhn, K., Richter, U., Meyer, E. H., Delannoy, E., de Longevialle, A. F., O'Toole, N., Borner, T., Millar, A. H., Small, I. D., Whelan, J. 2009. Phage-type RNA polymerase RPOTmp performs gene-specific transcription in mitochondria of *Arabidopsis thaliana*. Plant Cell 21:2762–2779.

Kurihara-Yonemoto, S., Handa, H. 2001. Low temperature affects the processing pattern and RNA editing status of the mitochondrial *cox*2 transcripts in wheat. Curr. Genet. 40:203–208.

Lambowitz, A. M., Zimmerly, S. 2004. Mobile group II introns. Ann. Rev. Genet. 38:1–35.

Lang, B. F., Laforest, M. J., Burger, G. 2007. Mitochondrial introns: a critical view. Trends Genet. 23:119–125.

Lange, H., Sement, F. M., Canaday, J., Gagliardi, D. 2009. Polyadenylation-assisted RNA degradation processes in plants. Trends Plant Sci. 14:497–504.

Lehmann, K., Schmidt, U. 2003. Group II introns: structure and catalytic versatility of large natural ribozymes. Crit. Rev. Biochem. Mol. Biol. 38:249–303.

Li, L., Wang, B., Liu, Y., Qiu, Y. L. 2009. The complete mitochondrial genome sequence of the hornwort *Megaceros aenigmaticus* shows a mixed mode of conservative yet dynamic evolution in early land plant mitochondrial genomes. J. Mol. Evol. 68:665–678.

Li-Pook-Than, J., Bonen, L. 2006. Multiple physical forms of excised group II intron RNAs in wheat mitochondria. Nucl. Acids Res. 34:2782–2790.

Li-Pook-Than, J., Carrillo, C., Bonen, L. 2004. Variation in mitochondrial transcript profiles of protein-coding genes during early germination and seedling development in wheat. Curr. Genet. 46:374–380.

Li-Pook-Than, J., Carrillo, C., Niknejad, N., Calixte, S., Crosthwait, J., Bonen, L. 2007. Relationship between RNA splicing and exon editing near intron junctions in wheat mitochondria. Physiol. Plant. 129:23–33.

Lippok, B., Brennicke, A., Wissinger, B. 1994. Differential RNA editing in closely related introns in *Oenothera* mitochondria. Mol. Gen. Genet. 243:39–46.

Long, M. B., Jones, J. P., Sullenger, B. A., Byun, J. 2003. Ribozyme-mediated revision of RNA and DNA. J. Clin. Invest. 112:312–318.

Luban, C., Beutel, M., Stahl, U., Schmidt, U. 2005. Systematic screening of nuclear encoded proteins involved in the splicing metabolism of group II introns in yeast mitochondria. Gene 354:72–79.

Malek, O., Knoop, V. 1998. *Trans*-splicing group II introns in plant mitochondria: the complete set of *cis*-arranged homologs in ferns, fern allies, and a hornwort. RNA 4:1599–1609.

Marienfeld, J. R., Newton, K. J. 1994. The maize NCS2 abnormal growth mutant has a chimeric *nad*4-*nad*7 mitochondrial gene and is associated with reduced complex I function. Genetics 138: 855–863.

Martin, W., Koonin, E. V. 2006. Introns and the origin of nucleus-cytosol compartmentalization. Nature 440:41–45.

Matthes, A., Schmidt-Gattung, S., Kohler, D., Forner, J., Wildrum, S., Raabe, M., Urlaub, H., Binder, S. 2007. Two DEAD-box proteins may be part of RNA-dependent high-molecular-mass protein complexes in *Arabidopsis* mitochondria. Plant Physiol. 145:1637–1646.

Medina, M., Collins, A. G., Takaoka, T. L., Kuehl, J. V., Boore, J. L. 2006. Naked corals: skeleton loss in Scleractinia. Proc. Natl. Acad. Sci. U.S.A. 103:9096–9100.

Merendino, L., Perron, K., Rahire, M., Howald, I., Rochaix, J. D., Goldschmidt-Clermont, M. 2006. A novel multifunctional factor involved in *trans*-splicing of chloroplast introns in *Chlamydomonas*. Nucl. Acids Res. 34:262–274.

Michel, F., Ferat, J. L. 1995. Structure and activities of group II introns. Ann. Rev. Biochem. 64:435–461.

Miyata, Y., Sugita, C., Maruyama, K., Sugita, M. 2008. RNA editing in the anticodon of tRNA-Leu(CAA) occurs before group I intron splicing in plastids of a moss *Takakia lepidozioides* S. Hatt. & Inoue. Plant Biol. 10:250–255.

Mohr, G., Lambowitz, A. M. 2003. Putative proteins related to group II intron reverse transcriptase/maturases are encoded by nuclear genes in higher plants. Nucl. Acids Res. 31:647–652.

Moreno, J. I., Buie, K. S., Price, R. E., Piva, M. A. 2009. Ccm1p/Ygr150cp, a pentatricopeptide repeat protein, is essential to remove the fourth intron of both *COB* and *COX1* pre-mRNAs in *Saccharomyces cerevisiae*. Curr. Genet. 55:475–484.

Nakagawa, N., Sakurai, N. 2006. A mutation in At-nMat1a, which encodes a nuclear gene having high similarity to group II intron maturase, causes impaired splicing of mitochondrial NAD4 transcript and altered carbon metabolism in Arabidopsis thaliana. Plant Cell. Physiol. 47:772–783.

Nielsen, H., Johansen, S. D. 2009. Group I introns: moving in new directions. RNA Biol. 6:1–9.

Notsu, Y., Masood, S., Nishikawa, T., Kubo, N., Akiduki, G., Nakazono, M., Hirai, A., Kadowaki, K. 2002. The complete sequence of the rice (*Oryza sativa* L.) mitochondrial genome: frequent DNA sequence acquisition and loss during the evolution of flowering plants. Mol. Genet. Genom. 268:434–445.

Oda, K., Yamato, K., Ohta, E., Nakamura, Y., Takemura, M., Nozato, N., Akashi, K., Kanegae, T., Ogura, Y., Kohchi, T., Ohyama, K. 1992. Gene organization deduced from the complete sequence of liverwort *Marchantia polymorpha* mitochondrial DNA: a primitive form of plant mitochondrial genome. J. Mol. Biol. 223:1–7.

Ogihara, Y., Yamazaki, Y., Murai, K., Kanno, A., Terachi, T., Shiina, T., Miyashita, N., Nasuda, S., Nakamura, C., Mori, N., Takumi, S., Murata, M., Futo, S., Tsunewaki, K. 2005. Structural dynamics of cereal mitochondrial genomes as revealed by complete nucleotide sequencing of the wheat mitochondrial genome. Nucl. Acids Res. 33:6235–6250.

Ohyama, K., Takemura, M. 2008. Molecular evolution of mitochondrial introns in the liverwort *Marchantia polymorpha*. Proc. Jpn. Acad. Ser. B 84:17–22.

Pla, M., Mathieu, C., de Paepe, R., Chetrit, P., Vedel, F. 1995. Deletion of the last two exons of the mitochondrial *nad7* gene results in lack of the NAD7 polypeptide in *Nicotiana sylvestris* CMS mutant. Mol. Gen. Genet. 248:79–88.

Qiu, Y. L., Palmer, J. D. 2004. Many independent origins of *trans*-splicing of a plant mitochondrial group II intron. J. Mol. Evol. 59:80–89.

Ran, J. H., Gao, H., Wang. X. Q. 2009. Fast evolution of the retroprocessed mitochondrial *rps*3 gene in Conifer II and further evidence for the phylogeny of gymnosperms. Mol Phylogen. Evol. 54(1):136–149.

Robart, A. R., Montgomery, N. K., Smith, K. L., Zimmerly, S. 2004. Principles of 3′ splice site selection and alternative splicing for an unusual group II intron from *Bacillus anthracis*. RNA 10:854–862.

Rochaix, J. D., Malnoe, P. 1978. Anatomy of the chloroplast ribosomal DNA of *Chlamydomonas reinhardii*. Cell 15:661–670.

Schmitz-Linneweber, C., Small, I. 2008. Pentatricopeptide repeat proteins: a socket set for organelle gene expression. Trends Plant Sci. 13:663–670.

Schock, I., Gregan, J., Steinhauser, S., Schweyen, R., Brennicke, A., Knoop, V. 2000. A member of a novel *Arabidopsis thaliana* gene family of candidate Mg^{2+} ion transporters complements a yeast mitochondrial group II intron-splicing mutant. Plant J. 24:489–501.

Sharp, P. A. 1991. Five easy pieces. Science 254:663.

Signorovitch, A. Y., Buss, L. W., Dellaporta, S. L. 2007. Comparative genomics of large mitochondria in placozoans. PLoS Genet. 3:e13.

Simon, D. M., Clarke, N. A.C., McNeil, B. A., Johnson, I., Pantuso, D., Dai, L., Chai, D., Zimmerly, S. 2008. Group II introns in eubacteria and archaea: ORF-less introns and new varieties. RNA 14:1704–1713.

Staudinger, M., Kempken, F. 2003. Electroporation of isolated higher-plant mitochondria: transcripts of an introduced *cox*2 gene, but not an *atp*6 gene, are edited *in organello*. Mol. Gen. Genom. 269:553–561.

Stupar, R. M., Lilly, J. W., Town, C. D., Cheng, Z., Kaul, S., Buell, C. R., Jiang, J. 2001. Complex mtDNA constitutes an approximate 620-kb insertion on Arabidopsis thaliana chromosome 2: implication of potential sequencing errors caused by large-unit repeats. Proc. Natl. Acad. Sci. U.S.A. 98:5099–5103.

Sutton, C. A., Conklin, P. L., Pruitt, K. D., Hanson, M. R. 1991. Editing of pre-mRNAs can occur before *cis*- and *trans*-splicing in Petunia mitochondria. Mol. Cell. Biol. 11:4274–4277.

Terasawa, K., Odahara, M., Kabeya, Y., Kikugawa, T., Sekine, Y., Fujiwara, M., Sato, N. 2007. The mitochondrial genome of the moss *Physcomitrella patens* sheds new light on mitochondrial evolution in land plants. Mol. Biol. Evol. 24:699–709.

Thompson, M. D., Copertino, D. W., Thompson, E., Favreau, M. R., Hallick, R. B. 1995. Evidence for the late origin of introns in chloroplast genes from an evolutionary analysis of the genus *Euglena*. Nucl. Acids Res. 23:4745–4752.

Toor, N., Keating, K. S., Pyle, A. M. 2009. Structural insights into RNA splicing. Curr. Opin. Struct. Biol. 19:260–266.

Toro, N., Jimenez-Zurdo, J. I., Garcia-Rodriguez, F. M. 2007. Bacterial group II introns: not just splicing. FEMS Microbiol. Rev. 31:342–358.

Unseld, M., Marienfeld, J. R., Brandt, P., Brennicke, A. 1997. The mitochondrial genome of *Arabidopsis thaliana* contains 57 genes in 366,924 nucleotides. Nat. Genet. 15:57–61.

Valles, Y., Halanych, K. M., Boore, J. L. 2008. Group II introns break new boundaries: presence in a bilaterian's genome. PLosOne 3:e1488.

Vicens, Q., Cech, T. R. 2006. Atomic level architecture of group I introns revealed. Trends Biochem. Sci. 31:41–51.

Vogel, J., Borner, T. 2002. Lariat formation and a hydrolytic pathway in plant chloroplast group II intron splicing. EMBO J. 21:3794–3803.

Wissinger, B., Schuster, W., Brennicke, A., 1991. *Trans* splicing in *Oenothera* mitochondria: *nad1* mRNAs are edited in exon and *trans*-splicing group II intron sequences. Cell 65:473–482.

Witkowski, J. A. 1988. The discovery of 'split' genes: a scientific revolution. Trends Biochem. Sci. 13:110–113.

Won, H., Renner, S. S. 2003. Horizontal gene transfer from flowering plants to *Gnetum*. Proc. Natl. Acad. Sci. U.S.A. 100:10824–10829.

Yang, A. J., Mulligan, R. M. 1991. RNA editing intermediates of *cox2* transcripts in maize mitochondria. Mol. Cell. Biol. 11:4278–4281.

Chapter 7
RNA Editing in Higher Plant Mitochondria

Anika Bruhs and Frank Kempken

Abstract RNA editing is a process in which the nucleotide sequence of a transcript is changed compared to its DNA template. RNA editing includes the insertion, deletion, or conversion of nucleotides. Many different kinds of RNA editing are known and apparently evolved separately from each other. Examples are: mitochondrial RNA editing in trypanosomes and plants, A-to-I editing, and apolipoprotein B RNA editing in certain mammalian cells. In higher plant mitochondria, RNA editing is mostly a conversion of cytidine residues to uracil, and there are several hundred editing sites. In this chapter we give an overview of plant mitochondrial RNA editing, its consequences for translation, the current knowledge of its mechanism, and some ideas on its evolution.

Keywords Higher plant mitochondria • RNA editing • Pentatricopeptid repeat proteins • In vitro and in organello systems

Abbreviations

CMS Cytoplasmic male sterility
MEF Mitochondrial RNA editing factor
NDH NAD(P)H dehydrogenase
PPR Pentatricopeptide repeat

7.1 Introduction

RNA editing is a process in which the nucleotide sequence of a transcript is changed compared to its DNA template. RNA editing includes the insertion, deletion, or conversion of nucleotides. As such, this process is different from RNA cleavage

F. Kempken (✉)
Botanisches Institut und Botanischer Garten, Abteilung für Botanische Genetik und Molekularbiologie, Christian-Albrechts-Universität zu Kiel, Olshausenstr. 40, D-24098, Kiel, Germany
email: fkempken@bot.uni-kiel.de

F. Kempken (ed.), *Plant Mitochondria*, Advances in Plant Biology 1,
DOI 10.1007/978-0-387-89781-3_7, © Springer Science+Business Media, LLC 2011

(see Chap. 5) or RNA splicing (see Chap. 6), which remove parts of a primary transcript but do not change the nucleotide sequence. Many different kinds of RNA editing are known and apparently evolved separately from each other. Different types of RNA editing with their major features are given in Table 7.1. Mitochondrial RNA editing in trypanosomes, A-to-I editing, and apolipoprotein B RNA editing in certain mammalian cells appear to be the best characterized cases (examples see Box 7.1).

Twenty years ago, RNA editing in higher plant mitochondria was simultaneously detected by three laboratories (Covello and Gray 1989; Gualberto et al. 1989; Hiesel et al. 1989), and later in plastids (Hoch et al. 1991). Higher plant mitochondria and plastids exhibit the same type of RNA editing, i.e., cytidine to uracil RNA editing and rare uracil to cytidine editing (reviewed in Shikanai 2006). Editing is thought to occur by deamination of cytidine (Fig. 7.1). However, uracil to cytidine is much more common in some ferns and mosses (Rüdinger et al. 2009, see Chap. 1). Though sharing the same mode of RNA editing, the number of RNA editing sites varies significantly in higher plant mitochondria and plastids. Higher plant mitochondria have several hundred editing sites (Table 7.2), but plastids have only about 30 editing sites (summarized in Zeng et al. 2007).

In this chapter we give an overview of plant mitochondrial RNA editing, its consequences for translation, the current knowledge regarding its mechanism, and some ideas on its evolution.

7.2 Extent and Consequences of RNA Editing in Higher Plant Mitochondria

7.2.1 Extent of RNA Editing

Up to 1,000 editing sites have been observed in the mitochondrial transcriptomes of the gymnosperm *Cycas taitungensis* (Chaw et al. 2008) and the lycophyte *Isoetes engelmannii* (Grewe et al. 2009). Higher plant mitochondria have roughly 500 editing sites (see Table 7.2); for example, *Arabidopsis thaliana* has 456 editing sites, of which 441 are in coding sequences (Giege and Brennicke 1999), *Oryza sativa* has 491 mitochondrial editing sites (Notsu et al. 2002), and *Brassica napus* has 427 editing sites (Handa 2003). These many sites pose a problem for editing site annotation, which was recognized by Volker Knoop and coworkers (Rüdinger et al. 2009). A new website called PREPACT (http://www.prepact.de) offers prediction, analysis, annotation, and graphical display of RNA editing sites (Lenz et al. 2009).

Though most editing sites occur in protein coding sequences, only a few editing sites are present in intronic sequences, and it has been observed that unspliced RNA is often only partially edited (Sutton et al. 1991; Yang and Mulligan 1991). Sometimes editing is a prerequisite for RNA splicing (Binder et al. 1992; Börner et al. 1995). RNA editing is not restricted to mRNA sequences, but may also occur in tRNA (Maréchal-Drouard et al. 1993, 1996; Binder et al. 1994).

Table 7.1 Types of RNA editing

Mode of editing	Organism	Organelle, compartment, or cell type	Template RNA	Original references
Cotranscriptional				
Insertion or deletion of guanosines	Paramyxovirus (RNA virus)	Cytoplasm	Viral	Cattaneo et al. 1989; Jacques et al. 1994; Thomas et al. 1988
Posttranscriptional				
Added/deleted uracil residues	Trypanosomes	Mitochondria	Protein encoding	Benne et al. 1986
Insertion of cytidines	*Physarum polycephalum*	Mitochondria	Protein encoding, rRNA	Mahendran et al. 1994; Mahendran et al. 1991
Adenine to inosine	Mammals, avian, Xenopus, Drosophila, squid	Nuclear	Double-stranded protein coding, noncoding	Bass and Weintraub 1988; Polson et al. 1991; Wagner et al. 1989
Cytidine to uracil editing	Mammals	Nuclear, intestine	Apolipoprotein B	Chen et al. 1987; Powell et al. 1987
Cytidine to uracil editing	Plants	Mitochondria	Protein encoding, tRNA, rRNA	Covello and Gray 1989; Gualberto et al. 1989; Hiesel et al. 1989
Cytidine to uracil editing	Plants	Plastids	Protein encoding	Hoch et al. 1991

Box 7.1 Other examples of RNA editing mechanisms

RNA editing in trypanosome mitochondria (Benne 1994, 1996; Benne et al. 1986) is characterized by the insertion or deletion of uridine residues from the mRNA template (Madison-Antenucci et al. 2002; Simpson et al. 2003). This change is achieved by employing a large number of guide RNAs (gRNAs) and catalyzed by a multienzyme complex that is not yet fully characterized. A subcomplex that is able to edit mRNAs *in vitro* has been isolated and is called the 20 S editosome (Stuart et al. 2005). RNA editing occurs in a 3′–5′ direction. The gRNAs subsequently bind the unedited mRNA template (Blum and Simpson 1990). Each gRNA indicates the position of an endonucleolytic cleavage point in the RNA and contains the information for the removal or insertion of uridine residues. The editing process contains several steps, including cleavage of the RNA template, addition of uridine residues by a terminal uridylyl transferase or removal of uridine residues by an exonuclease, and ligation of the RNA by a RNA ligase (Stuart et al. 2005). Recently, a 3D structural analysis of the core editing complex from *Leishmania tarentolae* mitochondria was published (Li et al. 2009).

Apolipoprotein B mRNA editing (reviewed in Wedekind et al. 2003) occurs at two positions within the *apoB* mRNA. At position 6,666, a glutamine codon is edited to a stop codon (CAA→UAA), and editing at position 6,802 causes a change in an amino acid (Thr→Ile) (Lau et al. 1991; Smith and Sowden 1996). In most mammals, including humans, *apoB* is edited in the intestine only, but there are some mammals in which *apoB* is also edited in the liver (Wedekind et al. 2003). Editing is performed by Apobec-1, a zinc-dependent cytidine deaminase (Teng et al. 1993). However, site-specific editing requires interaction with additional specificity factors (Dance et al. 2002). Interestingly, activation-induced deaminase (AID), a protein related to Apobec-1, is essential for immunoglobulin gene somatic hypermutation and class switch DNA recombination (Muramatsu et al. 2000).

7.2.2 Consequences of RNA Editing

Basically, all higher plant mRNAs coding proteins are edited with the exception of some chimeric ORF found in sterile male cytoplasm (Ward and Levings 1991, see Chap. 18). Most of the editing sites in higher plant mitochondria are found in protein coding regions. Thus, RNA editing can alter the amino acid composition of the protein and appears to be an essential mechanism for gene expression in plant organelles. RNA editing is particularly required to generate the conserved and biologically functional protein sequence. There is strong evidence that RNA editing occurs at functionally conserved sites; for example, editing sites in *atp6* have been

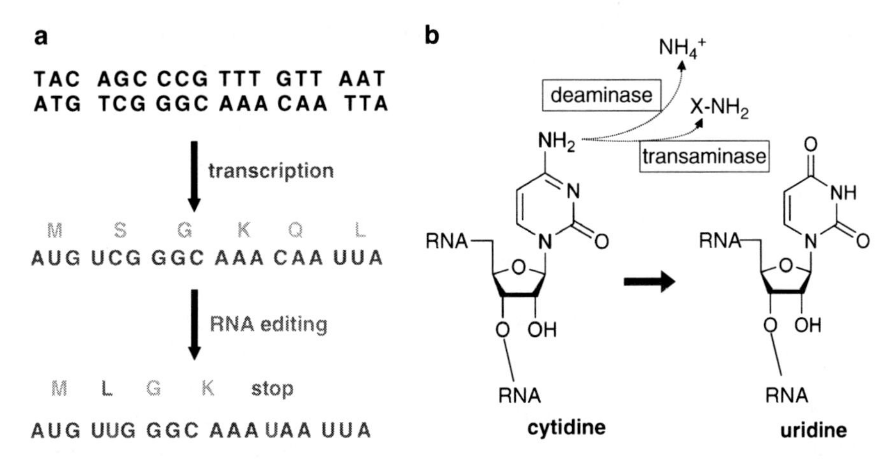

Fig. 7.1 RNA editing in plant organelles. (**a**) RNA editing converts two cytidines to uridines, which leads to amino acid changes and the creation of a stop codon. (**b**) A transaminase or deaminase presumably catalyzes the editing reaction (**b**: Modified from Takenaka et al. 2008)

Table 7.2 Extent of RNA editing in higher plant mitochondria

Organism	Genome size (kb)	Number of genes	Editing sites	References
Arabidopsis thaliana	367	55	441	Unseld et al. 1997
Zea mays	570	58	?	Clifton et al. 2004
Brassica napus	222	53	427	Handa 2003
Oryza sativum	491	56	491	Notsu et al. 2002
Sugar beet (TK81-0)	369	52	~370	Kubo et al. 2000
(TK81-MS)	501	52	~370	Satoh et al. 2006
Nicotiana tabacum	431	60	?	Sugiyama et al. 2005

shown to occur at functionally relevant positions (Kempken et al. 1998). In the *cox2* mRNA sequence, the editing of one codon changes a coding from threonine to methionine, creating an essential domain for copper binding (Gualberto et al. 1989; Maier et al. 1996). However, a surprisingly high number of viable mutations were recently identified with a loss of editing at specific sites (Takenaka and Brennicke 2009). This observation may indicate that only some editing sites are essential for the function of the encoded protein, whereas others are not. In some cases, RNA editing is necessary to generate a translation start codon (Chapdelaine and Bonen 1991; Hoch et al. 1991), to create a stop codon and thereby the conserved C-terminus of the protein (Kempken et al. 1991; Wintz and Hanson 1991; Kugita et al. 2003), or to remove a stop codon (Hiesel et al. 1994). Evidence also indicates a connection between RNA editing and cytoplasmic male sterility: (1) Sterile male *Sorghum bicolor* anthers exhibit a strong reduction in *atp6* RNA editing, which is gradually restored in F1 and F2 plants (Howad and Kempken 1997; Howad et al. 1999; Pring et al. 1999), but this effect is less pronounced when analyzing plants grown under natural conditions (Pring and Tang 2001); and (2) truncation of a

CMS-associated reading frame has been observed, which was due to the generation of a stop codon by RNA editing (Gallagher et al. 2002).

Not all RNA editing sites in coding regions lead to a change in the codon capacity. Editing sites in the third codon position are often silent, i.e., they do not lead to a change in the encoded amino acid (Schuster et al. 1990; Kempken et al. 1995). These sites are often not fully edited; only some RNAs are edited at these sites. However, partial editing is not limited to silent editing sites; there are many editing sites that lead to a changed amino acid but still exhibit partial editing (Pring et al. 1993; Brennicke et al. 1999). In fact, editing efficiency varies from site to site and, sometimes, partially edited transcripts may be more abundant than fully edited RNAs (Schuster et al. 1990; Lu et al. 1996). Plant mitochondria do not appear to possess a quality control for defective or partially edited RNAs (see Holec et al. 2008 for a review). Nevertheless, there must be some kind of mechanism to prevent partially edited or unedited transcripts from being translated, as this can have severe consequences. In fact, only peptides from fully edited *atp6* transcripts have been observed to accumulate in petunia (Lu and Hanson 1994). Likewise, in potato, only NAD9 peptides from fully edited transcripts accumulate, though a significant portion of partially edited *nad9* transcripts are found in various tissues (Grohmann et al. 1994). These results suggest a pretranslational selection between edited and incompletely edited mRNAs in plant mitochondria. However, polymorphic RPS12 polypeptides have been detected when analyzing *rps12* RNA editing in maize and petunia and its consequences for translation (Lu et al. 1996; Phreaner et al. 1996). In some cases, partially edited transcripts can be translated and the protein product can accumulate to detectable levels, suggesting that incompletely edited transcripts can encode more than one gene product in plant mitochondria and possibly provide a physiological or regulatory function of RNA editing.

7.3 Functional Analysis of the Higher Plant RNA Editing Mechanism

7.3.1 *The Use of In vitro and In Organello Systems*

To investigate the mechanism of recognition and editing of RNA editing sites, *in vitro* (Hirose and Sugiura 2001; Takenaka and Brennicke 2003; Nakajima and Mulligan 2005; Takenaka et al. 2007; Verbitskiy et al. 2008) and *in organello* approaches have been successfully established for mitochondria and plastids (reviewed in Takenaka et al. 2008). Until now, *in organello* systems for mitochondria from *Triticum aestivum* (Farré and Araya 2001), *Solanum tuberosum* (Choury et al. 2005), *Zea mays,* and *Sorghum bicolor* (Staudinger and Kempken 2003) were successfully set up. De novo transcription and RNA processing has been observed *in organello* using biotinylated UTP, and the *in organello* system has been shown to faithfully reflect the *in planta* situation. Transcripts that are fully edited *in planta* are fully edited *in organello*,

whereas those that are partially edited *in planta* are also partially edited in the *in organello* system (Staudinger and Kempken 2003; Staudinger and Kempken 2004). More intriguing is that, when the *A. thaliana cox2* gene is introduced into maize mitochondria via electroporation, the *cox2* transcript is spliced and approximately 100% RNA editing observed at almost all sites, while RNA editing in *in vitro* systems exhibits a lower activity of about 7% (Neuwirt et al. 2005). The use of this *in organello* system also confirmed that RNA editing and transcription are independent of each other (Hinrichsen et al. 2009), which had previously been suggested by the use of an *in vitro* system in which partially edited *atp4* transcripts in an *in vitro* system were found to be a substrate for further RNA editing (Verbitskiy et al. 2006). These findings indicate that RNA editing may indeed be a posttranscriptional process.
The use of *in organello* and *in vitro* systems allowed the size and localization of regions essential for the recognition and RNA editing of single editing sites to be identified. Using an *in organello* approach in wheat, a region essential for the editing of *cox2* editing sites C259 and C77 was characterized. In each case, a region of 16 nucleotides upstream and six nucleotides downstream was shown to be sufficient for recognition (Farré et al. 2001; Choury et al. 2004).

An *in vitro* approach with mitochondrial protein extract emphasized a region of 20 nucleotides upstream as essential for the first editing position of the pea *atp9* transcript. However, for efficient editing, an upstream sequence of 40 nucleotides is required (Takenaka et al. 2004). This observation was confirmed in an *in vitro Brassica oleracea* var. *botrytis* system, which also identified one nucleotide downstream of the *atp9* editing site as essential (Neuwirt et al. 2005). Furthermore, the 5′ region of an editing site has been shown to be important for editing another position 50–70 nucleotides downstream (van der Merwe et al. 2006).

The high number of editing sites found in mitochondria, which are partly densely packed within only some 50 genes, raises the question of whether these sites are addressed independently, collectively, or consecutively. The results of an analysis of *in organello* RNA editing in maize mitochondria using different chimeric Sorghum-maize *atp6* gene constructs suggest the presence of an upstream binding domain for a primary transcript-specific editing factor (Staudinger et al. 2005). All 19 RNA editing sites of the Sorghum *atp6* transcript were not edited in maize mitochondria, despite a great extent of sequence similarity. However, in the presence of the maize *atp6* 5′ UTR and 5′ pre-piece sequence, all editing sites were found to be partially edited. This situation is somewhat reminiscent of previous observations in CMS Sorghum (Howad and Kempken 1997). Considering the influence on every single editing site of the transcript, the presence of a region essential for editing machinery recognition of the transcript seems to be likely. This hypothesis is supported by *in vitro* RNA editing analysis showing that specific trans-factors seem to recognize neighboring editing sites, even if there is no sequence similarity between the essential *cis* regions (van der Merwe et al. 2006). The addition of multiple *cis* elements leads to strongly enhanced, up to 80%, *in vitro* editing, leading to the idea that there could be a kind of cooperative *cis* effect between the specific recognition sequences, which promotes the functional and effective assembly of the yet unknown RNA editosome (Verbitskiy et al. 2008). The effect of *cis*-acting enhancement decreases with increasing distance between the repeated specificity signals, suggesting a distance-dependent cooperation. To summarize, these results suggest that *cis/trans* interactions can be effective over larger distances and affect more than one editing event.

7.3.2 What Can Be Learned from Plastid RNA Editing?

Because higher plant mitochondrial and plastid RNA editing share some characteristics, we will also cover plastid editing in this section. However, the organelles are not capable of editing RNA molecules from the other organelle. Though it has long been established that mitochondrial sequences in transgenic tobacco are not edited (Sutton et al. 1995), plant mitochondria were shown only recently to not be able to edit plastid transcripts (Bolle et al. 2007).

In the case of several plastid RNA editing sites, some *cis*-acting sequences have been identified that are similar in size and location to *cis*-acting sequences of mitochondrial sites (Miyamoto et al. 2002; Hayes et al. 2006). In transgenic tobacco plants, a region of 22 nucleotides, 16 nucleotides upstream and five nucleotides downstream of an editing site in the *psbL* transcript, has been demonstrated to be sufficiently large for recognition (Chaudhuri and Maliga 1996). In a similar approach, Bock et al. (1996) were able to identify a region 12 nucleotides upstream and two nucleotides downstream of two adjacent editing sites in the *ndh* transcript, providing at least partial editing.

These results for the *ndhB* and *psbL* editing positions were later confirmed in an *in vitro* tobacco system (Hirose and Sugiura 2001). In similar approaches, regions localized −20 to +6 nucleotides and −15 to +5 nucleotides around the editing site were found to be sufficient for editing the tobacco *rpoB* and *ndhF* transcript (Reed et al. 2001; Sasaki et al. 2006). Thus, the recognition of mitochondrial and plastid editing sites occurs through similar *cis* element motifs, of which a number have been characterized in recent years. In contrast, the enzymes involved in the catalysis of the editing reaction have not been identified thus far. However, several proteins required for the editing process at specific sites have been identified.

Biochemical analysis of plastid extracts using UV cross-linking led to the isolation of protein factors that specifically bind to *cis* elements at plastid editing sites. Proteins 25, 56, and 70 kDa in size bind to *cis* elements of *psbL*, *psbE*, and *petB* editing sites, respectively (Hirose and Sugiura 2001; Miyamoto et al. 2002), and a 95 kDa protein binds two editing sites, one each in *ndhB* and *ndhF* transcripts (Kobayashi et al. 2008). Furthermore, using a genetic approach based on screening *A. thaliana* mutants with decreased NDH activity, three plastid factors essential for editing have been identified. Two of these nuclear encoded proteins, CRR4 and CRR21, seem to be essential for editing at the *ndh*D-1 and *ndh*D-2 editing sites (Kotera et al. 2005; Okuda et al. 2007), whereas the third factor, CLB19, is required for editing two sites located in plastid transcripts *rpoA* and *clpP* (Chateigner-Boutin et al. 2008).

7.3.3 Mitochondrial RNA Editing Factors

The first nuclear encoded factors for RNA editing in mitochondria were recently identified. So far, four such trans-factors, MEF1, MEF11, OGR1, and MEF9, have been described, all of which appear to recognize and bind several RNA motifs at at least three different editing sites, except MEF9 (Kim et al. 2009; Takenaka 2009; Verbitskiy et al. 2009; Zehrmann et al. 2009). In mitochondria, MEF1 was the first editing factor to be characterized (Zehrmann et al. 2009). A total of three sites, *rps4*-956, *nad7*-963, and *nad2*-1,160, are affected by this nuclear encoded protein; however, though editing occurs, decreased and even absent RNA editing is tolerated at all three sites, which led to the conclusion that the amino acid changes may be silent for the respective protein functions. MEF11 is also involved in RNA editing at multiple sites in *A. thaliana*: *cox3*-422, *nad4*-124, and *ccb203*-344. Though the editing sites *cox3*-422 and *ccb203*-344 each alter amino acids that are conserved in the respective proteins, the *nad4*-124 site is silent. Studies on mutants of *A. thaliana* also suggest additional function of the protein MEF11 because of the tolerance to lovastatin (Verbitskiy et al. 2009). In contrast to MEF1 and MEF11, which both affect three editing sites, and OGR1, which is responsible for seven sites (Kim et al. 2009), MEF9 appears to be an exception because it is associated with a single editing site, *nad7*-200 (Takenaka 2009). A pentatricopeptide repeat protein (PPR) of the P subfamily was recently found to be correlated with partial editing of the *rps3* transcript (Doniwa et al. 2010).

All specific *trans* factors identified thus far are PPRs, a family with about 450 members in *A. thaliana* that are composed of a variable number of repeats of roughly 35 amino acids and additional C-terminal domains (Small and Peeters 2000). Based on the nature of their PPR motifs, the plant PPR proteins can be separated into two major subfamilies and four smaller subclasses based on their characteristic C-terminal motifs (Lurin et al. 2004). Most of these PPR proteins are predicted to be targeted to mitochondria and/or plastids (Lurin et al. 2004).

Members of the PPR family are known to be involved in many posttranscriptional processes in mitochondria, as well as in plastids. Some of the proteins are implicated in the processing of multicistronic pre-mRNAs, the stabilization of pre-tRNAs, and the translation of specific genes in plastids (Delannoy et al. 2007; Schmitz-Linneweber and Small 2008). However, the molecular mechanisms of the biochemical reactions are still unknown.

Among the PPR proteins required for editing in plastids and mitochondria, some contain C-terminal E and DYW regions, whereas others lack the additional DYW regions (Kotera et al. 2005; Okuda et al. 2006, 2007; Chateigner-Boutin et al. 2008; see Chap. 8 for more details). Because several PPR proteins that are necessary for RNA editing contain E-extensions but no DYW domain, the hypothesis of a catalytic function linking the described DYW domain to the editing mechanism (Salone et al. 2007) appears to be unlikely. Instead, the identified editing factors are proposed to be specificity factors that interact directly or indirectly with one or more additional proteins, including one with enzymatic activity for the deamination step (Okuda et al. 2007; Chateigner-Boutin et al. 2008). Notably, the actual enzymatic mechanism for changes of the nucleotide identity has not been characterized, though base exchange and nucleotide excision can be excluded (Rajasekhar and Mulligan 1993; Yu and Schuster 1995) and deamination or transamination are still possible mechanisms (see Fig. 7.1; see Takenaka et al. 2008 for details).

Although the first proteins have been identified, how the specificity of binding is achieved is not known. Preliminary data indicate that there may be insufficient specificity for PPR proteins to distinguish between editing and nonediting sites (Zehrmann et al. 2009). Analysis of the phylogenetic distribution of *cox2* RNA editing sites using available sequence data has revealed the presence of mono- and dicotyledonous-specific RNA editing sites (Bolle and Kempken 2006). Both the mono- and dicotyledonous *cox2* transcripts have been shown to be efficiently edited in cauliflower and maize mitochondrial *in organello* systems, respectively, even for RNA editing sites not present in the endogenous *cox2* sequences (Bolle and Kempken 2006). The binding specificity of PPR proteins appears to be limited;

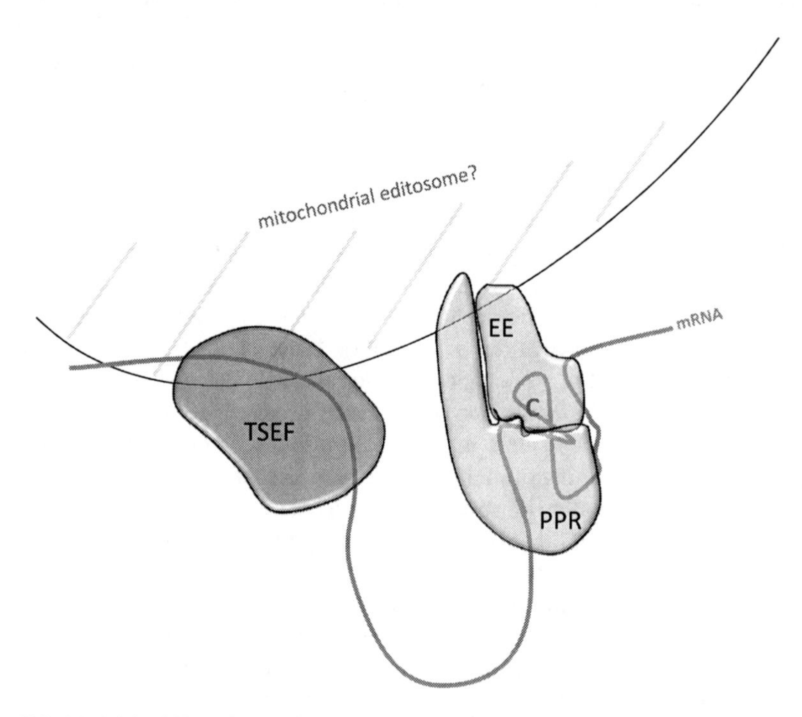

Fig. 7.2 Model for RNA editing. *C* cytosine to be edited, *EE* (unknown) editing enzyme, *PPR*, pentatricopeptide repeat protein as described by Zehrmann et al. (2009), *TSEF* transcript-specific editing factor based on work on Sorghum *atp6* mRNA in maize mitochondrial *in organello* approach (Staudinger et al. 2005). A secondary or tertiary RNA structure is proposed for the recognition of editing sites (According to Bolle and Kempken 2006)

thus, the editing information may be contained in the transcript itself, a complex secondary or tertiary structure that may support proper PPR protein binding. In this scenario, there would be no need for strong conservation of site-specific *trans*-acting factors because the ability to edit foreign editing sites would be based on the structural formation of the transcript. Following this hypothesis, RNA sequences extremely different from one another may also form similar structures recognized by a limited number of *trans*-factors. This concept is included in the provided model for RNA editing (see Fig. 7.2).

7.4 Evolution of RNA Editing

As shown in Table 7.1, many organisms employ complex and energy demanding systems to edit RNA sequences. The apparent question raised by this situation is why such complex systems exist instead of relying on DNA-based genetic information, as is the case in most organisms. Of course this question has no simple answer. One obvious function of RNA editing is to provide an additional level of gene expression regulation, and indeed there are examples of this kind. A single editing site in the mRNA of the apolipoprotein B gene (Box 7.1) leads to the formation of a stop codon in the coding region. Consequently, two forms of apolipoprotein, 48 kDa (apo-B48) and 100 kDa (apo-B100), are generated in the intestine. Another well characterized example of this kind is represented by A-to-I editing in the human brain. In contrast, higher plant mitochondrial RNA editing does not seem to generate different polypeptides, or at least that appears to be the general rule. As described in Sect. 7.2, there is only one example (*rps12*) of a heterologous population of differentially edited mRNAs being translated and a population of different proteins accumulating (Lu et al. 1996; Phreaner et al. 1996). Although the number of studies is limited, the *rps12* protein may be an exception to the general rule. Consequently, gene regulation does not appear to be the driving force for the establishment of RNA editing in plant mitochondria. Instead, Gray and Covello (1993) proposed a hypothesis for the establishment of plant mitochondrial RNA editing based on genetic drift. According to this hypothesis, RNA editing results from the correction of spontaneously occurring DNA mutations by the presence of enzymatic activity capable of correcting the DNA mutation on the RNA level. Later, this originally rather unspecific process was assumed to have been spread by genetic drift. The selective pressure would be provided by the correction of functionally essential coding information (Gray and Covello 1993).

A problem in this model is that we must assume that the editing activity is rather unspecific in the beginning, as specificity factors are most certainly a later edition. This assumption would imply that such editing activity would not only correct DNA mutations on the RNA level, but provide an additional source of mutation on the RNA level by editing at the wrong positions. Indeed, ectopic expression of the

Apobec-1 editing enzyme in mouse liver leads to an increase in liver carcinoma (Yamanaka et al. 1996), highlighting this problem.

However, Mulligan et al. (2007) provided a potential clue about this paradox. According to their analysis, editable cytidines are characterized by the presence of a short adjacent base pair motif (YYCGN) lacking cytidine residues that are never edited in plant mitochondria (Mulligan et al. 2007). This sequence motif might be the remnant of an ancient specific editing system that became established because it provided a selective force to reduce the number of undesired editing events. If RNA editing arose from the correction of otherwise lethal, or at least fitness reducing, mutations, the editing sites in present day mitochondrial transcriptomes should reflect this original situation. In fact, the deviation of amino acid sequences deduced for plant mitochondrial DNA from conserved animal and fungal mitochondrial protein sequences originally led to the analysis of the mitochondrial transcriptome and the discovery of plant mitochondrial RNA editing.

Recently, a number of conserved editing sites were found to be dispensable for plant fitness (Takenaka and Brennicke 2009). This truly surprising result questions the functional relevance of RNA editing, at least with respect to the correction of lethal mutations. This is not to say that none of the editing sites are essential. Some editing sites occur at positions that are known to be essential due to the presence of human mitochondrial mutations that cause severe diseases (Kempken et al. 1998) and, therefore, have to be assumed to be essential. When comparing all known human mitochondrial DNA mutations in the 13 mitochondrial protein coding genes to their plant mRNA counterparts, only four positions correlate between human mitochondrial DNA mutations and 140 known RNA editing sites, whereas approximately 81 known mutations in human mtDNA result in lost or reduced function (Kempken, unpublished data). The data from Takenaka and Brennicke (2009), together with the comparison described here, strongly suggest that only a few editing sites are truly functionally essential. Yet, there are hundreds of editing sites in higher plants and even more in primitive plant species. Many of these editing sites are even conserved, at least among mono- or dicot species.

Is there a solution to this complex question? A combination of two scenarios is possible. At the beginning, there were a few functionally relevant mutations in a sequence context that allowed for RNA editing. Subsequently, editing was genetically fixed during evolution and additional specificity factors (PPR proteins) were acquired. The reduction in unspecific editing sites may have served as a driving force for acquiring these additional factors. Later, the number of PPR proteins increased. A recent study (O'Toole et al. 2008) found that there is evidence for one or more waves of retrotransposition leading to an increase in PPR genes in flowering plants. Moreover, the number of PPR proteins correlates with differences in RNA editing between three species (O'Toole et al. 2008). One may assume that mutations, possibly even due to the process of reverse transcription, led to modified binding properties of the amplified PPR proteins. Consequently, the number of editing sites may have changed slowly as only those PPR mutations persisted with beneficial editing properties. In this scenario, conservation of editing sites would occur, even if not all of them are truly essential.

7.5 Concluding Remarks

Due to the establishment of *in vitro* and *in organello* systems, and the use of ecotype-specific differences in RNA editing, we have already identified important clues toward establishing a model of plant mitochondrial RNA editing. Most importantly, PPR proteins have been established as being required for editing specific sites. However, there are still a number of open questions, among which are the following:

- So far, the sequence-specificity of PPR proteins does not appear to be sufficient for editing site recognition. How is specific editing site binding achieved?
- The enzymatic activity that actually converts cytidine to uracil has not yet been identified. Is it a deaminase or transaminase? Is there a specific enzyme for each organelle, or do plastids and mitochondria share the same enzyme?
- How the translational machinery avoids translation of unedited or partially edited transcripts, or avoids the accumulation of proteins from unedited or partially edited transcripts, is not known. What is the regulatory mechanism?

Hopefully, one or all of these questions will soon be addressed.

Acknowledgments The laboratory work of the authors is funded by the German Research Association (D.F.G.). We thank Mrs. Stefanie Joanna Bollmann for help with comparing plant editing sites and human mitochondrial mutations.

Glossary

Biotinylated UTP: A modified nucleotide carrying a biotin group, which allows for the specific isolation of RNAs using streptavidin.

Cytoplasmic male sterility: A condition that leads to the inability to produce fertile pollen.

Editosome: A complex of proteins and RNA molecules able to edit RNA molecules; so far only described in trypanosomes.

Genetic drift: The change in the relative frequency at which a gene variant occurs in a population due to random sampling and chance.

RNA editing: A process that changes the nucleotide sequence of a transcript compared to its DNA template.

Silent RNA editing: An editing site at a third position in a codon that does not lead to a change in the amino acid sequence.

Transcriptome: Set of all RNA molecules produced in an organelle, cell, or organism.

References

Bass, B. L., Weintraub, H. 1988. An unwinding activity that covalently modifies its double-stranded RNA substrate. Cell 55:1089–1098.

Benne, R. 1994. RNA editing in trypanosomes. Eur. J. Biochem. 221:9–23.

Benne, R. 1996. RNA editing. The long and the short of it. Nature 380:391–392.

Benne, R., van den Burg, J., Brakenhoff, J. P. J., Sloof, P., van Boom, J. H., Tromp, M. C. 1986. Major transcript of the frameshifted *coxII* gene from trypanosome mitochondria contains four nucleotides that are not encoded in the DNA. Cell 46:819–826.

Binder, S., Marchfelder, A., Brennicke, A., Wissinger, B. 1992. RNA editing in trans-splicing intron sequences of *nad2* mRNAs in Oenothera mitochondria. J. Biol. Chem. 267:7615–7623.

Binder, S., Marchfeld, A., Brennicke, A. 1994. RNA editing of tRNAPhe and tRNACys in mitochondria of *Oenothera berteriana* is initiated in precursor molecules. Mol. Gen. Genet. 244:67–74.

Blum, B., Simpson, L. 1990. Guide RNAs in kinetoplastid mitochondria have a nonencoded 3′ Oligo (U)-tail involved in recognition of the preedited region. Cell 62:391–397.

Bock, R., Hermann, M., Kössel, H. 1996. *In vivo* dissection of *cis*-acting determinants for plastid RNA editing. EMBO J. 15:5052–5059.

Bolle, N., Kempken, F. 2006. Mono- and dicotyledonous plant-specific RNA editing sites are correctly edited in both in organello systems. FEBS Lett. 580:4443–4448.

Bolle, N., Hinrichsen, I., Kempken, F. 2007. Plastid mRNAs are neither spliced nor edited in maize and cauliflower mitochondrial in organello systems. RNA 13:2061–2065.

Börner, G. V., Mörl, M., Wissinger, B., Brennicke, A., Schmelzer, C. 1995. RNA editing of a group II intron in Oenothera as a prerequisite for splicing. Mol. Gen. Genet. 246:739–744.

Brennicke, A., Marchfelder, A., Binder, S. 1999. RNA editing. FEMS Microbiol. Rev. 23:297–316.

Cattaneo, R., Kaelin, K., Baczko, K., Billeter, M. A. 1989. Measles virus editing provides an additional cysteine rich protein. Cell 56:759–764.

Chapdelaine, Y., Bonen, L. 1991. The wheat mitochondrial gene for subunit *I* of the NADH dehydrogenase complex: a trans-splicing model for this gene-in-pieces. Cell 65:465–472.

Chateigner-Boutin, A.-L., Ramos-Vega, M., Guevara-García, A., Andrés, C., de la Luz Gutiérrez-Nava, M., Cantero, A., Delannoy, E., Jiménez, L. F., Lurin, C., Small, I., León, P. 2008. CLB19, a pentatricopeptide repeat protein required for editing of rpoA and clpP chloroplast transcripts. Plant J. 56:590–602.

Chaudhuri, S., Maliga, P. 1996. Sequences directing C to U editing of the plastid *psbL* mRNA are located within a 22 nucleotide segment spanning the editing site. EMBO J. 15:5958–5964.

Chaw, S. M., Shih, A. C., Wang, D., Wu, Y. W., Liu, S. M., Chou, T. Y. 2008. The mitochondrial genome of the gymnosperm *Cycas taitungensis* contains a novel family of short interspersed elements, Bpu sequences, and abundant RNA editing sites. Mol. Biol. Evol. 25:603–615.

Chen, S. H., Habib, G., Yang, C. Y., Gu, Z. W., Lee, B. R., Weng, S. A., Silbermann, S. R., Cai, S. J., Deslypere, J. P., Rosseneu, M., Gotto, A. M. Jr, Li, W. H., Chan, L. 1987. Apolipoprotein B-48 is the product of messenger RNA with an organ-specific in-frame translation codon. Science 328:363–366.

Choury, D., Farre, J. C., Jordana, X., Araya, A. 2004. Different patterns in the recognition of editing sites in plant mitochondria. Nucleic Acids Res. 32:6397–6406.

Choury, D., Farré, J. C., Jordana, X., Araya, A. 2005. Gene expression studies in isolated mitochondria: *Solanum tuberosum rps10* is recognized by cognate potato but not by the transcription, splicing and editing machinery of wheat mitochondria. Nucleic Acids Res. 33:7058–7065.

Clifton, S. W., Minx, P., Fauron, C. M., Gibson, M., Allen, J. O., Sun, H., Thompson, M., Barbazuk, W. B., Kanuganti, S., Tayloe, C., Meyer, L., Wilson, R. K., Newton, K. J. 2004. Sequence and comparative analysis of the maize NB mitochondrial genome. Plant Physiol. 136:3486–3503.

Covello, P. S., Gray, M. W. 1989. RNA editing in plant mitochondria. Nature 341:662–666.

Dance, G. S., Sowden, M. P., Cartegni, L., Cooper, E., Krainer, A. R., Smith, H. C. 2002. Two proteins essential for apolipoprotein B mRNA editing are expressed from a single gene through alternative splicing. J. Biol. Chem. 277:12703–12709.

Delannoy, E., Stanley, W. A., Bond, C. S., Small, I. D. 2007. Pentatricopeptide repeat (PPR) proteins as sequence-specificity factors in post-transcriptional processes in organelles. Biochem. Soc. Trans. 35:1643–1647.

Doniwa, Y., Ueda, M., Ueta, M., Wada, A., Kadowaki, K. I., Tsutsumi, N. 2010. The involvement of a PPR protein of the P subfamily in partial RNA editing of an Arabidopsis mitochondrial transcript. Gene 454(1–2):39–46.

Farré, J.-C., Araya, A. 2001. Gene expression in isolated plant mitochondria: high fidelity of transcription, splicing and editing of a transgene product in electroporated organelles. Nucleic Acids Res. 29:2484–2491.

Farré, J. C., Leon, G., Jordana, X., Araya, A. 2001. *cis* recognition elements in plant mitochondrion RNA editing. Mol. Cell Biol. 21:6731–6737.

Gallagher, L. J., Betz, S. K., Chase, C. D. 2002. Mitochondrial RNA editing truncates a chimeric open reading frame associated with S male-sterility in maize. Curr. Genet. 42:179–184.

Giege, P., Brennicke, A. 1999. RNA editing in Arabidopsis mitochondria effects 441 C to U changes in ORFs. Proc. Natl. Acad. Sci. U.S.A. 96:15324–15329.

Gray, M. W., Covello, P. S. 1993. RNA editing in plant mitochondria and chloroplasts. FASEB J. 7:64–71.

Grewe, F., Viehoever, P., Weisshaar, B., Knoop, V. 2009. A trans-splicing group I intron and tRNA-hyperediting in the mitochondrial genome of the lycophyte *Isoetes engelmannii*. Nucleic Acids Res. 37:5093–5104.

Grohmann, L., Thieck, O., Herz, U., Schröder, W., Brennicke, A. 1994. Translation of *nad9* mRNAs in mitochondria from *Solanum tuberosum* is restricted to completely edited transcripts. Nucleic Acids Res. 22:3304–3311.

Gualberto, J. M., Lamattina, L., Bonnard, G., Weil, J. H., Grienenberger, J. M. 1989. RNA editing in wheat mitochondria results in the conservation of protein sequences. Nature 341:660–662.

Handa, H. 2003. The complete nucleotide sequence and RNA editing content of the mitochondrial genome of rapeseed (*Brassica napus* L.): comparative analysis of the mitochondrial genomes of rapeseed and *Arabidopsis thaliana*. Nucleic Acids Res. 31:5907–5916.

Hayes, M. L., Reed, M. L., Hegemann, C. E., Hanson, M. R. 2006. Sequence elements critical for efficient RNA editing of a tobacco chloroplast transcript *in vivo* and *in vitro*. Nucleic Acids Res. 34:3742–3754.

Hiesel, R., Wissinger, B., Schuster, W., Brennicke, A. 1989. RNA editing in plant mitochondria. Science 246:1632–1634.

Hiesel, R., Combettes, B., Brennicke, A. 1994. Evidence for RNA editing in mitochondria of all major groups of land plants except the Bryophyta. Proc. Natl. Acad. Sci. U.S.A. 91:629–633.

Hinrichsen, I., Bolle, N., Paun, L., Kempken, F. 2009. RNA processing in plant mitochondria is independent of transcription. Plant Mol. Biol. 70:663–668.

Hirose, T., Sugiura, M. 2001. Involvement of a site-specific *trans*-acting factor and a common RNA-binding protein in the editing of chloroplast mRNAs: development of a chloroplast *in vitro* RNA editing system. EMBO J. 20:1144–1152.

Hoch, B., Maier, R. M., Appel, K., Igloi, G. L., Kössel, H. 1991. Editing of a chloroplast mRNA by creation of an initiation codon. Nature 353:178–180.

Holec, S., Lange, H., Canaday, J., Gagliardi, D. 2008. Coping with cryptic and defective transcripts in plant mitochondria. Biochim. Biophys. Acta 1779:566–573.

Howad, W., Kempken, F. 1997. Cell-type specific loss of *atp6* RNA editing in cytoplasmic male sterile *Sorghum bicolor*. Proc. Natl. Acad. Sci. U.S.A. 94:11090–11095.

Howad, W., Tang, H. V., Pring, D. R., Kempken, F. 1999. Nuclear genes from Tx CMS maintainer lines are unable to maintain *atp6* RNA editing in any anther cell-type in the *Sorghum bicolor* A3 cytoplasm. Curr. Genet. 36:62–68.

Jacques, J. P., Hausmann, S., Kolakofsky, D. 1994. Paramyxovirus mRNA editing leads to G deletions as well as insertions. EMBO J. 13:5496–5503.

Kempken, F., Mullen, J. A., Pring, D. R., Tang, H. V. 1991. RNA editing of sorghum mitochondrial *atp6* transcripts changes 15 amino acids and generates a carboxy-terminus identical to yeast. Curr. Genet. 20:417–422.

Kempken, F., Höfken, G., Pring, D. R. 1995. Analysis of silent RNA editing sites in *atp6* transcripts of *Sorghum bicolor*. Curr. Genet. 27:555–558.

Kempken, F., Howad, W., Pring, D. R. 1998. Mutations at specific *atp6* codons which cause human mitochondrial diseases also lead to male sterility in a plant. FEBS Lett. 441:159–160.

Kim, S.-R., Yang, J.-I., Moon, S., Ryu, C.-H., An, K., Yim, J., An, G. 2009. Rice OGR1 encodes a pentatricopeptide repeat-DYW protein and is essential for RNA editing in mitochondria. Plant J. 59:738–749.

Kobayashi, Y., Matsuo, M., Sakamoto, K., Wakasugi, T., Yamada, K., Obokata, J. 2008. Two RNA editing sites with cis-acting elements of moderate sequence identity are recognized by an identical site-recognition protein in tobacco chloroplasts. Nucleic Acids Res. 36:311–318.

Kotera, E., Tasaka, M., Shikanai, T. 2005. A pentatricopeptide repeat protein is essential for RNA editing in chloroplasts. Nature 433:326–330.

Kubo, T., Nishizawa, S., Sugawara, A., itchoda, N., Estiati, N., Mikami, T. 2000. The complete nucleotide sequence of the mitochondrial genome of sugar beet (*Beta vulgaris* L.) reveals a novel gene for tRNA (Cys) (GCA). Nucleic Acids Res. 28:2571–2576.

Kugita, M., Yamamoto, Y., Fujikawa, T., Matsumoto, T., Yoshinaga, K. 2003. RNA editing in hornwort chloroplasts makes more than half the genes functional. Nucleic Acids Res. 31:2417–2423.

Lau, P. P., Xiong, W. J., Zhu, H. J., Chen, S. H., Chan, L. 1991. Apolipoprotein B mRNA editing is an intranuclear event that occurs posttranscriptionally coincident with splicing and polyadenylation. J. Biol. Chem. 266:20550–20554.

Lenz, H., Rüdinger, M., Volkmar, U., Fischer, S., Herres, S., Grewe, F., Knoop, V. 2009. Introducing the plant RNA editing prediction and analysis computer tool PREPACT and an update on RNA editing site nomenclature. Curr. Genet. 56(2):189–201. Epub ahead of print.

Li, F., Ge, P., Hui, W. H., Atanasov, I., Rogers, K., Guo, Q., Osato, D., Falick, A. M., Zhou, Z. H., Simpson, L. 2009 Structure of the core editing complex (L-complex) involved in uridine insertion/deletion RNA editing in trypanosomatid mitochondria. Proc. Natl. Acad. Sci. U.S.A. 106:12306–12310.

Lu, B., Hanson, M. R. 1994. A single homogeneous form of ATP6 protein accumulates in Petunia mitochondria despite the presence of differentially edited *atp6* transcripts. Plant Cell 6:1955–1968.

Lu, B., Wilson, R. K., Phreaner, C. G., Mulligan, R. M., Hanson, M. R. 1996. Protein polymorphism generated by differential RNA editing of a plant mitochondrial *rps12* gene. Mol. Cell Biol. 16:1543–1549.

Lurin, C., Andrés, C., Aubourg, S., Bellaoui, M., Bitton, F., Bruyère, C., Caboche, M., Debast, C., Gualberto, J., Hoffmann, B., Lecharny, A., Le Ret, M., Martin-Magniette, M.-L., Mireau, H., Peeters, N., Renou, J.-P., Szurek, B., Taconnat, L., Small, I. 2004. Genome-wide analysis of Arabidopsis pentatricopeptide repeat proteins reveals their essential role in organelle biogenesis. Plant Cell 16:2089–2103.

Madison-Antenucci, S., Grams, J., Hajduk, S. L. 2002. Editing machines: the complexities of trypanosome RNA editing. Cell 108:435–438.

Mahendran, R., Spottswood, M. S., Miller, D. L. 1991. RNA editing by cytidine insertion in mitochondria of *Physarum polycephalum*. Nature 349:434–438.

Mahendran, R., Spottswood, M. S., Ghate, A., Ling, M., Jeng, K., Miller, D. L. 1994. Editing of the mitochondrial small subunit rRNA in *Physarum polycephalum*. EMBO J. 13:232–240.

Maier, R. M., Zeltz, P., Kössel, H., Bonnard, G., Gualberto, J. M., Grienenberger, J. M. 1996. RNA editing in plant mitochondria and chloroplasts. Plant Mol. Biol. 32:343–365.

Maréchal-Drouard, L., Ramamonjisoa, D., Cosset, A., Weil, J. H., Dietrich, A. 1993. Editing corrects mispairing in the acceptor stem of bean and potato mitochondrial phenylalanine transfer RNA. Nucleic Acids Res. 21:4909–4914.

Maréchal-Drouard, L., Kumar, R., Remacle, C., Small, I. 1996. RNA editing of larch mitochondrial $tRNA^{HIS}$ precursors is a prerequisite for processing. Nucleic Acids Res. 24:3229–3234.

Miyamoto, T., Obokata, J., Sugiura, M. 2002. Recognition of RNA editing sites is directed by unique proteins in chloroplasts: biochemical identification of *cis*-acting elements and *trans*-acting factors involved in RNA editing in tobacco and pea chloroplasts. Mol. Cell Biol. 22:6726–6734.

Mulligan, R. M., Chang, K. L., Chou, C. C. 2007. Computational analysis of RNA editing sites in plant mitochondrial genomes reveals similar information content and a sporadic distribution of editing sites. Mol. Biol. Evol. 24:1971–1981.

Muramatsu, M., Kinoshita, K., Fagarasan, S., Yamada, S., Shinkai, Y., Honjo, T. 2000. Class switch recombination and hypermutation require activation-induced cytidine deaminase (AID), a potential RNA editing enzyme. Cell 102:553–563.

Nakajima, Y., Mulligan, R. M. 2005. Nucleotide specificity of the RNA editing reaction in pea chloroplasts. J Plant Physiol. 162:1347–1354.

Neuwirt, J., Takenaka, M., van der Merwe, J. A., Brennicke, A. 2005. An *in vitro* RNA editing system from cauliflower mitochondria: editing site recognition parameters can vary in different plant species. RNA 11:1563–1570.

Notsu, Y., Masood, S., Nishikawa, T., Kubo, N., Akiduki, G., Nakazone, M., Hirai, A., HKadowaki, K. 2002. The complete sequence of the rice (*Oryza sativa* L.) mitochondrial genome: frequent DNA sequence acquisition and loss during the evolution of flowering plants. Mol. Gen. Genom. 268:434–445.

Okuda, K., Nakamura, T., Sugita, M., Shimizu, T., Shikanai, T. 2006. A pentatricopeptide repeat protein is a site recognition factor in chloroplast RNA editing. J. Biol. Chem. 281:37661–37667.

Okuda, K., Myouga, F., Motohashi, R., Shinozaki, K., Shikanai, T. 2007. Conserved domain structure of pentatricopeptide repeat proteins involved in chloroplast RNA editing. Proc. Natl. Acad. Sci. U.S.A. 104:8178–8183.

O'Toole, N., Hattori, M., Andres, C., Iida, K., Lurin, C., Schmitz-Linneweber, C., Sugita, M., Small, I. 2008. On the expansion of the pentatricopeptide repeat gene family in plants. Mol. Biol. Evol. 25:1120–1128.

Phreaner, C. G., Williams, M. A., Mulligan, R. M. 1996. Incomplete editing of *rps12* transcripts results in the synthesis of polymorphic polypeptides in plant mitochondria. Plant Cell 8:107–117.

Polson, A. G., Crain, P. F., Pomerantz, S. C., McCloskey, J. A., Bass, B. L. 1991. The mechanism of adenosine to inosine conversion by the double-stranded RNA unwinding/modifying activity: a high-performance liquid chromatography-mass spectrometry analysis. Biochemistry 30:11507–11514.

Powell, L. M., Willis, S. C., Pease, R. J., Edwards, Y. H., Knott, T. J., Scott, J. 1987. A novel form of tissue-specific RNA processing produces apolipoprotein-B48 in intestine. Cell 50:831–840.

Pring, D. R., Tang, H. V. 2001. Mitochondrial *atp6* transcript editing during microgametogenesis in male-sterile sorghum. Curr. Genet. 39:371–376.

Pring, D. R., Brennicke, A., Schuster, W. 1993. RNA editing gives a new meaning to the genetic information in mitochondria and chloroplasts. Plant Mol. Biol. 21:1163–1170.

Pring, D. R., Tang, H. V., Howad, W., Kempken, F. 1999. A unique two-gene gametophytic male sterility system in sorghum involving a possible role of RNA editing in fertility restoration. J Hered 90:386–393.

Rajasekhar, V. K., Mulligan, R. M. 1993. RNA editing in plant mitochondria: a-phosphate is retained during C-to-U conversion in mRNAs. Plant Cell 5:1843–1852.

Reed, M. L., Peeters, N. M., Hanson, M. R. 2001. A single alteration 20 nt 5′ to an editing target inhibits chloroplast RNA editing *in vivo*. Nucleic Acids Res. 29:1507–1513.

Rüdinger, M., Funk, H. T., Rensing, S. A., Maier, U. G., Knoop, V. 2009. RNA editing: only eleven sites are present in the *Physcomitrella patens* mitochondrial transcriptome and a universal nomenclature proposal. Mol. Gen. Genom. 281:473–481.

Salone, V., Rüdinger, M., Polsakiewicz, M., Hoffmann, B., Groth-Malonek, M., Szurek, B., Small, I., Knoop, V., Lurin, C. 2007. A hypothesis on the identification of the editing enzyme in plant organelles. FEBS Lett. 581:4132–4138.

Sasaki, T., Yukawa, Y., Wakasugi, T., Yamada, K., Sugiura, M. 2006. A simple *in vitro* RNA editing assay for chloroplast transcripts using fluorescent dideoxynucleotides: distinct types of sequence elements required for editing of ndh transcripts. Plant J. 47:802–810.

Satoh, M., Kubo, T., Mikami, T. 2006. The Owen mitochondrial genome in sugar beet (*Beta vulgaris* L.): possible mechanisms of extensive rearrangements and the origin of the mitotype-unique regions. Theor Appl Genet 113:477–484.

Schmitz-Linneweber, C., Small, I. 2008. Pentatricopeptide repeat proteins: a socket set for organelle gene expression. Trends Plant Sci. 13:663–670.

Schuster, W., Wissinger, B., Unseld, M., Brennicke, A. 1990. Transcripts of the NADH-dehydrogenase subunit *3* gene are differentially edited in Oenothera mitochondria. EMBO J. 9:263–269.

Shikanai, T. 2006. RNA editing in plant organelles: machinery, physiological function and evolution. Cell Mol. Life Sci. 63:698–708.

Simpson, L., Sbicego, S., Aphasizhev, R. 2003. Uridine insertion/deletion RNA editing in trypanosome mitochondria: a complex business. RNA 9:265–276.

Small, I. D., Peeters, N. 2000. The PPR motif – a TPR-related motif prevalent in plant organellar proteins. Trends Biochem. Sci. 25:46–47.

Smith, H. C., Sowden, M. P. 1996. Base-modification mRNA editing through deamination-the good, the bad and the unregulated. Trends Genet. 12:418–424.

Staudinger, M., Kempken, F. 2003. Electroporation of isolated higher-plant mitochondria: transcripts of an introduced *cox2* gene, but not an *atp6* gene, are edited in organello. Mol. Gen. Genom. 269:553–561.

Staudinger, M., Kempken, F. 2004. In organello editing of mitochondrial *atp9*, *cox2*, and *nad9* transcripts. Endocytobiosis Cell Res. 15:551–560.

Staudinger, M., Bolle, N., Kempken, F. 2005. Mitochondrial electroporation and in organello RNA editing of chimeric *atp6* transcripts. Mol. Gen. Genom. 273:130–136.

Stuart, K. D., Schnaufer, A., Ernst, N. L., Panigrahi, A. K. 2005. Complex management: RNA editing in trypanosomes. Trends Biochem. Sci. 30:97–105.

Sugiyama, Y., Watase, Y., Nagase, M., Makita, N., Yagura, S., Hirai, A., Sugiura, M. 2005. The complete nucleotide sequence and multipartite organization of the tobacco mitochondrial genome: comparative analysis of mitochondrial genomes in higher plants. Mol. Gen. Genom. 272:603–615.

Sutton, C. A., Conklin, P. L., Pruitt, K. D., Hanson, M. R. 1991. Editing of pre-mRNAs can occur before *cis*- and *trans*-splicing in Petunia mitochondria. Mol. Cell Biol. 11:4274–4277.

Sutton, C. A., Zoubenko, O. V., Hanson, M. R., Maliga, P. 1995. A plant mitochondrial sequence transcribed in transgenic tobacco chloroplasts is not edited. Mol. Cell Biol. 15:1377–1381.

Takenaka, M. 2009. MEF9, an E subclass PPR protein, is required for an RNA editing event in the nad7 transcript in mitochondria of Arabidopsis thaliana. Plant Physiol. 109:151–175.

Takenaka, M., Brennicke, A. 2003. *In vitro* RNA editing in pea mitochondria requires NTPor dNTP, suggesting involvement of an RNA helicase. J. Biol. Chem. 278:47526–47533.

Takenaka, M., Brennicke, A. 2009. Multiplex single-base extension typing to identify nuclear genes required for RNA editing in plant organelles. Nucleic Acids Res. 37:e13.

Takenaka, M., Neuwirt, J., Brennicke, A. 2004. Complex *cis*-elements determine an RNA editing site in pea mitochondria. Nucleic Acids Res. 32:4137–4144.

Takenaka, M., Verbitskiy, D., van der Merwe, J. A., Zehrmann, A., Plessmann, U., Urlaub, H., Brennicke, A. 2007. *In vitro* RNA editing in plant mitochondria does not require added energy. FEBS Lett. 581:2743–2747.

Takenaka, M., Verbitskiy, D., van der Merwe, J. A., Zehrmann, A., Brennicke, A. 2008. The process of RNA editing in plant mitochondria. Mitochondrion 8:35–46.

Teng, B. B., Burant, C. F., Davidson, N. O. 1993. Molecular cloning of an apolipoprotein B messenger RNA editing protein. Science 260:1816–1818.

Thomas, S. M., Lamb, R. A., Paterson, R. G. 1988. Two mRNAs that differ by two nontemplated nucleotides encode the amino coterminal proteins P and V of the paramyxovirus SV5. Cell 54:891–902.

Unseld, M., Marienfeld, J. R., Brandt, P., Brennicke, A. 1997. The mitochondrial genome of *Arabidopsis thaliana* contains 57 genes in 366,924 nucleotides. Nat. Genet. 15:57–61.

van der Merwe, J. A., Takenaka, M., Neuwirt, J., Verbitskiy, D., Brennicke, A. 2006. RNA editing sites in plant mitochondria can share cis-elements. FEBS Lett. 580:268–272.

Verbitskiy, D., Takenaka, M., Neuwirt, J., van der Merwe, J. A., Brennicke, A. 2006. Partially edited RNAs are intermediates of RNA editing in plant mitochondria. Plant J 47:408–416.

Verbitskiy, D., van der Merwe, J. A., Zehrmann, A., Brennicke, A., Takenaka, M. 2008. Multiple specificity recognition motifs enhance plant mitochondrial RNA editing *in vitro*. J. Biol. Chem. 283:24374–24381.

Verbitskiy, D., Zehrmann, A., van der Merwe, J. A., Brennicke, A., Takenaka, M. 2009. The PPR protein encoded by the LOVASTATIN INSENSITIVE 1 gene is involved in RNA editing at three sites in mitochondria of Arabidopsis thaliana. Plant J. 61:446–455.

Wagner, R. W., Smith, J. E., Cooperman, B. S., Nishikura, K. 1989. A double-stranded RNA unwinding activity introduces structural alterations by means of adenosine to inosine conversions in mammalian cells and Xenopus eggs. Proc. Natl. Acad. Sci. U.S.A. 86:2647–2651.

Ward, G. C., Levings, C. S., III 1991. The protein-encoding gene T-*urf13* is not edited in maize mitochondria. Plant Mol. Biol. 17:1083–1088.

Wedekind, J. E., Dance, G. S., Sowden, M. P., Smith, H. C. 2003. Messenger RNA editing in mammals: new members of the APOBEC family seeking roles in the family business. Trends Genet. 19:207–216.

Wintz, H., Hanson, M. R. 1991. A termination codon is created by RNA editing in the petunia mitochonrial *atp9* gene transcript. Curr. Genet. 19:61–64.

Yamanaka, S., Poksay, K. S., Driscoll, D. M., Innerarity, T. L. 1996. Hyperediting of multiple cytidines of apolipoprotein B mRNA by APOBEC-1 requires auxiliary protein(s) but not a mooring sequence motif. J. Biol. Chem. 271:11506–11510.

Yang, A. J., Mulligan, R. M. 1991. RNA editing intermediates of *cox2* transcripts in maize mitochondria. Mol. Cell Biol. 11:4278–4281.

Yu, W., Schuster, W. 1995. Evidence for a site-specific cytidine deamination reaction involved in C-to-U RNA editing of plant mitochondria. J. Biol. Chem. 270:18227–18233.

Zehrmann, A., Verbitskiy, D., van der Merwe, J. A., Brennicke, A., Takenaka, M. 2009. A DYW domain-containing pentatricopeptide repeat protein is required for RNA editing at multiple sites in mitochondria of Arabidopsis thaliana. Plant Cell 21:558–567.

Zeng, W. H., Liao, S. C., Chang, C. C. 2007. Identification of RNA editing sites in chloroplast transcripts of *Phalaenopsis aphrodite* and comparative analysis with those of other seed plants. Plant Cell Physiol. 48:362–368.

Chapter 8
RNA-Binding Proteins Required for Chloroplast RNA Processing

Reimo Zoschke, Christiane Kupsch, and Christian Schmitz-Linneweber

Abstract Chloroplasts and mitochondria both evolved from endosymbionts and therefore share many physiologic, biochemical, and genetic features. This chapter reviews a selection of chloroplast RNA-binding proteins, specifically those from land plants that meet the following criteria: (1) the encoding gene is known; (2) an association with RNA has been demonstrated; and (3) the protein does not have any (known) catalytic activity. This set of RNA-binding proteins predominantly consists of members from five protein families: the pentatricopeptide repeat (PPR) protein family; the chloroplast ribonucleoprotein (cpRNP) family; the chloroplast ribosome maturation (CRM) and RNA splicing protein family; the Whirly (Why) protein family; and the plant organelle RNA recognition (PORR) protein family. The functions of representative members of these families are summarized, and their protein–RNA interactions are discussed.

Keywords Chloroplast • RNA binding • RNA-protein interaction • RNA processing • Pentatricopetide • Ribonucleoprotein

8.1 Introduction

The mitochondria and chloroplasts of land plants share a number of unusual features that distinguish them from other organelles. They contain their own genome; they are capable of protein production; and they both reproduce by binary fission. In addition, both organelles boast a tremendously complex RNA metabolism that sets them apart from their bacterial ancestors as well as animal mitochondria. Their primary transcripts, which are polycistronic more often in chloroplasts than in plant mitochondria, are heavily processed (Barkan and Goldschmidt-Clermont 2000). Most undergo endonucleolytic cleavage, splicing, single-base editing, exonucleolytic trimming, and the 3′-terminal addition of ribonucleotides.

C. Schmitz-Linneweber (✉)
Institute of Biology, Humboldt University of Berlin, Chausseestr. 117, 10115 Berlin, Germany
e-mail: smitzlic@rz.hu-berlin.de

F. Kempken (ed.), *Plant Mitochondria*, Advances in Plant Biology 1,
DOI 10.1007/978-0-387-89781-3_8, © Springer Science+Business Media, LLC 2011

Descriptive studies on the individual processing steps began in earnest in the early 1980s (e.g., Strittmatter and Kossel 1984). Although this work is far from complete, we currently have a good estimate of the number of RNA processing events necessary for selected chloroplast and mitochondrial operons. This large number of processing steps in exemplary operons seems to accurately represent the overall complexity of organellar RNA metabolism (for example, the *psbB* operon; Barkan 1988; Westhoff and Hermann 1988).

Over the past 15 years, the field's focus has shifted from descriptive studies on organellar RNA processing toward efforts aimed at identifying the underlying processing machinery. Naturally, this machinery includes RNA-binding proteins, which critically function to: (1) confer specificity to individual processing events (i.e., through the recognition of specific RNA sequence elements); and (2) support processing steps by altering or stabilizing RNA conformations. At present, however, the mechanistic and structural details underlying the actions of individual organellar RNA-binding proteins are largely unknown. In addition, it is not yet clear whether organellar RNA processing is a critical regulator of gene expression, or whether it simply represents some particularly weird excess of evolutionary tinkering (Maier et al. 2008). In this chapter, we do not attempt to add to the discussion on the "why" behind organellar RNA processing. Rather, we try to summarize the classes of RNA-binding proteins that have been identified in chloroplasts and review selected individual family members in terms of their function(s) in the maturation of their RNA ligands. We restrict our discussion to RNA-binding proteins without any (known) catalytic activity, and consider ribonucleases, the ribosome, and polymerases to be beyond the scope of this chapter. Furthermore, we exclusively discuss proteins from angiosperms, and do not address more distantly related photosynthetic organisms (e.g., green algae). Finally, we discuss only proteins for which the underlying genes have been identified, and for which associations with RNA have been supported by direct experimental evidence (i.e., beyond mere genetic defects in organellar RNA metabolism in corresponding mutants). The RNA-binding protein families presented herein are encoded in the nucleus and have some members that localize to the mitochondrion, while others localize to the chloroplast. Since many more RNA-binding proteins have been analyzed to date from chloroplasts than from mitochondria, we herein focus on the chloroplast RNA-binding proteins (summarized in Table 8.1). Given the similarities in the endosymbiotic evolutionary backgrounds and expression systems of chloroplasts and plant mitochondria; however, we believe that information gained from chloroplast RNA-binding proteins could be instrumental to improve our understanding of mitochondrial RNA-binding proteins.

8.2 The Pentatricopeptide Repeat Proteins

The pentatricopeptide repeat (PPR) protein family as a whole was discovered relatively late, especially given that it is a remarkably large protein family comprising more than 450 members in *Arabidopsis* (Aubourg et al. 2000; Small and Peeters 2000; Lurin et al. 2004; Rivals et al. 2006). Prior to the identification of this family,

Table 8.1 Chloroplast RNA-binding protein families with members that have been shown to associate with RNA

Protein family	Number[a]	Functions	Phylogenetic origin[b]	Selected members[c]	Association with RNA shown	
					In vivo[d]	*In vitro*
PPR **P**entatricopeptide **R**epeat	450	Stability (Beick et al. 2008)	Eukaryotic	ZmPPR5	RIP-Chip (Beick et al. 2008)	EMSA (Williams-Carrier et al. 2008)
		Stability and translation (Pfalz et al. 2009)		ZmPPR10	RIP-Chip (Pfalz et al. 2009)	EMSA (Pfalz et al. 2009)
		Splicing (Schmitz-Linneweber et al. 2006)		ZmPPR4	RIP-Chip (Schmitz-Linneweber et al. 2006)	–
		Processing (Meierhoff et al. 2003)		AtHCF152	–	UV-cross-link (Nakamura et al. 2003)
		Editing (Kotera et al. 2005)		AtCRR4	–	EMSA (Okuda et al. 2006)
		Translation and processing/stability (Fisk et al. 1999)		ZmCRP1	RIP-Chip (Schmitz-Linneweber et al. 2005)	EMSA (Williams-Carrier et al. 2008)
cpRNP **C**hloro**p**last **R**ibo**n**ucleo**p**rotein	9	Editing, stability (Nakamura et al. 2001; Tillich et al. 2009)	Eukaryotic, (Embryophyta)	NtCP31	Co-IP Northern (Nakamura et al. 1999)	RNA affinity-column (Li and Sugiura 1991)
		–		NtCP33	Co-IP Northern (Nakamura et al. 1999)	RNA affinity-column (Li and Sugiura 1991)
		Stability (Nakamura et al. 2001)		NrCP29	Co-IP Northern (Nakamura et al. 1999)	RNA affinity-column (Li and Sugiura 1991)
		3′ processing (Schuster and Gruissem 1991)		SoCP28	–	UV-cross-link (Lisitsky et al. 1995)
PORR **P**lant **O**rganelle **R**NA **R**ecognition	17	Splicing (Kroeger et al. 2009)	Eukaryotic, (Embryophyta)	ZmWTF	RIP-Chip (Kroeger et al. 2009)	EMSA (Kroeger et al. 2009)

(continued)

Table 8.1 (continued)

Protein family	Number[a]	Functions	Phylogenetic origin[b]	Selected members[c]	Association with RNA shown	
					In vivo[d]	*In vitro*
CRM Chloroplast RNA Splicing and Ribosome Maturation	16	Splicing (Ostheimer et al. 2003)	Prokaryotic	ZmCAF1	RIP-Chip (Schmitz-Linneweber et al. 2005)	–
		Splicing (Ostheimer et al. 2003)		ZmCAF2	Co-IP slot blot (Ostheimer et al. 2003)	–
		Splicing (Jenkins, et al. 1997)		ZmCRS1	RIP-Chip (CS and Barkan, unpublished)	Footprinting, filter binding, EMSA (Ostersetzer et al. 2005; Barkan et al. 2007; Keren et al. 2008)
		Splicing (Asakura and Barkan 2007)		ZmCFM2	RIP-Chip (Asakura and Barkan 2007)	–
		Splicing (Asakura et al. 2008)		ZmCFM3	RIP-Chip (Asakura et al. 2008)	–
CRM-associated factors	2	Splicing (Jenkins et al. 1997)	Prokaryotic	ZmCRS2		(Ostheimer et al. 2006)
		Splicing (Watkins et al. 2007)	Prokaryotic	ZmRNC1	RIP-Chip (Watkins et al. 2007)	Filter binding, EMSA (Watkins et al. 2007)
Whirly	1	Splicing, chromosome integrity (Prikryl et al. 2008; Marechal et al. 2009)	Eukaryotic (Viridiplantae)	ZmWhy1	RIP-Chip (Prikryl et al. 2008)	Filter binding, EMSA (Prikryl et al. 2008)

[a] The number of genes from *Arabidopsis thaliana* classified into each family is indicated

[b] The closest relative in eukaryotes or the ancestor of the relevant organelle (i.e., prokaryotes)

[c] The first two letters are the initials of the species names: *Zm*, *Zea mays*; *At*, *Arabidopsis thaliana*; *So*, *Spinacia oleracea*; *Nt*, *Nicotiana tabacum*

[d] If multiple methods were used to determine the RNA targets, the most comprehensive experiment is cited (e.g., an RIP-Chip experiment is cited over a Co-IP or slot-blot analysis)

individual PPR genes, such as *pet309* in yeast, *cya-5* in *Neurospora crassa*, and *crp1* in maize, had been described as having functions in specific steps during the modification and/or translation of organellar RNAs (Manthey and McEwen 1995; Coffin et al. 1997; Fisk et al. 1999). However, the family itself was not elucidated until after the publication of the complete *Arabidopsis* genome sequence (Aubourg et al. 2000; Small and Peeters 2000; Lurin et al. 2004).

8.2.1 Structure

The PPR proteins are characterized by a degenerate motif of 35 amino acids that can be repeated up to 30 times with in a single protein. The motif is related to the tetratricopeptide repeat (TPR) motif in prokaryotes (Box 8.1). Lurin et al. found that about

Box 8.1 The pentatricopeptide repeat (PPR) motif

The standard PPR motif found in all eukaryotes is 35 amino acids long. It usually appears in a tandem "P repeat" that may comprise up to 30 consecutive units (called P-repeat; see also Fig. 8.1 and Small and Peeters 2000). The motif is degenerate and belongs to a group of helical-hairpin repeat motifs that are characterized by two to three α-helical elements connected by short linkers. Other members of this motif group are the heat, Arm, Puf, PORR, and TPR repeats.

The PPR motif consists of two alpha-helical elements connected by a short linker. The crystal structures of related TPR motifs show that the two helices fold back onto each other in a U-like fashion. Stacking of consecutive TPR "Us" leads to the formation of a superstructure that itself is helical due to a pitch between adjacent repeat units (Das et al. 1998; Blatch and Lassle 1999; Wilson et al. 2005). It is believed that the PPR domains form a similar structure. Crucial between-motif differences are found in helix A, which forms the inner surface of the superhelix. Among the PPRs, helix A is typically rich in hydrophilic side chains and has a highly charged edge. This surface is believed to make contact with RNA, although it is not yet known whether this contact is restricted to bases or includes the phosphorous backbone (as suggested by the many charges in this area).

Land plants harbor a distinct type of PPR protein that contains not only the standard 35 amino acid repeat, but also shorter and a longer versions of the repeat (designated S and L repeats, respectively; Fig. 8.1). These are characteristically found to alternate as "PLS triplets." Subtypes of this class of PPR proteins contain additional C-terminal domains called E (extended) and DYW motifs, which currently do not have any function assigned to them, although both are often found in PPR proteins involved in RNA editing.

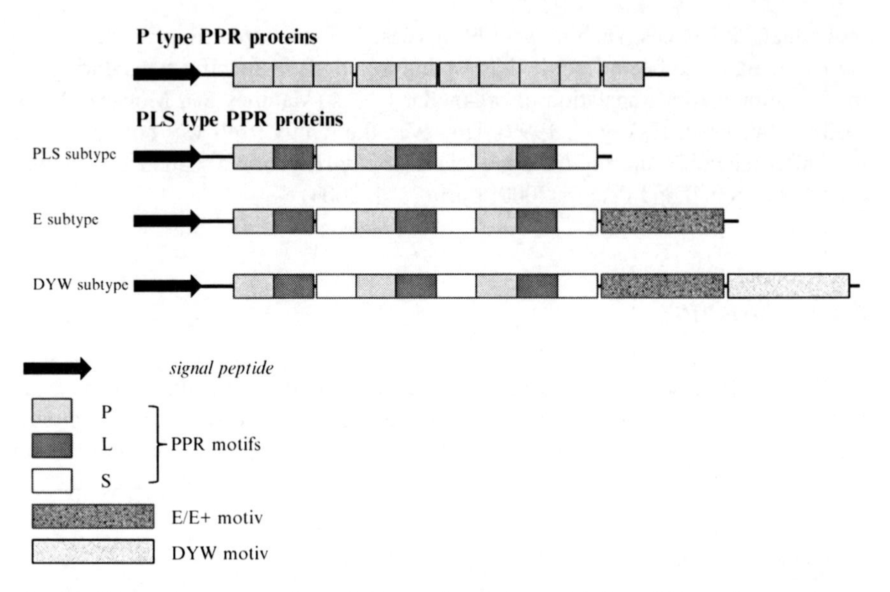

Fig. 8.1 Schematic representation of PPR-protein subclasses . P-Type PPR proteins contain tandem repeats of the classical 35 amino acid repeat unit. In PLS-PPR-Proteins, these canonical PPR motifs alternate with L- and S-motives. Note that PPR domains do not necessarily follow each other without a spacer (*thick line*). Most PPR proteins contain an N-terminal signal sequence for organellar import (*arrow*). Most PLS-type PPRs have in addition C-terminal extensions of unknown function

half of the PPR proteins in *Arabidopsis thaliana* showed tandem arrays of canonical repeats (or P-repeats; Lurin et al. 2004); such proteins are called P-type PPRs. The other family members additionally contain shorter (designated S, mostly 31 amino acids in length) and/or longer (designated L and usually 36 amino acids in length) versions. These alternative motifs are usually organized together with a canonical P repeat to form the so-called "PLS triplets." Although we currently do not have structural data allowing us to distinguish P-type versus PLS-type PPR proteins, recent evidence suggests that there may be a functional differentiation (see below).

Most PPR proteins consist almost entirely of PPR domains. However, in the PLS subfamily, additional domains can be found at the C terminus; these are called the E (for extended) and DYW (named after three conserved amino acids) domains. At the N terminus, PPR proteins usually carry a signal peptide that directs the protein into either the chloroplasts or the mitochondria. No PPR protein has yet been identified as having dual targeting, and only one member, GRP23, has been suggested to localize outside of these DNA-carrying organelles (Ding et al. 2006). Only a few family members contain other protein domains with assigned functions. For example, PPR4 has an N-terminal RNA recognition motif (RRM) domain (RNA recognition motif; Schmitz-Linneweber et al. 2006), OTP51 has an endonuclease domain (de Longevialle et al. 2008), and nine PPR proteins in *Arabidopsis* carry C-terminal small *mutS* related (SMR) domains (Ian Small, personal communication 2008).

8.2.2 Evolution

PPR proteins are found in all eukaryotes, but their numbers vary drastically between different eukaryotic groups. They are particularly numerous in land plants (e.g., 450 family members in *Arabidopsis* and 477 in rice; Lurin et al. 2004), whereas only a handful of PRP family members are typically found in yeast, animals, and algae (Schmitz-Linneweber and Small 2008). Basal land plant genomes, such as that of the moss, *Physcomitrella patens*, contain more than 100 PPR-encoding genes (O'Toole et al. 2008), and the fern-ally, *Selaginella moellendorffii*, is estimated to have more than 600 such genes (http://wiki.genomics.purdue.edu/index.php/PPR_gene_family). The expansion of the PPR family seems to have been at least partly due to the emergence of the plant-specific PLS-type PPRs, which have been shown to be the main source for organellar RNA-editing factors (Hammani et al. 2009). A correlation has been identified between the number of editing sites and the number of PLS-type PPR genes in a given land-plant genome (Salone et al. 2007), suggesting that the acquisition and spread of editing in plant organellar genomes might have been a driving force for PPR family expansion.

A common theme in large gene families is their rapid lineage-specific evolution through gene duplication, loss, conversion, etc. Therefore, it was surprising when phylogenetic reconstructions using all annotated rice and *Arabidopsis* PPR proteins revealed that this family had an extremely high percentage (>80%) of orthologous gene pairs (ÓToole et al. 2008) between the two plants. In other words, the PPR family (at least in angiosperms) appears to show exquisite evolutionary stability, suggesting that most of the PPR genes are vitally important for basic functions of plant organelles. In the future, it will be interesting to determine whether this conservation extends to basal embryophytes (e.g., *Selaginella*). Indeed, initial analyses have suggested that a partial set of the PPR-encoding gene in the bryophyte *Physcomitrella* is orthologous to angiosperm PPR genes (ÓToole et al. 2008).

Although we have some understanding of the evolutionary trends of land-plant PPR family members, their origin has not yet been fully elucidated. Given that PPR proteins localize predominantly in the mitochondria or chloroplasts, it would seem reasonable to hypothesize that they arose from the prokaryotic endosymbionts. However, analyses of various cyanobacterial and α-proteobacterial genomes failed to uncover any PPR domain-encoding reading frames (Lurin et al. 2004). In fact, with the exception of rare events of horizontal gene transfer (Lurin et al. 2004), prokaryotes do not harbor PPR domain-encoding sequences. This suggests that the PPR proteins may actually have a eukaryotic origin. It has been proposed that the PPR motif evolved from the related tetratricopeptide repeat (TPR) motif (Small and Peeters 2000), which can be found in all three domains of life (Blatch and Lassle 1999; D'Andrea and Regan 2003). The predicted structures for these two motifs are generally similar. The TPR repeat surface typically exposes fewer charged and polar amino acids compared to the PPR repeat, but some TPR proteins deviate from this rule and are capable of interacting with RNA (D'Andrea and Regan 2003; Sane et al. 2005), suggesting that the functional gap between PPRs and TPRs is not dramatic.

8.2.3 Functions of PPR Proteins

The majority of PPR proteins have an N-terminal signal peptide that directs (and is required for) their localization to mitochondria or chloroplasts (Lurin et al. 2004; O'Toole et al. 2008). Consistent with this localization, almost all characterized PPR proteins play a functional role within these organelles (Delannoy et al. 2007; Schmitz-Linneweber and Small 2008). As predicted from the high degree of conservation seen even between PPR proteins from distantly related species, most PPRs are nonredundant and essential. Analysis of PPR null mutants has shown that most of them confer strong phenotypes, including frequent deficiencies in early embryonic development (Lurin et al. 2004; Cushing et al. 2005; Schmitz-Linneweber and Small 2008). These mutants display highly specific defects in the processing of individual organellar transcripts, which has provided important insights into the functions of various PPR proteins.

An important function of the P-type PPR proteins seems to be stabilizing RNAs by protecting transcripts against both exo- and endonuclease digestion (compare Chap. 5). A particularly well-studied example is maize PPR10, which plays a role in protecting the plastid *atpH* and *psaJ* transcripts against exonucleolytic degradation (Pfalz et al. 2009). PPR10 associates with a specific sequence element of no more than 30 nucleotides found in the intercistronic *atpI-atpH* and *psaJ-rpl33* regions; when endonucleolytic cleavage opens the way for exonucleases, they are stopped at the PPR10-RNA complex (Pfalz et al. 2009). In the light of this work, many previous studies on pure PPRs need to be reevaluated, as a large proportion of pure PPRs may actually function as exonuclease blockers. For example, this could be true for the function of *Arabidopsis* PGR3 (a plastid protein) in stabilizing transcripts from the *petL* operon (Yamazaki et al. 2004), maize CRP1 in potentially stabilizing *petD/petB* mRNAs (Fisk et al. 1999), or *Arabidopsis* HCF152 in affecting transcripts from the *psbB* operon (Meierhoff et al. 2003). Since a large proportion of organellar 5′ ends are protected by PPR "roadblocks," it is conceivable that the blocking of exonucleases might be a particular ancient function of the PPR proteins. This hypothesis is supported by three observations: First, such PPR blocks seem to exist in mosses, as evidenced by work on PpPPR_38 from *Physcomitrella patens* (Hattori and Sugita 2009). Second, the blocking of exonucleases is not likely to require interactions with other factors and thus could have been acquired at an early stage of evolution after the PPRs "learned" to bind to RNA. And third, this task seems to fall to the original P-type PPRs, not the more derived members of the PLS class. Thus, it seems possible that the RNA-protecting ability of the PPRs was an early acquisition, with additional functions evolving thereafter.

Interestingly, the binding of maize PPR10 to RNA not only increases the target transcript's half-life, it also affects the translational efficiency of the *atpH* mRNA by a yet-unknown mechanism (Pfalz et al. 2009). Similarly, maize CRP1 was shown to be important for translation of the *psaC* and *petA* mRNAs in chloroplasts (Fisk et al. 1999; Schmitz-Linneweber et al. 2005). These findings suggest that the protection of the 5′ terminus of an mRNA could be linked to its translation. It has been speculated that PPRs associated with the 5′ ends of messages may support the recruitment of ribosomes (Pfalz et al. 2009) or dissolve obstructing secondary structures in the 5′-UTR. If this is true, a correlation might exist between the distance from the PPR-bound 5′ end of a transcript and the start codon and its impact on translation. In the future, it will be exciting to examine the importance of the PPR-binding site's location relative to the locations of other functional sites within a given mRNA.

In addition to protecting transcript ends, the evidence suggests that PPR proteins also protect internal sequence elements. For example, maize PPR5 was shown to associate with a short region within the *trnG*-UCC intron and prevents degradation of the unspliced precursor, presumably by protecting a sequence stretch against endonucleolytic attack (Beick et al. 2008). The protected region was found to resemble an RNase E entry site, which is typically a highly AU-rich sequence limited by stem-loop structures (McDowall et al. 1995; Schein et al. 2008). Although the nature of the nuclease responsible for the degradation process has not yet been established, AU-rich chloroplast RNAs would be expected to have numerous potential cleavage sites for endonucleases. It remains to be seen whether other PPR proteins function similarly to PPR5.

In addition to the above-described functions, numerous PPRs play roles in splicing (Schmitz-Linneweber and Small 2008). Many of these proteins serve mitochondrial introns, but several are also required for the splicing of chloroplast mRNAs (for splicing, see also Chap. 6). The latter group includes *Arabidopsis* OTP51, which supports splicing of the *ycf3* reading frame (de Longevialle et al. 2008), *Arabidopsis* HCF152, which affects *petB* splicing (Meierhoff et al. 2003), maize PPR4, which is essential for trans-splicing of the *rps12* mRNA (Schmitz-Linneweber et al. 2006), and *Physcomitrella* PpPPR_38, which is required for splicing of the *clpP* mRNA (Hattori et al. 2007). It is not yet known how these PPRs facilitate the splicing process, but speculatively, they help resolve or support structure elements in the larger group II intron superstructure.

When the PPR proteins were initially recognized as forming a large protein family in land plants, it was suggested that many of the members could play roles in organellar RNA editing (Small and Peeters 2000). This turned out to be true, at least for the PLS-type PPRs. The first RNA-editing factor discovered, CRR4, belonged to this PPR-subfamily (Kotera et al. 2005; Okuda et al. 2007). Since then, 14 other PPRs (all belonging to the PLS subfamily) have been found to have RNA-editing functions; they serve 22 of the 34 known editing sites in *Arabidopsis* (Hammani et al. 2009; and references therein). Currently, there is some debate regarding whether or not the DYW domain of the PLS-PPR proteins confers the catalytic activity required for RNA editing (see also Chap. 7). In addition, it is not yet clear whether PPR-mediated RNA editing plays a regulatory role in organelles.

Indeed, it is interesting to speculate on the regulatory relevance of the chloroplast (and mitochondrial) PPR proteins. Although most PPRs are essential for the specific RNA processing steps they mediate, it is largely unclear whether they participate in the regulation of gene expression under natural conditions. Good evidence for a regulatory role is only available for the *Arabidopsis* PPR protein, genome uncoupled 1 (GUN1), which was identified in a screen for *Arabidopsis* mutants that no longer transmit signals from the chloroplast to the nucleus (Susek et al. 1993; Koussevitzky et al. 2007), i.e., the so-called "retrograde signal" (Woodson and Chory 2008; Kleine et al. 2009). When chloroplast development is compromised, the retrograde signal switches off the expression of a specific set of nuclear genes, many of which are required for photosynthesis. In GUN1 mutants, the retrograde signal is still transmitted even if the chloroplast is severely damaged (e.g., by blockage of chloroplast translation, or oxidative stress, Susek et al. 1993; Koussevitzky et al. 2007). In order to understand the nature of the signal, it will be important to identify the targets of GUN1. In this respect, it is interesting that GUN1 is one of the few PPRs that has an additional protein domain of known function: It contains an SMR-domain. This domain type is found in eight other PPR proteins and is known to be involved in DNA-repair and recombination processes in bacteria (Moreira and Philippe 1999; Kunkel and Erie 2005). As GUN1 is localized to chloroplast nucleoids that contain chloroplast DNA (Koussevitzky et al. 2007), it is likely that GUN1 and related proteins will turn out to be DNA-binding proteins. Future work will be required to confirm this, and to further determine whether this binding is sequence-specific and relevant to the generation or transduction of the retrograde signal.

8.2.4 RNA Recognition by PPR Proteins

Most of the studies on PPR proteins have described their functions in RNA metabolism solely based on genetic evidence; actual associations with target RNAs have been described only for a handful of PPRs. Of the chloroplast-localized PPRs, only CRP1, PPR4, PPR5, and PPR10 have been demonstrated to be associated with RNA in vivo (Schmitz-Linneweber et al. 2005, 2006; Beick et al. 2008; Pfalz et al. 2009). These studies utilized coimmunoprecipitation techniques, which cannot discriminate between direct and indirect interactions. Biochemical analyses have shown that HCF152, PPR5, PPR10, CRR4, CRP1, and PpPPR_381specifically and directly recognize their RNA ligands (Nakamura et al. 2003; Okuda et al. 2006; Williams-Carrier et al. 2008; Hattori and Sugita 2009; Pfalz et al. 2009). However, in the absence of structural data, the direction of this interaction remains unclear. A common sequence motif has been identified in two target transcripts of CRP1 and PPR10 (Pfalz et al. 2009), whereas no simple sequence similarities have been identified among the binding sites of the PPRs that serve several RNA-editing sites, such as CLB19, CRR28, and OTP84 (Chateigner-Boutin et al. 2008; Hammani et al. 2009; Okuda et al. 2009). Recently, however, it was shown that the consensus sequence recognized by the editing PPR proteins might be more degenerate than previously anticipated. In order to achieve specificity, the editing factors may only distinguish purine nucleotides from pyrimidine nucleotides, and may be capable of uniquely distinguishing among the four nucleotides only at some positions within their binding sites (Hammani et al. 2009). If this is the case, such a strategy would allow the recognition of multiple target sites and would be consistent with the observation that some PPRs function in the processing of multiple RNA species. Clearly, additional detailed biochemical and structural investigations will be necessary if we hope to fully understand the PPR–RNA interaction.

8.3 The Plant Organellar RNA Recognition (PORR) Family

Recently, a novel RNA-binding domain specific to land plants was discovered and termed the plant organellar RNA recognition (PORR) domain (Konishi and Sugiyama 2006; Kroeger et al. 2009). PORR domains have been found in 17 *Arabidopsis* proteins. Almost all of them have orthologs in rice, suggesting that the PORR family may show the same high degree of conservation seen in the PPR protein family. PORR proteins are exclusively found in land plants, and most members are predicted to be targeted to either mitochondria or chloroplasts. Similar to the PPR domain, the PORR domain seems to provide an RNA-binding surface that is made of an array of consecutive alpha-helical elements (see also Box 8.1).

The founding member of this family, termed WTF ("What's this factor?"), was originally detected as a protein that copurified with chloroplast splice factors (Kroeger et al. 2009). WTF has been shown to directly interact with the RNase III-like protein, RNC1, and to reciprocally copurify from chloroplast extracts with CRM proteins (CAF1, CAF2, CFM3, see below; Kroeger et al. 2009). Consistent with the function of these factors in splicing, WTF also associates in vivo with about a dozen chloroplast group II introns (Kroeger et al. 2009), and genetic analyses have shown that WTF supports the splicing of most of its target introns (Kroeger et al. 2009). A second member of the family, RPD1, was initially described as functioning in the development of root hair primordia (Konishi and Sugiyama 2003, 2006). The protein is predicted to be localized to mitochondria (Kroeger et al. 2009), and analysis of null mutants showed that RPD1 is essential for embryonic development (Konishi and Sugiyama 2006). Many factors involved in mitochondrial gene expression are essential for embryogenesis. Therefore, it seems plausible that RPD1 may bind RNA and facilitate an essential RNA processing step, such as the splicing of a critical mitochondrial gene. Further studies on PORR proteins will be required to determine whether RPD1 and other family members are indeed predominantly involved in organellar RNA metabolism.

8.4 Chloroplast Ribonucleoproteins

8.4.1 Structure, Evolution, and RNA Targets

The first nuclear genes identified as encoding chloroplast-localized RNA-binding proteins were the chloroplast ribonucleoproteins (cpRNPs) (Li and Sugiura 1990). The cpRNPs are not only the most abundant RNA-binding proteins in chloroplasts, their abundance level is comparable to that of the plastid ribosomal proteins (Nakamura et al. 2001). In *Arabidopsis*, eight cpRNPs have been described to date. Each cpRNP contains a twin RNA recognition motif (RRM) (Li and Sugiura 1990; Ye et al. 1991; Ye and Sugiura 1992). Although RRM motifs are found throughout all three domains of life, the closest relatives of the cpRNPs are the eukaryotic RRM proteins of the heterogeneous nuclear ribonucleoprotein (hnRNP) class (Maruyama et al. 1999). Like hnRNPs, cpRNPs bind multiple RNAs, including tRNA precursors and mRNAs. The cpRNPs appear to bind intron-containing tRNAs, but not intron-free tRNAs (Nakamura et al. 2001), and they are not found together with translated (mature) RNA (Nakamura et al. 2001).

8.4.2 Expression

Whereas few remarkable expression patterns have been identified for PPR-encoding genes, possibly because their expression levels are typically low (an exception is the LOJ PPR gene that is expressed in lateral organ junctions; Prasad et al. 2005), cpRNPs are highly regulated on different levels. The accumulation of *cpRNP* transcripts is modulated by multiple signals, including light (Li and Sugiura 1990; Schuster and Gruissem 1991; Churin et al. 1999; Raab et al. 2006). For example, the abundance levels of *Arabidopsis CP29A* and *CP29B* transcripts may change as much as 200-fold after a light stimulus (Wang et al. 2006). CP29A expression is affected not only by singular changes in irradiance, but also by circadian signals and the presence of stress-induced hormones, such as abscisic acid (Raab et al. 2006). Indeed, the genes encoding CP29A and most other cpRNPs are strongly repressed early during leaf senescence (van der Graaff et al. 2006).

At the protein level, cpRNPs are extensively modified. These modifications include light-induced N-terminal acetylation (Wang et al. 2006), ADP-ribosylation after pathogen attack (Fu et al. 2007), and (perhaps the best known process) protein phosphorylation and acetylation (Kanekatsu et al. 1993, 1995; Lisitsky and Schuster 1995; Wang et al. 2006; Kleffmann et al. 2007). For example, spinach CP28 is phosphorylated at serine 22 after light irradiation (Lisitsky and Schuster 1995), diminishing the affinity of CP28 for RNA *in vitro* (Lisitsky and Schuster 1995). Recently, phosphorylation sites in CP29A, CP29B and CP33B proteins have been mapped in large-scale screens for phosphopeptides in *Arabidopsis* (Sugiyama et al. 2008; Reiland et al. 2009). The results from such screens will greatly facilitate further studies into the functional ramifications of the posttranslational modifications of cpRNPs.

8.4.3 Functions

The cpRNPs are multifunctional proteins that appear to be involved in various steps during chloroplast RNA metabolism. For example, spinach CP28 is involved in the 3′-end processing of four plastid mRNAs (Schuster and Gruissem 1991), while tobacco cpRNPs positively influence the stability of chloroplast mRNAs *in vitro* (Nakamura et al. 2001; Tillich et al. 2009). Tobacco CP31 contributes to the editing of three sites in vitro (Hirose and Sugiura 2001), and a close relative of tobacco CP31 in *Arabidopsis*, CP31A, was recently shown to be involved in editing 13 sites in vivo (Tillich et al. 2009). The editing events at some of these sites depended on the presence of an additional CP31 paralogue, CP31B, and initial experiments suggest that these two cpRNPs act cooperatively (Tillich et al. 2009). The cpRNPs are known to be present in large ribonucleoprotein complexes that are likely to contain other RNA-binding proteins. Interaction studies and proteomic analyses of such complexes will be required to establish an interaction map of cpRNPs. Given their intriguing expression patterns and posttranscriptional modifications, it seems safe to say that the cpRNPs are the most likely candidates to represent a set of chloroplast RNA-binding proteins that are both mechanistically required for RNA processing and involved in regulating chloroplast gene expression in response to external signals.

Although, the cpRNPs have long been scrutinized in the context of their functions in chloroplast RNA metabolism, recent data on the cpRNPs have come from an unexpected and seemingly unrelated line of research. Surprisingly, *Arabidopsis* CP31A was found to be a target for posttranslational modification by bacterial pathogens (Fu et al. 2007). Although it is presently unclear whether this has any functional relevance for the action of cpRNPs in plant-pathogen interactions, it is intriguing that other lines of evidence suggest a role for cpRNPs in pathogenesis as well. For example, CP29 has been shown to be a repressor of the gene encoding the nuclear-pathogenesis-related protein, PR10a (Boyle and Brisson 2001; Gonzalez-Lamothe et al. 2008), and gel-shift experiments have suggested that CP29 directly interacts with the promoter region of PR10, at least in vitro (Boyle and Brisson 2001). Additionally, a dual localization of CP29 to the chloroplast and the nucleus has been shown (Gonzales-Lamothe et al. 2008). This is surprising for two reasons: First, cpRNPs typically have a much higher affinity for RNA than for DNA (Li and Sugiura 1991); and second, cpRNPs have been shown to reside in the chloroplast (although a nuclear localization has not been excluded) (Li and Sugiura 1990; Ohta et al. 1995; Nakamura et al. 2001). Thus, if these new observations prove to be true in vivo, the functional potential of cpRNPs will have to be reevaluated. Although cpRNPs have not yet been shown to have a transcriptional function within chloroplasts, it remains possible that cpRNPs may also play a role in chloroplast-RNA metabolism. Future studies should examine whether cpRNPs are found in the nucleus, and (if so) whether they also function in nuclear RNA maturation.

8.5 Whirly Proteins

Similar to some cpRNPs, the members of an unrelated protein family, the Whirly proteins, also show dual localization to the organelles and nucleus, and bind both RNA and DNA (Prikryl et al. 2008). The Whirly proteins form two orthologous groups in land plants (Desveaux et al. 2005). Among these, potato Whirly 1 (StWhy1) has been shown to positively regulate nuclear PR10a gene expression and to bind specific sequence elements in the PR10 promoter in vitro (Desveaux et al. 2000). Furthermore, StWhy1 preferentially binds single-stranded DNA and acts as a transcriptional activator (Desveaux et al. 2000). Work on the *Arabidopsis* ortholog, AtWhy1, demonstrated that this protein is necessary for pathogen-defense responses (Desveaux et al. 2004). Different methods have shown that Why1 proteins from potato, *Arabidopsis* and maize all localize to chloroplasts (Desveaux et al. 2000, 2005; Prikryl et al. 2008). The plastid functions of *Arabidopsis* and maize Why1 have been studied in greater detail, and maize Why1 has been shown to bind nonspecifically to chloroplast DNA and in addition to a specific set of intron-containing RNAs (Prikryl et al. 2008). The impact of maize Why1 mutants on chloroplast transcription and DNA content was minimal, but there was a notable effect on *atpF* mRNA splicing and ribosomal RNA biogenesis (Prikryl et al. 2008). In *Arabidopsis*, the combined loss of paralogues AtWhy1 and AtWhy3 leads to instabilities in the plastid genome and eventually to the development of albino tissues (Marechal et al. 2009). The effects of these mutations on RNA metabolism have not yet been analyzed. Collectively, these findings seem to indicate that Why1 functions in various aspects of chloroplast nucleic acid metabolism. In the future, it will be exciting to investigate whether there is a functional link between its effects on DNA and RNA. Also, it will be important to determine the function of AtWhy2, which has been shown to localize to mitochondria and bind nonspecifically to mitochondrial DNA (Marechal et al. 2008).

8.6 Chloroplast RNA Splicing and Ribosome Maturation (CRM) Proteins and Associated Factors

8.6.1 Evolution and Structure of the CRM Proteins

The CRM protein family is probably the best understood RNA-binding protein family in chloroplasts. It is one of the few families for which information on the function of individual members has been gathered in different model systems, including maize, rice, and *Arabidopsis* (Asakura and Barkan 2006; Asakura et al. 2008). The CRM domain was formerly called the CRS1-YhbY domain, after the founding members of the CRM protein family in plants and bacteria, respectively (Till et al. 2001; Ostheimer et al. 2003; Barkan et al. 2007). This domain is of ancient bacterial origin, and already existed prior to the archeae/eubacteria split (Barkan et al. 2007). CRM domains are found as single copies in bacteria, while plant CRM proteins may have one to four CRM domains (Barkan et al. 2007). The CRM domain has loose similarity to the widespread KH RNA-binding domain, although it is unclear whether this is due to a phylogenetic relationship or convergent evolution (Barkan et al. 2007). The ancient single-CRM proteins are associated with ribosomes and are required for ribosomal maturation. In the plant lineage, at least one single-CRM protein localizes to the nucleolus, suggesting that this function may have been conserved among single-CRM members in plants (Barkan et al. 2007). In contrast, most of the 14 orthologous groups of multiple-CRM proteins in plants appear to have evolved other functions, most notably in organellar RNA splicing (Barkan et al. 2007).

8.6.2 Plant CRM Proteins Are Polyvalent Splicing Factors

CRM proteins are the dominant players in chloroplast RNA splicing. The five CRM proteins (CAF1, CAF2, CRS1, CFM2, and CFM3) have been demonstrated to associate with chloroplast group II introns (Ostheimer et al. 2003; Asakura and Barkan 2006, 2007; Asakura et al. 2008), while CFM3 additionally associates with the single group I intron present in angiosperm chloroplasts (Asakura et al. 2008). Most CRM proteins are essential splicing factors; unlike the splicing PPRs proteins, they are responsible for multiple target introns. This has been demonstrated not only genetically, but also through the utilization of novel ribonomic techniques in which the immunoprecipitation of CRM proteins was coupled with the identification of RNA ligands on tiling microarrays (Ostheimer et al. 2003; Asakura and Barkan 2006, 2007; Asakura et al. 2008). The CRM domain itself is likely to be the adaptor for RNA interaction, as indicated by observations that isolated recombinant CRM domains can bind RNA in vitro (Keren et al. 2008).

CAF1 and CAF2, which are closely related paralogues with two CRM domains each, are required for different sets of chloroplast subgroup IIB introns. CAF1 serves seven introns, while CAF2 has diverged to accommodate four others (Ostheimer et al. 2003); only the first intron of the chloroplast *ycf3* mRNA requires both proteins. A similarly rapid divergence of paralogous CRM proteins is also seen in the case of CRS1 and CFM2. CRS1 is the only CRM protein for which only one target intron has been identified, the *atpF* intron (Till et al. 2001; Schmitz-Linneweber et al. 2005). In contrast, its paralog, CFM2, is required for the splicing of four different introns (Asakura and Barkan 2007). These data on CRM paralogs seem to suggest that there is a strong selection against the overlapping of target introns. However, when we consider all of the CRM proteins analyzed to date, overlapping target ranges appear to be more the rule than the exception. For example, CAF2 and CFM2 each serve *ycf3* intron 1 and the *ndhB* intron; CAF1 and CFM1 are required for introns in *petG*, *trnG*-UCC, *rps16*, and *rpl16*; and CAF2 and CFM3 both support the splicing of *ndhB* and *petB* (summarized in Asakura et al. 2008).

8.6.3 CRM Ribonucleoprotein Particles

Based on the above-described overlapping of intron target ranges, it was speculated that the CRM proteins may be found together on their shared RNA targets. Indeed, several lines of evidence indicate that this is the case. First, density gradient centrifugation has shown that all analyzed CRM proteins are present in heterogeneous RNA-protein complexes of 500–700 kDa in size (e.g., Asakura et al. 2008). Second, coimmunoprecipitation analyses have demonstrated that several CRM proteins associate simultaneously with their shared intron ligands. For example, CFM3, CAF1, and CAF2 form part of a larger ribonucleoprotein particle, while CFM2 can be found together with CAF1 and CAF2 (Asakura and Barkan 2007; Asakura et al. 2008). Given the nonredundant nature of splicing, it has to be assumed that the CRM protein-binding sites on introns will differ from one another. Thus, it will be a future challenge to determine whether CRM proteins positioned adjacent on their targets act independently or cooperatively.

8.6.4 CRM-Associated Factors

The complexes that assemble on introns contain not only CRM proteins, but also several additional factors, including the splicing factors, WTF (see above), CRS2, and RNC1, which have all been shown to coprecipitate with CRM proteins (Ostheimer et al. 2003; Watkins et al. 2007; Kroeger et al. 2009). RNC1 is only found in plants and contains two ribonuclease III (RNase III) domains (Watkins et al. 2007). As the name implies, this domain usually confers ribonucleolytic activity (e.g., in bacterial RNase III proteins or eukaryotic Dicer enzymes). Notably, however, several of the amino acids crucial for catalysis are absent from the RNase III domains of RNC1. Consistent with this, RNC1 does not show any evidence of endonuclease activity (Watkins et al. 2007). RNC1 also does not exhibit binding specificity in vitro. However, it specifically impacts the splicing of ten chloroplast introns in vivo, and coimmunoprecipitation experiments have shown that it is exclusively associated with many (but not all) chloroplast introns (Watkins et al. 2007). This makes RNC1 the most versatile splicing factor described to date. The discrepancy between the in vitro binding data and the apparent in vivo specificity suggests that other proteins (possibly including CAF1 and CAF2) may be required to confer this binding specificity. Mutation experiments have shown that RNC1 is essential for the splicing of some introns, whereas others show only minor decreases in splicing efficiency following mutation of RNC1. In addition, at least three introns associate with RNC1 but do not depend on it for splicing (Watkins et al. 2007). It will be interesting to determine whether the latter associations represent evolutionary "experiments," or whether RNC1 has other physiologically relevant activities that are independent of splicing.

A second non-CRM protein found in CRM-ribonucleoprotein particles is CRS2, which is a peptidyl-tRNA hydrolase (PTH) homologue that is essential for splicing among all but one of the subgroup IIB introns in the chloroplast (Jenkins et al. 1997; Jenkins and Barkan 2001). Similar to the case of RNC1, CRS2 has lost its ancestral catalytic activity, as indicated by the maize gene failing to rescue an *Escherichia coli* PTH mutant (Jenkins and Barkan 2001). Also, CRS2 lacks several conserved amino acids that are important for the activity of the *E. coli* enzyme. Apparently, only its RNA-binding capability has been retained in plants, although

CRS2 alone does not bind RNA with high affinity. It has therefore been proposed that, similar to the case of RNC1, the CAF proteins recruit CRS2 to its target introns (Ostheimer et al. 2003). Both CAF1 and CAF2 have been shown to directly interact with CRS2 in a yeast-two-hybrid assay and in in vitro cofractionation studies (Ostheimer et al. 2003). The protein domains required for this interaction have been mapped in detail by researchers who solved the x-ray structure of CRS2 and dissected the CRS2-binding site on the CAF proteins using an in vitro deletion mutant analysis (Ostheimer et al. 2005, 2006). Although the precise function of CRS2 remains to be determined, it is believed to drive the folding of the intron into its catalytically active conformation. Here, as for most of the chloroplast splicing factors discussed above, future studies should include detailed analyses of the effect of protein-intron binding and precise mapping of the binding site.

8.6.5 CRS1: Highlighting the Molecular Mechanism Behind CRM Domain Functions

CRS1, which is required for splicing of the *atpF* mRNA, is the only land plant organelle splicing factor for which we have a more detailed understanding of the underlying molecular function (Jenkins et al. 1997). *In vitro* binding studies using recombinant CRS1 and deletion mutants of its *atpF* target intron, together with hydroxy radical footprinting studies on intronic conformation, have shown that CRS1 binds with high affinity and specificity to sequence elements in intron domains I and IV (Ostersetzer et al. 2005). Studies on the binding characteristics of the three CRM domains in CRS1 have demonstrated that each CRM domain mediates attachment to a different *atpF* binding site (Keren et al. 2008). These binding sites are not conserved in other group II introns, accounting for the intronic specificity of CRS1 (Ostersetzer et al. 2005). The binding of CRS1 reorganizes the intronic structure not only at the binding site, but also at the catalytic core (Ostersetzer et al. 2005), suggesting that CRS1 may actively promote the folding of the intron target. It will be a major future challenge to determine whether related CRM proteins have similar functions, and to understand whether different CRM proteins influence each other at their shared RNA targets. Also, given that several CRM proteins are predicted to be targeted to the mitochondria (Barkan et al. 2007) and at least one CRM protein (CFM3) is dually localized (Asakura et al. 2008), the data from chloroplast CRM proteins may prove useful in guiding research on their mitochondrial relatives.

8.7 Orphan Chloroplast RNA-Binding Proteins

There are numerous nuclear-encoded proteins that affect chloroplast RNA metabolism but do not have any recognizable RNA-binding domain and cannot be classified to any particular protein family by other means. For most of these proteins, an association with RNA has not yet been established. Some examples of these proteins include the DAG-like (DAL), pale cress (PAC), and HCF145 proteins from *Arabidopsis*, which seem to affect the maturation of chloroplast rRNAs or mRNAs, as judged by analyses of corresponding mutants with defects in chloroplast development (Meurer et al. 1998; Bisanz et al. 2003; Lezhneva and Meurer 2004). However, it should be noted that such effects may be secondarily caused by blocks in chloroplast biogenesis, and therefore should not be taken as exclusive evidence for RNA-binding activity (Schmitz-Linneweber et al. 2006; Beick et al. 2008).

Direct RNA binding has, however, been demonstrated for one protein outside the larger families discussed above: Atab2. This protein was isolated based on its homology to the *Chlamydomonas* chloroplast RNA-binding protein, Tab2 (Barneche et al. 2006), and comparative studies demonstrated that both Tab2 and Atab2 are required for the translation of essential components of photosystem I (Dauvillee et al. 2003; Barneche et al. 2006). In addition, Atab2 is also required for the accumulation of photosystem II subunits (Barneche et al. 2006). Atab2 binds A/U-rich stretches in different chloroplast 5′-UTRs, and Atab2 mutants show broad decreases in the translation rates for chloroplast mRNAs that encode subunits of photosystems I and II (Barneche et al. 2006). This suggests that Atab2 plays a general role in translation. Further experiments have demonstrated that Atab2 is not found in polysomes, suggesting that it is likely to participate in the initiation of translation rather than elongation (Barneche et al. 2006). Intriguingly, Atab2 expression is under the control of photoreceptors, specifically the blue-light absorbing cryptochromes (Barneche et al. 2006), making Atab2 one of the very few RNA-binding proteins that can translate external signals into a posttranscriptional response.

Chlamydomonas Tab2 has 38% identity with Atab2 and can almost completely complement a null mutant of Atab2 (Dauvillee et al. 2003; Barneche et al. 2006). This surprising conservation of the sequence and function of an RNA-binding protein between green algae and land plants is extremely rare and suggests the involvement of this protein in an ancient and highly important mechanism. It is intriguing that such an unprecedented conservation applies to translation, as this is arguably the most poorly understood step in chloroplast RNA metabolism. In the future we may find that key light-dependent regulatory mechanisms responsible for determining the output of chloroplast gene expression are actually centered around translation. In the near future, the identification and characterization of further regulatory factors essential for the translation of mRNAs for major photosynthetic proteins will be undoubtedly one of the most exciting tasks in organelle biology.

8.8 Outlook

The large number of chloroplast RNA-binding proteins identified to date accurately reflects the complexity of the associated RNA processing events. The functions of these proteins are diverse and cover all of the events presently known to occur during the lifecycle of a chloroplast RNA following transcription (Fig. 8.2).

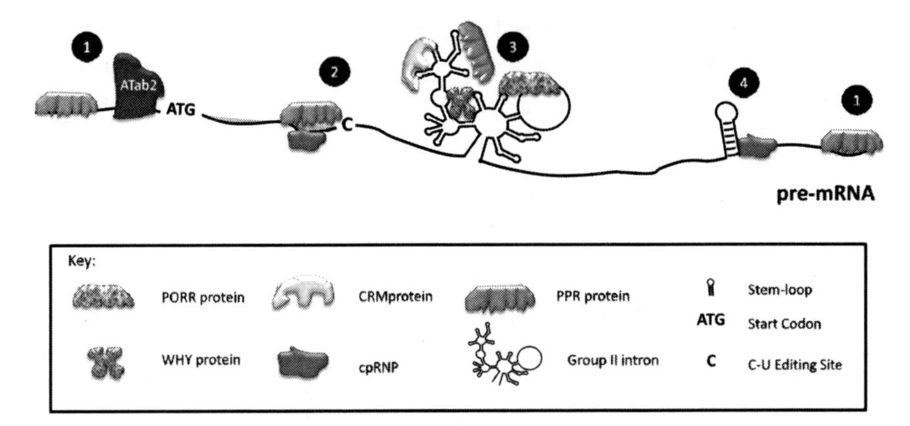

Fig. 8.2 Functions and target regions of the RNA-binding protein families in chloroplasts. This figure summarizes the chloroplast RNA-binding protein functions that have been well supported by genetic and biochemical experiments, including binding data. The mRNA schematic (*black line*) contains a start codon, an editing site, a group II intron, and a 3′ stem loop. (1) Transcript stability and translation are influenced by the binding of PPR proteins to the 5′-UTR and (in the case of stability) the 3′ ends of mRNAs. (4) The transcript stability of many chloroplast mRNAs is believed to be affected by the generation of 3′-terminal stem-loops, and the cpRNP, cp28, has been demonstrated to affect this processing. (2) C-to-U RNA editing is impacted by PPR proteins (which bind specific editing sites) and by cpRNPs (in a more general sense). The binding of PPR proteins to short cis-elements immediately upstream of RNA-editing sites is required for C-to-U processing, (3) RNA splicing. Group II introns form intricate secondary and tertiary structures that are targeted by many RNA-binding proteins; most of these proteins presumably play roles in generating of a correct intron fold. The list of proteins known to interact with group II introns and required for splicing includes a number of PPR and CRM proteins as well as the PORR protein, WTF, and the whirly protein, Why1

The PPR proteins are highly specific RNA-binding proteins that serve one to three RNAs. Most members of this family seem to fulfill a single function, such as the editing function of CRR4 (Kotera et al. 2005) and the splicing function of OTP51 (de Longevialle et al. 2008). However, emerging evidence suggests that some PPR proteins (e.g., PPR10 and CRP1) could be required for several steps in RNA metabolism and could affect both RNA stability and translation (Fisk et al. 1999; Pfalz et al. 2009). The other RNA-binding protein families tend to be more versatile in their target ranges; the CRM proteins, PORR proteins, Whirly proteins, and cpRNPs attach to multiple RNAs and are required for multiple processing steps. So far, the roles of the CRM and PORR family members all appear to be related to splicing; any additional functions await discovery. The cpRNPs and Whirly proteins, in contrast, have been associated with multiple functions: The cpRNPs assist in RNA editing and impact RNA stability, while the Whirly family members are important for splicing and genomic stability. However, although we can make these relatively broad generalizations with regard to the chloroplast RNA-binding protein families, the field is still largely in its hunter-and-gatherer phase, with novel factors being collected and enumerated but little headway being made in understanding the underlying molecular mechanisms.

We are just beginning to understand the composition of ribonucleoprotein particles on chloroplast RNAs, which are far more complex than initially thought, with multiple proteins being present on a given transcript at the same time. Several CRM proteins have been shown to coprecipitate with each other and with other RNA-binding proteins, such as CRS2, RNC1, WTF, and PPR5 (see, for example, Kroeger et al. 2009). However, it is not yet clear whether these coprecipitations are due to direct interactions, or if the proteins are simply tethered along the same RNA molecules. In addition, little is known about the exact RNA sequences bound by these factors, and we know virtually nothing about the kinetics within these particles. For example, is there a defined series of events, a succession of entries and exits of RNA-binding proteins, such as that seen in nuclear ribonucleoproteins (e.g., the spliceosome and nuclear mRNA export particles, Kohler and Hurt 2007; Rino and Carmo-Fonseca 2009)?

Another wide-open field for research on chloroplast RNA-binding proteins is the question of their suborganellar localization. How are the transcriptional, posttranscriptional, and translational machineries spatially organized? Given that only processed mRNAs are translated efficiently (Barkan 1988), there must be a mechanism capable of separating unprocessed RNAs from ribosomes. This could potentially be achieved by differentiating the localizations of processed versus unprocessed RNAs. Alternatively, RNA-binding proteins could mark unprocessed RNAs and block them against translation initiation. Hopefully, the application of new microscopic techniques capable of going beyond the Abbé limit in vivo will be instrumental in answering these questions in the near future.

Finally, it will be intriguing to investigate the connection between transcription and posttranscriptional processing. Of the three RNA polymerases active in the chloroplasts (Liere and Börner 2007), are any capable of recruiting processing factors (like nuclear RNA polymerase II)? Are there protein–protein contacts between polymerases and RNA-binding proteins? To date, a multitude of potential RNA processing factors, including some PPR proteins, have been copurified from crude extracts of chloroplast RNA polymerases (Pfalz et al. 2006). Interaction studies (e.g., using tagged versions of chloroplast RNA polymerases) should facilitate the identification of links between transcription and RNA processing in chloroplasts. Such studies could also help delineate the set of transcripts produced by different RNA polymerases. In combination with genetic analyses using the available mutants for all three RNA polymerases, such work will improve our understanding of whether the transcripts from the three polymerases differ in their downstream processing fates.

Glossary

Arm-repeat: Helical hairpin repeat protein named after the armadillo protein. The repeat is about 40 amino acids long and repeated in tandem. It was first identified in the *Drosophila melanogaster* segment polarity gene armadillo.

Co-IP and slot-blot: A technique related to RIP-Chip. However, instead of using a microarray for detection, co-precipitated RNA is blotted on a membrane using a vacuum manifold. DNA probes (e.g., PCR products) are labeled radioactively and hybridized to the membrane similar to an RNA gel blot hybridization experiment.

EMSA: The electrophoretic mobility shift assay (EMSA), also referred to as the gel retardation assay or gel shift assay, is a common technique used to characterize protein–RNA interactions (also done with DNA). EMSA takes advantage of the fact that complexes of protein and RNA migrate through a nondenaturing polyacrylamide gel more slowly than free RNA fragments. The assay is carried out by incubating a purified protein with labeled RNA containing the putative protein binding site and nonspecific competitor RNA.

Filter binding assay: Filter binding assays measure affinities between RNA (or DNA) and protein molecules using a double layer of different filters. The top filter is made of nitrocellulose and binds proteins but not nucleic acids. However, if RNA is sticking to the protein it will be retained on this filter already. The second, bottom filter is a highly positively charged nylon membrane that will bind all nucleic acids that have passed the first filter. The exact amount of RNA sticking to each of the two filters can be quantified if the RNA was radioactively labeled, for example. The ratio of bound to unbound RNA together can be correlated with protein input and competitor RNA in the initial binding reaction. Among others, the data can be used to determine the binding constant for protein–RNA interactions.

HEAT-repeat: Four proteins with simple, iterated helical-hairpin repeat domains gave rise to this acronym (Huntingtin, elongation factor 3 [EF3], protein phosphatase 2A [PP2A], and PI3-kinase [TOR1]). The domain is often found in proteins involved in transport processes.

Hydroxyradical footprinting: Hydroxyradical footprinting is a technique to determine the binding site of an RNA binding protein on its RNA target (also done with DNA-binding proteins). The recombinant purified protein is incubated with its cognate target and attacked with hydroxyl radicals. The radicals will destroy single-stranded RNA by attacking the phosphodiester connections between ribonucleotides in single-stranded areas.

KH-domain: The K homology (KH) domain was first identified in the human heterogeneous nuclear ribonucleoprotein (hnRNP) K. It is a wide-spread RNA binding motif of around 70 amino acids and is often found in multiple copies within one protein.

Puf-repeat: Named after the human Pumilio protein, the Puf proteins contain a C-terminal domain that consist of repeated helical hairpin repeat domains, the Puf domains. Puf proteins regulate various aspects of development by controlling mRNA stability and translation through sequence-specific interactions, often with sequence elements in the 3′ untranslated region of target mRNAs.

Ribonucleoprotein (RNP): Ribonucleoprotein (RNP) originally referred to proteins localized to the nucleus and associated with RNA. A more recent definition includes all particles that combine ribonucleic acid and protein together.

RIP-Chip: RNA-coimmunoprecipitation and chip analysis refers to a technique that couples immunoprecipitation of an RNP with subsequent analysis of the bound RNA using a microarray. This allows detecting targets of an RNA binding protein across an entire genome.

SMR-Domain: The small MutS-related (SMR) domain has been implicated in recombination and repair processes. It is related to the C-terminal part of the bacterial MutS2 domain and is found in bacteria as well as eukaryotes.

UV-cross-linking: A technique related to EMSAs, where protein–nucleic acid complexes are irradiated with ultraviolet light. This causes covalent bonds to form between the nucleic acid and proteins that are in close contact with the nucleic acid. The protein–RNA adduct is often treated with RNAses to remove non-cross-linked RNA. For detection, the RNA is often labeled and can be visualized in a denaturing polyacrylamide gel.

References

Asakura, Y., Barkan, A. 2006. *Arabidopsis* orthologs of maize chloroplast splicing factors promote splicing of orthologous and species-specific group II introns. Plant Physiol 142:1656–1663.

Asakura, Y., Barkan, A. 2007. A CRM domain protein functions dually in group I and group II intron splicing in land plant chloroplasts. Plant Cell 19:3864–3875.

Asakura, Y., Bayraktar, O. A., Barkan, A. 2008. Two CRM protein subfamilies cooperate in the splicing of group IIB introns in chloroplasts. RNA 14:2319–2332.

Aubourg, S., Boudet, N., Kreis, M., Lecharny, A. 2000. In *Arabidopsis thaliana*, 1% of the genome codes for a novel protein family unique to plants. Plant Mol Biol. 42:603–613.

Barkan, A. 1988. Proteins encoded by a complex chloroplast transcription unit are each translated from both monocistronic and polycistronic mRNAs. EMBO J 7:2637–2644.

Barkan, A., Goldschmidt-Clermont, M. 2000. Participation of nuclear genes in chloroplast gene expression. Biochimie 82:559–572.

Barkan, A., Klipcan, L., Ostersetzer, O., Kawamura, T., Asakura, Y., Watkins, K. P. 2007. The CRM domain: an RNA binding module derived from an ancient ribosome-associated protein. RNA 13:55–64.

Barneche, F., Winter, V., Crevecoeur, M., Rochaix, J. D. 2006. ATAB2 is a novel factor in the signalling pathway of light-controlled synthesis of photosystem proteins. EMBO J 25:5907–5918.

Beick, S., Schmitz-Linneweber, C., Williams-Carrier, R., Jensen, B., Barkan, A. 2008. The pentatricopeptide repeat protein PPR5 stabilizes a specific tRNA precursor in maize chloroplasts. Mol Cell Biol. 28(17):5337–47.

Bisanz, C., Begot, L., Carol, P., Perez, P., Bligny, M., Pesey, H., Gallois, J. L., Lerbs-Mache, S., Mache, R. 2003. The *Arabidopsis* nuclear DAL gene encodes a chloroplast protein which is required for the maturation of the plastid ribosomal RNAs and is essential for chloroplast differentiation. Plant Mol Biol. 51:651–663.

Blatch, G., Lassle, M. 1999. The tetratricopeptide repeat: a structural motif mediating protein-protein interactions. Bioessays 21:932–939.

Boyle, B., Brisson, N. 2001. Repression of the defense gene PR-10a by the single-stranded DNA binding protein SEBF. Plant Cell 13:2525–2537.

Chateigner-Boutin, A. L., Ramos-Vega, M., Guevara-Garcia, A., Andres, C., de la Luz Gutierrez-Nava, M., Cantero, A., Delannoy, E., Jimenez, L. F., Lurin, C., Small, I., Leon, P. 2008. CLB19, a pentatricopeptide repeat protein required for editing of rpoA and clpP chloroplast transcripts. Plant J. 56:590–602.

Churin, Y., Hess, W., Börner, T. 1999. Cloning and characterization of three cDNAs encoding chloroplast RNA binding proteins from barley: differential regulation of expression by light and plastid development. Curr Genet. 36:173–181.

Coffin, J. W., Dhillon, R., Ritzel, R. G., Nargang, F. E. 1997. The *Neurospora crassa cya-5* nuclear gene encodes a protein with a region of homology to the *Sacharomyces cerevisiae* PET309 protein and is required in a post-transcriptional step for the expression of the mitochondrially encoded COXI protein. Curr Genet. 32:273–280.

Cushing, D. A., Forsthoefel, N. R., Gestaut, D. R., Vernon, D. M. 2005. *Arabidopsis* emb175 and other ppr knockout mutants reveal essential roles for pentatricopeptide repeat (PPR) proteins in plant embryogenesis. Planta 221:424–436.

D'Andrea, L. D., Regan, L. 2003. TPR proteins: the versatile helix. Trends Biochem Sci. 28:655–662.

Das, A. K., Cohen, P. T. W., Barford D. 1998. The structure of the tetratricopeptide repeats of protein phosphatase 5: implications for TPR-mediated protein-protein interactions. EMBO J 17:1192–1199.

Dauvillee, D., Stampacchia, O., Girard-Bascou, J., Rochaix, J.-D. 2003. Tab2 is a novel conserved RNA binding protein required for translation of the chloroplast *psaB* mRNA. EMBO J 22:6378–6388.

de Longevialle, A. F., Hendrickson, L., Taylor, N. L., Delannoy, E., Lurin, C., Badger, M., Millar, A. H., Small, I. 2008. The pentatricopeptide repeat gene OTP51 with two LAGLIDADG motifs is required for the cis-splicing of plastid ycf3 intron 2 in *Arabidopsis thaliana*. Plant J. 56:157–168.

Delannoy, E., Stanley, W. A., Bond, C. S., Small, I. D. 2007. Pentatricopeptide repeat (PPR) proteins as sequence-specificity factors in post-transcriptional processes in organelles. Biochem Soc Trans 35:1643–1647.

Desveaux, D., Despres, C., Joyeux, A., Subramaniam, R., Brisson, N. 2000. PBF-2 is a novel single-stranded DNA binding factor implicated in PR-10a gene activation in potato. Plant Cell 12:1477–1489.

Desveaux, D., Marechal, A., Brisson, N. 2005. Whirly transcription factors: defense gene regulation and beyond. Trends Plant Sci 10:95–102.

Desveaux, D., Subramaniam, R., Despres, C., Mess, J. N., Levesque, C., Fobert, P. R., Dangl, J. L., Brisson, N. 2004. A "Whirly" transcription factor is required for salicylic acid-dependent disease resistance in *Arabidopsis*. Dev Cell 6:229–240.

Ding, Y. H., Liu, N. Y., Tang, Z. S., Liu, J., Yang, W. C. 2006. *Arabidopsis* GLUTAMINE-RICH PROTEIN23 is essential for early embryogenesis and encodes a novel nuclear PPR motif protein that interacts with RNA polymerase II subunit III. Plant Cell 18:815–830.

Fisk, D. G., Walker, M. B., Barkan A. 1999. Molecular cloning of the maize gene *crp1* reveals similarity between regulators of mitochondrial and chloroplast gene expression. EMBO J 18:2621–2630.

Fu, Z. Q., Guo, M., Jeong, B. R., Tian, F., Elthon, T. E., Cerny, R. L., Staiger, D., Alfano, J. R. 2007. A type III effector ADP-ribosylates RNA-binding proteins and quells plant immunity. Nature 447:284–288.

Gonzalez-Lamothe, R., Boyle, P., Dulude, A., Roy, V., Lezin-Doumbou, C., Kaur, G. S., Bouarab, K., Despres, C., Brisson, N. 2008. The transcriptional activator Pti4 is required for the recruitment of a repressosome nucleated by repressor SEBF at the potato PR-10a gene. Plant Cell 20:3136–3147.

Hammani, K., Okuda, K., Tanz, S. K., Chateigner-Boutin, A. L., Shikanai, T., Small I. 2009. A study of new *Arabidopsis* chloroplast RNA editing mutants reveals general features of editing factors and their target sites. Plant Cell 21:3686–3699.

Hattori, M., Miyake, H., Sugita M. 2007. A Pentatricopeptide repeat protein is required for RNA processing of clpP Pre-mRNA in moss chloroplasts. J Biol Chem 282:10773–10782.

Hattori, M., Sugita, M. 2009. A moss pentatricopeptide repeat protein binds to the 3′ end of plastid clpP pre-mRNA and assists with mRNA maturation. FEBS J 276:5860–5869.

Hirose, T., Sugiura, M. 2001. Involvement of a site-specific *trans*-acting factor and a common RNA-binding protein in the editing of chloroplast mRNAs: development of a chloroplast in vitro RNA editing system. EMBO J 20:1144–1152.

Jenkins, B. D., Barkan, A. 2001. Recruitment of a peptidyl-tRNA hydrolase as a facilitator of group II intron splicing in chloroplasts. EMBO J 20:872–879.

Jenkins, B. D., Kulhanek, D. J., Barkan, A. 1997. Nuclear mutations that block group II RNA splicing in maize chloroplasts reveal several intron classes with distinct requirements for splicing factors. Plant Cell 9:283–296.

Kanekatsu, M., Ezumi, A., Nakamura, T., Ohtsuki, K. 1995. Chloroplast ribonucleoproteins (RNPs) as phosphate acceptors for casein kinase II: purification by ssDNA-cellulose column chromatography. Plant Cell Physiol 36:1649–1656.

Kanekatsu, M., Munakata, H., Furuzono, K., Ohtsuki, K. 1993. Biochemical characterization of a 34 kDa ribonucleoprotein (p34) purified from the spinach chloroplast fraction as an effective phosphate acceptor for casein kinase II. FEBS Lett 335:176–180.

Keren, I., Klipcan, L., Bezawork-Geleta, A., Kolton, M., Shaya, F., Ostersetzer-Biran, O. 2008. Characterization of the molecular basis of group II intron RNA recognition by CRS1-CRM domains. J Biol Chem 283:23333–23342.

Kleffmann, T., von Zychlinski, A., Russenberger, D., Hirsch-Hoffmann, M., Gehrig, P., Gruissem, W., Baginsky, S. 2007. Proteome dynamics during plastid differentiation in rice. Plant Physiol 143:912–923.

Kleine, T., Voigt, C., Leister, D. 2009. Plastid signalling to the nucleus: messengers still lost in the mists? Trends Genet 25:185–192.

Kohler, A., Hurt, E. 2007. Exporting RNA from the nucleus to the cytoplasm. Nat Rev Mol Cell Biol 8:761–773.

Konishi, M., Sugiyama, M. 2006. A novel plant-specific family gene, ROOT PRIMORDIUM DEFECTIVE 1, is required for the maintenance of active cell proliferation. Plant Physiol 140:591–602.

Konishi, M., Sugiyama, M. 2003. Genetic analysis of adventitious root formation with a novel series of temperature-sensitive mutants of *Arabidopsis thaliana*. Development 130:5637–5647.

Kotera, E., Tasaka, M., Shikanai, T. 2005. A pentatricopeptide repeat protein is essential for RNA editing in chloroplasts. Nature 433:326–330.

Koussevitzky, S., Nott, A., Mockler, T. C., Hong, F., Sachetto-Martins, G., Surpin, M., Lim, J., Mittler, R., Chory, J. 2007. Signals from chloroplasts converge to regulate nuclear gene expression. Science 316:715–719.

Kroeger, T. S., Watkins, K. P., Friso, G., van Wijk, K. J., Barkan, A. 2009. A plant-specific RNA-binding domain revealed through analysis of chloroplast group II intron splicing. Proc Natl Acad Sci USA 106:4537–4542.

Kunkel, T. A., Erie, D. A. 2005. DNA mismatch repair. Annu Rev Biochem 74:681–710.

Lezhneva, L., Meurer, J. 2004. The nuclear factor HCF145 affects chloroplast psaA-psaB-rps14 transcript abundance in *Arabidopsis thaliana*. Plant J 38:740–753.

Li, Y. Q., Sugiura, M. 1991. Nucleic acid-binding specificities of tobacco chloroplast ribonucleoproteins. Nucleic Acids Res 19:2893–2896.

Li, Y. Q., Sugiura, M. 1990. Three distinct ribonucleoproteins from tobacco chloroplasts: each contains a unique amino terminal acidic domain and two ribonucleoprotein consensus motifs. EMBO J 9:3059–3066.

Liere, K., Börner, T. 2007. Transcription and transcriptional regulation in plastids. in R. Bock, ed. pp. 121–174. Cell and Molecular Biology of Plastids. Springer: New York.

Lisitsky, I., Liveanu, V., Schuster, G. 1995. RNA-binding characteristics of a ribonucleoprotein from spinach chloroplast. Plant Physiol 107:933–941.

Lisitsky, I., Schuster, G. 1995. Phosphorylation of a chloroplast RNA-binding protein changes its affinity to RNA. Nucleic Acids Res. 23:2506–2511.

Lurin, C., Andres, C., Aubourg, S., Bellaoui, M., Bitton, F., Bruyere, C., Caboche, M., Debast, C., Gualberto, J., Hoffmann, B., Lecharny, A., Le Ret, M., Martin-Magniette, M. L., Mireau, H., Peeters, N., Renou, J. P., Szurek, B., Taconnat, L., Small, I. 2004. Genome-wide analysis of *Arabidopsis* pentatricopeptide repeat proteins reveals their essential role in organelle biogenesis. Plant Cell 16:2089–2103.

Maier, U.-G., Bozarth, A., Funk, H., Zauner, S., Rensing, S., Schmitz-Linneweber, C., Börner, T., Tillich, M. 2008. Complex chloroplast RNA metabolism: just debugging the genetic programme? BMC Biol 6:36.

Manthey, G. M., McEwen, J. E. 1995. The product of the nuclear gene *PET309* is required for translation of mature mRNA and stability or production of intron-containing RNAs derived from the mitochondrial *COX1* locus of *Saccharomyces cerevisiae*. EMBO J 14:4031–4043.

Marechal, A., Parent, J. S., Sabar, M., Veronneau-Lafortune, F., Abou-Rached, C., and Brisson, N. 2008. Overexpression of mtDNA-associated AtWhy2 compromises mitochondrial function. BMC Plant Biol 8:42.

Marechal, A., Parent, J. S., Veronneau-Lafortune, F., Joyeux, A., Lang, B. F., Brisson, N. 2009. Whirly proteins maintain plastid genome stability in *Arabidopsis*. Proc Natl Acad Sci USA 106:14693–14698.

Maruyama, K., Sato, N., Ohta, N. 1999. Conservation of structure and cold-regulation of RNA-binding proteins in cyanobacteria: probable convergent evolution with eukaryotic glycine-rich RNA-binding proteins. Nucleic Acids Res 27:2029–2036.

McDowall, K. J., Kaberdin, V. R., Wu, S. W., Cohen, S. N., Lin-Chao, S. 1995. Site-specific RNase E cleavage of oligonucleotides and inhibition by stem-loops. Nature 374:287–290.

Meierhoff, K., Felder, S., Nakamura, T., Bechtold, N., Schuster, G. 2003. HCF152, an *Arabidopsis* RNA binding pentatricopeptide repeat protein involved in the processing of chloroplast *psbB-psbT-psbH-petB-petD* RNAs. Plant Cell 15:1480–1495.

Meurer, J., Grevelding, C., Westhoff, P., Reiss, B. 1998. The PAC protein affects the maturation of specific chloroplast mRNAs in *Arabidopsis thaliana*. Mol Gen Genet 258:342–351.

Moreira, D., Philippe, H. 1999. Smr: a bacterial and eukaryotic homologue of the C-terminal region of the MutS2 family. Trends Biochem Sci 24:298–300.

Nakamura, T., Meierhoff, K., Westhoff, P., Schuster, G. 2003. RNA-binding properties of HCF152, an *Arabidopsis* PPR protein involved in the processing of chloroplast RNA. Eur J Biochem 270:4070–4081.

Nakamura, T., Ohta, M., Sugiura, M., Sugita, M. 1999. Chloroplast ribonucleoproteins are associated with both mRNAs and intron-containing precursor tRNAs. FEBS Lett 460:437–441.

Nakamura, T., Ohta, M., Sugiura, M., Sugita, M. 2001. Chloroplast ribonucleoproteins function as a stabilizing factor of ribosome-free mRNAs in the stroma. J Biol Chem 276:147–152.

ÓToole, N., Hattori, M., Andres, C., Iida, K., Lurin, C., Schmitz-Linneweber, C., Sugita, M., Small, I. 2008. On the expansion of the pentatricopeptide repeat gene family in plants. Mol Biol Evol 25:1120–1128.

Ohta, M., Sugita, M., Sugiura, M. 1995. Three types of nuclear genes encoding chloroplast RNA-binding proteins (cp29, cp31, and cp33) are present in *Arabidopsis thaliana*: presence of cp31 in chloroplasts and its homologue in nuclei/cytoplasms. Plant Mol Biol 27:529–539.

Okuda, K., Chateigner-Boutin, A. L., Nakamura, T., Delannoy, E., Sugita, M., Myouga, F., Motohashi, R., Shinozaki, K., Small, I., Shikanai, T. 2009. Pentatricopeptide repeat proteins with the DYW motif have distinct molecular functions in RNA editing and RNA cleavage in *Arabidopsis* chloroplasts. Plant Cell 21:146–156.

Okuda, K., Myouga, F., Motohashi, R., Shinozaki, K., Shikanai, T. 2007. Conserved domain structure of pentatricopeptide repeat proteins involved in chloroplast RNA editing. Proc Natl Acad Sci USA 104:8178–8183.

Okuda, K., Nakamura, T., Sugita, M., Shimizu, T., Shikanai, T. 2006. A pentatricopeptide repeat protein is a site recognition factor in chloroplast RNA editing. J Biol Chem 281:37661–37667.

Ostersetzer, O., Cooke, A. M., Watkins, K. P., Barkan, A. 2005. CRS1, a chloroplast group II intron splicing factor, promotes intron folding through specific interactions with two intron domains. Plant Cell 17:241–255.

Ostheimer, G. J., Hadjivasiliou, H., Kloer, D. P., Barkan, A., Matthews, B. W. 2005. Structural analysis of the group II intron splicing factor CRS2 yields insights into its protein and RNA interaction surfaces. J Mol Biol 345:51–68.

Ostheimer, G. J., Rojas, M., Hadjivassiliou, H., Barkan, A. 2006. Formation of the CRS2-CAF2 group II intron splicing complex is mediated by a 22-amino acid motif in the COOH-terminal region of CAF2. J Biol Chem 281:4732–4738.

Ostheimer, G. J., Williams-Carrier, R., Belcher, S., Osborne, E., Gierke, J., Barkan, A. 2003. Group II intron splicing factors derived by diversification of an ancient RNA-binding domain. EMBO J 22:3919–3929.

Pfalz, J., Bayraktar, O. A., Prikryl, J., Barkan, A. 2009. Site-specific binding of a PPR protein defines and stabilizes 5′ and 3′ mRNA termini in chloroplasts. EMBO J 28:2042–2052.

Pfalz, J., Liere, K., Kandlbinder, A., Dietz, K. J., Oelmuller, R. 2006. pTAC2, -6, and -12 are components of the transcriptionally active plastid chromosome that are required for plastid gene expression. Plant Cell 18:176–197.

Prasad, A. M., Sivanandan, C., Resminath, R., Thakare, D. R., Bhat, S. R., Srinivasan, R. 2005. Cloning and characterization of a pentatricopeptide protein encoding gene (LOJ) that is specifically expressed in lateral organ junctions in *Arabidopsis thaliana*. Gene 353:67–79.

Prikryl, J., Watkins, K. P., Friso, G., Wijk, K. J., Barkan, A. 2008. A member of the Whirly family is a multifunctional RNA- and DNA-binding protein that is essential for chloroplast biogenesis. Nucleic Acids Res 36:5152–5165.

Raab, S., Toth, Z., de Groot, C., Stamminger, T., Hoth, S. 2006. ABA-responsive RNA-binding proteins are involved in chloroplast and stromule function in *Arabidopsis* seedlings. Planta 224:900–914.

Reiland, S., Messerli, G., Baerenfaller, K., Gerrits, B., Endler, A., Grossmann, J., Gruissem, W., Baginsky, S. 2009. Large-scale *Arabidopsis* phosphoproteome profiling reveals novel chloroplast kinase substrates and phosphorylation networks. Plant Physiol 150:889–903.

Rino, J., Carmo-Fonseca, M. 2009. The spliceosome: a self-organized macromolecular machine in the nucleus? Trends Cell Biol 19:375–384.

Rivals, E., Bruyere, C., Toffano-Nioche, C., Lecharny, A. 2006. Formation of the *Arabidopsis* pentatricopeptide repeat family. Plant Physiol 141:825–839.

Salone, V., Rudinger, M., Polsakiewicz, M., Hoffmann, B., Groth-Malonek, M., Szurek, B., Small, I., Knoop, V., Lurin, C. 2007. A hypothesis on the identification of the editing enzyme in plant organelles. FEBS Lett 581:4132–4138.

Sane, A. P., Stein, B., Westhoff, P. 2005. The nuclear gene HCF107 encodes a membrane-associated R-TPR (RNA tetratricopeptide repeat)-containing protein involved in expression of the plastidial psbH gene in *Arabidopsis*. Plant J 42:720–730.

Schein, A., Sheffy-Levin, S., Glaser, F., Schuster, G. 2008. The RNase E/G-type endoribonuclease of higher plants is located in the chloroplast and cleaves RNA similarly to the *E. coli* enzyme. RNA 14:1057–1068.

Schmitz-Linneweber, C., Small, I. 2008. Pentatricopeptide repeat proteins: a socket set for organelle gene expression. Trends Plant Sci 13:663–670.

Schmitz-Linneweber, C., Williams-Carrier, R., Barkan, A. 2005. RNA immunoprecipitation and microarray analysis show a chloroplast Pentatricopeptide repeat protein to be associated with the 5′ region of mRNAs whose translation it activates. Plant Cell 17:2791–2804.

Schmitz-Linneweber, C., Williams-Carrier, R., Williams, P., Kroeger, T., Vichas, A., Barkan, A. 2006. A pentatricopeptide repeat protein binds to and facilitates the *trans*-splicing of the maize chloroplast *rps12* pre-mRNA. Plant Cell 18:2650–2663.

Schuster, G., Gruissem, W. 1991. Chloroplast mRNA 3′ end processing requires a nuclear-encoded RNA-binding protein. EMBO J 10:1493–1502.

Small, I., Peeters, N. 2000. The PPR motif – a TPR-related motif prevalent in plant organellar proteins. Trends Biochem Sci 25:46–47.

Strittmatter, G., Kossel, H. 1984. Cotranscription and processing of 23S, 4.5S and 5S rRNA in chloroplasts from *Zea mays*. Nucleic Acids Res 12:7633–7647.

Sugiyama, N., Nakagami, H., Mochida, K., Daudi, A., Tomita, M., Shirasu, K., Ishihama, Y. 2008. Large-scale phosphorylation mapping reveals the extent of tyrosine phosphorylation in *Arabidopsis*. Mol Syst Biol 4:193.

Susek, R. E., Ausubel, F. M., Chory, J. 1993. Signal transduction mutants of *Arabidopsis* uncouple nuclear CAB and RBCS gene expression from chloroplast development. Cell 74:787–799.

Till, B., Schmitz-Linneweber, C., Williams-Carrier, R., Barkan, A. 2001. CRS1 is a novel group II intron splicing factor that was derived from a domain of ancient origin. RNA 7:1227–1238.

Tillich, M., Hardel, S. L., Kupsch, C., Armbruster, U., Delannoy, E., Gualberto, J. M., Lehwark, P., Leister, D., Small, I. D., Schmitz-Linneweber, C. 2009. Chloroplast ribonucleoprotein CP31A is required for editing and stability of specific chloroplast mRNAs. Proc Natl Acad Sci USA 106:6002–6007.

van der Graaff, E., Schwacke, R., Schneider, A., Desimone, M., Flugge, U. I., Kunze, R. 2006. Transcription analysis of *Arabidopsis* membrane transporters and hormone pathways during developmental and induced leaf senescence. Plant Physiol 141:776–792.

Wang, B. C., Wang, H. X., Feng, J. X., Meng, D. Z., Qu, L. J., Zhu, Y. X. 2006. Post-translational modifications, but not transcriptional regulation, of major chloroplast RNA-binding proteins are related to *Arabidopsis* seedling development. Proteomics 6:2555–2563.

Watkins, K. P., Kroeger, T. S., Cooke, A. M., Williams-Carrier, R. E., Friso, G., Belcher S. E., van Wijk, K. J., Barkan, A. 2007. A ribonuclease III domain protein functions in group ii intron splicing in maize chloroplasts. Plant Cell 19:2606–2623.

Westhoff, P., Hermann, R. G. 1988. Complex RNA maturation in chloroplasts. The psbB operon from spinach. Eur J Biochem 171:551–564.

Williams-Carrier, R., Kroeger, T., Barkan, A. 2008. Sequence-specific binding of a chloroplast pentatricopeptide repeat protein to its native group II intron ligand. RNA 14:1930–1941.

Wilson, C. G., Kajander, T., Regan, L. 2005. The crystal structure of NlpI. A prokaryotic tetratricopeptide repeat protein with a globular fold. Febs J 272:166–179.

Woodson, J. D., Chory, J. 2008. Coordination of gene expression between organellar and nuclear genomes. Nat Rev Genet 9:383–395.

Yamazaki, H., Tasaka, M., Shikanai, T. 2004. PPR motifs of the nucleus-encoded factor, PGR3, function in the selective and distinct steps of chloroplast gene expression in *Arabidopsis*. Plant J 38:152–163.

Ye, L., Sugiura, M. 1992. Domains required for nucleic acid binding activities in chloroplast ribonucleoproteins. Nucleic Acids Res 20:6275–6279.

Ye, L. H., Li, Y. Q., Fukami-Kobayashi, K., Go, M., Konishi, T., Watanabe, A., Sugiura, M. 1991. Diversity of a ribonucleoprotein family in tobacco chloroplasts: two new chloroplast ribonucleoproteins and a phylogenetic tree of ten chloroplast RNA-binding domains. Nucleic Acids Res 19:6485–6490.

Part III
Translation & Import

Chapter 9
The Plant Mitochondrial Proteome Composition and Stress Response: Conservation and Divergence Between Monocots and Dicots

Shaobai Huang, A. Harvey Millar, and Nicolas L. Taylor

Abstract Mitochondria play a key role in the synthesis of ATP, nucleotides, vitamins, cofactors, and the metabolism of amino acids and lipids. Here we compare the proteomes of mitochondria from monocot and dicot plants and highlight the conservation of the PPR and MSCF families and the mitochondrial electron transfer chain protein complex I. We also examine the divergence of mitochondrial preprotein cleavage during import and succinate dehydrogenase. New insights into the common stress responses of monocot and dicot plant mitochondria is also revealed by reanalysis of previous proteomic studies from a mitochondrial context. A large number of challenges remain in further characterizing the mitochondrial proteomes of model plants including the development of methods to produce highly pure mitochondria from more species, fractionation of mitochondria to allow further analysis, and deeper proteome analysis of whole digests or mitochondrial fractions. In addition, the need to identify contaminating proteins through the use of quantitative methods that can exclude contaminants during data analysis and better mechanisms of collating these large datasets with relevant information for use by other researchers are required. Some insights, directions, and methodology currently utilized in our lab for meeting these challenges are discussed.

Keywords Arabidopsis • Rice • Mitochondria • Proteome • Quantitative proteomics • Stress

Abbreviations

AMT	Accurate Mass Time
AUQA	Absolute quantitation

N.L. Taylor (✉)
ARC Centre of Excellence in Plant Energy Biology, M316, The University of Western Australia, 35 Stirling Hwy, 6009, Crawley, Western Australia, Australia
e-mail: ntaylor@cyllene.uwa.edu.au

F. Kempken (ed.), *Plant Mitochondria*, Advances in Plant Biology 1,
DOI 10.1007/978-0-387-89781-3_9,

BN-PAGE	Blue native-polyacrylamide gel electrophoresis
DIGE	Differential in gel electrophoresis
E1 subunit	2-Oxo acid dehydrogenase
E2 subunit	Dihydrolipoamide dehydrogenase
E3 subunit	Dihydrolipoamide acetyltransferase
ETC	Electron transfer chain
emPAI	Exponentially modified protein abundance index
FACS	Fluorescence activated cell sorting
GDC	Glycine decarboxylase complex
HSP70	Heat shock protein (70 kDa)
ICAT	Isotope-coded affinity tags
IDDA	Intelligent data dependent acquisition
IEF	Isoelectric focusing
iTRAQ	Isobaric tags for relative and absolute quantitation
LC-MS	Liquid chromatography – mass spectrometry
L-Protein	Dihydrolipoamide acetyltransferase of GDC
MACS	Magnetic activated cell sorting
MSCF	Mitochondrial substrate carrier family
MM	Molecular Mass
MPP	Mitochondrial processing peptidase
MS/MS	Tandem mass spectrometry
MS	Mass spectrometry
OGDC	2-Oxoglutarate dehydrogenase complex
PDC	Pyruvate dehydrogenase complex
PPDB	Plant proteome database
P-Protein	PLP dependant amino acid decarboxylase
PTM	Post translational modification
QCE	Quantitative contamination exclusion
ROS	Reactive oxygen species
RPD	The rice proteome database
RP-HPLC	Reverse phase – high performance liquid chromatography
SDH	Succinate dehydrogenase
SHMT	Serine hydroxyl-methyl transferase
SILAC	Stable isotope labeling with amino acids in cell culture
SRM	Selected reaction monitoring
SUBA	The SUB-cellular localization of *Arabidopsis* database (http://suba.plantenergy.uwa.edu.au)
TCA	Tricarboxylic acid
TMT	Tandem mass tags
XIC	Extracted ion chromatogram
ZE-FFE	Free-flow electrophoresis in zone electrophoresis.

9.1 Introduction

Mitochondria not only play a role in ATP production via respiratory oxidation of organic acids and the transfer of electrons to O_2, but also in the synthesis of nucleotides, metabolism of amino acids and lipids, and synthesis of vitamins and cofactors. They also participate in the multiorganellar photorespiratory pathway and provide organic acid intermediates for cellular biosynthesis. Such complicated coordination of multiple biochemical processes requires many hundreds of different proteins working together in protein complexes and in catalytic pathways, and is localized across two membrane systems and several aqueous spaces. To understand the diverse biochemical processes in mitochondria, it is important to know their protein constituents and the localization of these components within the mitochondria. The plant mitochondrial proteome has been estimated to contain as many as 2,000-3,000 different nuclear gene products, alongside the approximately 40 proteins from the mitochondrial genome (Kruft et al. 2001; Millar et al. 2001). In addition, it is likely that a substantial number of individual proteins also undergo various posttranslational modifications (PTM) (Millar et al. 2005). With the availability of annotated genome sequences for the model dicot *Arabidopsis* and monocot rice, one indirect approach to determine the plant mitochondrial proteome is based on localization prediction tools using the N-terminal portions of protein sequences (e.g., MitoprotII, TargetP, iPsort, Predotar), but the fidelity of these predictions can be as low as 50–60% when compared with experimentally proven datasets (Heazlewood et al. 2004; Huang et al. 2009a). The direct experimental analysis of the plant mitochondrial proteome has predominated in dicot species such as *Arabidopsis* (Kruft et al. 2001; Millar et al. 2001; Heazlewood et al. 2004) and pea (Bardel et al. 2002). In contrast, a limited number of reports of mitochondrial proteome analysis in monocot species such as rice (Heazlewood et al. 2003a; Huang et al. 2009a) and maize (Hochholdinger et al. 2004) have begun. A range of proteins that have been claimed to be mitochondrial proteins in these studies have also been claimed in other cellular organelles from mass spectrometry studies and/or by independent experiments. While some dual claims have been proven to be dual-targeted proteins, the majority simply arise from contamination. This highlights the crucial importance of achieving high purity organelle isolations for organellar proteomics and the requirement to utilize techniques that remove contaminants. Combining the traditional differential and gradient centrifugation techniques that are based on size and density of mitochondria with techniques that separate mitochondria from other plant organelles using free-flow electrophoresis in zone electrophoresis mode (ZE-FFE) (Table 9.1) has allowed a deeper and more comprehensive analysis of both the *Arabidopsis* organellar proteomes (Eubel et al. 2007) and rice mitochondrial proteome (Huang et al. 2009a).

The adoption of quantitative methods by many researchers to exclude contaminating proteins is increasing and leading to further improvement in the fidelity of both the mitochondrial proteomes and proteomes of other organelles. These techniques have included both gel-based software-driven analyses and differential in-gel electrophoresis (DIGE), as well as non-gel based, spectral counting and isobaric tagging (Fig. 9.1, Box 9.1, Table 9.1). This has lead to a number of proteins

Table 9.1 Purification of organelles to remove contaminants using quantitative analysis

Separation method	Quantitative analysis	References
Centrifugation (Percoll or sucrose gradient etc.) separation based on size and density of the organelles	Enzymatic markers and western blot. The enzymatic markers and western blot are applied to indicate the relative purity of organelles after separation.	Millar et al. (2001); Heazlewood et al. (2004); Reumann et al. (2007)
	Purity was determined by identification of chloroplastic complex after 2D blue-native/Tricine SDS-PAGE	Kruft et al. (2001)
FFE separation based on surface charge of the organelles	DIGE (Box 9.1)	Eubel et al. (2007); Eubel et al. (2008); Huang et al. (2009a)
	Spectral Counting (Box 9.1)	
Profiling of organelle membrane density gradients	ICAT-labeling (Precursor Quantitation, Box 9.1)	Dunkley et al. (2006)

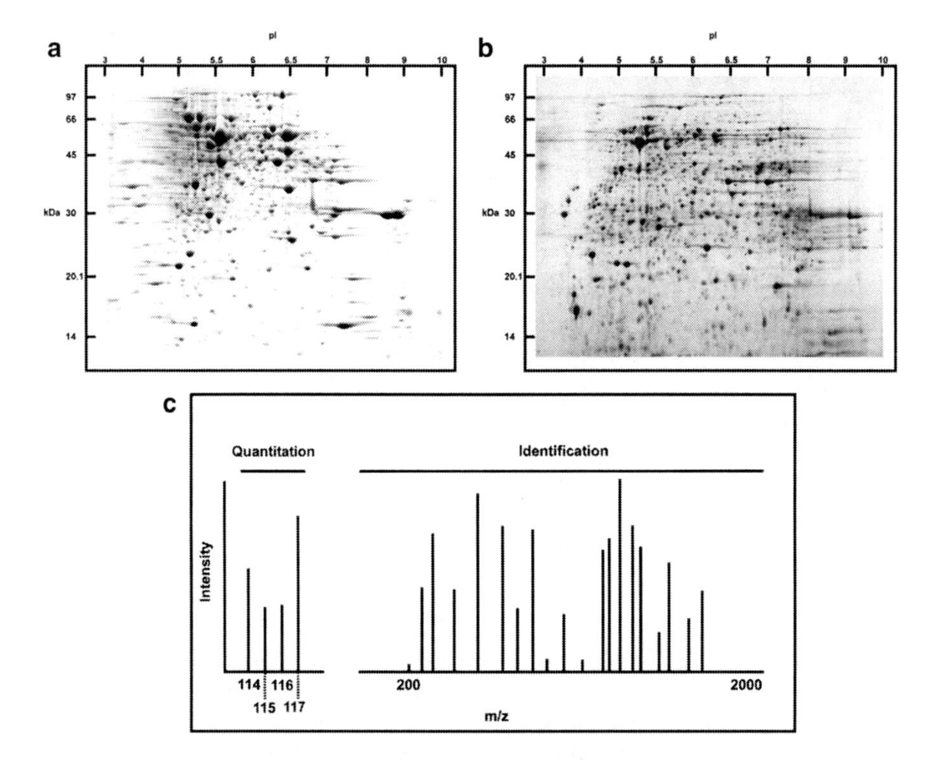

Fig. 9.1 Representative 2D gels of *Arabidopsis thaliana* cell culture mitochondria (**a**) and *Oryza sativa* shoot mitochondria (**b**) separated by IEF pH 3-10 in the first dimension and SDS-PAGE in the second dimension. (**c**) A representation of an MS/MS spectrum obtained during a reporter quantitation experiment (see Box 9.1). The known *m/z* tags (114–117) are used for quantitation of four samples and the remaining spectrum is used for protein identification

Box 9.1 Quantitative mass spectrometry techniques

Gel Based

Software Based Comparisons

This type of analysis, requires multiple 2D gels to be stained and scanned before analysis by a software package such as Same Spots or ImageQuant. Typically the software will align the gel images, normalize gels, detect spots, and then provide statistical comparison based on spot intensity. Spots of interest are then excised from either multiple gels or from a preparatory gel for identification by mass spectrometry.

Differential in Gel Electrophoresis (DIGE)

Proteins to be characterized are analyzed in a pairwise manner in the same gel through the use of fluorescent labels. Prelabeled proteins (two samples) are separated by 2D gel electrophoresis and then scanned using a fluorescence capable imager. Software (e.g., Decyder or Same Spots will align the

(continued)

Box 9.1 (continued)

gel images, normalize gels, detect spots, and then provide statistical comparison based on spot intensity and an internal standard. Spots of interest are then excised from a preparatory gel for identification by mass spectrometry.

Gel-Free

Accurate Mass Times (AMTs)

In this technique extracted ion chromatograms (XIC) for peptides are obtained with a defined mass and retention time during LC-MS. These AMTs are then compared across sample runs to compare peptides abundance directly by XIC peak area. The identification of AMTs of interest is then carried out on subsequent runs by tandem mass spectrometry. This approach can be used against multiple sample runs for quantitation.

Spectral Counting

This involves assessing the number of peptide matches to a particular protein as a measure of abundance. For example if protein A is identified by 35 peptides in sample A and by 70 in sample B then it is more abundant in sample B. The validity of this technique is limited to large datasets including multiple sample runs. A further refinement of this technique is known as exponentially modified protein abundance index (emPAI) where the relative abundance is estimated from a ratio of all possible proteotypic peptides versus those actually found.

Precursor Quantitation

This approach is similar to AMTs in that the peak area of an XIC for a peptide is used for quantitation. However instead of quantifying across two separate runs, the samples are run together and differentiated by a (usually small) mass shift. Quantitation is carried out in a pair wise manner, one sample is labeled with a mass shift reagent while the other remains unlabeled, samples are then combined and run together. Techniques that utilize this approach include ^{15}N labeling, ^{18}O labeling, ICAT, SILAC, and AQUA.

Reporter Quantitation

In this approach quantitation is based on the relative intensities of fragment peaks at known *m/z* values within a tandem mass spectra. Samples (up to 8) are labeled with an isobaric tag that has a known (and consistent across all samples) mass shift during MS. When a peptide is selected for MS/MS the labels fragment to yield ions at known *m/z* values and represent the abundances used for quantitation (see Fig. 9.1). In parallel to quantitation, other fragment ions are used for identification. Techniques that utilize this approach include iTRAQ, TMT, and ExacTag.

Selected Reaction Monitoring (*SRM*)

Selected reaction monitoring also measures peak area of an ion XIC to determine quantitation. However, unlike both AMTs and precursor quantitation, the parent peptide is fragmented and quantitation is carried out on a fragment ion of the parent peptide. Samples are run in parallel, and prior knowledge of the prototypic peptides of a protein and their fragmentation spectra is required.

previously reported as dual-targeted being excluded as contaminants (Eubel et al. 2003; Dunkley et al. 2006; Huang et al. 2009a). In addition these quantitative techniques have allowed the effects of various environmental stresses to be assessed in both *Arabidopsis* and rice including salinity (Kim et al. 2005; Ndimba et al. 2005; Yan et al. 2005; Chitteti and Peng 2007; Jiang et al. 2007; Chen et al. 2009), cold (Bae et al. 2003; Cui et al. 2005; Yan et al. 2006; Hashimoto and Komatsu 2007; Komatsu et al. 2009; Lee et al. 2009), heat (Lee et al. 2007; Palmblad et al. 2008), drought (Ndimba et al. 2005; Ali and Komatsu 2006), and Cd stress (Sarry et al. 2006; Ahsan et al. 2007). While very few direct analyses of isolated mitochondria have been carried out (Sweetlove et al. 2002; Millar et al. 2004; Chen et al. 2009), a number of whole-plant proteomic studies have revealed some mitochondrial proteins that quantitatively change following exposure to stress. These studies tend to identify highly abundant cellular proteins such as those involved in glycolysis, other carbon metabolism, the tricarboxylic acid cycle, and photosynthesis, and only reveal a few highly abundant mitochondrial proteins.

The recently expanded list of identified mitochondrial proteins from model monocot rice (Huang et al. 2009a) makes it feasible for the first time to compare it with the extensively analyzed *Arabidopsis* mitochondrial proteome to yield insights into the similarities and differences between monocot and dicot mitochondrial proteomes. Furthermore, insights have recently been gained into the presequence targeting signals of mitochondrial-targeted proteins in both rice and *Arabidopsis* using high resolution mass spectrometry (MS) data (Huang et al. 2009b).

In this chapter, we will discuss the conservation and divergence of the mitochondrial proteomes of the model monocot rice and the model dicot *Arabidopsis*. This will span elements involved in protein targeting and import that define the biogenesis of mitochondria, through to analysis of the steady-state proteome composition, and finally to the modification of the proteome during plant abiotic stress. Finally, we will propose some possible future directions in research into plant mitochondrial proteomes, especially in the application of MS/MS quantitative techniques to allow accurate quantification of proteins and modified peptides during different treatments/tissues.

9.2 Overall Comparisons of Monocot and Dicot Mitochondrial Proteomes

Two in-depth studies of a monocot and a dicot plant mitochondrial proteome have produced comparable numbers of identified mitochondrial proteins: 416 in *Arabidopsis* (Heazlewood et al. 2004) and 322 in rice (Huang et al. 2009a). These provide the first opportunity for a real comparison across a significant phylogenetic divide within the plant kingdom. Nearly 80% of the proteins identified in rice mitochondria have orthologs in *Arabidopsis* and are listed as mitochondrial proteins in the subcellular localization of *Arabidopsis* (SUBA) database. The number of proteins involved in energy production and metabolism are similar in rice and *Arabidopsis* mitochondrial datasets (Fig. 9.2) and the corresponding proteins are

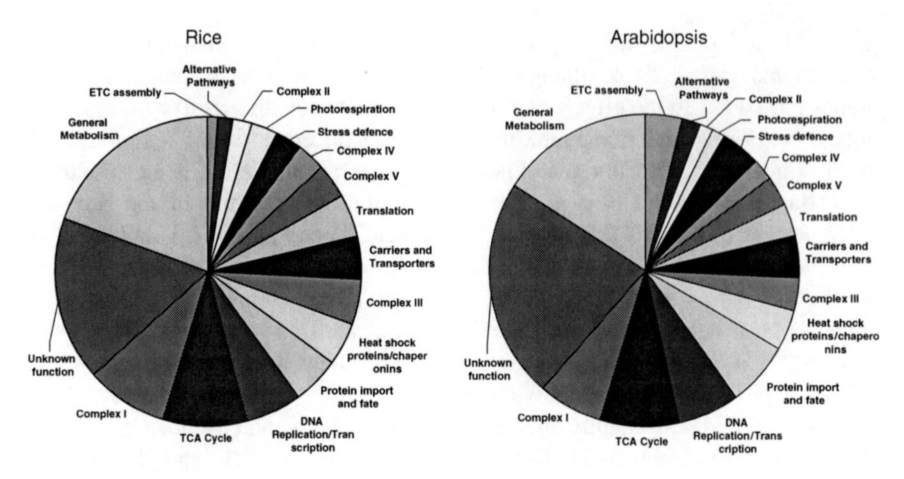

Fig. 9.2 Functional distribution of the 322 rice mitochondrial proteins (Huang et al. 2009a) and 416 *Arabidopsis* mitochondrial proteins (Heazlewood et al. 2004)

also largely conserved (Huang et al. 2009a). This conservation suggests that rice and *Arabidopsis* have very similar components in the electron transport chain complexes and tricarboxylic acid (TCA) cycle for energy production and metabolism. Similarly, proteins involved in ETC assembly, signaling, stress defense, carriers and transporters, and protein import/fate were also observed to be conserved between the rice and *Arabidopsis* mitochondrial datasets (Heazlewood et al. 2004; Huang et al. 2009a). Although the proportion of unknown proteins in rice was smaller than in *Arabidopsis* (Huang et al. 2009a), 49 out of 55 rice proteins had clear *Arabidopsis* orthologs, and 22 of them have been identified as *Arabidopsis* mitochondrial proteins. This indicates a substantial conservation of even unknown function proteins in mitochondria between the two model species. While it is not surprising at some level that monocot and dicots show quite similar mitochondrial proteomes, some specific examples that follow show the highly conserved nature of components of the mitochondrion, while others demonstrate a surprising divergence.

9.3 Specific Examples of Conservation of Plant Mitochondrial Proteome and Function Between Monocots and Dicots

Mitochondrial biogenesis driven from the organelle genome involves RNA editing (see also Chap. 7) and RNA processing (see also Chap. 5), and in both species these are governed by a large family of PPR proteins (see also Chap. 8). Several hundred are predicted to be located in plant mitochondria, and most form interleaved pairs in phylogenetic analysis when rice and *Arabidopsis* PPR families are analyzed together (O'Toole et al. 2008). This suggests a broad pair-wised conservation of function of PPRs in specific RNA editing and processing roles in mitochondria both dicots and monocots. Biogenesis driven by nuclear-encoded components of mitochondria is depended on recognition and import of cytosolic proteins. Comparison of presequences for specific proteins between rice and *Arabidopsis* shows that while amino acid sequence is poorly conserved, there is a remarkable conservation of presequence length for phylogenetically related proteins (Huang et al. 2009b), this may reflect a key conserved element in the efficiency of import or processing for specific types of plant mitochondrial proteins.

The transport of substrates and products between mitochondria and the cytosol requires carriers in the inner mitochondrial membrane. Both rice and *Arabidopsis* nuclear genomes contain families of mitochondrial carrier proteins of approximately 50 members. Phylogenetic studies reveal family members have an interleaved relationship showing conservation of specific carriers and their likely functions between rice and *Arabidopsis* (Taylor et al. 2010). Mass spectrometry has even identified that the abundant carriers in both *Arabidopsis* and rice are phylogenetically paired proteins within the carrier family (Taylor et al. 2010).

Given the fundamental function of the plant mitochondrion for ATP production coupled to the respiratory oxidation of organic acids and the transfer of electrons to O_2, it is expected that plant mitochondrial respiratory complexes from different species will have conserved subunit composition. For example, conservation of Complex I (NADH:ubiquinone oxidoreductase, a respiratory chain component with a large multi-subunit structure containing 30–40 subunits) between *Arabidopsis* and rice is expected and has been experimentally proved. By resolving Complex I using BN-PAGE separation followed by MS identification, a series of 30 different proteins from *Arabidopsis* and 24 proteins from rice have been identified (Heazlewood et al. 2003b). In these datasets, there are 14 highly conserved Complex I subunits found in other eukaryotes, such as in mammals and yeast (Heazlewood et al. 2003b). In addition to these, a series of plant-specific nuclear-encoded complex I-associated subunits have been identified in both *Arabidopsis* and rice, including a series of gamma carbonic anhydrase-like subunits and a range of small proteins of unknown function (Heazlewood et al. 2003b). This conservation of Complex I has also been reinforced by the recent finding that 28 Complex I proteins identified in rice mitochondria (Huang et al. 2009a) had clear orthologs in *Arabidopsis* (Heazlewood et al. 2004).

9.4 Specific Examples of Divergence in Plant Mitochondrial Proteome Monocots and Dicots

Despite the conservation of fundamental mitochondrial processes, plant mitochondrial function does show diversity between rice and *Arabidopsis*. Here again we can highlight examples from biogenesis to functional composition of respiratory chain machinery.

Plant mitochondrial proteins synthesized in the cytosol are recognized by receptors on the organelle surface termed Tom20, translocated through membrane(s), and the targeting signal is cleaved after import in mitochondria (Perry et al. 2006; Lister et al. 2007). There are significant differences in Tom20 between *Arabidopsis* and rice, the former having four copies in its genome and the latter only one copy (Lister et al. 2007). The diversification of Tom20 sequences in *Arabidopsis* potentially allows more variation in mitochondrial targeting signal peptides than the single Tom20 will allow in rice. This is supported by the observations that (1) presequences are longer but less basic and hydrophilic in *Arabidopsis* than in rice, and (2) a significantly lower proportion of four major amino acids (R, A, L and S) in the first ten N-terminal amino acids in presequences of *Arabidopsis* than those in rice (Huang et al. 2009b). This may have resulted from more selective pressure for a signal in presequences for rice mitochondrial proteins than in *Arabidopsis* orthologs. Proteins targeted to the mitochondrial matrix have their mitochondrial presequences cleaved after import by the mitochondrial processing peptidase (MPP), through what is presumed to be a conserved cleavage motif (Zhang et al. 2001). Recently, variations in motif(s) of the cleavage sites between *Arabidopsis* and rice mitochondrial matrix proteins have also been observed (Huang et al. 2009b). For example, the class II cleavage site that contains a conserved arginine at the -3 position (-3R) represents 55% and 58% of total identified cleavage sites of *Arabidopsis* and rice proteins, respectively (Huang et al. 2009b). In this group of class II cleavage sites, sequence logo analysis indicated that Phe (F) in the -1 position dominated in *Arabidopsis* while in rice, Y, F, L were evenly distributed (Fig. 9.3). Such observations suggested that the MPPs might have a slightly different cutting motif in *Arabidopsis* that in rice.

Complex II (succinate:ubiquinone oxidoreductase) has a central role in mitochondrial metabolism as a component of both the electron transport chain and the

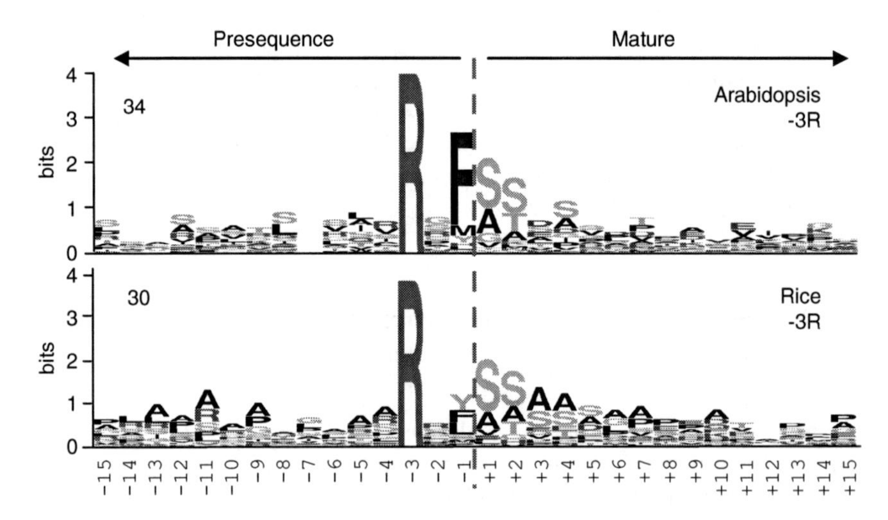

Fig. 9.3 Consensus sequences of the sites of rice and *Arabidopsis* mitochondrial presequence cleavage in group II (-3R). The red line represents the cleavage site by mitochondrial processing peptidase (Adapted from Huang et al. 2009b)

TCA cycle; however, higher plants have a divergent Complex II when compared with mammalian forms (Burger et al. 1996). The subunits of SDH1 and SDH2 of Complex II have similar sequences, but the membrane anchor proteins SDH3 and SDH4 are greatly diversified at the sequence level (Burger et al. 1996). Furthermore, Complex II in plants has four additional plant specific subunits (Eubel et al. 2003; Millar et al. 2004). It has been demonstrated that mitochondrial Complex II from *Arabidopsis* and rice differ significantly in several aspects: (1) The rice complex II is significantly less stable when resolved by native electrophoresis and activity staining when compared to *Arabidopsis* (Huang et al. 2010). (2) The *Arabidopsis* complex II contains 8 subunits while homologs for only 7 have been found in the rice genome and while the subunits SDH 1 and 2 subunits display high levels of amino acid identity, the remainder of the subunits are not well conserved at a sequence level, indicating significant divergence (Huang et al. 2010). (3) The succinate-dependent O_2 consumption and SDH activity of isolated *Arabidopsis* mitochondria is substantially stimulated by ATP, but a much more minor effect of ATP has been observed for the rice enzyme (Huang et al. 2010) indicating variation in its biochemical regulation between the two plants.

Divergence in mitochondrial functions and composition between monocots and dicots has also been previously reported, before the proteomes from both rice and *Arabidopsis* were available. For example, monocots including rice lack the constitutive type 2 *alternative oxidase* genes responsible for the high alternative oxidase activity of various dicot species, including *Arabidopsis* (Considine et al. 2002). Also, monocots lack the double lipoyl-domain acetyl transferase (E2) subunits of pyruvate dehydrogenase that is found in dicot plants, but retain the single lipoyl-domain proteins that are more similar to the E2 enzymes found in yeast (Millar et al. 1999).

9.5 Plant Mitochondrial Proteome Responses to Abiotic Stress

To understand if conservation or divergence in biogenesis and composition is likely to be maintained following the influence of environmental factors and plant life-history, we have assessed the similarities and differences between the monocot and dicot mitochondrial proteomes in their response to environmental stresses. This was probed by analyzing a range of proteome analysis reports investigating stresses in *Arabidopsis* (Table 9.2) and rice (Table 9.3), both at the whole tissue/organ level and at the mitochondrial level. To date, these datasets have not been collectively interpreted from a mitochondrial perspective. By collecting from these publications all proteins that are known to be located in mitochondria based on localization data from other proteomic studies, we were able to create a list of mitochondrial proteins that change in abundance during stress in *Arabidopsis* and rice. From these studies we extracted a redundant set of 100 mitochondrial stress-responding proteins: 62 in *Arabidopsis* and 38 in rice. Previous analysis has shown that in whole tissue/organ stress publications from *Arabidopsis*, the chloroplastic, mitochondrial, and peroxisomal

Table 9.2 The localization of mitochondrial proteins identified in whole tissue/organ and mitochondrial specific proteomic studies of environmental stress in *Arabidopsis*

Salt stress		Fold change 150 mMNaCl			
AGI #	Putative identity	6 h	48 h	Mitochondrial publication	Stress publication
At1g02930	Glutathione *S*-transferase (ATGST1)	0.68 ± 0.14	2.15 ± 0.81	Sweetlove et al. (2002); Heazlewood et al. (2004)	Jiang et al. (2007)
At1g09210	Calreticulin 2 (CRT2)	0.61 ± 0.17	1.62 ± 0.19	Sweetlove et al. (2002); Heazlewood et al. (2004)	Jiang et al. (2007)
At1g47260	Carbonic anhydrase gamma subunit	0.63 ± 0.12	0.7 ± 0.10	Millar et al. (2001); Heazlewood et al. (2003); Brugiere et al. (2004); Heazlewood et al. (2004); Meyer et al. (2007)	Jiang et al. (2007)
At1g51980	Metallo endopeptidase	0.35 ± 0.18	0.91 ± 0.23	Millar et al. (2001); Werhahn and Braun (2002); Brugiere et al. (2004); Heazlewood et al. (2004); Lister et al. (2004); Ito et al. (2006)	Jiang et al. (2007)
At1g53240	Malate dehydrogenase	1.08 ± 0.12	1.52 ± 0.23	Kruft et al. (2001); Millar et al. (2001); Sweetlove et al. (2002); Brugiere et al. (2004); Heazlewood et al. (2004)	Jiang et al. (2007)
At1g56340	Calreticulin 1 (CRT1)	0.57 ± 0.10	1.53 ± 0.10	Sweetlove et al. (2002); Heazlewood et al. (2004)	Jiang et al. (2007)
At2g05710	Aconitate hydratase	0.96 ± 0.18	0.59 ± 0.11	Kruft et al. (2001); Millar et al. (2001); Sweetlove et al. (2002); Heazlewood et al. (2004); Ito et al. (2006)	Jiang et al. (2007)
At2g20360	Hypothetical protein	1.77 ± 0.22	0.81 ± 0.14	Heazlewood et al. (2003); Brugiere et al. (2004); Heazlewood et al. (2004); Meyer et al. (2007)	Jiang et al. (2007)
At2g47510	Fumarase	0.63 ± 0.20	0.88 ± 0.16	Millar et al. (2001); Sweetlove et al. (2002); Heazlewood et al. (2004)	Jiang et al. (2007)
At3g04120	Glyceraldehyde-3-phosphate dehydrogenase	0.55 ± 0.08	0.46 ± 0.11	Sweetlove et al. (2002); Heazlewood et al. (2004)	Jiang et al. (2007)
At3g10920	Mn-SOD	1.33 ± 0.35	3.71 ± 1.08	Kruft et al. (2001); Millar et al. (2001); Eubel et al. (2003); Herald et al. (2003); Heazlewood et al. (2004)	Jiang et al. (2007)

At3g52300	Putative protein	1.25 ± 0.18	0.61 ± 0.12	Kruft et al. (2001); Millar et al. (2001); Heazlewood et al. (2003); Brugiere et al. (2004); Heazlewood et al. (2004)	Jiang et al. (2007)
At3g53870	Ribosomal protein S3a	0.83 ± 0.17	0.64 ± 0.07	Heazlewood et al. (2004); Peltier et al. (2006)	Jiang et al. (2007)
At4g11600	Glutathione peroxidase 6 (ATGPX6)	0.4 ± 0.17	2.61 ± 1.14	Brugiere et al. (2004)	Jiang et al. (2007)
At5g07440	Glutamate dehydrogenase 2 (GDH2)	0.63 ± 0.14	1.61 ± 0.18	Millar et al. (2001); Heazlewood et al. (2004); Ito et al. (2006)	Jiang et al. (2007)
At5g09590	HSC70-2	0.62 ± 0.20	0.32 ± 0.24	Millar et al. (2001); Heazlewood et al. (2004); Ito et al. (2006)	Jiang et al. (2007)
At5g10860	Unknown protein	0.75 ± 0.16	1.52 ± 0.11	Kruft et al. (2001); Millar et al. 2001; Heazlewood et al. 2004)	Jiang et al. (2007)
At5g20080	Cytochrome-b5 reductase/ oxidoreductase	0.92 ± 0.16	0.58 ± 0.14	Kruft et al. 2001; Heazlewood et al. (2004)	Jiang et al. (2007)
At5g47030	ATP synthase delta chain,	1.53 ± 0.26	1.39 ± 0.30	Kruft et al. (2001); Werhahn and Braun (2002); Heazlewood et al. (2003); Brugiere et al. (2004); Heazlewood et al. (2004)	Jiang et al. (2007)
		200 mM NaCl			
		6 h			
At1g02930	GST	1.49		Sweetlove et al. (2002); Heazlewood et al. (2004)	Ndimba et al. (2005)
At1g48030	Dihydrolipoamide dehydrogenase	1.14		Kruft et al. (2001); Millar et al. (2001); Heazlewood et al. (2004); Ito et al. (2006)	Ndimba et al. (2005)
At2g05710	Aconitate hydratase	1.2		Kruft et al. (2001); Millar et al. (2001); Sweetlove et al. (2002); Heazlewood et al. (2004); Ito et al. (2006)	Ndimba et al. (2005)
At3g02090	MPP beta subunit	1.22		Millar et al. (2001); Werhahn and Braun (2002); Brugiere et al. (2004); Heazlewood et al. (2004); Lister et al. (2004)	Ndimba et al. (2005)
At3g12580	Heat-shockprotein70, putative	0.79		Ito et al. (2006)	Ndimba et al. (2005)
At3g62530	Putative protein	0.79		Herald et al. (2003); Heazlewood et al. (2004)	Ndimba et al. (2005)

(continued)

Table 9.2 (continued)

		200 mM NaCl		
		6 h		
At4g37910	DnaK-type chaperoneF20D10.30	1.14	Millar et al. (2001); Brugiere et al. (2004); Heazlewood et al. (2004); Ito et al. (2006)	Ndimba et al. (2005)
At5g07440	Glutamate dehydrogenase 2	1.11	Kruft et al. (2001); Millar et al. (2001); Sweetlove et al. (2002); Werhahn and Braun (2002); Heazlewood et al. (2003); Heazlewood et al. (2004)	Ndimba et al. (2005)

		Fold change		
Cold		6°C		
AGI #	Putative identity	7 day	Mitochondrial publication	Stress publication
At1g61980	Hypothetical protein	2.40	Heazlewood et al. (2004)	Bae et al. (2003)
At3g04120	Glyceraldehyde 3-phosphate dehydrogenase	2.80	Sweetlove et al. (2002); Heazlewood et al. (2004)	Bae et al. (2003)

		Fold change		
Cd stress		0.5–200 μM CdCl		
AGI #	Putative identity	24 h	Mitochondrial publication	Stress publication
At1g02930	GST, phi family	3.3	Sweetlove et al. (2002); Heazlewood et al. (2004)	Sarry et al. (2006)
At1g13440	Glyceraldehyde-3-phosphate dehydrogenase (GAPC),	4.7	Sweetlove et al. (2002); Heazlewood et al. (2004)	Sarry et al. (2006)
At1g48030	Lipoamide dehydrogenase (LPD1)	2.9	Kruft et al. (2001); Millar et al. (2001); Heazlewood et al. (2004); Ito et al. (2006)	Sarry et al. (2006)
At1g53240	Malate dehydrogenase	2	Kruft et al. (2001); Millar et al. (2001); Sweetlove et al. (2002); Heazlewood et al. (2004)	Sarry et al. (2006)
At2g05710	Aconitate hydratase	1.9	Kruft et al. (2001); Millar et al. (2001); Sweetlove et al. (2002); Heazlewood et al. (2004); Ito et al. (2006)	Sarry et al. (2006)
At2g44350	Citrate synthase	1.7	Millar et al. (2001); Heazlewood et al. (2004); Ito et al. (2006)	Sarry et al. (2006)
At3g04120	Glyceraldehyde-3-phosphate dehydrogenase	4.9	Sweetlove et al. (2002); Heazlewood et al. (2004)	Sarry et al. (2006)

At3g06050	Peroxiredoxin	1.3	Kruft et al. (2001); Sweetlove et al. (2002); Heazlewood et al. (2004)	Sarry et al. (2006)
At3g12580	HSP 70	1.2	Ito et al. (2006)	Sarry et al. (2006)
At3g17240	Lipoamide dehydrogenase (LPD2)	3.7	Kruft et al. (2001); Millar et al. (2001); Heazlewood et al. (2004); Taylor et al. (2004); Ito et al. (2006)	Sarry et al. (2006)
At3g22200	4-Aminobutyrate aminotransferase	1.7	Sweetlove et al. (2002); Heazlewood et al. (2004)	Sarry et al. (2006)
At3g23990	HSP 60	3.9	Millar et al. (2001); Brugiere et al. (2004); Heazlewood et al. (2004); Ito et al. (2006)	Sarry et al. (2006)
At3g48000	Aldehyde dehydrogenase	1.9	Millar et al. (2001); Heazlewood et al. (2004); Ito et al. (2006)	Sarry et al. (2006)
At3g55440	Triosephosphate isomerize	2.8	Heazlewood et al. (2004)	Sarry et al. (2006)
At4g26970	Aconitate hydratase	1.6	Kruft et al. (2001); Millar et al. (2001); Heazlewood et al. (2004)	Sarry et al. (2006)
At5g07440	Glutamate dehydrogenase 2	2.1	Millar et al. (2001); Heazlewood et al. (2004); Ito et al. (2006)	Sarry et al. (2006)
At5g09590	HSP70	1.8	Millar et al. (2001); Heazlewood et al. (2004); Ito et al. (2006)	Sarry et al. (2006)
At5g18170	Glutamate dehydrogenase 1	1.4	Kruft et al. (2001); Millar et al. (2001); Heazlewood et al. (2004); Ito et al. (2006)	Sarry et al. (2006)
At5g63400	Adenylate kinase	3.4	Millar et al. (2001); Brugiere et al. (2004); Heazlewood et al. (2004)	Sarry et al. (2006)

(continued)

Table 9.2 (continued)

		Fold change		
Heat		38°C		
AGI #	Putative identity	4 h	Mitochondrial publication	Stress publication
At3g12580	Hsp 70	2.5	Ito et al. (2006)	Palmblad et al. (2008)
At5g09590	Hsc 70-5*d*	1.5	Millar et al. (2001); Heazlewood et al. (2004); Ito et al. (2006)	Palmblad et al. (2008)
At5g56030	Hsp 81-2*d*	1.3	Ito et al. (2006)	Palmblad et al. (2008)

		Fold change		
Drought		400 mM Sorbitol		
AGI #	Putative identity	6 h	Mitochondrial publication	Stress publication
At1g53240	Malate dehydrogenase	1.16	Kruft et al. (2001); Millar et al. (2001); Sweetlove et al. (2002); Brugiere et al. (2004); Heazlewood et al. (2004)	Ndimba et al. 2005)
At1g53240	Malate dehydrogenase	1.26	Kruft et al. (2001); Millar et al. (2001); Sweetlove et al. (2002); Brugiere et al. (2004); Heazlewood et al. (2004)	Ndimba et al. (2005)
At2g20420	Succinyl-CoA ligase beta subunit	1.34	Kruft et al. (2001); Millar et al. (2001); Sweetlove et al. (2002); Heazlewood et al. (2004); Ito et al. (2006)	Ndimba et al. (2005)

At3g02090	MPP beta subunit,	1.15	Millar et al. (2001); Werhahn and Braun (2002); Brugiere et al. (2004); Heazlewood et al. (2004); Lister et al. (2004)	Ndimba et al. (2005)
At3g10920	Mn-SOD	1.18	Kruft et al. (2001); Millar et al. (2001); Eubel et al. (2003); Herald et al. (2003); Heazlewood et al. (2004)	Ndimba et al. (2005)
At3g15020	Malate dehydrogenase	1.14	Millar and Heazlewood (2003); Heazlewood et al. (2004)	Ndimba et al. (2005)
At3g22200	Aminotransferase	1.16	Sweetlove et al. (2002); Heazlewood et al. (2004)	Ndimba et al. (2005)
At3g48000	Aldehyde dehydrogenase,	1.19	Millar et al. (2001); Heazlewood et al. (2004); Ito et al. (2006)	Ndimba et al. (2005)
At4g02930	Elongation factor Tu	1.16	Kruft et al. (2001); Millar and Heazlewood (2003); Heazlewood et al. (2004); Ito et al. (2006)	Ndimba et al. (2005)
At4g02930	Elongation factor Tu	1.11	Kruft et al. (2001); Millar and Heazlewood (2003); Heazlewood et al. (2004); Ito et al. (2006)	Ndimba et al. (2005)

(continued)

Table 9.2 (continued)

		Fold change		
Drought		400 mM Sorbitol		
AGI #	Putative identity	6 h	Mitochondrial publication	Stress publication
At5g08690	ATP synthase beta chain	1.21	Kruft et al. (2001); Millar et al. (2001); Sweetlove et al. (2002); Werhahn and Braun (2002); Heazlewood et al. (2003); Heazlewood et al. (2004)	Ndimba et al. (2005)

AGI #; gene accession number; Putative identity, TAIR protein name; Fold change, the fold change in protein abundance reported under each stress: values >1 indicate and increase in protein abundance and value <1 indicate a decrease in abundance; Mitochondrial publication, the proteomic publication which have claimed protein location; Stress publication, the stress publication that showed changes in protein abundance

Table 9.3 The localization of mitochondrial proteins identified in whole tissue/organ and mitochondrial specific proteomic studies of environmental stress in rice

Salt stress		Fold change 500 mM NaCl		
AGI #	Putative identity	2 h	Mitochondrial publication	Stress publication
Os02g53420.1	Mitochondrial heat shock protein 70 (HSP70)	≥2	Heazlewood et al. (2003); Huang et al. (2009a)	Chen et al. (2009)
Os01g49190.1	ATP synthase beta subunit	≤0.5	Huang et al. (2009a)	Chen et al. (2009)
Os03g27290.1	Cytochrome c oxidase subunit 6B-1	≤0.5	Heazlewood et al. (2003); Huang et al. (2009a)	Chen et al. (2009)
		150 mM NaCl		
AGI #	Putative identity	72 h	Mitochondrial publication	Stress publication
Os03g27290.1	Cytochrome *c* oxidase subunit 6B-1	≥30	Heazlewood et al. (2003); Huang et al. (2009a)	Yan et al. (2005)
		50 mM + 75 mM NaCl		
AGI #	Putative identity	7d + 7d	Mitochondrial publication	Stress publication
Os01g22520.1	Dihydrolipoyl dehydrogenase (E3)	~2	Heazlewood et al. (2003); Huang et al. (2009a)	Dooki et al. (2006)
Os12g07820.1	OsAPx6 – Ascorbate peroxidase	~1.5	Huang et al. (2009a)	Dooki et al. (2006)
Os06g02600.1	DAG protein	~1.7	Huang et al. (2009a)	Dooki et al. (2006)

(continued)

Table 9.3 (continued)

		130 mM NaCl			
AGI #	Putative identity	4d		Mitochondrial publication	Stress publication
Os06g40940.1	Glycine dehydrogenase 2 (P-protein)	1.3		Huang et al. (2009a)	Kim et al. (2005)
Os01g51410.1	Glycine dehydrogenase 2 (P-protein)	1.5		Huang et al. (2009a)	Kim et al. (2005)
Os01g22520.1	Dihydrolipoyl dehydrogenase (E3)	1.5		Heazlewood et al. (2003); Huang et al. (2009a)	Kim et al. (2005)
Os03g52840.1	Serine hydroxymethyl transferase	1.5		Heazlewood et al. (2003); Huang et al. (2009a)	Kim et al. (2005)
		150 mM NaCl	150 mM NaCl		
AGI #	Putative identity	10 h	24 h	Mitochondrial publication	Stress publication
Os05g47980.1	ATP synthase beta chain		0.27	Heazlewood et al. (2003); Komatsu and Tanaka (2005); Huang et al. (2009a)	Chitteti and Peng (2007)
Os01g22520.1	Dihydrolipoyl dehydrogenase (E3)		2.4	Heazlewood et al. (2003); Huang et al. (2009a)	Chitteti and Peng (2007)
		Fold change			
Cold		5°C			
AGI #	Putative identity	48 h		Mitochondrial publication	Stress publication
Os05g47980.1	ATP synthase beta chain	>0.5		Heazlewood et al. (2003); Komatsu and Tanaka (2005); Huang et al. (2009a)	Komatsu et al. (2009)

		10°C				
AGI #	Putative identity	24 h	72 h		Mitochondrial publication	Stress publication
Os08g09200.1	Aconitate hydratase	2.08	4.06		Huang et al. (2009a)	Lee et al. (2009)
Os08g09200.1	Aconitate hydratase	2.8	4.15		Huang et al. (2009a)	Lee et al. (2009)
Os08g09200.1	Aconitate hydratase	4.26	7.73		Huang et al. (2009a)	Lee et al. (2009)
Os01g51410.1	Glycine dehydrogenase 2 (P-protein)	2.93	4.36		Huang et al. (2009a)	Lee et al. (2009)
Os01g51410.1	Glycine dehydrogenase 2 (P-protein)	2.17	3.07		Huang et al. (2009a)	Lee et al. (2009)
		15°C	10°C	5°C		
AGI #	Putative identity	24 h			Mitochondrial publication	Stress publication
Os08g09200.1	Aconitate hydratase	1.72 ± 0.37	2.67 ± 0.30	2.17 ± 0.12	Huang et al. (2009a)	Cui et al. (2005)
Osm1g00580.1	ATP synthase F0 subunit 1	1.32 ± 0.18	1.45 ± 0.21	1.85 ± 0.28	Heazlewood et al. (2003); Huang et al. (2009a)	Cui et al. (2005)
Os06g15990.1	Aldehyde dehydrogenase	1.42 ± 0.05	1.69 ± 0.15	1.54 ± 0.18	Heazlewood et al. (2003); Komatsu and Tanaka (2005); Huang et al. (2009a)	Cui et al. (2005)
Os03g50540.1	NADH-ubiquinone oxidoreductase 75 kDa subunit	2.63 ± 0.18	3.15 ± 0.19	2.79 ± 0.08	Heazlewood et al. (2003); Huang et al. (2009a)	Cui et al. (2005)

(continued)

Table 9.3 (continued)

		5°C		
AGI #	Putative identity	48 h	Mitochondrial publication	Stress publication
Os12g13380.1	Adenylate kinase A	~1.4	Heazlewood et al. (2003); Komatsu and Tanaka (2005); Huang et al. (2009a)	Hashimoto and Komatsu (2007)

		6°C		
AGI #	Putative identity	24 h	Mitochondrial publication	Stress publication
Os01g51410.1	Glycine dehydrogenase 2 (P-protein)	≤0.66	Huang et al. (2009a)	Yan et al. (2006)
Os08g09200.1	Aconitate hydratase	≤0.66	Huang et al. (2009a)	Yan et al. (2006)
Os03g50540.1	NADH-ubiquinone oxidoreductase 75 kDa subunit	≤0.66	Heazlewood et al. (2003); Huang et al. (2009a)	Yan et al. (2006)
Os10g32550.1	Chaperonin CPN60-1	≤0.66	Huang et al. (2009a)	Yan et al. (2006)
Os01g51410.1	Glycine dehydrogenase 2 (P-protein)	≥1.5	Huang et al. (2009a)	Yan et al. (2006)
Os04g32330.1	Dihydrolipoyllysine-residue succinyltransferase (OGDC Subunit)	≥1.5	Heazlewood et al. (2003); Huang et al. (2009a)	Yan et al. (2006)
Os12g13380.1	Adenylate kinase A	≥1.5	Heazlewood et al. (2003); Komatsu and Tanaka (2005); Huang et al. (2009a)	Yan et al. (2006)

Cd Stress		Fold change 800 µM CdCl 4d		Mitochondrial publication	Stress publication
Os05g47980.1	ATP synthase beta chain	2.7		Heazlewood et al. (2003); Komatsu and Tanaka (2005); Huang et al. (2009a)	Ahsan et al. (2007)

Heat		Fold change 42°C			
AGI #	Putative identity	12 h	24 h	Mitochondrial publication	Stress publication
Os06g40940.1	Glycine dehydrogenase 2 (P-protein)	2.00 ± 0.20	1.19 ± 0.20	Huang et al. (2009a)	Lee et al. (2007)
Os06g40940.1	Glycine dehydrogenase 2 (P-protein)	2.18 ± 0.25	1.40 ± 0.40	Huang et al. (2009a)	Lee et al. (2007)
Os01g51410.1	Glycine dehydrogenase 2 (P-protein)	2.39 ± 0.40	1.94 ± 0.25	Huang et al. (2009a)	Lee et al. (2007)
Os03g58040.1	Glutamate dehydrogenase	1.71 ± 0.10	1.73 ± 0.23	Heazlewood et al. (2003); Huang et al. (2009a)	Lee et al. (2007)
Os02g50620.1	Pyruvate dehydrogenase E1 component alpha subunit	1.52 ± 0.24	1.79 ± 0.06	Huang et al. (2009a)	Lee et al. (2007)

(continued)

Table 9.3 (continued)

Drought		Fold change No water		
AGI #	Putative identity	14d	Mitochondrial publication	Stress publication
Os03g52840.1	Serine hydroxymethyl transferase	1.8	Heazlewood et al. (2003); Huang et al. (2009a)	Ali and Komatsu (2006)

AGI #, Gene accession number; Putative identity, OSA protein name; Fold change, the fold change in protein abundance reported under each stress: values >1 indicate an increase in protein abundance and values <1 indicate a decrease in abundance; Mitochondrial publication, the proteomic publication which have claimed protein location; Stress publication, the stress publication that showed changes in protein abundance proteins represent 56% of the proteins observed to change in abundance, whereas these proteins only represent 17% of the total *Arabidopsis* proteome (Taylor et al. 2009). This observation does not necessarily suggest that proteins from these organelles are more responsive to environmental stresses than proteins located in other parts of the cell. After all, protein abundance is a major factor in identifications from whole cell studies. However, it does show that changes in organelle proteins are of major significance in understanding the current data on stress response.

From the data tabulation it is clear that certain groups of proteins appear to be consistently observed to change in abundance under a variety of stress conditions (see Tables 9.2 and 9.3); however, due to the limited number of mitochondrial proteins identified and the limited number of studies under some stress conditions, it is unwise at this stage to attempt to identify proteins that respond to multiple stresses or only to a single stress. Also it should be noted that due to the majority of studies not being focussed on mitochondria, we are likely not identifying the largest stress responsive mitochondrial proteins but simply highly abundant mitochondrial proteins that change in abundance. Nevertheless it is clear that a number of mitochondrial proteins that change in abundance are common across all stresses and across the monocot–dicot divide. These include HSP70s, which are shown to decrease under salt stress in both plants and increase under Cd stress and heat in *Arabidopsis*. While initially this might appear contradictory, it is complicated by the fact that in most studies it is often assumed that the appearance of a particular protein spot on a gel is due to synthesis, and that the loss of a protein spot is due to degradation, without taking into account the addition of PTMs or the appearance or disappearance of degradation products. Many of the environmental stress publications to date do not include data on the theoretical molecular mass (MM) of a protein and the experimentally derived MM from the gel. Many examples exist in the literature where chemical stresses in other plant species lead to the formation of degradation products and shifts in pI by PTMs (Sweetlove et al. 2002; Taylor et al. 2002). Nevertheless it certainly appears that HSP70 has a role in the stress response of plant mitochondria. Also many components of the large multi-subunit enzyme complexes of PDC, OGDC, and GDC seem to increase in response to stress in both plants. For example, protein spots containing GDC P protein have been shown to generally increase in salt, cold, and heat (with a bias towards rice studies), whereas protein spots containing the common E3 (L-Protein of GDC) subunit are seen to increase under salt and Cd stress and the E1 subunits of OGDC and PDC increase in cold and heat, respectively. Also protein spots containing SHMT, linked to photorespiratory function and thus to GDC function, appear to increase during stress. These proteins have been shown previously to be particularly sensitive mitochondrial targets during stress (Taylor et al. 2002). Another well-known stress sensitive target is also found in this analysis; protein spots containing the TCA cycle aconitase are seen to increase in salt, Cd stress, and cold stress. This protein is known to be sensitive to the reactive oxygen species (ROS) hydrogen peroxide that is known to increase during environmental stress. Several other ROS protective enzymes were also seen to change across the stress regimes and monocot dicot divide. This includes increases in ascorbate peroxidase under salt stress (~1.5 fold).

This protein also highlights another major problem with some of the stress publication seen in this study. There is a lack of numerical data to support the extent of the changes observed, as in some cases only a graph is supplied in the manuscript and no precise numbers are supplied, not even as supplemental material; this makes comparisons of changes very difficult and has resulted in the approximate values (~) seen in Tables 9.2 and 9.3. Along with ascorbate peroxidase, glutathione peroxidase was seen to decrease under salt stress and Mn-SOD was seen to increase under the same stress.

This preliminary examination of the stress responses of the monocot, rice and the dicot, *Arabidopsis* provides us with the first protein level investigation of common changes that occur in plant mitochondria exposed to varying environmental conditions. Although limited in both the number of mitochondrial proteins identified and the number of stress studies, it does provide a valuable insight for future examination of the environmental stress response of the plant mitochondria and the potential for translation of mitochondrial-targeted biotechnology approaches between plant species.

9.6 Future Directions

We have discussed the conservation and divergence of the mitochondrial proteomes of the model monocot rice and model dicot *Arabidopsis*. A large number of challenges remain in further characterizing the mitochondrial proteomes of model plants to better represent the diversity of mitochondria in plant genera. These challenges include the development of methods to produce highly pure mitochondria from more species, fractionation of mitochondria to allow further analysis, and deeper proteome analysis of whole digests or mitochondrial fractions. Coupled to these method developments will be the need to identify contaminating proteins through the use of quantitative methods that can exclude contaminants during data analysis. Finally, better mechanisms of collating these large datasets with relevant information for use by other researchers are required. Some insights, directions, and methodology currently utilized in our lab for meeting these challenges are presented as follows.

9.6.1 Excluding Contaminants by Quantitative Analysis

Quantitative comparison combined with the improved mitochondrial purification techniques is crucial for removal of contaminants as mentioned in the Sect. 9.1. The quantitative contamination exclusion (QCE) approaches employ quantitative techniques (discussed later) and samples of different levels of purity that are run in parallel. After protein identification of multiple complex mixture samples, proteins that are enriched in the higher purity samples are confirmed as being mitochondrial, while those of a higher abundance in the less pure samples are excluded as proteins likely to be contaminants. We have employed this method in a number of studies and it has allowed us to confidently assign proteins to the rice mitochondrion and *Arabidopsis* peroxisome (Eubel et al. 2008; Huang et al. 2009a).

In the future, developments in other fields of research may be applied to enrich mitochondria from plants to increase their purity, target cell specificity or allow analysis of smaller samples. For example labeling of mitochondria with fluorescence tags and then isolating them from plant extracts using FACS (fluorescence-activated cell sorting) or MACS (magnetic activated cell sorting) opens a range of opportunities for mitochondrial proteomics through the isolation of mitochondria from a diverse range of tissue or cell types. Historically, the study of plant mitochondria has been from a fairly select group of plant species and tissue types, mostly dicots; however, even this set is much larger than the restricted proteome studies to date. The advent of new genome and transcriptome sequencing projects such as wheat, soybean, corn, grape, *Sorghum bicolor,* and *Medicago truncatula* coupled to organelle isolation techniques such as density gradient centrifugation, ZE-FFE, FACS, or MACS will soon open a wide array of plant species for mitochondrial proteome analysis, providing the chance to undertake much wider surveys of the respiratory apparatus in plants.

9.6.2 In-depth Identification of Mitochondrial Proteins

There are multiple ways to identify more proteins in mitochondria by reducing sample complexity. Firstly, the mitochondria can be fractionated using the membrane and soluble structure, yielding soluble matrix, inner and outer membranes, and the intermembrane space. However, the likely existence of contamination from other organelles will also complicate the analysis of compartmented mitochondrial proteomes, and an additional level of complexity is obtained while trying to confirm subcellular localization. We have had good success separating membrane fractions away from the soluble components of mitochondria to focus on this sub mitochondrial set of proteins (Millar and Heazlewood 2003; Taylor, N. L. et al., 2010). A second approach is to separate the mitochondrial proteins based on physical characteristics. This can be achieved by off-line techniques such as off-gel electrophoresis, gel filtration, or preliminary FPLC or HPLC fractionation of proteins followed by RP-HPLC-MS of peptides. However, with the majority of discovery proteomics research now occurring by LC-MS, avoiding or subsequently removing detergents, urea, and other solubilization components is essential. In our lab we have separated intact proteins by off-gel electrophoresis based on protein pI before a cleanup to remove MS incompatible components. This has allowed a significant increase in the total number of proteins that can be identified in a given sample (Taylor, N. L. et al., 2010). All of these techniques use intact proteins during their separation, followed by trypsin digestion; however following trypsin digests the generated peptides can be fractionated and then analyzed by RP-HPLC-MS. One major draw back of this approach is that many of the generated fractions can contain peptides from the same protein and reconstructing these data sets from large MS analysis files is challenging. Once a set of reduced complexity samples has been obtained, depth of MS analysis can also be modified by the methods employed in shotgun peptide analysis. In the context of gel-based proteomics of mitochondrial extracts, and in its most simple form, this can involve running separate preparative gel that may be overloaded for high abundance proteins but provides enough protein to identify lower abundance proteins. Alternatively multiple gels can be run and excised spots pooled during extraction and digestion to increase the amount of protein available for identification. Techniques can also be employed to increase identification of low abundance proteins in RP-HPLC-MS analyses. The most common method is to simply undertake multiple separations and analyses of the same samples and combined the data together. An increase in depth is achieved simply due to the random nature of automatic selection of different peptides for MS/MS fragmentation. This often results in modest but significant increases in the depth of proteome analysis. A further development of this process, sometimes referred to as intelligent data dependant acquisition (IDDA), involves running a sample multiple times, but for the following run either the peptides that were identified in the previous run or all the ions that were selected for MS/MS are excluded from subsequent analysis. This forces a new set of lower abundance ions to be selected during the subsequent analyses. When this process is repeated several times and the datasets combined, a significant increase in the depth of proteome analysis is possible. We are currently using this strategy to gain depth in analysis of plant mitochondrial proteomes and gaining substantial increases in identifications of low abundance mitochondrial proteins (Taylor, N. L. et al., 2010).

9.6.3 Refining Quantitative Analysis of Proteome Differences of Biologic Consequence

Quantitative analysis is not only crucial for removal of contaminants but also for insight into proteome differences of biological treatments. Combining information gained from proteome mining and quantitative changes in protein abundance improve both the validity of the data and the biological insights, especially for proteins of unknown or unclear function. Classically, gel-based proteomics approaches (see Box 9.1) dominate the quantitative studies based on changes in spots abundance after silver or Coomassie staining, but with variations among different gels. The advantage of DIGE analysis, even with relatively high infrastructure costs for image and software analysis, is that it minimizes gel to gel variations and statistically rigorous studies can be completed using fewer gels (Tonge et al. 2001). All gel-based techniques for quantitative analysis have their merits, including their simplicity in both experimental approach and downstream data analysis, but they have limitation of resolution of spot numbers with high abundance. On the other hand, quantitative analysis can be achieved by the gel-free analysis (see Box 9.1) of complexes mixtures of peptides, allowing a wider range of proteins from a sample and to detect a wider dynamic range of proteins contained in a sample. When this is combined with either Isotope-Coded Affinity Tags (ICAT), Isobaric Tags for Relative and Absolute Quantitation (iTRAQ), or Tandem Mass Tags (TMT), it allows the relative quantitation between samples. In our lab we have implemented iTRAQ relative quantitation and have seen the benefits of this techniques when compared to identical samples analyzed by either spectral counting or DIGE, including the confident assignment of more subtle quantitative changes as well as the detection of mitochondrial proteins normally excluded by gels (Taylor, N. L. et al., 2010). In the future, this will provide us with a robust method for analyzing the plant mitochondrial proteome quantitatively. However this untargeted approach relies on being able to detect labeled peptides, and therefore, in our experience, generally only identifies higher abundance mitochondrial proteins even when combined with methods discussed above to dig deeper into proteomes. An alternative is selected reaction monitoring (SRM), a mass spectrometry technique that allows targeted quantitation of unlabeled known proteotypic peptides using a triple quadrupole mass spectrometry. This technique is able to detect lower abundance peptides in a complex mixture due to it multiple targeted approach to the known peptide. Therefore, SRM analysis allows the filling of the gaps in shotgun type gel based and/or labeling MS/MS approaches, by providing information on target proteins or peptide relevants to the biology under investigation. We have begun the use of this approach to study quantitative changes in rice mitochondrial carrier proteins during seed germination (Taylor, N. L. et al., 2010).

9.6.4 Database Development and Access

Much of the information that is being collected by discovery or shotgun proteomics of plant mitochondria is currently being stored in databases. The (SUBA) database (www.plantenergy.uwa.edu.au/suba2/) provides information gathered from mitochondrial studies alongside other cellular organelles. This provides a valuable resource when determining if a protein observed to change during environmental stress has been identified before in published studies and if there is evidence for a mitochondrial localization. A similar database PPDB (Plant Proteome Database, http://ppdb.tc.cornell.edu/) includes information gathered from *Arabidopsis* and maize, with a focus on plastids but expanding to provide a range of data relevant for mitochondrial researchers, and for rice The Rice Proteome Database (RPD, http://gene64.dna.affrc.go.jp/RPD/) is available with various information on subcellular proteomes from rice. Several other databases exist that provide MS/MS spectral data for plant proteins, including mitochondrial proteins, that could be used to design SRM transitions for targeted proteome analysis; these include AtProteome (Baerenfaller et al. 2008) and ProMEX (Hummel et al. 2007).

Acknowledgments This work was supported by the Australian Research Council (ARC) (CE0561495) to AHM. NLT is supported as an ARC Australian Post-doctoral Fellow and AHM as an ARC Australian Professorial Fellow.

Glossary

PPR: Pentatricopeptide repeat proteins are a very large family in *Arabidopsis* containing at least 450 proteins. They are characterized by tandemly repeated, degenerate, 35 amino acid motifs and have been proposed to be sequence-specific RNA binding proteins. They play a role in posttranscriptional processes in organelles and PPR proteins have been implicated in RNA editing, RNA processing, RNA splicing and translational activation.

PTM: Post translational modifications are chemical alterations that occur to the amino acid backbone of a protein after translation, these may include phosphorylation, methylation, acetylation or nitrosylation.

Proteome: The complete profile of proteins expressed in a given tissue, cell, or biological system at a given time.

Proteotypic peptide: A peptide (preferably unique) that results from an enzymatic digestion of a protein and can be detected by mass spectrometry and that can be used as a proxy for protein presence and/or abundance.

ROS: Reactive oxygen species are small reactive molecules that contain an oxygen atom. They can be highly reactive due to the presence of unpaired valence shell electrons or simply unstable sets of electron pairs, and are formed as a natural byproduct of the metabolism of oxygen.

References

Ahsan, N., Lee, S. H., Lee, D. G., Lee, H., Lee, S. W., Bahk, J. D., Lee, B. H. 2007. Physiological and protein profiles alternation of germinating rice seedlings exposed to acute cadmium toxicity. C R Biol 330:735–746.

Ali, G. M., Komatsu, S. 2006. Proteomic analysis of rice leaf sheath during drought stress. J Proteome Res 5:396–403.

Bae, M. S., Cho, E. J., Choi, E. Y., Park, O. K. 2003. Analysis of the *Arabidopsis* nuclear proteome and its response to cold stress. Plant J 36:652–663.

Baerenfaller, K., Grossmann, J., Grobei, M. A., Hull, R., Hirsch-Hoffmann, M., Yalovsky, S., Zimmermann, P., Grossniklaus, U., Gruissem, W., Baginsky, S. 2008. Genome-scale proteomics reveals *Arabidopsis thaliana* gene models and proteome dynamics. Science 320:938–941.

Bardel, J., Louwagie, M., Jaquinod, M., Jourdain, A., Luche, S., Rabilloud, T., Macherel, D., Garin, J., Bourguignon, J. 2002. A survey of the plant mitochondrial proteome in relation to development. Proteomics 2:880–898.

Brugiere, S., Kowalski, S., Ferro, M., Seigneurin-Berny, D., Miras, S., Salvi, D., Ravanel, S., d'Herin, P., Garin, J., Bourguignon, J., Joyard, J., Rolland, N. 2004. The hydrophobic proteome of mitochondrial membranes from *Arabidopsis* cell suspensions. Phytochemistry 65:1693–1707.

Burger, C., Lang, F., Reith, M., Gray, W. 1996. Genes encoding the same three subunits of respiratory complex II are present in the mitochondrial DNA of two phylogenetically distinct eukaryotes. Proc Natl Acad Sci USA 93:2328–2332.

Chen, X., Wang, Y., Li, J., Jiang, A., Cheng, Y., Zhang, W. 2009. Mitochondrial proteome during salt stress-induced programmed cell death in rice. Plant Physiol Biochem 47:407–415.

Chitteti, B. R., Peng, Z. 2007. Proteome and phosphoproteome differential expression under salinity stress in rice (*Oryza sativa*) roots. J Proteome Res 6:1718–1727.

Considine, M. J., Holtzappfel, R. C., Day, D. A., Whelen, J., Millar, A. H. 2002. Molecular distinction between alternative oxidase from monocots and dicots. Plant Physiol 129:949–953.

Cui, S., Huang, F., Wang, J., Ma, X., Cheng, Y., Liu, J. 2005. A proteomic analysis of cold stress responses in rice seedlings. Proteomics 5:3162–3172.

Dooki, A. D., Mayer-Posner, F. J., Askari, H., Zaiee, A. A., Salekdeh, G. H. 2006. Proteomic responses of rice young panicles to salinity. Proteomics 6:6498–6507.

Dunkley, T. P. J., Hester, S., Shadforth, I. P., Runions, J., Weimar, T., Hanton, S. L., Griffin, J. L., Bessant, C., Brandizz,i F., Hawes, C., Watson, R. B., Dupree, P., Lilley, K. S. 2006. Mapping the *Arabidopsis* organelle proteome. PNAS 103:6518–6523.

Eubel, H., Jansch, L., Braun, H. P. 2003. New insights into the respiratory chain of plant mitochondria. Supercomplexes and a unique composition of complex II. Plant Physiol 133:274–286.

Eubel, H., Lee, C. P., Kou, J., Meyer, E. H., Taylor, N. L., Millar, A. H. 2007. Free flow electrophoresis for purification of plant mitochondria by surface charge. Plant J 52:583–594.

Eubel, H., Meyer, E. H., Taylor, N. L., Bussell, J. D., O'Toole, N., Heazlewood, J. L., Castleden, I., Small, I. D., Smith, S. M., Millar, A. H. 2008. Novel proteins, putative membrane transporters, and an integrated metabolic network are revealed by quantitative proteomic analysis of *Arabidopsis* cell culture peroxisomes. Plant Physiol 148:1809–1829.

Hashimoto, M., Komatsu, S. 2007. Proteomic analysis of rice seedlings during cold stress. Proteomics 7:1293–1302.

Heazlewood, J. L., Howell, K. A., Millar, A. H. 2003b. Mitochondrial complex I from *Arabidopsis* and rice: orthologs of mammalian and fungal components coupled with plant-specific subunits. Biochim Biophys Acta 1604:159–169.

Heazlewood, J. L., Howell, K. A., Whelan, J., Millar, A. H. 2003a. Towards an analysis of the rice mitochondrial proteome. Plant Physiol 132:230–242.

Heazlewood, J. L., Tonti-Filippini, J. S., Gout, A. M., Day, D. A., Whelan, J., Millar, A. H. 2004. Experimental analysis of the *Arabidopsis* mitochondrial proteome highlights signaling and regulatory components, provides assessment of targeting prediction programs, and indicates plant-specific mitochondrial proteins. Plant Cell 16:241–256.

Herald, V. L., Heazlewood, J. L., Day, D. A., Millar, A. H. 2003. Proteomic identification of divalent metal cation binding proteins in plant mitochondria. FEBS Lett 537:96–100.

Hochholdinger, F., Guo, L., Schnable, P. S. 2004. Cytoplasmic regulation of the accumulation of nuclear-encoded proteins in the mitochondrial proteome of maize. Plant J 37:199–208.

Huang, S., Taylor, N. L., Narsai, R., Eubel, H., Whelan, J., Millar, A. H. 2009a. Experimental analysis of the rice mitochondrial proteome, its biogenesis, and heterogeneity. Plant Physiol 149:719–734.

Huang, S., Taylor, N. L., Narsai, R., Eubel, H., Whelan, J., Millar, A. H. 2010. Functional and composition differences between mitochondrial complex II in *Arabidopsis* and rice are correlated with the complex genetic history of the enzyme. Plant Mol Biol 72:.31–342.

Huang, S., Taylor, N. L., Whelan, J., Millar, A. H. 2009b. Refining the definition of plant mitochondrial presequences through analysis of sorting signals, N-terminal modifications, and cleavage motifs. Plant Physiol 150:1272–1285.

Hummel, J., Niemann, M., Wienkoop, S., Schulze, W., Steinhauser, D., Selbig, J., Walther, D., Weckwerth, W. 2007. ProMEX: a mass spectral reference database for proteins and protein phosphorylation sites. BMC Bioinform 8:216.

Ito, J., Heazlewood, J. L., Millar, A. H. 2006. Analysis of the soluble ATP-binding proteome of plant mitochondria identifies new proteins and nucleotide triphosphate interactions within the matrix. J. Proteome Res 5:459–3469.

Jiang, Y., Yang, B., Harris, N. S., Deyholos, M. K. 2007. Comparative proteomic analysis of NaCl stress-responsive proteins in *Arabidopsis* roots. J. Exp. Bot. 58:3591–3607.

Kim, D. W., Rakwal, R., Agrawal, G. K., Jung, Y. H., Shibato, J., Jwa, N. S., Iwahashi, Y., Iwahashi, H., Kim, D. H., Shim, Ie. S., Usui, K. 2005. A hydroponic rice seedling culture model system for investigating proteome of salt stress in rice leaf. Electrophoresis 26:4521–4539.

Komatsu, S., Tanaka, N. 2005. Rice proteome analysis: a step toward functional analysis of the rice genome. Proteomics 5:938–949.

Komatsu, S., Yamada, E., Furukawa, K. 2009. Cold stress changes the concanavalin A-positive glycosylation pattern of proteins expressed in the basal parts of rice leaf sheaths. Amino Acids 36:115–123.

Kruft, V., Eubel, H., Jansch, L., Werhahn, W., Braun, H. P. 2001. Proteomic approach to identify novel mitochondrial proteins in *Arabidopsis*. Plant Physiol 127:1694–1710.

Lee, D. G., Ahsan, N., Lee, S. H., Kang, K. Y., Bahk, J. D., Lee, I. J., Lee, B. H. 2007. A proteomic approach in analyzing heat-responsive proteins in rice leaves. Proteomics 7:3369–3383.

Lee, D. G., Ahsan, N., Lee, S. H., Lee, J. J., Bahk, J. D., Kang, K. Y., Lee, B. H. 2009. Chilling stress-induced proteomic changes in rice roots. J. Plant Physiol 166:1–11.

Lister, R., Carrie, C., Duncan, O., Ho, L. H. M., Howell, K. A., Murcha, M. W., Whelan, J. 2007. Functional definition of outer membrane proteins involved in preprotein import into mitochondria. Plant Cell 19:3739–3759.

Lister, R., Chew, O., Lee, M. N., Heazlewood, J. L., Clifton, R., Parker, K. L., Millar, A. H., Whelan, J. 2004. A transcriptomic and proteomic characterization of the *Arabidopsis* mitochondrial protein import apparatus and its response to mitochondrial dysfunction. Plant Physiol 134:777–789.

Meyer, E. H., Heazlewood, J. L., Millar, A. H. 2007. Mitochondrial acyl carrier proteins in *Arabidopsis thaliana* are predominantly soluble matrix proteins and none can be confirmed as subunits of respiratory Complex I. Plant Mol Biol 64:319–327.

Millar, A. H., Heazlewood, J. L. 2003. Genomic and proteomic analysis of mitochondrial carrier proteins in *Arabidopsis*. Plant Physiol 131:443–453.

Millar, A. H., Heazlewood, J. L., Kristensen, B. K., Braun, H. P., Moller, I. M. 2005. The plant mitochondrial proteome. Trends Plant Sci 10:36–43.

Millar, A. H., Leaver, C. J., Hill, S. A. 1999. Characterization of the dihydrolipoamide acetyltransferase of the mitochondrial pyruvate dehydrogenase complex from potato and comparisons with similar enzymes in diverse plant species. Eur J Biochem 264:973–981.

Millar, A. H., Sweetlove, L. J., Giege, P., Leaver, C. J. 2001. Analysis of the *Arabidopsis* mitochondrial proteome. Plant Physiol 127:1711–1727.

Millar, A. H. , Trend, A. E., Heazlewood, J. L. 2004. Changes in the mitochondrial proteome during the anoxia to air transition in rice focus around cytochrome-containing respiratory complexes. J Biol Chem 279:39471–39478.

Ndimba, B. K., Chivasa, S., Simon, W. J., Slabas, A. R. 2005. Identification of *Arabidopsis* salt and osmotic stress responsive proteins using two-dimensional difference gel electrophoresis and mass spectrometry. Proteomics 5:4185–4196.

O'Toole, N., Hattori, M., Andres, C., Iida, K., Lurin, C., Schmitz-Linneweber, C., Sugita, M., Small, I. 2008. On the expansion of the pentatricopeptide repeat gene family in plants. Mol Biol Evol 25:1120–1128.

Palmblad, M., Mills, D. J., Bindschedler, L. V. 2008. Heat-shock response in *Arabidopsis thaliana* explored by multiplexed quantitative proteomics using differential metabolic labeling. J Proteome Res 7:780–785.

Peltier, J. B., Cai, Y., Sun, Q., Zabrouskov, V., Giacomelli, L., Rudella, A., Ytterberg, A. J., Rutschow, H., van Wijk, K. J. 2006. The oligomeric stromal proteome of *Arabidopsis thaliana* chloroplasts. Mol Cell Proteomics 5:114–133.

Perry, A. J., Hulett, J. M., Likic, V. A., Lithgow, T., Gooley, P. R. 2006. Convergent evolution of receptors for protein import into mitochondria. Curr Biol 16:221–229.

Reumann, S., Babujee, L., Ma, C., Wienkoop, S., Siemsen, T., Antonicelli, G. E., Rasche, N., Luder, F., Weckwerth, W., Jahn, O. 2007. Proteome analysis of *Arabidopsis* leaf peroxisomes reveals novel targeting peptides, metabolic pathways, and defense mechanisms. Plant Cell 19:3170–3193.

Sarry, J. E., Kuhn, L., Ducruix, C., Lafaye, A., Juno, C., Hugouvieux, V., Jourdain, A., Bastien, O., Fievet, J. B., Vailhen, D., Amekraz, B., Moulin, C., Ezan, E., Garin, J., Bourguignon, J. 2006. The early responses of *Arabidopsis thaliana* cells to cadmium exposure explored by protein and metabolite profiling analyses. Proteomics 6:2180–2198.

Sweetlove, L. J., Heazlewood, J. L., Herald, V., Holtzapffel, R., Day, D. A., Leaver, C. J., Millar, A. H. 2002. The impact of oxidative stress on *Arabidopsis* mitochondria. Plant J 32:891–904.

Taylor, N.L., Day, D. A., Millar, A. H. 2002. Environmental stress causes oxidative damage to plant mitochondria leading to inhibition of glycine decarboxylase. J Biol Chem 277:42663–42668.

Taylor, N. L., Heazlewood, J. L., Day, D. A., Millar, A. H. 2004. Lipoic acid-dependent oxidative catabolism of alpha-keto acids in mitochondria provides evidence for branched-chain amino acid catabolism in *Arabidopsis*. Plant Physiol 134:838–848.

Taylor, N. L., Tan, Y. F., Jacoby, R. P., Millar, A. H. 2009. Abiotic environmental stress induced changes in the *Arabidopsis thaliana* chloroplast, mitochondria and peroxisome proteomes. J Proteomics 72:367–378.

Taylor, N. L., Howell, K. A., Heazlewood, J. L., Tan, T. Y., Narsai, R., Huang, S., Whelan, J., Millar, A. H. 2010. Analysis of the rice mitochondrial carrier family reveals anaerobic accumulation of a basic amino acid carrier involved in arginine metabolism during seed germination. Plant Physiol 154:691–704.

Tonge, R., Shaw, J., Middleton, B., Rowlinson, R., Rayner, S., Young, J., Pognan, F., Hawkins, E., Currie, I., Davison, M. 2001. Validation and development of fluorescence two-dimensional differential gel electrophoresis proteomics technology. Proteomics 1:377–396.

Werhahn, W., Braun, H. P. 2002. Biochemical dissection of the mitochondrial proteome from *Arabidopsis thaliana* by three-dimensional gel electrophoresis. Electrophoresis 23:640–646.

Yan, S., Tang, Z., Su, W., Sun, W. 2005. Proteomic analysis of salt stress-responsive proteins in rice root. Proteomics 5: 235–244.

Yan, S. P., Zhang, Q. Y., Tang, Z. C., Su, W. A., Sun, W. N. 2006. Comparative proteomic analysis provides new insights into chilling stress responses in rice. Mol Cell Proteomics 5:484–496.

Zhang, X-P., Sjoling, S., Tanudji, M., Somogyi, L., Andreu, D., Eriksson, L. E. G., Graslund, A., Whelan, J., Glaser, E. 2001. Mutagensis and computer modelling approach to study determinants for recognition of signal peptides by the mitochondrial processing peptidase. Plant J 27:427–438.

Chapter 10
Import of RNAs into Plant Mitochondria

Anne-Marie Duchêne, Samira El Farouk-Ameqrane, François Sieber, and Laurence Maréchal-Drouard

Abstract The plant mitochondrial genetic system has to provide about 35 polypeptides, which are essential for cell survival as they are components of the respiratory chain or contribute to its biogenesis. Thus, an active mitochondrial translation system is an absolute requisite. With the exception of a few ribosomal proteins, all protein factors involved in plant mitochondrial translation are nuclear-encoded and imported via the classical protein import channel. At the RNA level, the three ribosomal RNAs are encoded by the mitochondrial genome. In contrast, the transfer RNA (tRNA) population encoded by the plant mitochondrial genome is not sufficient to decode the 61 sense codons of the universal genetic code used by plant mitochondria. It is now well established that to compensate for the lack of tRNAs, several nuclear-encoded tRNAs used by the cytosolic translation machi-nery are also found in the mitochondrion. In this review, evolutionary aspects and functions of imported tRNAs are presented. Then, the basic questions on the tRNA mitochondrial import selectivity, regulation, targeting, and translocation in plants are discussed and compared to what has been discovered in tRNA mitochondrial import in evolutionary divergent organisms.

Keywords Transfer RNA • Codon usage • Translation • Aminoacyl-tRNA synthetase • TOM • UDAC

Abbreviations

aaRS	Aminoacyl-tRNA synthetase
cp-like	Chloroplast-like
eEF1α	Elongation factor 1A
RIC	RNA import complex
TOM	Translocase of the outer mitochondrial membrane
VDAC	Voltage-dependent anion channel

L. Maréchal-Drouard (✉)
Institut de Biologie Moléculaire des Plantes, Unité Propre de Recherche du CNRS, associated with Strasbourg University, 12 rue du Général Zimmer 67084, Strasbourg cedex, France
e-mail: Laurence.drouard@ibmp-cnrs.unistra.fr

F. Kempken (ed.), *Plant Mitochondria*, Advances in Plant Biology 1,
DOI 10.1007/978-0-387-89781-3_10, © Springer Science+Business Media, LLC 2011

10.1 Introduction

Biogenesis of plant mitochondria requires the import of up to 2,000 nuclear-encoded proteins (see Chaps. 11 and 12), but the plant mitochondrial genome still encodes about 35 protein genes (e.g., Unseld et al. 1997). Plant mitochondrial translation is essential for the synthesis of proteins involved in oxidative phosphorylation, cytochrome c biogenesis, and protein synthesis. Maintenance, integrity, and efficient expression of the mitochondrial genome are thus fundamental not only in plants but for eukaryotic organisms in general. With the exception of a few ribosomal proteins, all the protein components (the other ribosomal proteins, initiation and elongation factors, and aminoacyl-tRNA synthetases) of the mitochondrial translation machinery are nuclear-encoded and imported into the mitochondria via the general protein import pathway. At the RNA level, the 3 ribosomal RNAs (5S rRNA, 18S rRNA, and 26S rRNA) are encoded by the plant mitochondrial genome. Furthermore, a complete set of tRNAs has to be present in the organelle to allow mitochondrial translation. In higher plants, only part of the required tRNA population is encoded by the mitochondrial genome (http://gobase.bcm.umontreal.ca/). In 1988, the presence of cytosolic tRNA species into bean mitochondria was experimentally proved (Maréchal-Drouard et al. 1988). Indeed, it is now well admitted that the missing tRNAs are encoded by nuclear genes and transported from the cytosol into the organelle (Maréchal-Drouard et al. 1990; Duchêne and Maréchal-Drouard 2001; Glover et al. 2001). This import of tRNAs is not plant-specific, but is rather occurring in a broad variety of organisms (Salinas et al. 2008; Alfonzo and Söll 2009). Mechanisms governing RNA targeting into mitochondria remain poorly understood, and it is only recently that they started to be unraveled.

The present review describe our knowledge in plant tRNA mitochondrial import. It focuses on evolutionary aspects and on the function of imported tRNAs. The most recent data concerning the basic questions on the tRNA mitochondrial import selectivity, regulation, targeting, and translocation are presented. Finally our present knowledge in the plant field is briefly compared to what has been discovered in tRNA import into mitochondria of other evolutionary divergent organisms.

10.2 What Are the Imported tRNAs in Photosynthetic Organisms?

10.2.1 The Number and Identity of Mitochondrial-Encoded tRNAs Change from One Plant Species to Another

Several mitochondrial genomes are now completely sequenced (http://gobase.bcm.umontreal.ca/), in green algae, red algae, bryophytes, and angiosperms, allowing the identification of mitochondrial *trn* genes. Some examples are given in Table 10.1. With the notable exception of *Chlamydomonas reinhardtii*, 17–29 *trn*

Table 10.1 Example of sequenced mitochondrial genomes in photosynthetic organisms

Organism	Abbrev.	Accession number	Genome length (bp)	Number of mito *trn* genes	Number of amino acid with no corresponding mito *trn* genes	References
Chondrus crispus	Chon	Z47547	25836	23	1	Leblanc et al. (1995)
Porphyra purpurea	Porp	AF114794	36753	24	1	Burger et al. (1999)
Chlamydomonas reinhardtii	Chla	CRU03843	15758	3	17	Boer and gray (1988); Michaelis et al. (1990)
Ostreococcus tauri	Ostr	CR954200	44237	28	0	Robbens et al. (2007)
Prototheca wickerhamii	Prot	PWU02970	55328	25	0	Wolff et al. (1994)
Marchantia polymorpha	Marc	MPOMTCG	186609	29	0	Oda et al. (1992a)
Physcomitrella patens	Phys	AB251495	105340	24	1	Terasawa et al. (2007)
Arabidopsis thaliana	Arab	Y08501	366924	23	6	Unseld et al. (1997)
Brassica napus	Bras	AP006444	221853	17	6	Handa (2003)
Beta vulgaris	Beta	BA000024	501020	29	5	Kubo et al. (2000)
Nicotiana tabacum	Nico	BA000042	430597	22	5	Sugiyama et al. (2005)
Oryza sativa	Oryz	BA000029	490520	20	6	Notsu et al. (2002)
Sorghum bicolor	Sorg	DQ984518	468628	18	6	
Tripsacum dactyloides	Trip	DQ984517	704100	18	6	
Triticum aestivum	Trit	AP008982	452528	25	7	Ogihara et al. (2005)
Zea mays strain NB	Zea	AY506529	569630	22	6	Clifton et al. (2004)

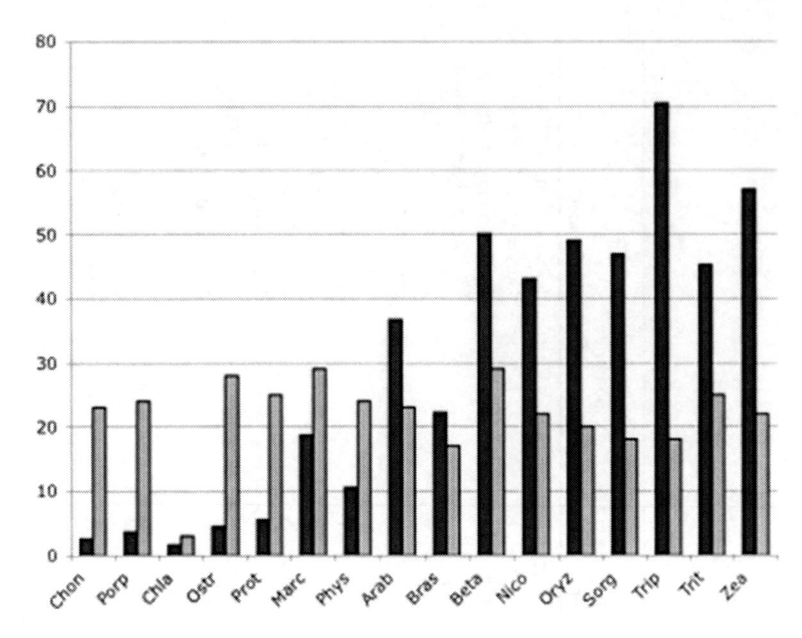

Fig. 10.1 The number of mitochondrial tRNA genes is independent on the genome length. The abbreviations used are those given in Table 10.1. On the histogram, black bars give the length of mitochondrial genome (×10 kb) and grey bars, the number of mitochondrial tRNA genes

genes have been identified in these mitochondrial genomes. This number is independent of the genome length, although angiosperms harbor the largest mitochondrial genomes reported so far (Fig. 10.1). The number of *trn* genes seems sufficient or nearly sufficient for mitochondrial translation in alga (except for *C. reinhardtii*) and bryophytes, since *trn* genes were identified for nearly each amino acid. But it is clearly insufficient in angiosperms where *trn* genes for 5–7 amino acids are missing (Table 10.1) and should be compensated by tRNA import. To be noted, not only the number but also the identity of mitochondrial-encoded tRNAs change from one plant species to the other (Fig. 10.2).

It is worth mentioning that another aspect of mitochondrial genome evolution is the acquisition of genes with another origin. Numerous *trn* genes of plastidial origin have been identified in angiosperm mitochondrial genomes. For example the native genes for $tRNA^{His}$ and $tRNA^{Asn}$ were lost in all angiosperms studied so far and replaced by cp-like genes (Fey et al. 1997), except in wheat where the cp-like *trnH* gene itself was lost and the corresponding tRNA replaced by cytosolic $tRNA^{His}$ imported into mitochondria (Glover et al. 2001; Fig. 10.2). Another example is that of $tRNA^{Cys}$ in *B. vulgaris* mitochondria. The native *trnC1*-GCA is a pseudogene and another *trnC2*-GCA gene of unknown origin is expressed in mitochondria (Kubo et al. 2000).

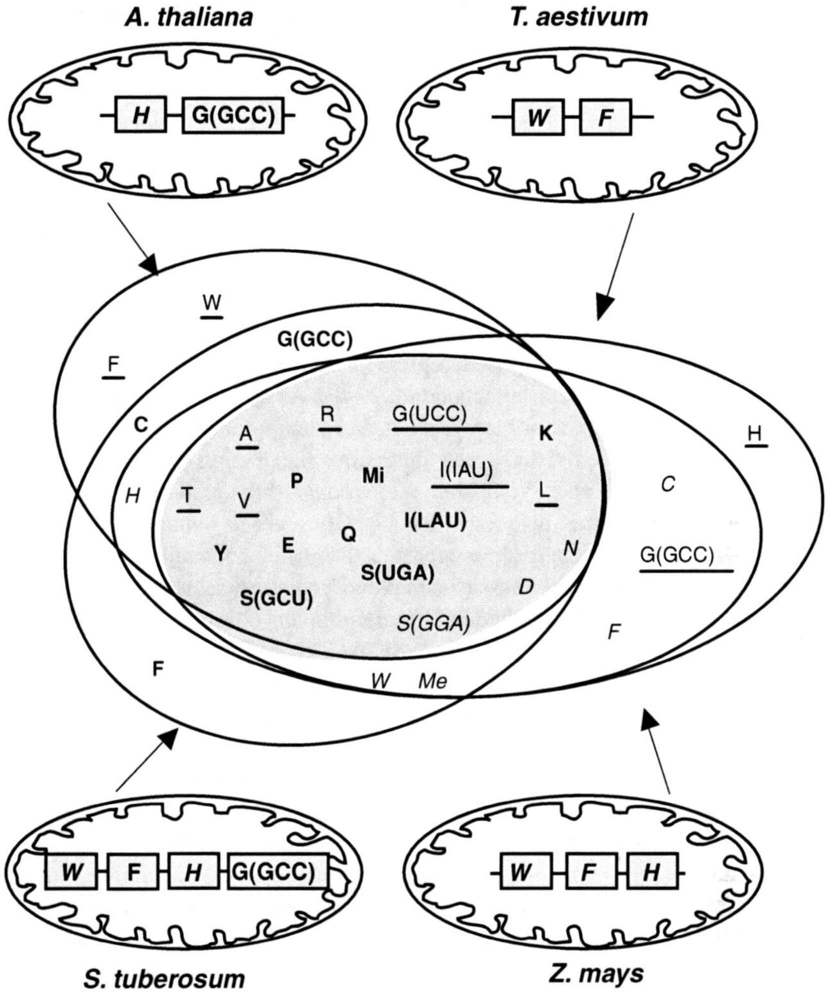

Fig. 10.2 The number and identity of imported tRNA species differ from one plant to the other. The extent of mitochondrial localization varies from one tRNA species to the other. In the middle of the figure, the mitochondrial tRNA population present in two dicotyledonous (*Arabidopsis thaliana* and *Solanum tuberosum*) and two monocotyledonous (*Triticum aestivum* and *Zea mays*) plants is presented. The tRNA population present in the mitochondria of each plant is circled. The tRNAs common to the four plant species is under grey background. On each side of the figure and for each plant, the case of four tRNAs (W, F, H and G(GCC)) is described. The one letter amino acid code is used to represent tRNAs. When necessary, the anticodon is between brackets. Native and chloroplast-like mitochondrially-encoded tRNAs are written in bold and in italics respectively. The imported tRNA species are underlined

10.2.2 Some Cytosolic tRNAs are Imported into Mitochondria

To compensate the missing *trn* genes, the corresponding tRNAs are expected to be imported from the cytosol. However, it is difficult to evaluate the exact number of missing tRNAs for various reasons:

1. Although rare, annotation mistakes cannot be excluded. For example *trnH* is not indicated in *Arabidopsis thaliana* mitochondrial sequence although it is present and expressed (Marienfeld et al. 1999), and *trnI*-CAU (the C being probably post-transcriptionally modified into lysidine L) cannot be found in *Marchantia polymorpha* sequence, probably because it is annotated as *trnM* (Oda et al. 1992b).
2. The minimal number of tRNA isoacceptors for each amino acid is not known. In the extensively studied human mitochondria, 22 tRNAs appear sufficient for translation, with two isoacceptors for Leu and Ser, and a unique one for the 18 other amino acids. However in the red and green algae *Chondrus crispus*, *Porphyra purpurea*, *Ostreococcus tauri*, and *Prototheca wickerhamii* (http://gobase.bcm.umontreal.ca/), two isoacceptors are also found for Arg, Gly, and Met (annotated as Met-e for elongator and Met-I for initiator), suggesting that a minimal number of 25 tRNAs is required for mitochondrial translation in photosynthetic organisms. This is also the minimal number of tRNAs found in *C. reinhardtii* and *O. tauri* chloroplasts (http://gobase.bcm.umontreal.ca/) (Maul et al. 2002; Robbens et al. 2007).
3. In most cases, the expression of *trn* genes is not known and difficult to predict because of the absence of a promoter consensus sequence. For example in *A. thaliana*, contrary to the other higher plants studied so far, cp-like *trnW* is present but not anymore expressed, and the cytosolic $tRNA^{Trp}$ is imported into mitochondria (Duchêne and Maréchal-Drouard 2001; Fig. 10.2).
4. Only a few plant mitochondrial tRNAs have been sequenced at the RNA level. Consequently, the posttranscriptional modifications of nucleotides are not known. When localized in the anticodon, they can affect the decoding capacity of tRNAs. One of the best examples is the recognition of the AUA isoleucine codon by a tRNA encoded by a mitochondrial tRNA gene possessing a CAT (methionine) anticodon. The C present at the first position of the anticodon of this tRNA is posttranscriptionally modified in a derivative of Lysidine (L*), thus permitting both to be aminoacylated with isoleucine (and not methionine) and to pair with AUA isoleucine codon and not with AUG methionine codon (Weber et al. 1990).

Therefore tRNA import can be anticipated only when mitochondrial *trn* genes for one amino acid are missing (Table 10.1), giving a minimal number of imported tRNAs. But this number is clearly underestimated, as shown by the extensive studies performed in *C. reinhardtii* and *Triticum aestivum*. With an absolute need of 25 tRNAs (see above), at least 22 and 10 cytosolic tRNAs should be respectively imported into *C. reinhardtii* and *T. aestivum* mitochondria, while 31 and 16 were identified. Among them, 5 and 4 cytosolic $tRNA^{Leu}$ were respectively detected in *C. reinhardtii* and *T. aestivum* mitochondria (Glover et al. 2001; Vinogradova et al. 2009). Such differences could be correlated to the mitochondrial translation requirements and to the presence of modified nucleotides restricting the codon-anticodon recognition (see previous text).

10.2.3 tRNA Import and Evolution

Gene loss and tRNA import acquisition have occurred very early during plant evolution. A very ancient acquisition of import is that of tRNA^{Ile} in land plants. In the algae *P. wickerhamii*, 2 *trnI* genes with GAU and CAU anticodons were found in mitochondrial genome, allowing the reading of the 3 Ile codons. In *C. crispus*, *P. purpurea*, but also in *Mesostigma viride* mitochondrial genomes (Turmel et al. 2002), only one *trnI*-GAU gene is present. How is AUA codon read? tRNA^{Ile} import can be imagined. However, a unique *trnI*-GAU seems sufficient for chloroplastic translation in *O. tauri* (Robbens et al. 2007). In land plants (bryophytes and angiosperms) the *trnI*-GAU gene was lost and only the *trnI*-CAU gene is present (C being converted into L* after posttranscriptional modification). Like *Escherichia coli* tRNA^{Ile}(LAU), plant mitochondrial tRNA^{Ile}(L*AU) is supposed to decode only AUA codons, and tRNA import seems necessary for reading AUU and AUC codons. Indeed, import of a tRNA^{Ile}(IAU) (A edited into inosine I) was shown in *M. polymorpha* (Akashi et al. 1996), in *T. aestivum* (Glover et al. 2001) and is likely to occur in *Solanum tuberosum* (Maréchal-Drouard et al. 1990) (Table 10.2).

Comparing the different genomes shows that *trn* genes for Ala, Arg, Leu, Thr, and Val were lost in all known angiosperm mitochondrial genomes. Import of the corresponding cytosolic tRNAs was shown in *T. aestivum* (except for tRNA^{Thr}), *Zea mays*, *S. tuberosum* (Table 10.2), (Maréchal-Drouard et al. 1990; Kumar et al. 1996; Glover et al. 2001). Interestingly, several mitochondrial tRNA genes that are still present in angiosperms are missing in the gymnosperm larch and the corresponding cytosolic tRNAs are imported in this plant species (Kumar et al. 1996). In the same range of idea, in green algae, the number of missing mitochondrial tRNA genes vary from zero (e.g., in *P. wickerhamii*) to most of them (e.g., in *C. reinhardtii*).

Differences between dicots and monocots are also observed, for example for tRNA^{Gly}. Two *trnG* genes are found in most algae and bryophytes. One of these (*trnG*-UCC) was lost in all angiosperms. The second *trnG* (GCC anticodon) is still present in mitochondrial genomes of dicots but was lost in all monocots mitochondrial genomes. Import of tRNA^{Gly}(UCC) but also of tRNA^{Gly}(CCC) was shown in the dicots, tobacco and potato (Brubacher-Kauffmann et al. 1999; Salinas et al. 2005). Import of both tRNA^{Gly}(UCC) and tRNA^{Gly} (GCC) was shown in wheat and maize (Kumar et al. 1996; Glover et al. 2001; Table 10.2). A more recent loss is that of *trnF* in Brassicaceae (*A. thaliana* and *Brassica napus*) (Chen et al. 1997). Last, loss of gene and import of the corresponding tRNA can occur in individual species, for example tRNA^{Trp} in *A. thaliana* (Duchêne and Maréchal-Drouard 2001; Table 10.2).

No complete sequence of a mitochondrial genome from a gymnosperm is available. However, based on experimental evidence, at least 11 nuclear-encoded tRNAs are imported into larch mitochondria and some of them were never reported to be so in any plant species, including angiosperms (Kumar et al. 1996).

Table 10.2 Loss of mitochondrial *trn* genes and import of nuclear-encoded tRNA into mitochondria

				Other missing *trn* genes or loss of *trn* expression	Demonstration of the corresponding tRNA import	Reference
Rhodophytes			*C. crispus*	T		
			P. purpurea	T		
Chlorophytes			*C. reinhardtii*	all but 3	31	Vinogradova et al. (2009)
			O. tauri	–		
			P. wickerhamii	–		
Bryophytes			*M. polymorpha*	–	I(IAU)	Akashi et al. (1996)
			P. patens	N		
Angiosperms	Dicots	Brassicales	*A. thaliana*	W, M-e	W	Duchêne and Maréchal-Drouard (2001)
					F	Chen et al. (1997)
Loss oftrnA,R,L,T,V and 1 *trnG*		Loss of *trnF*				
			B. napus	–		
		Caryophyllales	*B. vulgaris*	–		
		Solanales	*N. tabacum*	–	V	Delage et al. (2003b)
					G	Salinas et al. (2005)
			S. tuberosum[a]	nd	ARLTI	Maréchal-Drouard et al. (1990)
					G	Brubacher-Kauffmann et al. (1999)
					V	Kumar et al. (1996)
	Monocots		*O. sativa*	–		
			S. bicolor	–		
	Loss of the 2nd *trnG*		*T. dactyloides*	–		
			T. aestivum	H	A R L V G H I(IAU)	Glover et al. (2001)
			Z. mays strain NB	–	A R L T V G	Kumar et al. (1996)

nd not determined

[a]*Solanum tuberosum* mitochondrial genome is not completely sequenced, but tRNA import into mitochondria has been extensively studied in this organism

Altogether, it is quite clear that each imported cytosolic tRNA can be acquired by mitochondria independently at different times during plant evolution and that these acquisitions are mostly correlated with the loss of the corresponding mitochondrial tRNA genes. In *C. reinhardtii*, the mitochondrial genome only encodes three mitochondrial tRNA genes (coding for a $tRNA^{Trp}$, a $tRNA^{Met}$ and a $tRNA^{Gln}$), and 31 cytosolic tRNAs are imported into the mitochondria of this green alga (Vinogradova et al. 2009). In higher plants, the number and identity of the tRNA species imported into mitochondria is flexible. Taken together, it seems that there is no apparent restriction on which tRNA can or cannot be imported into plant mitochondria.

10.3 Why Are tRNAs Imported?

Transfer RNAs, as adapter molecules mediating the translation of the RNA alphabet into the protein alphabet, represent key elements of all translation machineries. If the minimal number of tRNAs (see previous) is not expressed from a mitochondrial genome, then it is tempting to speculate that the imported cytosolic tRNA species will be useful to compensate this incomplete set. In that case, the import of nuclear-encoded tRNAs into mitochondria will be a prerequisite for mitochondrial protein synthesis. Indeed, with a few possible exceptions discussed later, this is exactly the case. For example, in *A. thaliana*, mitochondrial tRNA genes corresponding to six amino acids have been lost during evolution and the corresponding cytosolic tRNAs are now found in the mitochondria (Duchêne and Maréchal-Drouard 2001). A similar situation has been described in wheat or in potato (Maréchal-Drouard et al. 1990; Glover et al. 2001), and in principle there is a quite strong correlation between the missing mitochondrially-encoded tRNAs and the imported nuclear-encoded tRNAs. As already reported previously for $tRNAs^{Ile}$ and $tRNAs^{Gly}$, there are some tRNA isoacceptors corresponding to the same amino acid that are either encoded by the mitochondrial DNA or imported from the cytosol. Here again, the main hypothesis is that the imported tRNA isoacceptors are essential for reading of specific codons that are not efficiently recognized by mitochondrially-encoded tRNA isoacceptors. In most cases, this remains to be experimentally proven. In particular, most modified nucleotides involved in codon/anticodon recognitions are not known for plant mitochondrial tRNAs and thus the potential redundancy of some tRNAs remains to be established. In this range of idea, we can wonder why plant mitochondria do need to import a $tRNA^{Gly}$ with a CCC anticodon, as the nuclear-encoded imported $tRNA^{Gly}$ with a UCC anticodon is, according to the conventional wobble rules, a priori able to read both GGA and GGG codons. In that case we cannot exclude that a hyper- or hypo-modification of the first position of anticodon restricts the codon reading pattern of a given tRNA. Another example of apparent redundancy has been described in *M. polymorpha* mitochondria, where nuclear-encoded $tRNA^{Val}$(AAC) and mitochondrially-encoded $tRNA^{Val}$(UAC) coexist (Akashi et al. 1998). Although we cannot exclude that a few tRNAs play extra functions, a differential codon recognition profile is the most probable explanation for the presence of apparently redundant tRNAs. This is reminiscent of what was recently shown in yeast mitochondria, where the nuclear-encoded $tRNA^{Lys}$(CUU) is essential under high temperature to decode the AAG codon which is no longer recognized by the $tRNA^{Lys}$(UUU) encoded by the mitochondrial genome (Kamenski et al. 2007). Finally, from an evolutionary point of view and to keep an active mitochondrial translation machinery, it would make sense if the acquisition of a new cytosolic tRNA in mitochondria precedes the loss of the corresponding mitochondrial tRNA gene. Thus, in theory all intermediary steps must be visible. In particular, we must find plant mitochondria that are able to import a nuclear-encoded tRNA even though the mitochondrial counterpart is still expressed.

In *E. coli*, in *S. cerevisiae,* and in chloroplasts it has been shown that the tRNA population is adapted to the codon usage (Pfitzinger et al. 1987). Thus we can wonder whether, in addition to tRNA import specificity, the extent of mitochondrial localization of each imported tRNAs varies so as to adjust the concentration to the need of the mitochondrial translation machinery. In agreement with this hypothesis, an analysis of the expression and sublocalization of the nuclear-encoded tRNA species in *C. reinhardtii* recently showed that indeed, in the mitochondria of this green alga where the codon usage is highly biased, the mitochondrial localization of tRNA isoacceptors is tightly linked to both the nuclear and the mitochondrial codon usages. Consequently, the extent of mitochondrial localization for the nuclear-encoded tRNAs is highly variable and ranges from 0% to 98% (Vinogradova et al. 2009; Fig. 10.3). In higher plants, cytosolic and mitochondrial codon usages are quite similar (http://www.kazusa.or.jp/codon/) and finding such correlation is more difficult, if there is any. Nevertheless, according to 2D gel quantification performed on *S. tuberosum* mitochondrial tRNA fraction, $tRNAs^{Leu}$ are the most abundant imported tRNAs and leucine codons are the most often used mitochondrial codons. By contrast, in plant mitochondria valine and threonine codons are 2–3 times less frequent than leucine codons and $tRNAs^{Val}$ and $tRNAs^{Thr}$ were found to be 2–3 times less abundant than $tRNAs^{Leu}$ (L. Maréchal-Drouard, personal communication). These variations can represent another example of optimization of the plant mitochondrial tRNA population to obtain an efficient mitochondrial protein synthesis. Further experiments will be required to confirm this hypothesis and more importantly to understand how this goal can be achieved.

10.4 Use of Imported tRNA by the Mitochondrial Translational Apparatus

As fidelity of translation is dependent on the accuracy of the aminoacylation reaction, it is generally admitted that a strong co-evolution must exist between aminoacyl-tRNA synthetases (aaRSs) and their cognate tRNAs. In plant mitochondria, this question, which is particularly complex because different tRNAs with different origins coexist, has been addressed in *A. thaliana* (Duchêne et al. 2005; Duchêne et al. 2009). Cytosolic tRNAs corresponding to nine amino acids are expected to be imported into *A. thaliana* mitochondria (Tables 10.2 and 10.3). All their cognate

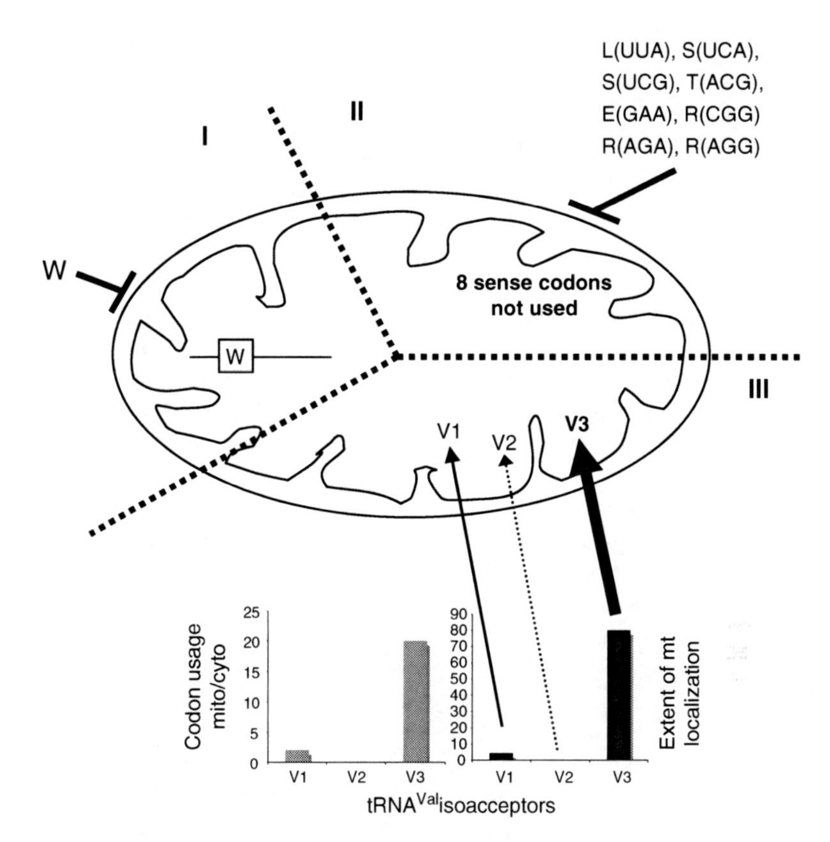

Fig. 10.3 The extent of mitochondrial localization varies from one tRNA species to the other in *Chlamydomona reinhardtii*. The biological significance for the differential import of cytosolic tRNAs into *C. reinhardtii* mitochondria is presented. **I** A $tRNA^{Trp}$ expressed from a mitochondrial gene exists and there is no import of cytosolic $tRNA^{Trp}$, **II** In *C. reinhardtii* mitochondria, the codon distribution is highly biased and 8 sense codons are not used, the corresponding cytosolic tRNAs were not found in mitochondria, **III** the extent of mitochondrial localization (given for the three $tRNA^{Val}$ isoacceptors, V1: V(AAC), V2:V(CAC) and V3: V(UAC)) is linked not only to the frequency of the cognate codon in mitochondria but also to its frequency in the cytosol (for example, see the codon usage mito/cyto ratio given for the $tRNA^{Val}$ isoacceptor family)

Table 10.3 Imported tRNAs and their cognate aaRSs in *A. thaliana* mitochondria

	Mitochondrial aaRSs	
Imported tRNAs	Cyto-mito cytosolic-like	Mito-chloro bacterial-like
$tRNA^{Ala}$	Yes (1)	Yes
$tRNA^{Gly}$ (5)	Yes (2)	Yes (3)
$tRNA^{Val}$	Yes	Yes
$tRNA^{Thr}$	Yes	Yes
$tRNA^{Leu}$	Yes	No
$tRNA^{Ile}$ (5)	No	Yes
$tRNA^{Phe}$	No	Yes (4)
$tRNA^{Trp}$	No	Yes

Although imported, $tRNA^{Arg}$ is not indicated because the mitochondrial ArgRS is not clearly identified. (1) Also targeted to chloroplasts; (2) inactive in mitochondria; (3) active in mitochondria; (4) similar to other mitochondrial PheRSs; (5) coexist with mitochondrial-encoded tRNA isoacceptors (Box 10.1)

Box 10.1 RNA import in other compartments or organisms

Several chloroplast genomes of different land plants have been completely sequenced (http://gobase.bcm.umontreal.ca/). Unlike plant mitochondria, it is generally assumed that the chloroplast genome encodes all tRNA species needed for protein synthesis and that chloroplasts of photosynthetic plants do not import tRNAs from the cytosol. By contrast, the reduced, but still transcribed chloroplast genome of the nonphotosynthetic plant, *Epifagus virginiana*, lacks at least 11 tRNA genes (Wolfe et al. 1992). Translation of plastidial mRNAs in this obligate parasite is therefore likely to involve cytosol-imported tRNAs, suggesting that in some extreme cases, tRNA import may also occurs in plastids.

Transport of tRNAs from the cytosol to the mitochondria is not restricted to plants but is a quite general phenomenon (e.g., Schneider and Maréchal-Drouard 2000; Salinas et al. 2008; Alfonzo and Söll 2009). It ranges from the import of only one cytosolic $tRNA^{Lys}$ in the opossum *Didelphis virginiana* to all tRNAs in some protozoans such as *Trypanosoma brucei* or *Leishmania tarentolae*. In the yeast *Saccharomyces cerevisiae*, the cytosolic $tRNA^{Lys}$(CUU) is imported. It must be noted that in this organism as well as in *Homo sapiens*, the mitochondrial import of cytosolic $tRNA^{Gln}$ has also been reported (Rinehart et al. 2005; Rubio et al. 2008). However, recent data on the formation of Gln-$tRNA^{Gln}$ via the indirect transamidation pathway in the mitochondria of these two organisms weaken either the efficient mitochondrial import of cytosolic $tRNA^{Gln}$ or their involvement in mitochondrial translation (Frechin et al. 2009; Nagao et al. 2009). Whereas the natural import of tRNA is still under debate in human mitochondria, the import of 5S ribosomal RNA (rRNA) has been clearly demonstrated (Entelis et al. 2001), but the function of this rRNA, which was not found so far associated to mitochondrial ribosomes remains to be determined. In *S. cerevisiae*, the imported $tRNA^{Lys}$(CUU), which at first glance appears to be redundant with the mitochondrially-encoded $tRNA^{Lys}$(UUU), is essential in the conditional adaptation of mitochondrial protein synthesis under heat stress condition. In all the other organisms where tRNA import has been experimentally proven, the role of the imported tRNAs in the mitochondrion is very likely to participate also to the mitochondrial protein synthesis, as they usually compensate the lack of mitochondrial tRNA genes. This is particularly obvious in the case of protozoans such as *T. brucei*. No tRNA gene is encoded on the *T. brucei* mitochondrial genome and all required tRNAs are imported. Two tRNAs ($tRNA^{Sec}$ and $tRNA^{Meti}$) remain in the cytosol as they are not necessary to get an active mitochondrial translation machinery (Bouzaidi-Tiali et al. 2007).

Whereas the transport of RNA into mitochondria appears to be a common and essential process for the biogenesis of the organelle in evolutionary divergent organisms, the question of the mechanism(s) involved is still not solved yet.

(continued)

Box 10.1 (continued)

As in plant mitochondria, it seems that in protozoans, such as *Leishmania* and *T. brucei*, the mechanism is independent of the protein import pathway and can be reproduced in vitro in the absence of cytosolic factors (Schneider and Maréchal-Drouard 2000; Salinas et al. 2008; Alfonzo and Söll 2009). However, this does not exclude that in vivo, cytosolic factors are required. One example is the crucial role played by the cytosolic translation elongation factor eEF1α in targeting cytosolic tRNAs to the surface of mitochondria in *T. brucei* (Bouzaidi-Tiali et al. 2007). In contrast to plants, the homologue of VDAC is not necessary for tRNA import in vivo in *T. brucei* (Pusnik et al. 2009). One of the most studied tRNA mitochondrial import mechanism is that of *Leishmania*. In this organism, an RNA Import Complex (RIC) constituted of a dozen of subunits (including the α subunit of the F1-ATP synthase, subunit 6b, and the Rieske protein of complex III of the respiratory chain and subunit 6 of complex IV) has been identified on the inner mitochondrial membrane (Mukherjee et al. 2007). However, RIC has not been found so far in any other mitochondria of protozoan and the Rieske protein was shown not to be involved in tRNA mitochondrial import in *T. brucei* (Paris et al. 2009). Besides the existence of pathways that are not directly dependent on the protein import channel, a second mechanism utilizing the protein import pathway has been described for tRNA mitochondrial import. In *S. cerevisiae*, the imported cytosolic $tRNA^{Lys}$(CUU) is first aminoacylated by the cytosolic lysyl-tRNA synthetase and then targeted to the surface of mitochondria by a shuttle, the glycolytic enzyme enolase. Finally, the tRNA is taken up by the precursor form of the mitochondrial lysyl-tRNA synthetase before its translocation through the protein import channel (Tarassov et al. 1995; Entelis et al. 2006). Taken together, evidence for distinct tRNA import machineries seems to emerge. However, due to controversial data, the use of different approaches (in particular in vitro and in vivo approaches are not per se comparable), and a still limited number of studies, it is probably too soon to draw conclusions. Further genetic and biochemical studies will be required to delineate how divergent these tRNA import mechanisms are. Finally, up to now, only the import of abundant RNA molecules (tRNA and 5s rRNA) has been clearly demonstrated. Is there any restriction for the import of this type of RNAs into mitochondria? To answer this question, one of the next challenges will be to look for the import of other types of RNAs present as minor species.

aaRSs were shown to be dual targeted to another compartment, either the cytosol or chloroplasts (Table 10.3). The cytosolic-mitochondrial aaRSs are cytosolic enzymes that have acquired a mitochondrial targeting signal during evolution. The chloroplastic-mitochondrial enzymes present no similarities with mitochondrial aaRS in other organisms (except for PheRS) but strong similarities with bacterial aaRSs, in particular with cyanobacterial aaRSs in half cases (Duchêne et al. 2009).

Among the imported tRNA species found in *A. thaliana* mitochondria, $tRNA^{Leu}$ is recognized by a cyto-mito cytosolic-like LeuRS. By contrast, $tRNA^{Ile}$, $tRNA^{Phe}$ and $tRNA^{Trp}$ are recognized by chloro-mito bacterial type enzymes. The situation is even more complex for $tRNA^{Ala}$, $tRNA^{Gly}$, $tRNA^{Val}$ and $tRNA^{Thr}$ since in these cases two a priori redundant aaRSs are found in mitochondria, one cyto-mito and one chloro-mito. The role of these enzymes in mitochondria is an open question. A specialization is suggested in the case of GlyRSs, because the cyto-mito GlyRS has no apparent aminoacylation activity within the organelle, this activity being assured by the chloro-mito GlyRS (Duchêne et al. 2001). Therefore the tRNA/aaRS coevolution is not evident for imported tRNAs into plant mitochondria.

10.5 Mechanism of tRNA Mitochondrial Import

In addition to the question of selectivity and regulation of tRNA import presented above, there are two other major aspects that need to be discussed when talking about the mechanism of tRNA mitochondrial import. First, we need to understand how some cytosolic tRNAs exported from the nucleus escape the channeling to the cytosolic translation machinery to be targeted to the surface of mitochondria. Second, we need to characterize the tRNA mitochondrial membrane import channel(s) and to identify the protein factors constituting the channel(s).

Though plant tRNA sequences are highly conserved, the number and the identity of imported tRNAs greatly vary from one plant to the other. Thus it is quite difficult to understand how different plants regulate the import of identical tRNAs. Using in vivo approaches, different nucleotides or domains were shown to be essential for the import of a few cytosolic tRNAs: position 73 in $tRNA^{Ala}$, the anticodon and the D-arm in $tRNA^{Val}$, and $tRNA^{Gly}$ (Dietrich et al. 1996; Delage et al. 2003b; Salinas et al. 2005; Fig. 10.4A). In a few cases (e.g., position 73 in $tRNA^{Ala}$, the anticodon in $tRNA^{Val}$), mutations of these nucleotides prevent tRNA import in vivo as well as the recognition by the cognate aaRS. However, this is not always true. For example, mutations in the D-loop of $tRNA^{Val}$ or in the anticodon or D-loop of $tRNA^{Gly}$ lead to in vivo tRNA import inhibition although the recognition by the cognate aaRS is not affected. This shows that import determinants carried by cytosolic tRNAs can be distinct from the identity elements required for recognition by the cognate aaRS. Furthermore, recognition of a tRNA by an imported aaRS is not sufficient to permit its import into the organelle (Brubacher-Kauffmann et al. 1999; Mireau et al. 2000). An elegant hypothesis can be that if a cytosolic aaRS recognizes and aminoacylates a tRNA, then this tRNA will be channeled to the cytosolic translation machinery, but if a tRNA is recognized by the precursor form of a mitochondrial aaRS, then it will be addressed to the mitochondria. However, the confirmation that aaRSs play a crucial role either in the targeting and/or in the translocation of tRNAs through the double mitochondrial membrane is still waiting for appropriate experiments.

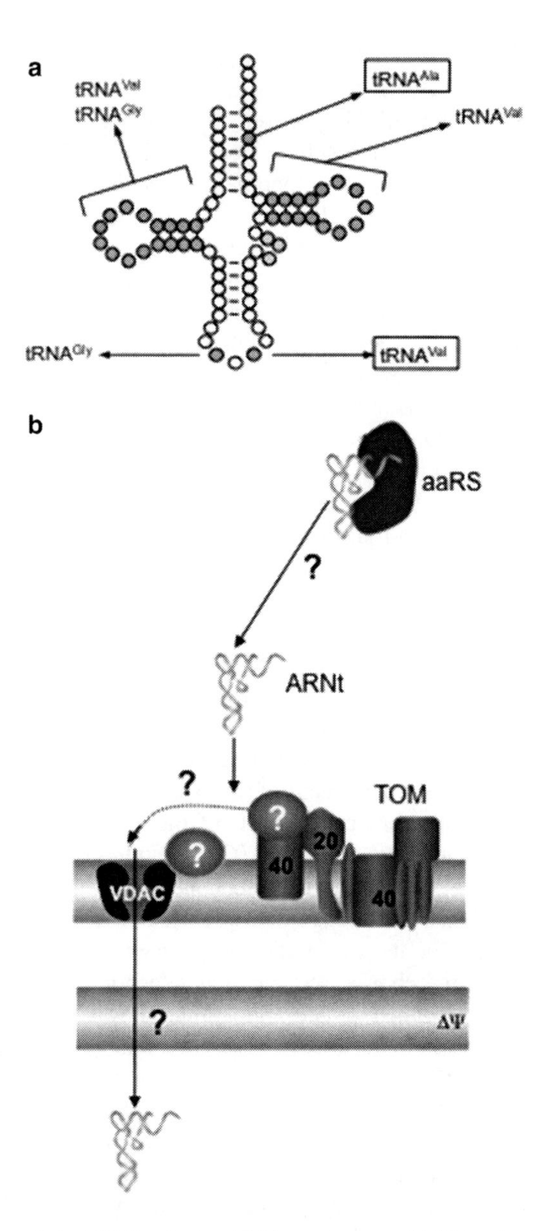

Fig. 10.4 Overview of plant tRNA mitochondrial import process. (**a**) The major import determinants (grey dots) identified in mature $tRNA^{Ala}$, $tRNA^{Val}$ and $tRNA^{Gly}$ are presented. Their identification is based on the abolition of mitochondrial import of mutated version of tRNAs expressed in transgenic tobacco plants. When the import determinant is also an identity element for the recognition by the cognate aaRS, the tRNA is surrounded. (**b**) Model of tRNA import into higher plant mitochondria. The hypothesis that aaRSs can play a role in the tRNA targeting and/or translocation steps is presented with a question mark. As shown by in vitro import experiments, the implication of TOM20 and TOM40, two proteins of the translocase of the outer membrane involved in the protein import pathway, and of VDAC, the voltage dependent anion channel for the translocation step is presented. The involvement of additional protein factors at the level of the outer membrane is indicated by grey ovals with question marks. How the translocation step through the mitochondrial inner membrane is achieved is still unknown. This model does not preclude the existence of other pathways to import tRNAs into plant mitochondria

Altogether, tRNA mitochondrial import selectivity appears complex in plants and we can wonder whether other levels of regulation do exist. In particular, the potential influence of the set of mitochondrial-encoded tRNA species on the identity of the imported tRNAs will need to be studied. Finally, we cannot also exclude the possibility that tRNAs, once translocated, are differentially degraded inside the mitochondrion and that their steady-state level results from both their import efficiency and their stability within the organelle.

Once the cytosolic tRNAs are bound to the surface of mitochondria, as a final step in the import process they need to be translocated through the two mitochondrial membranes to reach the mitochondrial matrix. With a size of 25 kDa, tRNAs represent large hydrophilic and polyanionic macromolecules and, as for proteins, an active import process involving several protein factors, is expected to occur. By contrast to the well-known and versatile set of machineries and mechanisms involved in protein mitochondrial import (Chacinska et al. 2009), it is only during the last decade that tRNA import pathways began to be elucidated. Still a lot of questions remain; this is particularly true in the plant field where only scarce data exist. An in vitro system set up with isolated potato mitochondria gave the first clues (Fig. 10.4b; Delage et al. 2003a). It showed that tRNA import requires both a membrane potential and ATP. This later requirement constitutes a feature common to all tRNA import systems described so far (Salinas et al. 2008; Alfonzo and Söll 2009). In vitro tRNA import was achieved in the absence of any added cytosolic protein factors and is independent of the protein import channel. However, it is worth noting that two major components of the TOM (for translocase of the outer mitochondrial membrane) complex (TOM20 and TOM40) are important for the binding of tRNAs at the surface of mitochondria (Salinas et al. 2006). Under these in vitro conditions, the voltage-dependent anion channel (VDAC), which plays a key role in metabolite transport, is implicated in the tRNA-translocation step through the outer mitochondrial membrane. When in vitro import of different cytosolic tRNAs into isolated potato mitochondria was tested, the selectivity observed did not always reflect the in vivo situation. For example, the potato cytosol-specific $tRNA^{Gly}$(GCC) was as efficiently imported into isolated mitochondria as the potato imported $tRNA^{Gly}$(UCC), (Salinas et al. 2005). Among the different hypotheses to explain this discrepancy, one easiest explanation is that cytosolic factors involved in the selectivity of the process in vivo are missing in vitro. We cannot also exclude that different pathways are used by plant mitochondria to import different tRNAs.

Glossary

Aminoacyl-tRNA synthetase: enzyme that catalyzes the amino acid attachment at the 3′ extremity of a tRNA molecule.

Angiosperm: the flowering plants.

Bryophyte: embryophytes (often called "land plants") that are nonvascular. They neither have flowers nor produce seeds and reproduce via spores.

Cytochromec: a small heme (an iron atom contained in a large heterocyclic ring called porphyrin) protein associated with the inner mitochondrial membrane that is an essential component for the electron transport chain.

D-arm: in the secondary structure of a "classical" tRNA molecule, the D-arm is usually a 4 bp stem ending in a loop that often contains dihydrouridine.

Enolase: a metalloenzyme of the glycolysis that converts the 2-phosphoglycerate to phosphoenolpyruvate.

Gymnosperm: a group of spermatophyte seed-bearing plants with ovules on scales of a cone or similar structure.

Lysidine: a derivative of cytidine in which the carbonyl is replaced by the amino acid lysine. Lysidine typically occurs in the anticodon of a tRNA molecule.

Oxidative phosphorylation: in mitochondria, a metabolic pathway that uses energy released by the oxidation of NADH and succinate to produce ATP.

Selenocysteine: an amino acid. It has a structure similar to cysteine, but with an atom of selenium taking the place of the sulfur. Often abbreviated Sec.

Translocase of the outer mitochondrial membrane: a protein complex located in the outer mitochondrial membrane of mitochondria and involved in the translocation of nuclear-encoded proteins from the cytosol to the inter membrane space of mitochondria.

References

Akashi, K., Sakurai, K., Hirayama, J., Fukuzama, H., Ohyama, K. 1996. Occurence of nuclear-encoded $tRNA^{Ile}$ in mitochondria of the liverwort *Marchantia polymorpha*. Curr. Genet. 30:181–185.

Akashi, K., Takenaka, M., Yamaoka, S., Suyama, Y., Fukuzawa, H., Ohyama, K. 1998. Coexistence of nuclear DNA-encoded tRNAVal(AAC) and mitochondrial DNA-encoded tRNAVal(UAC) in mitochondria of a liverwort *Marchantia polymorpha*. Nucleic Acids Res. 26:2168–2172.

Alfonzo, J. D., Söll, D. 2009. Mitochondrial tRNA import-the challenge to understand has just begun. Biol. Chem. 390:717–722.

Boer, P. H., Gray, M. W. 1988. Transfer RNA genes and the genetic code in *Chlamydomonas reinhardtii* mitochondria. Curr. Genet. 14:583–590.

Bouzaidi-Tiali, N., Aeby, E., Charriere, F., Pusnik, M., Schneider, A. 2007. Elongation factor 1a mediates the specificity of mitochondrial tRNA import in T. brucei. EMBO. J. 26:4302–4312.

Brubacher-Kauffmann, S., Maréchal-Drouard, L., Cosset, A., Dietrich, A., Duchêne, A. M. 1999. Differential import of nuclear-encoded tRNAGly isoacceptors into solanum *Tuberosum mitochondria*. Nucleic Acids Res. 27:2037–2042.

Burger, G., Saint-Louis, D., Gray, M. W., Lang, B. F. 1999. Complete sequence of the mitochondrial DNA of the red alga *Porphyra purpurea*. Cyanobacterial introns and shared ancestry of red and green algae. Plant Cell. 11:1675–1694.

Chacinska, A., Koehler, C. M., Milenkovic, D., Lithgow, T., Pfanner, N. 2009. Importing mitochondrial proteins: machineries and mechanisms. Cell 138:628–644.

Chen, H. C., Viry-Moussaid, M., Dietrich, A., Wintz, H. 1997. Evolution of a mitochondrial tRNA PHE gene in A. thaliana: import of cytosolic tRNA PHE into mitochondria. Biochem. Biophys. Res. Commun. 237:432–437.

Clifton, S. W., Minx, P., Fauron, C. M., Gibson, M., Allen, J. O., Sun, H., Thompson, M., Barbazuk, W. B., Kanuganti, S., Tayloe, C., Meyer, L., Wilson, R. K., Newton, K. J. 2004. Sequence and comparative analysis of the maize NB mitochondrial genome. Plant Physiol. 136:3486–3503.

Delage, L., Dietrich, A., Cosset, A., Maréchal-Drouard, L. 2003a. In vitro import of a nuclearly encoded tRNA into mitochondria of *Solanum tuberosum*. Mol. Cell. Biol. 23:4000–4012.

Delage, L., Duchêne, A. M., Zaepfel, M., Maréchal-Drouard, L. 2003b. The anticodon and the D-domain sequences are essential determinants for plant cytosolic tRNA(Val) import into mitochondria. Plant J. 34:623–633.

Dietrich, A., Maréchal-Drouard, L., Carneiro, V., Cosset, A., Small, I. 1996. A single base change prevents import of cytosolic tRNA(Ala) into mitochondria in transgenic plants. Plant J. 10:913–918.

Duchêne, A. M., Maréchal-Drouard, L. 2001. The chloroplast-derived *trnW* and *trnM-e* genes are not expressed in *Arabidopsis* mitochondria. Biochem. Biophys. Res. Comm. 285:1213–1216.

Duchêne, A. M., Peeters, N., Dietrich, A., Cosset, A., Small, I. D., Wintz, H. 2001. Overlapping destinations for two dual targeted glycyl-tRNA synthetases in *Arabidopsis thaliana* and *Phaseolus vulgaris*. J. Biol. Chem. 276:15275–15283.

Duchêne, A. M., Giritch, A., Hoffmann, B., Cognat, V., Lancelin, D., Peeters, N. M., Zaepfel, M., Maréchal-Drouard, L., Small, I. D. 2005. Dual targeting is the rule for organellar aminoacyl-tRNA synthetases in *Arabidopsis thaliana*. Proc. Natl. Acad. Sci. U.S.A. 102:16484–16489.

Duchêne, A. M., Pujol, C., Maréchal-Drouard, L. 2009. Import of tRNAs and aminoacyl-tRNA synthetases into mitochondria. Curr. Genet. 55:1–18.

Entelis, N. S., Kolesnikova, O. A., Dogan, S., Martin, R. P., Tarassov, I. A. 2001. 5 S rRNA and tRNA import into human mitochondria. Comparison of in vitro requirements. J. Biol. Chem. 276:45642–45653.

Entelis, N., Brandina, I., Kamenski, P., Krasheninnikov, I. A., Martin, R. P., Tarassov, I. 2006. A glycolytic enzyme, enolase, is recruited as a cofactor of tRNA targeting toward mitochondria in *Saccharomyces cerevisiae*. Genes Dev. 20:1609–1620.

Fey, J., Dietrich, A., Cosset, A., Desprez, T., Maréchal-Drouard, L. 1997. Evolutionary aspects of "chloroplast-like" trnN and trnH expression in higher-plant mitochondria. Curr. Genet. 32:358–360.

Frechin, M., Senger, B., Braye, M., Kern, D., Martin, R. P., Becker, H. D. 2009. Yeast mitochondrial Gln-tRNA(Gln) is generated by a GatFAB-mediated transamidation pathway involving Arc1p-controlled subcellular sorting of cytosolic GluRS. Genes Dev. 23:1119–1130.

Glover, K. E., Spencer, D. F., Gray, M. W. 2001. Identification and structural characterization of nucleus-encoded transfer RNAs imported into wheat mitochondria. J. Biol. Chem. 276:639–648.

Handa, H. 2003. The complete nucleotide sequence and RNA editing content of the mitochondrial genome of rapeseed (*Brassica napus L.*): comparative analysis of the mitochondrial genomes of rapeseed and *Arabidopsis thaliana*. Nucleic Acids Res. 31:5907–5916.

Kamenski, P., Kolesnikova, O., Jubenot, V., Entelis, N., Krasheninnikov, I. A., Martin, R. P., Tarassov, I. 2007. Evidence for an adaptation mechanism of mitochondrial translation via tRNA import from the cytosol. Mol. Cell. 26:625–637.

Kubo, T., Nishizawa, S., Sugawara, A., Itchoda, N., Estiati, A., Mikami, T. 2000. The complete nucleotide sequence of the mitochondrial genome of sugar beet (*Beta vulgaris L.*) reveals a novel gene for tRNA(Cys)(GCA). Nucleic Acids Res. 28:2571–2576.

Kumar, R., Maréchal-Drouard, L., Akama, K., Small, I. 1996. Striking differences in mitochondrial tRNA import between different plant species. Mol. Gen. Genet. 252:404–411.

Leblanc, C., Boyen, C., Richard, O., Bonnard, G., Grienenberger, J. M., Kloareg, B. 1995. Complete sequence of the mitochondrial DNA of the rhodophyte *Chondrus crispus* (Gigartinales). Gene content and genome organization. J. Mol. Biol. 250:484–495.

Maréchal-Drouard, L., Weil, J. H., Guillemaut, P. 1988. Import of several tRNAs from the cytoplasm into the mitochondria in bean *Phaseolus vulgaris*. Nucleic Acids Res. 16:4777–4788.

Maréchal-Drouard, L., Guillemaut, P., Cosset, A., Arbogast, M., Weber, F., Weil, J. H., Dietrich, A. 1990. Transfer RNAs of potato (*Solanum tuberosum*) mitochondria have different genetic origins. Nucleic Acids Res. 18:3689–3696.

Marienfeld, J., Unseld, M., Brennicke, A. 1999. The mitochondrial genome of Arabidopsis is composed of both native and immigrant information. Trends Plant Sci. 4:495–502.

Maul, J. E., Lilly, J. W., Cui, L., dePamphilis, C. W., Miller, W., Harris, E. H., Stern, D. B. 2002. The *Chlamydomonas reinhardtii* plastid chromosome: islands of genes in a sea of repeats. Plant Cell 14:2659–2679.

Michaelis, G., Vahrenholtz, C., Pratje, E. 1990. Mitochondrial DNA of *Chlamydomonas reinhardtii*: the gene for apocytochrome b and the complete functional map of the 15.8 kb DNA. Mol. Gen. Genet. 223:211–216.

Mireau, H., Cosset, A., Maréchal-Drouard, L., Fox, T. D., Small, I. D., Dietrich, A. 2000. Expression of Arabidopsis thaliana mitochondrial alanyl-tRNA synthetase is not sufficient to trigger mitochondrial import of tRNAAla in yeast. J. Biol. Chem. 275:13291–13296.

Mukherjee, S., Basu, S., Home, P., Dhar, G., Adhya, S. 2007. Necessary and sufficient factors for the import of transfer RNA into the kinetoplast mitochondrion. EMBO Rep. 8:589–595.

Nagao, A., Suzuki, T., Katoh, T., Sakaguchi, Y. 2009. Biogenesis of glutaminyl-mt tRNAGln in human mitochondria. Proc. Natl. Acad. Sci. U.S.A. 106:16209–16214.

Notsu, Y., Masood, S., Nishikawa, T., Kubo, N., Akiduki, G., Nakazono, M., Hirai, A., Kadowaki, K. 2002. The complete sequence of the rice (Oryza sativa L.) mitochondrial genome: frequent DNA sequence acquisition and loss during the evolution of flowering plants. Mol. Genet. Genomics 268:434–445.

Oda, K., Yamato, K., Ohta, E., Nakamura, Y., Takemura, M., Nozato, N., Akashi, K., Kanegae, T., Ogura, Y., Kohchi, T., et al. 1992a. Gene organization deduced from the complete sequence of liverwort *Marchantia polymorpha* mitochondrial DNA. A primitive form of plant mitochondrial genome. J. Mol. Biol. 223:1–7.

Oda, K., Yamato, K., Ohta, E., Nakamura, Y., Takemura, M., Nozato, N., Akashi, K., Ohyama, K. 1992b. Transfer RNA genes in the mitochondrial genome from a liverwort, Marchantia polymorpha: the absence of chloroplast-like tRNAs. Nucleic Acids Res. 20:3773–3777.

Ogihara, Y., Yamazaki, Y., Murai, K., Kanno, A., Terachi, T., Shiina, T., Miyashita, N., Nasuda, S., Nakamura, C., Mori, N., Takumi, S., Murata, M., Futo, S., Tsunewaki, K. 2005. Structural dynamics of cereal mitochondrial genomes as revealed by complete nucleotide sequencing of the wheat mitochondrial genome. Nucleic Acids Res. 33:6235–6250.

Paris, Z., Rubio, M. A., Lukes, J., Alfonzo, J. D. 2009. Mitochondrial tRNA import in *Trypanosoma brucei* is independent of thiolation and the Rieske protein. RNA 15:1398–1406.

Pfitzinger, H., Guillemaut, P., Weil, J. H., Pillay, D. T. N. 1987. Adjustment of the tRNA population to the codon usage in chloroplasts. Nucleic Acids Res. 15:1377–1386.

Pusnik, M., Charriere, F., Maser, P., Waller, R. F., Dagley, M. J., Lithgow, T., Schneider, A. 2009. The single mitochondrial porin of *Trypanosoma brucei* is the main metabolite transporter in the outer mitochondrial membrane. Mol. Biol. Evol. 26:671–680.

Rinehart, J., Krett, B., Rubio, M. A., Alfonzo, J. D., Söll, D. 2005. *Saccharomyces cerevisiae* imports the cytosolic pathway for Gln-tRNA synthesis into the mitochondrion. Genes Dev. 19:583–592.

Robbens, S., Derelle, E., Ferraz, C., Wuyts, J., Moreau, H., Van de Peer, Y. 2007. The complete chloroplast and mitochondrial DNA sequence of *Ostreococcus tauri*: organelle genomes of the smallest eukaryote are examples of compaction. Mol. Biol. Evol. 24:956–968.

Rubio, M. A., Rinehart, J. J., Krett, B., Duvezin-Caubet, S., Reichert, A. S., Söll, D., Alfonzo, J. D. 2008. Mammalian mitochondria have the innate ability to import tRNAs by a mechanism distinct from protein import. Proc. Natl. Acad. Sci. U.S.A. 105:9186–9191.

Salinas, T., Schaeffer, C., Maréchal-Drouard, L., Duchêne, A. M. 2005. Sequence dependence of tRNA(Gly) import into tobacco mitochondria. Biochimie. 87:863–872.

Salinas, T., Duchêne, A. M., Delage, L., Nilsson, S., Glaser, E., Zaepfel, M., Maréchal-Drouard, L. 2006. The voltage-dependent anion channel, a major component of the tRNA import machinery in plant mitochondria. Proc. Natl. Acad. Sci. U.S.A. 103:18362–18367.

Salinas, T., Duchêne, A. M., Maréchal-Drouard, L. 2008. Recent advances in tRNA mitochondrial import. Trends Biochem. Sci. 33:320–329.

Schneider, A., Maréchal-Drouard, L. 2000. Mitochondrial tRNA import: are there distinct mechanisms? Trends Cell Biol. 10:509–513.

Sugiyama, Y., Watase, Y., Nagase, M., Makita, N., Yagura, S., Hirai, A., Sugiura, M. 2005. The complete nucleotide sequence and multipartite organization of the tobacco mitochondrial genome: comparative analysis of mitochondrial genomes in higher plants. Mol. Genet. Genomics 272:603–615.

Tarassov, I., Entelis, N., Martin, R. P. 1995. An intact protein translocating machinery is required for mitochondrial import of a yeast cytoplasmic tRNA. J. Mol. Biol. 245:315–323.

Terasawa, K., Odahara, M., Kabeya, Y., Kikugawa, T., Sekine, Y., Fujiwara, M., Sato, N. 2007. The mitochondrial genome of the moss *Physcomitrella patens* sheds new light on mitochondrial evolution in land plants. Mol. Biol. Evol. 24:699–709.

Turmel, M., Otis, C., Lemieux, C. 2002. The complete mitochondrial DNA sequence of *Mesostigma viride* identifies this green alga as the earliest green plant divergence and predicts a highly compact mitochondrial genome in the ancestor of all green plants. Mol. Biol. Evol. 19:24–38.

Unseld, M., Marienfeld, J. R., Brandt, P., Brennicke, A. 1997. The mitochondrial genome of *Arabidopsis thaliana* contains 57 genes in 366,924 nucleotides. Nat. Genet. 15:57–61.

Vinogradova, E., Salinas, T., Cognat, V., Remacle, C., Maréchal-Drouard, L. 2009. Steady-state levels of imported tRNAs in Chlamydomonas mitochondria are correlated with both cytosolic and mitochondrial codon usages. Nucleic Acids Res. 37:1521–1528.

Weber, F., Dietrich, A., Weil, J. H., Maréchal-Drouard, L. 1990. A potato mitochondrial isoleucine tRNA is coded for by a mitochondrial gene possessing a methionine anticodon. Nucleic Acids Res. 18:5027–5030.

Wolfe, K. H., Morden, C. W., Palmer, J. D. 1992. Function and evolution of a minimal plastid genome from a nonphotosynthetic parasitic plant. Proc. Natl. Acad. Sci. U.S.A. 89:10648–10652.

Wolff, G., Plante, I., Lang, B. F., Kuck, U., Burger, G. 1994. Complete sequence of the mitochondrial DNA of the chlorophyte alga *Prototheca wickerhamii*. Gene content and genome organization. J. Mol. Biol. 237:75–86.

Chapter 11
Protein Import into Plant Mitochondria

Elzbieta Glaser and James Whelan

Abstract The presence of plastids in plant cells requires a higher level of precursor recognition by the mitochondrial protein import apparatus than in nonplant organisms. Although the plant presequences display the overall features observed in yeast and mammals, they are generally longer and more hydrophilic. Most of them are highly organelle specific, but some have ambiguous targeting specificity delivering a protein to both mitochondria and chloroplasts. Many components of plant protein import apparatus appear different to that in yeast and mammalian systems. The three outer membrane mitochondrial proteins characterized to play role as receptors in plants – Tom20, OM64, and metaxin – are plant specific. However, the channel forming units of the TOM and SAM complexes, Tom40 and Sam50, respectively, are orthologous to these components in yeast. While components of the MIA and TIM complexes also display high levels of orthology, functional studies indicate divergences in function and mechanism. Differences exist also in terms of intraorganellar localization of proteolytic events, e.g., the location of the mitochondrial processing peptidase, MPP, involved in removing targeting signals is different, whereas the function and location of the presequence protease, PreP, degrading targeting peptides, is well conserved. Overall, although the protein import machinery of mitochondria from all organisms appears to have coopted and uses the channel forming subunits from the endosymbiont that gave rise to mitochondria, there is a greater diversity in plant components in comparison to those from nonplant species.

Keywords Targeting peptides • Dual targeting • Mitochondrial protein import • Import machinery • TOM • TIM • Precursor processing • Presequence degradation • PreP

E. Glaser (✉) and J. Whelan (✉)
Department of Biochemistry and Biophysics, Stockholm University,
SE-106 91, Stockholm, Sweden
e-mail: e_glaser@dbb.su.se

F. Kempken (ed.), *Plant Mitochondria*, Advances in Plant Biology 1,
DOI 10.1007/978-0-387-89781-3_11,

Abbreviations

AIP	Arylhydrocarbon receptor-interacting protein
cTP	Chloroplast targeting peptide
dTP	Dual targeting peptide
ERV1	Essential for respiration and viability
Hot13p	Helper of small Tms
IDE	Insulin degrading enzyme
MIA	Mitochondrial intermembrane assembly machinery
MPP	Mitochondrial processing peptidase
MSF	Mitochondrial import stimulating factor
mTP	mitochondrial targeting peptide
OMP85	Outer membrane protein 85
Oxa1p	Cytochrome oxidase assembly
PAM	Presequence assisted motor
PBF	Presequence binding factor
PRAT	Preprotein and amino acid transporter
PreP	Presequence degrading peptidase
SAM	Sorting and assembly machinery of the outer membrane
TF	Targeting Factor
TIM	Translocase of the inner membrane
TOB	Topogenesis of β-barrel proteins
TOM	Translocase of the outer membrane
UCP	Uncoupling protein

11.1 Introduction

Although the coding capacity of plant mitochondrial genomes is greater than that observed in fungi and mammals, it still is limited to approximately 60 proteins. Thus the majority of the estimated 2,000 proteins located in mitochondria are encoded by nuclear located genes where they are transcribed (see also Chap. 9). mRNA is exported to the cytosol where translation occurs and proteins are imported into mitochondria (Millar et al. 2005). Although there is still some debate if protein import into mitochondria is cotranslational or posttranslational, the mitochondrial machinery responsible for the import of proteins is the same irrespective of the kinetic link between translation and translocation into mitochondria (Neupert and Herrmann 2007). Protein uptake into mitochondria is achieved by the combined action of a number of multisubunit membrane-bound protein complexes termed translocases, in combination with a number of soluble components in the cytosol, intermembrane space, and matrix such as peptidases and molecular chaperones. Together this mitochondrial import apparatus must specifically recognize mitochondrial proteins from the total protein complement being translated in the cytosol and achieve the unidirectional transport of these proteins into one of four mitochondrial compartments, and in the case of the outer and inner membranes achieve the correct orientation or protein topology. The recognition event is dependent on targeting signals that interact specifically with mitochondrial receptors. Most of the mitochondrial proteins contain an N-terminal cleavable targeting signal called a presequence; however, many proteins, especially hydrophobic membrane proteins have an internal noncleavable signal (Neupert and Herrmann 2007).

The import of proteins into mitochondria can be described by four main import pathways: the general import pathway, the carrier import pathway, the sorting and assembly pathway of β-barrel proteins into the outer membrane, and the mitochondrial intermembrane space import and assembly pathway (Pfanner and Geissler 2001; Neupert and Herrmann 2007; Chacinska et al. 2009). The general import pathway describes the import of proteins that contain the N-terminal targeting signals. These proteins are recognized by receptors in the TOM complex (translocase of the outer membrane) and are subsequently passed to the TIM23 complex (translocase of the inner membrane). They may be imported into the matrix, stop-transfer or conservatively sorted into the inner membrane (Hartl et al. 1989). Some proteins of the outer membrane may also use this pathway but are "stopped" at the outer membrane (Shore et al. 1995). The carrier import pathway describes the import for proteins of the mitochondrial carrier family into the inner membrane. They are initially recognized by receptors on the TOM complex, and then transferred to the TIM22 complex with the aid of small Tim proteins in the intermembrane space. The import of β-barrel proteins into the outer membrane is described by the sorting and assembly machinery (SAM) of the outer membrane. This pathway is also referred to as the topogenesis of β-barrel proteins (TOB) (Paschen et al. 2003). These proteins interact with the TOM complex and then are passed to the SAM complex with the aid of the small Tim proteins in the intermembrane space. Import of proteins into the intermembrane space is described by the mitochondrial intermembrane space import and assembly machinery (MIA), where proteins are passed from the TOM complex to a Mia40 protein, and subsequently undergo oxidative protein folding.

After the import into mitochondria, both precursors and targeting peptides are exposed to different proteolytic events. Precursor proteins are proteolytically processed by processing proteases, the mature proteins are additionally trimmed by intermediate proteases and the free targeting peptides are degraded by the mitochondrial presequence degrading protease (Neupert 1997; Stahl et al. 2002; Vogtle et al. 2009).

The sequencing of the Arabidopsis genome followed closely by the sequence of other plant genomes has given greater insight into many processes in plant biology in general (Flavell 2009). The understanding of the process of protein import into plant mitochondria has also been greatly accelerated by the availability of plant genome sequences. First, the genome sequenced allowed the first complete overview of the protein import apparatus in plants as it allowed the knowledge that had been gained in the forefront models for protein import into mitochondria, *Saccharomyces cerevisiae,*and *Neurospora crassa,* to be applied and utilized in plant systems. Second, it made it possible to identify mitochondrial precursor proteins and analyze targeting signals with respect to both targeting properties and processing determinants. Third, the enabling technologies that accompanied the sequencing of genomes allowed various genetic approaches that had been so successfully used in fungal systems to be applied to plants; in particular, the reverse genetic approach has been most widely used in plants.

11.2 Mitochondrial Precursor Proteins and Cytosolic Factors

Most mitochondrial proteins are synthesized in the cytosol on free polyribosomes as precursor proteins equipped with N-terminal extensions functioning as targeting signals. In vivo and in vitro studies showed that these precursors are imported into mitochondria posttranslationally. However, it has also been proposed that import might be cotranslational (Lithgow 2000) as there exist mechanisms that direct mRNAs of some mitochondrial proteins (Stein et al. 1994) to the surface of mitochondria, where they are translated on ribosomes that are bound to the outer membrane (Corral-Debrinski et al. 2000). The newly synthesized precursor proteins interact in the cytosol with different cytosolic factors and are protected from degradation and aggregations by molecular chaperones, such as Hsp70 and Hsp90, that also contribute to maintenance of an import competent conformation of the precursors. Several cytosolic factors such as MSF (Hachiya et al. 1993), PBF (Murakami and Mori 1990), and TF (Ono and Tuboi 1990), have been shown to interact with mitochondrial targeting peptides and were proposed to guide the precursors to organellar import receptors. However, the importance of these factors for import has not been confirmed in vivo. Using two-hybrid screening, Yano et al. (2003)identified arylhydrocarbon receptor-interacting protein (AIP) mediating precursor protein binding to the mitochondrial receptor Tom20. Depletion of AIP by RNA interference impaired the import (Yano et al. 2003). There is an *Arabidopsis* homologue of AIP that shows 45% sequence similarity (27% identity) to the mammalian AIP. A puzzling observation is that mitochondrial precursor proteins synthesized in wheat germ extract (WGE) fail to be imported into mitochondria, whereas the same precursors synthesized in rabbit reticulocyte lysate are easily imported (Dessi et al. 2003). A series of studies points to the conclusion that the import incompetence of WGE-synthesized precursors is a result of interaction of WGE inhibitory factors with the mature portion of precursor proteins (Dessi et al. 2003). There also exists a WGE factor that inhibits chloroplast protein import (Schleiff et al. 2002).

11.3 Mitochondrial Targeting Signals

The targeting signals of the mitochondrial proteins are most often present as cleavable presequences at the *N*-terminus, but many precursors lack the i*N*-terminal extensions and contain internal targeting signals (Box 11.1). Targeting signals are necessary and

Box 11.1

Mitochondrial N-terminal targeting signals are generally specific; however, some of them, called dual targeting peptides are ambiguous and target the precursor protein to both mitochondria and chloroplasts.

Most mitochondrial targeting signals are localized as N-terminal cleavable extensions in precursor proteins targeting the proteins to the mitochondrial matrix. There are also internal targeting signals involved in import of membrane proteins. The N-terminal mitochondrial targeting peptides, mTPs, vary in length from 18 to117 amino acids with an average length of 42–50 amino acids. They have high content of hydrophobic, hydroxylated, and positively charged amino acids and a low abundance of negatively charged residues (Zhang and Glaser 2002; Bhushan et al. 2006; Huang et al. 2009). mTPs are predicted to form amphiphilic α-helices that are important for import into the mitochondria. mTPs are very similar in amino acid composition to the chloroplastic targeting peptides, cTPs, with the main difference in the N-terminal portion of the targeting peptides, where arginine is overrepresented in mTPs, whereas serine and proline are overrepresented in cTPs. Despite this high similarity, organellar protein import is generally very specific (Whelan and Glaser 1997). However, there are 51 proteins known to date that are encoded by a single gene, translated in the cytosol as a single precursor protein and imported into both mitochondria and chloroplasts using the same ambiguous targeting peptide, the dual targeting peptide, dTP (Peeters and Small 2001). The overall properties of dTPs resemble characteristics of mTPs and cTPs with some differences in quantity and distribution of amino acids observed, such as significant increase in phenylalanine, leucine, and serine and decrease in acidic residues and glycine. There is a high content of serine and an intermediary content of argining in the N-terminal portion of the dTPs in comparison to mTPs and cTPs (Pujol et al. 2007; Berglund et al. 2009; Carrie et al. 2009). There is no general rule for how the determinants for dual targeting are distributed within dTPs but in most cases the N-terminal portion was shown to be essential for import into both organelles, and in a few cases a domain structure was observed (Pujol et al. 2007; Berglund et al. 2009).

sufficient to direct proteins to the organelle. They can also mediate targeting of non-native passenger proteins (de Castro Silva Filho et al. 1996; Westermann and Neupert 2000; Duby et al. 2001b). The N-terminal mitochondrial targeting peptides (mTPs) recognize import receptors and target precursor proteins across the outer and inner mitochondrial membrane to the matrix (Glaser et al. 1998). They can also mediate import into the intermembrane space if the matrix-targeting sequence is followed by an additional hydrophobic signal sequence arresting the precursor in transit in the inner membrane. After the proteolytic cleavage of the signal sequence, the mature protein will be released to the IMS (Burri et al. 2005).

11.3.1 N-Terminal Targeting Signals

The features of plant mTPs have been analyzed using experimentally determined cleavage sites and a global sequence collection from the mitochondrial proteome of *A. thaliana* and *Oryza sativa* (Zhang and Glaser 2002; Heazlewood et al. 2004; Bhushan et al. 2006;Huang et al. 2009). The length of mTPs in different plant species varies substantially, from 18 to 117 residues with an average length of 42–50 amino acids residues. The great majority (74–80%) of the mTPs are in the range of 20–70 residues long. The overall amino acid composition of mTPs shows that there is high content of 33–35% hydrophobic (L, A, F, I, V), 22–23% hydroxylated (S, T), and 14–15% positively charged (R, K) amino acid residues,and a very low abundance of acidic amino acids. Four dominant amino acids (S, R, A, and L) constitute approximately 50% of all the residues. Furthermore, proline and glycine are well represented and constitute about 11% of the total content. Average pI values for plant mTPs are very high, 11.61–11.84; they are higher than for yeast mTPs (11.32). In comparison to non-plant mTPs, plant mTPs are about 9–20 amino acid residues longer and contain about two- to five-fold more serine residues (Glaser et al. 1998; Zhang and Glaser 2002; Huang et al. 2009). Interestingly, calculation of the hydrophobicity values (−1.30 to 1.39) for the first 10 amino acid residues in plant mTPs showed that the N-terminal sequences were hydrophilic (more hydrophilic than in yeast).

mTPs have the potential to form amphiphilic α-helices (von Heijne 1986) that are important for import. The predicted overall helical content has been estimated to 29–40% for different species. Eighty-four percent and ninety percent of Arabidopsis and rice mTPs, respectively, form an α-helix within the first 10 amino acid residues. Some of the mTPs have two or more α-helices (Huang et al. 2009). The NMR structure of a few mTPs was determined, e.g., of the *Nicotiana plumbaginifolia* $F_1\beta$, and shown to form amphiphilic α-helices in membranelike environment (Moberg et al. 2004). The NMR structure of the cytosolic domain of rat Tom20 (Abe et al. 2000; Muto et al. 2001) in complex with mTPs revealed that the hydrophobic side of the amphipathic α-helix of mTPs is located in a hydrophobic groove of the Tom20. Interaction of plant mTPs with Arabidopsis Tom20 was shown to involve not only hydrophobic, but also ionic residues as determined from structural studies by NMR (Perry et al. 2006) Nearly all mTPs (97%) contain Hsp70 binding motifs, a hydrophobic stretch of five amino acid residues flanked by positively charged residues (Zhang and Glaser 2002) that is important for the interaction with the mitochondrial Hsp70.

Comparison of plant mTPs to chloroplast targeting peptides (cTPs) (Zhang and Glaser 2002; Bhushan et al. 2006; Huang et al. 2009) showed a remarkable simil-arity in the overall amino acid composition. The detailed analysis showed a decrease in the amount of arginine and an increased in the amount of proline and serine in the N-terminal portion of the cTPs in comparison to mTPs (Bhushan et al. 2006; Huang et al. 2009). pI was lower for cTPs (11.27) than for mTPs. The N-terminal portion of cTPs(1–10) was substantially less hydrophilic (hydrophobicity value of 0.2) than the corresponding portion of mTPs(1–10). Furthermore, cTPs were shown to form mostly random coils and the predicted α-helical structure is much lower than for the mTPs. Only 30% of Arabidopsis cTPs form an α-helix within the first 10 amino acid residues (Huang et al. 2009). The Presence of helical elements in cTPs has been found by NMR studies (Wienk et al. 1999; Bruce 2000).

11.3.2 Internal Targeting Signals

Internal targeting signals have been found for proteins of the outer membrane, hydrophobic carrier proteins of the inner membrane, and for a class of IMS proteins (Brix et al. 1999, 2000;Habib et al. 2003;Horie et al. 2003). These internal signals are noncleavable, not well defined, lack consistent patterns, and may be distributed all over the protein (Rapaport 2003). Recent MS analysis of mitochondrial precursors identified 25 proteins in rice and 18 proteins in Arabidopsis that lacked N-terminal presequences (Huang et al. 2009). These proteins include porins, OMP85, small TIMs, adenylate kinase, UCP3, and some subunits of the respiratory complexes I, III, and IV. Some plant carrier proteins, such as adenine nucleotide carrier and phosphate carrier from maize, contain a cleavable N-terminal region, whereas oxoglutarate malate carrier does not contain a cleavable extension. It has been shown that the N-terminal extensions are not necessary for import of these proteins into the inner membrane; however, in the presence of these sequences the insertion into the inner membrane was much more efficient (Murcha et al. 2005b).

11.4 Dual Targeting Signals

Generally, import of proteins into mitochondria and chloroplasts is considered to be organelle specific. However, there exist a group of 51 proteins known to date that are encoded by a single gene in the nucleus, translated in the cytosol as a single translation product and targeted to both chloroplasts and mitochondria. These proteins contain an ambiguous N-terminal signal peptide referred to as a dual targeting peptide (dTP) that is recognized by both mitochondrial and chloroplast receptors (Peeters and Small 2001; Silva-Filho 2003; see Box 11.1). In addition, there are at least another four proteins in plants that are encoded by a single gene, but contain different targeting peptides due to alternative translation (for reviews see Peeters and Small 2001; Silva-Filho 2003; Carrie et al. 2009). Out of the 51 dually targeted proteins to mitochondria and chloroplasts, 18 are aminoacyl-tRNA synthetases (aaRS) (Duchene et al. 2005; Carrie et al. 2009; Berglund et al. 2009a). Other dual-targeted enzymes are related to protection against oxidative stress, DNA synthesis and processing, cellular protein folding, and turnover and energy conversion (Carrie et al. 2009). The number of dual targeted proteins is constantly growing, and based on prediction program (ATP, ambiguous targeting predictor) there might be as many as 523 dual targeted proteins in Arabidopsis (Mitschke et al. 2009).

Determinants for dual targeting are not fully understood. dTPs are very similar to mTPs and cTPs in overall amino acid composition, and it is difficult to identify specific features for dTPs. It has been proposed that the information for organellar targeting can be organized in domains, as in RNA polymerase RpoT:2 (Hedtke et al. 2000), glutathione reductase (GR) (Rudhe et al. 2002b), presequence protease (PreP) (Bhushan et al. 2003) or spread all through the targeting peptide (Chew et al. 2003) or associated with amino acid residues in the N-terminal portion of the presequence (Pujol et al. 2007). Furthermore, expression of the 5′ untranslated region (UTR) upstream of the ATG start codon has been also shown to be involved in generating a dTP (Christensen et al. 2005). Sequence analysis of dTPs using statistical methods and Sequence Logos revealed the overall significant increase in phenylalanines, leucines, and serines and a decrease in acidic amino acids and glycine in dTPs in comparison to single mTPs and cTPs (Pujol et al. 2007; Berglund et al. 2009b). The N-terminal portion of dTPs has significantly more serines than mTPs. The amount of arginines is similar as in mTPs, but significantly higher than in cTPs. dTPs are about 50–60 amino acid residues long (Berglund et al. 2009b). The lack of an amphiphilic α-helix has been observed (Peeters and Small 2001); however, in some cases amphiphilic α-helix is present with a hydrophilic side formed by a polar residue, serine (Pujol et al. 2007; Berglund et al. 2009b). It has been suggested that the dTPs might be intermediary in character in comparison to mTPs and cTPs (Peeters and Small 2001; Berglund et al. 2009b). Characterization of recognition determinants of aaRS-dTPs, in which aaRS-dTPs GFP fusion proteins were imported in vitro in single and dual import system (Rudhe et al. 2002a) as well as in vivo showed essential role of the N-terminal portion of dTPs (about 20 amino acids) for import into both organelles. Furthermore, for some aaRS-dTPs, a domain structure was demonstrated that additionally suggested important function of the N-terminal portion of dTPs for import into both or into a single organelle (Berglund et al. 2009a). This was in agreement with a study where mitochondrial and chloroplastic targeting peptides fused in tandem showed that the position of the targeting sequence is important, and that the N-terminal portion has the greatest importance for the final location (de Castro Silva Filho et al. 1996). It is worth noting that the dual targeting ability of dTPs has been shown to be affected by the nature of the mature protein (Chew et al. 2003). Furthermore, investigations of determinants for processing of the glutathione reductase (GR) dTP in mitochondria by MPP and in chloroplasts by the stromal processing protease (SPP) (Richter and Lamppa 2002) revealed that the recognition of the processing site differed between MPP and SPP and that the dual targeting peptide of GR has similar determinants for processing by MPP as mTPs, whereas processing by SPP showed a low level of sensitivity to single mutations and likely involves recognition of the physiochemical properties of the sequence in the vicinity of cleavage rather than a requirement for specific amino acid residues (Rudhe et al. 2004).

11.5 An In-Silico Picture of the Protein Import Machinery in Plants

Using the protein import apparatus of yeast and *N. crassa* as a template a virtual picture of the plant protein import apparatus can be obtained (Fig. 11.1). Although this approach identifies many of the components of the protein import apparatus in plants, clearly some critical components are missing or not identified. Overall it is evident while the membrane located channel forming subunits of the membrane bound translocases are well conserved, similarity with other components, most notably the outer membrane receptors is lacking (Lister et al. 2005). Thus Tom40, Sam50 (Tob50), Tim17, 23, and 22, and Oxa 1p can all be clearly identified. Additional components of the TIM23 complex, in particular Tim50 and Tim44 can also be identified. Pam16, 18, HSP70, and Mge1 can also be identified; however, caution needs to be exercised with these components as it is not clear which are mitochondrial, some may be located in other HSP70 complexes in other locations in the cell. In the case of Tom40, Sam50 (Tob50) and Oxa 1p they appear similar to

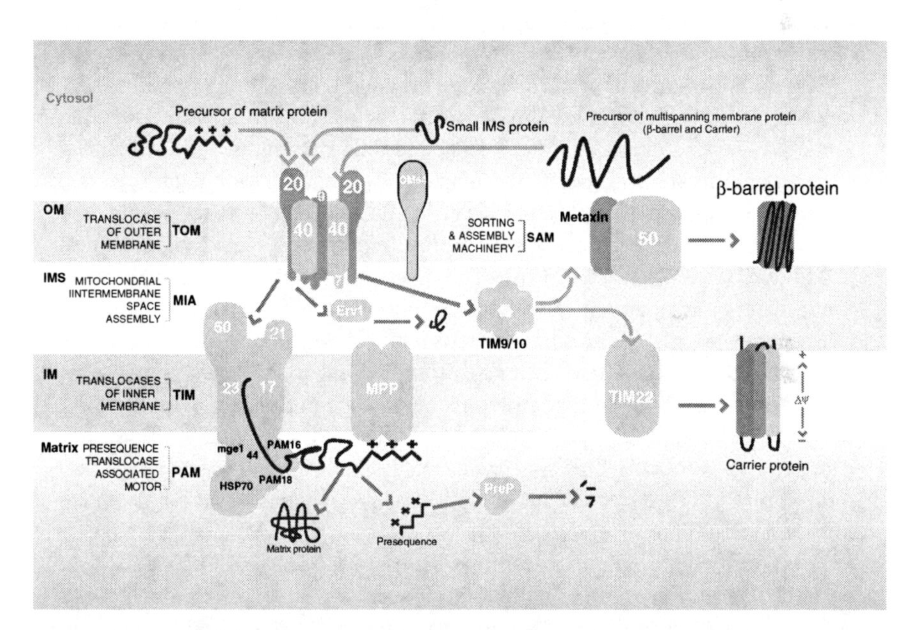

Fig. 11.1 Overview of protein import pathways and machinery in plant mitochondria. The import of proteins into mitochondria is achieved by the combined action of a number of multi-subunit membrane embedded protein complexes in combinations with a variety of soluble components in the intermembrane space and matrix. All proteins imported into mitochondria interact with the TOM complex and are sorted to different intraorganelle locations. Sorting of proteins to various compartments is achieved by interactions with a variety of other proteins complexes. Indicated in green are components that are orthologous to components identified in yeast, the exception being OM64 (*outlined in black*) that displays high sequence identity to a component of the outer membrane envelope of chloroplasts. *Indicated in red* are components that are not orthologous to components in yeast or mammalian systems. Although Metaxin has been identified by limited sequence similarity with mammalian Metaxin, it is not clear if they are orthologous

their fungal (and mammalian) counterparts and they are likely to fulfil the same functions, although it is possible that mechanistic details differ. Thus with the plant Tom40 it lacks the tryptophan residue at position 243 that has been shown to be critical for sorting proteins to the TIM23 complex (Gabriel et al. 2003).A similar situation exists with the channel or pore forming subunits of the inner membrane translocases. In yeast Tim17, 23, and 22 are encoded by single genes and belong to a larger family of proteins called preprotein and amino acid transporters (PRAT) (Rassow et al. 1999). However in Arabidopsis there are 17 genes that encode PRAT proteins and at least 24 in rice (Murcha et al. 2007). Ten of the 17 genes in Arabidopsis encode proteins targeted to mitochondria, the others encode outer envelope proteins of the chloroplast (OEP) similar to OEP16 that was first characterized in pea (Pohlmeyer et al. 1997). Of these ten proteins, 3 encode Tim17-like proteins, three Tim23-like proteins, and 2 encode Tim22-like proteins. A protein encoded at

Box 11.2

Plant mitochondrial protein import receptors are evolutionarily distinct to other organisms

It is a first reaction to propose that the reason that the protein import receptors of plant mitochondria are different fromthat in yeast and mammals is due to the presence of plastids in plant cells, and thus this requires mitochondrial receptors with greater specificity than that present in nonplant systems. However this ignores the facts that:(1) protein import receptors for plastids are fundamentally different than mitochondrial import receptors from yeast and plants (Soll and Schleiff 2004).(2) The plant import receptors recognize the same features as nonplant mitochondrial import receptors, as mitochondrial targeting peptides are quite similar across wide phylogenetic gaps (see above). Furthermore, structural studies suggest that both plant and yeast Tom20 proteins form remarkably similar structures, even though they are evolutionarily distinct (Lister and Whelan 2006; Perry et al. 2006).(3) Finally, the plastid endosymbiosis followed the mitochondrial symbiosis by approximately 500 million to 1 billion years (Dyall et al. 2004). This suggests that either plants lost the "yeast"-like receptor components after the plastid endosymbiosis, or after the mitochondrial endosymbiosis divergence occurred. Cells that went on to evolve into yeast and mammalian-like cells were on one branch while cells that went on to form plants and trypanosomes are on another branch. Thus the nonorthologous nature of outer membrane protein receptors between yeast (and mammals) and plants may represent an early divergence of the eukaryotic lineage, rather than have been driven by a protein sorting problem that likely arose only 500 million to 1 billion years later. The presence of the protein import receptor, OM64, on the outer membrane of plant mitochondria that displays a high level of sequence identity with the plastid import receptor Toc64 is consistent with this sequence of events (Chew et al. 2004).

the loci At2g42210 is found in complex I (Meyer et al. 2008). The function of a PRAT protein encoded at the chromosomal locus At3g25120 is unknown. Additionally a PRAT protein encoded at the chromosomal locus At5g24650 appears to be targeted to mitochondria ad plastids (Murcha et al. 2007). Arabidopsis and rice both contain single genes for components of the MIA pathway, Mia40 and Erv1 (Lister et al. 2003). A chaperone protein Hot13p chaperone that is required for the correct import and assembly of Mia40 in yeast has also been reported to be present in plants (Mesecke et al. 2008), but the protein sequence identity is limited to a small region in a protein with over 600 amino acids and thus it is unlikely that the plant protein is orthologous to the yeast protein (Carrie et al. 2010). It is particularly notable that none of the outer membrane receptor components of the plant mitochondrial protein import apparatus are orthologous to yeast (or mammalian) proteins. Thus even though a protein labelled Tom20 is present in the plant TOM complex, it is not orthologous to yeast Tom20 (Boxes 11.2 and 11.3) (Likic et al. 2005). Likewise no orthologue to yeast Tom70 can be identified in the Arabidopsis genome (Chan et al. 2006).

Box 11.3

The protein import machinery of plant differs in many respects compared to the model system in yeast (see Chap. 12 for more details)

Although it is often stated that the process of protein import into mitochondria is well conserved between organisms, there are many differences in the processes in plants compared to yeast (and mammalian) systems. In fact differences in all import pathways have been observed.

TOM complex: Receptor components Tom20, 22, and 70 of yeast not present in plants. Additional components such as OM64 present. Only Tom40 and Tom7 are orthologous to yeast components.

SAM complex: Apart from SAM50 the other components of the yeast complex are not present in plants, namely SAM35, 37, mdm10, and Min1 are not identified in plants. The relationship of plant metaxin to SAM35 or 37 is unclear.

MIA: In contrast to yeast Erv1 is the only essential component in plants, whereasMia40 is present in plants but not essential as observed in yeast.

TIM17/23 complex:Although all components identified in yeast are present in plants, they differ in that Tim17 is longer in plants and is exposed on the outer membrane. Thus Tim17 appears to have additional roles(s) in plants.

TIM22: Biochemical reconstitution of import of carrier proteins into the inner membrane has only been reported in plants. Furthermore, plants lack Tim54 and Tim12 that are present in the TIM22 complex in yeast.

MPP: Is an integral part of the cytochrome bc1 complex in plants compared to a matrix location in yeast.

11.6 Functional Studies on the Mitochondrial Protein Import Apparatus of Plants

11.6.1 The Outer Membrane

The outer membrane is the site of recognition of mitochondrial proteins from the total cytosolic pool. The TOM complex achieves three functions, it recognizes the mitochondrial precursors, translocates them across the outer membrane and passes them to the next machine in the protein import apparatus. The plant TOM complex was first characterized from potato by biochemical approaches (Jansch et al. 1998). The purified TOM complex consists of Tom40 that is orthologous to Tom40 in yeast, and Tom20, 9, 5, 6, and 7. Tom40 and Tom7 of the TOM complex are orthologous to the yeast and mammalian complexes (Jansch et al. 1998; Lister et al. 2005). A 70-kDa protein was reported in this isolation but not identified (Jansch et al. 1998). This 70-kDa protein may be similar to OM64 that has been shown to be located on the outer mitochondrial membrane (Chew et al. 2004), but in Arabidopsis it is not associated with the Tom complex under conditions where the other subunits are still observed in the Tom complex (Lister et al. 2007). Tom40 is a relatively abundant protein in plants as it can be readily visualized on 2D-PAGE gels from total mitochondrial preparations from Arabidopsis and rice (Milla et al. 2001; Howell et al. 2006).

Initial functional studies on the receptor components of the TOM complex in plants suggested that the plant Tom20 acted as a receptor for proteins imported via the general and carrier import pathways (Heins and Schmitz 1996), and subsequent studies in Arabidopsis inactivating Tom20 by genetic means supports these findings (Lister et al. 2007). A functional TOM complex can form with all Tom20 proteins inactivated (Duncan and Whelan, unpublished data), and even though the rate of import is substantially reduced for a variety of proteins in Arabidopsis with all Tom20 protein inactivated, plant growth is minimally affected (Lister et al. 2007). Inactivating one or two Tom20 proteins does have some differential effect on the rate of import of different precursor proteins, suggesting that different Tom20 isoforms may have different affinities for different precursor proteins (Lister et al. 2007). Other functional studies on components of the TOM complex show that inactivation of Tom40 is lethal (Lister et al. 2007). Also while plants lack an orthologue to Tom22 in yeast, that is considered a central convergence point for precursor proteins before they are translocated into the Tom40 channel (van Wilpe et al. 1999), plants do contain a Tom9 subunit (Jansch et al. 1998; Lister et al. 2003). Complementation studies have shown that a plant Tom9 can complement a yeast *tom22* mutant, if the receptor component of the yeast Tom22 is attached to Tom9 (Macasev et al. 2004). Thus plant Tom9 appears to be functionally equivalent to yeast Tom22 except for the receptor domain functions of Tom22.

Two additional receptor components have been proposed for plant mitochondria, mtOM64, and metaxin (Lister et al. 2007). mtOM64 is a protein that is located on the outer mitochondrial membrane that displays high sequence identity to the plastid protein import receptor/outer envelope component Toc64 (Chew et al. 2004). Inactivation of mtOM64 results in a decreased rate of import for only one precursor protein, a plant-specific subunit of the ATP synthase complex (Lister et al. 2007). Notably Toc64 cannot compete with mtOM64 for binding of mitochondrial precursor proteins. Plant metaxin was initially identified by limited sequence identity with mammalian metaxin, originally proposed to affect the import of proteins into mammalian mitochondria (Armstrong et al. 1997). Mammalian metaxin displays limited sequence identity to yeast Sam37, whose role in this complex is not fully elucidated to date. Inactivation of metaxin in Arabidopsis leads to severe developmental abnormalities and infertility (Lister et al. 2007). Analysis of these plants reveal that metaxin in plants plays a role in importing β-barrel protein via the SAM complex as in the absence of metaxin transcript abundance for both porin and Tom40, both β-barrel proteins of the outer mitochondrial membrane increased several fold. Furthermore there is a large accumulation of un-imported porin in the cytosol (Lister et al. 2007).

11.6.2 The Intermembrane Space

There are only a few studies published to date examining the role of intermembrane space protein involved in protein import in plants. One study carried out a biochemical reconstitution of the import of carrier proteins into the intermembrane space with outer membrane ruptured mitochondria (Lister et al. 2002). It showed that the import of carrier proteins could be stimulated in a zinc-dependent manner by the addition of a fraction of proteins from the intermembrane space that was enriched for the small Tim proteins, Tim 9 and 10. This indicated that the carrier import pathway in plants is similar to that described in yeast where small intermembrane space proteins chaperone the carrier proteins from the outer to the inner membrane. Notably this reconstitution assay has never been reported in yeast, and along with the absence of Tim54 and 12 in plants (and other organisms) it suggest that some differences exist with the import of carrier proteins in plants. It is notable that many carrier proteins, but not all, in plants have N-terminal cleavable extensions (Winning et al. 1992). These are not essential for import, but seem to increase the efficiency of import (Murcha et al. 2005b). In the case of the extension of the adenine nucleotide translocator and phosphate translocator, the extension is removed in two steps, the first processing by MPP and the second in the intermembrane space by an unidentified peptidase (Murcha et al. 2004).

Although Arabidopsis contains distinct orthologues to both Mia40 and Erv1 in yeast, functional studies reveal that there are significant differences in how this pathway operates in plants compared to yeast (Carrie et al. 2010). Mia40 is an essential protein in yeast, but deletion in Arabidopsis has no phenotypic affects, thus Mia40 is not an essential component of the disulfide relay system of plant mitochondria. Notably trypanosomes lack a gene encoding Mia40 and they also import carrier proteins and contain small Tim proteins, indicating that import of small Tim proteins can occur without Mia40. Both plants and trypanosomes contain a different arrangement of cysteine pairs in their Erv1 proteins that may fulfil the role of Mia40, similar to related proteins in the endoplasmic reticulum (Carrie et al. 2010). Furthermore in plants Mia40 is a dual targeted protein, also targeted to peroxisomes, suggesting oxidative protein folding in plant peroxisomes (Carrie et al. 2010).

11.6.3 The Inner Membrane

The inner membrane is the site of three translocases, TIM23 that imports proteins via the general import pathway (Pfanner and Geissler 2001; Neupert and Herrmann 2007), TIM22 that imports protein via the carrier import pathway (Pfanner and Geissler 2001; Neupert and Herrmann 2007), and Oxa 1p that is a translocase conserved from the bacterial endosymbiotic ancestor of mitochondria that is responsible for inserting proteins into the inner membrane that are synthesized within mitochondria (Hell et al. 2001; Luirink et al. 2001). Apart from investigations showing that Arabidopsis Oxa1p and a member of the PRAT family of proteins can complement a yeast *oxa1p* and *tim22* mutants, respectively, only the TIM23 complex in plants has been investigated experimentally (Sakamoto et al. 2000; Murcha et al. 2007).

Although the TIM23 complex is well conserved at a sequence level to that in yeast, functionally it differs to some degree. First, the three genes that encode Tim17-like proteins in Arabidopsis differ from yeast in that two encode protein that are 218 and 243 amino acids rather than the 158 amino acids of the yeast Tim17 protein (Murcha et al. 2007). The third gene encodes a protein of only 133 amino acids, and essentially lacks the fourth transmembrane region (Murcha et al. 2007). Functional studies show that the longer protein can only complement a yeast *tim17* mutant if the extensions are removed. Thus the extensions on the Arabidopsis appear to carry out additional functions. Characterization of the role of the extension on one of the Arabidopsis Tim17 proteins revealed that it was exposed on the outside of the outer membrane (Murcha et al. 2005a). Rupture of the outer membrane followed by protein import revealed that antibodies to the extension inhibited import of proteins via the general import pathway, but not via the carrier pathway. These studies reveal a "receptor"-like role for the extension on Arabidopsis Tim17 compared to yeast Tim17. Notably in yeast the N-terminal region of Tim23 is inserted into the outer membrane and increases the kinetics of import (Donzeau et al. 2000). Expression analysis at the level of transcript reveals that only one gene, *Tim17–2,* is widely expressed (Murcha et al. 2007); thus, the role of the other two Tim17 proteins is unclear. Given that Arabidopsis *Tim17–3* encodes a protein of only 133 amino acids that lacks the fourth transmembrane region typical of PRAT proteins, it may be a pseudogene.

The three genes encoding Tim23-like proteins in Arabidopsis encode proteins that display relatively high levels of sequence identity to yeast Tim23. The predicted Arabidopsis Tim23 proteins lack two features compared to yeast Tim23

(Murcha et al. 2007). First, they do not contain a predicted leucine zipper region at the N-terminal region that inserts into the outer membrane in yeast Tim23 (Donzeau et al. 2000). Second, two of the three Arabidopsis Tim23 proteins lack one of the consensus amino acid residues within the PRAT domain. This amino acid must be changed to the consensus for the Arabidopsis proteins to complement a yeast *tim23* mutant (Murcha et al. 2007).

11.7 Processing of Precursor Proteins and Degradation of Targeting Peptides

Several proteases have been identified to participate in the processing and trimming of precursor proteins and proteolysis of targeting peptides inside mitochondria. Mitochondrial processing peptidase (MPP) and inner membrane peptidase (IMP) catalyze cleavage of targeting peptides from precursor proteins (Neupert 1997; Gakh et al. 2002). After the cleavage, the intermediate mature proteins can be additionally trimmed by the mitochondrial intermediate peptidase (MIP; Isaya et al. 1992) or a recently identified intermediate cleaving peptidase (Icp55; Naamati et al. 2009; Vogtle et al. 2009) in the matrix or a Rhomboid protease in the membrane (McQuibban et al. 2003). The cleaved targeting peptides have been shown to be degraded by a novel peptidasome, presequence protease (PreP; Stahl et al. 2002). All these proteases are independent of ATP and differ in function from ATP-dependent proteases, such as the AAA, Lon, and Clp proteases, that participate in quality control of miss-folded and damaged organellar proteins and in assembly of the maturated proteins by regulation of the stoichiometric amounts of polypeptides in protein complexes (for a review, see Koppen and Langer 2007).

11.7.1 Processing of Precursor Proteins

The first step in processing of precursor proteins is the endoproteolytic cleavage of targeting peptides catalyzed by the mitochondrial processing peptidase (MPP). MPP is a metalloendopeptidase (for reviews, see Neupert 1997; Glaser and Dessi 1999) consisting of two structurally related subunits, α-MPP and β-MPP (about 50 kDa each), which cooperate in processing. The catalytic site is located on the β-MPP subunit and constitutes an inverted zinc-binding motif ($HXXEHX_{74-76}E$), which classifies the protease to the pitrilysin family (subfamily M16B). In mammals and yeast MPP is localized to the matrix. In *N. crassa,* β-MPP was found to be partially integrated into the cytochrome bc_1 complex of the respiratory chain as a Core 1 protein of (Schulte et al. 1989). In contrast to non plant sources, plant MPP activity resides in the mitochondrial inner membrane and MPP is completely integrated into the bc_1 complex of the respiratory chain as Core1 and Core 2 proteins (Box 11.3). It has been shown in several higher plant species such as potato, spinach, wheat (Braun et al. 1992; Glaser et al. 1994; Glaser and Dessi 1999) and also in lower plants, the staghorn fern *Platycerium bifurcatum*, and the horsetail *Equisetum arvense* (Brumme et al. 1998) indicating that integration of MPP into the bc_1 complex of the respiratory chain is a general feature for plants. However, despite the fact that the MPP/bc_1 complex is bifunctional, the processing activity is not dependent on the electron transfer (Emmermann et al. 1993; Eriksson et al. 1996).A connection between import and processing has been reported (Whelan et al. 1996); however, it has also been shown that the targeting peptide cleavage occurs after precursor translocation, indicating that processing catalyzed by the MPP/bc_1 complex in plants and import are independent events (Dessi et al. 2000).

Analysis of the cleavage sites indicated loosely defined cleavage motifs for a majority (80%) of precursor proteins containing an arginine either at position −2 (−2R) (R-X↓X-S-T/S-T) or −3 (−3R) (R-X-F/Y↓A/S-T/S/A) relative to the cleavage site (Zhang and Glaser 2002; Huang et al. 2009), however the occurrence of the −3R motif has been recently revised by identification of Icp55 (see below). There is also a third group of precursors lacking a conserved arginine (no R) in the C-terminal portion of the targeting peptide (Zhang and Glaser 2002). Recent studies in both Arabidopsis and rice revealed a motif of (F/Y↓S/A) for this group of proteins (Huang et al. 2009). Several reports also showed that the processing depends not only on the consensus upstream to the cleavage site but also on the downstream residues (Tanudji et al. 1999; Duby et al. 2001a; Rudhe et al. 2004) and flexible linkers within the targeting peptide (Waltner and Weiner 1995). 3D structure of the recombinant yeast MPP in complex with synthetic targeting peptides of the −2R group revealed binding properties of the targeting sequences to MPP (Taylor et al. 2001) and the recognition sites for the −2R and the +1 aromatic residue were observed. The authors suggested that the presequences adopt context-dependent conformations through mitochondrial import and processing, helical for recognition by mitochondrial import machinery and extended for cleavage by MPP (Taylor et al. 2001).

Recent global analysis of the mitochondrial N-terminal proteome in yeast (Vogtle et al. 2009) identified a novel intermediate peptidase Icp55 that catalyses trimming of the precursor protein at the N-terminus removing a single amino acid after MPP cleavage. The Icp55 protease showed specificity for tyrosine, phenylalanine or leucine, amino acid residues highly abundant in −3R cleavage motif revealing that −3R is not a MPP recognition motif but it indicates a second processing site. The authors proposed that Icp55 converts the instable form of an intermediate protein into a stable protein (Vogtle et al. 2009). Naamati et al. (2009) reported that Icp55 can remove three amino acids from the N-terminus. Furthermore, the authors showed that Icp55 is distributed to both mitochondria and the nucleus and that the nuclear Icp55 is processed in mitochondria (Naamati et al. 2009). Icp55 is a member of aminopeptidase P (APP) family of metalloproteases and is peripherally attached to the matrix side of the mitochondrial inner membrane (Rawlings et al. 2008).

Mitochondrial Intermediate Peptidase, MIP has been known for many years as a protease cleaving off a characteristic octapeptide sequence (F/L/I)XX(T/S/G)XXXX downstream of the MPP cleavage site (Branda and Isaya 1995; Schneider et al. 1998). MIP is a thiol-dependent metallopeptidase that belongs to the thimet (thiol and metal-dependent) oligopeptidase family (Rawlings et al. 2008). It has been identified as a monomer ofapproximately 75 kDa in the matrix of fungi and mammals. MIP contains a metal-binding site, HEXXH, and is enriched in cysteine residues. Twelvesubstrates for the MIP protease have been found in yeast (Branda and Isaya 1995; Vogtle et al. 2009). Plant MIP has not been thoroughly studied. The *A. thaliana* MIP homologue is predicted to be localized to the mitochondrial matrix, however, in contrast to nonplant species, it harbors an expanded C-terminal domain that is predicted to form a trans-membrane segment. An mutant with a T-DNA insertion in the gene of MIP was not lethal in *A. thaliana*. No overrepresentation of the −10R motif or the characteristic octapeptide has been found in the recent global proteome studies of the mitochondrial processing in Arabidopsis (Huang et al. 2009).

Some proteins destined for the intermembrane space have a bipartite N-terminal signal sequence consisting of a MPP specific targeting signal followed by a hydrophobic stop-transfer sorting signal that is cleaved after the MPP cleavage by inner membrane peptidase, IMP. MPP cleaves the first part of the presequence from the precursor that is immobilized in the TIM complex and the cleavage of the stop-transfer signal releases protein to the IMS. IMP consists of two catalytic subunits, Imp1 and Imp2 anchored to the inner membrane via an N-terminal membrane-spanning domain. The C-terminal domain of IMP faces the IMS and carries the catalytic site characterized by a conserved Ser-Lys dyad (Chen et al. 1999). Imp1 and Imp2 belong to the type-I signal peptidase (SP) family of proteases. Homologues of both Imp1 and Imp2 have been found in *A. thaliana*.

Another novel mitochondrial integral inner membrane protease is a Rhomboid protease, Pcp1, a serine protease with a catalytic site composed of Ser-His-Asn located within the hydrophobic membrane-spanning segment. The Rhomboid protease has been shown to cleave and generate mature proteins after cleavage of the selected target precursors, cytochrome c peroxidase (Ccp1) or dynamin-like GTPase (Mgm1), by the m-AAA protease and MPP (McQuibban et al. 2003). Rhomboid protease regulates also mitochondrial membrane remodeling indicating an important role of this protease in regulation of membrane biogenesis (McQuibban et al. 2003). Five rhomboid-like proteins in *A. thaliana* were predicted to be mitochondrially targeted, however only AtRBL12 was shown experimentally to be imported into mitochondria. Complementation of the yeast lacking a functional copy of mitochondrial rhomboid with AtRBL12 indicated that this plant protease does not recognize the yeast Ccp1 or Mgm1, implying that plant mitochondrial rhomboids function in a specific manner and thus differ from their yeast and mammal counterparts (Kmiec-Wisniewska et al. 2008).

11.7.2 Degradation of Targeting Peptides

The cleaved, free targeting peptides are potentially toxic to the integrity and function of mitochondria as they can penetrate membranes, dissipate membrane potential, and uncouple respiration (Hugosson et al. 1994). Therefore they have to be degraded or removed from the organelle. A novel organellar peptidasome called the presequence protease (PrePP) responsible for the degradation of targeting peptides in mitochondria and also in chloroplasts has been identified (Stahl et al. 2002; Bhushan et al. 2003; Moberg et al. 2003).

PreP was originally isolated from the potato mitochondrial matrix and identified by mass spectrometric analysis, ESI-MS/MS (Stahl et al. 2002) in the *A. thaliana* database as a novel organellar zinc metalloprotease containing an inverted zinc binding motif (HXXEH) belonging to the pitrilysin protease family (subfamily M16C). There are two isoforms of PreP in *A. thaliana*, *At*PreP1 and *At*PreP2, containing 85 amino acids long N-terminal cleavable targeting peptides. In vitro and in vivoimport studies of *At*PreP fusion constructs with GFP (green fluorescent protein), demonstrated that both isoenzymes are dually targeted to both mitochondria and chloroplasts using an ambiguous targeting signal (Bhushan et al. 2003, 2005). Furthermore, the targeting peptide of *At*PreP1 was shown to be organized in domains with an N-terminal domain required for the mitochondrial import and the C-terminal domain sufficient for the chloroplast import. Studies of a single and a double *At*PreP knockout mutants showed that although both *At*PreP1 and *At*PreP2 are expressed in all tissues, *At*PreP1 is expressed to a much higher level than *At*PreP2. The single *At*PreP1 and the double knockout mutant resulted in a chlorotic phenotype, especially during early plant development. The mutant plants exhibited also a slower growth rate. The accumulated biomass was 40% lower all through the development in the mutant plants in comparison to wild-type. Both mitochondria and chloroplasts exhibited altered morphology. Chloroplasts contained less grana stacking and less starch granules; chlorophyll a and b content was diminished. Mitochondria were variable in size, partially uncoupled, and the respiratory rates were lower (Nilsson Cederholm et al. 2009). These results demonstrated the importance of PreP for efficient organellar functions and normal plant growth and development.

Both *At*PreP1 and *At*PreP2 were shown to cleave unstructured peptides that are in the range of 10–65 amino acid residues, but not small proteins. Detailed substrate specificity studies showed that both proteases have preference for positively charged amino acids in the P_1' position and small-uncharged residues or serine residues in the P_1 position. Despite the fact that there were similarities in substrate specificity, mapping of the cleavage sites showed unique cleavage patterns that were context- and structure-dependent for both *At*PreP isoforms (Stahl et al. 2005).

The 3D crystal structure of the recombinant inactive mutant *At*PreP1 E80Q has been solved at 2.1 Å resolution (Johnson et al. 2006). It was the first closed conformation of a protease from the pitrilysin family. The structure revealed that *At*PreP creates bowl-shaped halves, connected by a hinge region. These two halves create a large internal chamber of about 10 000 Å^3, where the active site resides. Interestingly, the active site is not only created by the inverted zinc-binding motif ($H_{77}XXE_{80}H_{81}$) with a distal Glu_{177} located in the N-terminal portion of the enzyme, but also R_{848} and Y_{854} located in the C-terminal part of the enzyme, at a distance of almost 800 amino acid residues from the zinc-binding motif are essential for the catalysis.

The degradation occurs only when the proteolytic chamber is closed but locking the enzyme in a closed conformation in cysteine double mutants under oxidizing condition inhibits proteolytic activity. Therefore, a mechanism involving hinge-bending motions causing opening and closing of the enzyme in response to substrate binding has been proposed (Johnson et al. 2006). The crystal structure revealed also two non-catalytic Mg^{2+} binding sites. Mutation studies demonstrated that one of these sites located inside the proteolytic chamber close to the active site was essential for the enzyme activity (Backman et al. 2009).

Homologues of PreP are present in all species except archaea. The *Saccharomyces cerevisiae* homologue of *At*PreP, Mop112/Cym1 has been reported to be localized to the IMS in contrast to plants and mammalian mitochondria, where PreP is found in the matrix (Kambacheld et al. 2005; Falkevall et al. 2006). Interestingly, *At*PreP complements MOP112 deleted yeast strains suggesting functional conservation of the yeast and plant homologues. Revisiting the submitochondrial localization of Mop112 revealed matrix localization of the endogenous Mop112/Cym1 in yeast, as has been previously reported for the plant and human homologue (Alikhani et al. unpublished).

The human PreP homologue, hPreP is localized in the mitochondrial matrix. Beside the degradation of targeting peptides, it has been also shown to degrade amyloid beta peptides [Aβ40, Aβ42, and Arctic Aβ40 (E22G)] associated with Alzheimer's disease (AD). Immunologic studies showed that hPreP is the sole protease responsible for degradation of Aβ in mitochondria (Falkevall et al. 2006). hPreP is a functional analogue of insulin degrading enzyme (IDE) that also has been implied in AD, as it degrades Aβ before insoluble plaques are formed. Unlike IDE, PreP cannot degrade insulin. PreP and IDE are structurally very similar, but IDE harbors an exosite in the catalytic chamber, which is hypothesized to unfold small proteins (Shen et al. 2006). The corresponding site is absent in the PreP structure, making PreP incapable ofdegrading folded small proteins. This fact can make PreP a better candidate compared to IDE for clearing up Aβ since it cannot degrade an important regulating protein such as insulin. Molecular homology model of hPreP has been created based on the solved 3D structure of *At*PreP (Johnson et al. 2006). Surprisingly, two cysteines (Cys90 in the N-terminal portion and Cys527 located in the hinge region) in the homology model of hPreP were discovered to be in close vicinity to each other. These cysteines are conserved in all mammalian PreP sequences and can form a disulphide bridge under oxidizing conditions deactivating the enzyme. These findings are highly interesting and indicate a possible inhibition of hPreP under elevated ROS production in mitochondria that is implicated in AD, and might therefore be of physiologic importance in AD (Glaser and Alikhani 2009). Furthermore, functional analysis of single nucleotide polymorphism variants of hPreP (hPreP-SNPs) showed a dramatically decreased activity for the hPreP(A525D) variant with a mutation situated in the hinge region (Pinho et al. 2010).This region is hypothesized to be of importance for the opening and closing of the proteolytic chamber. Hence, we believe that uncommon substitution in hPreP may contribute to less efficient clearance of Aβ and other toxic peptides in mitochondria, which thereby may contribute to mitochondrial dysfunctions.

11.8 Future Issues

- Sorting process and the role of cytosolic factors in the sorting between mitochondria and chloroplasts is not fully understood.
- There is a growing numberof dual targeting proteins to mitochondria and chloroplasts, and it will be challenging to identify them and understand their import pathway and regulation of import to both mitochondria and chloroplasts.
- Plant mitochondrial outer membrane proteome has the potential to identify novel receptor components.
- Orthologous components may have additional or different functions in one species compared to another, e.g., Tim17 and Mia40. On the other hand non-orthologous components may have similar functions, e.g. Tom20 from plants and yeast.
- Novel proteases have been identified and their importance to mitochondrial biogenesis and morphology is intensively studied.

Acknowledgments This work was supported by a grant from The Swedish Research Council to E.G. J.W. was supported by a grant from the Wenner-Gren Foundation.

Glossary

Ambiguous targeting signal: A targeting signal that directs a protein to more than one location in a cell.

Amphiphilic (amphipathic): With reference to targeting signals refers to the fact that they contain polar and apolar properties.

Dual-targeting: A targeting process in which a protein is authentically targeted and accumulates in more than one location in a cell.

Intraorganelle sorting: The requirement to sort a protein to the correct location inside the organelle once it has been targeted to that organelle. The default targeting signal for mitochondria targets protein to the matrix and thus all proteins that are not located in the matrix must be stopped on their ways to the matrix (stop-transfer) or retargeted from the matrix to their final location (conservative sorting).

Mis-targeting: A process in which a protein is targeted to a location other than that where it is normally found. Mis-targeting can be observed in some in vitro uptake assays, in vivo targeting assays that tag proteins with artificial passenger proteins, when using heterologous systems or when using chimeric constructs.

Molecular chaperone: A protein that binds transiently to unfolded domains of other proteins, preventing them for misfolding and conferring an import competent conformation.

Passenger protein: A protein that when linked to a presequence will be directed to mitochondria. Passenger proteins have properties that allow them to be easily visualized, such as fluorescence (GFP) or can be manipulated to fold or unfold under altered experimental (DHFR).

Precursor processing: Proteolytic cleavage of the presequence from a precursor protein by mitochondrial processing peptidases (MPP, IMP, ICP55, MIP) resulting in a mature protein and a free presequence that is further degraded by presequence protease, PreP.

Presequence: N-terminal cleavable targeting signals that direct proteins to mitochondria.

Transit peptide: N-terminal cleavable targeting signals that direct proteins to chloroplasts.

Translocase: A multi-subunit membrane bound protein complex that mediates the recognition and insertion of proteins into or across a membrane.

References

Abe, Y., Shodai, T., Muto, T., Mihara, K., Torii, H., Nishikawa, S., Endo, T., Kohda, D. 2000. Structural basis of presequence recognition by the mitochondrial protein import receptor tom20. Cell 100(5):551–560.

Armstrong, L. C., Komiya, T., Bergman, B. E., Mihara, K., Bornstein, P. 1997. Metaxin is a component of a preprotein import complex in the outer membrane of the mammalian mitochondrion. J. Biol. Chem. 272(10):6510–6518.

Backman, H. G., Pessoa, J., Eneqvist, T., Glaser, E. 2009. Binding of divalent cations is essential for the activity of the organellar peptidasome in arabidopsis thaliana, atprep. FEBS Lett. 583(17):2727–2733.

Berglund, A. K., Pujol, C., Duchene, A. M., Glaser, E. 2009a. Defining the determinants for dual targeting of amino acyl-trna synthetases to mitochondria and chloroplasts. J. Mol. Biol. 393(4):803–814.

Berglund, A. K., Spanning, E., Biverstahl, H., Maddalo, G., Tellgren-Roth, C., Maler, L., Glaser, E. 2009b. Dual targeting to mitochondria and chloroplasts: Characterization of thr-trna synthetase targeting peptide. Mol. Plant 2(6):1298–1309.

Bhushan, S., Lefebvre, B., Stahl, A., Wright, S. J., Bruce, B. D., Boutry, M., Glaser, E. 2003. Dual targeting and function of a protease in mitochondria and chloroplasts. EMBO Rep. 4(11):1073–1078.

Bhushan, S., Stahl, A., Nilsson, S., Lefebvre, B., Seki, M., Roth, C., McWilliam, D., Wright, S. J., Liberles, D. A., Shinozaki, K., Bruce, B. D., Boutry, M., Glaser, E. 2005. Catalysis, subcellular localization, expression and evolution of the targeting peptides degrading protease, atprep2. Plant Cell Phsyiol. 46(6):985–996.

Bhushan, S., Kuhn, C., Berglund, A. K., Roth, C., Glaser, E. 2006. The role of the n-terminal domain of chloroplast targeting peptides in organellar protein import and miss-sorting. FEBS Lett. 580(16):3966–3972.

Branda, S. S., Isaya. G. 1995. Prediction and identification of new natural substrates of the yeast mitochondrial intermediate peptidase. J. Biol. Chem. 270(45):27366–27373.

Braun, H. P., Emmermann, M., Kruft, V., Schmitz, U. K. 1992. The general mitochondrial processing peptidase from potato is an integral part of cytochrome c reductase of the respiratory chain. EMBO J. 11(9):3219–3227.

Brix, J., Rudiger, S., Bukau, B., Schneider-Mergener, J., Pfanner, N. 1999. Distribution of binding sequences for the mitochondrial import receptors tom20, tom22, and tom70 in a presequence-carrying preprotein and a non-cleavable preprotein. J. Biol. Chem. 274(23):16522–16530.

Brix, J., Ziegler, G. A., Dietmeier, K., Schneider-Mergener, J., Schulz, G. E., Pfanner, N. 2000. The mitochondrial import receptor tom70: Identification of a 25 kda core domain with a specific binding site for preproteins. J. Mol. Biol. 303(4):479–488.

Bruce, B. D. 2000. Chloroplast transit peptides: Structure, function and evolution. Trends Cell Biol. 10(10):440–447.

Brumme, S., Kruft, V., Schmitz, U. K., Braun, H. P. 1998. New insights into the co-evolution of cytochrome c reductase and the mitochondrial processing peptidase. J. Biol. Chem. 273(21):13143–13149.

Burri, L., Strahm, Y., Hawkins, C. J., Gentle, I. E., Puryer, M. A., Verhagen, A., Callus, B., Vaux, D., Lithgow, T. 2005. Mature diablo/smac is produced by the imp protease complex on the mitochondrial inner membrane. Mol. Biol. Cell 16(6):2926–2933.

Carrie, C., Giraud, E., Duncan, O., Hsu, L., Wang, Y., Clifton, R., Murcha, M. W., Filipovska, A., Rackham, O., Vrielink, A., Whelan, J. 2010. Conserved and novel functions for Arabidopsis MIA40 in assembly of proteins in mitochondria and peroxisomes. J Biol. Chem. (PMID: 20829360).

Carrie, C., Giraud, E., Whelan, J. 2009. Protein transport in organelles: Dual targeting of proteins to mitochondria and chloroplasts. FEBS J. 276(5):1187–1195.

Chacinska, A., Koehler, C. M., Milenkovic, D., Lithgow, T., Pfanner, N. 2009. Importing mitochondrial proteins: Machineries and mechanisms. Cell 138(4):628–644. doi:S0092-8674(09)00967-2 [pii].10.1016/j.cell.2009.08.005.

Chan, N. C., Likic, V. A., Waller, R. F., Mulhern, T. D., Lithgow, T. 2006. The c-terminal tpr domain of tom70 defines a family of mitochondrial protein import receptors found only in animals and fungi. J. Mol. Biol. 358(4):1010–1022. doi:S0022-2836(06)00277-4 [pii]10.1016/j.jmb.2006.02.062.

Chen, X., Van Valkenburgh, C., Fang, H., Green, N. 1999. Signal peptides having standard and nonstandard cleavage sites can be processed by imp1p of the mitochondrial inner membrane protease. J. Biol. Chem. 274(53):37750–37754.

Chew, O., Rudhe, C., Glaser, E., Whelan, J. 2003. Characterization of the targeting signal of dual-targeted pea glutathione reductase. Plant Mol. Biol. 53(3):341–356.

Chew, O., Lister, R., Qbadou, S., Heazlewood, J. L., Soll, J., Schleiff, E., Millar, A. H., Whelan, J. 2004. A plant outer mitochondrial membrane protein with high amino acid sequence identity to a chloroplast protein import receptor. FEBS Lett. 557(1–3):109–114. doi:S0014579303014571 [pii].

Christensen, A. C., Lyznik, A., Mohammed, S., Elowsky, C. G., Elo, A., Yule, R., Mackenzie, S. A. 2005. Dual-domain, dual-targeting organellar protein presequences in arabidopsis can use non-aug start codons. Plant Cell 17(10):2805–2816.

Corral-Debrinski, M., Blugeon, C., Jacq, C. 2000. In yeast, the 3′ untranslated region or the presequence of atm1 is required for the exclusive localization of its mrna to the vicinity of mitochondria. Mol. Cell Biol. 20(21):7881–7892.

de Castro Silva Filho, M., Chaumont, F., Leterme, S., Boutry, M. 1996. Mitochondrial and chloroplast targeting sequences in tandem modify protein import specificity in plant organelles. Plant Mol. Biol. 30(4):769–780.

Dessi, P., Rudhe, C., Glaser, E. 2000. Studies on the topology of the protein import channel in relation to the plant mitochondrial processing peptidase integrated into the cytochrome bc1 complex. Plant J. 24(5):637–644.

Dessi, P., Pavlov, P. F., Wallberg, F., Rudhe, C., Brack, S., Whelan, J., Glaser, E. 2003. Investigations on the in vitro import ability of mitochondrial precursor proteins synthesized in wheat germ transcription-translation extract. Plant Mol. Biol. 52(2):259–271.

Donzeau, M., Kaldi, K., Adam, A., Paschen, S., Wanner, G., Guiard, B., Bauer, M. F., Neupert, W., Brunner, M. 2000. Tim23 links the inner and outer mitochondrial membranes. Cell 101(4):401–412. doi:S0092-8674(00)80850-8 [pii].

Duby, G., Degand, H., Boutry, M. 2001a. Structure requirement and identification of a cryptic cleavage site in the mitochondrial processing of a plant f1-atpase beta-subunit presequence. FEBS Lett. 505(3):409–413.

Duby, G., Oufattole, M., Boutry, M. 2001b. Hydrophobic residues within the predicted n-terminal amphiphilic alpha-helix of a plant mitochondrial targeting presequence play a major role in in vivo import. Plant J. 27(6):539–549.

Duchene, A. M., Giritch, A., Hoffmann, B., Cognat, V., Lancelin, D., Peeters, N. M., Zaepfel, M., Marechal-Drouard, L., Small, I. D. 2005. Dual targeting is the rule for organellar aminoacyl-trna synthetases in arabidopsis thaliana. Proc. Natl. Acad. Sci. U.S.A. 102(45):16484–16489.

Dyall, S. D., Brown, M. T., Johnson, P. J. 2004. Ancient invasions: From endosymbionts to organelles. Science 304(5668):253–257. doi:10.1126/science.1094884, 304/5668/253 [pii].

Emmermann, M., Braun, H. P., Arretz, M., Schmitz, U. K. 1993. Characterization of the bifunctional cytochrome c reductase-processing peptidase complex from potato mitochondria. J. Biol. Chem. 268(25):18936–18942.

Eriksson, A. C., Sjoling, S., Glaser, E. 1996. Characterization of the bifunctional mitochondrial processing peptidase (mpp)/bc1 complex in spinacia oleracea. J. Bioenerg. Biomembr. 28 (3):285–292.

Falkevall, A., Alikhani, N., Bhushan, S., Pavlov, P. F., Busch, K., Johnson, K. A., Eneqvist, T., Tjernberg, L., Ankarcrona, M., Glaser, E. 2006. Degradation of the amyloid beta-protein by the novel mitochondrial peptidasome, prep. J. Biol. Chem. 281(39):29096–29104.

Flavell, R. 2009. Role of model plant species. Methods Mol. Biol. 513:1–18. doi:10.1007/978-1-59745-427-8_1.

Gabriel, K., Egan, B., Lithgow, T. 2003. Tom40, the import channel of the mitochondrial outer membrane, plays an active role in sorting imported proteins. EMBO J. 22 (10):2380–2386. doi:10.1093/emboj/cdg229.

Gakh, O., Cavadini, P., Isaya, G. 2002. Mitochondrial processing peptidases. Biochim. Biophys. Acta 1592(1):63–77.

Glaser, E., Alikhani, N. 2009. The organellar peptidasome, prep: A journey from arabidopsis to Alzheimer's disease. Biochim. Biophys. Acta 1797(6–7):1076–1080.

Glaser, E., Dessi, P. 1999. Integration of the mitochondrial-processing peptidase into the cytochrome bc1 complex in plants. J. Bioenerg. Biomembr. 31(3):259–274.

Glaser, E., Eriksson, A., Sjoling, S. 1994. Bifunctional role of the bc1 complex in plants. Mitochondrial bc1 complex catalyses both electron transport and protein processing. FEBS Lett. 346(1):83–87.

Glaser, E., Sjoling, S., Tanudji, M., Whelan, J. 1998. Mitochondrial protein import in plants. Signals, sorting, targeting, processing and regulation. Plant Mol. Biol. 38(1–2):311–338.

Habib, S. J., Vasiljev, A., Neupert, W., Rapaport, D. 2003. Multiple functions of tail-anchor domains of mitochondrial outer membrane proteins. FEBS Lett. 555(3):511–515.

Hachiya, N., Alam, R., Sakasegawa, Y., Sakaguchi, M., Mihara, K., Omura, T. 1993. A mitochondrial import factor purified from rat liver cytosol is an atp-dependent conformational modu lator for precursor proteins. EMBO J. 12(4):1579–1586.

Hartl, F. U., Pfanner, N., Nicholson, D. W., Neupert, W. 1989. Mitochondrial protein import. Biochim. Biophys. Acta 988(1):1–45. doi:0304-4157(89)90002-6 [pii].

Heazlewood, J. L., Tonti-Filippini, J. S., Gout, A. M., Day, D. A., Whelan, J., Millar, A. H. 2004. Experimental analysis of the arabidopsis mitochondrial proteome highlights signaling and regulatory components, provides assessment of targeting prediction programs, and indicates plant-specific mitochondrial proteins. Plant Cell 16(1):241–256.

Hedtke, B., Borner, T., Weihe, A. 2000. One RNA polymerase serving two genomes. EMBO Rep. 1(5):435–440.

Heins, L., Schmitz, U. K. 1996. A receptor for protein import into potato mitochondria. Plant J. 9(6):829–839.

Hell, K., Neupert, W., Stuart, R. A. 2001. Oxa1p acts as a general membrane insertion machinery for proteins encoded by mitochondrial DNA. EMBO J. 20(6):1281–1288. doi:10.1093/emboj/20.6.1281.

Horie, C., Suzuki, H., Sakaguchi, M., Mihara, K. 2003. Targeting and assembly of mitochondrial tail-anchored protein tom5 to the tom complex depend on a signal distinct from that of tail-anchored proteins dispersed in the membrane. J. Biol. Chem. 278(42):41462–41471.

Howell, K. A., Millar, A. H., Whelan, J. 2006. Ordered assembly of mitochondria during rice germination begins with pro-mitochondrial structures rich in components of the protein import apparatus. Plant Mol. Biol. 60(2):201–223. doi:10.1007/s11103-005-3688-7.

Huang, S., Taylor, N. L., Whelan, J., Millar, A. H. 2009. Refining the definition of plant mitochondrial presequences through analysis of sorting signals, n-terminal modifications, and cleavage motifs. Plant Physiol.150(3):1272–1285.

Hugosson, M., Andreu, D., Boman, H. G., Glaser, E. 1994. Antibacterial peptides and mitochondrial presequences affect mitochondrial coupling, respiration and protein import. Eur. J. biochem. FEBS 223(3):1027–1033.

Isaya, G., Kalousek, F., Rosenberg, L. E. 1992. Sequence analysis of rat mitochondrial intermediate peptidase: Similarity to zinc metallopeptidases and to a putative yeast homologue. Proc. Natl. Acad. Sci. U.S.A. 89(17):8317–8321.

Jansch, L., Kruft, V., Schmitz, U. K., Braun, H. P. 1998. Unique composition of the preprotein translocase of the outer mitochondrial membrane from plants. J. Biol. Chem. 273(27):17251–17257.

Johnson, K. A., Bhushan, S., Stahl, A., Hallberg, B. M., Frohn, A., Glaser, E., Eneqvist, T. 2006. The closed structure of presequence protease prep forms a unique 10,000 angstroms3 chamber for proteolysis. EMBO J. 25(9):1977–1986.

Kambacheld, M., Augustin, S., Tatsuta, T., Muller, S., Langer, T. 2005. Role of the novel metallopeptidase mop112 and saccharolysin for the complete degradation of proteins residing in different subcompartments of mitochondria. J. Biol. Chem. 280(20):20132–20139. doi:M500398200 [pii]10.1074/jbc.M500398200.

Kmiec-Wisniewska, B., Krumpe, K., Urantowka, A., Sakamoto, W., Pratje, E., Janska, H. 2008. Plant mitochondrial rhomboid, atrbl12, has different substrate specificity from its yeast counterpart. Plant Mol. Biol. 68(1–2):159–171.

Koppen, M., Langer, T. 2007. Protein degradation within mitochondria: Versatile activities of aaa proteases and other peptidases. Crit. Rev. Biochem. Mol. Biol. 42(3):221–242.

Likic, V. A., Perry, A., Hulett, J., Derby, M., Traven, A., Waller, R. F., Keeling, P. J., Koehler, C. M., Curran, S. P., Gooley, P. R., Lithgow, T. 2005. Patterns that define the four domains conserved in known and novel isoforms of the protein import receptor tom20. J. Mol. Biol. 347(1):81–93. doi:S0022-2836(05)00005-7 [pii]10.1016/j.jmb.2004.12.057.

Lister, R., Whelan, J. 2006. Mitochondrial protein import: Convergent solutions for receptor structure. Curr. Biol. 16(6):R197–199. doi:S0960-9822(06)01174-2 [pii]10.1016/j.cub.2006.02.024.

Lister, R., Mowday, B., Whelan, J., Millar, A. H. 2002. Zinc-dependent intermembrane space proteins stimulate import of carrier proteins into plant mitochondria. Plant J. 30(5):555–566. doi:1316 [pii].

Lister, R., Murcha, M. W., Whelan, J. 2003. The mitochondrial protein import machinery of plants (mpimp) database. Nucleic Acids Res. 31(1):325–327.

Lister, R., Hulett, J. M., Lithgow, T., Whelan, J. 2005. Protein import into mitochondria: Origins and functions today (review). Mol. Membr. Biol. 22(1–2):87–100.

Lister, R., Carrie, C., Duncan, O., Ho, L. H., Howell, K. A., Murcha, M. W., Whelan, J. 2007. Functional definition of outer membrane proteins involved in preprotein import into mitochondria. Plant Cell 19(11):3739–3759. doi:tpc.107.050534 [pii] 10.1105/tpc.107.050534.

Lithgow, T. 2000. Targeting of proteins to mitochondria. FEBS Lett.476(1–2):22–26. doi:S0014-5793(00)01663-X [pii].

Luirink, J., Samuelsson, T., de Gier, J. W. 2001. Yidc/oxa1p/alb3: Evolutionarily conserved mediators of membrane protein assembly. FEBS Lett. 501(1):1–5. doi:S0014-5793(01)02616-3 [pii].

Macasev, D., Whelan, J., Newbigin, E., Silva-Filho, M. C., Mulhern, T. D., Lithgow, T. 2004. Tom22′, an 8-kda trans-site receptor in plants and protozoans, is a conserved feature of the tom complex that appeared early in the evolution of eukaryotes. Mol. Biol. Evol. 21(8):1557–1564. doi:10.1093/molbev/msh166[pii].

McQuibban, G. A., Saurya, S., Freeman, M. 2003. Mitochondrial membrane remodelling regulated by a conserved rhomboid protease. Nature 423(6939):537–541.

Mesecke, N., Bihlmaier, K., Grumbt, B., Longen, S., Terziyska, N., Hell, K., Herrmann, J. M. 2008. The zinc-binding protein hot13 promotes oxidation of the mitochondrial import receptor mia40. EMBO Rep. 9(11):1107–1113. doi:embor2008173 [pii].

Meyer, E. H., Taylor, N. L., Millar, A. H. 2008. Resolving and identifying protein components of plant mitochondrial respiratory complexes using three dimensions of gel electrophoresis. J. Proteome Res. 7(2):786–794. doi:10.1021/pr700595p.

Milla, A. H., Sweetlove, L. J., Giege, P., Leaver, C. J. 2001. Analysis of the arabidopsis mitochondrial proteome. Plant Physiol. 127(4):1711–1727.

Millar, A. H., Heazlewood, J. L., Kristensen, B. K., Braun, H. P., Moller, I. M. 2005. The plant mitochondrial proteome. Trends Plant Sci. 10(1):36–43. doi:S1360-1385(04)00273-0 [pii]10.1016/j.tplants.2004.12.002.

Mitschke, J., Fuss, J., Blum, T., Hoglund, A., Reski, R., Kohlbacher, O., Rensing, S. A. 2009. Prediction of dual protein targeting to plant organelles. New Phytol. 183(1):224–235.

Moberg, P., Stahl, A., Bhushan, S., Wright, S. J., Eriksson, A., Bruce, B. D., Glaser, E. 2003. Characterization of a novel zinc metalloprotease involved in degrading targeting peptides in mitochondria and chloroplasts. Plant J. 36(5):616–628.

Moberg, P., Nilsson, S., Stahl, A., Eriksson, A. C., Glaser, E., Maler, L. 2004. Nmr solution structure of the mitochondrial f1beta presequence from nicotiana plumbaginifolia. J. Mol. Biol. 336(5):1129–1140.

Murakami, K., Mori, M. 1990. Purified presequence binding factor (pbf) forms an import-comp-etent complex with a purified mitochondrial precursor protein. EMBO J. 9(10):3201–3208.

Murcha, M. W., Elhafez, D., Millar, A. H., Whelan, J. 2004. The n-terminal extension of plant mitochondrial carrier proteins is removed by two-step processing: the first cleavage is by the mitochondrial processing peptidase. J. Mol. Biol. 344 (2):443–454. doi:S0022-2836(04) 01201-X [pii]10.1016/j.jmb.2004.09.045.

Murcha, M. W., Elhafez, D., Millar, A. H., Whelan, J. 2005a. The c-terminal region of tim17 links the outer and inner mitochondrial membranes in arabidopsis and is essential for protein import. J. Biol. Chem. 280 (16):16476–16483. doi:M413299200 [pii] 10.1074/jbc.M413299200.

Murcha, M. W., Millar, A. H., Whelan, J. 2005b. The n-terminal cleavable extension of plant carrier proteins is responsible for efficient insertion into the inner mitochondrial membrane. J. Mol. Biol. 351 (1):16–25. doi:S0022-2836(05)00649-2 [pii] 10.1016/j.jmb.2005.06.004.

Murcha, M. W., Elhafez, D., Lister, R., Tonti-Filippini, J., Baumgartner, M., Philippar, K., Carrie, C., Mokranjac, D., Soll, J., Whelan, J. 2007. Characterization of the preprotein and amino acid transporter gene family in arabidopsis. Plant Physiol. 143(1):199–212. doi:pp.106.090688 [pii]10.1104/pp.106.090688.

Muto, T., Obita, T., Abe, Y., Shodai, T., Endo, T., Kohda, D. 2001. Nmr identification of the tom20 binding segment in mitochondrial presequences. J. Mol. Biol. 306(2):137–143.

Naamati, A., Regev-Rudzki, N., Galperin, S., Lill, R., Pines, O. 2009. Dual targeting of nfs1 and discovery of its novel processing enzyme, icp55. J. Biol. Chem. 284(44):30200–30208.

Neupert, W. 1997. Protein import into mitochondria. Annu. Rev. Biochem. 66:863–917.

Neupert, W., Herrmann, J. M. 2007. Translocation of proteins into mitochondria. Annu. Rev. Biochem. 76:723–749. doi:10.1146/annurev.biochem.76.052705.163409.

Nilsson Cederholm, S., Backman, H. G., Pesaresi, P., Leister, D., Glaser, E. 2009. Deletion of an organellar peptidasome prep affects early development in arabidopsis thaliana. Plant Mol. Biol. 71(4–5):497–508.

Ono, H., Tuboi, S. 1990. Purification and identification of a cytosolic factor required for import of precursors of mitochondrial proteins into mitochondria. Arch. Biochem. Biophys. 280(2):299–304.

Paschen, S. A., Waizenegger, T., Stan, T., Preuss, M., Cyrklaff, M., Hell, K., Rapaport, D., Neupert, W. 2003. Evolutionary conservation of biogenesis of beta-barrel membrane proteins. Nature 426(6968):862–866. doi:10.1038/nature02208nature02208 [pii].

Peeters, N., Small, I. 2001. Dual targeting to mitochondria and chloroplasts. Biochim. Biophys. Acta 1541(1–2):54–63.

Perry, A. J., Hulett, J. M., Likic, V. A., Lithgow, T., Gooley, P. R. 2006. Convergent evolution of receptors for protein import into mitochondria. Curr. Biol. 1 (3):221–229. doi:S0960-9822(05)01596-4 [pii]10.1016/j.cub.2005.12.034.

Pfanner, N., Geissler, A. 2001. Versatility of the mitochondrial protein import machinery. Nat. Rev. Mol. Cell Biol. 2(5):339–349. doi:10.1038/3507300635073006 [pii].

Pinho, C. M., Bjork, B. F., Alikhani, N., Backman, H. G., Eneqvist, T., Fratiglioni, L., Glaser, E., Graff, C. 2010. Genetic and biochemical studies of snps of the mitochondrial a beta-degrading protease, hprep. Neurosci. Letts. 469(2):204–208.

Pohlmeyer, K., Soll, J., Steinkamp, T., Hinnah, S., Wagner, R. 1997. Isolation and characterization of an amino acid-selective channel protein present in the chloroplastic outer envelope membrane. Proc. Natl. Acad. Sci. U.S.A. 94(17):9504–9509.

Pujol, C., Marechal-Drouard, L., Duchene, A. M. 2007. How can organellar protein n-terminal sequences be dual targeting signals? In silico analysis and mutagenesis approach. J. Mol. Biol. 369(2):356–367.

Rapaport, D. 2003. Finding the right organelle. Targeting signals in mitochondrial outer-membrane proteins. EMBO Rep. 4(10):948–952.

Rassow, J., Dekker, P. J., van Wilpe, S., Meijer, M., Soll, J. 1999. The preprotein translocase of the mitochondrial inner membrane: Function and evolution. J. Mol. Biol. 286(1):105–120. doi:S0022-2836(98)92455-X [pii]10.1006/jmbi.1998.2455.

Rawlings, N. D., Morton, F. R., Kok, C. Y., Kong, J., Barrett, A. J. 2008. Merops: The peptidase database. Nucleic Acids Res. 36(Database issue):D320–325.

Richter, S., Lamppa, G. K. 2002. Determinants for removal and degradation of transit peptides of chloroplast precursor proteins. J. Biol. Chem. 277(46):43888–43894.

Rudhe, C., Chew, O., Whelan, J., Glaser, E. 2002a. A novel in vitro system for simultaneous import of precursor proteins into mitochondria and chloroplasts. Plant J. 30(2):213–220.

Rudhe, C., Clifton, R., Whelan, J., Glaser, E. 2002b. N-terminal domain of the dual-targeted pea glutathione reductase signal peptide controls organellar targeting efficiency. J. Mol. Biol. 324(4):577–585.

Rudhe, C., Clifton, R., Chew, O., Zemam, K., Richter, S., Lamppa, G., Whelan, J., Glaser, E. 2004. Processing of the dual targeted precursor protein of glutathione reductase in mitochondria and chloroplasts. J. Mol. Biol. 343(3):639–647.

Sakamoto, W., Spielewoy, N., Bonnard, G., Murata, M., Wintz, H. 2000. Mitochondrial localization of atoxa1, an arabidopsis homologue of yeast oxa1p involved in the insertion and assembly of protein complexes in mitochondrial inner membrane. Plant Cell Physiol. 41(10):1157–1163.

Schleiff, E., Motzkus, M., Soll, J. 2002. Chloroplast protein import inhibition by a soluble factor from wheat germ lysate. Plant Mol. Biol. 50(2):177–185.

Schneider, G., Sjoling, S., Wallin, E., Wrede, P., Glaser, E., von Heijne, G. 1998. Feature-extraction from endopeptidase cleavage sites in mitochondrial targeting peptides. Proteins 30(1):49–60.

Schulte, U., Arretz, M., Schneider, H., Tropschug, M., Wachter, E., Neupert, W., Weiss, H. 1989. A family of mitochondrial proteins involved in bioenergetics and biogenesis. Nature 339(6220):147–149.

Shen, Y., Joachimiak, A., Rosner, M. R., Tang, W. J. 2006. Structures of human insulin-degrading enzyme reveal a new substrate recognition mechanism. Nature 443(7113):870–874.

Shore, G. C., McBride, H. M., Millar, D. G., Steenaart, N. A., Nguyen, M. 1995. Import and insertion of proteins into the mitochondrial outer membrane. Eur. J. Biochem. FEBS 227(1–2):9–18.

Silva-Filho, M. C. 2003.One ticket for multiple destinations: Dual targeting of proteins to distinct subcellular locations. Curr. Opin. Plant Biol. 6(6):589–595.

Soll, J., Schleiff, E. 2004. Protein import into chloroplasts. Nat. Rev. Mol. Cell Biol. 5(3):198–208. doi:10.1038/nrm1333nrm1333 [pii].

Stahl, A., Moberg, P., Ytterberg, J., Panfilov, O., Brockenhuus Von Lowenhielm, H, Nilsson, F., Glaser, E. 2002. Isolation and identification of a novel mitochondrial metalloprotease (prep) that degrades targeting presequences in plants. J. Biol. Chem. 277(44):41931–41939.

Stahl, A., Nilsson, S., Lundberg, P., Bhushan, S., Biverstahl, H., Moberg, P., Morisset, M., Vener, A., Maler, L., Langel, U., Glaser, E. 2005. Two novel targeting peptide degrading proteases, preps, in mitochondria and chloroplasts, so similar and still different. J. Mol. Biol. 349(4):847–860.

Stein, I., Peleg, Y., Even-Ram, S., Pines, O. 1994. The single translation product of the fum1 gene (fumarase) is processed in mitochondria before being distributed between the cytosol and mitochondria in saccharomyces cerevisiae. Mol. Cell Biol. 14(7):4770–4778.

Tanudji, M., Sjoling, S., Glaser, E., Whelan, J. 1999. Signals required for the import and processing of the alternative oxidase into mitochondria. J. Biol. Chem. 274(3):1286–1293.

Taylor, A. B., Smith, B. S., Kitada, S., Kojima, K., Miyaura, H., Otwinowski, Z., Ito, A., Deisenhofer, J. 2001. Crystal structures of mitochondrial processing peptidase reveal the mode for specific cleavage of import signal sequences. Structure 9(7):615–625.
van Wilpe, S., Ryan, M. T., Hill, K., Maarse, A. C., Meisinger, C., Brix, J., Dekker, P. J., Moczko, M., Wagner, R., Meijer, M., Guiard, B., Honlinger, A., Pfanner, N. 1999. Tom22 is a multifunctional organizer of the mitochondrial preprotein translocase. Nature 401(6752):485–489. doi:10.1038/46802.
Vogtle, F. N., Wortelkamp, S., Zahedi, R. P., Becker, D., Leidhold, C., Gevaert, K., Kellermann, J., Voos, W., Sickmann, A., Pfanner, N., Meisinger, C. 2009. Global analysis of the mitochondrial n-proteome identifies a processing peptidase critical for protein stability. Cell 139(2):428–439.
von Heijne, G. 1986. Mitochondrial targeting sequences may form amphiphilic helices. EMBO J. 5(6):1335–1342.
Waltner, M., Weiner, H. 1995. Conversion of a nonprocessed mitochondrial precursor protein into one that is processed by the mitochondrial processing peptidase. J. Biol. Chem. 270(44):26311–26317.
Westermann, B., Neupert, W. 2000. Mitochondria-targeted green fluorescent proteins: Convenient tools for the study of organelle biogenesis in saccharomyces cerevisiae. Yeast (Chichester, England) 16(15):1421–1427.
Whelan, J., Tanudji, M. R., Smith, M. K., Day, D. A. 1996. Evidence for a link between translocation and processing during protein import into soybean mitochondria. Biochim. Biophys. Acta 1312(1):48–54.
Wienk, H. L., Czisch, M., de Kruijff, B. 1999. The structural flexibility of the preferredoxin transit peptide. FEBS Lett. 453(3):318–326.
Winning, B. M., Sarah, C. J., Purdue, P. E., Day, C. D., Leaver, C. J. 1992. The adenine nucleotide translocator of higher plants is synthesized as a large precursor that is processed upon import into mitochondria. Plant J. 2(5):763–773.
Yano, M., Terada, K., Mori, M. 2003. Aip is a mitochondrial import mediator that binds to both import receptor tom20 and preproteins. J. Cell Biol. 16(1):45–56.
Zhang, X. P., Glaser, E. 2002. Interaction of plant mitochondrial and chloroplast signal peptides with the hsp70 molecular chaperone. Trends Plant Sci. 7(1):14–21.

Chapter 12
Mitochondrial Protein Import in Fungi and Animals

D. Becker, U.K. Seibold, and W. Voos

Abstract Mitochondria are essential constituents of virtually all eukaryotic cells. They represent the major production site of chemical energy in the form of ATP, and house a number of essential biosynthetic pathways. In recent years, mitochondria have been identified as key components of signaling pathways. As endosymbiotic organisms, they contain their own genome as well as an independent and fully functional endogenous protein biosynthesis machinery. However, the coding capacity of the mitochondrial genome is restricted to a few polypeptides, i.e., eight in the baker`s yeast Saccharomyces cerevisiae and thirteen in the human genome. Hence, most mitochondrial proteins – recent proteomic studies predict a total number of at least 500-800 different polypeptides – are encoded in the nuclear genome and have to be synthesized as precursors at cytosolic ribosomes. In this review mitochondrial protein import in fungi and animals is discussed.

Keywords Mitochondria • Protein import • Presequence receptor • Membrane translocase • Import motor • Chaperone

Abbreviations

AAA	ATPases associated with a wide variety of cellular activities
AAC	ATP/ADP carrier
EM	Electron microscopy
GIP	General import pore
HSP	Heat shock protein
IM	Inner membrane
IMS	Intermembrane space

W. Voos (✉)
Institut für Biochemie und Molekularbiologie, Universität Bonn, Nussallee 11, D-53115 Bonn, Germany
e-mail: wolfgang.voos@uni-bonn.de

F. Kempken (ed.), *Plant Mitochondria*, Advances in Plant Biology 1,
DOI 10.1007/978-0-387-89781-3_12,

MCF	Metabolite carrier family
MIA	Mitochondrial intermembrane space assembly
MIP	Mitochondrial intermediate peptidase
MPP	Mitochondrial processing peptidase
NMR	Nuclear magnetic resonance
OM	Outer membrane
PAM	Presequence translocase-associated motor
SAM	Sorting and assembly machinery
TIM	Translocase of the inner mitochondrial membrane
TOB	Topogenesis of β-barrel proteins
TOM	Translocase of the outer mitochondrial membrane
$\Delta\psi$	Electric inner membrane potential
TPR	Tetratricopeptide repeat
VDAC	Voltage-dependent anion channel

12.1 Introduction

Mitochondria are essential constituents of virtually all eukaryotic cells. They represent the major production site of chemical energy in the form of ATP, and house a number of essential biosynthetic pathways (see also Chaps. 13 and 14). In recent years, mitochondria have been identified as key components of signaling pathways, e.g., apoptosis (McBride et al. 2006). Among the different organelles present in fungal and animal cells, mitochondria are unique due to their prokaryotic ancestry. As endosymbiotic organisms, they contain their own genome as well as independent and fully functional endogenous protein biosynthesis machinery. However, the coding capacity of the mitochondrial genome is restricted to a few polypeptides, i.e., eight in the baker`s yeast *Saccharomyces cerevisiae* and thirteen in the human genome (see also Chaps. 1 and 9). Hence, most mitochondrial proteins – recent proteomic studies predict a total number of at least 500–800 different polypeptides – are encoded in the nuclear genome and have to be synthesized as precursors at cytosolic ribosomes (Sickmann et al. 2003). The mitochondrial precursor proteins ("preproteins") are imported into the organelle after their translation has been initiated in the cytosol (Neupert and Herrmann 2007; Chacinska et al. 2009). Please note that plant mitochondrial protein import is specifically discussed in Chap. 11. In the majority of cases, translation seems to be even completed prior to the mitochondrial import process. Concurrent translation and transport events are conceivable, but except for a few examples, these processes are not as tightly coupled as in the case of nascent polypeptide chains destined for the endoplasmic reticulum (ER). Given the complex biosynthetic process of mitochondrial proteins, several important prerequisites are necessary to guarantee an efficient transport to their final destinations. First, a signal has to be provided to target the precursor proteins to their correct (submitochondrial) localization. Second, due to the impermeability of biologic membranes for polypeptides, proteinaceous translocase complexes are required to accomplish their transport across or into the mitochondrial double-membrane system. Third, newly imported polypeptide chains have to acquire their enzymatic activity by folding to the native conformation and, in most cases, by assembly into the active enzyme complexes.

Similar to other transport processes for proteins across biologic membranes, the import of mitochondrial preproteins follows several general principles. Correct targeting of the newly synthesized precursor proteins is ensured by signal sequences that contain the information for their specific intramitochondrial localization. The targeting motifs are recognized by dedicated receptor proteins on the surface of the outer mitochondrial envelope, thereby initiating translocation. Membrane translocation of precursor polypeptides is mediated by specialized proteinaceous pore structures that are integrated into the two mitochondrial phospholipid membranes. Importantly, in order to maintain ion gradients across the mitochondrial membrane system, these pore structures provide only narrow and tightly regulated openings. Consequently, the mitochondrial import process for polypeptides requires the preproteins to be in an unfolded or only partially folded conformational state. The energy needed to fuel protein translocation into mitochondria is derived from two different sources, the hydrolysis of ATP and the electric potential across the inner membrane. Subsequent folding of a mitochondrial protein into its native state or its assembly into multi-subunit enzyme complexes requires the activity of further (proteinaceous) components. A large number of general chaperone proteins as well as specific assembly factors have been described to drive or assist these maturation processes.

The mitochondrial protein biogenesis process is determined by the complex structure of the organelle. Mitochondria can be divided into at least four subcompartments: the outer membrane (OM), the intermembrane space (IMS), the inner membrane (IM), and the matrix. Protein import into each of these subcompartments depends on individual pathways suited to the specific biologic or biochemical demands. The picture is further complicated by the fact that some subcompartments, in particular the inner membrane, utilize multiple translocation pathways. In many cases, pathways to specific subcompartments share some components at early stages, but segregate in the later steps. Furthermore, different preproteins may have individual requirements for specific components of the translocation machinery. Taken together, in spite of significant progress in the analysis of mitochondrial import during the past decades, lots of mechanistic details still remain to be clarified. The transport pathways into mitochondria can be best described by relating to the so-called translocase complexes, large proteinaceous machineries that confer preprotein recognition, provide the pore structures and mediate transport across or into the two lipid bilayers (Pfanner and Chacinska 2002; Endo et al. 2003). An overview of the translocation pathways into mitochondria is depicted in Fig. 12.1. The following paragraph gives a short summary on the mechanisms and machineries of the different pathways.

The OM contains two separate translocase complexes (Becker et al. 2008b). Virtually all nuclear-encoded preproteins enter mitochondria via the TOM (translocase of the outer mitochondrial membrane) complex, which exposes receptor components to the

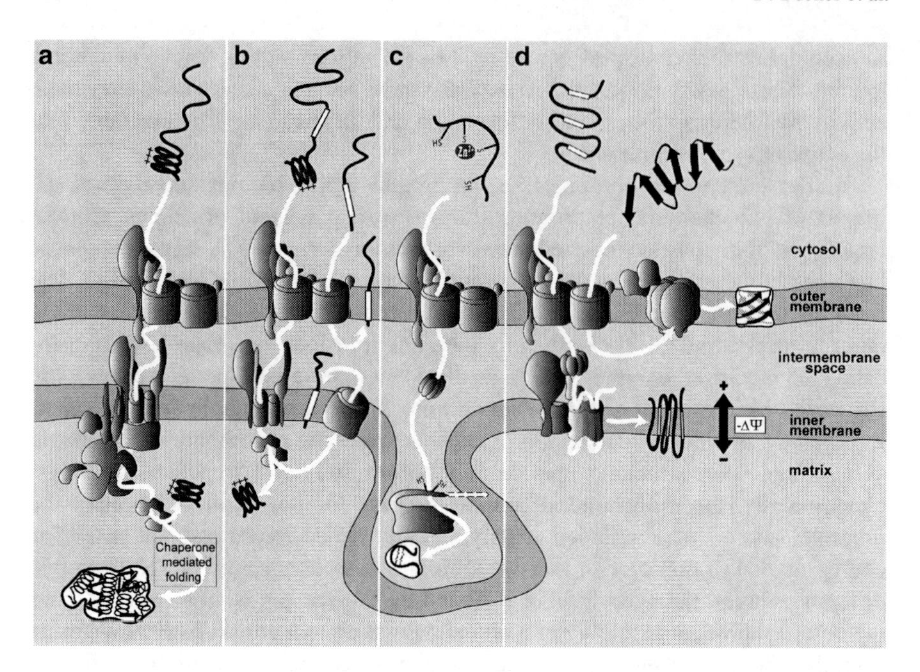

Fig. 12.1 Schematic overview of mitochondrial protein translocation pathways. Indicated are the mitochondrial subcompartments: outer membrane (*OM*), intermembrane space (*IMS*), inner membrane (*IM*), and matrix (*MA*). Several specific preprotein translocase complexes that are integrated into the phospholipid membranes confer the import of cytosolic precursor proteins to their final intramitochondrial localization. An electric potential across the inner membrane ($\Delta\psi$), generated by the respiratory chain, is required for polypeptide transport processes across the inner membrane. (**a**) The translocation pathway for soluble, presequence-containing preproteins into the mitochondrial matrix, comprising the outer membrane TOM complex, the inner membrane TIM23 complex, the import motor complex PAM, and the processing peptidase MPP (for details see text). (**b**) Insertion pathways of monotopic membrane proteins into the outer and inner membranes using the TOM and TIM23 protein complexes. (**c**) The disulfide-relay system responsible for the transport and folding of a subset of intermembrane space proteins containing intramolecular disulfide bonds (MIA pathway). (**d**) The insertion of hydrophobic β-barrel proteins into the outer membrane via the TOM and SAM complexes (*right*) and the insertion of polytopic inner membrane proteins of the metabolite carrier family into the inner membrane requiring the TOM and TIM22 protein complexes (*left*)

cytosol and provides the pore for crossing of the outer membrane. Integral OM proteins can either insert directly into the membrane, or they require the assistance of individual TOM components. The OM contains an additional protein complex, termed SAM/TOB (sorting and assembly machinery/topogenesis of β-barrel proteins), dedicated to the insertion of the abundant β-barrel membrane proteins that form large open-pore structures for metabolites in the OM. The import of proteins into the IMS does not follow a unified principle. Some proteins, like the abundant cytochrome *c*, seem to directly cross the OM, while others utilize parts of the matrix-targeting pathway in combination with the so-called stop-transfer mechanism (Glick et al. 1992a). In the latter case, detachment of the soluble IMS protein from the inner membrane is accomplished by a processing event. A genuine IMS pathway, named after its core component Mia40 (mitochondrial intermembrane space import and assembly), is represented by the recently described disulfide relay system, in which some soluble cysteine-containing preproteins traverse the outer membrane via TOM, but are then directly bound and folded by consecutive redox reactions in the IMS (Hell 2008). Most integral membrane proteins of the IM, in particular those belonging to the abundant metabolite carrier family (MCF), also utilize the TOM pore but are then transferred to the TIM22 (translocase of the inner membrane) complex in the inner membrane that governs their insertion into the IM (Koehler et al. 1999b). The best-studied pathway to date is the import of soluble proteins into the mitochondrial matrix (van der Laan et al. 2010). Here, hydrophilic preproteins containing N-terminal presequences first cross the OM via the TOM complex and are subsequently inserted into the IM via TIM23 (also termed presequence translocase). Full translocation into the matrix is accomplished with the help of the import motor complex, also termed PAM (presequence translocase-associated import motor). After completion of translocation into the matrix, the N-terminal targeting sequences are usually removed by dedicated processing peptidases. In some cases, polypeptides containing hydrophobic presequences segregate from the matrix transport route and instead, are directly integrated into the IM (Box 12.1).

12.2 The Multitude of Mitochondrial Targeting Signals

Proteomic as well as an impressive number of individual studies have revealed the amazing diversity of mitochondrial targeting sequence elements that reflect the complexity and versatility of the mitochondrial import machinery (Sickmann et al. 2003; Pagliarini et al. 2008; Elstner et al. 2009). The picture becomes even more complicated, as in some cases targeting seems to be dictated by structural elements rather than specific primary sequence motifs.

The most common mitochondrial targeting signals are cleavable N-terminal extensions, termed presequences (Cavalier-Smith 2006). They are found in precursor pro-

Box 12.1 Processing of mitochondrial precursor proteins

Upon arrival in the matrix, most N-terminal presequences are proteolytically removed by the heterodimeric metallopeptidase MPP (mitochondrial processing peptidase). In fungi and mammals, MPP is found in the matrix, whereas in plant mitochondria, it resides on the matrix-exposed side of the inner mitochondrial membrane (Taylor et al. 2001; Gakh et al. 2002). Interestingly, the two subunits of MPP (α- and β-MPP) are homologous to the core components (core I and II) of the respiratory chain complex III (cytochrome bc_1 complex) and in plants, MPP has been shown to constitute an integral part of the bc_1

(continued)

Box 12.1 (continued)

complex (Gakh et al. 2002; see also Chap. 13). Thus, the two MPP subunits may have evolved from gene duplication of the respiratory chain subunits. The importance of MPP-mediated cleavage is, among others, highlighted by its involvement in the pathology of Friedreich's ataxia, a progressive neurodegenerative disorder caused by mutations in the human frataxin homologue, a mitochondrial protein involved in cellular iron homeostasis (Schapira 1999; Patel and Isaya 2001).

Precursor proteins with more complex N-terminal sorting signals, directing them to the IMS first undergo MPP processing in the matrix, followed by cleavage of the sorting sequence, most often by the inner membrane peptidase (IMP), localized at the outer surface of the inner membrane. Other two-step processing events in the matrix include the mitochondrial intermediate peptidase (MIP; alternatively termed Oct1), which catalyzes preprotein maturation by removal of an octapeptide after MPP processing. Typical Oct substrates are iron/sulfur-containing proteins as well as nuclear-encoded respiratory chain subunits (Gakh et al. 2002). A very recent study resulted in the identification of another intermediate peptidase in the yeast *S. cerevisiae*, Icp55 (intermediate cleaving peptidase). Icp55 removes one further N-terminal residue in a subset of MPP-generated cleavage intermediates (Vögtle et al. 2009). The systematic analysis of the N-termini generated by Icp55 and Oct1 provided the first experimental indication for the existence of the N-end rule in mitochondria (Vögtle et al. 2009). Formerly known for bacteria and the eukaryotic cytosol, the N-end rule connects the N-terminal amino acid with the half-life of a protein differentiating between stabilizing and destabilizing residues (Mogk et al. 2007; Varshavsky 2008). In a few special cases, other inner membrane and matrix proteases such as the AAA proteases iAAA (ATPases associated with various cellular activities, intermembrane space-exposed) and mAAA (matrix-exposed) as well as members of the rhomboid-like proteases (PARL in mammals; Pcp1/Rbd1 in *S. cerevisiae*), membrane-integrated serine proteases that catalyze intramembraneous substrate cleavage are involved in presequence processing (Esser et al. 2002; Herlan et al. 2003; Duvezin-Caubet et al. 2007).

teins destined for the mitochondrial matrix as well as in some inner membrane and intermembrane space proteins (Pfanner and Chacinska 2002). Typical presequences are 20–80 residues in length and, although not conserved in their primary sequence, they share the tendency to form amphipathic α-helices with a positive net charge (Roise and Schatz 1988; von Heijne et al. 1989). Most presequence-containing inner membrane and a number of intermembrane space proteins contain so-called bipartite targeting signals, i.e., the N-terminal targeting sequence is followed by a hydrophobic sorting signal. This sorting signal ensures their correct localization to the respective mitochondrial subcompartment. Former studies applying presequence prediction programs on the mitochondrial proteome suggested approximately 45–50% of mitochondrial precursor proteins to contain cleavable N-terminal extensions (Sickmann et al. 2003; Prokisch et al. 2006; Habib et al. 2007). Very recent analysis of the yeast proteome, however, indicates that up to 70% of mitochondrial proteins are synthesized with presequences (Vögtle et al. 2009). Hence, the presequence pathway seems to represent the major pathway of mitochondrial protein import.

A substantial number of mitochondrial proteins are synthesized without cleavable presequences. Instead, they carry (multiple) targeting signals distributed throughout the mature proteins. This remarkably heterogeneous group includes virtually all outer membrane proteins studied so far, a large number of intermembrane space proteins, and (polytopic) inner membrane proteins as well as a few matrix proteins. These internal targeting signals are of highly diverse nature and less well characterized as the presequences described above.

1. In the mitochondrial outer membrane, α-helical proteins are the predominant membrane-integral proteins (Burri et al. 2006; Zahedi et al. 2006). The bulk of them contain single α-helical transmembrane domains that are often located either at the N-terminus ("signal-anchor" function, see below) or close to the C-terminus ("tail anchor"). In either case, these proteins are mainly exposed to the cytosol with a small portion present in the IMS. Targeting information is found within the transmembrane domain (TMD) and in positive residues flanking it (Rapaport 2003, 2005; Becker et al. 2009b). However, consensus sequences are not apparent. Thus, targeting seems to be governed by structural elements rather than specific primary sequences. Likewise, the import modes of α-helical outer membrane proteins seem to be diverse and remain to be established in many cases. While the TOM complex or at least the TOM channel has been suggested to play a role for some of them, a considerable number of signal-anchored proteins does not seem to require the TOM complex at all for outer membrane insertion, as shown for the outer membrane isoform of cytochrome b_5 reductase (Mcr1 in *S. cerevisiae*) and the SAM component Sam37 (Habib et al. 2005; Meineke et al. 2008). Insertion of another subset of signal-anchored proteins has been shown to depend on the integral outer membrane protein Mim1 (mitochondrial import) (Becker et al. 2008a; Hulett et al. 2008). The insertion of tail-anchored proteins is likewise poorly understood, and conflicting data concerning the involvement of the TOM complex have been published, e.g., for integration of different members of the Bcl-2 protein family (Becker et al. 2008b, 2009b). Furthermore, a so-far unique case is represented by the polytopic outer membrane protein peripheral benzodiazepine receptor (PBR), which solely depends on Tom70 for outer membrane insertion (Otera et al. 2007). Recent studies additionally suggest the lipid composition of the outer membrane to play a role for the selectivity of tail-anchor integration (Kemper et al. 2008).
2. Several outer membrane proteins are inserted into the lipid phase via multiple membrane-spanning antiparallel β-strands forming cylindrically shaped structures (Schulz 2000; Wimley 2003). Major mitochondrial representatives of these highly rigid and thermodynamically stable β-barrel proteins are porin

(voltage-dependent anion channel in mammals, VDAC), the most abundant outer membrane protein, and the protein-conducting TOM component Tom40 (Paschen et al. 2005). Since β-barrels are exclusively found in the outer membranes of mitochondria and chloroplasts and in Gram-negative bacteria, they support the endosymbiotic origin of these eukaryotic organelles (Schleiff and Soll 2005; Dolezal et al. 2006). Only recently, a targeting signal for mitochondrial β-barrel insertion was identified by sequence comparison of different β-barrel proteins and mutational analysis of the yeast Tom40 protein (Kutik et al. 2008). Characteristic features of this signal sequence, which locates to the last predicted transmembrane β-strand, are a large polar residue, an invariant glycine, and two bulky hydrophobic residues. This signature motif is both necessary and sufficient for selective recognition of a β-barrel precursor at the level of the SAM complex and it is conserved across the entire eukaryotic kingdom.

3. Additional non-cleavable targeting information was recently identified in a subset of intermembrane space proteins. Typical representatives are of relatively small size (~8–22 kDa) and contain highly conserved cysteine residues often arranged in twin CX_3C or twin CX_9C motifs (Milenkovic et al. 2007; Sideris and Tokatlidis 2007; Milenkovic et al. 2009). Specific recognition and import of these cysteine-rich proteins involves the transient formation of disulfide bonds and additionally requires a hydrophobic residue as part of the signal sequence (Banci et al. 2009; Milenkovic et al. 2009).
4. Finally, the inner membrane contains a large number of polytopic proteins predominantly synthesized as mature-sized precursors. Among them are the numerous members of the mitochondrial carrier family (at least 35 in *S. cerevisiae* and more than 45 members in humans) that mediate the exchange of metabolites, nucleotides, and cofactors between the cytosol and mitochondria across the tightly sealed inner membrane (Palmieri et al. 1996). Despite the diversity of different MCF members, they share several common features; they are nuclear-encoded inner membrane proteins of approximately 300 amino acids divided into three similarly sized tandem repeats (modules) that presumably originated from gene duplication. Each of these modules folds into two transmembrane helices that are interconnected by hydrophilic loops. The most abundant mitochondrial MCF member is the ADP/ATP carrier (AAC or ANT for adenine nucleotide translocator) (Klingenberg 2008). However, despite considerable efforts, no clear consensus sequence for the selective targeting of carrier proteins to the inner mitochondrial membrane has been identified so far. It appears that targeting information governing outer membrane translocation and guidance through the intermembrane space is present in each of the three modules mentioned above. Cooperation between these domains and concomitant interaction of the carrier precursor with several receptor binding sites has been reported to be required for full translocation efficiency (Endres et al. 1999; Brix et al. 2000; Wiedemann et al. 2001). Insertion into the inner membrane is mainly dictated by targeting information present in the carboxy-terminal module as shown for the yeast dicarboxylate carrier (Brix et al. 1999; Brandner et al. 2005). Additional N-terminal cleavable presequences as found in the mitochondrial phosphate carrier, as well as in the citrate carrier, seem to be dispensable for targeting, but

are rather required for specific and efficient chaperone interaction mainly with those of the heat-shock protein 70 class (see below; Zara et al. 1992). Besides, Tim23 and Tim22, polytopic inner membrane proteins containing four transmembrane segments follow the carrier transport route as well. Studies on the import of Tim23 suggested the N-terminal domain of the protein as well as transmembrane helices one and four to be necessary for targeting and import (Davis et al. 2007). Moreover, positive charges in helix-connecting loops seem to be required for insertion into the inner membrane (Davis et al. 1998).

12.3 Crossing the Outer Membrane: The TOM Complex

The TOM complex represents the central entry gate for virtually all nuclear-encoded mitochondrial proteins studied so far. It is composed of seven different subunits which have been most intensively studied in the unicellular model organisms *Neurospora crassa* and *S. cerevisiae,* i.e., in fungal systems: The two rather loosely attached receptor proteins Tom20 and Tom70, the central receptor Tom22, the only essential component Tom40, which forms a cation-selective, voltage-gated protein-conducting channel through the outer membrane, and the so-called small subunits Tom5, Tom6, and Tom7 (numbers refer to the approximate molecular mass of the respective subunit in kDa) (Pfanner et al. 2004). By now, the mammalian counterparts for all of these components have been identified (for review see Hoogenraad et al. 2002; Kato and Mihara 2008).

Both Tom20 and Tom70 are integrated into the outer membrane via single N-terminal transmembrane domains and expose globular domains to the cytosol. Although the two receptors display partially overlapping substrate specificity and have also been shown to interact with each other, Tom20 is mainly required for the initial recognition of presequence-containing preproteins, whereas Tom70 serves as major receptor for the inner membrane carrier precursors (Brix et al. 2000; Igura et al. 2005; Saitoh et al. 2007). The nuclear magnetic resonance (NMR) structure of the Tom20 receptor domain from rat in complex with a presequence peptide showed that Tom20 contains a hydrophobic groove on its surface that accommodates the hydrophobic side of the amphiphilic helix formed by a presequence (Abé et al. 2000). Crystallographic analysis as well as cross-linking studies suggested the yeast Tom70 to dimerize (Wu and Sha 2006; Mills et al. 2009). However, there is evidence for a functional role of the monomeric structure, as well. This discrepancy led to the suggestion of an interconversion between the two forms as a fundamental aspect of Tom70 function. Additionally, the crystal structure of the cytosolic domain of Tom70 revealed a highly conserved putative binding pocket for substrate proteins containing a high proportion of hydrophobic residues in its interior with some charged residues at the opening (Wu and Sha 2006). So far, little is known about the selective recognition of other precursor types at the level of the TOM. The biogenesis of Porin has been shown to involve Tom20 (Schleiff et al. 1999; Krimmer et al. 2001), whereas both Tom20 and Tom70 have been implicated in mitochondrial targeting of Tom40 and Tom22 (Model et al. 2001). Tom20 and Tom70 pass on their cargo to the C-tail anchored receptor Tom22, which provides docking sites for the two surface receptors and assists in the insertion of precursor proteins into the pore-forming β-barrel protein Tom40 (Bolliger et al. 1995). Additionally, Tom22 plays a prominent role in the biogenesis, assembly, and stability of the TOM complex (van Wilpe et al. 1999; Wiedemann et al. 2003). Together with Tom40 and the tail-anchored small Tom proteins, it constitutes the so-called general import pore (GIP, also termed the TOM core complex; Meisinger et al. 2001; Model et al. 2001). Transfer of precursor proteins from Tom22 to the Tom40 channel involves the Tom5 subunit (Dietmeier et al. 1997). Besides, Tom5 promotes the assembly of the TOM complex while Tom6 and Tom7 play antagonistic roles in the assembly and stability of the TOM (Hönlinger et al. 1996; Dekker et al. 1998; Model et al. 2001). Hence, the TOM complex acts as a dynamically organized molecular machine allowing for efficient protein translocation across the outer membrane.

Electrophysiologic measurements and cryo-electron microscopy (EM) of the purified *S. cerevisiae* TOM complex revealed near-threefold symmetry with three elliptical pores of about 2.5 × 1.5 nm, allowing for the translocation of polypeptides in an α-helical conformation, and even for the simultaneous translocation of two α-helices (Ahting et al. 2001; Becker et al. 2005; Model et al. 2008). The GIP complex, by contrast, displays only two pores (Model et al. 2008). Each of these pores is likely formed by one Tom40 monomer. Alternatively, two or more monomers might contribute β-strands to one large barrel. Intriguingly, even though the TOM complex serves as the general import site for mitochondrial precursor proteins, it promotes different modes of membrane translocation: (1) Presequence-containing precursors traverse the channel as linear polypeptide chains with their N-termini ahead (Rassow et al. 1990). (2) Noncleavable carrier precursors of the inner membrane are most probably translocated in a loop formation such that their midsection reaches the intermembrane space first while both N- and C- termini are still on the cytosolic face of the outer membrane (Wiedemann et al. 2001). (3) A recently introduced model suggests the insertion of some outer membrane proteins to proceed at the protein-lipid interphase, i.e., independently of the interior of the TOM channel (Gabriel et al. 2003; Ahting et al. 2005). This mechanism seems plausible as it circumvents the thermodynamically unfavorable opening of a β-barrel required for lateral release into the surrounding membrane.

After passage through the outer membrane, the import routes of the different precursor proteins diverge in dependence of their targeting information. Presequence-containing precursor proteins are bound by the intermembrane space domain of Tom22 ($Tom22_{IMS}$) that serves as a *trans*-binding site (Bolliger et al. 1995; Moczko et al. 1997; Komiya et al. 1998).

12.4 Sorting and Assembly in the Outer Membrane: The SAM/TOB Complex

The sorting and assembly machinery (SAM; Kozjak et al. 2003), alternatively termed TOB complex (topogenesis of β-barrel proteins) (Paschen et al. 2003) mediates the integration of β-barrel-containing proteins into the outer mitochondrial

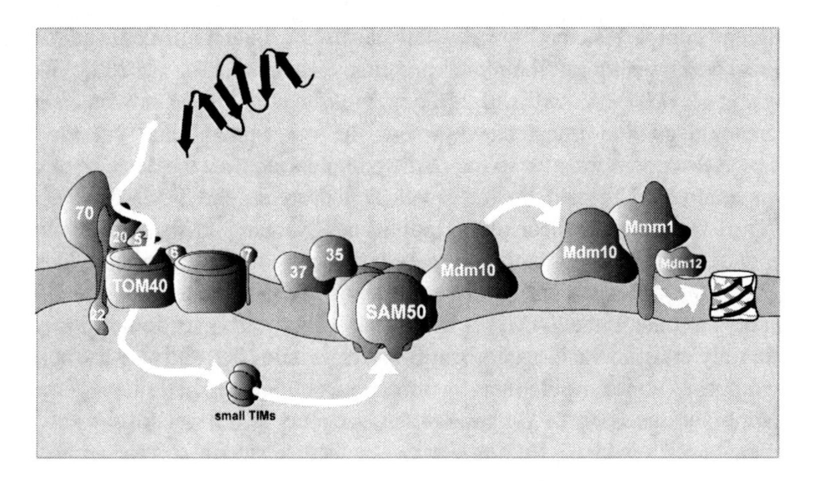

Fig. 12.2 Summary of the biogenesis of β-barrel proteins that are inserted into the mitochondrial outer membrane (SAM pathway). The hydrophobic precursor proteins are recognized by membrane receptors and transported across the outer membrane through the TOM40 pore to the SAM complex via a putative soluble intermediate in the intermembrane space. The SAM components Sam50, Sam37, and Sam35 together with the outer membrane proteins Mdm10, Mdm12, and Mmm1 assist in the insertion into the outer membrane and assembly to the native structure

membrane (Fig. 12.2). A wealth of information regarding the players and mechanisms of this pathway has accumulated ever since the first SAM/TOB component has been identified.

Three proteins have been reported to constitute the SAM core complex in yeast: the two essential components Sam50 (Tob55) and Sam35 (Tob/Tom38) as well as the non-essential Sam37, the first component to be identified (Mas/Tom37) (Kozjak et al. 2003; Paschen et al. 2003; Milenkovic et al. 2004; Waizenegger et al. 2004). The β-barrel protein Sam50 represents the core of the SAM machinery and it is conserved from bacteria to humans. Its bacterial orthologue Omp85 (outer membrane protein) is involved in the insertion of β-barrel proteins into the outer membrane of gram-negative bacteria (Dolezal et al. 2006). The C-terminal β-barrel domain of Sam50 forms a channel, while the α-helical N-terminus features a so-called POTRA (polypeptide transport-associated) domain (Kim et al. 2007). Both Sam35 and Sam37 are peripherally attached to the cytosolic face of the outer membrane. Sam35 has recently been proposed to be involved in recognition of the β-barrel targeting signal (see below; Kutik et al. 2008). The third subunit, Sam37, was suggested to play a role in substrate release from the SAM complex (Chan and Lithgow 2008). Since besides Tom40, Sam35, and Sam37 are the only essential β-barrel proteins in yeast mitochondria, unraveling the precise functions of the two SAM components poses an interesting challenge for the future. The two genes metaxin 1 and 2 most likely represent the mammalian counterparts of Sam37 and Sam35, respectively (Milenkovic et al. 2004; Kozjak-Pavlovic et al. 2007). Depletion of the human metaxin 2 gene resulted in selective deficiencies of Tom40 and VDAC import. Moreover, though human Sam50 and the metaxins seem to be in different complexes, their steady-state levels are interconnected, indicating a functional relationship of the three proteins (Humphries et al. 2005; Kozjak-Pavlovic et al. 2007). As outlined in the previous section, SAM substrates are initially recognized and translocated across the outer membrane via the TOM complex. Subsequent transfer to the SAM complex requires the activity of one of the two complexes formed by the small Tim proteins, the Tim9-Tim10, or the Tim8-Tim13 complex, respectively (Hoppins and Nargang 2004; Wiedemann et al. 2004). These soluble intermembrane space complexes display chaperone-like properties, i.e., they prevent premature folding and/or aggregation of the β-barrel precursors en route to the SAM. As will be subject of a later section of this review, they not only assist in the biogenesis of β-barrel proteins but also play a crucial role in the import of noncleavable inner membrane precursors. Exactly how maturation and membrane insertion of β-barrel proteins occurs is not yet fully established. However, based on latest findings, a two-step mechanism for recognition and subsequent insertion of β-barrel precursors into the outer membrane has been proposed (Kutik et al. 2008): Initial docking of the β-signal located in the last β-strand of the precursor requires the peripheral membrane protein Sam35, the predicted β-signal receptor. Intriguingly, β-signal binding induces a conformational change that not only results in the stimulation of the SAM channel, but in an increase in pore size as well. Thus, several β-strands can be inserted into the hydrophilic proteinaceous environment at the same time. Subsequent membrane integration depends on conserved glycine residues within the β-signal as well as further β-strands. As suggested by Kutik and colleagues, Sam35, a peripheral membrane protein located at the cytosolic face of the outer membrane, is located in a hydrophilic, proteinaceous environment by close association with multiple Sam50 molecules forming a cavity. Release of the β-barrel protein from the SAM might thus be accomplished by local rearrangements within the complex and does not require lateral opening of the Sam50 β-barrel domain.

In addition to the constituents of the SAM core complex, the SAM holo complex contains a fourth subunit, Mdm10 (mitochondrial distribution and morphology). Mdm10 does not seem to participate in the general β-barrel biogenesis pathway; instead, it is specifically required for the assembly of Tom40 (Meisinger et al. 2004). Two other proteins, Mdm12 and Mmm1 (maintenance of mitochondrial morphology), have been implicated in the general β-barrel import pathway (Meisinger et al. 2007). Additionally, Mdm10, together with Mdm12, Mmm1, and Mdm34, is involved in the maintenance of mitochondrial morphology and DNA structure (Boldogh et al. 2003; Youngman et al. 2004). Mdm10 is thus present in two distinct outer membrane complexes. Intriguingly, a very recent study discovered the MDM proteins to be involved in ER-mitochondria contacts (Kornmann et al. 2009). Unlike formerly published, Kornmann et al. showed that Mmm1 is not a mitochondrial outer membrane protein, but instead localizes to the ER membrane. Mmm1 and Mdm10 directly interact to form the core of the so-called ERMES (endoplasmic reticulum-mitochondria encounter structure) complex that functions as a molecular "zipper" connecting ER and mitochondria, presumably for interorganelle calcium and phospholipid exchange (Kornmann et al. 2009).

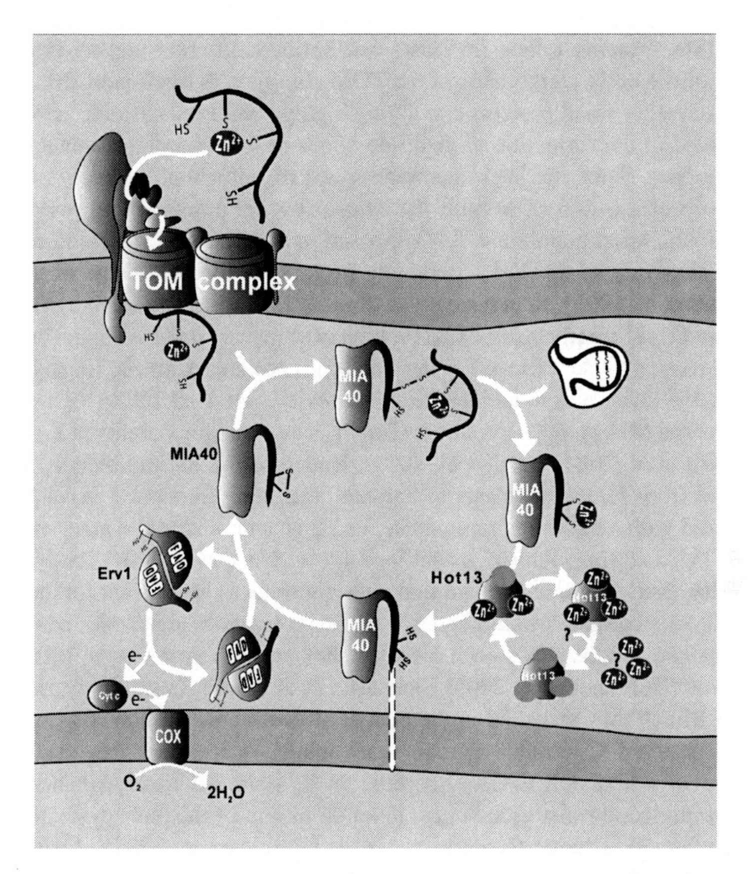

Fig. 12.3 Summary of the transport and folding reactions of disulfide-containing preproteins of the intermembrane space (MIA pathway). After traversing the outer membrane via the TOM complex, the preproteins undergo a covalent linkage to Mia40 via disulfide bridging. During and after release from Mia40, internal disulfide bonds are formed and the final folding state of the preprotein is acquired. Re-oxidation of Mia40 SH-groups proceeds via Hot13 and Erv1

Surprisingly, the SAM complex does not exclusively promote the insertion of β-barrel proteins into the outer membrane. A recent study revealed its participation in the biogenesis of the tail-anchored α-helical TOM components Tom22 as well as Tom5, 6, and 7 (Stojanovski et al. 2007). Thus, strictly speaking, the alias TOB is no longer appropriate.

12.5 Redox-Regulated Import into the Intermembrane Space: The MIA Pathway

The MIA system represents the most recently identified mitochondrial protein import machinery. It drives redox-regulated import and maturation of a subset of soluble intermembrane space proteins (Fig. 12.3). As outlined in a previous section, soluble IMS proteins follow (at least) two entirely different sorting routes after crossing of the outer membrane via the TOM complex. A fraction of IMS proteins consists of rather small proteins containing highly conserved cysteine residues that are involved in the formation of disulfide bonds and/or binding of metal ions like zinc or copper. Since the intermembrane space of mitochondria freely exchanges small molecules (~2–6 kDa) with the cytosol by the pore-forming porins of the outer mitochondrial membrane, it is supposed to be a reducing environment. Thus, the first discovery of disulfide-containing proteins in the IMS came as a surprise (Tsukihara et al. 1995). Representatives of these IMS residents, such as the copper chaperone Cox17 and the related Cox19, both of which are involved in the biogenesis of cytochrome *c* oxidase as well as the small Tim proteins Tim8, 9, 10, and 13 were the first MIA substrates to be identified (for review, see Hell 2008).

The central MIA component, Mia40/Tim40, is essential for viability in *S. cerevisiae* (Chacinska et al. 2004; Naoé et al. 2004). Underscoring its importance, Mia40 is conserved from fungi and plants to animals, including humans. Fungal Mia40 is synthesized with a bipartite presequence, i.e., it is anchored to the inner membrane via the TIM23 sorting pathway, as outlined above (Naoé et al. 2004; Terziyska et al. 2005). However, membrane integration does not seem to be relevant for the activity of Mia40. In accordance, homologues in higher eukaryotes are synthesized without a presequence and, as was shown recently, they are imported via the MIA system themselves (Hofmann et al. 2005; Chacinska et al. 2008). Intriguingly, the import mode of Mia40 thus switched in the course of evolution. All homologues share a highly conserved C-terminal domain containing six invariant cysteine residues organized in a CPC-CX_9C-CX_9C pattern. In its oxidized form, Mia40 contains three intramolecular disulfide bonds, deletion of any of the three cysteine pairs in the yeast Mia40 is lethal (Naoé et al. 2004; Grumbt et al. 2007). Apart from a single exception (see below), all MIA substrates identified so far contain either a highly conserved twin CX_3C or twin CX_9C motif (Hell 2008). Most of our knowledge on the MIA pathway was deduced from studies employing the small Tim proteins as model substrates. Crystallographic analysis of the heterohexameric Tim9-Tim10 complex revealed two disulfide bonds within each subunit that are important for the structural integrity of the complex (Webb et al. 2006). As soon as they emerge on the IMS side of the outer membrane, newly imported MIA substrates interact with Mia40 via transient disulfide bonding (Chacinska et al. 2004; Milenkovic et al. 2007; Sideris and Tokatlidis 2007). Reorganization of these bonds finally results in the formation of intramolecular disulfides within the Mia40 client proteins, which are thereby released. Based on the dissection of the import and assembly of small Tim proteins, a "folding-trap" mechanism has been proposed as import driving force (Lutz et al. 2003; Müller et al. 2008): Mia40 substrates cross the outer membrane in an unfolded reduced state. This state seems to be stabilized by zinc ions in the cytosol (Lu and Woodburn 2005). Mia40-catalyzed oxidative folding prevents backward diffusion, and thus traps the proteins in the IMS. Oxidation of precursor proteins results in the reduction of cysteine residues in Mia40, thereby converting the enzyme into its inactive reduced form. Reoxidation of Mia40 requires the activity of the FAD-dependent sulfhydryl oxidase Erv1 (essential for respiration and vegetative growth) (Mesecke et al. 2005; Rissler et al. 2005). Erv1 homologues have been found in mitochondria of all eukaryotic kingdoms (Sevier et al. 2005; Coppock and Thorpe 2006). Of note, the mammalian Erv1 homologue ALR (augmenter of liver regeneration) promotes the regeneration of the liver; however, mechanistic details of this effect await further research (Pawlowski and Jura 2006). Interestingly, although it contains a twin CX_2C motif instead of the twin CX_3C or CX_9C motifs found in all formerly identified MIA substrates, Erv1 itself depends on the MIA system for import (Gabriel et al. 2007). Thus, the substrate spectrum of the Mia40-Erv1 disulfide relay system seems to be broader than hitherto anticipated. Electron transfer from Mia40 to the Erv1 prosthetic group FAD results in the regeneration of the oxidized functional form of Mia40. At least in vitro, Erv1 is capable of transferring electrons directly to molecular oxygen, thereby generating hydrogen peroxide (Dabir et al. 2007). However, in vivo, it has been shown to shuttle electrons to cytochrome *c*, which in turn feeds them into the cytochrome *c* oxidase (complex IV) to generate water from molecular oxygen (Allen et al. 2005; Bihlmaier et al. 2007; Dabir et al. 2007). This connection of the Mia40-Erv1 disulfide relay system to the respiratory chain of mitochondria circumvents the production of potentially deleterious reactive oxygen species. Secondly, reoxidation of Mia40 via cytochrome *c* as electron acceptor allows for a more efficient regeneration of the enzyme even at low oxygen levels. The viability of yeast cells even under anaerobic conditions suggests the existence of an alternative final electron acceptor besides oxygen. An additional MIA component, the cysteine-rich Hot13 (helper of Tim) protein has been suggested to bind zinc ions, thereby keeping Mia40 in a zinc-free state and thus, promoting Erv1-mediated Mia40 oxidation into the active form (Curran et al. 2004; Mesecke et al. 2008).

Interestingly, the mitochondrial disulfide relay system shares basic principles with the protein oxidation systems found in the bacterial periplasm (DsbA/DsbB; for review see Nakamoto and Bardwell 2004) and in the endoplasmic reticulum of eukaryotes (Ero1/PDI; for review, see Tu and Weissman 2004), i.e., the introduction of disulfides in newly synthesized proteins and the accompanying feeding of electrons in an electron-accepting system is a widely conserved principle. Unraveling the precise mechanisms and evolutionary origins of the different systems represents thus an interesting field of research for the future.

12.6 Insertion into the Inner Membrane Via the TIM22 Complex: The Carrier Pathway

The TIM22 machinery is exclusively dedicated to the insertion of multi-spanning hydrophobic inner membrane proteins (Fig. 12.4), including those of the large family of mitochondrial solute carriers as well as components of inner membrane protein translocases (Koehler 2004; Rehling et al. 2004). Similar to the TOM and SAM machineries, the *S. cerevisiae* TIM22 complex is made up of a core module and auxiliary subunits. The TIM22 core complex consists of three integral membrane

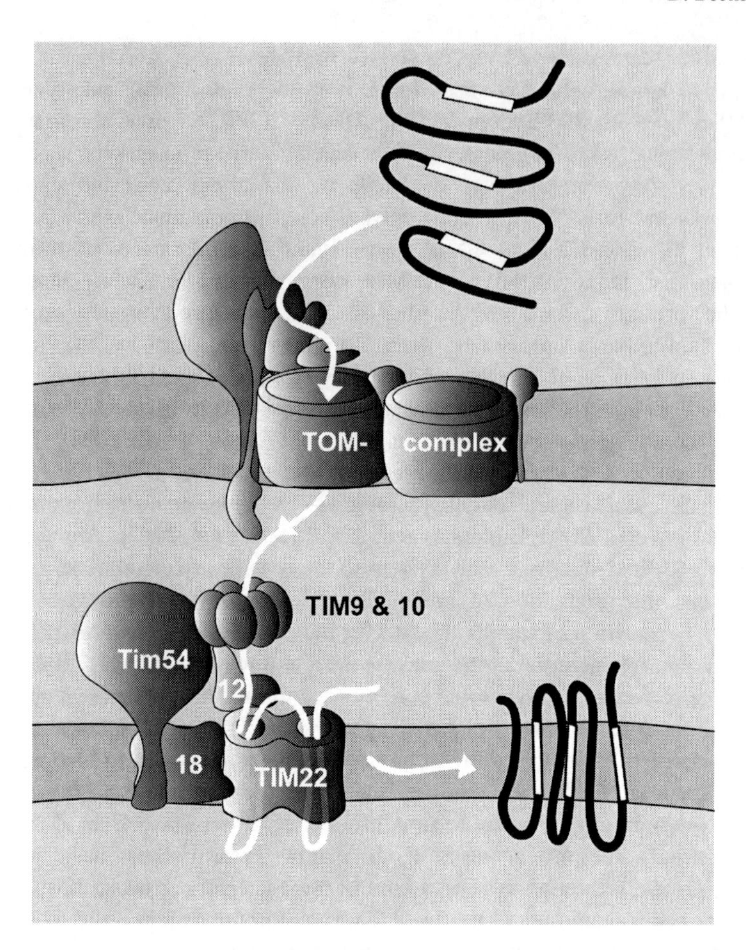

Fig. 12.4 Import and inner membrane insertion of polytopic membrane proteins of the metabolite carrier family containing internal signal sequences (TIM22 pathway). After transport through the outer membrane via the TOM complex, the preproteins are stabilized by chaperone-like interactions with members of the small Tim family (Tim9, Tim10, and Tim12) in the intermembrane space. The insertion into the inner membrane is accomplished by the TIM22 complex and driven by the inner membrane potential

proteins, Tim22, Tim54, and Tim18 (Rehling et al. 2004). Additional components are the peripherally attached small subunits Tim9, Tim10, and Tim12 (Rehling et al. 2004). Tim22 represents the only essential TIM22 component in yeast. As mentioned in a previous section, import of Tim22 is mediated by the TIM22 complex but, different from most other proteins following the carrier transport route studied so far, Tim22 is directed to the Tom20 receptor of the outer membrane (Kurz et al. 1999). According to electrophysiologic measurements, Tim22 forms a cation-selective voltage-activated channel that both recognizes its substrates and inserts them into the inner membrane utilizing the membrane potential $\Delta\psi$ as the only external energy source. Depending on the open state, the inner diameter of the Tim22 pore was estimated to range from approximately 1.1 to 1.8 nm. Thus, in the fully open state, it is sufficiently large for the simultaneous insertion of two tightly packed α-helices, allowing for the proposed mechanism of protein insertion in a loop (or "hairpin") conformation (Kovermann et al. 2002). The existence of two pores per TIM22 complex, as revealed by single-particle cryoelectron microscopy, led to the alternative term twin-pore translocase (Rehling et al. 2003). While initially reported to be essential, the Tim54 subunit has been shown to be dispensable for the survival of yeast cells (Kovermann et al. 2002). Its molecular function in the TIM22 complex remains poorly understood. Interestingly, a link between Tim54 and the assembly of the yeast iAAA protease Yme1 of the inner membrane into a functional complex has been reported (Hwang et al. 2007). Furthermore, a recent study suggests Tim54 to connect the TIM22 core module with the auxiliary unit by interacting with Tim10 (Wagner et al. 2008). The import of Tim54 likewise represents an exception to the rule: It preferentially interacts with Tom70 for targeting to the outer membrane, yet it follows the TIM23 pathway (see below) (Kerscher et al. 1997; Kurz et al. 1999); additionally, the assembly of Tim54 into the translocase complex requires IMS components usually involved in the carrier pathway (Wagner et al. 2008). Tim18 seems neither required for TIM22 function nor for the viability of yeast cells. Its function is still elusive; however, a role for Tim18 in the assembly of the translocase complex has been suggested (Kerscher et al. 2000; Koehler et al. 2000; Wagner et al. 2008). As described for the biogenesis of β-barrel containing outer membrane proteins, the small Tim proteins Tim9 and Tim10 form a soluble complex that prevents aggregation and/or misfolding of hydrophobic precursor proteins in the aqueous environment of the intermembrane space (Koehler et al. 1998b Vasiljev et al. 2004; Davis et al. 2007). In yeast, both Tim9 and Tim10 are essential for viability and the complex they form represents the major intermembrane space chaperone. Together with the homologous Tim12, they assemble into another complex that associates with the TIM22 core module and faces the intermembrane space (Koehler et al. 1998a; Sirrenberg et al. 1998; Webb et al. 2006). The importance of intermembrane space chaperoning by the small Tim complexes is emphasized by the so-called deafness dystonia syndrome (also termed Mohr-Tranebjaerg syndrome), a neurodegenerative disorder caused by mutations in the human Tim8 homologue (deafness-dystonia peptide 1, DDP1). Besides, it represents the first clinical syndrome primarily associated with mitochondrial protein import (Koehler et al. 1999a; Roesch et al. 2002). The Tim8-Tim13 complex has been shown to escort some noncarrier inner membrane proteins to the TIM22 complex; however, it appears not to be involved in the carrier biogenesis as such (Davis et al. 2000; Paschen et al. 2000). In higher eukaryotes, both the structural organization and mechanistic features of the TIM22 machinery differ from those of the fungal complex in several aspects. While a human Tim22 homologue with approximately 40% sequence similarity as well as one humanTim9 homologue and two human Tim10 homologues (Tim10a and Tim10b) have been identified, the yeast Tim54 and Tim18 appear not to be present in higher eukaryotes (Bauer et al. 1999b). Furthermore, a clear-cut Tim12 seems to be absent in mammals and other eukaryotes except *S. cerevisiae*. Instead, the Tim10b variant might represent the functional counterpart of yeast Tim12 in humans (Bauer et al. 1999b; Mühlenbein et al. 2004). As in yeast, the small Tims assemble into two different complexes, the Tim9-Tim10a and the Tim9-Tim10a-Tim10b complexes. However, both of these complexes are quantitatively associated with the inner mitochondrial membrane, as shown in human mitochondria (Mühlenbein et al. 2004). Thus, a different mechanism must have evolved to ensure import competence of hydrophobic carrier proteins en route to the TIM22 complex.

As indicated in a previous section of this review, intense research on the biogenesis of several MCF members – in particular, the generation of stable translocation intermediates by artificial arrest of the precursors – allowed for the experimental dissection of the import and assembly reaction into five sequential stages (for review, see Rehling et al. 2004). Due to their hydrophobic nature, newly synthesized MCF family members require the activity of cytosolic factors to guide them to the mitochondrial surface receptors. The role of these cytosolic components in protein targeting to mitochondria is only partially understood. However, considerable knowledge has accumulated on the involvement of chaperones from the Hsp70 and Hsp90 classes in the import of carrier precursors (Young et al. 2003; Zara et al. 2009). As shown in yeast, carrier precursors interact with cytosolic Hsp70 and docking of Hsp70 onto the Tom70 receptor is required for productive preprotein recognition; furthermore, the cooperative recruitment of three Tom70 dimers upon binding to the AAC precursor was reported (Wiedemann et al. 2001). In mammals, both Hsp90 and Hsp70 associate with preproteins in a multi-chaperone complex for delivery to the mitochondrial surface receptor. The N-terminal tetratricopeptide repeat (TPR) motifs ("clamp" domain) within Tom70 are thought to mediate chaperone interactions (Wu and Sha 2006; Schlegel et al. 2007; Mills et al. 2009). Subsequent transfer of the precursors to Tom70 requires cytosolic ATP. Whether ATPase cycling solely serves to dissociate the chaperone-precursor complex or additionally advances preprotein insertion into the TOM channel – as indicated by studies employing Hsp90-specific inhibitors (Young et al. 2003; Fan et al. 2006) – is currently under debate. In any case, the precursor is passed on to the Tom40 pore into which it is inserted, supposedly in a loop conformation (Wiedemann et al. 2001; Curran et al. 2002). Interaction with the essential Tim9-Tim10 complex is not only required for guidance of the hydrophobic precursor proteins through the intermembrane space, but also for the completion of outer membrane translocation and dissociation from the TOM complex (Truscott et al. 2002; Vasiljev et al. 2004). Subsequent insertion of the carrier precursor into the Tim22 channel strictly depends on the membrane potential $\Delta\psi$ that plays a dual role: It exerts an electrophoretic force on positive charges found in hydrophilic loops of the precursor, and it contributes to the activation of the voltage-dependent Tim22 channel (Kovermann et al. 2002; Pebay-Peyroula et al. 2003). Maximal gating, however, requires both the membrane potential and the presence of a specific targeting signal, suggesting a cooperative effect on the channel (Kovermann et al. 2002). Intriguingly, the two pores present in the TIM22 complex are functionally coupled, as demonstrated in electrophysiologic measurements. If a membrane potential in the physiologic range is applied, the presence of a model peptide affects the two pores differently; one closes, while the other one is simultaneously activated (Rehling et al. 2003). Based on the analysis of the targeting signal-mediated gating properties displayed by the twin-pore, the three carrier modules were suggested to insert in a sequential manner (Rehling et al. 2004).

The third module is assumed to be of particular importance for this process since it contains an internal targeting signal (Brix et al. 2000; Kovermann et al. 2002). Indeed, insertion of the other two modules was shown to depend on the presence of the third module (Endres et al. 1999; Brandner et al. 2005). Subsequent lateral release of the carrier protein into the lipid phase occurs by an as yet unknown mechanism. Based on the structural analysis of AAC from different species, it has been widely accepted for a long time that carrier proteins form functional homodimers although different models were also proposed (Nury et al. 2008 and references therein). Given that the determination of assembly states is difficult in the case of membrane proteins, the functional organization of carrier proteins remains a controversial issue.

Hence, the highly complex course of events governing TIM22-mediated membrane integration of a carrier precursor and its assembly into the functional state await further biochemical and biophysical analysis. Another open question relates to the barrier function of the inner membrane and how it is maintained in the absence of a precursor: The electrochemical proton gradient generated by the respiratory chain activity is crucial for the cell's energy metabolism as it provides the energy for the bulk of cellular ATP synthesis. Yet, how an uncontrolled flux of ions through the inner membrane is prevented at the level of the TIM22 machinery is currently unknown.

12.7 Import into the Matrix: The TIM23-PAM Pathway

Complete translocation of precursor polypeptide chains to the matrix compartment represents a complicated process that comprises several translocase machineries with specific enzymatic activities (Fig. 12.5). As in the case of most preproteins, recognition of targeting signals and transport through the outer membrane is mediated by the TOM complex. Translocation across the inner membrane involves the tightly regulated cooperation of the TOM complex with another translocase in the inner membrane, the TIM23 complex, which provides the translocation pore through the inner membrane.

12.7.1 The TIM23 Complex

Generally, the TIM23 complex mediates the import of all presequence-containing precursor proteins into the mitochondrial matrix. In addition, some TIM23 substrates contain bipartite targeting signals that direct them either to the intermembrane space or the inner membrane. Thus, the TIM23 machinery catalyzes the transport of a highly diverse set of precursor proteins into three different mitochondrial compartments. The first TIM23 components were identified almost two decades ago; however, recent research has significantly broadened our knowledge on the formerly unexpected complex reaction cycle that governs the import of presequence-containing precursor proteins. To meet the different demands of its various substrates, the presequence translocase (transiently) interacts with three different mitochondrial systems: the TOM complex of the outer membrane, the import motor located at the matrix side of the inner mitochondrial membrane, and complexes III

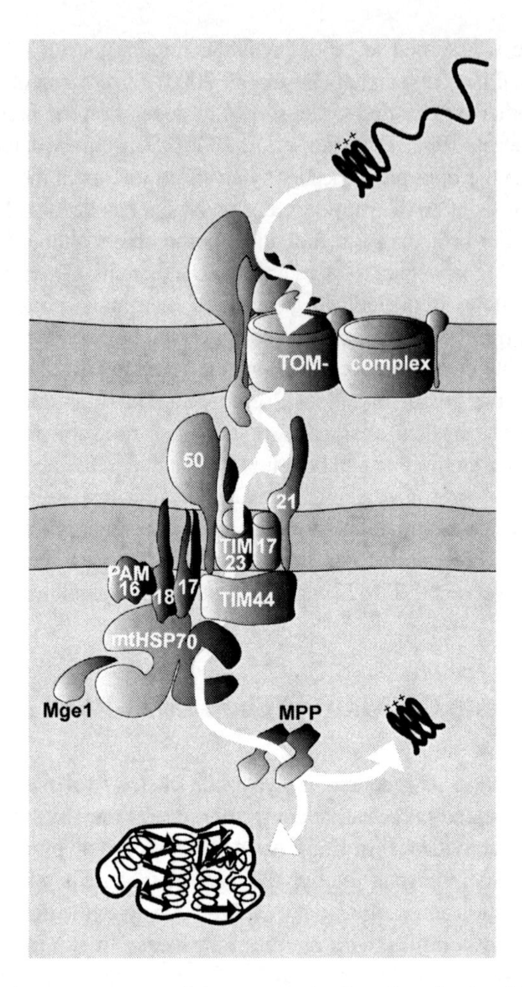

Fig. 12.5 Import of presequence-containing preproteins into the mitochondrial matrix (TIM23-PAM pathway). The positively charged N-terminal signals of the preproteins are recognized by TOM receptors and transferred to the TIM23 inner membrane complex via Tim50 and Tim21. Insertion of the N-terminus into the inner membrane is driven by the membrane potential, while complete translocation requires the hydrolysis of ATP by the import motor complex PAM, consisting of the chaperone mtHsp70 as the core component that is anchored to the inner membrane via Tim44. The activity of mtHsp70 is tightly regulated by the accessory proteins Pam 18, Pam16, and Mge1. The presequences are removed by the matrix processing peptidase MPP

and IV of the respiratory chain. Furthermore, while the membrane potential $\Delta\psi$ is sufficient to drive the initial steps of inner membrane translocation, completion of matrix import depends on ATP as second energy source. Hence, the import of presequence-containing precursor proteins clearly represents the most complicated mitochondrial protein translocation pathway.

The core of the TIM23 complex is made up of three integral inner membrane proteins that are all essential for the viability of yeast cells: Tim23, Tim17, and Tim50 (Wiedemann et al. 2004). More recent studies resulted in the identification of an additional, membrane-integral subunit, the nonessential Tim21 (Chacinska et al. 2005; Mokranjac et al. 2005). Tim23 and Tim17 as well as the channel-forming subunit of the carrier translocase, Tim22, exhibit sequence similarity and share the same topology, suggesting that the two TIM complexes have evolved from a common ancestor (Dolezal et al. 2006). Tim23 forms a cation-selective voltage-gated channel through the inner membrane (Truscott et al. 2001). With an estimated pore diameter of approximately 1.3 nm, it is large enough to accommodate one polypeptide chain in an α-helical conformation. The soluble intermembrane space domain of Tim23 mediates the selectivity of the channel, i.e., it fulfills receptor function (Bauer et al. 1996; Alder et al. 2008). As shown by electrophysiologic measurements, both functions, presequence recognition and channel activity, are in fact intimately linked. In presence of a membrane potential of sufficient magnitude, peptide binding to the soluble N-terminal domain induces the opening of the pore formed by the C-terminal domain (Truscott et al. 2001). Hence, the membrane potential $\Delta\psi$ drives the initial steps of precursor translocation across the inner membrane by exerting an electrophoretic force on the positive charges present in N-terminal presequences on the one hand, and by activating the Tim23 channel on the other. As to which part of the Tim23 protein forms the actual voltage sensor is currently not defined. Tim23 and Tim17 interact via their hydrophobic membrane-integral domains and this association is critical for the functionality of the translocase (Ryan et al. 1998; Alder et al. 2008). However, in contrast to Tim23, the function of Tim17 remains rather elusive. Its soluble N-terminal portion contains several highly conserved negative charges that have been implicated in initial steps of the translocation reaction by affecting the gating properties of the Tim23 channel (Meier et al. 2005). Similar results were obtained in recent electrophysiologic studies (Martinez-Caballero et al. 2007). Furthermore, Tim17 was shown to play an active role in inner membrane sorting on the one hand, and in the recruitment of the PAM complex (presequence translocase-associated motor) on the other (see below; Chacinska et al. 2005).

Tim50 spans the inner membrane via a single transmembrane domain. Its C-terminus protrudes into the intermembrane space and contacts the soluble part of Tim23 (Geissler et al. 2002; Mokranjac et al. 2003). In the absence of a precursor protein, Tim50 induces the closure of the Tim23 channel, thus maintaining the permeability barrier of the inner membrane (Meinecke et al. 2006). This effect is selectively overridden by presequences, i.e., Tim50 and presequences act in an antagonistic manner with regard to the channel gating properties. Additionally, Tim50 exhibits receptor function by mediating the initial (membrane potential-insensitive) contact of a presequence-containing precursor polypeptide to the TIM23 complex. As described in a previous section of this review, presequence-containing precursor polypeptides are bound by the intermembrane space domain of Tom22 ($Tom22_{IMS}$) as they emerge from the Tom40 channel. The concomitant interaction of $Tim50_{IMS}$ with the presequence results in an intermediate state of the translocation reaction, in which the translocases of both outer and inner membranes are indirectly linked via the precursor protein (Chacinska et al. 2003; Alder et al. 2008; Mokranjac et al. 2009). Even before the identification of TOM and TIM translocases, matrix-destined precursor proteins *in transit* were demonstrated to span both inner and outer mitochondrial membranes (Rassow et al. 1989). A super-complex containing core components of both TOM and TIM23 was isolated by artificial arrest of the precursor polypeptide inside the translocation channels. This import intermediate represents a stage in which the N-terminus of the preprotein protrudes into the matrix while the C-terminus still resides on the cytosolic face of the outer membrane (Dekker et al. 1997; Chacinska et al. 2003). However, a direct physical contact between components of the two translocases could not be detected in these studies. Only recently, such a direct interaction involving Tim21 could be demonstrated (Chacinska et al. 2005; Mokranjac et al. 2005). The soluble inter-membrane space domain of Tim21 ($Tim21_{IMS}$) interacts with $Tom22_{IMS}$, which results in the release of the presequence from Tom22 followed by transfer of the precursor to the Tim23 channel. Thus, Tim50 and Tim21 seem to have antagonistic functions by competing for the presequence bound to $Tom22_{IMS}$.

Homologues of the TIM23 constituents have also been found in higher eukaryotes, including humans: A single human gene encodes for Tim23 and two genes encoding Tim17 have been identified (Bömer et al. 1996; Bauer et al. 1999a). Both Tim17 isoforms are ubiquitously expressed and they each form separate complexes with Tim23; however, the relevance for this is not clear. Representing the first mouse model of defective mitochondrial protein import, homozygous *tim23* knock-out mice have been reported to be inviable and heterozygous mutants displayed a neurological phenotype and a markedly reduced lifespan (Ahting et al. 2009). Additionally, a human Tim50 homologue displaying phosphatase activity was identified (Guo et al. 2004). Downregulation of human Tim50 has been demonstrated to sensitize human cells to apoptotic stimuli by increasing cytochrome *c* release. This is in consistence with the essential role of yeast Tim50 in maintaining the permeability barrier of the inner membrane. A possible Tim21 homologue was first identified by large-scale cDNA analysis in CD34+ hematopoietic stem/progenitor cells and later in the context of the "Mammalian Gene Collection" project (Zhang et al. 2000; Gerhard et al. 2004). On the basis of sequence analysis, the "Tim21-like" protein is assumed to be a cleavable inner membrane protein, however, neither on its structural nor its functional characterization has been reported so far.

12.7.2 The Import Motor (PAM)

In the case of presequence-containing inner membrane or intermembrane space precursor proteins, both the import modes and energetic requirements are controversially discussed (see below). However, the completion of import into the matrix is clearly dependent on the cooperation of the presequence translocase with the ATP-powered PAM complex (Shariff et al. 2004; Krayl et al. 2007).

The central component of the import motor complex is the essential mtHsp70 (mitochondrial Hsp70, Ssc1 in yeast) that binds to the incoming precursor polypeptide and promotes its forward movement in an ATP-dependent fashion (Voos and Pfanner 2001; Voos and Röttgers 2002). Generally, the activity of Hsp70-type chaperones is based on mutual allosteric control between the N-terminal nucleotide-binding domain (NBD) and the C-terminal peptide-binding domain (PBD). In the ATP-bound state of the enzyme, the PBD is in an "open" conformation, allowing for the rapid binding and release of substrate peptides. ATP hydrolysis catalyzed by the NBD induces a "closed" conformation of the PBD, i.e., the substrate peptide is trapped in the PBD with high affinity (for review see Erbse et al. 2004). Vice versa, peptide binding to the PBD enhances ATPase activity in the NBD (Mayer et al. 2000). Additionally, the activity of Hsp70 chaperones is influenced by specific partner proteins, termed cochaperones. The translocation-specific activity of mtHsp70 is tightly regulated by five cochaperones: The essential Pam18 (Tim14) subunit is an integral membrane protein belonging to the so-called J-domain protein family (Mokranjac et al. 2003; Truscott et al. 2003). J-proteins interact with Hsp70 chaperones via their conserved J-domain, thereby stimulating the low intrinsic ATPase activity (Mayer and Bukau 2005). The likewise essential nucleotide exchange factor Mge1 (mitochondrial GrpE homolog), a soluble matrix protein, enhances ATPase activity by catalyzing the release of ADP, thus allowing for a new round of ATP binding and hydrolysis (Voos et al. 1994; Miao et al. 1997; Weiss et al. 2002). Apart from these common Hsp70 cochaperones, the import motor contains (at least) three additional proteins: Tim44, an essential peripheral membrane protein, provides a dynamic (ATP-sensitive) mtHsp70 binding site, thereby recruiting the otherwise soluble mtHsp70 to the exit site of the translocation channel (Voos et al. 1996; D'Silva et al. 2004). Furthermore, Tim44 binds to precursor polypeptides *in transit* as soon as they emerge from the Tim23 channel and might thus serve as a membrane anchor for the generation of a driving force according to the "power stroke" model (see below; Voisine et al. 1999; Matouschek et al. 2000; Krayl et al. 2007). The J-like protein Pam16 (Tim18) shares sequence homology with Pam18, however, it does not contain the highly conserved J-domain signature motif (Frazier et al. 2004; D'Silva et al. 2005). Accordingly, it does not stimulate mtHsp70 activity; instead, it plays a regulative role by antagonizing Pam18 as confirmed by the recently available crystal structure of the heterodimer formed by Pam18 and Pam16 (Li et al. 2004; Mokranjac et al. 2006). Furthermore, the recently identified membrane-integral Pam17 indirectly regulates mtHsp70 activity by promoting the association of the Pam16-Pam18 dimer to the PAM (van der Laan et al. 2005). Thus, the PAM complex represents one of the most highly sophisticated Hsp70 systems known to date. For many years now, the molecular mechanism underlying the translocation-specific activity of mtHsp70 has been under debate. Currently, there are two major mechanistic models to describe the role of the mtHsp70 motor in unfolding and importing precursor proteins into the matrix: The "Brownian ratchet" ("trapping" or "holding") and the "Power stroke" ("pulling") model (Voos and Röttgers 2002; Sousa and Lafer 2006; Tomkiewicz et al. 2007). In the former, mtHsp70 solely biases the random Brownian motion of an unfolded precursor protein: As soon as the polypeptide chain arrives on the matrix side of the inner membrane, it is bound by mtHsp70, thereby preventing its retrograde movement in the translocase channel. Substrate binding enhances the ATPase activity of mtHsp70, which in turn induces the high-affinity state of the enzyme, i.e., the preprotein becomes trapped in the peptide-binding domain. Repeated cycles of precursor binding and release finally result in the completion of matrix import. Unfolding of the precursor protein and its movement across the membranes is exclusively based on spontaneous reversible movement (hysteresis) and Tim44 solely serves to increase the local mtHsp70 concentration at the exit site of the Tim23 channel.

In contrast, the power stroke model postulates a force to be exerted on the precursor protein *in transit*: Conformational changes in mtHsp70 induced by ATP binding and hydrolysis translate into a power stroke that is imposed on the polypeptide chain. The preprotein is thus actively pulled into the matrix. Concomitant binding of mtHsp70 to a membrane anchor is a crucial prerequisite for this model. As mtHsp70 was shown to simultaneously bind to Tim44 and to a peptide substrate, Tim44 was suggested to fulfill this anchor function (Horst et al. 1996).

A considerable number of studies have demonstrated that the trapping model is applicable on the import of unfolded or loosely folded precursor proteins. However, the analysis of conditional mtHsp70 mutants demonstrated that trapping and unfolding of a precursor protein by mtHsp70 represent two distinct functions (Voisine et al. 1999). Furthermore, the ability of mtHsp70 to actively pull precursor proteins into the matrix compartment was shown to be tightly coupled to the viability of yeast cells (Voos et al. 1996; Voisine et al. 1999; Becker et al. 2009a). Meanwhile, it seems likely that both mechanisms contribute to the import process; as to which extent one or the other mode predominates, seems to depend on the intrinsic properties of a given preprotein and the stage of the translocation reaction (Wilcox et al. 2005; Krayl et al. 2007). Initially, spontaneous fluctuations might enable an N-terminal portion of the precursor to reach the matrix compartment. (Additionally, the negatively charged mitochondrial matrix exerts an electrophoretic effect on positive charges present in the presequence.) Binding of mtHsp70 to the incoming precursor would then prevent its backward diffusion by a trapping mechanism. In the case of a folded precursor or folded domains in the C-terminal portion of the precursor, continuation and completion of import would require the pulling activity driven by ATP-induced conformational changes in mtHsp70.

The interplay between the import motor and the presequence translocase is mediated by a complex network of protein-protein contacts: Pam18 interacts with Tim17 in a reaction depending on Pam16 (Chacinska et al. 2005; D'Silva et al. 2008), Pam17 interacts with Tim23 (Popov-Celeketic et al. 2008), and Tim44 plays a role in the association of the PAM complex with the presequence translocase (Hutu et al. 2008). Interestingly, recent analysis has additionally revealed a connection between the presequence pathway and the respiratory chain: Tim21 as well as the Pam16-Pam18 module were found to transiently interact with supercomplexes of the respiratory chain complexes III (cytochrome bc_1) and IV (cytochrome *c* oxidase) in the course of the translocation reaction (van der Laan et al. 2006; Wiedemann et al. 2007). The physiologic relevance of these interactions has not been analyzed so far. However, it seems reasonable to suppose that the particularly high electrochemical gradient in the direct vicinity of proton-pumping machineries supports the membrane potential-dependent process of precursor insertion into the membrane.

The first mammalian PAM component to be identified was Tim44. Interestingly, it was found to be highly overexpressed in diabetic newborn mice that additionally displayed altered cristae (Wada and Kanwar 1998). Upregulation of Tim44 as a response to elevated glucose levels might serve to enhance the import of nuclear-encoded proteins involved in oxidative phosphorylation. Additionally, Tim44 from rat was shown to interact with matrix Hsp70 in an ATP-dependent manner; however, in contrast to fungal Tim44, the mammalian homologue is mainly found in the soluble matrix fraction of mitochondria (Ishihara and Mihara 1998; Bauer et al. 1999a). Furthermore, a human homologue to the bacterial GrpE, HMGE (human mitochondrial GrpE homologue) has been identified and characterized (Choglay et al. 2001). Finally, the human Pam18 and Pam16 homologues (DNAJC19 and Magmas, respectively) have been identified. As their yeast counterparts, DNAJC19 (DNAJ domain containing) and Magmas (mitochondria-associated granulocyte macrophage CSF signaling molecule) form heterodimeric complexes with each other and they have been shown in vitro to regulate mtHsp70 activity in the same manner as their yeast homologues (Elsner et al. 2009). Besides, DNAJC19 and Magmas are of particular interest as they are associated with several human disorders: A novel autosomal recessive disorder termed "DCMA syndrome" (dilated cardiomyopathy with ataxia) has been assigned to mutations in the DNAJC19 gene (Davey et al. 2006). The clinical syndrome resembles that of the Barth syndrome, which is caused by defective cardiolipin remodeling in the mitochondrial inner membrane as a result of mutations in the tafazzin gene (TAZ1/G4.5) (Schlame and Ren 2006). Magmas is suspected to be involved in increased rates of anaerobic metabolism, resistance to apoptosis, and other cellular alterations characteristic of cancer cells (Jubinsky et al. 2001).

12.8 Folding and Assembly into Active Enzyme Complexes

Mitochondrially synthesized as well as all newly imported proteins enter the matrix compartment in an unfolded or only partially folded state, respectively. In order to achieve their native and functional conformation, the nascent polypeptides have to pass through a dedicated chaperone-catalyzed folding pathway. In this context, different from its involvement in the translocation process, the major Hsp70 chaperone in the mitochondrial matrix, mtHsp70 (or Ssc1 in yeast), acts as a soluble matrix protein and it is assisted by a different J-domain cochaperone, termed Mdj1 (Horst et al. 1997). Here, the action of mtHsp70 closely resembles a classical chaperone function, very similar to its bacterial relatives. The second Hsp70-type chaperone in yeast mitochondria, Ssq1, seems to be exclusively involved in protein folding reactions. Together with its J-domain cochaperone Jac1, Ssq1 is necessary for cold-shock survival and the folding and/or assembly of Fe/S-cluster containing proteins like the Rieske FeS protein of the cytochrome bc_1 complex (Vickery and Cupp-Vickery 2007).

Box 12.2 Controversial issues concerning the Tim23-PAM pathway mechanism

Concerning the molecular mechanism of the Tim23-PAM pathway, divergent models have been proposed. Due to the complex nature of the import process and technical difficulties in its characterization, these discrepancies have not been completely resolved to date.

1. One problematic issue arises due to the fact that the same translocase complex assists transport of some precursors through the inner membrane, while other preproteins are inserted into the phospholipid bilayer. Recently, an interesting model was proposed as to how the TIM23 complex switches between membrane integration and matrix translocation. According to this model, the TIM23 complex exists in (at least) two distinct forms: a Tim21-containing but PAM-free complex that specifically catalyzes inner membrane sorting ("TIM23^{SORT}") and a Tim21-free but PAM-bound form promoting import into the matrix ("TIM23^{MOTOR}") (Chacinska et al. 2005; van der Laan et al. 2007; Wiedemann et al. 2007). Tim21 thus functions as a "molecular switch": recruitment of Tim21 results in dissociation of the PAM complex; vice versa, release of Tim21 promotes the association of PAM with the TIM23 complex. Tim17 might play a crucial role in the switch between the two forms, as it is both involved in protein sorting to the inner membrane and interacts with Pam18 (Chacinska et al. 2005). It is thus tempting to speculate that Tim17 functions as a sensor that recognizes the targeting properties of incoming precursors and either promotes membrane insertion or the recruitment of the PAM complex via interacting with Pam18. This model was corroborated by successful membrane integration of an inner membrane protein into cardiolipin-containing proteoliposomes containing the TIM23^{SORT} constituents Tim23, Tim17, Tim50, and Tim21 (van der Laan et al. 2007). However, the physiologic relevance of these experiments needs to be addressed in future studies using intact mitochondria. Besides, the only model substrate employed by van der Laan et al., cytochrome c_1, was formerly shown to be imported into intact mitochondria independent of intramitochondrial ATP (Wachter et al. 1992). In contrast, the import of several other presequence-containing precursors destined for the inner membrane or the intermembrane space has been reported to require matrix ATP (Glick et al. 1993; Gärtner et al. 1995). Furthermore, in coisolation studies, Tim21 was found in association with the presequence translocase actively translocating preproteins into the matrix, and with PAM components (Mokranjac et al. 2005; Tamura et al. 2006; Popov-Celeketic et al. 2008). Hence, in contrast to the model outlined above, it was proposed that the PAM complex associates permanently with the presequence translocase TIM23. Yet, these apparently controversial views are not necessarily incompatible and a unified model to describe the multistep TIM23-PAM reaction cycle has recently been proposed (Chacinska et al. 2009).

(continued)

Box 12.2 (continued)

2. Another controversial issue relates to the import route followed by precursor proteins containing bipartite targeting signals: According to the "stop-transfer" model, the hydrophobic sorting signal mediates the arrest of the precursor inside the Tim23 channel followed by lateral release into the lipid phase (by an as-yet unknown mechanism). The inner membrane protein cytochrome b_2 as well as the IMS resident cytochrome c_1 have been demonstrated to follow the stop-transfer route (Glick et al. 1992b; Chacinska et al. 2005; Neupert and Herrmann 2007). The "conservative sorting" principle was originally described for the Rieske iron/sulfur cluster protein. According to this model, bipartite signal-containing preproteins are completely imported into the matrix and subsequently inserted into the membrane via a route conserved from the bacterial ancestor of mitochondria (Hartl et al. 1986). The identification of Oxa1 as the core of a mitochondrial export machinery finally allowed for the verification of the conservative sorting model (Hell et al. 1998). Oxa1 represents a homologue of the bacterial export component YidC and mediates the cotranslational membrane insertion of mitochondria-encoded proteins (Kuhn et al. 2003). Additionally, it seems to mediate the export of some nuclear-encoded proteins after their import into the matrix (Neupert and Herrmann 2007). Thus, there is experimental evidence for both routes, i.e., the two models do not seem to exclude each other.

Similar to the situation in bacteria, the folding reactions of some mitochondrial matrix proteins to their native conformation additionally require the assistance of the multi-subunit chaperonin complex, composed of the proteins Hsp60/Hsp10 (Voos and Röttgers 2002), the relatives of the GroEL and GroES proteins in *E. coli*. Interestingly, as in the case of mtHsp70, loss of Hsp60 function is lethal in yeast, reflecting the importance of this enzyme in the protein maturation process. In humans, a point mutation in Hsp60 (V72I) has been described to be associated with the neurodegenerative disorder hereditary spastic paraplegia (Hansen et al. 2002). In its native structure, the Hsp60 complex consists of a tetradecameric double ring system, often referred to as a "double doughnut" structure. Each of the two heptameric Hsp60 rings provides a hydrophobic cavity for the folding of a single polypeptide with a size of up to 50 kDa molecular mass. For efficient folding the cavity is closed by a single heptameric ring of Hsp10 molecules (Höhfeld and Hartl 1994). Closure of the reaction chamber provides a protected environment for the folding of the polypeptide substrate in an ATP-dependent manner (Walter 2002). Substrate dissociation from the *cis* side of the folding complex is initiated by ATP binding to the opposite ring, concomitant with the binding of a new substrate protein to this (*trans*) ring (Hartl and Hayer-Hartl 2009). The mitochondrial Hsp60 complex has also been described to function as a single ring-shaped chaperone, resulting in the hypothesis of a dynamic equilibrium between single and double ring structures in the case of mitochondrial Hsp60 (Levy-Rimler et al. 2002).

Some folding reactions may also require the assistance of peptidyl-prolyl cis-trans isomerases. In yeast, Cpr3 is the only mitochondrially localized member of this enzyme family known so far. By in vitro import experiments using artificial substrate proteins, Cpr3 was shown to contribute to the folding of newly imported polypeptides (Matouschek et al. 1995; Rassow et al. 1995). Although the details of the reaction cycle await further research, the human homologue of Cpr3, Cyclophilin D, has been shown to be involved in the opening of the mitochondrial transition pore, which eventually results in apoptotic cell death (Zorov et al. 2009).

Glossary

Chaperone: Proteins that prevent irregular interactions of their substrate polypeptides based on the nonspecific binding to exposed hydrophobic segments. In many cases, substrate affinity of chaperones is regulated by nucleotide binding.

Import motor or PAM (presequence translocase-associated motor): ATP-consuming translocation system attached to the inner face of the inner mitochondrial membrane with the core component mtHsp70 that drives import of precursor proteins into the matrix compartment.

MCF (metabolite carrier family): Family of strongly hydrophobic proteins integrated into the inner mitochondrial membrane that confer exchange of metabolites between the cytosol and the mitochondrial matrix compartment.

MIA (mitochondrial intermembrane space import and assembly): Disulfide-relay system in the intermembrane space that mediates the import and folding of cysteine-rich precursor proteins in the intermembrane space.

Peptidyl-prolyl *cis-trans* isomerase: Enzyme that catalyzes the *cis-trans* isomeration of peptide bonds involving the amino acid proline.

Processing: Specific removal of N-terminal targeting sequences during the mitochondria import process by specific endoproteases.

Receptor: Protein integrated into the mitochondrial outer membrane that interacts specifically with polypeptides containing mitochondrial targeting signals, thereby initiating the import process.

SAM/TOB (sorting and assembly machinery/topogenesis of β-barrel proteins): Multi-subunit protein complex that mediates the insertion and assembly of (mainly) β-barrel precursor proteins into the outer mitochondrial membrane.

TIM22 (translocase of the inner mitochondrial membrane with 22-kDa-core protein): Multi-subunit protein complex that mediates the insertion of (mainly polytopic) mitochondrial precursor proteins into the inner mitochondrial membrane in dependency of the electric potential $\Delta\psi$ across the inner membrane.

TIM23 (translocase of the inner mitochondrial membrane with 23-kDa-core protein; also termed presequence translocase): Multi-subunit protein complex that mediates the translocation of mitochondrial precursor proteins across or into the inner mitochondrial membrane in dependency of the electric potential $\Delta\psi$ across the inner membrane. For matrix import, TIM23 cooperates with PAM.

TPR (tetratricopeptide repeat) motif: Degenerate protein sequence of approximately 34 amino acids containing the consensus residues -W-LG-Y-A-F-A-P-. The sequence occurs in tandem arrays that arrange into a highly conserved tertiary structure. Acts as protein-protein interaction module.

Translocase complex: Membrane-integrated multi-subunit protein complex that mediates the translocation of mitochondrial precursor proteins across the respective mitochondrial membrane.

References

Abé, Y., Shodai, T., Muto, T., Mihara, K., Torii, H., Nishikawa, S., Endo, T., Kohda, D. 2000. Cell 100:551–560.
Ahting, U., Thieffry, M., Engelhardt, H., Hegerl, R., Neupert, W., Nussberger, S. 2001. J. Cell Biol. 153:1151–1160.
Ahting, U., Waizenegger, T., Neupert, W., Rapaport, D. 2005. J. Biol. Chem. 280:48–53.
Ahting, U., Floss, T., Uez, N., Schneider-Lohmar, I., Becker, L., Kling, E., Iuso, A., Bender, A., de Angelis, M. H., Gailus-Durner, V., Fuchs, H., Meitinger, T., Wurst, W., Prokisch, H., Klopstock, T. 2009. Biochim. Biophys. Acta 1787:371–376.
Alder, N. N., Jensen, R. E., Johnson, A. E. 2008. Cell 134:439–450.
Allen, S., Balabanidou, V., Sideris, D. P., Lisowsky, T., Tokatlidis, K. 2005. J. Mol. Biol. 353:937–944.
Banci, L., Bertini, I., Cefaro, C., Ciofi-Baffoni, S., Gallo, A., Martinelli, M., Sideris, D. P., Katrakili, N., Tokatlidis, K. 2009. Nat. Struct. Mol. Biol. 16:198–206.
Bauer, M. F., Sirrenberg, C., Neupert, W., Brunner, M. 1996. Cell 87:33–41.
Bauer, M. F., Gempel, K., Reichert, A. S., Rappold, G. A., Lichtner, P., Gerbitz, K. D., Neupert, W., Brunner, M., Hofmann, S. 1999a. J. Mol. Biol. 289:69–82.
Bauer, M. F., Rothbauer, U., Muhlenbein, N., Smith, R. J., Gerbitz, K., Neupert, W., Brunner, M., Hofmann, S. 1999b. FEBS Lett. 464:41–47.
Becker, L., Bannwarth, M., Meisinger, C., Hill, K., Model, K., Krimmer, T., Casadio, R., Truscott, K. N., Schulz, G. E., Pfanner, N., Wagner, R. 2005. J. Mol. Biol. 353:1011–1020.
Becker, T., Pfannschmidt, S., Guiard, B., Stojanovski, D., Milenkovic, D., Kutik, S., Pfanner, N., Meisinger, C., Wiedemann, N. 2008a. J. Biol. Chem. 283:120–127.
Becker, T., Vogtle, F. N., Stojanovski, D., Meisinger, C. 2008b. Biochim. Biophys. Acta 1777:557–563.
Becker, D., Krayl, M., Strub, A., Li, Y., Mayer, M. P., Voos, W. 2009a. J. Biol. Chem. 284:2934–2946.
Becker, T., Gebert, M., Pfanner, N., van der Laan, M. 2009b. Curr. Opin. Cell Biol. 21:484–493.
Bihlmaier, K., Mesecke, N., Terziyska, N., Bien, M., Hell, K., Herrmann, J. M. 2007. J. Cell Biol. 179:389–395.
Boldogh, I. R., Nowakowski, D. W., Yang, H. C., Chung, H., Karmon, S., Royes, P., Pon, L. A. 2003. Mol. Biol. Cell 14:4618–4627.
Bolliger, L., Junne, T., Schatz, G., Lithgow, T. 1995. EMBO J. 14:6318–6326.
Bömer, U., Rassow, J., Zufall, N., Pfanner, N., Meijer, M., Maarse, A. C. 1996. J. Mol. Biol. 262:389–395.

Brandner, K., Rehling, P., Truscott, K. N. 2005. J. Biol. Chem. 280:6215–6221.
Brix, J., Rudiger, S., Bukau, B., Schneider-Mergener, J., Pfanner, N. 1999. J. Biol. Chem. 274:16522–16530.
Brix, J., Ziegler, G. A., Dietmeier, K., Schneider-Mergener, J., Schulz, G. E., Pfanner, N. 2000. J. Mol. Biol. 303:479–488.
Burri, L., Vascotto, K., Gentle, I. E., Chan, N. C., Beilharz, T., Stapleton, D. I., Ramage, L., Lithgow, T. 2006. FEBS J. 273:1507–1515.
Cavalier-Smith, T. 2006. Proc. Biol. Sci. 273:1943–1952.
Chacinska, A., Rehling, P., Guiard, B., Frazier, A. E., Schulze-Specking, A., Pfanner, N., Voos, W., Meisinger, C. 2003. EMBO J. 22:5370–5381.
Chacinska, A., Pfannschmidt, S., Wiedemann, N., Kozjak, V., Sanjuan Szklarz, L. K., Schulze-Specking, A., Truscott, K. N., Guiard, B., Meisinger, C., Pfanner, N. 2004. EMBO J. 23:3735–3746.
Chacinska, A., Lind, M., Frazier, A. E., Dudek, J., Meisinger, C., Geissler, A., Sickmann, A., Meyer, H. E., Truscott, K. N., Guiard, B., Pfanner, N., Rehling, P. 2005. Cell 120:817–829.
Chacinska, A., Guiard, B., Muller, J. M., Schulze-Specking, A., Gabriel, K., Kutik, S., Pfanner, N. 2008. J. Biol. Chem. 283:29723–29729.
Chacinska, A., Koehler, C. M., Milenkovic, D., Lithgow, T., Pfanner, N. 2009. Cell 138:628–644.
Chan, N. C., Lithgow, T. 2008. Mol. Biol. Cell 19:126–136.
Choglay, A. A., Chapple, J. P., Blatch, G. L., Cheetham, M. E. 2001. Gene 267:125–134.
Coppock, D. L., Thorpe, C. 2006. Antioxid. Redox Signal 8:300–311.
Curran, S. P., Leuenberger, D., Schmidt, E., Koehler, C. M. 2002. J. Cell. Biol. 158:1017–1027.
Curran, S. P., Leuenberger, D., Leverich, E. P., Hwang, D. K., Beverly, K. N., Koehler, C. M. 2004. J. Biol. Chem. 279:43744–43751.
D'Silva, P., Liu, Q., Walter, W., Craig, E. A. 2004. Nat. Struct. Mol. Biol. 11:1084–1091.
D'Silva, P. R., Schilke, B., Walter, W., Craig, E. A. 2005. Proc. Natl. Acad. Sci. U.S.A. 102:12419–12424.
D'Silva, P. R., Schilke, B., Hayashi, M., Craig, E. A. 2008. Mol. Biol. Cell 19:424–432.
Dabir, D. V., Leverich, E. P., Kim, S. K., Tsai, F. D., Hirasawa, M., Knaff, D. B., Koehler, C. M. 2007. EMBO J. 26:4801–4811.
Davey, K. M., Parboosingh, J. S., McLeod, D. R., Chan, A., Casey, R., Ferreira, P., Snyder, F. F., Bridge, P. J., Bernier, F. P. 2006. J. Med. Genet. 43:385–393.
Davis, A. J., Ryan, K. R. and Jensen, R. E. 1998. Mol. Biol. Cell 9:2577–2593.
Davis, A. J., Sepuri, N. B., Holder, J., Johnson, A. E., Jensen, R. E. 2000. J. Cell. Biol. 150:1271–1282.
Davis, A. J., Alder, N. N., Jensen, R. E., Johnson, A. E. 2007. Mol. Biol. Cell 18:475–486.
Dekker, P. J., Martin, F., Maarse, A. C., Bomer, U., Muller, H., Guiard, B., Meijer, M., Rassow, J., Pfanner, N. 1997. EMBO J. 16:5408–5419.
Dekker, P. J., Ryan, M. T., Brix, J., Muller, H., Honlinger, A., Pfanner, N. 1998. Mol. Cell Biol. 18:6515–6524.
Dietmeier, K., Hönlinger, A., Bömer, U., Dekker, P. J., Eckerskorn, C., Lottspeich, F., Kübrich, M., Pfanner, N. 1997. Nature 388:195–200.
Dolezal, P., Likic, V., Tachezy, J., Lithgow, T. 2006. Science 313:314–318.
Duvezin-Caubet, S., Koppen, M., Wagener, J., Zick, M., Israel, L., Bernacchia, A., Jagasia, R., Rugarli, E. I., Imhof, A., Neupert, W., Langer, T., Reichert, A. S. 2007. Mol. Biol. Cell 18:3582–3590.
Elsner, S., Simian, D., Iosefson, O., Marom, M., Azem, A. 2009. Int. J. Mol. Sci. 10:2041–2053.
Elstner, M., Andreoli, C., Klopstock, T., Meitinger, T., Prokisch, H. 2009. Meth. Enzymol. 457:3–20.
Endo, T., Yamamoto, H., Esaki, M. 2003. J. Cell Sci. 116:3259–3267.
Endres, M., Neupert, W., Brunner, M. 1999. EMBO J. 18:3214–3221.
Erbse, A., Mayer, M. P., Bukau, B. 2004. Biochem. Soc. Trans. 32:617–621.
Esser, K., Tursun, B., Ingenhoven, M., Michaelis, G., Pratje, E. 2002. J. Mol. Biol. 323:835–843.

Fan, A. C., Bhangoo, M. K., Young, J. C. 2006. J. Biol. Chem. 281:33313–33324.
Frazier, A. E., Dudek, J., Guiard, B., Voos, W., Li, Y., Lind, M., Meisinger, C., Geissler, A., Sickmann, A., Meyer, H. E., Bilanchone, V., Cumsky, M. G., Truscott, K. N., Pfanner, N., Rehling, P. 2004. Nat. Struct. Mol. Biol. 11:226–233.
Gabriel, K., Egan, B., Lithgow, T. 2003. EMBO J. 22:2380–2386.
Gabriel, K., Milenkovic, D., Chacinska, A., Muller, J., Guiard, B., Pfanner, N., Meisinger, C. 2007. J. Mol. Biol. 365:612–620.
Gakh, O., Cavadini, P., Isaya, G. 2002. Biochim. Biophys. Acta 1592:63–77.
Gärtner, F., Voos, W., Querol, A., Miller, B. R., Craig, E. A., Cumsky, M. G., Pfanner, N. 1995. J. Biol. Chem. 270:3788–3795.
Geissler, A., Chacinska, A., Truscott, K. N., Wiedemann, N., Brandner, K., Sickmann, A., Meyer, H. E., Meisinger, C., Pfanner, N., Rehling, P. 2002. Cell 111:507–518.
Gerhard, D. S., Wagner, L., Feingold, E. A., et al. 2004. Genome Res. 14:2121–2127.
Glick, B. S., Beasley, E. M., Schatz, G. 1992a. Trends Biochem. Sci. 17:453–459.
Glick, B. S., Brandt, A., Cunningham, K., Muller, S., Hallberg, R. L., Schatz, G. 1992b. Cell 69:809–822.
Glick, B. S., Wachter, C., Reid, G. A., Schatz, G. 1993. Prot. Sci. 2:1901–1917.
Grumbt, B., Stroobant, V., Terziyska, N., Israel, L., Hell, K. 2007. J. Biol. Chem. 282:37461–37470.
Guo, Y., Cheong, N., Zhang, Z., De Rose, R., Deng, Y., Farber, S. A., Fernandes-Alnemri, T., Alnemri, E. S. 2004. J. Biol. Chem. 279:24813–24825.
Habib, S. J., Waizenegger, T., Lech, M., Neupert, W., Rapaport, D. 2005. J. Biol. Chem. 280:6434–6440.
Habib, S. J., Neupert, W., Rapaport, D. 2007. Meth. Cell Biol. 80:761–781.
Hansen, J. J., Durr, A., Cournu-Rebeix, I., Georgopoulos, C., Ang, D., Nielsen, M. N., Davoine, C. S., Brice, A., Fontaine, B., Gregersen, N., Bross, P. 2002. Am. J. Hum. Genet. 70:1328–1332.
Hartl, F. U., Hayer-Hartl, M. 2009. Nat. Struct. Mol. Biol. 16:574–581.
Hartl, F. U., Schmidt, B., Wachter, E., Weiss, H., Neupert, W. 1986. Cell 47:939–951.
Hell, K. 2008. Biochim. Biophys. Acta 1783:601–609.
Hell, K., Herrmann, J. M., Pratje, E., Neupert, W., Stuart, R. A. 1998. Proc. Natl. Acad. Sci. U.S.A. 95:2250–2255.
Herlan, M., Vogel, F., Bornhovd, C., Neupert, W., Reichert, A. S. 2003. J. Biol. Chem. 278:27781–27788.
Hofmann, S., Rothbauer, U., Muhlenbein, N., Baiker, K., Hell, K., Bauer, M. F. 2005. J. Mol. Biol. 353:517–528.
Höhfeld, J., Hartl, F. U. 1994. J. Cell. Biol. 126:305–315.
Hönlinger, A., Bömer, U., Alconada, A., Eckerskorn, C., Lottspeich, F., Dietmeier, K., Pfanner, N. 1996. EMBO J. 15:2125–2137.
Hoogenraad, N. J., Ward, L. A., Ryan, M. T. 2002. Biochim. Biophys. Acta 1592:97–105.
Hoppins, S. C., Nargang, F. E. 2004. J. Biol. Chem. 279:12396–12405.
Horst, M., Oppliger, W., Feifel, B., Schatz, G., Glick, B. S. 1996. Protein Sci. 5:759–767.
Horst, M., Oppliger, W., Rospert, S., Schonfeld, H. J., Schatz, G., Azem, A. 1997. EMBO J. 16:1842–1849.
Hulett, J. M., Lueder, F., Chan, N. C., Perry, A. J., Wolynec, P., Likic, V. A., Gooley, P. R., Lithgow, T. 2008. J. Mol. Biol. 376:694–704.
Humphries, A. D., Streimann, I. C., Stojanovski, D., Johnston, A. J., Yano, M., Hoogenraad, N. J., Ryan, M. T. 2005. J. Biol. Chem. 280:11535–11543.
Hutu, D. P., Guiard, B., Chacinska, A., Becker, D., Pfanner, N., Rehling, P., van der Laan, M. 2008. Mol. Biol. Cell 19:2642–2649.
Hwang, D. K., Claypool, S. M., Leuenberger, D., Tienson, H. L., Koehler, C. M. 2007. J. Cell. Biol. 178:1161–1175.
Igura, M., Ose, T., Obita, T., Sato, C., Maenaka, K., Endo, T., Kohda, D. 2005. Acta Crystallogr. Sect. F Struct. Biol. Cryst. Commun. 61:514–517.
Ishihara, N., Mihara, K. 1998. J. Biochem. 123:722–732.
Jubinsky, P. T., Messer, A., Bender, J., Morris, R. E., Ciraolo, G. M., Witte, D. P., Hawley, R. G., Short, M. K. 2001. Exp. Hematol. 29:1392–1402.

Jubinsky, P. T., Short, M. K., Mutema, G., Morris, R. E., Ciraolo, G. M., Li, M. 2005. J. Mol. Histol. 36:69–75.
Kato, H., Mihara, K. 2008. Biochem. Biophys. Res. Commun. 369:958–963.
Kemper, C., Habib, S. J., Engl, G., Heckmeyer, P., Dimmer, K. S., Rapaport, D. 2008. J. Cell Sci. 121:1990–1998.
Kerscher, O., Holder, J., Srinivasan, M., Leung, R. S., Jensen, R. E. 1997. J. Cell Biol. 139:1663–1675.
Kerscher, O., Sepuri, N. B., Jensen, R. E. 2000. Mol. Biol. Cell 11:103–116.
Kim, S., Malinverni, J. C., Sliz, P., Silhavy, T. J., Harrison, S. C., Kahne, D. 2007. Science 317:961–964.
Klingenberg, M. 2008. Biochim. Biophys. Acta 1778:1978–2021.
Koehler, C. M. 2004. Annu. Rev. Cell. Dev. Biol. 20:309–335.
Koehler, C. M., Jarosch, E., Tokatlidis, K., Schmid, K., Schweyen, R. J., Schatz, G. 1998a. Science 279:369–373.
Koehler, C. M., Merchant, S., Oppliger, W., Schmid, K., Jarosch, E., Dolfini, L., Junne, T., Schatz, G., Tokatlidis, K. 1998b. EMBO J. 17:6477–6486.
Koehler, C. M., Leuenberger, D., Merchant, S., Renold, A., Junne, T., Schatz, G. 1999a. Proc. Natl. Acad. Sci. U.S.A. 96:2141–2146.
Koehler, C. M., Merchant, S., Schatz, G. 1999b. Trends Biochem. Sci. 24:428–432.
Koehler, C. M., Murphy, M. P., Bally, N. A., Leuenberger, D., Oppliger, W., Dolfini, L., Junne, T., Schatz, G., Or, E. 2000. Mol. Cell. Biol. 20:1187–1193.
Komiya, T., Rospert, S., Koehler, C., Looser, R., Schatz, G., Mihara, K. 1998. EMBO J. 17:3886–3898.
Kornmann, B., Currie, E., Collins, S. R., Schuldiner, M., Nunnari, J., Weissman, J. S., Walter, P. 2009. Science 325:477–481.
Kovermann, P., Truscott, K. N., Guiard, B., Rehling, P., Sepuri, N. B., Muller, H., Jensen, R. E., Wagner, R., Pfanner, N. 2002. Mol. Cell 9:363–373.
Kozjak, V., Wiedemann, N., Milenkovic, D., Lohaus, C., Meyer, H. E., Guiard, B., Meisinger, C., Pfanner, N. 2003. J. Biol. Chem. 278:48520–48523.
Kozjak-Pavlovic, V., Ross, K., Benlasfer, N., Kimmig, S., Karlas, A., Rudel, T. 2007. EMBO Rep 8:576–582.
Krayl, M., Lim, J. H., Martin, F., Guiard, B., Voos, W. 2007. Mol. Cell. Biol. 27:411–425.
Krimmer, T., Rapaport, D., Ryan, M. T., Meisinger, C., Kassenbrock, C. K., Blachly-Dyson, E., Forte, M., Douglas, M. G., Neupert, W., Nargang, F. E., Pfanner, N. 2001. J. Cell Biol. 152:289–300.
Kuhn, A., Stuart, R., Henry, R., Dalbey, R. E. 2003. Trends Cell. Biol. 13:510–516.
Kurz, M., Martin, H., Rassow, J., Pfanner, N., Ryan, M. T. 1999. Mol. Biol. Cell 10:2461–2474.
Kutik, S., Stojanovski, D., Becker, L., Becker, T., Meinecke, M., Kruger, V., Prinz, C., Meisinger, C., Guiard, B., Wagner, R., Pfanner, N., Wiedemann, N. 2008. Cell 132:1011–1024.
Levy-Rimler, G., Bell, R. E., Ben-Tal, N., Azem, A. 2002. FEBS Lett. 529:1–5.
Li, Y., Dudek, J., Guiard, B., Pfanner, N., Rehling, P., Voos, W. 2004. J. Biol. Chem. 279:38047–38054.
Lu, H., Woodburn, J. 2005. J. Mol. Biol. 353:897–910.
Lutz, T., Neupert, W., Herrmann, J. M. 2003. EMBO J. 22:4400–4408.
Martinez-Caballero, S., Grigoriev, S. M., Herrmann, J. M., Campo, M. L., Kinnally, K. W. 2007. J. Biol. Chem. 282:3584–3593.
Matouschek, A., Rospert, S., Schmid, K., Glick, B. S., Schatz, G. 1995. Proc. Natl. Acad. Sci. U.S.A. 92:6319–6323.
Matouschek, A., Pfanner, N., Voos, W. 2000. EMBO Rep. 1:404–410.
Mayer, M. P., Bukau, B. 2005. Cell Mol. Life Sci. 62:670–684.
Mayer, M. P., Rüdiger, S., Bukau, B. 2000. Biol. Chem. 381:877–885.
McBride, H. M., Neuspiel, M., Wasiak, S. 2006. Curr. Biol. 16:R551–560.
Meier, S., Neupert, W., Herrmann, J. M. 2005. J. Biol. Chem. 280:7777–7785.
Meinecke, M., Wagner, R., Kovermann, P., Guiard, B., Mick, D. U., Hutu, D. P., Voos, W., Truscott, K. N., Chacinska, A., Pfanner, N., Rehling, P. 2006. Science 312:1523–1526.

Meineke, B., Engl, G., Kemper, C., Vasiljev-Neumeyer, A., Paulitschke, H., Rapaport, D. 2008. FEBS Lett. 582:855–860.
Meisinger, C., Ryan, M. T., Hill, K., Model, K., Lim, J. H., Sickmann, A., Müller, H., Meyer, H. E., Wagner, R., Pfanner, N. 2001. Mol. Cell Biol. 21:2337–2348.
Meisinger, C., Rissler, M., Chacinska, A., Szklarz, L. K., Milenkovic, D., Kozjak, V., Schonfisch, B., Lohaus, C., Meyer, H. E., Yaffe, M. P., Guiard, B., Wiedemann, N., Pfanner, N. 2004. Dev. Cell 7:61–71.
Meisinger, C., Pfannschmidt, S., Rissler, M., Milenkovic, D., Becker, T., Stojanovski, D., Youngman, M. J., Jensen, R. E., Chacinska, A., Guiard, B., Pfanner, N., Wiedemann, N. 2007. EMBO J. 26:2229–2239.
Mesecke, N., Terziyska, N., Kozany, C., Baumann, F., Neupert, W., Hell, K., Herrmann, J. M. 2005. Cell 121:1059–1069.
Mesecke, N., Bihlmaier, K., Grumbt, B., Longen, S., Terziyska, N., Hell, K., Herrmann, J. M. 2008. EMBO Rep. 9:1107–1113.
Miao, B., Davis, J. E., Craig, E. A. 1997. J. Mol. Biol. 265:541–552.
Milenkovic, D., Kozjak, V., Wiedemann, N., Lohaus, C., Meyer, H. E., Guiard, B., Pfanner, N., Meisinger, C. 2004. J. Biol. Chem. 279:22781–22785.
Milenkovic, D., Gabriel, K., Guiard, B., Schulze-Specking, A., Pfanner, N., Chacinska, A. 2007. J. Biol. Chem. 282:22472–22480.
Milenkovic, D., Ramming, T., Muller, J. M., Wenz, L. S., Gebert, N., Schulze-Specking, A., Stojanovski, D., Rospert, S., Chacinska, A. 2009. Mol. Biol. Cell 20:2530–2539.
Mills, R. D., Trewhella, J., Qiu, T. W., Welte, T., Ryan, T. M., Hanley, T., Knott, R. B., Lithgow, T., Mulhern, T. D. 2009. J. Mol. Biol. 388:1043–1058.
Moczko, M., Bomer, U., Kubrich, M., Zufall, N., Honlinger, A., Pfanner, N. 1997. Mol. Cell Biol. 17:6574–6584.
Model, K., Meisinger, C., Prinz, T., Wiedemann, N., Truscott, K. N., Pfanner, N., Ryan, M. T. 2001. Nat. Struct. Biol. 8:361–370.
Model, K., Meisinger, C., Kuhlbrandt, W. 2008. J. Mol. Biol. 383:1049–1057.
Mogk, A., Schmidt, R., Bukau, B. 2007. Trends Cell. Biol. 17:165–172.
Mokranjac, D., Paschen, S. A., Kozany, C., Prokisch, H., Hoppins, S. C., Nargang, F. E., Neupert, W., Hell, K. 2003. EMBO J. 22:816–825.
Mokranjac, D., Popov-Celeketic, D., Hell, K., Neupert, W. 2005. J. Biol. Chem. 280:23437–23440.
Mokranjac, D., Bourenkov, G., Hell, K., Neupert, W., Groll, M. 2006. EMBO J. 25:4675–4685.
Mokranjac, D., Sichting, M., Popov-Celeketic, D., Mapa, K., Gevorkyan-Airapetov, L., Zohary, K., Hell, K., Azem, A., Neupert, W. 2009. Mol. Biol. Cell 20:1400–1407.
Mühlenbein, N., Hofmann, S., Rothbauer, U., Bauer, M. F. 2004. J. Biol. Chem. 279:13540–13546.
Müller, J. M., Milenkovic, D., Guiard, B., Pfanner, N., Chacinska, A. 2008. Mol. Biol. Cell 19:226–236.
Nakamoto, H., Bardwell, J. C. 2004. Biochim. Biophys. Acta 1694:111–119.
Naoé, M., Ohwa, Y., Ishikawa, D., Ohshima, C., Nishikawa, S., Yamamoto, H., Endo, T. 2004. J. Biol. Chem. 279:47815–47821.
Neupert, W., Herrmann, J. M. 2007. Annu. Rev. Biochem. 76:723–749.
Nury, H., Manon, F., Arnou, B., le Maire, M., Pebay-Peyroula, E., Ebel, C. 2008. Biochemistry 47:12319–12331.
Otera, H., Taira, Y., Horie, C., Suzuki, Y., Suzuki, H., Setoguchi, K., Kato, H., Oka, T., Mihara, K. 2007. J. Cell. Biol. 179:1355–1363.
Pagliarini, D. J., Calvo, S. E., Chang, B., Sheth, S. A., Vafai, S. B., Ong, S. E., Walford, G. A., Sugiana, C., Boneh, A., Chen, W. K., Hill, D. E., Vidal, M., Evans, J. G., Thorburn, D. R., Carr, S. A., Mootha, V. K. 2008. Cell 134:112–123.
Palmieri, F., Bisaccia, F., Capobianco, L., Dolce, V., Fiermonte, G., Iacobazzi, V., Indiveri, C., Palmieri, L. 1996. Biochim. Biophys. Acta 1275:127–132.
Paschen, S. A., Rothbauer, U., Kaldi, K., Bauer, M. F., Neupert, W., Brunner, M. 2000. EMBO J. 19:6392–6400.

Paschen, S. A., Waizenegger, T., Stan, T., Preuss, M., Cyrklaff, M., Hell, K., Rapaport, D., Neupert, W. 2003. Nature 426:862–866.
Paschen, S. A., Neupert, W., Rapaport, D. 2005. Trends Biochem. Sci. 30:575–582.
Patel, P. I., Isaya, G. 2001. Am. J. Hum. Genet. 69:15–24.
Pawlowski, R., Jura, J. 2006. Mol. Cell Biochem. 288:159–169.
Pebay-Peyroula, E., Dahout-Gonzalez, C., Kahn, R., Trezeguet, V., Lauquin, G. J., Brandolin, G. 2003. Nature 426:39–44.
Pfanner, N., Chacinska, A. 2002. Biochim. Biophys. Acta 1592:15–24.
Pfanner, N., Wiedemann, N., Meisinger, C., Lithgow, T. 2004. Nat. Struct. Mol. Biol. 11:1044–1048.
Popov-Celeketic, D., Mapa, K., Neupert, W., Mokranjac, D. 2008. EMBO J. 27:1469–1480.
Prokisch, H., Andreoli, C., Ahting, U., Heiss, K., Ruepp, A., Scharfe, C., Meitinger, T. 2006. Nucl. Acids Res. 34:D705–711.
Rapaport, D. 2003. EMBO Rep. 4:948–952.
Rapaport, D. 2005. J. Cell. Biol. 171:419–423.
Rassow, J., Guiard, B., Wienhues, U., Herzog, V., Hartl, F.-U., Neupert, W. 1989. J. Cell. Biol. 109:1421–1428.
Rassow, J., Hartl, F. U., Guiard, B., Pfanner, N., Neupert, W. 1990. FEBS Lett. 275:190–194.
Rassow, J., Mohrs, K., Koidl, S., Barthelmess, I. B., Pfanner, N., Tropschug, M. 1995. Mol. Cell. Biol. 15:2654–2662.
Rehling, P., Model, K., Brandner, K., Kovermann, P., Sickmann, A., Meyer, H. E., Kuhlbrandt, W., Wagner, R., Truscott, K. N., Pfanner, N. 2003. Science 299:1747–1751.
Rehling, P., Brandner, K., Pfanner, N. 2004. Nat. Rev. Mol. Cell. Biol. 5:519–530.
Rissler, M., Wiedemann, N., Pfannschmidt, S., Gabriel, K., Guiard, B., Pfanner, N., Chacinska, A. 2005. J. Mol. Biol. 353:485–492.
Roesch, K., Curran, S. P., Tranebjaerg, L., Koehler, C. M. 2002. Hum. Mol. Genet. 11:477–486.
Roise, D., Schatz, G. 1988. J. Biol. Chem. 263:4509–4511.
Ryan, K. R., Leung, R. S., Jensen, R. E. 1998. Mol. Cell Biol. 18:178–187.
Saitoh, T., Igura, M., Obita, T., Ose, T., Kojima, R., Maenaka, K., Endo, T., Kohda, D. 2007. EMBO J. 26:4777–4787.
Schapira, A. H. 1999. Biochim. Biophys. Acta 1410:99–102.
Schlame, M., Ren, M. 2006. FEBS Lett. 580:5450–5455.
Schlegel, T., Mirus, O., von Haeseler, A., Schleiff, E. 2007. Mol. Biol. Evol. 24:2763–2774.
Schleiff, E., Soll, J. 2005. EMBO Rep. 6:1023–1027.
Schleiff, E., Silvius, J. R., Shore, G. C. 1999. J. Cell Biol. 145:973–978.
Schulz, G. E. 2000. Curr. Opin. Struct. Biol. 10:443–447.
Sevier, C. S., Kadokura, H., Tam, V. C., Beckwith, J., Fass, D., Kaiser, C. A. 2005. Prot. Sci. 14:1630–1642.
Shariff, K., Ghosal, S., Matouschek, A. 2004. Biophys. J. 86:3647–3652.
Sickmann, A., Reinders, J., Wagner, Y., Joppich, C., Zahedi, R., Meyer, H. E., Schonfisch, B., Perschil, I., Chacinska, A., Guiard, B., Rehling, P., Pfanner, N., Meisinger, C. 2003. Proc. Natl. Acad. Sci. U.S.A. 100:13207–13212.
Sideris, D. P., Tokatlidis, K. 2007. Mol Microbiol 65:1360–1373.
Sirrenberg, C., Endres, M., Folsch, H., Stuart, R. A., Neupert, W., Brunner, M. 1998. Nature 391:912–915.
Sousa, R., Lafer, E. M. 2006. Traffic 7:1596–1603.
Stojanovski, D., Guiard, B., Kozjak-Pavlovic, V., Pfanner, N., Meisinger, C. 2007. J. Cell Biol. 179:881–893.
Tamura, Y., Harada, Y., Yamano, K., Watanabe, K., Ishikawa, D., Ohshima, C., Nishikawa, S., Yamamoto, H., Endo, T. 2006. J. Cell Biol. 174:631–637.
Taylor, A. B., Smith, B. S., Kitada, S., Kojima, K., Miyaura, H., Otwinowski, Z., Ito, A., Deisenhofer, J. 2001. Structure 9:615–625.
Terziyska, N., Lutz, T., Kozany, C., Mokranjac, D., Mesecke, N., Neupert, W., Herrmann, J. M., Hell, K. 2005. FEBS Lett. 579:179–184.

Tomkiewicz, D., Nouwen, N., Driessen, A. J. 2007. FEBS Lett. 581:2820–2828.
Truscott, K. N., Kovermann, P., Geissler, A., Merlin, A., Meijer, M., Driessen, A. J., Rassow, J., Pfanner, N., Wagner, R. 2001. Nat. Struct. Biol. 8:1074–1082.
Truscott, K. N., Wiedemann, N., Rehling, P., Muller, H., Meisinger, C., Pfanner, N., Guiard, B. 2002. Mol. Cell. Biol. 22:7780–7789.
Truscott, K. N., Voos, W., Frazier, A. E., Lind, M., Li, Y., Geissler, A., Dudek, J., Muller, H., Sickmann, A., Meyer, H. E., Meisinger, C., Guiard, B., Rehling, P., Pfanner, N. 2003. J. Cell Biol. 163:707–713.
Tsukihara, T., Aoyama, H., Yamashita, E., Tomizaki, T., Yamaguchi, H., Shinzawa-Itoh, K., Nakashima, R., Yaono, R., Yoshikawa, S. 1995. Science 269:1069–1074.
Tu, B. P., Weissman, J. S. 2004. J. Cell Biol. 164:341–346.
van der Laan, M., Chacinska, A., Lind, M., Perschil, I., Sickmann, A., Meyer, H. E., Guiard, B., Meisinger, C., Pfanner, N., Rehling, P. 2005. Mol. Cell Biol. 25:7449–7458.
van der Laan, M., Wiedemann, N., Mick, D. U., Guiard, B., Rehling, P., Pfanner, N. 2006. Curr. Biol. 16:2271–2276.
van der Laan, M., Meinecke, M., Dudek, J., Hutu, D. P., Lind, M., Perschil, I., Guiard, B., Wagner, R., Pfanner, N., Rehling, P. 2007. Nat. Cell Biol. 9:1152–1159.
van der Laan, M., Hutu, D. P., Rehling, P. 2010. Biochim. Biophys. Acta 732–739.
van Wilpe, S., Ryan, M. T., Hill, K., Maarse, A. C., Meisinger, C., Brix, J., Dekker, P. J., Moczko, M., Wagner, R., Meijer, M., Guiard, B., Honlinger, A., Pfanner, N. 1999. Nature 401:485–489.
Varshavsky, A. 2008. Nat. Struct. Mol. Biol. 15:1238–1240.
Vasiljev, A., Ahting, U., Nargang, F. E., Go, N. E., Habib, S. J., Kozany, C., Panneels, V., Sinning, I., Prokisch, H., Neupert, W., Nussberger, S., Rapaport, D. 2004. Mol. Biol. Cell 15: 1445–1458.
Vickery, L. E., Cupp-Vickery, J. R. 2007. Crit. Rev. Biochem. Mol. Biol. 42:95–111.
Vögtle, F. N., Wortelkamp, S., Zahedi, R. P., Becker, D., Leidhold, C., Gevaert, K., Kellermann, J., Voos, W., Sickmann, A., Pfanner, N., Meisinger, C. 2009. Cell 139:428–439.
Voisine, C., Craig, E. A., Zufall, N., von Ahsen, O., Pfanner, N., Voos, W. 1999. Cell 97:565–574.
von Heijne, G., Steppuhn, J., Herrmann, R. G. 1989. Eur J. Biochem. 180:535–545.
Voos, W., Pfanner, N. 2001. Front. Mol. Biol. 37:61–89.
Voos, W., Röttgers, K. 2002. Biochim. Biophys. Acta 1592:51–62.
Voos, W., Gambill, B. D., Laloraya, S., Ang, D., Craig, E. A., Pfanner, N. 1994. Mol. Cell. Biol. 14:6627–6634.
Voos, W., von Ahsen, O., Müller, H., Guiard, B., Rassow, J., Pfanner, N. 1996. EMBO J. 15:2668–2677.
Wachter, C., Schatz, G., Glick, B. S. 1992. EMBO J. 11:4787–4794.
Wada, J., Kanwar, Y. S. 1998. Proc. Natl. Acad. Sci. U.S.A. 95:144–149.
Wagner, K., Gebert, N., Guiard, B., Brandner, K., Truscott, K. N., Wiedemann, N., Pfanner, N., Rehling, P. 2008. Mol. Cell. Biol. 28:4251–4260.
Waizenegger, T., Habib, S. J., Lech, M., Mokranjac, D., Paschen, S. A., Hell, K., Neupert, W., Rapaport, D. 2004. EMBO Rep. 5:704–709.
Walter, S. 2002. Cell Mol. Life Sci. 59:1589–1597.
Webb, C. T., Gorman, M. A., Lazarou, M., Ryan, M. T., Gulbis, J. M. 2006. Mol. Cell 21:123–133.
Weiss, C., Niv, A., Azem, A. 2002. Protein Expr. Purif. 24:268–273.
Wiedemann, N., Pfanner, N., Ryan, M. T. 2001. EMBO J. 20:951–960.
Wiedemann, N., Kozjak, V., Chacinska, A., Schonfisch, B., Rospert, S., Ryan, M. T., Pfanner, N., Meisinger, C. 2003. Nature 424:565–571.
Wiedemann, N., Frazier, A. E., Pfanner, N. 2004. J. Biol. Chem. 279:14473–14476.
Wiedemann, N., van der Laan, M., Hutu, D. P., Rehling, P., Pfanner, N. 2007. J. Cell Biol. 179:1115–1122.
Wilcox, A. J., Choy, J., Bustamante, C., Matouschek, A. 2005. Proc. Natl. Acad. Sci. U.S.A. 102:15435–15440.
Wimley, W. C. 2003. Curr. Opin. Struct. Biol. 13:404–411.
Wu, Y., Sha, B. 2006. Nat. Struct. Mol. Biol. 13:589–593.
Young, J. C., Hoogenraad, N. J., Hartl, F. U. 2003. Cell 112:41–50.

Youngman, M. J., Hobbs, A. E., Burgess, S. M., Srinivasan, M., Jensen, R. E. 2004. J. Cell Biol. 164:677–688.

Zahedi, R. P., Sickmann, A., Boehm, A. M., Winkler, C., Zufall, N., Schonfisch, B., Guiard, B., Pfanner, N., Meisinger, C. 2006. Mol. Biol. Cell 17:1436–1450.

Zara, V., Palmieri, F., Mahlke, K., Pfanner, N. 1992. J. Biol. Chem. 267:12077–12081.

Zara, V., Ferramosca, A., Robitaille-Foucher, P., Palmieri, F., Young, J. C. 2009. Biochem. J. 419:369–375.

Zhang, Q. H., Ye, M., Wu, X. Y., Ren, S. X., Zhao, M., Zhao, C. J., Fu, G., Shen, Y., Fan, H. Y., Lu, G., Zhong, M., Xu, X. R., Han, Z. G., Zhang, J. W., Tao, J., Huang, Q. H., Zhou, J., Hu, G. X., Gu, J., Chen, S. J., Chen, Z. 2000. Genome Res. 10:1546–1560.

Zorov, D. B., Juhaszova, M., Yaniv, Y., Nuss, H. B., Wang, S., Sollott, S. J. 2009. Cardiovasc. Res. 83:213–225.

Part IV
Biochemistry, Regulation & Function

Chapter 13
Biogenesis and Supramolecular Organization of the Oxidative Phosphorylation System in Plants

Elina Welchen, Jennifer Klodmann, and Hans-Peter Braun

Abstract Oxidative phosphorylation (OXPHOS) is the major process to generate ATP in most eukaryotes. The site of this process is the mitochondrion. Structural basis for OXPHOS is the presence of five large multi-subunit complexes within the inner mitochondrial membrane (termed complexes I–V), the soluble intermembrane space-localized protein cytochrome c and the lipid ubiquinone, all together also referred to as the "OXPHOS system." In plants, additional so-called "alternative" components are part of this system. The OXPHOS system is formed by more than 100 different proteins as well as numerous prosthetic groups like FeS clusters, heme groups, and copper ions. Here, we summarize recent insights into the assembly of the OXPHOS system. Furthermore, findings on its supramolecular organization are summarized and discussed.

Keywords Oxidative phosphorylation system (OXPHOS) • Supercomplex • Respirasome • Biogenesis • Assembly • BN-SDS PAGE

Abbreviations

AA	Ascorbic acid
ACP	Acyl carrier protein
ADP	Adenosine diphosphate
ATP	Adenosine triphosphate
CA	Carbonic anhydrase (also see Glossary)
COX	Cytochrome c oxidase

H.-P. Braun (✉)
Institute for Plant Genetics, Leibniz Universität Hannover, Herrenhäuser Str. 2, D-30419, Hannover, Germany
e-mail: braun@genetik.uni-hannover.de

F. Kempken (ed.), *Plant Mitochondria*, Advances in Plant Biology 1,
DOI 10.1007/978-0-387-89781-3_13,

EM	Electron microscopy
FAD	Flavin adenine dinucleotide
Fe-S cluster	Iron-sulfur cluster
FMN	Flavin mononucleotide
IMS	Intermembrane space
L-GalL	L-galactono-1,4-lactone
L-GalLDH	L-galactono-1,4-lactone dehydrogenase
MPP	Mitochondrial processing peptidase (also see Glossary)
mRNA	Messenger RNA
NADH	Nicotinamide adenine dinucleotide
NADPH	Nicotinamide adenine dinucleotide phosphate
OSCP	Oligomycin sensitive conferring protein
OXPHOS	Oxidative phosphorylation (also see Glossary)
PAGE	Polyacrylamide gelelectrophoresis
RNAi	RNA interference
ROS	Reactive oxygen species
SDH	Succinate dehydrogenase
UTR	Untranslated region
UV	Ultraviolet

13.1 Introduction

Mitochondria are the power stations of the cell. Decisive events for cellular survival take place within mitochondria. In addition to their essential role in the aerobic respiration, they are required for other important metabolic and signaling processes such as lipid metabolism, amino acid biosynthesis, calcium signaling, and the control of the redox status as well as the fate of the entire cell. All these processes have to be coordinated with the rest of the cell in accordance with different stages of the cell cycle and with respect to environmental demands. Mitochondria produce ATP via a process termed oxidative phosphorylation (OXPHOS). The OXPHOS system of plants differs from those of most animals because it includes additional so-called "alternative" electron transport components and participates in light-dependent metabolic pathways like photorespiration. Therefore, activities of plant mitochondria have to be coordinated with the ones of chloroplasts (see also Chap. 15).

Both mitochondria and chloroplasts are semiautonomous organelles that have retained a genome throughout evolution. Consequently, mitochondrial biogenesis requires the expression of two separate genomes. About 30 mitochondrial proteins are encoded by the mitochondrial genome, several of which represent components of the OXPHOS system (Unseld et al. 1997). The remaining approximately 2,000 mitochondrial proteins (see also Chap. 9), including most components of the OXPHOS system, are nuclear encoded and have to be posttranslationally imported into the organelle (Mackenzie and McIntosh 1999; Adams and Palmer 2003; see also Chaps. 11 and 12). Therefore, precise communication mechanisms are necessary for the biogenesis of respiratory complexes (Goffart and Wiesner 2003; Welchen and Gonzalez 2006; Gonzalez et al. 2007). Adding even greater complexity to this stage, the respiratory protein complexes constitute supercomplexes in the mitochondrial internal membrane (Dudkina et al. 2006, 2008; Acín-Pérez et al. 2008; Peters et al. 2008). Loss-of-function analyses revealed that mutations within genes encoding subunits of an individual OXPHOS complex affect assembly of another OXPHOS complex, most likely because supercomplex formation is important for protein complex stabilization (Acín-Pérez et al. 2004, 2008; Ugalde et al. 2004a; Diaz et al. 2006; Li et al. 2007). In this chapter we summarize insights into (1) the assembly processes for OXPHOS complexes, and (2) the supramolecular organization of the OXPHOS system. Results basically refer to mitochondria of plants, but also include background information obtained for animal and fungal systems.

13.2 Composition and Biogenesis of Mitochondrial OXPHOS Complexes

13.2.1 Composition and Biogenesis of Complex I

Complex I (NADH-ubiquinone oxidoreductase) under most conditions is the main entrance point for electrons into the respiratory chain. It transfers two electrons from NADH to ubiquinone. Coupled to this process four protons are translocated across the inner mitochondrial membrane. Complex I has a molecular mass of about 1,000 kDa and is composed of more than 40 distinct protein subunits, one FMN cofactor, and eight Fe-S clusters. Complex I assembly has been studied first in the model fungus *Neurospora crassa* (Schulte 2001). The proteins that constitute complex I are encoded by two genomes: Nine have a mitochondrial genetic origin in higher plants, while the remaining subunits are encoded in the nucleus. Recently, composition of complex I has been investigated intensively in humans (Ugalde et al. 2004b; Vogel et al. 2007; Remacle et al. 2008), in the vascular plants *Arabidopsis thaliana* and *Oryza sativa* (Heazlewood et al. 2003a, b; Meyer et al. 2008) and in the green algae *Chlamydomonas reinhardtii* (Cardol et al. 2004, 2005). In Arabidopsis, 42 distinct complex I subunits were described by a novel 3D BN/SDS/SDS PAGE system coupled to tandem mass spectrometry (Meyer et al. 2008). In general, mitochondrial complex I particles from all organisms investigated have many subunits in common, including 14 orthologues to the prokaryotic enzyme that constitute the "core" subunits of the complex (Cardol et al. 2005). The function of most of the extra subunits in mitochondria is yet unclear, but it is speculated that they might be important for enzyme stabilization or protection of the complex toward reactive oxygen species (ROS) (Vogel et al. 2007).

In addition, complex I includes some specific extra subunits in plants which partially integrate side functions into this respiratory complex. Among them is a group of subunits structurally related to an archaebacterial gamma-type carbonic anhydrase (γCA) (Parisi et al. 2004; Perales et al. 2004). The γCAs are zink-containing enzymes that catalyze the interconversion of CO_2 and HCO_3^-. Five structurally related γCAs are present in complex I of Arabidopsis and at least three in rice (Heazlewood et al. 2003a). The Arabidopsis γCAs are attached to the hydrophobic membrane domain of mitochondrial complex I on its matrix-exposed side and form an extra domain characteristic for this respiratory complex in plants (Dudkina et al. 2005a; Sunderhaus et al. 2006; Peters et al. 2008). The functional role of these subunits within plant mitochondria is currently unknown, but they are essential for assembly of complex I and therefore have to be considered to represent integral components (Perales et al. 2005). Ectopic overexpression of the γCA2 protein in Arabidopsis causes male sterility in transgenic plants (Villarreal et al. 2009). Although direct evidence for γCA activity of complex I in plants is still lacking the present data indicate that the complex I-integrated γCAs are involved in mitochondrial HCO_3^- formation to allow efficient recycling of inorganic carbon for chloroplastic CO_2 fixation (Braun and Zabaleta 2007).

Another specific protein of plant mitochondrial complex I is L-galactono-1,4-lactone dehydrogenase (L-GalLDH), an enzyme involved in the conversion of L-galactono-1,4-lactone (L-GalL) to ascorbic acid (AA). It was demonstrated before that this enzyme is attached to the inner mitochondrial membrane (Bartoli et al. 2000) and its binding to complex I was first reported by Heazlewood et al. (2003a). The addition of rotenone, a specific inhibitor of complex I, in the presence of pyruvate and malate blocks the synthesis of AA in mitochondria isolated from Arabidopsis leaves, suggesting that respiration can control ascorbate synthesis in plants. It is speculated that L-GalLDH activity is modulated by the redox state of complex I in Arabidopsis (Millar et al. 2003). Using an RNAi approach, it recently was shown that L-GalLDH, in addition to ascorbate synthesis, also plays an important role in the regulation of cell growth-related processes in plants (Alhagdow et al. 2007). Recent data also indicate a role of this protein for assembly of complex I (Pineau et al. 2008).

In *N. crassa* and mammalian complex I, a subunit homologous to an acyl carrier protein (ACP) involved in fatty acid and lipoic acid synthesis was reported to be associated with complex I (Brandt 2006 and citations therein). In *Arabidopsis thaliana*, three small ACPs are present in mitochondria, but in contrast to the situation in *N. crassa* and mammals no significant amounts of these proteins are associated to complex I (Millar et al. 2005; Meyer et al. 2007).

Complex I assembly so far is only partially understood. Some insights into this process were provided by the analysis of complex I mutants in maize, tobacco, and Arabidopsis and by searching for homologues of the assembly factors characterized for other model systems. Results were recently summarized in two extensive reviews (Vogel et al. 2007; Remacle et al. 2008). Similar to bacteria, eukaryotic complex I has an L-like shape and is composed of two arms: a hydrophilic "peripheral arm" exposed to the matrix side that contains the Fe-S clusters and a FMN center; and a hydrophobic "membrane arm" embedded in the inner mitochondrial membrane. Studies based on *N. crassa* mutants revealed that assembly of the full complex is proceeded by the interaction of the peripheral arm with two subcomplexes of the membrane arm referred to as large and small intermediate (Guenebaut et al. 1997; Sazanov and Walker 2000; Schulte 2001). Two chaperones termed *complex I intermediate assembly proteins* (CIA30 and CIA84) are bound to the large membrane arm intermediate in Neurospora. Disruption of either of the two CIA genes leads to a specific block of complex I assembly (Kuffner et al. 1998). Human orthologues of the CIA proteins have been identified (Gabaldon et al. 2005), but a functional role for complex I assembly only has been demonstrated for CIA30 (Dunning et al. 2007). CIA30 homologues also were described for several other organisms by a bioinformatic screen (Cardol et al. 2005). The Arabidopsis genome codes for three CIA30-like proteins which additionally comprise an N-terminal NAD(P)H binding site of unknown function (Schulte 2001).

13.2.2 Composition and Biogenesis of Complex II

Complex II or succinate-ubiquinone oxidoreductase is the smallest OXPHOS complex, but has a central role in mitochondrial metabolism because it is involved in both the tricarboxylic (TCA) acid cycle and the mitochondrial electron transport chain. Complex II catalyzes the oxidation of succinate to fumarate and the reduction of ubiquinone to ubiquinol. The activity of this complex is not coupled to the generation of an electrochemical gradient of protons across the membrane to drive the synthesis of ATP (Horsefield et al. 2004). Complex II is the simplest constituent of the electron transport chain. In bacteria and heterotrophic eukaryotes it contains only four subunits and five cofactors: one flavin adenine dinucleotide (FAD), three Fe-S clusters, and one b-type heme (Yankovskaya et al. 2003). Two of the four proteins protrude into the mitochondrial matrix and are anchored in the mitochondrial membrane by the other two proteins which are small and hydrophobic. The larger one of the hydrophilic proteins, termed SDH1, has a covalently bound FAD molecule. The other hydrophilic component, SDH2, is an iron-sulfur protein responsible for electron transport from FAD to the membrane domain of complex II. Three non-heme Fe-S clusters are involved in this process (Scheffler 1998). The two hydrophobic proteins (SDH3 and SDH4) constitute the ubiquinone binding site and contain a *b*-type heme of unknown function (Cecchini 2003; Yankovskaya et al. 2003; Horsefield et al. 2006).

In animals and fungi, all of complex II subunits are encoded in the nucleus (Scheffler 1998), whereas *sdh* genes have been identified in the mitochondrial genomes of the zooflagellate *Reclinomonas americana* (*sdh*2, *sdh*3 and *sdh*4) (Burger et al. 1996) and in the liverwort *Marchantia polymorpha* (*sdh*3 and *sdh*4) (Oda et al. 1992). On the contrary, all the functional genes encoding complex II proteins are encoded in the nuclear genome in Arabidopsis (Unseld et al. 1997) because the sequences for complex II genes found in the mitochondrial genome of higher plants all turned out to represent pseudogenes (Giegé et al. 1998). In higher plants, complex II comprises four extra subunits of unknown function (SDH5, SDH6, SDH7, SDH8) which might integrate secondary enzyme activities into the complex (Eubel et al. 2003; Millar et al. 2004).

In Arabidopsis, SDH1 and SDH3 are encoded by two and SDH2 even by three genes (designated *sdh1-1*, *sdh1-2*, *sdh*2-1, *sdh*2-2, *sdh*2-3, *sdh*3-*1*, and *sdh*3-2) (Figueroa et al. 2001, 2002; Huang et al. 2009a, b). The present data suggest that *sdh2-1* and *sdh2-2* exhibit similar expression patterns in all organs of adult plants

(Figueroa et al. 2001; Elorza et al. 2004), while *sdh2-3* is highly expressed in the embryo during the maturation phase of seed development (Elorza et al. 2006) and may have a specific role during the early stages of germination (Roschzttardtz et al. 2009). The *sdh1-1* and *sdh1-2* genes are very much conserved and expressed at high levels in roots and flowers, respectively (Figueroa et al. 2002). Analysis of Arabidopsis mutants revealed that SDH1-1 is especially important for gametophyte development (Leon et al. 2007).

Using a combined transcriptomic and proteomic approach, it recently was demonstrated that complex II from Arabidopsis and rice differ significantly in very important aspects as stability, number of subunits, and in the gene expression profiles of several gene orthologues (Huang et al. 2009b). The regulation of complex II gene expression therefore seems not to be conserved between monocot and dicot plants.

13.2.3 Composition and Biogenesis of Complex III

Ubiquinol produced by the action of complexes I and II is oxidized by complex III, also called cytochrome bc_1 complex or ubiquinol-cytochrome *c* oxidoreductase. The electrons from ubiquinol are transferred to cytochrome *c* and four protons are pumped for each pair of electrons passing through this multi-subunit complex (Tzagoloff 1995; Berry et al. 2000). Biogenesis of complex III was extensively studied in mammals and yeast, but unfortunately little is known about the corresponding process in plants. Complex III structure is highly conserved in different groups of eukaryotes. It functions as a homodimer of two identical units composed of 10–11 subunits each and has an overall molecular mass of about 500 kDa (Braun and Schmitz 1995a; Iwata et al. 1998; Dudkina et al. 2006). Its catalytic center is constituted by cytochrome *b*, cytochrome c_1, and the Rieske Fe-S protein. Cytochrome *b* is the only subunit encoded by mitochondrial DNA, whereas all the other subunits are of nuclear origin. The remaining seven to eight proteins do not have homologues in bacteria and their functions are largely unknown. The two largest of them, which are named "core proteins" because they originally were assumed to form the center of the complex, exhibit sequence similarity to the two subunits of the mitochondrial processing peptidase (MPP), a metallo-endopeptidase that specifically cleaves off presequences from several 100 mitochondrial precursor proteins. MPP is localized within the mitochondrial matrix in mammals and yeast. In contrast, higher plant MPP activity was shown to reside in the mitochondrial inner membrane where it is completely integrated into the bc_1 complex of the respiratory chain (Braun et al. 1992, 1995; Braun and Schmitz 1993,1995b; Eriksson et al. 1994). Indeed, the two core proteins of complex III from plants represent the MPP subunits in plants. The complex III therefore is bifunctional in this group of organisms. However, processing activity does not depend on the electron transfer activity of complex III (Glaser et al. 1994; Braun and Schmitz 1995b; Eriksson et al. 1996).

Accessory proteins that do not belong to the functional bc_1 complex are required for its assembly. In yeast, the Bcs1 chaperone is involved in the binding of the FeS protein to complex III. The FeS protein is believed to play a role in complex III dimerization. Moreover, mutations in two yeast proteins designated Cbp3 or Cbp4 cause severe structural defects in bc_1 assembly but no data are currently available on their specific role during assembly of this complex (Zara et al. 2009). Finally, complex III assembly also requires the correct insertion of its prosthetic groups which are essential for its function. In this respect, synthesis and maturation of *c*-type cytochromes are crucial steps (reviewed in Giegé 2007; Hamel et al. 2009).

13.2.4 Composition and Biogenesis of Complex IV

Complex IV or cytochrome *c* oxidase (COX) is the terminal enzyme of the mitochondrial electron transport chain. It belongs to the heme-copper *aa*3-type terminal oxidase superfamily and catalyzes the sequential transfer of electrons from reduced cytochrome *c* to di-oxygen, thereby generating water. This electron transfer reaction is coupled to electrogenic proton pumping across the inner mitochondrial membrane (Capaldi 1990; Barrientos et al. 2002; Herrmann and Funes 2005). The membrane embedded complex faces both the intermembrane space and the matrix, but slightly more emerges to the IMS. The biogenesis of COX is complicated because it consists of subunits encoded by two different genomes and it includes several prosthetic groups essential for its function, including two heme *a*, three copper ions, zinc, magnesium, and sodium ions (Carr and Winge 2003). There is little information about COX biogenesis in plants and most data were obtained by functional complementation of yeast mutants. The core of the enzyme complex is composed of three large and hydrophobic subunits (Cox1, Cox2, and Cox3) encoded by the mitochondrial genome (Capaldi 1990; Poyton and McEwen 1996; Adams and Palmer 2003). The amino acid sequences of these three proteins are highly conserved between different species. In prokaryotes the organization of the complex is simple because the counterparts of the three mitochondrial encoded subunits constitute the functional holo-enzyme. In eukaryotes, COX consists of 8–11 additional small peripheral subunits encoded by the nuclear genome (Millar et al. 2004; Khalimonchuk and Rödel 2005). The catalytic center of the eukaryotic enzyme is constituted by the mitochondrial encoded subunits which bind redox active metal centers (Fig. 13.1a). Cox1, the largest subunit of COX, is composed of 12 transmembrane domains and contains the heme-*a* and a bimetallic site composed of heme-*a3* and a copper ion (Cu_B). This protein appears to be the key subunit for COX assembly and participates in proton translocation via two pores, the D- and K-channels (Carr and Winge 2003; Khalimonchuk and Rödel 2005). The Cox2 subunit binds a binuclear Cu_A center which is exposed to the intermembrane space and constitutes the entry site for electrons which are delivered by cytochrome *c* (Poyton and McEwen 1996). Like Cox1, Cox3 also is a very hydrophobic protein spanning the inner mitochondrial membrane with seven transmembrane helices. It does not carry any prosthetic groups and is not directly involved

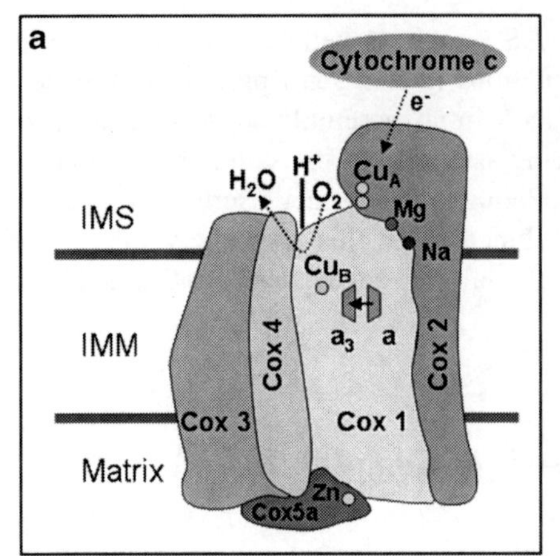

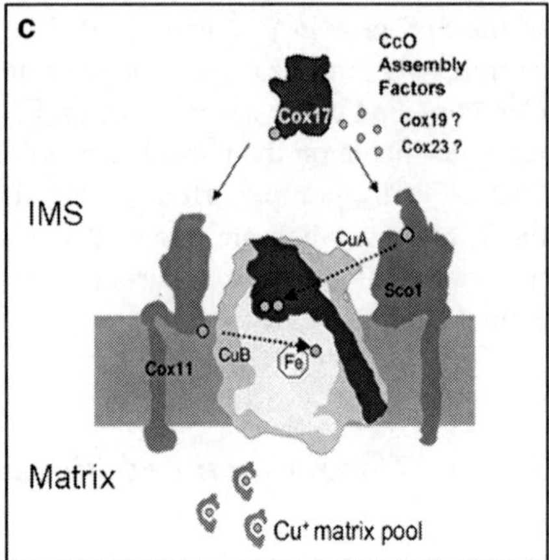

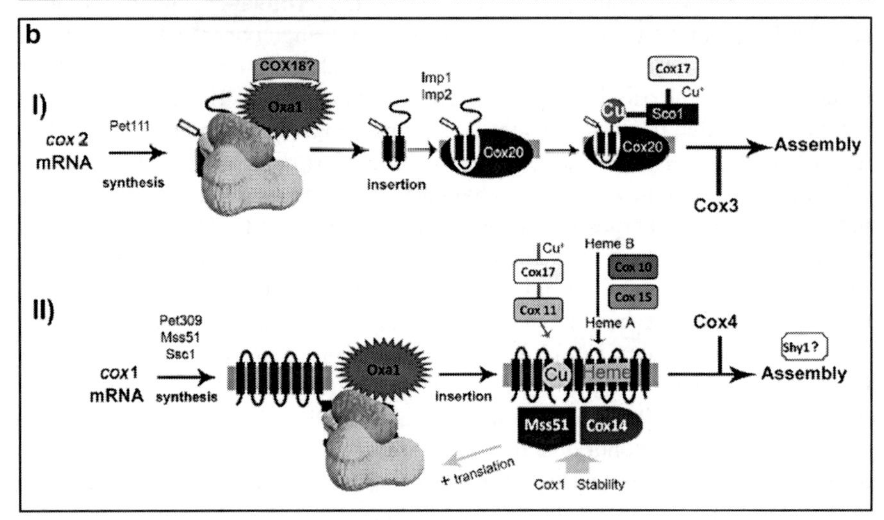

Fig. 13.1 Function and assembly of cytochrome c oxidase (complex IV). (**a**) Electron transfer within complex IV. Electrons are provided by the mobile intermembrane space protein cytochrome c and enter complex IV through the catalytic binuclear Cu_A center on the mitochondrial encoded Cox2 subunit. Final electron transfer steps and water reduction take place within Cox1 by its bimetallic center composed by the heme-aa_3 and the Cu_B. Cox1 also participates in proton translocation (*IMM*, Inner mitochondrial membrane; *IMS*, intermembrane space; modified from Khalimonchuk and Rödel 2005). (**b**) Model for the synthesis, maturation and assembly of the Cox1 and Cox2 subunits in *Saccharomyces cerevisiae*. Explanations are given in Sect. 13.2.4 (modified from Herrmann and Funes 2005). (**c**) Copper delivery to the catalytic subunits of complex IV in *Saccharomyces cerevisiae*: The metal chaperone Cox17 (and possibly further protein factors like Cox19 and Cox23) brings the Cu(I) atoms to the Cox11 and Sco1 proteins for its insertion into Cox1 and Cox2 subunits of the apoenzyme, respectively. Further explanations are given in Sect. 13.2.4

in proton translocation. Its specific role within COX so far is not known, but it was proposed that Cox3 is involved in the assembly and/or stability of COX (Meunier and Taanman 2002).

All nuclear encoded subunits are relatively small compared to the mitochondrial encoded proteins. Their molecular masses range from 5.4 to 14.9 kDa in yeast and from 5 to 17.1 kDa in mammals (Capaldi 1990). Despite of the fact that these subunits rather play an auxiliary role in the catalytic function of COX, studies based on yeast knock-out mutants have shown that they are indispensable for enzyme conformation and activity (Carr and Winge 2003). For instance, early stage COX assembly in mammals has been reported to depend on the Cox4 subunit (Nijtmans et al. 1998). In the last years it also has been proposed that several of the nuclear encoded COX subunits are important for the stability of the enzyme complex and for protecting the catalytic heart of the enzyme against ROS toxicity (Fontanesi et al. 2006).

Also in plants, COX includes several nuclear-encoded subunits, most of which have a molecular mass below 10 kDa. Using 2D Blue-native/SDS-PAGE (see Box 13.1 for introduction to this procedure), it was shown that complex IV of potato consist of 10 subunits and COX of *Arabidopsis thaliana* of even more subunits (Jänsch et al. 1996; Eubel et al. 2003; Millar et al. 2004). Several of the

Box 13.1 Characterization of the respiratory chain by Blue native PAGE

The most powerful procedure to comprehensively characterize the composition of the respiratory chain complexes and the ATP synthase complex is two-dimensional Blue-native/SDS PAGE (BN/SDS PAGE). The procedure was invented by Herman Schägger two decades ago (Schägger and von Jagow 1991). The starting point for BN PAGE is a purified protein fraction, e.g., from isolated mitochondria, which is directly incubated with a non-ionic detergent and with Coomassie blue. This already is the most important trick of the procedure. Normally, Coomassie-blue is used *after* an electrophoresis has been carried out to stain and thereby visualize proteins within a gel. However, in the case of BN/SDS PAGE, this incubation takes place *before* gel electrophoresis. Coomassie binds tightly to proteins and introduces a negative charge into them, because under the conditions used, it is a negatively charged compound. However, in contrast to SDS, which normally is used for charge introduction into proteins, Coomassie blue does not denature proteins and protein complexes. If Coomassie-pretreated protein fractions are separated on a polyacrylamide gel, resulting bands represent intact protein complexes. For the second gel dimension, stripes of the BN gel can be cut out and transferred horizontally onto a second gel dimension, which is carried out in the presence of SDS. Under these conditions, the protein complexes are dissected into their subunits, which are separated according to size. On a resulting two-dimensional gel, proteins belonging to a distinct protein complex form a vertical row of spots (see Fig. 13.2). BN/SDS PAGE was extensively used for the characterization of the respiratory chain. In combination with usage of gentle nonionic detergents, it also nicely allows to study its supramolecular structure (Eubel et al. 2003).

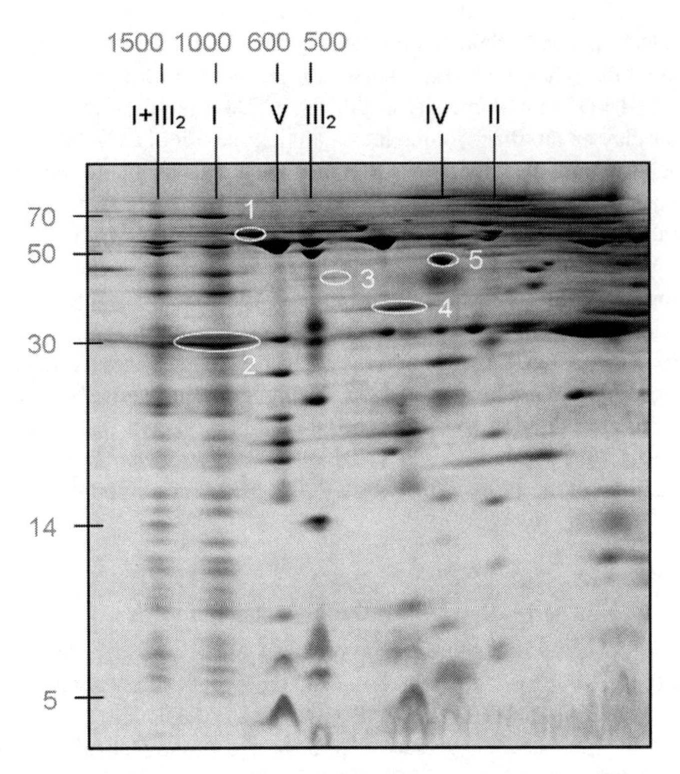

Fig. 13.2 Characterization of the respiratory chain of Arabidopsis mitochondria by 2D Blue native/SDS PAGE. Separation of the first gel dimension was from left to right (BN PAGE), separation of the second gel dimension (SDS PAGE) from top to bottom. The molecular masses of standard proteins (in kDa) are given in gray above the gel (first gel dimension) and left to the gel (second gel dimension). The protein complexes of the OXPHOS system are indicated in roman letters above the gel. Subunits belonging to other mitochondrial protein complexes are marked directly on the gel by white numbers. 1: HSP60 complex, 2: prohibitin complex, 3: glutamate dehydrogenase complex, 4: TOM complex, 5: formate dehydrogenase complex

nuclear encoded COX subunits of plants were first identified by homology searches using the sequences of COX genes from yeast and mammals, e.g., the Cox5b, Cox5c, Cox6a, and Cox6b proteins in potato, rice, and Arabidopsis (Nakagawa et al. 1990; Kadowaki et al. 1996; Ohtsu et al. 2001; Curi et al. 2003). In Arabidopsis it was found that transcript levels of nuclear genes encoding subunits 5b, 6a, and 6b increase when plants are incubated with carbohydrates, but this was not observed for the organelle-encoded cox2 gene (Welchen et al. 2002; Curi et al. 2003). Furthermore, several nuclear genes show increased expression in anthers and are induced by carbohydrates, light, UV-B, and hormones (Welchen et al. 2004; Comelli et al. 2009; Mufarrege et al. 2009). Some additional plant specific nuclear encoded COX subunits were detected by mass spectrometry, but their function currently is not known (Millar et al. 2004).

Biogenesis of a functional cytochrome *c* oxidase complex depends on the presence of all its structural subunits and more than two dozen COX-specific assembly factors (Fig. 13.1b). The multiple levels of regulation described to date involve the availability of subunits and assembly factors regulated at the transcriptional and translational levels, availability of cofactors, protein import into mitochondria, and protein insertion into the inner mitochondrial membrane. In the last years excellent reviews were published on COX biogenesis in humans and yeast (Herrmann and Funes 2005; Khalimonchuk and Rödel 2005; Fontanesi et al. 2006, 2008; Barrientos et al. 2009).

The three mitochondrial encoded COX subunits are synthesized on mitochondrial ribosomes which are associated with the inner mitochondrial membrane. The corresponding mRNAs are directed to the membrane by a set of specific membrane-bound translational activators that in yeast interact with the 5′-untranslated region (UTR) of the COX transcripts. Translation of Cox1 mRNA is activated by Pet309, Mss51, and Ssc1 (Fontanesi et al. 2010), translation of Cox2 by Pet111 and translation of Cox3 by Pet54, Pet122, and Pet494 (Khalimonchuk and Rödel 2005 and references therein).

Upon or during their synthesis the mitochondrial encoded COX subunits undergo a translocation/insertion step to attain their proper orientation in the inner mitochondrial membrane. Translocation of the Cox1, Cox2, and Cox3 subunits requires the function of Oxa1 that transiently interacts with nascent polypeptides during their translation. Oxa1 is required for cotranslational translocation and insertion of Cox2, while a distant homologue Cox18 (OXa2 in mammals) is necessary for the export of its C-terminal domain (Bonnefoy et al. 2009). Upon emerging in the IMS, the N-terminus of Cox2 is processed yielding the mature form of Cox2 that can be inserted into COX. The cleavage step is performed by Imp1/Imp2 protease which resides on the outer surface of inner mitochondrial membrane (Gakh et al. 2002). Following its translocation Cox2 associates with Cox20, a membrane-bound chaperone required for processing (Hell et al. 2000). In parallel, Cox1 insertion appears to depend on the translation activator Mss51 that interacts with Cox14 and facilitates membrane insertion of the nascent polypeptide. Another factor, Shy1, which also is necessary for COX assembly, may cooperate with the Cox1/Mss51/Cox14 complex during Cox1 insertion (Barrientos et al. 2009; Fontanesi et al. 2010). Finally, the Cox1 and Cox2 subunits are equipped with their respective cofactors on the basis of independent assembly lines. Prior to their assembly with the other COX subunits, these two subunits remain bound to specific chaperones, Mss51 for Cox1 and Cox20 for subunit Cox2, to prevent potentially harmful redox reactions of the unassembled subunits (Herrmann and Funes 2005).

Mitochondrial copper homeostasis, heme-*a* biosynthesis, and insertion of both prosthetic groups into the corresponding newly synthesized subunits forming the catalytic core of the enzyme involve several pathways that must be precisely coordinated (Fig. 13.1c). Copper is a metal essential for COX biogenesis and function (Cobine et al. 2006). In yeast, mitochondrial copper acquisition and insertion into the apoenzyme depends on two proteins, Cox11 and Sco1, that mediate Cu(I) insertion into Cox1 and Cox2, respectively (Glerum et al. 1996a; Hiser et al. 2000; Horng et al. 2004). Furthermore, additional metallochaperones were identified which have conserved amino acid sequences and bind Cu(I) by a characteristic

Cys-Cys-X-Cys domain. The first characterized member of this metallochaperone family was a small and hydrophilic protein termed Cox17 in yeast (Glerum et al. 1996b; Abajian et al. 2004; Horng et al. 2004). Involvement of Cox17 in copper homeostasis was originally proposed because the respiratory-deficient phenotype of a cox17 null mutant can be rescued by supplementation of copper to the medium (Glerum et al. 1996b). Yeast Cox17 is localized in two cellular compartments, the cytoplasm and the mitochondrial intermembrane space. This localization reflects its suggested function as a soluble copper chaperone, probably shuttling copper from the cytoplasm into the mitochondrial intermembrane space and finally transferring Cu(I) to the Cox11 and Sco1 proteins (Glerum et al. 1996a, Beers et al. 1997; Horng et al. 2004). Further proteins that are required for COX assembly contain a Cys-X_9-Cys motif: Cox19 (Nobrega et al. 2002; Rigby et al. 2007), Cox23 (Barros et al. 2004), Mia40 (Grumbt et al. 2007), Pet191 (Khalimonchuk et al. 2008), and Cmc1 (Horn et al. 2008). Albeit the precise role of these proteins is not completely known, they most likely mediate trafficking of copper within the mitochondrial intermembrane space (Leary et al. 2009).

In Arabidopsis, among the other proteins implicated in complex IV assembly that were mentioned above, homologues of Oxa1, Cox10, Cox11, Cox15, Cox17, Cox19, Cox23, Sco1, and Cmc1 are encoded by its nuclear genome. AtOxa1 as well as the AtCox17 and AtCox19 proteins can complement the corresponding yeast mutants (Hamel et al. 1997; Balandin and Castresana 2002; Attallah et al. 2007a,b). Moreover, AtCox19 proteins can be in vitro imported into isolated mitochondria and are afterwards attached to the inner mitochondrial membrane (Attallah et al. 2007a).

13.2.5 Composition and Biogenesis of Complex V

The mitochondrial ATP synthase, also named F_1F_0-ATP synthase or complex V, is located in the inner mitochondrial membrane together with the protein complexes of the respiratory chain. It is responsible for the terminal step of the OXPHOS process, the phosphorylation of ADP, by using the proton gradient across the inner mitochondrial membrane which is generated by the complexes I to IV. The general structure of complex V is highly conserved in both prokaryotic and eukaryotic organisms. Like the bacterial enzyme complex, mitochondrial F_1F_0-ATP synthase is composed of two domains, one within the inner mitochondrial membrane termed F_0 and the other within the mitochondrial matrix termed F_1. The two domains are linked via a central stalk and a peripheral stalk plus the OSCP (oligomycin sensitive conferring protein) accessory subunit (see Devenish et al. 2008 for review). F_1 is hydrophilic and can be removed from the mitochondrial inner membrane as a stable subcomplex. It consists of five different subunits designated α, β, γ, δ, and ε in 3:3:1:1:1 stoichiometry (Arnold et al. 1999; Velours and Arselin 2000). The F_0 domain consists of three different subunits in *E. coli*, termed a, b, and c, which occur in 1:1:12 stoichiometry. In mammals and yeast the F_0 domain contains a central core of the same subunits but with a slightly different stoichiometry of 1:2:9–12. Additionally, a series of associated proteins form part of eukaryotic F_1F_0 ATP synthase like the so-called IF1 protein (Stock et al. 1999; Velours and Arselin 2000). Recent data have led to a modified view about the ATP synthase structure; the catalytic hexagon $\alpha_3\beta_3$ is attached to the membrane-embedded F_0-a subunit via two stalks: the rotor stalk $\gamma\varepsilon c_{ring}$ and the "stator" stalk δb_2. Rotation of the c-subunit ring relative to the F_0-a subunit is critical for the proton translocation event (Weber and Senior 2003; Pickova et al. 2005).

The plant mitochondrial ATP synthase also displays the classical F_1 five subunit structure. Also, the OSCP (termed δ′ in plants) and IF1 subunits have been identified in the plant F_1F_0 ATP synthase complex (Jänsch et al. 1996; Heazlewood et al. 2003b). For the F_0 domain, a search of the Arabidopsis genome identified sequence orthologues for some of the known subunits identified in mammals and yeast. By using a proteomic approach, a series of nine known subunits were identified along with two additional new proteins unique to the F_1F_0 complex in Arabidopsis (Heazlewood et al. 2003b).

The assembly of complex V has not been investigated in plants but many insights were provided by the analysis of respiratory deficient mutants in yeast. In *S. cerevisiae*, several chaperone-like proteins are essential for the biogenesis of the ATP synthase: Atp10 (Ackerman and Tzagoloff 1990), Atp11 and Atp12 (Ackerman 2002), ATP22 (Helfenbein et al. 2003), Fmc1 (Lefebvre-Legendre et al. 2001), Atp23 (Zeng et al. 2007), Oxa1 (Jia et al. 2007), and others. Homologues for many of these proteins were identified in other organisms (Pickova et al. 2005). The mechanism of F_1 assembly seems to be conserved in all eukaryotic lineages investigated and involves the functions of Atp11 and Atp12, which mediate the formation of the $\alpha_3\beta_3$ core of the F_1 domain. In the absence of Atp11 and Atp12 the αβ hexamer is not formed and α and β subunits form large insoluble aggregates (Ackerman 2002). In contrast, composition of the F_0 domain as well as its assembly is more versatile. In yeast, the Atp10 and Atp22 proteins are essential for the formation of the F_0 part. Atp10p is required for the incorporation of the F_0-a subunit (Pickova et al. 2005). The metalloprotease Atp23, which is conserved from yeast to humans, seems to act as a chaperone in conjunction with Atp10 and mediates the association of "subunit 6" with "subunit 9" (= subunit c) ring during F_0 assembly (Zeng et al. 2007).

Furthermore, it recently was demonstrated that Oxa1, but not Cox18/Oxa2, directly supports the assembly of the membrane embedded F_0-sector. Oxa1 was found to physically interact with newly synthesized mitochondrial encoded Atp9 (subunit c) protein in a posttranslational manner to maintain the assembly competence of the Atp9-F_1 subcomplex for its association with Atp6 and to ensure the efficient assembly of the complex. Moreover, the Oxa1-ATP synthase association may also support the activity of Oxa1, either in a bioenergetics manner or by regulating the activity of Oxa1 as a translocase for the complex IV assembly, possibly to maintain a balance between the biogenesis levels of COX and ATP synthase complexes (Jia et al. 2007).

13.3 Supermolecular Organization of the Mitochondrial OXPHOS System

The protein complexes of the respiratory chain were biochemically first described in the early 1960s upon subfractionation of mitochondrial membranes from beef using bile salts like cholate or deoxycholate (reviewed in Hatefi 1985). In some cases, defined combinations of OXPHOS complexes were obtained, e.g., NADH – cytochrome c oxidoreductase, which represents a copurification of the complexes I and III (Hatefi et al. 1961; Fowler and Richardson 1963; Hatefi and Rieske 1967). Later, the five OXPHOS complexes also were biochemically purified by the use of detergents in combination with chromatographic procedures. Since they all represent stable protein particles easily separable from each other upon membrane solubilization, their separate existence under in vivo condition was suggested. This view was further supported by activity measurements of the respiratory protein complexes in inner membrane vesicles during lipid dilution experiments (Hackenbrock et al. 1986). According to the "fluid state model," the protein complexes of the respiratory chain independently move within the lipid bilayer of the inner mitochondrial membrane and electron transport is based on random collisions between them and the mobile electron carriers ubiquinone and cytochrome c (also called the "random collision model" of respiration). However, several other experimental observations rather indicated occurrence of stable interactions of respiratory protein complexes which led to the suggestion of an alternative "solid state model". These observations include, among others, the copurification of defined respiratory protein complexes (Hatefi et al. 1961; Fowler and Richardson 1963; Hatefi and Rieske 1967), high electron transfer activities of defined combinations of respiratory complexes during reconstitution experiments (Fowler and Hatefi 1961; Hatefi et al. 1962; Ragan and Heron 1978) and the results of noninvasive flux control experiments (Boumans et al. 1998; Genova et al. 2003; Bianchi et al. 2004). The presence of defined assemblies of OXPHOS complexes termed respiratory supercomplexes was recently very much supported by native gel electrophoresis procedures and by single particle electron microscopy. Using several experimental systems the following supercomplexes could be described: (1) dimeric complex III + complex I, (2) dimeric complex III + one or two copies of monomeric complex IV, (3) dimeric complex III + complex I + one to four copies of complex IV, and (4) dimeric ATP synthase (summarized in several recent reviews, e.g., Boekema and Braun 2007; Dudkina et al. 2008, 2010b; Vonck and Schäfer 2009). However, not all OXPHOS complexes form part of supramolecular structures at all time periods. Rather, separate OXPHOS complexes coexist with supercomplexes which are dynamically made and degraded depending on the physiologic state of mitochondria and the ambient cell (Fig. 13.3). Finally, recent data indicate interactions of supercomplexes forming even larger structures called "respiratory strings" or "respiratory megacomplexes" (Bultema et al. 2009; Wittig and Schägger 2009). Formation of these structures is assumed to represent a major determent for cristae structure of the inner mitochondrial membrane.

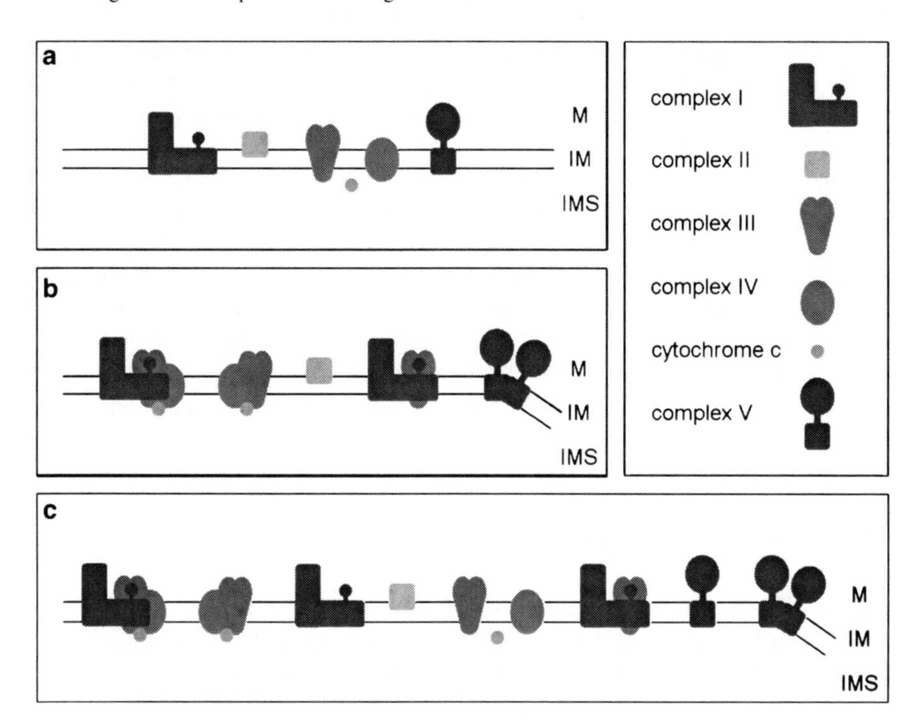

Fig. 13.3 Schematic model of the mitochondrial OXPHOS system. (**a**) Fluid-state model. Based on this model, the respiratory protein complexes are present separately and electron transfer takes place based on random collisions of the involved components. (**b**) Solid-state model: the OXPHOS complexes stably interact forming OXPHOS supercomplexes as described in the text. (**c**) Integrated model of the OXPHOS system. Singular OXPHOS complexes and supercomplexes coexist. Supercomplexes are dynamically formed and degraded (*IM*, Inner mitochondrial membrane; *IMS*, intermembrane space; *M*, matrix)

13.3.1 The I + III Supercomplex

Association of complexes I and III is supported by reconstitution and flux control experiments and by their copurification (Hatefi et al. 1961; Fowler and Richardson 1963; Hatefi and Rieske 1967; Boumans et al. 1998; Genova et al. 2003; Bianchi et al. 2004). Gentle solubilization of mitochondrial membranes using dodecylmaltoside, Triton X-100 or digitonin allowed resolving a supercomplex composed of monomeric complex I and dimeric complex III by Blue native PAGE for beef and several other organisms (Schägger and Pfeiffer 2000; Eubel et al. 2003). The I + III_2 supercomplex has a molecular mass of about 1,500 kDa and so far was not found to include additional subunits except the ones reported to be present in monomeric complexes I and dimeric III. A first structural characterization of the I + III_2 supercomplex was carried out by single particle electron microscopy (Dudkina et al. 2005a). For this procedure, tens of thousands of EM photographs are taken from a purified particle, classified according to orientation, and used for the calculation of average structures of very high resolution. Within the I + III_2 supercomplex, dimeric complex III is laterally associated to the membrane arm of complex I. The membrane arm is slightly bent around the complex III dimer. The peripheral arm of complex I and the core protein domain of complex III both protrude into the mitochondrial matrix but do not physically interact. The ubiquinol binding pocket of complex III and the assumed location of the ubiquinone binding site at complex I are in close proximity. However, it currently is not known whether electron transport within the I + III_2 supercomplex is based on direct ubiquinone/ubiquinol channeling.

13.3.2 The III + IV Supercomplex

Flux control experiments revealed that the respiratory chain of *Saccharomyces cerevisiae*, which lacks complex I, behaves like a single functional unit (Boumans et al. 1998). Presented results indicate tight interaction of complex III, complex IV and cytochrome c. Similar results were obtained upon membrane solubilization by detergents and separation of protein complexes by Blue native PAGE (Schägger and Pfeiffer 2000; Cruciat et al. 2000; Eubel et al. 2004). In beef, yeast and Arabidopsis, two major forms of III + IV supercomplexes occur which are composed of one complex III dimer and one or two monomers of complex IV. The supercomplexes have molecular masses of about 850 and 700 kDa. Recently, EM average structures were presented for yeast (Heinemeyer et al. 2007). The III_2 + IV_2 supercomplex has a very symmetric shape. The complex III dimer is in the center; the complex IV momomers bind to the complex III dimer at exactly opposite sites. Base on a structural model, which was calculated using the crystal structures available for complexes III and IV, the cytochrome c binding pockets on the surface of the two complexes are in close proximity allowing electron transport by a short rage ping-pong movement of cytochrome c. Indeed, some EM images show an additional protein density at the position of the binding pockets for cytochrome c (Heinemeyer et al. 2007). Therefore, the III + IV supercomplexes most likely form the scaffold for a very efficient electron transport mechanism mediated by the mobile electron carrier cytochrome c.

13.3.3 The I + III + IV Supercomplex

Even larger supercomplexes of greater than 1,500 kDa were described by Blue native PAGE analyses for beef, Arabidopsis, and several further organisms (Schägger and Pfeiffer 2000; Eubel et al. 2004). They were found to be composed of complex I, one copy of dimeric complex III, and one to four copies of monomeric complex IV. The term *respirasome* was suggested for these supercomplexes, because they autonomously can carry out respiration in the presence of the mobile electron transporters ubiquinol and cytochrome c (Schägger and Pfeiffer 2000). Low resolution EM structures were meanwhile generated for beef and potato (Schäfer et al. 2006; Dudkina et al. 2008). Based on modeling using crystal structures of singular OXPHOS complexes an internal architecture of the EM-derived supercomplex structures was suggested. Like within the I + III_2 supercomplex, dimeric complex III is bound to the membrane arm of complex I which is bend around complex III at the site of interaction. Complex IV is bound to complex III_2 and at the same time laterally attached to the tip of the membrane arm of complex I. The physiologic role of the respirasomes, which is discussed to be related to substrate and electron channeling, has to be further investigated.

13.3.4 The Dimeric ATP Synthase Supercomplex

A dimer of ATP synthase first was reported upon mild solubilization of mitochondrial membranes using nonionic detergents and their subsequent resolution by Blue native PAGE (Arnold et al. 1998, 1999). Meanwhile, these dimers also were described for many other organisms. Dimeric ATP synthase of yeast has a molecular mass of about 1,200 kDa and can be dissociated into monomers of 600 kDa upon high concentration detergent treatment. Interestingly, the ATP synthase dimers of yeast include some dimer-specific subunits, e.g., subunit g. If the gene encoding this subunit is deleted, dimers are not formed anymore (Arnold et al. 1998). At the same time, mitochondrial ultrastructure is very much altered which led to the proposal that cristae formation might be disturbed in the absence of dimers (Giraud et al. 2002; Paumard et al. 2002). A similar ultrastructural phenotype was reported upon in vivo cross-linking of the F_1 headpieces of ATP synthase complexes (Gavin et al. 2004). It therefore was postulated that interaction of the ATP synthase monomers within dimers in vivo rather takes place between the membrane-bound F_0 parts of ATP synthase complexes and that this interaction might cause a local bend in the inner mitochondrial membrane. EM average structures obtained for beef, the algae *Polytomella* and yeast indeed support this conclusion (Minauro-Sanmiguel et al. 2005; Dudkina et al. 2005b, 2006; Thomas et al. 2008). In all investigated organisms, the long axes of ATP synthase complexes form an angle of 30–90° within the dimers. The physiologic relevance of the ATP synthase dimers might therefore be to shape the inner mitochondrial membrane and allow cristae formation. Indeed, long rows of laterally associated ATP synthase dimers were observed using high-resolution EM techniques (Allen et al. 1989; Allen 1995; Strauss et al. 2008; Dudkina et al. 2010a). Formation of ATP synthase supercomplexes most likely is essential to maintain fully bioenergetically competent mitochondria, to allow the formation of respiratory supercomplexes and thereby to mediate fast metabolite/substrate channeling (Bornhövd et al. 2006; Wittig and Schägger 2009).

13.4 Regulation of the Biogenesis of the OXPHOS Complexes and Supercomplexes

The particular situation of respiratory enzymes arranged in a series of protein complexes and supercomplexes whose components are encoded by more than 100 genes localized on two different genomes requires a large degree of coordination with respect to biogenesis. Several efforts have been made to elucidate this process in eukaryotes but many questions remain to be answered. A significant number of studies have shown that mitochondrial respiration is modulated in plant cells according to developmental stages (Logan et al. 2001; Howell et al. 2006), in response to metabolic or environmental stimuli (Curi et al. 2003; Giegé et al. 2005; Welchen and Gonzalez 2005, 2006; Gonzalez et al. 2007; Mufarrege et al. 2009; Welchen et al. 2009) and in response to stress (Sweetlove et al. 2002; Attallah et al. 2007a,b; Ho et al. 2008; Comelli et al. 2009). Furthermore, several authors have also described changes in the expression of nuclear genes encoding mitochondrial proteins caused by signals generated in the organelle (Yu et al. 2001; Gomez Casati et al. 2002; Lister et al. 2004). This retrograde regulation is based on chemical cross-talk between the mitochondrion and the nucleus to adjust the expression of specific genes according to the conditions required within the organelle (see also Chaps. 14 and 16).

In the last years much was speculated about the possible coordination between mitochondria and nucleus to adjust the synthesis of different proteins forming part of the OXPHOS machinery. Investigations addressing this issue indicate that the subunits made within mitochondria are synthesized in excess and that their expression is not coordinated with the protein biosynthesis of their nuclear encoded counterparts (Giegé et al. 2005). In this way, the limiting factors during the assembly of new complexes are the nuclear components and the regulation of this process rather takes place at the posttranslational level. Nonetheless, expression of the genes encoding nuclear encoded OXPHOS subunits is regulated at the transcriptional level through the interaction of common regulatory elements present in their promoter regions with specific transcription factors (Welchen and Gonzalez 2006; Gonzalez et al. 2007; Comelli and Gonzalez 2009).

Almost as a general rule, nuclear genes for respiratory chain components show increased expression in anthers and are induced by carbohydrates (Zabaleta et al. 1998; Curi et al. 2003; Elorza et al. 2004; Welchen et al. 2004, 2009; Welchen and Gonzalez 2005; Mufarrege et al. 2009). This is in accordance with the increased number of mitochondria per cell needed for pollen maturation (Lee and Warmke 1979) and with increased respiration induced by up-regulated sugar metabolism (see also Chap. 18). Analyses of the promoter regions of different genes involved in OXPHOS assembly from both Arabidopsis and rice led to the discovery of a common regulatory element known as *site* II (TGGGCY) usually in more than one copy (Welchen and Gonzalez 2006). *Site* II elements have been previously described as regulator elements for expression in proliferating tissues (Tremousaygue et al. 2003). Mutagenesis studies allowed demonstrating involvement of *site* II elements in regulating nuclear gene expression in vivo, either at basal level or in response to several environmental conditions. They regulate gene expression in coordination with one or several additional regulatory elements also present in the promoter regions (Welchen and Gonzalez 2005; Comelli et al. 2009; Comelli and Gonzalez 2009; Mufarrege et al. 2009; Welchen et al. 2009).

A further level of complexity concerning OXPHOS gene regulation is added due to the fact that isoforms for most components of the respiratory chain occur in plants which most likely arose by numerous gene duplications during evolution. It was proposed that these gene duplications were the basis for gene alterations allowing "subfunctionalization" or "neofunctionalization" of genes to acquire more plasticity and new expression characteristics (Haberer et al. 2004; Duarte et al. 2006). There are several examples for specialization between genes related to the OXPHOS process (Elorza et al. 2004, 2006; Mufarrege et al. 2009; Welchen et al. 2009). Moreover, regarding to the effect of *site II* elements on the expression of duplicated genes in response to the availability of carbohydrates, an evolutionary model was recently postulated: One of the genes retained ancient patterns in the expression, while the other one incorporated novel regulatory elements at the promoter level that allowed a progressive divergence of regulatory mechanisms and the acquisition of new properties (Comelli and Gonzalez 2009).

However, despite all efforts to uncover the regulatory mechanisms that assure coordinated expression of the mitochondrial and the nuclear genomes for OXPHOS assembly, many aspects of this process still are not quite understood. This especially is true for supercomplex assembly. Does supercomplex formation require fully assembled respiratory complexes? Interestingly, genetic alterations causing loss of complexes III and IV seem to disturb assembly of complex I (Acín-Pérez et al. 2004; Diaz et al. 2006; Li et al. 2007). Therefore, supercomplex formation might be an essential step for building up a functional OXPHOS apparatus. Also, changes in complex I amount lead to reduced levels of complexes III and IV (Ugalde et al. 2004a). On the other hand, some mutations in complex I do not have a visible effect on the amount of any of the other OXPHOS complexes (Perales et al. 2005). Although not completely clarified, these findings seem to indicate that only some of the complex I subunits affect assembly of other OXPHOS complexes, most likely those which are at the interphases of OXPHOS complexes within the supercomplexes. Further interesting results on supercomplex assembly were obtained by analyzing the time course of incorporation of the 13 mitochondrial encoded OXPHOS subunits into respiratory complexes and supercomplexes. Experimental findings indicate that supercomplexes are structures formed by the specific and defined association of individual complexes and there is a temporal gap between the formation of the individual complexes and the one of the supercomplexes (Acín-Pérez et al. 2008).

13.5 Outlook

Within the last 10 years, several new insights were obtained into the assembly processes of the OXPHOS system and its supramolecular structure but at the same time many basic questions are still unresolved. For instance, transcription factors that regulate expression of nuclear genes encoding OXPHOS components are largely unknown. High-resolution structures are necessary to better understand the physiological role of OXPHOS supercomplexes. To elucidate the in vivo configuration of the OXPHOS system, noninvasive methods will be of great importance. Although of central impact for cellular metabolism, comprehensive knowledge on the OXPHOS system is not available today and will require large efforts in future basic research.

Glossary

Carbonic anhydrase (CA): Enzyme that catalyzes the interconversion of carbon dioxide and bicarbonate and protons.

Cofactor: A nonpeptide component that is bound to a protein and is essential for its biological activity; tightly bound cofactors are also called *prosthetic groups;* loosely attached cofactors are called *coenzymes.*

Fluid state model (also called the "random collision model"): According to this model, the respiratory chain complexes move independently within the inner mitochondrial membrane, and electron transfer is based on random collisions between them, ubiquinol, and cytochrome c.

Mitochondrial processing peptidase (MPP): Metallo-endopeptidase that cleaves off presequences from mitochondrial precursor proteins; in plants it is associated to complex III of the OXPHOS system.

OXPHOS complexes: Protein complexes constituting the OXPHOS system termed complex I (NADH:ubiquinone oxidoreductase), complex II (succinate: ubiquinone oxidoreductase), complex III (cytochrome bc_1 complex or ubiquinol-cytochrome *c* oxidoreductase), complex IV (cytochrome *c* oxidase), and complex V (ATP synthase or F_1F_0-ATP synthase).

OXPHOS supercomplex: Defined assembly of several OXPHOS complexes.

OXPHOS system: The oxidative phosphorylation system catalyzes O_2 dependent ATP formation. It is composed of the respiratory chain complexes and ATP synthase complex.

Respirasome: OXPHOS supercomplex that contains one copy of complex I, dimeric complex III and 1–4 copies of complex IV. It is able to autonomously carry out respiration in the presence of ubiquinol and cytochrome c.

Respiratory strings (also calles "megacomplexes"): Associations of OXPHOS supercomplexes.

Solid state model: According to this model, stable interactions between the respiratory chain complexes occur, thereby forming respiratory supercomplexes.

Substrate channeling or electron channeling: The optimized exchange of substrates or electrons between enzymes due to their assembly within protein complexes or supercomplexes.

References

Abajian, C., Yatsunyk, L. A., Ramirez, B. E., Rosenzweig, A. C. 2004. Yeast *cox17* solution structure and Copper(I) binding. J. Biol. Chem. 279:53584–53592.

Abdel-Ghany, S. E., Ye, H., Garifullina, G. F., Zhang, L., Pilon-Smits, E. A., Pilon, M. 2005. Iron–sulfur cluster biogenesis in chloroplasts. Involvement of the scaffold protein CpIscA. Plant Physiol. 138:161–172.

Acín-Pérez, R., Bayona-Bafaluy, M. P., Fernandez-Silva, P., Moreno-Loshuertos, R., Perez-Martos, A., Bruno, C., Moraes, C. T., Enriquez, J. A. 2004. Respiratory complex III is required to maintain complex I in mammalian mitochondria. Mol. Cell 13:805–815.

Acín-Pérez, R., Fernández-Silva, P., Peleato, M. L., Pérez-Martos, A., Enriquez, J. A. 2008. Respiratory active mitochondrial supercomplexes. Mol. Cell 32:529–539.

Ackerman, S. H. 2002. Atp11p and Atp12p are chaperones for F_1-ATPase biogenesis in mitochondria. Biochim. Biophys. Acta 1555:101–105.

Ackerman, S. H., Tzagoloff, A. 1990. *ATP10*, a yeast nuclear gene required for the assembly of the mitochondrial F_1–F_0 complex. J. Biol. Chem. 265:9952–9959.

Adams, K. L., Palmer, J. D. 2003. Evolution of mitochondrial gene content: gene loss and transfer to the nucleus. Mol. Phylogenet. Evol. 29:380–395.

Alhagdow, M., Mounet, F., Gilbert, L., Nunes-Nesi, A., Garcia, V., Just, D., Petit, J., Beauvoit, B., Alisdair R. Fernie, A. R., Rothan, C., Baldet, P. 2007. Silencing of the mitochondrial ascorbate synthesizing enzyme L-Galactono-1,4-lactone dehydrogenase affects plant and fruit development in tomato. Plant Physiol. 145:1408–1422.

Allen, R. D. 1995. Membrane tubulation and proton pumps. Protoplasma 189:1–8.

Allen, R. D., Schroeder, C. C., Fok, A. K. 1989. An investigation of mitochondrial inner membranes by rapid-freeze deep-etch techniques. J. Cell Biol. 108:2233–2240.

Arnold, I., Pfeiffer, K., Neupert, W., Stuart, R. A., Schägger, H. 1998. Yeast mitochondrial F_1F_0-ATP synthase exists as a dimer: identification of three dimer-specific subunits. EMBO J. 17:7170–7178.

Arnold, I., Pfeiffer, K., Neupert, W., Stuart, R. A., Schägger, H. 1999. ATP synthase of yeast mitochondria. Isolation of subunit j and disruption of the *ATP18* gene. J. Biol. Chem. 274:36–40.

Attallah, C. V., Welchen, E., Gonzalez, D. H. 2007a. The promoters of *Arabidopsis thaliana* genes At*COX17-1* and *-2*, encoding a copper chaperone involved in cytochrome *c* oxidase biogenesis, are preferentially active in roots and anthers and induced by biotic and abiotic stress. Physiol. Plant. 129:123–134.

Attallah, C. V., Welchen, E., Pujol, C., Bonnard, G., Gonzalez, D. H. 2007b. Characterization of *Arabidopsis thaliana* genes encoding functional homologues of the yeast metal chaperone Cox19p, involved in cytochrome *c* oxidase biogenesis. Plant Mol. Biol. 65:343–355.

Balandin, T., Castresana, C. 2002. At*COX17*, an Arabidopsis homolog of the yeast copper chaperone COX17. Plant Physiol. 129:1852–1857.

Barrientos, A., Barrios, M. H., Valnot, I., Rötig, A., Rustin, P., Tzagoloff, A. 2002. Cytochrome oxidase in health and disease. Gene 286:53–63.

Barrientos, A., Gouget, K., Horn, D., Soto, I. C., Fontanesi, F. 2009. Suppression mechanisms of COX assembly defects in yeast and human: insights into the COX assembly process. Biochim. Biophys. Acta 1793:97–107.

Barros, M. H., Johnson, A., Tzagoloff, A. 2004. COX23, a homologue of COX17, is required for cytochrome oxidase assembly. J. Biol. Chem. 279:31943–31947.

Bartoli, C. G., Pastori, G. M., Foyer, C. H. 2000. Ascorbate biosynthesis in mitochondria is linked to the electron transport chain between complexes III and IV. Plant Physiol. 123:335–344.

Beers, J., Glerum, D. M., Tzagoloff, A. 1997. Purification, characterization, and localization of yeast Cox17p, a mitochondrial copper shuttle. J. Biol. Chem. 272:33191–33196.

Berry, E. A., Guergova-Kuras, M., Huang, L. S., Crofts, A. R. 2000. Structure and function of cytochrome *bc* complexes. Annu. Rev. Biochem. 69:1005–1075.

Bianchi, C., Genova, M. L., Parenti Castelli, G., Lenaz, G. 2004. The mitochondrial respiratory chain is partially organized in a supercomplex assembly: kinetic evidence using flux control analysis. J. Biol. Chem. 279:36562–36569.

Boekema, E. J., Braun, H. P. 2007. Supramolecular structure of the mitochondrial Oxidative Phosphorylation System. J. Biol. Chem. 282:1–4.

Bonnefoy, N., Fiumera, H. L., Dujardin, G., Fox, T. D. 2009. Roles of Oxa1-related inner-membrane translocases in assembly of respiratory chain complexes. Biochim. Biophys. Acta 1793:60–70.

Bornhövd, C., Vogel, F., Neupert, W., Reichert, A. S. 2006. Mitochondrial membrane potential is dependent on the oligomeric state of F_1F_0-ATP synthase supracomplexes. J. Biol. Chem. 281:13990–13998.

Boumans, H., Grivell, L. A., Berden, J. A. 1998. The respiratory chain in yeast behaves as a single functional unit. J. Biol. Chem. 273:4872–4877.

Brandt, U. 2006. Energy converting NADH: quinone oxidoreductase (complex I). Annu. Rev. Biochem. 75:69–92.

Braun, H. P., Schmitz, U. K. 1993. Purification and sequencing of cytochrome *b* from potato reveal methionine cleavage of a mitochondrially encoded protein. FEBS Lett. 316:128–132.

Braun, H. P., Schmitz, U. K. 1995a. The bifunctional cytochrome *c* reductase/processing peptidase complex from plant mitochondria. J. Bioenerg. Biomembr. 27:423–436.

Braun, H. P., Schmitz, U. K. 1995b. Are the core proteins of the mitochondrial bc1 complex evolutionary relics of a processing peptidase? Trends Biochem. Sci. 20:171–175.

Braun, H. P., Zabaleta, E. 2007. Carbonic anhydrase subunits of the mitochondrial NADH dehydrogenase complex (complex I) in plants. Physiol. Plantarum 129:114–122.

Braun, H. P., Emmermann, M., Kruft, V., Schmitz, U. K. 1992. The general mitochondrial processing peptidase from potato is an integral part of cytochrome *c* reductase of the respiratory chain. EMBO J. 11:3219–3227.

Braun, H. P., Emmermann, M., Kruft, V., Bödicker, M., Schmitz, U. K. 1995. The general mitochondrial processing peptidase from wheat is integrated into the cytochrome bc_1 complex of the respiratory chain. Planta 195:396–402.

Bultema, J. B., Braun, H. P., Boekema, E. J., Kouril, R. 2009. Megacomplex organization of the oxidative phosphorylation system by structural analysis of respiratory supercomplexes from potato. Biochim. Biophys. Acta (Bioenergetics) 1787:60–67.

Burger, G., Lang, B. F., Reith, M., Gray, M. W. 1996. Genes encoding the same three subunits of respiratory complex II are present in the mitochondrial DNA of two phylogenetically distant eukaryotes. Proc. Natl. Acad. Sci. U.S.A. 93:2328–2332.

Capaldi, R. A. 1990. Structure and function of cytochrome *c* oxidase. Annu. Rev. Biochem. 59:569–596.

Cardol, P., Vanrobaeys, F., Devreese, B., Van Beeumen, J., Matagne, R. F., Remacle, C. 2004. Higher plant-like subunit composition of mitochondrial complex I from *Chlamydomonas reinhardtii*: 31 conserved components among eukaryotes. Biochim. Biophys. Acta 1658:212–224.

Cardol, P., Gonzalez-Halphen, D., Reyes-Prieto, A., Baurain, D., Matagne, R. F., Remacle, C. 2005. The mitochondrial oxidative phosphorylation proteome of *Chlamydomonas reinhardtii* deduced from the genome sequencing project. Plant Physiol. 137:447–459.

Carr, H. S., Winge, D. R. 2003. Assembly of cytochrome *c* oxidase within the mitochondrion. Acc. Chem. Res. 36:309–316.

Cecchini, G. 2003. Function and structure of complex II of the respiratory chain. Annu. Rev. Biochem. 72:77–109.

Cobine, P. A., Pierrel, F., Winge, D. R. 2006. Copper trafficking to the mitochondrion and assembly of copper metalloenzymes. Biochim. Biophys. Acta 1763:759–772.

Comelli, R. N., Gonzalez, D. H. 2009. Divergent regulatory mechanisms in the response of respiratory chain component genes to carbohydrates suggests a model for gene evolution after duplication. Plant Signal. Behav. 4:1–3.

Comelli, R., Viola, I., Gonzalez, D. H. 2009. Characterization of promoter elements required for expression and induction by sucrose of the Arabidopsis *COX5b-1* nuclear gene, encoding the zinc-binding subunit of cytochrome *c* oxidase. Plant Mol. Biol. 69:729–743.

Cruciat, C. M., Brunner, S., Baumann, F., Neupert, W., Stuart, R. A. 2000. The cytochrome bc1 and cytochrome c oxidase complexes associate to form a single supracomplex in yeast mitochondria. J. Biol. Chem. 275:18093–18098.

Curi, G. C., Welchen, E., Chan, R. L., Gonzalez, D. H. 2003. Nuclear and mitochondrial genes encoding cytochrome *c* oxidase subunits respond differently to the same metabolic factors. Plant Physiol. Biochem. 41:689–693.

Devenish, R. J., Prescott, M., Rodgers, A. J. 2008. The structure and function of mitochondrial F_1F_0-ATP synthases. Int. Rev. Cell. Mol. Biol. 267:1–58.

Diaz, F., Fukui, H., Garcia, S., Moraes, C. T. 2006. Cytochrome *c* oxidase is required for the assembly/stability of respiratory complex I in mouse fibroblasts. Mol. Cell Biol. 26:4872–4881.

Duarte, J. M., Cui, L., Wall, P. K., Zhang, Q., Zhang, X., Leebens-Mack, J., Ma, H., Altman, N., dePamphilis, C. W. 2006. Expression pattern shifts following duplication indicative of subfunctionalization and neofunctionalization in regulatory genes of Arabidopsis. Mol. Biol. Evol. 23:469–478.

Dudkina, N. V., Eubel, H., Keegstra, W., Boekema, E. J., Braun, H. P. 2005a. Structure of a mitochondrial supercomplex formed by respiratory chain complexes I and III. Proc. Natl. Acad. Sci. U.S.A. 102:3225–3229.

Dudkina, N. V., Heinemeyer, J., Keegstra, W., Boekema, E. J., Braun, H. P. 2005b. Structure of dimeric ATP synthase from mitochondria: an angular association of monomers induces the strong curvature of the inner membrane. FEBS Lett. 579:5769–5772.

Dudkina, N. V., Sunderhaus, S., Braun, H. P., Boekema, E. J. 2006. Characterization of dimeric ATP synthase and cristae membrane ultrastructure from Saccharomyces and Polytomella mitochondria. FEBS Lett. 580:3427–3432.

Dudkina, N. V., Sunderhaus, S., Boekema, E. J., Braun, H. P. 2008. The higher level of the oxidative phosphorylation system: mitochondrial supercomplexes. J. Bioenerg. Biomembr. 40:419–424.

Dudkina, N. V., Oostergetel, G. Lewejohann, D., Braun, H. P., Boekema, E. J. 2010a. Row-like organization of ATP synthase in intact mitochondria determined by cryo-electron tomography. Biochim. Biophys. Acta (Bioenergetics) 1797:272–277.

Dudkina, N. V., Kouřil, R., Peters, K., Braun, H. P., Boekema, E. J. 2010b. Structure and function of mitochondrial supercomplexes. Biochim. Biophys. Acta 1797:664–670.

Dunning, C. J., McKenzie, M., Sugiana, C., Lazarou, M., Silke, J., Connelly, A., Fletcher, J. M., Kirby, D. M., Thorburn, D. R., Ryan, M. T. 2007. Human CIA30 is involved in the early assembly of mitochondrial complex I and mutations in its gene cause disease. EMBO J. 26:3227–3237.

Elorza, A., León, G., Gómez, I., Mouras, A., Holuigue, L., Araya, A., Jordana, X. 2004. Nuclear *SDH2-1* and *SDH2-2* genes, encoding the iron-sulfur subunit of mitochondrial complex II in Arabidopsis, have distinct cell specific expression patterns and promoter activities. Plant Physiol. 136:4072–4087.

Elorza, A., Roschzttardtz, H., Gómez, I., Mouras, A., Holuigue, L., Araya, A., Jordana, X. 2006. A nuclear gene for the iron-sulfur subunit of mitochondrial complex II is specifically expressed during Arabidopsis seed development and germination. Plant Cell Physiol. 47:14–21.

Eriksson, A. C., Sjöling, S., Glaser, E. 1994. The ubiquinol cytochrome *c* oxidoreductase complex of spinach leaf mitochondria is involved in both respiration and protein processing. Biochim. Biophys. Acta 1186:221–231.

Eriksson, A. C., Sjöling, S., Glaser, E. 1996. Characterization of the bifunctional mitochondrial processing peptidase (MPP)/bc_1 complex in *Spinacia oleracea*. J. Bioenerg. Biomembr. 28:285–292.

Eubel, H., Jänsch, L., Braun, H. P. 2003. New insights into the respiratory chain of plant mitochondria. Supercomplexes and a unique composition of Complex II. Plant Physiol. 133:274–286.

Eubel, H., Heinemeyer, J., Braun, H. P. 2004. Identification and characterization of respirasomes in potato mitochondria. Plant Physiol. 134:1450–1459.

Figueroa, P., León, G., Elorza, A., Holuigue, L., Jordana, X. 2001. Three different genes encode the iron-sulfur subunit of succinate dehydrogenase in *Arabidopsis thaliana*. Plant Mol. Biol. 46:241–250.

Figueroa, P., León, G., Elorza, A., Holuigue, L., Araya, A., Jordana, X. 2002. The four subunits of mitochondrial respiratory complex II are encoded by multiple nuclear genes and targeted to mitochondria in *Arabidopsis thaliana*. Plant Mol. Biol. 50:725–734.

Fontanesi, F., Soto, I. C., Horn, D., Barrientos, A. 2006. Assembly of mitochondrial cytochrome *c*-oxidase, a complicated and highly regulated cellular process. Am. J. Physiol. Cell Physiol. 291:C1129–C1147.

Fontanesi, F., Soto, I. C., Barrientos, A. 2008. Cytochrome *c* oxidase biogenesis: new levels of regulation. IUBMB Life 60:557–568.

Fontanesi, F., Soto, I. C., Horn, D., Barrientos, A. 2010. Mss51 and Ssc1 facilitate translational regulation of cytochrome C oxidase biogenesis. Mol. Cell. Biol. 30:245–259.

Fowler, L. R., Hatefi, Y. 1961. Reconstitution of the electron transport system III. Reconstitution of DPNH oxidase, succinic oxidase, and DPNH succinic oxidase. Biochem. Biophys. Res. Commun. 5:203–208.

Fowler, L. R., Richardson, H. S. 1963. Studies on the electron transfer system. J. Biol. Chem. 238:456–463.

Gabaldon, T., Rainey, D., Huynen, M. A. 2005. Tracing the evolution of a large protein complex in the eukaryotes, NADH:ubiquinone oxidoreductase (complex I). J. Mol. Biol. 348:857–870.

Gakh, O., Cavadini, P., Isaya, G. 2002. Mitochondrial processing peptidases. Biochim. Biophys. Acta 1592:63–77.

Gavin, P. D., Prescott, M., Luff, S. E., Devenish, R. J. 2004. Cross-linking ATP synthase complexes in vivo eliminates mitochondrial cristae. J. Cell Sci. 117:2233–2243.

Genova, M. L., Bianchi, C., Lenaz, G. 2003. Structural organization of the mitochondrial respiratory chain. Ital. J. Biochem. 52:58–61.

Giegé, P. 2007. Mitochondrial respiratory complex biogenesis: communication, gene expression and assembly. In Plant Mitochondria, ed. D. C. Logan, pp. 141–170. Oxford, UK: Blackwell.

Giegé, P., Knoop, V., Brennicke, A. 1998. Complex II subunit 4 (*sdh4*) homologous sequences in plant mitochondrial genomes. Curr. Genet. 34:313–317.

Giegé, P., Sweetlove, L. J., Cognat, V., Leaver, C. J. 2005. Coordination of nuclear and mitochondrial genome expression during mitochondrial biogenesis in Arabidopsis. Plant Cell 17:1497–1512.

Giraud, M. F., Paumard, P., Soubannier, V., Vaillier, J., Arselin, G., Salin, B., Schaeffer, J., Brèthes, D., di Rago, P., Velours, J. 2002. Is there a relationship between the supramolecular organization of the mitochondrial ATP synthase and the formation of cristae? Biochim. Biophys. Acta 1555, 174–180.

Glaser, E., Eriksson, A. C., Sjöling, S. 1994. Bifunctional role of the bc1 complex in plants. Mitochondrial bc_1 complex catalyses both electron transport and protein processing. FEBS Lett. 346:83–87.

Glerum, D. M., Shtanko, A., Tzagoloff, A. 1996a. Characterization of *COX17*, a yeast gene involved in copper metabolism and assembly of cytochrome oxidase. J. Biol. Chem. 271:14504–14509.

Glerum, D. M., Shtanko, A., Tzagoloff, A. 1996b. *SCO1* and *SCO2* act as high copy suppressors of a mitochondrial copper recruitment defect in *Saccharomyces cerevisiae*. J. Biol. Chem. 271:20531–20535.

Goffart, S., Wiesner, R. J. 2003. Regulation and co-ordination of nuclear gene expression during mitochondrial biogenesis. Exp. Physiol. 88:33–40.

Gomez Casati, D. F., Busi, M. V., Gonzalez-Schaina, N., Mouras, A., Zabaleta, E., Araya, A. 2002. A mitochondrial dysfunction induces the expression of nuclear-encoded complex I genes in engineered male sterile *Arabidopsis thaliana*. FEBS Lett. 532:70–74.

Gonzalez, D. H., Welchen, E., Attallah, C. V., Comelli, R. N., Mufarrege, E. M. 2007. Transcriptional coordination of the biogenesis of the oxidative phosphorylation machinery in plants. Plant J. 51:105–116.

Grumbt, B., Stroobant, V., Terziyska, N., Israel, L., Hell, K. 2007. Functional characterization of Mia40p, the central component of the disulfide relay system of the mitochondrial intermembrane space. J. Biol. Chem. 282:37461–37470.

Guenebaut, V., Vincentelli, R., Mills, D., Weiss, H., Leonard, K. R. 1997. Three dimensional structure of NADH-dehydrogenase from *Neurospora crassa* by electron microscopy and conical tilt reconstruction. J. Mol. Biol. 265:409–418.

Haberer, G., Hindemitt, T., Meyers, B. C., Mayer, K. F. X. 2004. Transcriptional similarities, dissimilarities, and conservation of *cis*-elements in duplicated genes of Arabidopsis. Plant Physiol. 136:3009–3022.

Hackenbrock, C. R., Chazotte, B., Gupta, S. S. 1986. The random collision model and a critical assessment of the diffusion and collision in mitochondrial electron transport. J. Bioenerg. Biomembr. 18:331–368.

Hamel, P., Sakamoto, W., Wintz, H., Dujardin, G. 1997. Functional complementation of an oxa1-yeast mutation identifies an Arabidopsis thaliana cDNA involved in the assembly of respiratory complexes. Plant J. 12:1319–1327.

Hamel, P., Corvest, V., Giegé, P., Bonnard, G. 2009. Biochemical requirements for the maturation of mitochondrial c-type cytochromes. Biochim. Biophys. Acta 1793:125–138.

Hatefi, Y. 1985. The mitochondrial electron transport and oxidative phosphorylation system. Annu. Rev. Biochem. 54:1015–1069.

Hatefi, Y., Rieske, J. S. 1967. The preparation and properties of DPNH-cytochrome c reductase (complex I-III of the respiratory chain). Methods Enzymol. 10:225–231.

Hatefi, Y., Haavik, A. G., Jurtshuk, P. 1961. Studies on the electron transport system XXX. DPNH-cytochrome c reductase I. Biochim. Biophys. Acta 52:106–118.

Hatefi, Y., Haavik, A. G., Fowler, L. R., Griffiths, D. E. 1962. Studies on the electron transfer system. Reconstitution of the electron transfer system. J. Biol. Chem. 237:2661–2669.

Heazlewood, J. L., Howell, K. A., Millar, A. H. 2003a. Mitochondrial complex I from Arabidopsis and rice: orthologs of mammalian and fungal components coupled with plant-specific subunits. Biochim. Biophys. Acta 1604:159–169.

Heazlewood, J. L., Howell, K. A., Whelan, J., Millar, A. H. 2003b. Towards an analysis of the rice mitochondrial proteome. Plant Physiol. 132:230–242.

Heinemeyer, J., Braun, H. P., Boekema, E. J., Kuril, R. 2007. A structural model of the cytochrome *c* reductase/oxidase supercomplex from yeast mitochondria. J. Biol. Chem. 282:12240–12248.

Helfenbein, K. G., Ellis, T. P., Dieckmann, C. L., Tzagoloff, A. 2003. *ATP22*, a nuclear gene required for expression of the F_0 sector of mitochondrial ATPase in *Saccharomyces cerevisiae*. J. Biol. Chem. 278:19751–19756.

Hell, K., Tzagoloff, A., Neupert, W., Stuart, R. A. 2000. Identification of Cox20p, a novel protein involved in the maturation and assembly of cytochrome oxidase subunit 2. J. Biol. Chem. 275:4571–4578.

Herrmann, J. M., Funes, S. 2005. Biogenesis of cytochrome oxidase sophisticated assembly lines in the mitochondrial inner membrane. Gene 354:43–52.

Hiser, L., Di Valentin, M., Hamer, A. G., Hosler, J. P. 2000. Cox11p is required for stable formation of the Cu(B) and magnesium centers of cytochrome *c* oxidase. J. Biol. Chem. 275:619–623.

Ho, L. H. M., Giraud, E., Uggalla, V., Lister, R., Clifton, R., Glen, A., Thirkettle-Watts, D., Van Aken, O., Whelan, J. 2008. Identification of regulatory pathways controlling gene expression of stress-responsive mitochondrial proteins in Arabidopsis. Plant Physiol. 147:1858–1873.

Horn, D., Al-Ali, H., Barrientos, A. 2008. Cmc1p is a conserved mitochondrial twin CX_9C protein involved in cytochrome *c* oxidase biogenesis. Mol. Cell Biol. 28:4354–4364.

Horng, Y. C., Cobine, P. A., Maxfield, A. B., Carr, H. S., Winge, D. R. 2004. Specific copper transfer from the *Cox17* metallochaperone to both *Sco1* and *Cox11* in the assembly of yeast cytochrome *c* oxidase. J Biol. Chem. 279:35334–35340.

Horsefield, R., Iwata, S., Byrne, B. 2004. Complex II from a structural perspective. Curr. Protein Pept. Sci. 5:107–118.

Horsefield, R., Yankovskaya, V., Sexton, G., Whittingham, W., Shiomi, K., Ōmura, S., Byrne, B., Cecchini, G., Iwata, S. 2006. Structural and computational analysis of the quinone-binding site of complex II (succinate-ubiquinone oxidoreductase): a mechanism of electron transfer and proton conduction during ubiquinone reduction. J. Biol. Chem. 281:7309–7316.

Howell, K. A., Millar, A. H., Whelan, J. 2006. Ordered assembly of mitochondria during rice germination begins with promitochondrial structures rich in components of the protein import apparatus. Plant Mol. Biol. 60:201–223.

Huang, S., Taylor, N. L., Narsai, R., Eubel, H., Whelan, J., Millar, A. H. 2009a. Experimental analysis of the rice mitochondrial proteome, its biogenesis and heterogeneity. Plant Physiol. 149:719–734.

Huang, S., Taylor, N. L., Narsai, R. Eubel, H., Whelan, J., Millar, A. H. 2009b. Functional and composition differences between mitochondrial complex II in Arabidopsis and rice are correlated with the complex genetic history of the enzyme. Plant Mol. Biol. 72:331–342.

Iwata, S., Lee, J. W., Okada, K., Lee, J. K., Iwata, M., Rasmussen, B., Link, T. A., Ramaswamy, S., Jap, B. K. 1998. Complete structure of the 11-subunit bovine mitochondrial cytochrome bc_1 complex. Science 281:64–71.

Jänsch, L., Kruft, V., Schmitz, U. K., Braun, H. P. 1996. New insights into the composition, molecular mass and stoichiometry of the protein complexes of plant mitochondria. Plant J. 9:357–368.

Jia, L., Dienhart, M. K., Stuart, R. A. 2007. Oxa1 directly interacts with Atp9 and mediates its assembly into the mitochondrial F_1F_0-ATP synthase complex. Mol. Biol. Cell 18:1897–1908.

Kadowaki, K., Kubo, N., Ozawa, K., Hirai, A. 1996. Targeting presequence acquisition after mitochondrial gene transfer to the nucleus occurs by duplication of existing targeting signal. EMBO J. 15:6652–6661.

Khalimonchuk, O., Rödel, G. 2005. Biogenesis of cytochrome *c* oxidase. Mitochondrion 5:363–388.

Khalimonchuk, O., Rigby, K., Bestwick, M., Pierrel, F., Cobine, P. A., Winge, D. R. 2008. Pet191 is a cytochrome c oxidase assembly factor in Saccharomyces cerevisiae. Eukaryot. Cell 7:1427–1431.

Kuffner, R., Rohr, A., Schmiede, A., Krull, C., Schulte, U. 1998. Involvement of two novel chaperones in the assembly of mitochondrial NADH:ubiquinone oxidoreductase (complex I). J. Mol. Biol. 283:409–417.

Leary, S. C., Winge, D. R., Cobine, P. A. 2009. "Pulling the plug" on cellular copper: the role of mitochondria in copper export. Biochim. Biophys. Acta 1793:146–153.

Lee, S., Warmke, H. 1979. Organelle size and number in fertile and T-cytoplasmic male-sterile corn. Am. J. Bot. 66:141–148.

Lefebvre-Legendre, L., Vaillier, J., Benabdelhak, H., Velours, J., Slonimski, P. P., di Rago, J. P. 2001. Identification of a nuclear gene (FMC1) required for the assembly/stability of yeast mitochondrial F_1-ATPase in heat stress conditions. J. Biol. Chem. 276:6789–6796.

Leon, G., Holuigue, L., Jordana, X. 2007. Mitochondrial complex II is essential for gametophyte development in Arabidopsis. Plant Physiol. 143:1534–1546.

Li, Y., D'Aurelio, M., Deng, J-H., Park, J-S., Manfredi, G., Hu, P., Lu, J., Bai, Y. 2007. An assembled complex IV maintains the stability and activity of complex I in mammalian mitochondria. J. Biol. Chem. 282:17557–17562.

Lister, R., Chew, O., Lee, M. N., Heazlewood, J. L., Clifton, R., Parker, K. L., Millar, A. H., Whelan, J. 2004. A transcriptomic and proteomic characterization of the Arabidopsis mitochondrial protein import apparatus and its response to mitochondrial dysfunction. Plant Physiol. 134:777–789.

Logan, D. C., Millar, A. H., Sweetlove, L. J., Hill, S. A., Leaver, C. J. 2001. Mitochondrial biogenesis during germination in maize embryos. Plant Physiol. 125:662–672.

Mackenzie, S., McIntosh, L. 1999. Higher plant mitochondria. Plant Cell 11:571–586.

Meunier, B., Taanman, J. W. 2002. Mutations of cytochrome *c* oxidase subunits 1 and 3 in *Saccharomyces cerevisiae*: assembly defect and compensation. Biochim. Biophys. Acta. 1554:101–107.

Meyer, E. H., Heazlewood, J. L., Millar, A. H. 2007. Mitochondrial acyl carrier proteins in *Arabidopsis thaliana* are predominantly soluble matrix proteins and none can be confirmed as subunits of respiratory Complex I. Plant Mol. Biol. 64:319–327.

Meyer, E. H., Taylor, N. L., Millar, A. H. 2008. Resolving and identifying protein components of plant mitochondrial respiratory complexes using three dimensions of gel electrophoresis. J. Proteome Res. 2:786–794.

Millar, A. H., Mittova, V., Kiddle, G., Heazlewood, J. L., Bartoli, C. G., Theodoulou, F. L., Foyer, C. H. 2003. Control of ascorbate synthesis by respiration and its implications for stress responses. Plant Physiol. 133:443–447.

Millar, A. H., Eubel, H., Jänsch, L., Kruft, V., Heazlewood, J. L., Braun, H. P. 2004. Mitochondrial cytochrome *c* oxidase and succinate dehydrogenase complexes contain plant specific subunits. Plant Mol. Biol. 56:77–90.

Millar, A. H., Cronan, J. E., Fearnley, I. M., Walker, J. E. 2005. Mammalian mitochondria contain a soluble acyl carrier protein. FEBS Lett. 579:4892–4896.

Minauro-Sanmiguel, F., Wilkens, S., Garcia, J. J. 2005. Structure of dimeric mitochondrial ATP synthase: novel F_0 bridging features and the structural basis of mitochondrial cristae biogenesis. Proc. Natl. Acad. Sci. 102:12356–12358.

Mufarrege, E., Curi, G., Gonzalez, D. H. 2009. Common sets of promoter elements determine the expression characteristics of three Arabidopsis genes encoding isoforms of mitochondrial cytochrome c oxidase subunit 6b. Plant Cell Physiol. 50:1393–1399.

Nakagawa, T., Maeshima, M., Nakamura, K., Asahi, T. 1990. Molecular cloning of a cDNA for the smallest nuclear-encoded subunit of sweet potato cytochrome *c* oxidase: analysis with the cDNA of the structure and import into mitochondria of the subunit. Eur. J. Biochem. 191:557–561.

Nijtmans, L. G., Taanman, J. W., Muijsers, A. O., Speijer, D., Van den Bogert, C. 1998. Assembly of cytochrome-c oxidase in cultured human cells. Eur. J. Biochem. 254:389–394.

Nobrega, M. P., Bandeira, S. C., Beers, J., Tzagoloff, A. 2002. Characterization of *COX19*, a widely distributed gene required for expression of mitochondrial cytochrome oxidase. J. Biol. Chem. 277:40206–40211.

Oda, K., Yamato, K., Ohta, E., Nakamura, Y., Takemura, M., Nozato, N., Akashi, K., Kanegae, T., Ogura, Y., Kohchi, T., Ohyama, K. 1992. Gene organization deduced from the complete sequence of liverwort *Marchantia polymorpha* mitochondrial DNA. J. Mol. Biol. 223:1–7.

Ohtsu, K., Nakazono, M., Tsutsumi, N., A. Hirai, A. 2001. Characterization and expression of the genes for cytochrome *c* oxidase subunit VIb (COX6b) from rice and *Arabidopsis thaliana*. Gene 264:233–239.

Parisi, G., Perales, M., Fornasari, M. S., Colaneri, A., González-Schain, N., Gómez-Casati, D., Zimmermann, S., Brennicke, A., Araya, A., Ferry, J. G., Echave, J., Zabaleta, E. 2004. Gamma carbonic anhydrases in plant mitochondria. Plant Mol. Biol. 55:193–207.

Paumard, P., Vaillier, J., Coulary, B., Schaeffer, J., Soubannier, V., Mueller, D. M., Brethes, D., di Rago, J. P., Velours, J. 2002. The ATP synthase is involved in generating mitochondrial cristae morphology. EMBO J. 21:221–230.

Perales, M., Parisi, G., Fornasari, M. S., Colaneri, A., Villarreal, F., González-Schain, N., Echave, J., Gómez Casati, D., Braun, H. P., Araya, A., Zabaleta, E. 2004. Gamma carbonic anhydrase like complex interact with plant mitochondrial complex I. Plant Mol. Biol. 56:947–957.

Perales, M., Eubel, H. Heinemeyer, J., Colaneri, A., Zabaleta, E., Braun, H. P. 2005. Disruption of a nuclear gene encoding a mitochondrial gamma carbonic anhydrase reduces complex I and supercomplex I + III_2 levels and alters mitochondrial physiology in Arabidopsis. J. Mol. Biol. 350:263–277.

Peters, K., Dukina, N. V., Jänsch, L., Braun, H. P., Boekema, E. J. 2008. A structural investigation of complex I and I + III_2 supercomplex from *Zea mays* at 11–13 Å resolution: assignment of the carbonic anhydrase domain and evidence for structural heterogeneity within complex I. Biochim. Biophys. Acta 1777:84–89.

Pickova, A., Potocky, M., Houstek, J. 2005. Assembly factors of F_1F_0-ATP synthase across genomes. Proteins 59:393–402.

Pineau, B., Layoune, O., Danon, A., De Paepe, R. 2008. L-galactono-1,4-lactone dehydrogenase is required for the accumulation of plant respiratory complex I. J. Biol. Chem. 283:32500–32505.

Poyton, R. O., McEwen, J. E. 1996. Crosstalk between nuclear and mitochondrial genomes. Annu. Rev. Biochem. 65:563–607.

Ragan, C. I., Heron, C. 1978. The interaction between mitochondrial NADH-ubiquinone oxidoreductase and ubiquinol-cytochrome c oxidoreductase. Biochem J. 174:783–790.

Remacle, C., Barbieri, R., Cardol, P., Hamel, P. P. 2008. Eukaryotic complex I: functional diversity and experimental systems to unravel the assembly process. Mol. Genet. Genomics 280:93–110.

Rigby, K., Zhang, L., Cobine, P. A., George, G. N., Winge, D. R. 2007. Characterization of the cytochrome *c* oxidase assembly factor *Cox19* of *Saccharomyces cerevisiae*. J. Biol. Chem. 282:10233–10242.

Roschzttardtz, H., Fuentes, I., Vasquez, M., Corvalan, C., Leon, G., Gomez, I., Araya, A., Holuigue, L., Vicente-Carbajosa, J., Jordana, X. 2009. A nuclear gene encoding the iron-sulfur subunit of mitochondrial Complex II is regulated by B3 domain transcription factors during seed development in Arabidopsis. Plant Physiol. 150:84–95.

Sazanov, L. A., Walker, J. E. 2000. Cryo-electron crystallography of two subcomplexes of bovine complex I reveals the relationship between the membrane and peripheral arms. J. Mol. Biol. 302:455–464.

Schäfer, E., Seelert, H., Reifschneider, N. H., Krause, F., Dencher, N. A., Vonck, J. 2006. Architecture of active mammalian respiratory chain supercomplexes. J. Biol. Chem. 281:15370–15375.

Schägger, H., Pfeiffer, K. 2000. Supercomplexes in the respiratory chains of yeast and mammalian mitochondria. EMBO J. 19:1777–1783.

Schägger, H., von Jagow, G. 1991. Blue native electrophoresis for isolation of membrane protein complexes in enzymatically active form. Anal. Biochem. 199:223–231.

Scheffler, I. E. 1998. Molecular genetics of succinate:quinone oxidoreductase in eukaryotes. Progr. Nucl. Acid Res. Mol. Biol. 60:267–315.

Schulte, U. 2001. Biogenesis of respiratory complex I. J. Bioenerg. Biomembr. 33:205–212.

Stock, D., Leslie, A. G., Walker, J. E. 1999. Molecular architecture of the rotary motor in ATP synthase. Science 286:1700–1705.

Strauss, M., Hofhaus, G., Schröder, R. R., Kühlbrandt, W. 2008. Dimer ribbons of ATP synthase shape the inner mitochondrial membrane. EMBO J. 27:1154–1160.

Sunderhaus, S., Dudkina, N., Jänsch, L., Klodmann, J., Heinemeyer, J., Perales, M., Zabaleta, E., Boekema, E., Braun, H. P. 2006. Carbonic anhydrase subunits form a matrix-exposed domain attached to the membrane arm of mitochondrial complex I in plants. J. Biol. Chem. 281:6482–6488.

Sweetlove, L. J., Heazlewood, J. L., Herald, V., Holtzapffel, R., Day, D. A., Leaver, C. J., Millar, A. H. 2002. The impact of oxidative stress on Arabidopsis mitochondria. Plant J. 32:891–904.

Thomas, D., Bron, P., Weimann, T., Dautant, A., Giraud, M. F., Paumard, P., Salin, B., Cavalier, A., Velours, J., Brèthes, D. 2008. Supramolecular organization of the yeast F1Fo-ATP synthase. Biol. Cell. 100:591–601.

Tremousaygue, D., Garnier, L., Bardet, C., Dabos, P., Herve, C., Lescure, B. 2003. Internal telomeric repeats and 'TCP domain' protein-binding sites co-operate to regulate gene expression in *Arabidopsis thaliana* cycling cells. Plant J. 33:957–966.

Tzagoloff, A. 1995. Ubiquinol-cytochrome-c oxidoreductase from *Saccharomyces cerevisiae*. Methods Enzymol. 260:51–63.

Ugalde, C., Janssen, R. J., Van Den Heuvel, L. P., Smeitink, J. A., Nijtmans, L. G. 2004a. Differences in assembly or stability of complex I and other mitochondrial OXPHOS complexes in inherited complex I deficiency. Hum. Mol. Genet. 13:659–667.

Ugalde, C., Vogel, R., Huijbens, R., Van Den Heuvel, B., Smeitink, J., Nijtmans, L. 2004b. Human mitochondrial complex I assembles through the combination of evolutionary conserved modules: a framework to interpret complex I deficiencies. Hum. Mol. Genet. 13:2461–2472.

Unseld, M., Marienfeld, J. R., Brandt, P., Brennicke, A. 1997. The mitochondrial genome of Arabidopsis thaliana contains 57 genes in 366,924 nucleotides. Nat. Genet. 15:57–61.

Velours, J., Arselin, G. 2000. The *Saccharomyces cerevisiae* ATP synthase. J. Bioenerg. Biomembr. 32:383–390.

Villarreal, F., Martín, V., Colaneri, A., González-Schain, N., Perales, M., Martín, M., Lombardo, C., Braun, H. P., Bartoli, C., Zabaleta, E. 2009. Ectopic expression of mitochondrial gamma Carbonic Anhydrase 2 causes male sterility by anther indehiscence. Plant Mol. Biol. 70:471–485.

Vogel, R. O., Smeitink, J. A. M., Nijtmans, L. G. J. 2007. Human mitochondrial complex I assembly: a dynamic and versatile process. Biochim. Biophys. Acta-Bioenergetics 1767:1215–1227.

Vonck, J., Schäfer, E. 2009. Supramolecular organization of protein complexes in the mitochondrial inner membrane. Biochim. Biophys. Acta. 1793:117–24.

Weber, J., Senior, A. E. 2003. ATP synthesis driven by proton transport in F_1F_0-ATP synthase. FEBS Lett. 545:61–70.

Welchen, E., Gonzalez, D. H. 2005. Differential expression of the Arabidopsis cytochrome *c* genes *Cytc-1* and *Cytc-2*: evidence for the involvement of TCP-domain protein binding elements in anther- and meristem-specific expression of the *Cytc-1* gene. Plant Physiol. 139:88–100.

Welchen, E., Gonzalez, D. H. 2006. Overrepresentation of elements recognized by TCP-domain transcription factors in the upstream regions of nuclear genes encoding components of the mitochondrial oxidative phosphorylation machinery. Plant Physiol. 141:540–545.

Welchen, E., Chan, R. L., Gonzalez, D. H. 2002. Metabolic regulation of genes encoding cytochrome *c* and cytochrome *c* oxidase subunit Vb in Arabidopsis. Plant Cell Environ. 25:1605–1615.

Welchen, E., Chan, R. L., Gonzalez, D. H. 2004. The promoter of the Arabidopsis nuclear gene *COX5b-1*, encoding subunit 5b of the mitochondrial cytochrome *c* oxidase, directs tissue-specific expression by a combination of positive and negative regulatory elements. J. Exp. Bot. 55:1997–2004.

Welchen, E., Viola, I. L., Kim, H. J., Prendes, L. P., Comelli, R. N., Hong, J. C., Gonzalez, D. H. 2009. A segment containing a G-box and an ACGT motif confers differential expression characteristics and responses to the Arabidopsis *Cytc-2* gene, encoding an isoform of cytochrome *c*. J. Exp. Bot. 60:829–845.

Wittig, I., Schägger, H. 2009. Supramolecular organization of ATP synthase and respiratory chain in mitochondrial membranes. Biochim. Biophys. Acta 1787:672–680.

Yankovskaya, V., Horsefield, R., Törnroth, S., Luna-Chavez, C., Miyoshi, H., Léger, C., Byrne, B., Cecchini, G., Iwata, S. 2003. Architecture of succinate dehydrogenase and reactive oxygen species generation. Science 299:700–704.

Yu, J., Nickels, R., McIntosh, L. 2001. A genome approach to mitochondrial nuclear communication in Arabidopsis. Plant Physiol. Biochem. 39:345–353.

Zabaleta, E., Heiser, V., Grohmann, L., Brennicke, A. 1998. Promoters of nuclear-encoded respiratory chain complex I genes from *Arabidopsis thaliana* contain a region essential for anther/pollen-specific expression. Plant J. 15:49–59.

Zara, V., Conte, L., Trumpower, B. L. 2009. Biogenesis of the yeast cytochrome bc_1 complex. Biochim. Biophys. Acta 1793:89–96.

Zeng, X., Neupert, W., Tzagoloff, A. 2007. The metalloprotease encoded by ATP23 has a dual function in processing and assembly of subunit 6 of mitochondrial ATPase. Mol. Biol. Cell. 18:617–626.

Chapter 14
Mitochondrial Electron Transport and Plant Stress

Allan G. Rasmusson and Ian M. Møller

Abstract Due to the sessile nature of plants, it is crucial for their survival and growth that they can handle a constantly changing, and thus stressful, ambient environment by modifying their structure and metabolism. The central metabolism of plants is characterized by many alternative options for metabolic pathways, which allow a wide range of adjustments of metabolic processes in response to environmental variations. Many of the metabolic pathways in plants involve the processing of redox compounds and the use of adenylates. They converge at the mitochondrial electron transport chain (ETC) where redox compounds from carbon degradation are used for powering ATP synthesis. The standard ETC contains three sites of energy conservation in complexes I, III, and IV, which are in common with most other eukaryotes. However, the complexity of the plant metabolic system is mirrored in the ETC. In addition to the standard enzymes, plants have a large set of supplementary electron transport enzymes. Many of these, such as the external and internal NAD(P)H dehydrogenases, proline dehydrogenase, and glycerol-3-phosphate dehydrogenase, feed into the ubiquinone pool and they therefore bypass the first site of energy conservation in the ETC. The alternative oxidase provides a non-energy-conserving alternative to electron transport through complexes III and IV. There also appears to be a special coupling between specific NAD(P)H dehydrogenases and specific members of the alternative oxidase family. These additional enzymes therefore give a great flexibility in the type and origin of the substrate, the electron transport route(s) used, and the energy yield. At the same time special reactions, such as ascorbate biosynthesis, can take place. In this way, the mitochondrial ETC can mediate major adjustments in cellular metabolism that is important for cellular function under a great variety of stress conditions such as low temperature and drought.

A.G. Rasmusson (✉)
Department of Cell and Organism Biology, Lund University, Sölvegatan 35B, SE-223 62, Lund, Sweden
e-mail: allan.rasmusson@cob.lu.se

F. Kempken (ed.), *Plant Mitochondria*, Advances in Plant Biology 1,
DOI 10.1007/978-0-387-89781-3_14,

Keywords Alternative pathways • Amino acid degradation • Ascorbate synthesis • Carbon starvation • Dehydrogenases • Glycerol metabolism • NADH • NADPH NO • Oxidases • Oxidative stress • Protein oxidation • Reactive oxygen species • Redox metabolism • Respiratory chain

Abbreviations

AOX	Alternative oxidase
ETC	Electron transport chain
GalDH	L-galactono-1,4-lactone dehydrogenase
P5C	Delta-1-pyrroline-5-carboxylate
ROS	Reactive oxygen species
UCP	Uncoupling protein
UQ	Ubiquinone

14.1 Introduction

Plants respire just like all other aerobic organisms, i.e., they use the oxygen in the air to oxidize organic molecules to CO_2 and water. The energy released in the oxidation is initially mainly conserved in reducing equivalents (NADH, NADPH) and finally in ATP. These compounds are then used to drive all of the energy-requiring processes that collectively characterize "life" – transport, synthesis and degradation of cellular components, signaling, defense, growth, and reproduction. The final reactions in the respiratory oxidation process – the Krebs cycle (also called the citric acid cycle or the tricarboxylic acid cycle), the electron transport chain (ETC) and ATP synthesis – take place in the mitochondria and supply the cell not only with energy, but also with a variety of important carbon metabolites used in biosynthesis of many cellular components.

Because plants are sessile, it is of special importance for them to be able to adjust and acclimate to a wide range of conditions. This means that the energy metabolism has to be flexible and accommodate the changing requirements of the cell. The ETC of plant mitochondria is much more complex than that of mammalian mitochondria and contains a number of unique components including alternative NAD(P)H dehydrogenases and an alternative oxidase (AOX) that contribute significantly to the flexibility and versatility of plant metabolism. Additionally, a set of enzymes directly connects the ETC in plants to reactions where carbon metabolites are oxidized.

It is the purpose of this chapter to describe the properties of the ETC in plant mitochondria and the ways in which they help the plant cell respond to stressful changes in environmental conditions. We will first briefly describe the standard ETC present in (almost) all mitochondria. We will then treat the energy-bypass enzymes, NAD(P)H dehydrogenases and the AOX, that allow plant mitochondria to make the energy-balance adjustments important to adaptational responses. The formation of reactive oxygen species (ROS) is one of the first stress indications and we will outline the ways in which ROS can be formed by the ETC, cause damage and/or be used in intracellular signaling. Finally, we will describe a range of reactions linked to the plant ETC, which allow it to contribute to a wide range of metabolic pathways linked to cellular stress tolerance.

14.2 The Standard Electron Transport Chain

The ETC of plant mitochondria contains the same four multi-subunit complexes as mammalian mitochondria – complex I or the NADH dehydrogenase complex, Complex II or the succinate dehydrogenase complex, complex III or the cytochrome *bc*1 complex, and complex IV or cytochrome oxidase. Complexes I and II oxidize NADH and succinate, respectively, and pass the electrons along to a small mobile carrier, ubiquinone (UQ), which is lipophilic and diffuses in the lipid bilayer of the inner membrane. Whereas the electron transport complexes contain active centers that can bind single electrons, UQ can bind one or two electrons, and will also bind protons to the same extent. UQ passes the electrons on to complex III. From complex III the electrons are passed on by another mobile electron carrier, cytochrome *c*, which is located on the outer surface of the inner membrane, to cytochrome oxidase where they are used to reduce oxygen to water (Fig. 14.1a).

A more detailed description of the structure and biosynthesis of these complexes, as well as the supercomplexes they form, is found in Chap. 13. For the purpose of this chapter it suffices to mention that electron transport through complexes I, III, and IV (but not through complex II) is coupled to translocation of 4, 4, and 2 protons, respectively, per electron pair, from the inner matrix side of the inner membrane to the intermembrane space. This gives rise to an electrochemical proton gradient across the inner membrane such that the electrical gradient is about −180 mV (negative inside) and the pH is about 0.5 pH units higher in the matrix (about pH 8) than in the cytosol (pH 7.2–7.5). ATP is synthesized when protons pass back into the matrix through the ATP synthase (often called complex V although it is not part of the ETC and does not participate in electron transport) driven by the electrochemical gradient. The protons can alternatively pass through an uncoupling protein (UCP) also located in the inner membrane but not part of the ETC. This reaction leads to dissipation of the proton gradient without the formation of ATP (Vercesi et al. 2006). Thus, with regard to plant energy-balance adjustments, the UCP may have overlapping functions with the energy-bypass proteins, the NAD(P)H dehydrogenases and the AOX, in the ETC.

14.3 Mutations in Complexes I–IV

Relatively few plant mutants deficient in respiratory proton-translocating complexes have been isolated and investigated. A possible reason for this may be that the complexes have a vital function in mediating efficient respiratory energy

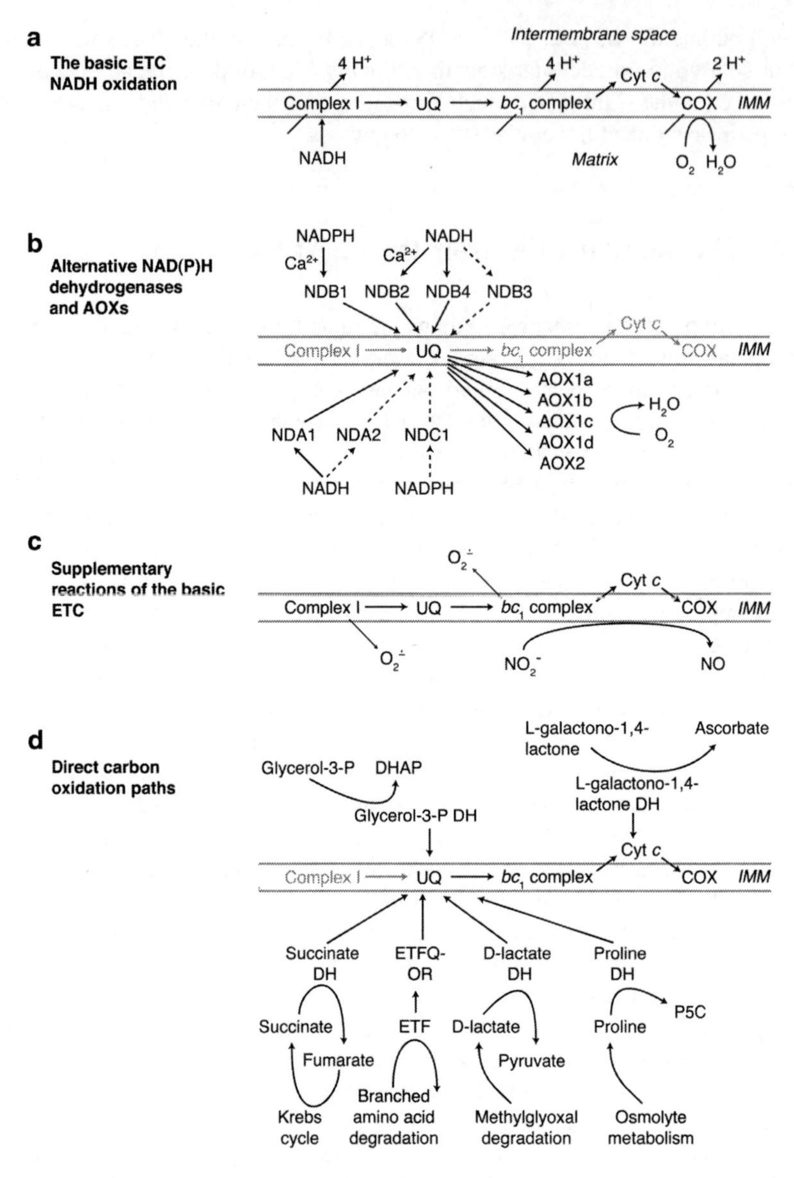

Fig. 14.1 The highly branched ETC in plant mitochondria. (**a**) The proton-pumping ETC complexes mediate oxidation of NADH, derived from the Krebs cycle, via UQ to oxygen. In this process, protons are pumped from the matrix to the intermembrane space, generating an electrochemical proton gradient across the inner mitochondrial membrane (IMM) that is used for ATP synthesis and transport. (**b**) In plants (and in many fungi and protists), alternative NAD(P)H dehydrogenases and AOXs bypass the proton-pumping complexes of main ETC. All of these enzymes are embedded in the IMM. (**c**) In addition to NADH oxidation with oxygen as acceptor, the basic ETC process will lead to some formation of superoxide by complex I and the *bc*1 complex. Under hypoxic or anoxic conditions, nitrite can be used as electron acceptor, leading to NO formation. The exact site of nitrite reduction is not known, but it appears to involve complexes III and IV.

conversion for supporting the cell and that their absence therefore decreases vitality. A number of subunits of the respiratory complexes are encoded by the mitochondrial genome and deletion mutants, where most mitochondrial DNA molecules in a plant lack genes for essential subunits of complexes I and IV have been analyzed (Marienfeld and Newton 1994; Gutierres et al. 1997; Newton et al. 2004; Noctor et al. 2004; Jiao et al. 2005). Especially, the CMSII mutant of *Nicotiana sylvestris*, lacking a functional complex I has been intensively studied. It survives most likely due to a high capacity of alternative NADH dehydrogenases, which are present in plant mitochondria (see below) (Sabar et al. 2000). However, the mutant displays a severe phenotype of male sterility, slow growth, leaf deformations, decreased photosynthesis, and a strongly perturbed metabolism and redox homeostasis (Dutilleul et al. 2003a, b, 2005). At the same time, the mutant displays increased in vivo respiration rate, increased levels of antioxidant enzymes and metabolites, insensitivity to drought stress, and a changed response to a bacterial elicitor (Boccara et al. 2001;

Box 14.1 Mitochondria play an important role in ascorbate biosynthesis

Ascorbate is a central antioxidant component of plant stress metabolism and an important vitamin in human nutrition (Noctor and Foyer 1998; Valpuesta and Botella 2004). It is synthesized via two different pathways. The last enzyme in one of the pathways, L-galactono-1,4-lactone dehydrogenase (GalDH) is part of the ETC in the inner membrane of *A. thaliana* mitochondria (Fig. 14.1d), where it is physically associated with complex I and required for its assembly. Though there is a second pathway for ascorbate synthesis, an Arabidopsis mutant deficient in the mitochondrial GalDH was not viable in the absence of ascorbate supplementation, illustrating the importance of the ETC-mediated ascorbate biosynthesis for cellular metabolism (Pineau et al. 2008).

GalDH donates electrons to cytochrome *c* in the ETC (Fig. 14.1d) (Bartoli et al. 2000; Heazlewood et al. 2003). Therefore, this activity may compete

(continued)

Fig. 14.1 (continued) (**d**) In parallel to the standard chain, several enzymes oxidize carbon metabolites directly, as part of degradation processes, and in one case, in synthesis (ascorbate, see Box 14.1). The enzymes are depicted feeding electrons into the ETC from the correct side of the IMM, but otherwise their physical location cannot be strictly inferred from the figure – some are soluble and others are embedded in the IMM. *Straight arrows* denote electron or proton transport and bent arrows conversions of metabolites. The enzyme complement depicted is based on the model plant Arabidopsis and uncertain connections are depicted by dashed lines. The different categories of ETC enzymes are displayed in separate panels for clarity, but are present together in the inner membrane and can theoretically all be connected via the mobile carriers UQ and cytochrome *c*. However, specific associations between dehydrogenase and oxidase pathways may exist, as discussed in this chapter, or between complexes as treated in Chap. 13. The major proton-pumping complexes are included throughout, but are shown in gray in panels where they are not directly involved in the processes depicted. COX, cytochrome oxidase; DH, dehydrogenase; ETF, electron-transfer flavoprotein; ETF-OR, electron-transfer flavoprotein:quinone oxidoreductase

Box 14.1 (continued)

with complex III for access to oxidized cytochrome *c*. This is consistent with the observation that overexpression of *AOX1a* in Arabidopsis leads to elevated ascorbate levels in leaves and higher ascorbate synthesis capacity in isolated mitochondria (Bartoli et al. 2006). The AOX oxidation of ubiquinol would act to decrease the reductive load at complex III and IV, allowing a faster flux of GalDH activity. Consistently, mitochondria from a tomato line suppressed for malate dehydrogenase showed elevated capacity for using Gal as a substrate for oxygen consumption (Nunes-Nesi et al. 2005). Of special interest in this respect is the observation that the ascorbate synthesis in leaves is modulated by photosynthetically active and phytochrome-modulating light (Yabuta et al. 2008; Bartoli et al. 2009), further supporting previously indicated interactions between the mitochondrial ETC and photosynthesis (see Chap. 15) and light signaling (Rasmusson and Escobar 2007). In isolated mitochondria, inhibition of complex I activity by rotenone during the simultaneous oxidation of Gal and NAD-linked Krebs cycle intermediates inhibits not only electron transport to oxygen, but also ascorbate synthesis. On the other hand, GalDH activity is high and unaffected by rotenone when Gal is oxidized together with succinate or external NADH as the substrate (Millar et al. 2003). This indicates functional interactions between the electron transport enzymes, but the molecular mechanism remains to be elucidated.

In the light of recent advances in our knowledge, it is clear that although GalDH probably only constitutes a small flux of electron transport in mitochondria, the flexibility of the mitochondrial ETC that is imposed by the AOX is likely to be very important for the maintenance of cellular levels of ascorbate.

Budar et al. 2003; Noctor et al. 2007). An Arabidopsis mutant for the nuclear *NDUFS4* gene, encoding a complex I subunit, was recently shown to have strong broadly overlapping phenotypic effects (Meyer et al. 2009). It is therefore clear that proton-pumping ETC complexes have a central importance for cellular metabolism and redox homeostasis, extending beyond the boundaries of the mitochondria. This is also evident from the additional involvement in ascorbate synthesis (Box 14.1).

14.4 Alternative NAD(P)H Dehydrogenases

In most eukaryotes, complex I is the major dehydrogenase for oxidation of NADH derived from mitochondrial matrix reactions like the Krebs cycle, to initially reduce UQ in the ETC. By pumping protons, complex I contributes to link this oxidation to ATP synthesis. However, in plant, fungal, and protist mitochondria there are additional type II NAD(P)H dehydrogenases that perform a similar reaction without energy conservation (Fig. 14.1b). These alternative NAD(P)H dehydrogenases are generally flavoproteins made up by a single polypeptide of 50–60 kDa that binds FAD as cofactor and is peripherally bound to the inner or outer surface of the inner mitochondrial membrane (Rasmusson et al. 2004, 2008).

Matrix NADH is oxidized by a type II NAD(P)H dehydrogenase reaction taking place on the inner surface of the inner mitochondrial membrane (Rasmusson and Møller 1991). In plants, this function is most likely performed by the NDA protein family (Rasmusson et al. 1999; Moore et al. 2003; Michalecka et al. 2003; Elhafez et al. 2006). Proteins of the homologous NDB family are localized at the outer surface of the inner mitochondrial membrane and are therefore responsible for oxidation of cytosolic NAD(P)H (Rasmusson et al. 1999; Elhafez et al. 2006). Arabidopsis and potato NDB1 oxidizes NADPH to UQ in a fully Ca^{2+}-dependent reaction, whereas the Arabidopsis NDB2 has been shown to be a Ca^{2+}-regulated NADH dehydrogenase and Arabidopsis NDB4 a Ca^{2+}-independent NADH dehydrogenase (Michalecka et al. 2004; Geisler et al. 2007). Arabidopsis additionally contain an *NDB3* gene and an *NDC1* gene, about which less is known. NDC1 may have a function as internal NADPH dehydrogenase in mitochondria, an activity that has been measured in mitochondrial membranes, but for which the enzyme has not yet been identified (Rasmusson and Møller 1991; Melo et al. 1996). *NDA*, *NDB*, and *NDC* type genes are present in both mono- and dicotyledons, but with variations in gene copy numbers within each gene family (Rasmusson et al. 2004).

The internal NADH dehydrogenase can directly bypass the complex I reaction, and thus function as an overflow path under conditions when complex I activity is restricted, e.g., due to a large electrochemical proton gradient. However, the reducing power carried by matrix NADH can also be exported by exchange of malate and oxaloacetate with the cytosol, where NADH is then reformed. Therefore, the external NADH dehydrogenases (NDB2 and NDB4), which oxidize cytosolic NADH, can also bypass the complex I reaction. In mitochondria isolated from different plant materials, there is a substantial variation in Ca^{2+} dependence for oxidation of external NADH (Cowley and Palmer 1978; Nash and Wiskich 1983; Møller 1997). This is likely due to variations in expression of the Ca^{2+}-dependent and -independent external NADH dehydrogenases. However, the physiologic importance of the Ca^{2+} dependence has not been determined.

The external NADPH dehydrogenase has been studied in transgenic *N. sylvestris* genotypes overexpressing and sense-suppressing the potato *NDB1* gene. The encoded enzyme was clearly active in both stems and leaves, as the NADP(H) reduction level in total extracts from both organs was modulated depending on the presence of the enzyme. Furthermore, in leaves the modulation was dependent on the light conditions (Liu et al. 2008, 2009). The variation in *NDB1*-expression in stem was associated with changes in the metabolite profile as well as expression of floral meristem identity genes and, as a likely consequence, a bolting time phenotype (Liu et al. 2009). A change in the NADP(H) reduction level may affect the hundreds of reactions of central and intermediary metabolism using NADPH. Direct importance of NADP(H) redox levels has been suggested for both abiotic and biotic stress (Allan et al. 2008; Scharte et al. 2009).

14.5 Alternative Oxidase

The main ETC of plant mitochondria is sensitive to cyanide, which inhibits complex IV. However, plants also have a cyanide-insensitive alternative pathway, catalyzed by the AOX, for this final step in respiration. The AOX is a di-iron protein that is embedded in the matrix leaflet of the inner mitochondrial membrane, where it oxidizes ubiquinol and transfers the electrons to oxygen to give water as the product (Huq and Palmer 1978). The polypeptide molecular mass is generally around 35 kDa for homologues in most plant species. In its least active form, it is a homodimer, which can be activated by reduction of a dithiol (formed by a cysteine residue in the N-terminal part of each monomer) to become a noncovalent dimer (Siedow and Umbach 2000; Moore et al. 2002; Berthold and Stenmark 2003). The thiol reduction is carried out by a mitochondrial thioredoxin (Gelhaye et al. 2004), using electrons derived from mitochondrial NADPH formed from carbon reactions associated with the Krebs cycle (Vanlerberghe et al. 1995; Møller and Rasmusson 1998). The AOX can be further activated if the reduced cysteine residue binds keto-acids, such as the respiratory substrate pyruvate, allowing AOX to efficiently compete with the bc_1-complex for ubiquinol (Millar et al. 1993, 1996). It is, however, not fully clear to what extent these regulatory changes take place in vivo (Millenaar and Lambers 2003).

AOXs has a wide distribution in nature, among plants, protists, some fungi, and lower animals (McDonald and Vanlerberghe 2004). In higher plants, AOX genes occur in two gene subfamilies, *AOX1* and *AOX2*. *AOX1* is present in monocotyledonous and dicotyledonous plants, while *AOX2* is only present in dicotyledonous plants (Considine et al. 2002; Karpova et al. 2002). The number of *AOX1* and *AOX2* homologues varies in dicotyledons. In soybean for example, two *AOX2* genes have been found, but only one *AOX1* gene (Thirkettle-Watts et al. 2003). In Arabidopsis, there are four *AOX1* genes *AOX1*a-d and one *AOX2* gene. The different gene homologues for AOX have individual expression patterns through development and in different organs. For example, distinct differences are observed for the Arabidopsis homologues (Clifton et al. 2006). Overall, however, most organs of investigated plants contain at least one active AOX.

The alternative oxidase is in many cases induced by stress conditions, which has been detected in different plant species and at the transcript, protein, and enzymatic capacity level, as well as by changes in vivo activity. The stress conditions include abiotic factors like low temperature (Vanlerberghe and McIntosh 1992), drought (Ribas-Carbo et al. 2005), and nutrient deficiency (Sieger et al. 2005), as well as biotic effectors and pathogen response intermediates (Lennon et al. 1997; Huang et al. 2002) and perturbation of metabolism by inhibitors (Djajanegara et al. 2002; Clifton et al. 2005) or in mutants lacking respiratory enzymes (Sabar et al. 2000; Karpova et al. 2002). There has, however, in several cases been a discrepancy between gene expression and in vivo activity, suggesting an intricate temporal control of the enzyme (Rasmusson et al. 2009). Nevertheless, studies of transgenic plants deficient in or suppressed for *AOX*-

genes, especially the stress-responsive *AOX1a*-genes in Arabidopsis and tobacco, have confirmed the importance of AOX function for growth at low temperature (Fiorani et al. 2005), tissue homeostasis at combined drought and elevated light stress (Giraud et al. 2008) and avoidance of programmed cell death induced by biotic stress signals (Ordog et al. 2002; Robson and Vanlerberghe 2002; Vanlerberghe et al. 2002).

AOX has been suggested to have a variety of roles, which are not mutually exclusive. Where no reference is given see (Vanlerberghe et al. 2009):

- To act as an overflow mechanism when the cells needs to turn over carbon compounds without the need to produce ATP.
- To act as a "survival protein" during disruptions in energy metabolism.
- To prevent ETC overreduction and in that way lower the rate of ROS production.
- To lower the oxygen concentration in the tissue to minimize ROS production (Gupta et al. 2009)
- To counteract short-term metabolic fluctuations (Rasmusson et al. 2009).

14.6 Coordination of Alternative NAD(P)H Dehydrogenases and the Alternative Oxidases

By analyzing microarray data from experiments in which cellular metabolism had been disturbed by a very wide range of treatments – abiotic stress, biotic stress, metabolic inhibitors, etc. – a clear pattern of coexpression (transcript numbers) has been observed between particular AOX1 homologues and NAD(P)H dehydrogenase homologues (Rasmusson et al. 2009). NDB2 is coexpressed with AOX1a (Clifton et al. 2005; Ho et al. 2007), NDB4 is coexpressed with AOX1c (Ho et al. 2007), NDA2 is coexpressed with AOX1d, and NDB3 is coexpressed with AOX1b (Rasmusson et al. 2009). Also in directed analyses using quantitative PCR technology, coordinated responses have been observed in response to various external cues (Clifton et al. 2005; Elhafez et al. 2006; Escobar et al. 2006).

The observed pairwise correlation between transcripts of NAD(P)H dehydrogenases and AOXs, comes with the caveat that there is, in general, a poor correlation between transcript levels and protein levels. For AOX, the situation is further compounded by the poor correlation between the amount of protein and the activity measured, as recently reviewed (Rasmusson et al. 2009). Thus, the marked pairwise correlation may not be reflected in a correlation between activities. However, pairwise coexpression at the protein levels have been observed in some cases (Svensson and Rasmusson 2001; Sieger et al. 2005; Escobar et al. 2006), and the RNA level correlation is so striking that it should have a biologic explanation. Possible reasons for pairwise expression of NAD(P)H dehydrogenases and AOXs include pairwise protein interactions, or a pairwise fit of kinetic parameters that could adapt an expressed enzyme pair to cell-specific levels in substrates and effectors. For example, the coexpression of *AOX1a* and *NDB2*, which is an external Ca^{2+}-stimulated

NADH dehydrogenase, may mediate increased nonenergy-conserving electron flux from cytosolic NADH to oxygen with very little ATP formation under conditions where the cytosolic concentrations of Ca^{2+} and NADH are relatively high. This may, for example, be the case at high glucose levels, where both genes are induced (Price et al. 2004). Given the thermodynamic ability of Type II NAD(P)H dehydrogenases to reduce UQ to very high levels, distinct compensatory control of the AOX in response to changing NADH dehydrogenase activity is likely to be essential for maintenance of a relatively constant UQ reduction level. This is especially important because overreduction may lead to elevated ROS production.

14.7 Formation of Reactive Oxygen Species and Its Consequences

Chloroplasts, peroxisomes, and mitochondria are major sites of ROS production in the plant cell (Foyer and Noctor 2009), but large amounts of ROS can also be produced at the plasma membrane during the oxidative burst associated with the defense response against invading pathogens and abiotic stress (Torres and Dangl 2005). The ROS concentration in plant cells is generally elevated under stress conditions, and this can lead to induction of programmed cell death events, if ROS is allowed to accumulate above cellular antioxidant capacities.

In the mitochondrial ETC, the two oxidases, cytochrome oxidase or AOX, interact with O_2 and transfer four electrons to reduce it to water. However, oxygen can also pick up a single electron at sites in complexes I and III (Fig. 14.1c). When this happens, superoxide (O_2^-) is formed. Superoxide can be transformed into H_2O_2 by superoxide dismutase, an enzyme present in the matrix, and H_2O_2 can, in turn, give rise to the very reactive hydroxyl radical ($HO^•$) if it interacts with free metal ions such as Fe^{2+} or Cu^+ (the so-called Fenton reaction). Plant mitochondria contain several enzyme systems, all using NADPH, dedicated to removing H_2O_2 from the matrix, as reviewed (Møller 2001; Møller et al. 2007).

It has been estimated that 1–5% of the oxygen consumed by mammalian mitochondria is diverted to ROS formation, depending on the conditions (Herrero and Barja 1997). It appears that plant mitochondria produce ROS at similar or higher rates, and that the rate depends on the reduction level of the ETC. This means that ROS formation will be high when input into the ETC increases relative to output, so that any metabolic change leading to increases in the formation of respiratory substrates or in a restriction in ETC capacity will lead to increased ROS production. By reoxidizing ubiquinol, the AOX can decrease the reduction level of the ETC and lower ROS production (Purvis and Shewfelt 1993; Maxwell et al. 1999; Møller 2001).

It is a general observation that ROS production increases whenever a plant cell is under stress and there is evidence that the mitochondria contribute to that increase. The stress-induced ROS formation can contribute to signaling which leads to activation/repression of nuclear genes, so-called retrograde signaling, but it can also damage cellular components (Neill et al. 2002; Gadjev et al. 2006; Rhoads et al. 2006; Møller et al. 2007; Foyer et al. 2009; Sweetlove and Møller 2009). H_2O_2 is relatively stable and can cross membranes through aquaporins (Bienert et al. 2007). This makes it a good candidate for retrograde signaling (however, see below). The subcellular localization of ROS formation can be studied by confocal laser microscopy using ROS-sensitive fluorescent dyes (Kristiansen et al. 2009), cytochemically or by using alamethicin (Szal et al. 2009), a peptide that forms channels through plant membranes (Johansson et al. 2004b; Matic et al. 2005). It is also possible to monitor the redox state of the mitochondria by transforming Arabidopsis with a redox-sensitive mitochondrially targeted green fluorescent protein (Schwarzländer et al. 2009).

14.7.1 Mitochondrial Protein Oxidation

ROS can interact with and modify/damage many cellular components, like nucleic acids, sugars, fatty acids, and proteins (Møller et al. 2007). Peroxidation of polyunsaturated fatty acids, which occurs at increased rates under oxidative stress conditions, can give rise to reactive breakdown products such as 4-hydroxy-2-nonenal (HNE), which in turn can form adducts with proteins. An example of this is the binding of HNE to the lipoic acid in the H-protein of glycine decarboxylase, which leads to the inhibition of the enzyme activity and breakdown of the protein, and thus inactivation of glycine oxidation (Taylor et al. 2002). It should be noted that although AOX activity can help lowering ROS production under stress conditions, the enzyme itself is also sensitive to inhibition by lipid peroxidation products

Table 14.1 Common ROS-induced modifications of mitochondrial proteins (Based on Halliwell and Gutteridge 2007)

Amino acid residue affected	Product(s)	Comment
Met	Methionine sulfoxide and methionine sulfone	First oxidation step is reversible
Cys	Cystine	Reversible
	Three-step oxidation to cysteic acid	First step is reversible
	Nitrosylation	Conjugation with NO
Trp	N-formylkynurenine and kynurenine	Studied in mammalian and plant mitochondria using mass spectrometry (Taylor et al. 2003; Møller and Kristensen 2004)
Arg, His, Lys, Pro, and Thr	Carbonyl group formed	Studied in plant mitochondria using dinitrophenylhydrazine (DNP) and anti-DNP antibodies (Kristensen et al. 2004)
Tyr	Bityrosine	

(Winger et al. 2005). UCP, which is stimulated by superoxide (Considine et al. 2003), may therefore be a more important ameliorator of severe oxidative stress (Fernie et al. 2004).

Proteins can react directly with ROS in a number of ways and the most important are listed in Table 14.1. Several studies have addressed the question of which mitochondrial proteins are oxidized/affected by oxidative stress (Karpova et al. 2002; Sweetlove et al. 2002; Taylor et al. 2003; Kristensen et al. 2004; O'Brien et al. 2004; Sieger et al. 2005; Møller and Kristensen 2006) and that has been reviewed in detail (Møller et al. 2007). It was concluded that many of the same proteins are modified in mitochondria from plants, yeast, and mammals when treated in similar ways. This indicates that these proteins are either more susceptible to oxidation, more exposed to ROS or both. However, most of these proteins are quite abundant and it is possible that this increases the risk of collision with ROS. The high abundance might also facilitate the detection of the oxidized form.

Since most of mitochondrial ROS is produced in complexes I and III located in the inner mitochondrial membrane one might expect most of the oxidized proteins to be at least membrane-associated if not actually subunits of these two respiratory complexes. The results of several investigations indicate that such damage is observed. However, even under unstressed conditions soluble proteins such as the Krebs cycle enzymes, which may not be physically close to the site of ROS synthesis, are affected (Sweetlove et al. 2002; Taylor et al. 2003; Johansson et al. 2004a; Kristensen et al. 2004; Møller and Kristensen 2006). Mitochondrial aconitase appears to be particularly susceptible to oxidative damage: It contains oxidized tryptophan in vivo (Taylor et al. 2003; Møller and Kristensen 2006) and carbonylation (Kristensen et al. 2004) and fragmentation products (Sweetlove et al. 2002) are detected after in vitro and in vivo oxidative stress, respectively. Damage to aconitase may lead to the release of iron and this has been suggested to lead to further ROS production via the Fenton reaction, i.e., a vicious circle (Ilangovan et al. 2006).

Damaged and redundant proteins are degraded, and mitochondria contain a number of different proteases that can mediate this (Adam et al. 2001). The half-life of plant mitochondrial proteins has been estimated to be 6–72 h (Sweetlove and Møller 2010). We do not yet know to what extent removal of oxidized proteins contribute to protein turnover, but the replacement of all mitochondrial proteins every 3–4 days is costly. It has been estimated that 2–20% of the ATP produced by a plant mitochondrion may be used for replacing its proteins (Møller et al. 2007).

The presence of oxidized proteins is a disadvantage to the living organism that is worth avoiding, as indicated by the observation that plants, animals and yeast have all found a way to minimize the transfer of oxidized proteins to the next generation, as reviewed (Møller et al. 2007). For instance, in mammals a small, metabolically relatively inactive subpopulation of "stem" mitochondria in each cell may specialize in DNA replication and division, while the metabolically active (and therefore more ROS-producing and ROS-damaged) mitochondria would not proliferate (Lemasters 2005).

Although H_2O_2 is a likely candidate for mediating retrograde regulation (see above), it lacks the specificity needed to regulate genes in response to stress in a specific compartment, i.e., H_2O_2 is the same irrespective of the source. It has therefore been speculated that peptides deriving from degradation of mitochondrial proteins could act as retrograde signals from the mitochondrion to the nucleus (Koppen and Langer 2007; Møller et al. 2007; Sweetlove and Møller 2010). Such peptides, especially if containing oxidized amino acids, could have a source-specificity that H_2O_2 lacks.

In summary, being a major generator of both ATP and ROS in all cells, both of which are central to stress response and stress tolerance, it is clear that the mitochondrial ETC and its maintenance has a central role in plant responses to stressful environmental factors.

14.8 Supplementary Pathways of Carbon Conversion Are Active Under Particular Stress Conditions

14.8.1 Respiration and Stress Tolerance

The ATP demand of the cell varies depending on plant species, plant organ, and external conditions. Plant mitochondria will also oxidize a great variety of respiratory substrates, including glycolytic pyruvate, Krebs cycle intermediates, photorespiratory glycine and other amino acids. Also, the outputs of activity through the respiratory carbon metabolism paths will vary. In addition to the default product CO_2, Krebs cycle intermediates are needed for nitrogen assimilation and other biosynthetic pathways, and in some cases for exudation to the soil (e.g., for chelating toxic ions). Therefore the ETC must adjust its composition and the rates through the different branches to the particular demands. Respiration is active in all cells at almost every stage of plant life, and is thus active under a wider range of external conditions (e.g., temperature variation and water status) than photosynthesis and provides ATP for stress tolerance mechanisms. In response to cold stress, leaf respiration will initially decline. Later, either acclimation of the present leaves or formation of new acclimated leaves will restore electron transport activity through the proton-pumping ETC complexes to maintain ATP synthesis and growth (Atkin et al. 2006). In germinating pea seeds, mitochondrial respiration was found to be active at temperatures between –3.5 and 40°C (Stupnikova et al. 2006). At the lower extreme, the ETC activity was mainly due to the external NADH dehydrogenase, which is clearly vital for seed germination under low temperature conditions. NDA1 expression was observed to decrease during cold stress of potato leaves, whereas complex I activity and capacities for external NAD(P)H dehydrogenases were unaffected (Svensson et al. 2002). Presumably cold treatment leads to a slower influx of NAD-linked substrates into the matrix and a resulting smaller need for additional NADH-oxidizing capacity. Induction of AOX is also a common response to cold treatment (Vanlerberghe and McIntosh 1992), further emphasizing the importance of balancing supplementary UQ-reducing and ubiquinol-oxidizing pathways during changes in temperature.

14.8.2 Mitochondrial D-Lactate Dehydrogenase Removes By products of Carbon Metabolism

Methyl glyoxal is a cytotoxic side product of metabolic pathways like threonine degradation, lipid peroxidation, and glycolysis, where a major nonenzymatic production by dephosphorylation of triose phosphates occurs (Kalapos 1999). In humans, methyl glyoxal has been linked to oxidative stress (Mukhopadhyay et al. 2007). The degradation of methyl glyoxal involves reaction with glutathione. After conversion of the formed hemithioacetal by glyoxalases, D-lactate is oxidized by a mitochondrial D-lactate dehydrogenase linked to the UQ pool. This enzyme has been found in plant mitochondria (Atlante et al. 2005; Engqvist et al. 2009). It is likely that degradation of methyl glyoxal has special importance under stress, because overexpression of glyoxalases of this pathway leads to increased tolerance to soil salinity and high Zn^{2+} concentrations (Singla-Pareek et al. 2003, 2006).

14.8.3 Electron-Transfer Flavoprotein: Quinone Oxidoreductase Mediates Degradation of Amino Acids Under Carbon Starvation

In leaves, the amount of starch displays cyclic changes during the diurnal cycle with an almost complete depletion taking place at the end of the night in several species, including Arabidopsis. As a consequence, extending the night even by only a few hours, leads to carbon starvation (Smith and Stitt 2007). During carbon starvation stress, the cell mobilizes nonsugar substrates for maintaining respiration. A major source are amino acids as indicated by the induction of glutamate dehydrogenase and enzymes degrading branched amino acids in darkness, including the electron-transfer flavoprotein:quinone oxidoreductase which mediates the reduction to the ETC (Fujiki et al. 2000). Consistently, there is a decreased dark survival of Arabidopsis mutants lacking the electron transfer flavoprotein and the enzyme oxidizing it (Ishizaki et al. 2005, 2006).

14.8.4 Glycerol Metabolism Involves an External Glycerol-3-Phosphate Dehydrogenase

Like in yeast and mammals, plant mitochondria are able to oxidize glycerol-3-phosphate from the cytosol via a UQ-reducing glycerol-3-phosphate dehydrogenase (Shen et al. 2003). In conjunction with a cytosolic NAD-reducing homologue, this may constitute a shuttle for NADH oxidation (Shen et al. 2006), or be involved in maintenance of glycerol and glycerol phosphate levels, as previously reviewed

(Rasmusson et al. 2008). An Arabidopsis mutant for the NAD-glycerol-3-phosphate dehydrogenase displayed changes in ROS levels and redox levels after ABA treatment indicating a potential involvement in stress metabolism (Shen et al. 2006). A further indication of potential stress involvement of glycerol metabolism is the observation that accumulation of glycerol leads to resistance to several stresses in Arabidopsis (Eastmond 2004).

14.8.5 Mitochondrial Proline Degradation Readjusts the Cellular Osmotic Potential upon Alleviation of Drought Stress

Plants are variable with respect to drought stress tolerance, and respiration is both a potential target for desiccation damage and involved in the adjustment to the stress. Total respiratory oxygen consumption is not affected by drought conditions that severely inhibit photosynthesis, but under severe water stress the activity of AOX will increase, leading to a lower ATP yield (Ribas-Carbo et al. 2005). A possible reason for the general drought-tolerance of respiration is that during drought stress and desiccation, the mitochondrial inner membrane is protected by a mitochondrial late-embryogenesis abundant protein (Tolleter et al. 2007). However, the change in relative energy yield of respiration also indicates that modifications in respiratory pathways contribute to plant drought acclimation.

A major tolerance mechanism for stresses linked to dehydration is the accumulation of compatible solutes that lower the osmotic potential without damaging cellular function. The imino acid proline is an abundant osmolyte in many plant species. Upon drought stress, proline is synthesized in the cytosol from glutamate via the intermediate delta-1-pyrroline-5-carboxylate (P5C), and accumulates in the cell (Hare et al. 1999). Under such conditions, mitochondrial electron transport is substantially decreased (Gibon et al. 2000). When water status returns to normal, the proline is oxidized by the mitochondrial proline dehydrogenase of the ETC, forming P5C (Fig. 14.1d) (Elthon and Stewart 1981; Rayapati and Stewart 1991; Mani et al. 2002; Di Martino et al. 2006). P5C is oxidized to glutamate by the NAD(P)-reducing P5C dehydrogenase in the mitochondrial matrix (Forlani et al. 1997), which provides further reductant for the ETC. This pathway has a regulatory role for proline levels, as indicated by the observation that the proline concentration in the plant is elevated in proline dehydrogenase mutants (Nanjo et al. 1999). Since large amounts of proline are degraded upon rehydration, this pathway likely constitutes a major electron transport flux, comparable to, e.g., photorespiratory mitochondrial oxygen consumption. Since proline additionally can induce programmed cell death in plant cells (Mani et al. 2002; Nanjo et al. 2003; Deuschle et al. 2004), it is vital that proline levels are rigorously regulated. This is mediated by a cytosolic-mitochondrial cycle of proline and P5C, which serves to minimize ROS production during proline degradation (Miller et al. 2009).

14.9 Electron Transport Regulation During Oxygen Deprivation

Plant roots are exposed to hypoxia, or in its extreme form, anoxia, during flooding. Under these conditions it is generally assumed that the mitochondrial ETC is severely limited by lack of electron acceptors (oxygen) and the plant cell switches to fermentation to provide the required ATP and recycle reducing equivalents. However, Stoimenova et al. (2007) showed that under anaerobic conditions mitochondria isolated from barley and rice roots oxidize NADH, NAPDH, and succinate with nitrite as the electron acceptor reducing it to NO. Inhibitor sensitivities indicated that ubisemiquinone in complex III is involved, similar to superoxide formation (Møller 2001). However, since KCN inhibited nitrite reduction more than myxothiazol did, also complex IV is likely to reduce nitrite (Fig. 14.1c) (Stoimenova et al. 2007). At present, the mechanism of NO removal is not clear. Avoidance of NO accumulation is important as NO could otherwise react with cysteine groups and cause nitrosylation (see above). NO is also considered to be an important messenger and its concentration must therefore be tightly regulated (Crawford and Guo 2005). Since NO is an inhibitor of complex IV, it has been suggested that the ETC can be autoregulatory via NO formation to adjust oxygen consumption under hypoxia, and thus to avoid complete anoxia (Borisjuk et al. 2007; Benamar et al. 2008). It has also been suggested that pyruvate accumulation under hypoxia may activate AOX and limit ROS production by decreasing the oxygen concentration (Gupta et al. 2009; Zabalza et al. 2009).

Reoxygenation following anoxia causes problems for plant cells because the initial high reduction level of all the ETC components gives rise to a burst of ROS production. In rice seedlings grown under anoxia the amount of complexes III and IV was low probably because of a downregulated heme biosynthesis. Biosynthesis of the complexes as well as AOX was induced within 24 h of return to oxygenated conditions (Millar et al. 2004), and it was emphasized that the AOX is "… also a heme-independent respiratory pathway." Thus, the AOX could be used in the immediate phase following conditions that prevent heme accumulation, such as anoxia, to provide a rapid flux increase to meet the requirements for aerobic respiration and prevent an overreduction of the respiratory chain.

14.10 Concluding Remarks

Given the plethora of biochemical pathways that intersect at the ETC in plant mitochondria it is clear that the plant ETC is far from the virtually linear pathway often depicted in textbooks of general biochemistry. The ETC of plant mitochondria consists of a number of UQ-reducing enzymes and enzyme complexes coupled to several ubiquinol-oxidizing enzymes and enzyme complexes via one mobile carrier, UQ. They are coupled in such a way that the energy yield is very flexible; anything from 0 to 10 protons pumped per two electrons is possible.

This highly flexible but complex system is involved in signaling and tolerance to plant stresses. The UQ reduction level reflects the balance between input and output, which must be carefully regulated by the cell because over-reduction leads to increased ROS production and possible damage. Also, the system must dynamically accommodate sudden changes in flux through ETC enzymes linked to specific synthesis and degradation reactions (e.g., proline degradation and ascorbate synthesis) (Box 14.1). By combining transgenic lesions and global analyses methods these questions are presently addressed.

We need to improve our understanding of the dynamics of electron transport in plant mitochondria in order to better predict the respiratory contribution to the whole plant carbon budget under the variable conditions that plants and their organs are exposed to. Such information is essential for improving our understanding of growth yield variations, the interaction between plants and the atmosphere, plant stress tolerance, as well as postharvest respiratory biomass losses.

Acknowledgments A.G.R. acknowledges financial support from the Swedish Research Council, and I.M.M. from the Faculty of Agricultural Sciences.

Glossary

Acclimation: A plant stress response, physiologic, and/or morphologic, that increases the ability of the plant to tolerate stress.

Alternative oxidase: A peripheral membrane protein that oxidizes ubiquinol to ubiquinone and reduces O_2 to H_2O, and which constitutes an energy-bypass.

Anoxia: The special case of hypoxia where O_2 is virtually absent.

Ascorbate: A major cellular antioxidant (vitamin c), synthesized by the electron transport chain in plants.

Compatible solute: A molecule that is synthesized in response to osmotic stress in so large amounts that the osmotic potential is lowered significantly without harming cellular constituents,

Electrochemical proton gradient: The combined effect of the difference in electrical charge and in proton concentration across a membrane, determining the energy change for protons transported across the membrane.

Electron transport chain: A set of redox enzymes linked by associated smaller molecules that allow electrons to pass from reduced substrates (e.g., NADH) to acceptors (e.g., O_2).

Energy-bypass: A protein that forms a parallel path for electron or proton transport, bypassing one or several of the sites for energy coupling between electron transport and ATP synthesis.

Glycine oxidation: The conversion of two glycines to serine, CO_2, and NH_4^+. This reaction is the major provider of NADH in the mitochondrial matrix in leaves of C3-plants in the light.

Hypoxia: The condition where the O_2 concentration is below a threshold level imposing a deficiency effect on the plant.

Male sterility: A defect in production of functional pollen often associated with changes in mitochondria.

NAD(P)H dehydrogenases: Membrane proteins that oxidize NADH to NAD^+, and/or NADPH to $NADP^+$ and reduce ubiquinone to ubiquinol. Plants contain type I (complex I) and type II NAD(P)H dehydrogenases, the latter being energy-bypasses to the former.

NO: Nitric oxide; a signaling molecule and inhibitor of complex IV.

Oxidative stress: A condition where reactive oxygen species accumulate to such high levels that they negatively affect cellular function.

Reactive oxygen species: ROS; oxygen-containing molecules that can damage cellular constituents by chemical reaction with them. Many ROS are radicals.

Reductant: A molecule (e.g., NADH, NADPH, malate, and reduced ferredoxin) that carries low redox potential electrons, which can be donated to another molecule via a redox reaction.

Retrograde regulation: The process where the functional status in an organelle is signaled to the nucleus and affects gene expression.

Ubiquinone: An electron and proton carrier located in the mitochondrial inner membrane, where it mediates electron and proton transport.

Uncoupling protein: A proton transporter and energy-bypass that allows protons exported by the electron transport chain to flow back into the matrix without passing the ATP synthase.

References

Adam, Z., Adamska, I., Nakabayashi, K., Ostersetzer, O., Haussuhl, K., Manuell, A., Zheng, B., Vallon, O., Rodermel, S. R., Shinozaki, K., Clarke, A. K. 2001. Chloroplast and mitochondrial proteases in Arabidopsis. A proposed nomenclature. Plant Physiol. 125:1912–1918.

Allan, W. L., Simpson, J. P., Clark, S. M., Shelp, B. J. 2008. Gamma-hydroxybutyrate accumulation in Arabidopsis and tobacco plants is a general response to abiotic stress: putative regulation by redox balance and glyoxylate reductase isoforms. J. Exp. Bot. 59:2555–2564.

Atkin, O. K., Loveys, B. R., Atkinson, L. J., Pons, T. L. 2006. Phenotypic plasticity and growth temperature: understanding interspecific variability. J. Exp. Bot. 57:267–281.

Atlante, A., de Bari, L., Valenti, D., Pizzuto, R., Paventi, G., Passarella, S. 2005. Transport and metabolism of D-lactate in Jerusalem artichoke mitochondria. Biochim. Biophys. Acta 1708:13–22.

Bartoli, C. G., Pastori, G. M., Foyer, C. H. 2000. Ascorbate biosynthesis in mitochondria is linked to the electron transport chain between complexes III and IV. Plant Physiol. 123:335–343.

Bartoli, C. G., Yu, J., Gomez, F., Fernandez, L., McIntosh, L., Foyer, C. H. 2006. Inter-relationships between light and respiration in the control of ascorbic acid synthesis and accumulation in *Arabidopsis thaliana* leaves. J. Exp. Bot. 57:1621–1631.

Bartoli, C. G., Tambussi, E. A., Diego, F., Foyer, C. H. 2009. Control of ascorbic acid synthesis and accumulation and glutathione by the incident light red/far red ratio in *Phaseolus vulgaris* leaves. FEBS Lett. 583:118–122.

Benamar, A., Rolletschek, H., Borisjuk, L., Avelange-Macherel, M. H., Curien, G., Mostefai, H. A., Andriantsitohaina, R., Macherel, D. 2008. Nitrite-nitric oxide control of mitochondrial respiration at the frontier of anoxia. Biochim. Biophys. Acta 1777:1268–1275.

Berthold, D. A., Stenmark, P. 2003. Membrane-bound diiron carboxylate proteins. Annu. Rev. Plant Biol. 54:497–517.

Bienert, G. P., Møller, A. L., Kristiansen, K. A., Schulz, A., Møller, I. M., Schjoerring, J. K., Jahn, T. P. 2007. Specific aquaporins facilitate the diffusion of hydrogen peroxide across membranes. J. Biol. Chem. 282:1183–1192.

Boccara, M., Boue, C., Garmier, M., De Paepe, R., Boccara, A. C. 2001. Infra-red thermography revealed a, role for mitochondria in pre-symptomatic cooling during harpin-induced hypersensitive response. Plant J. 28:663–670.

Borisjuk, L., Macherel, D., Benamar, A., Wobus, U., Rolletschek, H. 2007. Low oxygen sensing and balancing in plant seeds: a role for nitric oxide. New Phytol. 176:813–823.

Budar, F., Touzet, P., De Paepe, R. 2003. The nucleo-mitochondrial conflict in cytoplasmic male sterilities revisited. Genetica 117:3–16.

Clifton, R., Lister, R., Parker, K. L., Sappl, P. G., Elhafez, D., Millar, A. H., Day, D. A., Whelan, J. 2005. Stress-induced co-expression of alternative respiratory chain components in *Arabidopsis thaliana*. Plant Mol. Biol. 58:193–212.

Clifton, R., Millar, A. H., Whelan, J. 2006. Alternative oxidases in Arabidopsis: a comparative analysis of differential expression in the gene family provides new insights into function of non-phosphorylating bypasses. Biochim. Biophys. Acta 1757:730–741.

Considine, M. J., Holtzapffel, R. C., Day, D. A., Whelan, J., Millar, A. H. 2002. Molecular distinction between alternative oxidase from monocots and dicots. Plant Physiol. 129:949–953.

Considine, M. J., Goodman, M., Echtay, K. S., Laloi, M., Whelan, J., Brand, M. D., Sweetlove, L. J. 2003. Superoxide stimulates a proton leak in potato mitochondria that is related to the activity of uncoupling protein. J. Biol. Chem. 278:22298–22302.

Cowley, R. C., Palmer, J. M. 1978. Interaction of citrate and calcium in regulating oxidation of exogenous NADH in plant mitochondria. Plant Sci. Lett. 11:345–350.

Crawford, N. M., Guo, F. Q. 2005. New insights into nitric oxide metabolism and regulatory functions. Trends Plant Sci. 10:195–200.

Deuschle, K., Funck, D., Forlani, G., Stransky, H., Biehl, A., Leister, D., van der Graaff, E., Kunzee, R., Frommer, W. B. 2004. The role of Δ^1-pyrroline-5-carboxylate dehydrogenase in proline degradation. Plant Cell 16:3413–3425.

Di Martino, C., Pizzuto, R., Pallotta, M. L., De Santis, A., Passarella, S. 2006. Mitochondrial transport in proline catabolism in plants: the existence of two separate translocators in mitochondria isolated from durum wheat seedlings. Planta 223:1123–1133.

Djajanegara, I., Finnegan, P. M., Mathieu, C., McCabe, T., Whelan, J., Day, D. A. 2002. Regulation of alternative oxidase gene expression in soybean. Plant Mol. Biol. 50:735–742.

Dutilleul, C., Driscoll, S., Cornic, G., De Paepe, R., Foyer, C. H., Noctor, G. 2003a. Functional mitochondrial complex I is required by tobacco leaves for optimal photosynthetic performance in photorespiratory conditions and during transients. Plant Physiol. 131:264–275.

Dutilleul, C., Garmier, M., Noctor, G., Mathieu, C., Chétrit, P., Foyer, C. H., De Paepe, R. 2003b. Leaf mitochondria modulate whole cell redox homeostasis, set antioxidant capacity, and determine stress resistance through altered signaling and diurnal regulation. Plant Cell 15:1212–1226.

Dutilleul, C., Lelarge, C., Prioul, J. L., De Paepe, R., Foyer, C. H., Noctor, G. 2005. Mitochondria-driven changes in leaf NAD status exert a crucial influence on the control of nitrate assimilation and the integration of carbon and nitrogen metabolism. Plant Physiol. 139:64–78.

Eastmond, P. J. 2004. Glycerol-insensitive Arabidopsis mutants: *Gli*1 seedlings lack glycerol kinase, accumulate glycerol and are more resistant to abiotic stress. Plant J. 37:617–625.

Elhafez, D., Murcha, M. W., Clifton, R., Soole, K. L., Day, D. A., Whelan, J. 2006. Characterization of mitochondrial alternative NAD(P)H dehydrogenases in Arabidopsis: intraorganelle location and expression. Plant Cell Physiol. 47:43–54.

Elthon, T. E., Stewart, C. R. 1981. Sub-mitochondrial location and electron-transport characteristics of enzymes involved in proline oxidation. Plant Physiol. 67:780–784.

Engqvist, M., Drincovich, M. F., Flugge, U. I., Maurino, V. G. 2009. Two D-2-hydroxy-acid dehydrogenases in *Arabidopsis thaliana* with catalytic capacities to participate in the last reactions of the methylglyoxal and beta-oxidation pathways. J. Biol. Chem. 284:25026–25037.

Escobar, M. A., Geisler, D. A., Rasmusson, A. G. 2006. Reorganization of the alternative pathways of the Arabidopsis respiratory chain by nitrogen supply: opposing effects of ammonium and nitrate. Plant J. 45:775–788.

Fernie, A. R., Carrari, F., Sweetlove, L. J. 2004. Respiratory metabolism: glycolysis, the TCA cycle and mitochondrial electron transport. Curr. Opin. Plant Biol. 7:254–261.

Fiorani, F., Umbach, A. L., Siedow, J. N. 2005. The alternative oxidase of plant mitochondria is involved in the acclimation of shoot growth at low temperature. A study of Arabidopsis *AOX1a* transgenic plants. Plant Physiol. 139:1795–1805.

Forlani, G., Scainelli, D., Nielsen, E. 1997. Delta(1)-pyrroline-5-carboxylate dehydrogenase from cultured cells of potato – purification and properties. Plant Physiol. 113:1413–1418.

Foyer, C. H., Noctor, G. 2009. Redox regulation in photosynthetic organisms: signaling, acclimation, and practical implications. Antioxid Redox Signal. 11:861–905.

Foyer, C. H., Bloom, A., Queval, G., Noctor, G. 2009. Photorespiratory metabolism: genes, mutants, energetics, and redox signaling. Annu. Rev. Plant Biol. 60:455–484.

Fujiki, Y., Sato, T., Ito, M., Watanabe, A. 2000. Isolation and characterization of cDNA, clones for the E1 beta and E2 subunits of the branched-chain alpha-ketoacid dehydrogenase complex in Arabidopsis. J. Biol. Chem. 275:6007–6013.

Gadjev, I., Vanderauwera, S., Gechev, T. S., Laloi, C., Minkov, I. N., Shulaev, V., Apel, K., Inze, D., Mittler, R., Van Breusegem, F. 2006. Transcriptomic footprints disclose specificity of reactive oxygen species signaling in Arabidopsis. Plant Physiol. 141:436–445.

Geisler, D. A., Broselid, C., Hederstedt, L., Rasmusson, A. G. 2007. Ca^{2+}-binding and Ca^{2+}-independent respiratory NADH and NADPH dehydrogenases of *Arabidopsis thaliana*. J. Biol. Chem. 282:28455–28464.

Gelhaye, E., Rouhier, N., Gerard, J., Jolivet, Y., Gualberto, J., Navrot, N., Ohlsson, P. I., Wingsle, G., Hirasawa, M., Knaff, D. B., Wang, H., Dizengremel, P., Meyer, Y., Jacquot, J. P. 2004. A specific form of thioredoxin h occurs in plant mitochondria and regulates the alternative oxidase. Proc. Natl. Acad. Sci. U.S.A. 101:14545–14550.

Gibon, Y., Sulpice, R., Larher, F. 2000. Proline accumulation in canola leaf discs subjected to osmotic stress is related to the loss of chlorophylls and to the decrease of mitochondrial activity. Physiol. Plant. 110:469–476.

Giraud, E., Ho, L. H., Clifton, R., Carroll, A., Estavillo, G., Tan, Y. F., Howell, K. A., Ivanova, A., Pogson, B. J., Millar, A. H., Whelan, J. 2008. The absence of ALTERNATIVE OXIDASE1a in Arabidopsis results in acute sensitivity to combined light and drought stress. Plant Physiol. 147:595–610.

Gupta, K. J., Zabalza, A., van Dongen, J. T. 2009. Regulation of respiration when the oxygen availability changes. Physiol. Plant. 137:383–391.

Gutierres, S., Sabar, M., Lelandais, C., Chetrit, P., Diolez, P., Degand, H., Boutry, M., Vedel, F., de Kouchkovsky, Y., De Paepe, R. 1997. Lack of mitochondrial and nuclear-encoded subunits of complex I and alteration of the respiratory chain in *Nicotiana sylvestris* mitochondrial deletion mutants. Proc. Natl. Acad. Sci. U.S.A. 94:3436–3441.

Halliwell, B., Gutteridge, J. M. C. 2007. Free Radicals in Biology and Medicine. Oxford, UK: Oxford University Press.
Hare, P. D., Cress, W. A., van Staden, J. 1999. Proline synthesis and degradation: a model system for elucidating stress-related signal transduction. J. Exp. Bot. 50:413–434.
Heazlewood, J. L., Howell, K. A., Millar, A. H. 2003. Mitochondrial complex I from *Arabidopsis* and rice: orthologs of mammalian and fungal components coupled with plant-specific subunits. Biochim. Biophys. Acta 1604:159–169.
Herrero, A., Barja, G. 1997. Sites and mechanisms responsible for the low rate of free radical production of heart mitochondria in the long-lived pigeon. Mech. Ageing Dev. 98:95–111.
Ho, L. H., Giraud, E., Lister, R., Thirkettle-Watts, D., Low, J., Clifton, R., Howell, K. A., Carrie, C., Donald, T., Whelan, J. 2007. Characterization of the regulatory and expression context of an alternative oxidase gene provides insights into cyanide-insensitive respiration during growth and development. Plant Physiol. 143:1519–1533.
Huang, X., von Rad, U., Durner, J. 2002. Nitric oxide induces transcriptional activation of the nitric oxide-tolerant alternative oxidase in Arabidopsis suspension cells. Planta 215:914–923.
Huq, S., Palmer, J. M. 1978. Superoxide and hydrogen-peroxide production in cyanide resistant *Arum maculatum* mitochondria. Plant Sci. Lett. 11:351–358.
Ilangovan, G., Venkatakrishnan, C. D., Bratasz, A., Osinbowale, S., Cardounel, A. J., Zweier, J. L., Kuppusamy, P. 2006. Heat shock-induced attenuation of hydroxyl radical generation and mitochondrial aconitase activity in cardiac h9c2 cells. Am. J. Physiol. Cell Physiol. 290:C313–324.
Ishizaki, K., Larson, T. R., Schauer, N., Fernie, A. R., Graham, I. A., Leaver, C. J. 2005. The critical role of Arabidopsis electron-transfer flavoprotein: ubiquinone oxidoreductase during dark-induced starvation. Plant Cell 17:2587–2600.
Ishizaki, K., Schauer, N., Larson, T. R., Graham, I. A., Fernie, A. R., Leaver, C. J. 2006. The mitochondrial electron transfer flavoprotein complex is essential for survival of Arabidopsis in extended darkness. Plant J. 47:751–760.
Jiao, S. X., Thornsberry, J. M., Elthon, T. E., Newton, K. J. 2005. Biochemical and molecular characterization of photosystem I deficiency in the NCS6 mitochondrial mutant of maize. Plant Mol. Biol. 57:303–313.
Johansson, E., Olsson, O., Nyström, T. 2004a. Progression and specificity of protein oxidation in the life cycle of *Arabidopsis thaliana*. J. Biol. Chem. 279:22204–22208.
Johansson, F. I., Michalecka, A. M., Møller, I. M., Rasmusson, A. G. 2004b. Oxidation and reduction of pyridine nucleotides in alamethicin-permeabilised plant mitochondria. Biochem. J. 380:193–202.
Kalapos, M. P. 1999. Methylglyoxal in living organisms – chemistry, biochemistry, toxicology and biological implications. Toxicol. Lett. 110:145–175.
Karpova, O. V., Kuzmin, E. V., Elthon, T. E., Newton, K. J. 2002. Differential expression of alternative oxidase genes in maize mitochondrial mutants. Plant Cell 14:3271–3284.
Koppen, M., Langer, T. 2007. Protein degradation within mitochondria: versatile activities of aaa proteases and other peptidases. Crit. Rev. Biochem. Mol. Biol. 42:221–242.
Kristensen, B. K., Askerlund, P., Bykova, N. V., Egsgaard, H., Møller, I. M. 2004. Identification of oxidised proteins in the matrix of rice leaf mitochondria by immunoprecipitation and two-dimensional liquid chromatography-tandem mass spectrometry. Phytochemistry 65:1839–1851.
Kristiansen, K. A., Jensen, P. E., Møller, I. M., Schulz, A. 2009. Monitoring reactive oxygen species formation and localisation in living cells by use of the fluorescent probe cm-H(2)dcfda and confocal laser microscopy. Physiol. Plant. 136:369–383.
Lemasters, J. J. 2005. Selective mitochondrial autophagy, or mitophagy, as a targeted defense against oxidative stress, mitochondrial dysfunction, and aging. Rejuvenation Res. 8:3–5.
Lennon, A. M., Neuenschwander, U. H., Ribas-Carbo, M., Giles, L., Ryals, J. A., Siedow, J. N. 1997. The effects of salicylic acid and tobacco mosaic virus infection on the alternative oxidase of tobacco. Plant Physiol. 115:783–791.

Liu, Y. J., Norberg, F. E., Szilagyi, A., De Paepe, R., Åkerlund, H. E., Rasmusson, A. G. 2008. The mitochondrial external NADPH dehydrogenase modulates the leaf NADPH/NADP$^+$ ratio in transgenic *Nicotiana sylvestris*. Plant Cell Physiol. 49:251–263.

Liu, Y. J., Nunes-Nesi, A., Wallström, S. V., Lager, I., Michalecka, A. M., Norberg, F. E., Widell, S., Fredlund, K. M., Fernie, A. R., Rasmusson, A. G. 2009. A redox-mediated modulation of stem bolting in transgenic *Nicotiana sylvestris* differentially expressing the external mitochondrial NADPH dehydrogenase. Plant Physiol. 150:1248–1259.

Mani, S., Van De Cotte, B., Van Montagu, M., Verbruggen, N. 2002. Altered levels of proline dehydrogenase cause hypersensitivity to proline and its analogs in Arabidopsis. Plant Physiol. 128:73–83.

Marienfeld, J. R., Newton, K. J. 1994. The maize NCS2 abnormal growth mutant has a chimeric *nad4-nad7* mitochondrial gene and is associated with reduced complex I function. Genetics 138:855–863.

Matic, S., Geisler, D. A., Møller, I. M., Widell, S., Rasmusson, A. G. 2005. Alamethicin permeabilizes the plasma membrane and mitochondria but not the tonoplast in tobacco (*Nicotiana tabacum* L. Cv bright yellow) suspension cells. Biochem. J. 389:695–704.

Maxwell, D. P., Wang, Y., McIntosh, L. 1999. The alternative oxidase lowers mitochondrial reactive oxygen production in plant cells. Proc. Natl. Acad. Sci. U.S.A. 96:8271–8276.

McDonald, A., Vanlerberghe, G. 2004. Branched mitochondrial electron transport in the animalia: presence of alternative oxidase in several animal phyla. IUBMB Life 56:333–341.

Melo, A. M. P., Roberts, T. H., Møller, I. M. 1996. Evidence for the presence of two rotenone-insensitive NAD(P)H dehydrogenases on the inner surface of the inner membrane of potato tuber mitochondria. Biochim. Biophys. Acta 1276:133–139.

Meyer, E. H., Tomaz, T., Carroll, A. J., Estavillo, G., Delannoy, E., Tanz, S. K., Small, I. D., Pogson, B. J., Millar, A. H. 2009. Remodeled respiration in ndufs4 with low phosphorylation efficiency suppresses Arabidopsis germination and growth and alters control of metabolism at night. Plant Physiol. 151:603–619.

Michalecka, A. M., Svensson, Å. S., Johansson, F. I., Agius, S. C., Johanson, U., Brennicke, A., Binder, S., Rasmusson, A. G. 2003. Arabidopsis genes encoding mitochondrial type II NAD(P)H dehydrogenases have different evolutionary origin and show distinct responses to light. Plant Physiol. 133:642–652.

Michalecka, A. M., Agius, S. C., Møller, I. M., Rasmusson, A. G. 2004. Identification of a mitochondrial external NADPH dehydrogenase by overexpression in transgenic *Nicotiana sylvestris*. Plant J. 37:415–425.

Millar, A. H., Wiskich, J. T., Whelan, J., Day, D. A. 1993. Organic-acid activation of the alternative oxidase of plant mitochondria. FEBS Lett. 329:259–262.

Millar, A. H., Hoefnagel, M. H. N., Day, D. A., Wiskich, J. T. 1996. Specificity of the organic acid activation of alternative oxidase in plant mitochondria. Plant Physiol. 111:613–618.

Millar, A. H., Mittova, V., Kiddle, G., Heazlewood, J. L., Bartoli, C. G., Theodoulou, F. L., Foyer, C. H. 2003. Control of ascorbate synthesis by respiration and its implications for stress responses. Plant Physiol. 133:443–447.

Millar, A. H., Trend, A. E., Heazlewood, J. L. 2004. Changes in the mitochondrial proteome during the anoxia to air transition in rice focus around cytochrome-containing respiratory complexes. J. Biol. Chem. 279:39471–39478.

Millenaar, F. F., Lambers, H. 2003. The alternative oxidase: *in vivo* regulation and function. Plant Biol. 5:2–15.

Miller, G., Honig, A., Stein, H., Suzuki, N., Mittler, R., Zilberstein, A. 2009. Unraveling delta(1)-pyrroline-5-carboxylate-proline cycle in plants by uncoupled expression of proline oxidation enzymes. J. Biol. Chem. 284:26482–26492.

Møller, I. M. 1997. The oxidation of cytosolic NAD(P)H by external NAD(P)H dehydrogenases in the respiratory chain of plant mitochondria. Physiol. Plant. 100:85–90.

Møller, I. M. 2001. Plant mitochondria and oxidative stress: electron transport, NADPH turnover, and metabolism of reactive oxygen species. Annu. Rev. Plant Physiol. Plant Mol. Biol. 52:561–591.

Møller, I. M., Kristensen, B. K. 2004. Protein oxidation in plant mitochondria as a stress indicator. Photochem. Photobiol. Sci. 3:730–735.

Møller, I. M., Kristensen, B. K. 2006. Protein oxidation in plant mitochondria detected as oxidized tryptophan. Free Radic. Biol. Med. 40:430–435.

Møller, I. M., Rasmusson, A. G. 1998. The role of NADP in the mitochondrial matrix. Trends Plant Sci. 3:21–27.

Møller, I. M., Jensen, P. E., Hansson, A. 2007. Oxidative modifications to cellular components in plants. Annu. Rev. Plant Biol. 58:459–481.

Moore, A. L., Albury, M. S., Crichton, P. G., Affourtit, C. 2002. Function of the alternative oxidase: is it still a scavenger? Trends Plant Sci. 7:478–481.

Moore, C. S., Cook-Johnson, R. J., Rudhe, C., Whelan, J., Day, D. A., Wiskich, J. T., Soole, K. L. 2003. Identification of AtNDI1, an internal non-phosphorylating NAD(P)H dehydrogenase in Arabidopsis mitochondria. Plant Physiol. 133:1968–1978.

Mukhopadhyay, S., Sen, S., Majhi, B., Das, K. P., Kar, M. 2007. Methyl glyoxal elevation is associated with oxidative stress in rheumatoid arthritis. Free Radic. Res. 41:507–514.

Nanjo, T., Kobayashi, M., Yoshiba, Y., Kakubari, Y., Yamaguchi-Shinozaki, K., Shinozaki, K. 1999. Antisense suppression of proline degradation improves tolerance to freezing and salinity in *Arabidopsis thaliana*. FEBS Lett. 461:205–210.

Nanjo, T., Fujita, M., Seki, M., Kato, T., Tabata, S., Shinozaki, K. 2003. Toxicity of free proline revealed in an Arabidopsis T-DNA-tagged mutant deficient in proline dehydrogenase. Plant Cell Physiol. 44:541–548.

Nash, D., Wiskich, J. T. 1983. Properties of substantially chlorophyll free pea leaf mitochondria prepared by sucrose density gradient separation. Plant Physiol. 71:627–634.

Neill, S., Desikan, R., Hancock, J. 2002. Hydrogen peroxide signalling. Curr. Opin. Plant Biol. 5:388–395.

Newton, K. J., Gabay-Laughnan, S., De Paepe, R. 2004. Mitochondrial mutations in plants. In: eds. D. A. Day, A. H. Millar, J. Whelan. Plant Mitochondria: From Genome to Function, pp. 121–142. Dordrecht: Kluwer.

Noctor, G., Foyer, C. H. 1998. Ascorbate and glutathione: Keeping active oxygen under control. Annu. Rev. Plant Physiol. Plant Mol. Biol. 49:249–279.

Noctor, G., Dutilleul, C., De Paepe, R., Foyer, C. H. 2004. Use of mitochondrial electron transport mutants to evaluate the effects of redox state on photosynthesis, stress tolerance and the integration of carbon/nitrogen metabolism. J. Exp. Bot. 55:49–57.

Noctor, G., De Paepe, R., Foyer, C. H. 2007. Mitochondrial redox biology and homeostasis in plants. Trends Plant Sci. 12:125–134.

Nunes-Nesi, A., Carrari, F., Lytovchenko, A., Smith, A. M., Loureiro, M. E., Ratcliffe, R. G., Sweetlove, L. J., Fernie, A. R. 2005. Enhanced photosynthetic performance and growth as a consequence of decreasing mitochondrial malate dehydrogenase activity in transgenic tomato plants. Plant Physiol. 137:611–622.

O'Brien, K. M., Dirmeier, R., Engle, M., Poyton, R. O. 2004. Mitochondrial protein oxidation in yeast mutants lacking manganese-(mnsod) or copper- and zinc-containing superoxide dismutase (cuznsod): evidence that mnsod and cuznsod have both unique and overlapping functions in protecting mitochondrial proteins from oxidative damage. J. Biol. Chem. 279:51817–51827.

Ordog, S. H., Higgins, V. J., Vanlerberghe, G. C. 2002. Mitochondrial alternative oxidase is not a critical component of plant viral resistance but may play a role in the hypersensitive response. Plant Physiol. 129:1858–1865.

Pineau, B., Layoune, O., Danon, A., De Paepe, R. 2008. L-galactono-1,4-lactone dehydrogenase is required for the accumulation of plant respiratory complex I. J. Biol. Chem. 283:32500–32505.

Price, J., Laxmi, A., St. Martin, S. K., Jang, J. C. 2004. Global transcription profiling reveals multiple sugar signal transduction mechanisms in Arabidopsis. Plant Cell 16:2128–2150.

Purvis, A. C., Shewfelt, R. L. 1993. Does the alternative pathway ameliorate chilling injury in sensitive plant tissues. Physiol. Plant. 88:712–718.

Rasmusson, A. G., Escobar, M. A. 2007. Light and diurnal regulation of plant respiratory gene expression. Physiol. Plant. 129:57–67.

Rasmusson, A. G., Møller, I. M. 1991. NAD(P)H dehydrogenases on the inner surface of the inner mitochondrial membrane studied using inside-out submitochondrial particles. Physiol. Plant. 83:357–365.

Rasmusson, A. G., Svensson, A. S., Knoop, V., Grohmann, L., Brennicke, A. 1999. Homologues of yeast and bacterial rotenone-insensitive NADH dehydrogenases in higher eukaryotes: two enzymes are present in potato mitochondria. Plant J. 20:79–87.
Rasmusson, A. G., Soole, K. L., Elthon, T. E. 2004. Alternative NAD(P)H dehydrogenases of plant mitochondria. Annu. Rev. Plant Biol. 55:23–39.
Rasmusson, A. G., Geisler, D. A., Møller, I. M. 2008. The multiplicity of dehydrogenases in the electron transport chain of plant mitochondria. Mitochondrion 8:47–60.
Rasmusson, A. G., Fernie, A. R., van Dongen, J. T. 2009. Alternative oxidase: a defence against metabolic fluctuations? Physiol. Plant 137:371–382.
Rayapati, P. J., Stewart, C. R. 1991. Solubilization of a proline dehydrogenase from maize (*Zea mays* L) mitochondria. Plant Physiol. 95:787–791.
Rhoads, D. M., Umbach, A. L., Subbaiah, C. C., Siedow, J. N. 2006. Mitochondrial reactive oxygen species. Contribution to oxidative stress and interorganellar signaling. Plant Physiol. 141:357–366.
Ribas-Carbo, M., Taylor, N. L., Giles, L., Busquets, S., Finnegan, P. M., Day, D. A., Lambers, H., Medrano, H., Berry, J. A., Flexas, J. 2005. Effects of water stress on respiration in soybean leaves. Plant Physiol. 139:466–473.
Robson, C. A., Vanlerberghe, G. C. 2002. Transgenic plant cells lacking mitochondrial alternative oxidase have increased susceptibility to mitochondria-dependent and -independent pathways of programmed cell death. Plant Physiol. 129:1908–1920.
Sabar, M., De Paepe, R., de Kouchkovsky, Y. 2000. Complex I impairment, respiratory compensations, and photosynthetic decrease in nuclear and mitochondrial male sterile mutants of *Nicotiana sylvestris*. Plant Physiol. 124:1239–1249.
Scharte, J., Schon, H., Tjaden, Z., Weis, E., von Schaewen, A. 2009. Isoenzyme replacement of glucose-6-phosphate dehydrogenase in the cytosol improves stress tolerance in plants. Proc. Natl. Acad. Sci. U.S.A. 106:8061–8066.
Schwarzländer, M., Fricker, M. D., Sweetlove, L. J. 2009. Monitoring the in vivo redox state of plant mitochondria: effect of respiratory inhibitors, abiotic stress and assessment of recovery from oxidative challenge. Biochim. Biophys. Acta 1787:468–475.
Shen, W., Wei, Y., Dauk, M., Zheng, Z., Zou, J. 2003. Identification of a mitochondrial glycerol-3-phosphate dehydrogenase from *Arabidopsis thaliana*: evidence for a mitochondrial glycerol-3-phosphate shuttle in plants. FEBS Lett. 536:92–96.
Shen, W. Y., Wei, Y. D., Dauk, M., Tan, Y. F., Taylor, D. C., Selvaraj, G., Zou, J. T. 2006. Involvement of a glycerol-3-phosphate dehydrogenase in modulating the NADH/NAD^+ ratio provides evidence of a mitochondrial glycerol-3-phosphate shuttle in Arabidopsis. Plant Cell 18:422–441.
Siedow, J. N., Umbach, A. L. 2000. The mitochondrial cyanide-resistant oxidase: structural conservation amid regulatory diversity. Biochim. Biophys. Acta 1459:432–439.
Sieger, S. M., Kristensen, B. K., Robson, C. A., Amirsadeghi, S., Eng, E. W., Abdel-Mesih, A., Møller, I. M., Vanlerberghe, G. C. 2005. The role of alternative oxidase in modulating carbon use efficiency and growth during macronutrient stress in tobacco cells. J. Exp. Bot. 56:1499–1515.
Singla-Pareek, S. L., Reddy, M. K., Sopory, S. K. 2003. Genetic engineering of the glyoxalase pathway in tobacco leads to enhanced salinity tolerance. Proc. Natl. Acad. Sci. U.S.A. 100:14672–14677.
Singla-Pareek, S. L., Yadav, S. K., Pareek, A., Reddy, M. K., Sopory, S. K. 2006. Transgenic tobacco overexpressing glyoxalase pathway enzymes grow and set viable seeds in zinc-spiked soils. Plant Physiol. 140:613–623.
Smith, A. M., Stitt, M. 2007. Coordination of carbon supply and plant growth. Plant Cell Environ. 30:1126–1149.
Stoimenova, M., Igamberdiev, A. U., Gupta, K. J., Hill, R. D. 2007. Nitrite-driven anaerobic ATP synthesis in barley and rice root mitochondria. Planta 226:465–474.
Stupnikova, I., Benamar, A., Tolleter, D., Grelet, J., Borovskii, G., Dorne, A. J., Macherel, D. 2006. Pea seed mitochondria are endowed with a remarkable tolerance to extreme physiological temperatures. Plant Physiol. 140:326–335.

Svensson, Å. S., Rasmusson, A. G. 2001. Light-dependent gene expression for proteins in the respiratory chain of potato leaves. Plant J. 28:73–82.

Svensson, Å. S., Johansson, F. I., Møller, I. M., Rasmusson, A. G. 2002. Cold stress decreases the capacity for respiratory NADH oxidation in potato leaves. FEBS Lett. 517:79–82.

Sweetlove, L. J., Møller, I. M. 2009. Oxidation of proteins in plants – mechanisms and consequences. Adv. Bot. Res. 52:1–23.

Sweetlove, L. J., Heazlewood, J. L., Herald, V., Holtzapffel, R., Day, D. A., Leaver, C. J., Millar, A. H. 2002. The impact of oxidative stress on *Arabidopsis* mitochondria. Plant J. 32:891–904.

Szal, B., Lukawska, K., Zdolinska, I., Rychter, A. M. 2009. Chilling stress and mitochondrial genome rearrangement in the MSC16 cucumber mutant affect the alternative oxidase and antioxidant defense system to a similar extent. Physiol. Plant. 137:435–445.

Taylor, N. L., Day, D. A., Millar, A. H. 2002. Environmental stress causes oxidative damage to plant mitochondria leading to inhibition of glycine decarboxylase. J. Biol. Chem. 277:42663–42668.

Taylor, S. W., Fahy, E., Murray, J., Capaldi, R. A., Ghosh, S. S. 2003. Oxidative post-translational modification of tryptophan residues in cardiac mitochondrial proteins. J. Biol. Chem. 278:19587–19590.

Thirkettle-Watts, D., McCabe, T. C., Clifton, R., Moore, C., Finnegan, P. M., Day, D. A., Whelan, J. 2003. Analysis of the alternative oxidase promoters from soybean. Plant Physiol. 133:1158–1169.

Tolleter, D., Jaquinod, M., Mangavel, C., Passirani, C., Saulnier, P., Manon, S., Teyssier, E., Payet, N., Avelange-Macherel, M. H., Macherel, D. 2007. Structure and function of a mitochondrial late embryogenesis abundant protein are revealed by desiccation. Plant Cell 19:1580–1589.

Torres, M. A., Dangl, J. L. 2005. Functions of the respiratory burst oxidase in biotic interactions, abiotic stress and development. Curr. Opin. Plant Biol. 8:397–403.

Valpuesta, V., Botella, M. A. 2004. Biosynthesis of L-ascorbic acid in plants: new pathways for an old antioxidant. Trends Plant Sci. 9:573–577.

Vanlerberghe, G. C., McIntosh, L. 1992. Lower growth temperature increases alternative pathway capacity and alternative oxidase protein in tobacco. Plant Physiol. 100:115–119.

Vanlerberghe, G. C., Day, D. A., Wiskich, J. T., Vanlerberghe, A. E., Mcintosh, L. 1995. Alternative oxidase activity in tobacco leaf mitochondria – dependence on tricarboxylic-acid cycle-mediated redox regulation and pyruvate activation. Plant Physiol. 109:353–361.

Vanlerberghe, G. C., Robson, C. A., Yip, J. Y. H. 2002. Induction of mitochondrial alternative oxidase in response to a cell signal pathway down-regulating the cytochrome pathway prevents programmed cell death. Plant Physiol. 129:1829–1842.

Vanlerberghe, G. C., Cvetkovska, M., Wang, J. 2009. Is the maintenance of homeostatic mitochondrial signaling during stress a physiological role for alternative oxidase? Physiol. Plant. 137:392–406.

Vercesi, A. E., Borecky, J., de Godoy Maia, I., Arruda, P., Cuccovia, I. M., Chaimovich, H. 2006. Plant uncoupling mitochondrial proteins. Annu. Rev. Plant Biol. 57:383–404.

Winger, A. M., Millar, A. H., Day, D. A. 2005. Sensitivity of plant mitochondrial terminal oxidases to the lipid peroxidation product 4-hydroxy-2-nonenal (HNE). Biochem. J. 387:865–870.

Yabuta, Y., Maruta, T., Nakamura, A., Mieda, T., Yoshimura, K., Ishikawa, T., Shigeoka, S. 2008. Conversion of L-galactono-1,4-lactone to L-ascorbate is regulated by the photosynthetic electron transport chain in Arabidopsis. Biosci. Biotechnol. Biochem. 72:2598–2607.

Zabalza, A., van Dongen, J. T., Froehlich, A., Oliver, S. N., Faix, B., Gupta, K. J., Schmalzlin, E., Igal, M., Orcaray, L., Royuela, M., Geigenberger, P. 2009. Regulation of respiration and fermentation to control the plant internal oxygen concentration. Plant Physiol. 149:1087–1098.

Chapter 15
Interaction Between Chloroplasts and Mitochondria: Activity, Function, and Regulation of the Mitochondrial Respiratory System during Photosynthesis

Keisuke Yoshida and Ko Noguchi

Abstract Activities of the plant mitochondrial respiratory system are flexibly and drastically modulated in response to variations in natural conditions, such as the light environment. Accumulating studies addressing the interactions between chloroplasts and mitochondria have provided evidence for the critical importance of the mitochondrial respiratory system for optimal photosynthetic performance. In this organelle cross-talk, several unique reactions and components in the plant mitochondria, such as the photorespiratory metabolism and the non-proton-pumping respiratory pathways, play key roles. In this chapter, we review recent advances of understanding about the activity, function, and regulation of the mitochondrial respiratory system during photosynthesis.

Keywords Alternative oxidase • Organelle cross-talk • Photorespiration • Photosynthesis • Redox-shuttling

Abbreviations

AOX	Alternative oxidase
C	Carbon
COX	Cytochrome *c* oxidase
CP	Cytochrome pathway
GDC	Glycine decarboxylase complex
MRR	Mitochondrial retrograde regulation
N	Nitrogen
NDex	External type II NAD(P)H dehydrogenase
NDin	Internal type II NAD(P)H dehydrogenase
NDs	Type II NAD(P)H dehydrogenases

K. Yoshida (✉)
Chemical Resources Laboratory, Tokyo Institute of Technology, Nagatsuta 4259-R1-8, Midori-ku, 226-8503 Yokohama, Japan
e-mail: yoshida.k.ao@m.titech.ac.jp

F. Kempken (ed.), *Plant Mitochondria*, Advances in Plant Biology 1,
DOI 10.1007/978-0-387-89781-3_15,

PDC	Pyruvate dehydrogenase complex
ROS	Reactive oxygen species
TCA cycle	Tricarboxylic acid cycle
UCP	Uncoupling protein

15.1 Introduction

Mitochondria in higher plants, as well as other eukaryotes, play a crucial role in providing energy for a variety of cellular events. In addition to this well-established role, mitochondria fulfill a wide range of functions cooperatively with other subcellular compartments. To sustain these functions properly, metabolism and electron transport in the mitochondria must be regulated in a coordinated manner. Higher plants are often exposed to severe stresses, which leads to drastic changes in the cellular environment. Under such conditions, the regulation of the mitochondrial respiratory system seems to be more important. Especially, light intensity is the most variable environmental factor in plant habitats. It has been demonstrated that the respiratory system in plant mitochondria shows a flexible acclimation in response to changes in the light environment.

Studies addressing interactions between chloroplasts and mitochondria in illuminated leaves have attracted much attention of plant researchers. As summarized in Fig. 15.1a, multiple metabolic interactions between chloroplasts and mitochondria have been suggested to date. Using light energy, water and CO_2, chloroplasts produce carbohydrates in a series of photosynthetic reactions. A part of these carbohydrates is catabolized to generate ATP in the mitochondria. Even this fundamental metabolism can be considered as a form of interaction between chloroplasts and mitochondria. Several other examples of the metabolic interaction have been proposed, including (1) transport of reducing equivalents from the chloroplasts to the mitochondria by redox-shuttling machineries, (2) supply of carbon (C) skeletons for nitrogen (N) assimilation in the chloroplasts, and (3) photorespiratory metabolism. Detailed information about these interactions is described in other reviews (e.g., Gardeström et al. 2002; Raghavendra and Padmasree 2003; Nunes-Nesi et al. 2007b; Noguchi and Yoshida 2008). The physiologic importance of the organelle interplay is also supported by the co-localizations of chloroplasts and mitochondria (and peroxisomes) in cells of illuminated leaves (Fig. 15.1b; Islam et al. 2009). Furthermore, in addition to higher plants, a unicellular green alga *Chlamydomonas* possesses unique features of organelle cross-talk (Box 15.1). These facts highlight that communications between chloroplasts and mitochondria are indispensable for photosynthetic eukaryotes.

Within this research topic, much interest has been directed to the contribution of the mitochondrial respiratory system to photosynthesis. It has been revealed that the dysfunction of the mitochondrial respiratory system affects the photosynthetic efficiency and sometimes even biomass production. This suggests the prerequisite

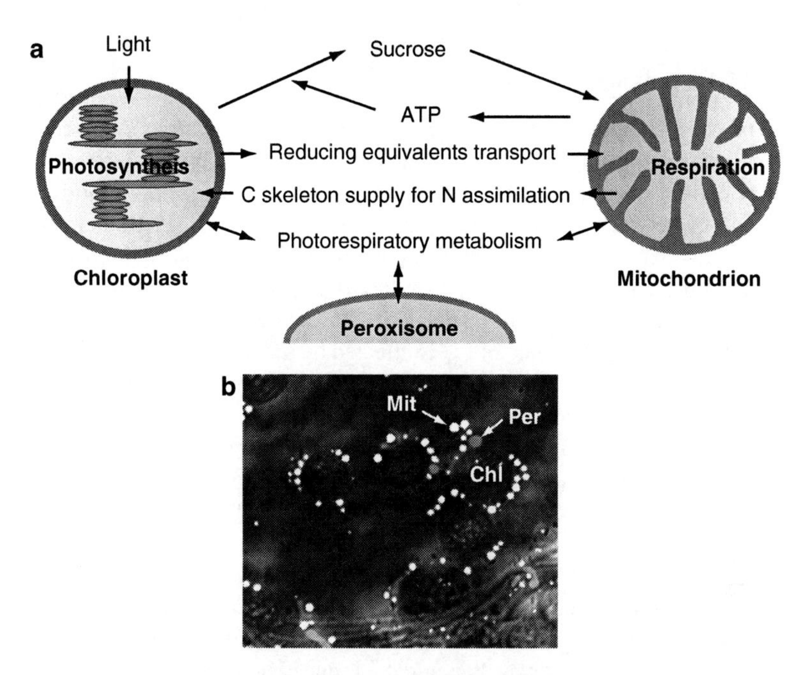

Fig. 15.1 Overview of organelle cross-talk. (**a**) Metabolic interactions between chloroplasts and mitochondria. *C* carbon, *N* nitrogen. (**b**) Visualization of chloroplasts, mitochondria and peroxisomes in green leaf tissue of *Arabidopsis* transformant by differential interference contrast microscopy. Mitochondria (*Mit*) were visualized by GFP, while peroxisomes (*Per*) were by RFP. Mitochondria and peroxisomes localize adjacent to the chloroplasts (*Chl*) (Image was kindly provided by Drs. K. Oikawa, S. Mano and Prof. M. Nishimura, National Institute for Basic Biology)

Box 15.1 Interaction between chloroplasts and mitochondria in *Chlamydomonas reinhardtii*

The modes of interaction between chloroplasts and mitochondria in *Chlamydomonas reinhardtii* are partly common to those in higher plants. For example, light-enhanced dark respiration (LEDR, see Glossary), which is often observed in higher plants, is also evident in *C. reinhardtii* (Xue et al. 1996). On the other hand, *C. reinhardtii* appears to possess several distinct aspects in the organelle cross-talk. One of such aspects is the remote control of photosynthetic genes by the mitochondrial respiratory chain (Matsuo and Obokata 2006). Under heterotrophic conditions, expression of nuclear-encoded photosynthetic genes is induced in response to an activation of the cytochrome pathway in the respiratory chain. As another example, mitochondria may contribute to the CO_2-concentrating mechanism by providing energy, which is beneficial for photosynthetic carbon assimilation (Huertas et al. 2002).

(continued)

Box 15.1 (continued)

The most exciting topic of organelle cross-talk in *C. reinhardtii* appears to be the interrelationship between the state transition of the photosystems and the mitochondrial respiration. State transition is a process balancing light excitation energy between the antennae systems of photosystem II (PSII) and photosystem I (PSI); light harvesting complex II (LHCII) is phosphorylated by thylakoid protein kinase depending on the redox state of plastoquinone (PQ), and subsequently moves from PSII to PSI (Rochaix 2007). It is known that, in *C. reinhardtii*, the redox state of PQ is affected by the mitochondrial respiration even in the dark. When intracellular ATP is depleted by the cessation of ATP supply from the mitochondria, stimulated glycolysis enhances nonphotochemical reduction of PQ through the chlororespiratory pathway (Peltier and Cournac 2002). *C. reinhardtii* mutants lacking several proton-pumping sites in the respiratory chain showed accelerated nonphotochemical PQ reduction, resulting in a high level of LHCII phosphorylation and preferential excitation of PSI by light (Cardol et al. 2003). As a result, the operation of cyclic electron transport around PSI was enhanced, which was functional for the restoration of ATP pool. When both mitochondrial respiration and state transition were impaired, photosynthetic efficiency was dramatically reduced (Cardol et al. 2009). These results highlight that the combined energetic contributions of the state transition and the mitochondrial respiration are beneficial for photosynthesis in *C. reinhardtii*. The tight connection of the state transition and the mitochondrial respiration is also exemplified by another result. The *C. reinhardtii* mutant *stm6*, identified on the basis of impaired state transition, has mutation in the *Moc1* gene whose product is targeted to the mitochondria and essential for expression of several respiratory complexes (Schönfeld et al. 2004).

While state transition in higher plants are of limited amplitude, that in *C. reinhardtii* plays a pivotal role in light-acclimation of this alga because as much as 80% of LHCII is mobile between two photosystems (Cardol et al. 2009). This difference may attribute to the distinct mode of organelle cross-talk in *C. reinhardtii*. In this cross-talk, ATP production by the phosphorylating respiratory pathway seems to be more important for photosynthesis. On the other hand, it is still unclear whether the nonphosphorylating pathways play significant roles in *C. reinhardtii*. It should be noted that the expression profiles of *AOX* genes in *C. reinhardtii* are somewhat different from those in higher plants; the promoter region in *AOX1* gene was unresponsive to stress (Baurain et al. 2003; Molen et al. 2006). In addition to the mitochondrial nonphosphorylating respiratory pathways, involvement of the chloroplastic chlororespiratory pathways in the organelle cross-talk and the light acclimation also should be studied. In *C. reinhardtii*, the plastid terminal oxidase (PTOX), a homologous component to AOX in the chloroplasts, may engage as a safety valve for preventing cellular over-reduction more favorably than the mitochondrial respiratory chain (Cournac et al. 2002). Notably, it was recently suggested that type II NAD(P)H dehydrogenase mediates chlororespiratory PQ reduction in *C. reinhardtii* (Jans et al. 2008).

roles of the mitochondria for optimal C assimilation. Considering the importance of such organelle communications, it is meaningful to understand activities and regulatory mechanisms of the mitochondrial respiratory system in the light. Here we mainly review the current understanding about the activity and function of the mitochondrial respiratory system (including the C/N metabolism in the matrix and the electron transport in the respiratory chain) in illuminated leaves, paying special attention to the interaction with the photosynthetic system. We also summarize recent progress on possible regulatory mechanisms of the respiratory system in the light at the transcriptional to posttranslational level.

15.2 Nonphotorespiratory C Metabolism in the Mitochondrial Matrix Under Light Conditions

C metabolism in the mitochondrial matrix (e.g., tricarboxylic acid [TCA] cycle) provides NADH and $FADH_2$, which are used for the electron transport and oxidative phosphorylation in the respiratory chain. During the oxidation of C compounds for the production of reducing equivalents, CO_2 is also released by four decarboxylating components (pyruvate dehydrogenase complex [PDC], NAD-dependent malic enzyme [NAD-ME], NAD-dependent isocitrate dehydrogenase [NAD-IDH], and 2-oxoglutarate dehydrogenase complex [OGDC]; Fig. 15.2). It has been

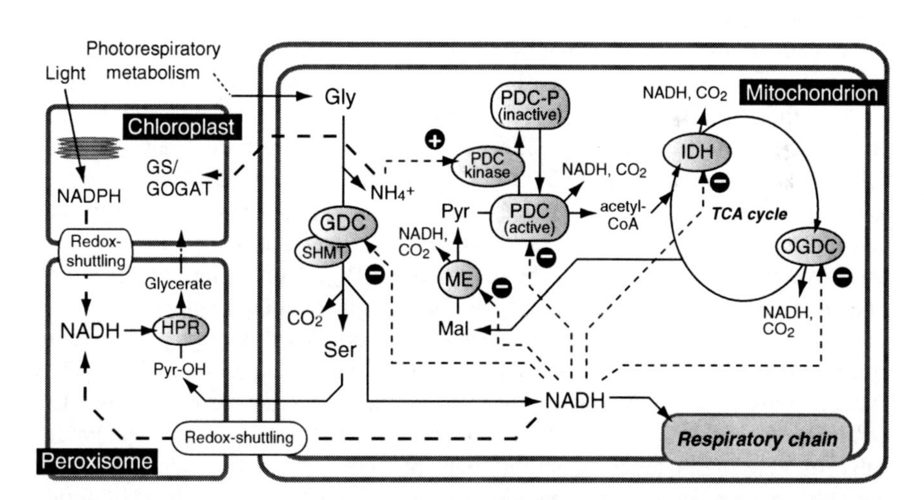

Fig. 15.2 Relationships between photorespiratory and nonphotorespiratory carbon metabolism in the mitochondria. Photorespiratory glycine (*Gly*) oxidation, a major process in the mitochondria in illuminated leaves, produces NH_4^+ and NADH. NH_4^+ stimulates pyruvate dehydrogenase complex (*PDC*) kinase, leading to an inactivation of PDC. A high $NADH/NAD^+$ ratio allosterically inhibits several dehydrogenases, such as PDC, NAD-malic enzyme (*ME*), NAD-isocitrate dehydrogenase (*IDH*) and 2-oxoglutarate dehydrogenase complex (*OGDC*). *GDC* Gly decarboxylase complex, *GS* glutamine synthetase, *GOGAT* glutamine-2-oxoglutarate aminotransferase, *HPR* hydroxypyruvate reductase, *Mal* malate, *PDC-P* phosphorylated PDC, *Pyr* pyruvate, *Pyr-OH* hydroxypyruvate, *Ser* serine, *SHMT* Ser hydroxymethyltransferase, *TCA cycle* tricarboxylic acid cycle

assumed that, in plants, the activity of C metabolism in the mitochondrial matrix is modulated during the day period. We outline the recent view of how metabolism in the mitochondrial matrix is affected in illuminated leaves.

To estimate the in vivo activity of C metabolism in the mitochondrial matrix in illuminated leaves, determination of the nonphotorespiratory CO_2 release rate in the light (R_{day}) has been performed for a long time. Precise determination of R_{day} is complicated because of the impairment of the direct measurement by other CO_2-mediated reactions such as photosynthetic CO_2 fixation and photorespiratory CO_2 release (Krömer 1995). Nevertheless, several methods for the determination of R_{day} (Laisk's method, Kok's method [for a review, see Atkin et al. 2000] and radiolabeling measurements using ^{14}C [Pärnik and Keerberg 2007]) demonstrated that R_{day} is generally lower (25–100%) than the CO_2 release rate in the dark (R_n).

What is the mechanism responsible for the lower rate in R_{day}? The most well-known factor is the inactivation of PDC in the light (Budde and Randall 1990; Tovar-Méndez et al. 2003). PDC is phosphorylated under light conditions, which leads to the reversible inactivation of PDC (see Fig. 15.2). This event is closely linked to photorespiration, because PDC kinase is activated by NH_4^+, a byproduct of the reaction catalyzed by the glycine decarboxylase complex (GDC). Furthermore, GDC generates a large amount of NADH in the mitochondria. An elevated NADH level would result in the allosteric inhibition of activities of several TCA cycle dehydrogenases (see Fig. 15.2; Noctor et al. 2007). Therefore, the repression of the TCA cycle activity in the light may also attribute to the lower rate of R_{day}.

Considering that the reaction catalyzed by PDC is an entry point to, and thereby controlling step of the TCA cycle, the light-dependent inhibition of PDC and the TCA cycle are likely to be linked. Nevertheless, it remains to be elucidated which one of these two factors is predominant for the lower rate of R_{day}. Tcherkez et al. (2005) reported intriguing results; by an unique technique using respiratory substrates labeled with ^{13}C at a specific position, they demonstrated that, the TCA cycle metabolism was reduced by 95% in the light, while the PDC reaction was lowered only by 27% (or less when CO_2 refixation by photosynthesis was considered). According to their result, it seems that a downregulation of the TCA cycle activity is the primary cause for the lower rate of R_{day}. Another important finding in their study was that, during the day period, the glucose degradation by glycolysis was strongly inhibited, likely resulting in a decreased flux of respiratory substrate into the TCA cycle. They also observed that the repression of the TCA cycle activity in the light was partly restored under photorespiratory (low CO_2) conditions (Tcherkez et al. 2008). Considering that the TCA cycle is enzymatically subjected to the feedback inhibition by photorespiratory byproduct (NADH), their observation appears to be paradoxical. In fact, GDC antisense lines of potato showed higher R_{day} than wild-type (Bykova et al. 2005), suggesting that the repression of the TCA cycle activity can be attributed, at least in part, to the effects of photorespiratory by-products (NH_4^+ and NADH). The mechanisms determining R_{day} may not be so simple, but there is little doubt that they strongly depend on photorespiratory metabolism, a major process in the mitochondria of illuminated leaves.

15.3 Photorespiration: The Impact on the Mitochondrial and Cellular Metabolism

Photorespiration, which is triggered by oxygenation of ribulose-1,5-bisphospate (RuBP) by Rubisco, is the best-known example in the organelle cross-talk. Because photorespiration leads to a substantial loss of net C gain, this pathway has often been believed to have a negative effect on the plant growth. On the other hand, several photorespiratory mutants have provided evidence for its necessity for plant survival under ambient CO_2 (for a review, see Foyer et al. 2009). As a physiologic function, photorespiration is thought to mitigate photoinhibition especially under high-light (HL) and drought (CO_2-limited) stresses by suppressing the over-reduction and overenergization states in the chloroplasts (Kozaki and Takeba 1996; Igamberdiev et al. 2001). Also, a complete photorespiratory cycle may be important for the recovery of 75% of C from phosphoglycolate, a product of RuBP oxygenation, and the removal of glyoxylate and hydroxypyruvate (Pyr-OH), the potent inhibitors of photosynthesis (Campbell and Ogren 1990; Givan and Kleczkowski 1992). It has also been suggested that, in the context of nitrate (NO_3^-) assimilation, photorespiration may be a beneficial reaction (Rachmilevitch et al. 2004). Several recent reviews (Keys and Leegood 2002; Oliver and Gardeström 2004; Foyer et al. 2009) are available for understanding the nature of photorespiration. Here, the consequences of mitochondrial NH_4^+ and NADH produced by GDC, a key enzyme of photorespiration in the mitochondria, are mainly focused.

Photorespiratory by-products, NH_4^+ and NADH, have a large impact on metabolism in the mitochondria (see Fig. 15.2). The efficient recycling of these by-products is essential for continuing photorespiratory metabolism. NH_4^+ diffuses to the chloroplasts and is re-fixed by the glutamine synthetase (GS)/glutamine-2-oxoglutarate aminotransferase (GOGAT) system in this organelle. Alternatively, NH_4^+ may be fixed within the mitochondria and transported to the chloroplasts safely by shuttle machineries (Taira et al. 2004; Linka and Weber 2005), although the localization of NH_4^+-fixing GS in the mitochondria is still controversial (Eubel et al. 2007). One molecule of glutamate (Glu) produced by the GS/GOGAT system is transported to the peroxisomes, where Glu serves as an amino donor for the conversion of glyoxylate to Gly. 2-oxoglutarate (2-OG) formed by Glu deamination is shuttled back into the chloroplasts. Thus, C and N cycling at a whole cell level is involved in the photorespiration. This proper cycling is essential for plant growth under ambient CO_2 environment (Igarashi et al. 2003).

Reoxidation of NADH is also required for efficient photorespiratory metabolism, because GDC, as well as PDC and the TCA cycle enzymes, are inhibited at a low NADH concentration (Oliver 1994; see Sect. 15.2). There are two major NADH-reoxidizing machineries; the respiratory chain in the mitochondria and the Pyr-OH reduction in the peroxisomes (see Fig. 15.2). Reoxidation by the respiratory chain and the following ATP production by oxidative phosphorylation may be helpful for sucrose synthesis, an ATP-requiring reaction in the cytosol (Gardeström and Wigge 1988). However, it seems that nonphosphorylating bypass pathways play a more dominant role in the efficient dissipation of NADH generated during photorespiration

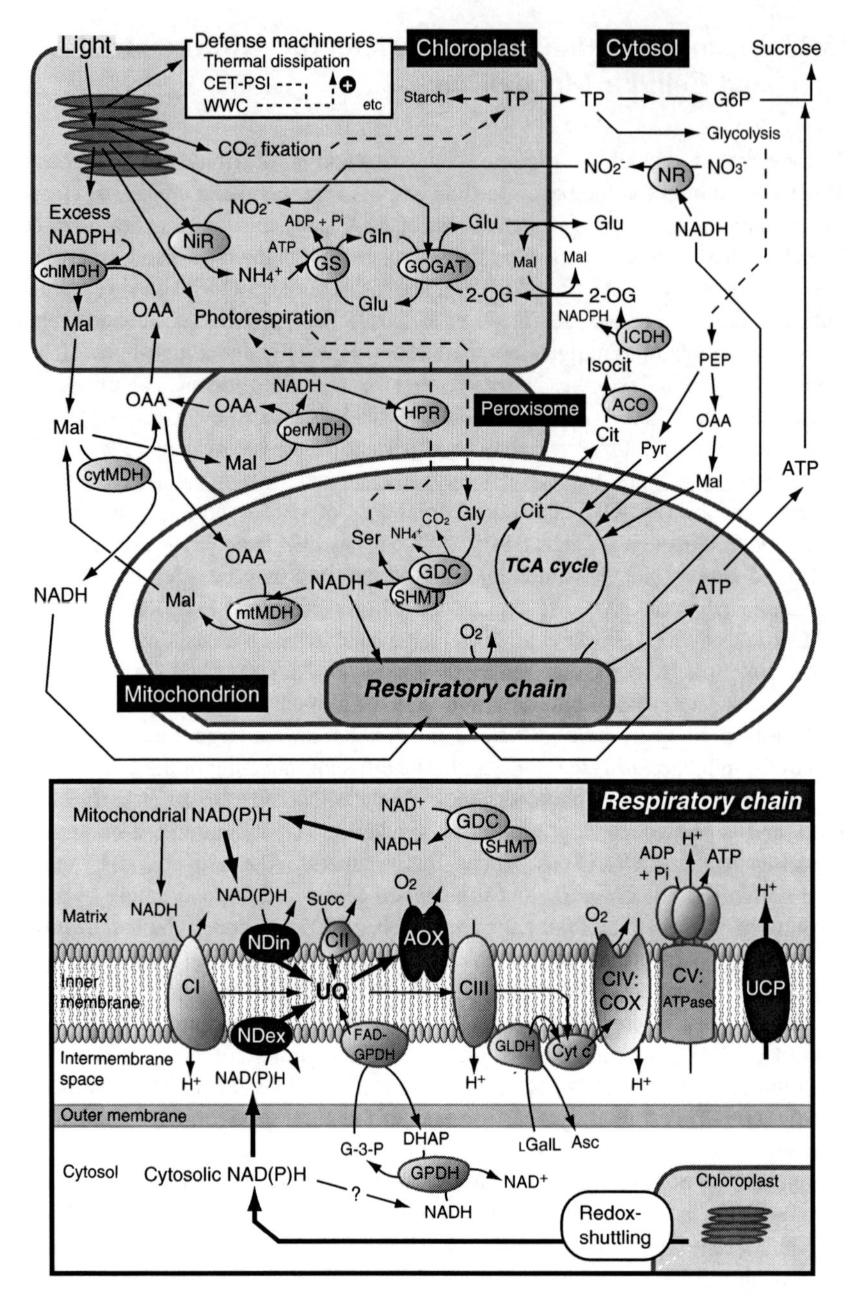

Fig. 15.3 Metabolic network between subcellular compartments in illuminated leaves. Detailed information is depicted in the main text. *ACO* aconitase, *AOX* alternative oxidase, *Asc* ascorbate, *CET-PSI* cyclic electron transport around photosystem I, *CI-V* complex I-V, *Cit* citrate, *COX* cytochrome *c* oxidase, *Cyt* c cytochrome *c*, *DHAP* dihydroxyacetone phosphate, $_{L}$*GalL* L-galactono-1,4-lactone, *GLDH* $_{L}$GalL dehydrogenase, *Gln* glutamine, *Glu* glutamate, *G-3-P* glycerol-3-phosphate, *GPDH* G-3-P dehydrogenase, *G6P* glucose-6-phosphate, *ICDH* NADP-isocitrate dehydrogenase, *Isocit*

(see Sects. 15.6 and 15.7). NADH that cannot be oxidized by the respiratory chain must be transported out of the mitochondria. Experiments using isolated mitochondria incubated in simulated in vivo conditions indicated that 25–50% of NADH produced by Gly oxidation is exported from the mitochondria (Krömer and Heldt 1991). Then, reducing equivalent is imported into the peroxisomes, where it is consumed in Pyr-OH reduction by Pyr-OH reductase (HPR). HPR needs the same amount of NADH as produced by GDC for the completion of the photorespiratory cycle. As a part of the NADH generated by GDC is oxidized by the respiratory chain, the rest required for HPR must be supplied from the chloroplasts. Transport of the reducing equivalents across the organelle envelope is facilitated by redox-shuttling machineries, such as the malate (Mal)-oxaloacetate (OAA) shuttle (Ebbighausen et al. 1985; Heineke et al. 1991; Flügge 1999; Scheibe 2004).

The impact of photorespiratory Gly oxidation is not limited to C and N metabolism. It will change the electron transport properties in the respiratory chain (see Sects. 15.6 and 15.7). Accumulating studies have disclosed the framework of the photorespiratory pathway and its interaction with the cellular C/N network, but further studies are needed to gain a complete picture. Recently, novel findings likely associated with photorespiration have been reported (e.g., glycolate oxidation in the mitochondria; Niessen et al. 2007, a cytosolic bypass pathway for HPR reaction; Timm et al. 2008, ferredoxin-dependent GOGAT and its interaction to serine hydroxymethyltransferase [SHMT]; Jamai et al. 2009). These findings imply that photorespiration potentially possesses more flexible metabolic pathways, leading to more complicated consequences for the cellular metabolism.

15.4 Does the TCA Cycle Have Physiologic Function(s) During Photosynthesis?

Light-dependent inhibition of the TCA cycle activity may give us the misleading idea that C metabolism in the TCA cycle does not play beneficial roles under light conditions. However, partial operation of the TCA cycle is necessary for sustaining de novo NH_4^+ assimilation in the light (Fig. 15.3). NH_4^+ assimilation is a light-dependent reaction in the chloroplasts and requires a C skeleton derived from the mitochondrial TCA cycle. Although C skeletons can be exported from the mitochondria as either citrate or 2-OG, the former is thought to be the primary pathway based on the analysis of metabolite-exchange activities using isolated spinach mitochondria (Hanning and Heldt 1993). The predominance of the citrate-export

Fig. 15.3 (continued) isocitrate, (*chl* or *cyt* or *mt* or *per*)*MDH* (chloroplastic or cytosolic or mitochondrial or peroxisomal) malate dehydrogenase, *NDex* external NAD(P)H dehydrogenase, *NDin* internal NAD(P)H dehydrogenase, *NiR* NO_2^- reductase, *NR* NO_3^- reductase, *OAA* oxaloacetate, *2-OG* 2-oxoglutarate, *PEP* phospho*enol*pyruvate, *TP* triose phosphate, *UCP* uncoupling protein, *UQ* ubiquinone, *WWC* water–water cycle. Other abbreviations are described in the caption of Fig. 15.2

pathway may be related to feedback inhibition of mitochondrial NAD-IDH in the light (Igamberdiev and Gardeström 2003). This idea is supported by the experimental fact that, while repression of citrate synthase (CS) in tomato leaves affected N assimilation (Sienkiewicz-Porzucek et al. 2008), the NAD-IDH mutation in *Arabidopsis* caused little effects (Lemaitre et al. 2007). Citrate is exported via the citrate-OAA shuttle (Hanning et al. 1999), followed by the conversion to 2-OG by cytosolic aconitase and NADP-dependent isocitrate dehydrogenase (ICDH) (Gálvez et al. 1999). 2-OG is imported to the chloroplasts via 2-OG/malate transporter (OMT; Taniguchi et al. 2002). OMT is indispensable for NH_4^+ assimilation and photorespiratory metabolism; antisense suppression of this transporter in tobacco caused a decrease in amino acid contents, suppression of photorespiratory metabolism and retardation of plant growth (Schneidereit et al. 2006). 2-OG imported to the chloroplasts is used as C skeleton for NH_4^+ assimilation by the GS/GOGAT system. 2-OG and glutamine (Gln) may regulate GS activity via transcriptional or post-translational control (Gálvez et al. 1999), suggesting a fine-regulated co-ordination of the availability of C skeleton and the synthesis of amino acid.

For net NH_4^+ assimilation, NO_3^- reduction via NO_3^- reductase (NR) in the cytosol and nitrite (NO_2^-) reductase (NiR) in the chloroplasts are required. It should be noted that NO_3^- reduction affects the properties of the respiratory electron transport via the consumption of cytosolic NADH (see Fig. 15.3; Escobar et al. 2006). This also highlights the close interaction between C/N metabolism and the mitochondrial respiratory system.

Fernie and co-workers have comprehensively assessed the role(s) of TCA cycle enzymes in illuminated leaves using transgenic tomato plants (Nunes-Nesi et al. 2008). Repression of the fumarase activity resulted in a slower rate of photosynthesis, which seemed to be a result of decreased stomatal function (Nunes-Nesi et al. 2007a). On the other hand, suppression of aconitase or NAD-dependent malate dehydrogenase (NAD-MDH) stimulated photosynthesis, which was linked to elevated sucrose synthesis activity or ascorbate-biosynthesis capacity, respectively (Carrari et al. 2003; Nunes-Nesi et al. 2005). Although decrease in citrate synthase (CS) or succinyl-CoA ligase did not affect apparent photosynthetic performance, profiles of cellular metabolism (e.g., amino acid composition) were greatly modified (Studart-Guimarães et al. 2007, Sienkiewicz-Porzucek et al. 2008). Taken together, the impact of the TCA cycle on photosynthesis and cellular metabolism is different depending on the enzymes involved. Therefore, when the role of the TCA cycle in illuminated leaves is debated, we may need to evaluate the individual reactions, rather than the overall TCA cycle flux.

15.5 Light-Dependent Response of the Respiratory Chain; Upregulation of the Alternative Oxidase

The respiratory chain (mitochondrial electron transport chain localized in the inner membrane, see Chap. 13) is the most intensively studied subject in the research area of plant mitochondria. One major feature of the plant respiratory chain is the presence of non-proton pumping alternative pathways (alternative oxidase [AOX] and type II NAD[P]H dehydrogenases [NDs], see Chap. 14). Especially, the physiologic roles of AOX have been intensively examined (for a recent review, see McDonald 2008). A classically-known significance of AOX is heat generation to facilitate pollination during floral development. Later, it has been suggested that AOX fulfills an important role in the maintenance of the mitochondrial and even cellular redox homeostasis especially under stress conditions (Maxwell et al. 1999; Umbach et al. 2005). As discussed below in detail, this AOX role is quite important under excess light conditions where much reducing equivalents are generated via the photosynthetic electron transport and the photorespiratory metabolism.

A large number of studies have shown that AOX transcript level and/or protein amount are increased by a variety of environmental stresses (see Chap. 14), including HL (e.g., Finnegan et al. 1997; Svensson and Rasmusson 2001; Yoshida and Noguchi 2009). As increases in AOX protein amount and capacity are sometimes not followed by elevated engagement of AOX (for a recent review, see Rasmusson et al. 2009), it seems to be more important to measure the in vivo electron flux to AOX. Using an O_2 isotope discrimination technique, Ribas-Carbo et al. (2000) demonstrated that the electron partitioning was shifted from the cytochrome pathway (CP) to AOX along with the light-dependent greening of soybean cotyledons. They also demonstrated that the in vivo engagement of AOX is higher in photosynthetic tissue than in other tissues (Ribas-Carbo et al. 1997). Light-dependent increase in the AOX activity is reflected in the activation state of AOX (Noguchi et al. 2005; see Sect. 15.9).

15.6 Function of AOX as a Dissipation System of Excess Cellular Reducing Equivalents

As discussed above, light-dependent AOX upregulation from transcript to in vivo activity levels has been repeatedly demonstrated. This raises the question of what is the physiologic function of AOX in illuminated leaves. In the 1980–1990s, the effects of chemical inhibition of the respiratory chain on photosynthetic efficiency and contents of some key metabolites were examined using leaf mesophyll protoplasts. The photosynthetic O_2 evolution was lowered by adding low concentrations of antimycin A (inhibitor of CP) or oligomycin (inhibitor of mitochondrial H^+-ATPase) (e.g., Krömer et al. 1993; Padmasree and Raghavendra 1999a). Based on the analyses of the ATP/ADP ratio and metabolite contents, it was suggested that the decrease in photosynthetic rate could be attributed to a restriction of sucrose synthesis as a result of ATP depletion (see Fig. 15.3). On the other hand, AOX inhibition similarly lowered photosynthetic rate, but distinctly affected the metabolite contents (Padmasree and Raghavendra 1999a, b). Notably, AOX inhibition increased the ratio of Mal/OAA, which is indicative of the cellular reduction level. We carried out simultaneous measurements of the photosynthetic O_2 evolution rate and electron transport efficiency in broad bean leaves, and examined effects of respiratory inhibitors on these parameters (Yoshida et al. 2006). AOX inhibition, but not CP inhibition, lowered photosynthetic efficiency under a wide range of light conditions. Bartoli et al. (2005) also reported a decrease in photosynthetic efficiency by AOX inhibition in drought-exposed wheat leaves.

The work with AOX inhibitors has been recently complemented by experiments using AOX-deficient plants. This may alleviate concerns about unexpected side-effects of AOX inhibitors (e.g., salicylhydroxamic acid [SHAM], Robinson et al. 1995), although it has been confirmed that such effects on photosynthesis in isolated intact chloroplasts are marginal (Padmasree and Raghavendra 1999a; Bartoli et al. 2005; Yoshida et al. 2006). Giraud et al. (2008) clearly demonstrated that a knockout mutant of *Arabidopsis AOX1a* resulted in a phenotypic difference (anthocyanin accumulation), lowered photosynthetic efficiency, and elevated reactive oxygen species (ROS) production after exposure to combined light and drought stress. They also observed that the cellular metabolite profiles and transcript abundances of several components (notably chloroplast-located antioxidant defense systems) were significantly modified.

Using *Arabidopsis* mutants defective in the cyclic electron transport around photosystem I (CET-PSI, Shikanai 2007; see Glossary), we examined the respiratory properties in conjunction with the redox state in the chloroplasts and the transport activity of reducing equivalents via the Mal-OAA shuttle (Yoshida et al. 2007). Under excess light conditions, AOX protein amount and capacity were upregulated concomitant with an accumulation of reducing equivalents in the chloroplasts and increases in the activities of enzymes involved in the Mal-OAA shuttle. Effects of HL or the CET-PSI mutation on the total respiratory rate and the maximal activity of cytochrome *c* oxidase (COX) were marginal, suggesting that AOX is preferentially up-regulated within two ubiquinol-oxidizing pathways by excess light. Other mutants of photosynthetic electron transport also showed little effects on the total respiratory rate (Armstrong et al. 2007).

Taken together, these findings suggest that AOX plays an important role as a dissipation system of excess cellular reducing equivalents generated by photochemical reactions in the chloroplasts (see Fig. 15.3). Furthermore, AOX seems to contribute to photorespiratory metabolism by supporting efficient oxidation of mitochondrial NADH (see Fig. 15.3; Igamberdiev et al. 1997). This idea is supported by a recent report demonstrating that *Arabidopsis aox1a* mutants showed a high Gly/serine (Ser) ratio in the light (Strodtkötter et al. 2009). AOX may be also involved in the control of ascorbate synthesis in the light (Bartoli et al. 2006). These roles of AOX may interact with the signal transduction to the nucleus, as AOX deficiency triggers the modification of transcript profiles of several genes (Vanlerberghe et al. 2009).

Because of uncoupling from the formation of a membrane potential, AOX can perform electron transport more efficiently without being restricted by cellular energy status (Vanlerberghe and Ordog 2002; Finnegan et al. 2004; McDonald 2008). Furthermore, de novo synthesis of AOX is likely to be attained more quickly and economically than the phosphorylating pathway requiring coordinated protein-complex assembly (Giegé et al. 2005). These two advantageous aspects of AOX would provide further plasticity of the photo-acclimation mechanisms in plants under excess light conditions. Even under other environmental stresses, the roles of AOX presented here may be beneficial because the photosynthetic apparatus is photo-damaged under such AOX-inducible conditions (Bartoli et al. 2005; Noguchi and Terashima 2006; Watanabe et al. 2008). In this sense, AOX becomes an essential accessory for stabilizing the autotrophic system of plants.

15.7 Function of Other Components in the Light

Several studies have implied that, in addition to AOX, other non-phosphorylating components in the respiratory chain play important roles in illuminated leaves. Uncoupling protein (UCP) dissipates the proton gradient across the mitochondrial inner membrane by facilitating re-entry of protons into the matrix (Vercesi et al. 2006). Using the *Arabidopsis ucp1* mutant, Sweetlove et al. (2006) demonstrated that UCP contributes to photosynthesis through efficient oxidation of Gly produced during photorespiratory metabolism. A similar role is suggested for AOX (see Sect. 15.6). The physiologic significance of this redundancy is unclear so far. It is known that the AOX activity is inhibited by 4-hydroxy-2-nonenal (HNE) produced as a consequence of lipid peroxidation, while the UCP activity is activated (Smith et al. 2004; Winger et al. 2005). Therefore, differential activation of AOX and UCP depending on the stress severity and the photorespiratory rate may be beneficial.

NDs are other candidates that efficiently oxidize the cellular reducing equivalents (see Fig. 15.3, Rasmusson et al. 2008). In the light, external NDs (NDex) may contribute to the oxidation of cytosolic reducing equivalents that are exported from the chloroplast. On the other hand, internal NDs (NDin) likely play a key role in oxidizing NADH generated by photorespiratory metabolism (Igamberdiev et al. 1997). In *Arabidopsis*, expression of *NDB2* (encoding NDex) often shows a coordinated light-induction pattern with *AOX1a* (see Sect. 15.8), implying cooperation between AOX and NDex to efficiently dissipate cytosolic NADH in illuminated leaves. However, there is no direct evidence showing that NDs actually contribute to efficient photosynthesis so far. According to Liu et al. (2008), overexpression of potato NDB1 (external Ca^{2+}-dependent NADPH dehydrogenase) in tobacco did not cause a significant change in chloroplast NADP-MDH activation state (an indicator of chloroplast redox state) and xanthophyll ratios. Further studies using mutants or transformants are needed to assess the involvement of NDs in optimal photosynthesis.

The plant material frequently used for the studies on the interaction between chloroplasts and mitochondria is the cytoplasmic male sterile mutant of tobacco (CMSII; Gutierres et al. 1997; Sabar et al. 2000). CMSII lacks functional complex I in the respiratory chain due to a deletion in the *nad7* region of the mitochondrial DNA (see Chap. 18). This mutant shows a decreased photosynthetic capacity, which is attributed to the disruption of the chloroplast redox balance (Dutilleul et al. 2003a). CMSII also possesses several physiologic characteristics, such as altered C/N metabolism and NAD(H) pool size (Dutilleul et al. 2005), impaired acclimation to HL conditions (Priault et al. 2006a), and decreased leaf internal conductance to CO_2 diffusion (Priault et al. 2006b). These results demonstrate that the NADH turnover by complex I is essential for maintaining a variety of cellular processes and thereby efficient photosynthesis. It was also demonstrated that the plant cell possesses flexibility to cope with an unexpected disturbance of mitochondrial electron transport by the induction of anaplerotic pathways (such as AOX and NDs), which enables plant survival (Dutilleul et al. 2003b). Similar observations were obtained in the mosaic mutant of cucumber (MSC16), which shows a decreased complex I capacity (Juszczuk et al. 2007).

Novel components recently found in the plant mitochondria may broaden the mechanisms of cross-talk between chloroplasts and mitochondria. A mitochondrial glycerol-3-phosphate (G-3-P) shuttle plays a crucial role to adjust cellular NADH/NAD^+ ratio by channeling cytosolic reducing equivalents to the respiratory chain (Shen et al. 2006). When plants are exposed to excess light, this shuttle may be involved in the oxidation of chloroplast-derived reducing equivalents (see Fig. 15.3). The electron transfer flavoprotein (ETF)/ETF ubiquinone oxidoreductase (ETFQO) electron transfer complex was suggested to be involved in the chlorophyll degradation pathway activated during dark-induced carbohydrate deprivation (Ishizaki et al. 2005, 2006). Although distinct from the mitochondrial role for maintaining cellular redox homeostasis, this can be represented as one of the interactions between chloroplasts and mitochondria.

15.8 How is the Respiratory Gene Expression Regulated in Illuminated Leaves?

Because the respiratory system plays an indispensable role in illuminated leaves, it is important to understand how the system is regulated under such situations. It is known that a number of genes encoding mitochondrial proteins are altered in their transcript levels in response to fluctuating light environments (Fig. 15.4; Rasmusson and Escobar 2007). For example, the expression of genes encoding respiratory chain components (e.g., *AOX1a*, *NDA1*, *NDB2*, *NDC1* in *Arabidopsis*) is subjected to light- or circadian-dependent regulation (Michalecka et al. 2003; Escobar et al. 2004; Elhafez et al. 2006; Yoshida and Noguchi 2009). The mechanistic insight into the regulation for such light-dependent responses is discussed here.

Thanks to several studies using cultured cells treated with specific respiratory inhibitors, the mitochondrial retrograde regulation (MRR) is well-known as machinery controlling respiratory gene expression (Rhoads and Subbaiah 2007). One of the key factors for MRR has been suggested to be ROS generated by the overreduction of the respiratory chain. Especially, *AOX* expression was sensitively induced by the impairment of the respiratory electron transport and resulting ROS production (e.g., Clifton et al. 2005; Yoshida and Noguchi 2009). So far, there is no direct evidence showing the involvement of MRR in the light-dependent induction of *AOX* expression. As speculated by the fact that illuminated leaves often exhibit a large transient increase in O_2 consumption when exposed to darkness (light-enhanced dark

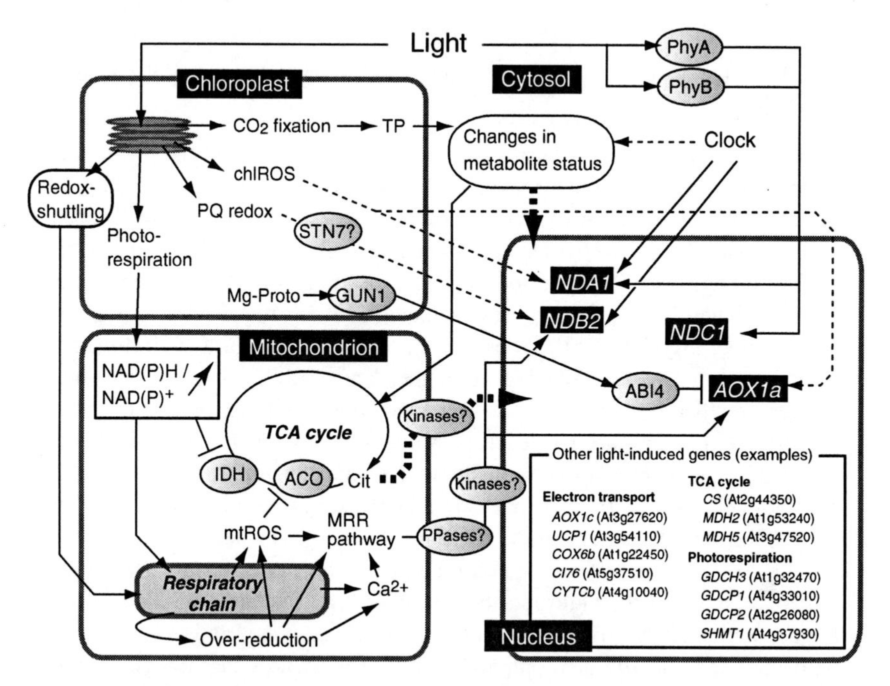

Fig. 15.4 Simplified scheme of the light-dependent regulations of respiratory gene expression in *Arabidopsis*. Signaling pathways denoted by dashed lines are still controversial. Detailed information is depicted in the main text. *Mg-Proto* Mg-protoporphyrin IX, *MRR* mitochondrial retrograde regulation, *PhyA(B)* phytochrome A(B), *PPase*s phosphatases, *PQ* plastoquinone, (*chl* or *mt*)*ROS* (chloroplastic or mitochondrial) reactive oxygen species. Other abbreviations are described in the captions of Figs. 15.2 and 15.3

respiration [LEDR]; Atkin et al. 2000), the respiratory chain is likely to be more reduced under light conditions possibly as a consequence of photorespiratory metabolism and redox-shuttling from the chloroplasts. Recently, we directly assessed the ubiquinone (UQ) redox state in plant tissues and confirmed that the respiratory chain is actually reduced in illuminated leaves (Yoshida et al. 2010). These findings indirectly support that *AOX* expression in illuminated leaves is regulated by MRR (see Fig. 15.4). One question is whether the signal(s) derived from other organelles also regulate *AOX* expression (Box 15.2). The incubation of the *Arabidopsis* leaf with inhibitors of chloroplastic electron transport chain (chlETC) or methyl viologen under light conditions did not result in a large increase in *AOX1a* transcript level (Yoshida and Noguchi 2009). In the light, the magnitude of ROS generation is much larger in the chloroplast than in the mitochondria (Foyer and Noctor 2003). Even under such conditions, mitochondrial ROS may be distinguished and registered by a local detection mechanism (Rhoads et al. 2006).

In *Arabidopsis*, expression of some respiratory genes in the light seems to be regulated in a distinct manner from MRR (see Fig. 15.4). It is known that *NDB2* shows a similar expression pattern as *AOX1a* under several stress conditions

Box 15.2 Does the chloroplast retrograde signal regulate respiratory gene expression?

Considering the flexible and drastic responses of the mitochondrial respiratory system to varying light environments, it is likely that the signals derived from the chloroplasts regulate expression of nucleus-encoded respiratory genes. Chloroplast-nucleus retrograde signaling mechanisms have been intensively studied more than mitochondrion-nucleus retrograde signaling. Several pathways mediated by reactive oxygen species (ROS; H_2O_2 and $^1O_2$), redox, and Mg-protoporphyrin IX (Mg-Proto) are known as the chloroplast retrograde signaling (for a review, see Woodson and Chory 2008), but, to date, it is unclear whether these chloroplast-originated signals regulate respiratory gene expression. In *Arabidopsis* leaves, expression of *NDA1* and *NDB2* was affected by the manipulation of the photosynthetic electron transport (Yoshida and Noguchi 2009). To conclude whether these genes are subjected to the chloroplast retrograde regulation, more mechanistic evidence is required.

The *Arabidopsis* leaf-variegated mutant *var2* is more sensitive to photo-oxidative stress in the chloroplasts than wild-type (Sakamoto 2006). While the capacities of tricarboxylic acid (TCA) cycle enzymes and cytochrome *c* oxidase (COX) in *var2* are comparable to those in wild-type, *AOX1a* expression was more induced in *var2* than in wild-type (Yoshida et al. 2008). Also, the photo-susceptible green sectors in *var2* leaves had higher levels of AOX protein than the white sectors lacking developed chloroplast. These results imply the involvement of chloroplastic photo-oxidative stress in the regulation of *AOX1a* expression.

Recently, the *Arabidopsis* transcription factor ABI4 was suggested to repress *AOX1a* expression under non-stress conditions (Giraud et al. 2009). It is known that ABI4 is acting downstream of Mg-Proto signaling pathway (Woodson and Chory 2008). This finding strongly suggests that *AOX1a* is, at least potentially, regulated by the chloroplast retrograde signaling (see Fig. 15.4). It may explain the lower level of AOX protein in *var2* white sectors compared to the green, although there is no experimental evidence that Mg-Proto accumulates in the undifferentiated plastids in *var2* white sectors.

Häusler et al. (2009) demonstrated that, using an *Arabidopsis* double mutant impaired in the starch synthesis and the triose phosphate (TP)/phosphate translocator (*adg1-1/tpt-1*), alterations of several properties in the chloroplasts (e.g., perturbation of the redox poise) lowered the *AOX1a* transcript level especially during dark period. Therefore, it is possible to consider that chloroplasts in this mutant sent some signals to repress *AOX1a* expression.

(Clifton et al. 2005). Also in our analysis, transcript levels of these genes showed a synchronized increase after HL treatment (Yoshida and Noguchi 2009). However, some treatments including incubation in high CO_2 and perturbation of chlETC affected expression of *AOX1a* and *NDB2* in a different way. Distinct responses of

AOX1a and *NDB2* are also supported by the comparative analysis of the promoter region in each gene (Ho et al. 2008). In etiolated seedlings, expression of *NDA1* and *NDC1* was suggested to be directly controlled by light via a photoreceptor (phytochrome)-mediated manner (Escobar et al. 2004). As phytochrome-mediated reactions are important mainly at the plant developmental stage (Franklin et al. 2005), it is unclear whether these regulations occur in mature leaves. Instead, *NDA1* expression appears to be altered by ROS generated in the chloroplasts in mature leaves (Yoshida and Noguchi 2009).

Furthermore, the fluctuation of metabolite levels during photosynthetic CO_2 fixation (e.g., accumulation of carbohydrate) seem to affect the respiratory gene expression (see Fig. 15.4; Price et al. 2004; Rasmusson and Escobar 2007; Yoshida and Noguchi 2009). Although further studies are needed to elucidate mechanisms of the respiratory gene expression in response to altered cellular C metabolism during photosynthesis, citrate may be involved in the signaling as a key molecule (Djajanegara et al. 2002; Gray et al. 2004; Yoshida and Noguchi 2009). If so, ROS-dependent inhibition of the aconitase activity (Verniquet et al. 1991) and feedback inhibition of the NAD-IDH activity (Igamberdiev and Gardeström 2003), likely resulting in citrate accumulation, may link to the citrate-dependent induction in the light (see Fig. 15.4).

It should be mentioned that the increase in the transcript abundance is not necessarily accompanied by elevated in vivo activity. For example, *Arabidopsis AOX1a* and *NDB2* coexpress in several situations, but this does not always lead to cooperation at the in vivo activity level. This is partly because the posttranslational regulations of these gene products are different; AOX1a is regulated by the redox state and pyruvate (Siedow and Umbach 2000), while NDB2 activity is Ca^{2+}-dependent (Geisler et al. 2007). Furthermore, some NDs are known to be targeted not only to the mitochondria but also to other organelles (Carrie et al. 2008). In order to understand the responses of the mitochondrial respiratory system to stress, it is important to assess the regulatory manner at the post-translational level and the in vivo activity, in addition to the transcriptional regulation.

15.9 Posttranslational Regulation: What are the Consequences of Changes in the Mitochondrial Redox Environment in the Light?

Compared with the information available on transcriptional responses, that of post-translational modifications of the mitochondrial respiratory system is limited, especially in the context of light-dependent regulation. One of the most dramatically fluctuating factors in the mitochondrial environment during photosynthesis is the redox state (Igamberdiev and Gardeström 2003). We have already explained that some mitochondrial enzymes are allosterically inhibited by a high NADH/NAD^+ ratio (see Sect. 15.2). Another possible consequence of the fluctuation in the mitochondrial redox state is the operation of the thioredoxin (Trx) system. The Trx system plays a pivotal role in sensing the local redox environment and regulating the activity of target enzymes (Buchanan and Balmer 2005; Hisabori et al. 2007). The Trx system was identified in the plant mitochondria (Laloi et al. 2001; Reichheld et al. 2005), as well as other subcellular compartments. In addition, the experimental strategies for screening Trx-target proteins have listed a variety of potential Trx-target proteins in the plant mitochondria (Balmer et al. 2004; Winger et al. 2007a; Martí et al. 2009). As one example, AOX is suggested to be a target protein of the mitochondrial Trx system; AOX is reduced and activated by the Trx system (Gelhaye et al. 2004). Therefore, the light-dependent conversion of the redox state of AOX (Noguchi et al. 2005) may be regulated by the mitochondrial Trx system.

Another likely posttranslational event of the mitochondrial proteins under excess light conditions is the ROS-derived modification. Reduction of the respiratory chain in the light (Yoshida et al. 2010) gives rise to ROS in the mitochondria (Sweetlove and Foyer 2004). ROS are highly reactive and cause rapid and deleterious oxidation of biomolecules such as proteins, lipid, and DNA. Sweetlove et al. (2002) treated *Arabidopsis* cultured cells with H_2O_2, menadione, or antimycin A to induce oxidative damage and identified damaged proteins in the mitochondria. The degraded proteins after oxidative damage included components related to the electron transport, the oxidative phosphorylation, and the TCA cycle. HNE, a product of ROS-induced lipid peroxidation, also causes deleterious effects on the proteins via forming HNE-Michael adducts (Millar and Leaver 2000). Especially, the enzymes containing lipoic acid as enzyme cofactor (e.g., GDC H-subunit) are readily modified and lowered in the activities by HNE (Taylor et al. 2002). HNE adduction also degrades some mitochondrial proteins (Winger et al. 2007b). HNE-dependent modifications are evident in plants under adverse environmental conditions, such as drought (Taylor et al. 2005).

Oxidative stress also leads to the carbonyl group formation (Kristensen et al. 2004), the oxidation of tryptophan (Møller and Kristensen 2006), and the tyrosine nitration (Turko et al. 2003) in mitochondrial proteins. In addition, several mitochondrial proteins are subjected to phosphorylation (Bykova et al. 2003). For future research, it is important to examine in an eco-physiologic context whether the post-translational regulations summarized here actually occur in illuminated leaves.

15.10 Conclusions and Perspectives

It is now obvious that the functions of the plant mitochondria are not confined to energy production alone. For supporting an optimal rate of photosynthesis, leaf mitochondria in the light play indispensable roles in the photorespiratory metabolism, the dissipation of excess cellular reducing equivalents, the ATP supply for sustaining sucrose synthesis and the C skeleton supply for N assimilation (see Figs. 15.1 and 15.3). The mitochondrial environment is dynamically modulated through metabolic interactions and redox exchange with the chloroplasts. Especially when plants are exposed to severe stress, unfavorable consequences in the mitochondria may occur (e.g., inhibition of metabolism, ROS generation and damage to the mitochondrial components). These deleterious effects lead to the depression of mitochondrial function as a support system of photosynthesis, which can have a negative impact on plant growth. Under excess light conditions, unique components in the plant mitochondria, AOX and other non-phosphorylating pathways, fulfill key roles in the flexible readjustment of the mitochondrial redox environment, in addition to having direct contributions to photosynthesis. Further studies using rapidly progressing experimental procedures, such as reverse genetics, microarrays, and proteomics, will advance our understanding about the light response of the mitochondrial respiratory system and relationship with other subcellular compartments (e.g., the chloroplast). Such efforts may lead to the identification of novel components essential for organelle cross-talk. Unique approaches, such as the in vitro reconstituted system of simulated organelle interplay (Raghavendra et al. 1998), are also worth trying to find new aspects of the organelle cross-talk mechanism. It is important, in parallel, to continue eco-physiologic studies paying more attention to the nature of plant behavior under varying natural conditions. For example, it is intriguing to examine whether mitochondrial physiologic roles during photosynthesis are variable depending on the leaf developmental/senescent stages (Keech et al. 2007).

Acknowledgments We acknowledge Drs. K. Oikawa, S. Mano, and Prof. M. Nishimura (National Institute for Basic Biology) for kindly providing the image shown in Fig. 15.1b. We also acknowledge Prof. W. Sakamoto (Okayama University), Drs. D. Tholen (CAS-MPG Partner Institute), and K. Iwabuchi (Osaka University) for helpful advice.

Glossary

Cyclic electron transport around photosystem I (CET-PSI): Photosynthetic electron transport depending solely on the photosystem I (PSI) photochemical reaction, not on the photosystem II (PSII). Because electrons are recycled from NAD(P)H or ferredoxin to plastoquinone (PQ), CET-PSI can generate ΔpH without accumulating NADPH.

Light-enhanced dark respiration (LEDR): A large transient increase in the respiratory rate detected during light-dark transfer in photosynthetic eukaryotes. LEDR reflects an increased supply of nonphotorespiratory substrates that accumulate during photosynthesis.

R_{day}: Nonphotorespiratory CO_2 release rate in the light.

Redox-shuttling: Reductant transport machinery across the organelle envelope. For example, Malate (Mal)-oxaloacetate (OAA) shuttle, Mal-aspartate (Asp) shuttle and triose phosphate (TP)-phosphoglycerate (PGA) shuttle are known.

References

Armstrong, A. F., Wardlaw, K. D., Atkin, O. K. 2007. Assessing the relationship between respiratory acclimation to the cold and photosystem II redox poise in *Arabidopsis thaliana*. Plant Cell Environ 30:1513–1522.

Atkin, O. K., Millar, A. H., Gardeström, P., Day, D. A. 2000. Photosynthesis, carbohydrate metabolism and respiration in leaves of higher plants. In Photosynthesis: Physiology and Metabolism, eds. R. C. Leegood, T. D. Sharkey, S. von Caemmerer, pp. 153–175. Dordrecht: Kluwer.

Balmer, Y., Vensel, W. H., Tanaka, C. K., Hurkman, W. J., Gelhaye, E., Rouhier, N., Jacquot, J.-P., Manieri, W., Schürmann, P., Droux, M., Buchanan, B. B. 2004. Thioredoxin links redox to the regulation of fundamental processes of plant mitochondria. Proc Natl Acad Sci USA 101:2642–2647.

Bartoli, C. G., Gomez, F., Gergoff, G., Guiamét, J. J., Puntarulo, S. 2005. Upregulation of the mitochondrial alternative oxidase pathway enhances photosynthetic electron transport under drought conditions. J Exp Bot 53:1269–1276.

Bartoli, C. G., Yu, J., Gómez, F., Fernández, L., McIntoch, L., Foyer, C. H. 2006. Inter-relationships between light and respiration in the control of ascorbic acid synthesis and accumulation in *Arabidopsis thaliana* leaves. J Exp Bot 57:1621–1631.

Baurain, D., Dinant, M., Coosemans, N., Metagne, R. F. 2003. Regulation of the alternative oxidase *Aox1* gene in *Chlamydomonas reinhardtii*. Role of the nitrogen source on the expression of a reporter gene under the control of the *Aox1* promoter. Plant Physiol 131:1418–1430.

Buchanan, B. B., Balmer, Y. 2005. Redox regulation: a broadening horizon. Annu Rev Plant Biol 56:187–220.

Budde, R. J. A., Randall, D. D. 1990. Pea leaf mitochondrial pyruvate dehydrogenase complex is inactivated *in vivo* in a light-dependent manner. Proc Natl Acad Sci USA 87:673–676.

Bykova, N. V., Egsgaard, H., Møller, I. M. 2003. Identification of 14 new phosphoproteins involved in important plant mitochondrial processes. FEBS Lett 540:141–146.

Bykova, N. V., Keerberg, O., Pärnik, T., Bauwe, H., Gardeström, P. 2005. Interaction between photorespiration and respiration in transgenic potato plants with antisense reduction in glycine decarboxylase. Planta 222:130–140.

Campbell, W. J., Ogren, W. L. 1990. Glyoxylate inhibition of ribulose bisphosphate carboxylase/oxygenase activation in intact, lysed and reconstituted chloroplasts. Photosynth Res 23:257–268.

Cardol, P., Alric, J., Girard-Bascou, J., Franck, F., Wollman, F.-A., Finazzi, G. 2009. Impaired respiration discloses the physiological significance of state transitions in *Chlamydomonas*. Proc Natl Acad Sci USA 106: 15979–15984.

Cardol, P., Gloire, G., Havaux, M., Remacle, C., Matagne, R., Franck, F. 2003. Photosynthesis and state transitions in mitochondrial mutants of *Chlamydomonas reinhardtii* affected in respiration. Plant Physiol 133:2010–2020.

Carrari, F., Nunes-Nesi, A., Gibon, Y., Lytovchenko, A., Loureiro, M. E., Fernie, A. R. 2003. Reduced expression of aconitase results in an enhanced rate of photosynthesis and marked shifts in carbon partitioning in illuminates leaves of wild species tomato. Plant Physiol 133:1322–1335.

Carrie, C., Murcha, M. W., Kuehn, K., Duncan, O., Barthet, M., Smith, P. M., Eubel, H., Meyer, E., Day, D. A., Millar, A. H., Whelan, J. 2008. Type II NAD(P)H dehydrogenases are targeted to mitochondria and chloroplasts or peroxisomes in *Arabidopsis thaliana*. FEBS Lett 582:3073–3079.

Clifton, R., Lister, R., Parker, K. L., Sappl, P. G., Elhafez, D., Millar, A. H., Day, D. A., Whelan, J. 2005. Stress-induced co-expression of alternative respiratory chain components in *Arabidopsis thaliana*. Plant Mol Biol 58:193–212.

Cournac, L., Latouche, G., Cerovic, Z., Redding, K., Ravenel, J., Peltier, G. 2002. In vivo interactions between photosynthesis, mitorespiration, and chlororespiration in *Chlamydomonas reinhardtii*. Plant Physiol 129:1921–1928.

Djajanegara, I., Finnegan, P. M., Mathieu, C., McCabe, T., Whelan, J., Day, D. A. 2002. Regulation of alternative oxidase gene expression in soybean. Plant Mol Biol 50:735–742.

Dutilleul, C., Driscoll, S., Cornic, G., De Paepe, R., Foyer, C. H., Noctor, G. 2003a. Functional mitochondrial complex I is required by tobacco leaves for optimal photosynthetic performance in photorespiratory conditions and during transients. Plant Physiol 131:264–275.

Dutilleul, C., Garmier, M., Noctor, G., Mathieu, C., Chétrit, P., Foyer, C. H., De Paepe, R. 2003b. Leaf mitochondria modulate whole cell redox homeostasis, set antioxidant capacity, and determine stress resistance through altered signaling and diurnal regulation. Plant Cell 15:1212–1226.

Dutilleul, C., Lelarge, C., Prioul, J.-L., De Paepe, R., Foyer, C. H., Noctor, G. 2005. Mitochondria-driven changes in leaf NAD status exert a crucial influence on the control of nitrate assimilation and the integration of carbon and nitrogen metabolism. Plant Physiol 139:64–78.

Ebbighausen, H., Jia, C., Heldt, H. W. 1985. Oxaloacetate translocator in plant mitochondria. Biochim Biophys Acta 810:184–199.

Elhafez, D., Murcha, M. W., Clifton, R., Soole, K. L., Day, D. A., Whelan, J. 2006. Characterization of mitochondrial alternative NAD(P)H dehydrogenase in *Arabidopsis*: intraorganelle location and expression. Plant Cell Physiol 47:43–54.

Escobar, M. A., Franklin, K. A., Svensson, A. S., Salter, M. G., Whitelam, G. C., Rasmusson, A. G. 2004. Light regulation of the *Arabidopsis* respiratory chain. Multiple discrete photoreceptor responses contribute to induction of type II NAD(P)H dehydrogenase genes. Plant Physiol 136:2710–2721.

Escobar, M. A., Geisler, D. A., Rasmusson, A. G. 2006. Reorganization of the alternative pathways of the *Arabidopsis* respiratory chain by nitrogen supply: opposing effects of ammonium and nitrate. Plant J 45:775–788.

Eubel, H., Lee, C. P., Kuo, J., Meyer, E. H., Taylor, N. L., Millar, A. H. 2007. Free-flow electrophoresis for purification of plant mitochondria by surface charge. Plant J 52:583–594.

Finnegan, P. M., Soole, K. L., and Umbach, A. L. 2004. Alternative mitochondrial electron transport proteins in higher plants. In Plant Mitochondria: From Genome to Function, eds. D. A. Day, A. H. Millar, J. Whelan, pp. 163–230. London: Kluwer.

Finnegan, P. M., Whelan, J., Millar, A. H., Zhang, Q., Smith, M. K., Wiskich, J. T., Day, D. A. 1997. Differential expression of the multigene family encoding the soybean mitochondrial alternative oxidase. Plant Physiol 114:455–466.

Flügge, U.-I. 1999. Phosphate translocators in plastids. Annu Rev Plant Physiol Plant Mol Biol 50:27–45.

Foyer, C. H., Bloom, A. J., Queval, G., Noctor, G. 2009. Photorespiratory metabolism: genes, mutants, energetics, and redox signaling. Annu Rev Plant Biol 60:455–484.

Foyer, C. H., Noctor, G. 2003. Redox sensing and signalling associated with reactive oxygen in chloroplasts, peroxisomes and mitochondria. Physiol Plant 119:355–364.

Franklin, K. A., Larner, V. S., Whitelam, G. C. 2005. The signal transducing photoreceptors of plants. Int J Dev Biol49:653–664.

Gálvez, S., Lancien, M., Hodges, M. 1999. Are isocitrate dehydrogenases and 2-oxoglutarate involved in the regulation of glutamate synthesis? Trends Plant Sci 4:484–490.

Gardeström, P., Igamberdiev, A. U., Raghavendra, A. S. 2002. Mitochondrial functions in the light and significance to carbon-nitrogen interactions. In Photosynthetic Nitrogen Assimilation and Associated Carbon and Respiratory Metabolism, eds. C. H. Foyer, G. Noctor, pp. 151–172. Dordrecht: Kluwer.

Gardeström, P., Wigge, B. 1988. Influence of photorespiration on ATP/ADP ratios in the chloroplasts, mitochondria, and cytosol, studied by rapid fractionation of barley (*Hordeum vulgare*) protoplasts. Plant Physiol 88:69–76.

Geisler, D. A., Broselid, C., Hederstedt, L., Rasmusson, A. G. 2007. Ca^{2+}-binding and Ca^{2+}-independent respiratory NADH and NADPH dehydrogenases of *Arabidopsis thaliana*. J Biol Chem 282:28455–28464.

Gelhaye, E., Rouhier, N., Gérard, J., Jolivet, Y., Gualberto, J., Navrot, N., Ohlsson, P.-I., Wingsle, G., Hirasawa, M., Knaff, D. B., Wang, H., Dizengremel, P., Meyer, Y., Jacquot, J.-P. 2004.

A specific form of thioredoxin *h* occurs in plant mitochondria and regulates the alternative oxidase. Proc Natl Acad Sci USA 101:14545–14550.

Giegé, P., Sweetlove, L. J., Cognat, V., Leaver, C. J. 2005. Coordination of nuclear and mitochondrial genome expression during mitochondrial biogenesis in *Arabidopsis*. Plant Cell 17:1497–1512.

Giraud, E., Ho, L. H. M., Clifton, R., Carroll, A., Estavillo, G., Tan, Y.-F., Howell, K. A., Ivanova, A., Pogson, B. J., Millar, A. H., Whelan, J. 2008. The absence of ALTERNATIVE OXIDASE1a in *Arabidopsis* results in acute sensitivity to combined light and drought stress. Plant Physiol 147:595–610.

Giraud, E., Van Aken, O., Ho, L. H. M., Whelan, J. 2009. The transcription factor ABI4 is a regulator of mitochondrial retrograde expression of *Alternative oxidase1a*. Plant Physiol 150:1286–1296.

Givan, C. V., Kleczkowski, L. A. 1992. The enzymic reduction of glyoxylate and hydroxypyruvate in leaves of higher plants. Plant Physiol 100:552–556.

Gray, G. R., Maxwell, D. P., Villarimo, A. R., McIntosh, L. 2004. Mitochondria/nuclear signaling of alternative oxidase gene expression occurs through distinct pathways involving organic acids and reactive oxygen species. Plant Cell Rep 23:497–503.

Gutierres, S., Saber, M., Lelandais, C., Chetrit, P., Diolez, P., Degand, H., Boutry, M., Vedel, F., de Kouchkovsky, Y., De Paepe, R. 1997. Lack of mitochondrial and nuclear-encoded subunits of complex I and alteration of the respiratory chain in *Nicotiana sulvestris* mitochondrial deletion mutants. Proc Natl Acad Sci USA 94: 3436–3441.

Hanning, I., Baumgarten, K., Schott, K., Heldt, H. W. 1999. Oxaloacetate transport into plant mitochondria. Plant Physiol 119:1025–1031.

Hanning, I., Heldt, H. 1993. On the function of mitochondrial metabolism during photosynthesis in spinach (*Spinacia oleracea* L.) leaves. Plant Physiol 103:1147–1154.

Häusler, R. E., Geimer, S., Kunz, H. H., Schmitz, J., Dörmann, P., Bell, K., Hetfeld, S., Guballa, A., and Flügge, U.-I. 2009. Chlororespiration and grana hyperstacking: how an *Arabidopsis* double mutant can survive despite defects in starch biosynthesis and daily carbon export from chloroplasts. Plant J 149:515–533.

Heineke, D., Riens, B., Grosse, H., Hoferichter, P., Peter, U., Flügge, U.-I., Heldt, H. W. 1991. Redox transfer across the inner chloroplast envelope membrane. Plant Physiol 95: 1131–1137.

Hisabori, T., Motohashi, K., Hosoya-Matsuda, N., Ueoka-Nakanishi, H., Romano, P. G. N. 2007. Towards a functional dissection of thioredoxin networks in plant cells. Photochem Photobiol 83:145–151.

Ho, L. H. M., Giraud, E., Uggalla, V., Lister, R., Clifton, R., Glen, A., Thirkettle-Watts, D., Van Aken, O., Whelan, J. 2008. Identification of regulatory pathways controlling gene expression of stress responsive mitochondrial proteins in *Arabidopsis*. Plant Physiol 147:1858–1873.

Huertas, I. E., Colman, B., Espie, G. S. 2002. Mitochondrial-driven bicarbonate transport supports photosynthesis in a marine microalga. Plant Physiol 130:284–291.

Igamberdiev, A. U., Gardeström, P. 2003. Regulation of NAD- and NADP-dependent isocitrate dehydrogenases by reduction levels of pyridine nucleotides in mitochondria and cytosol of pea leaves. Biochim Biophys Acta 1606:117–125.

Igamberdiev, A. U., Bykova, N. V., Gardeström, P. 1997. Involvement of cyanide-resistant and rotenone-insensitive pathways of mitochondrial electron transport during oxidation of glycine in higher plants. FEBS Lett 412:265–269.

Igamberdiev, A. U., Bykova, N. V., Lea, P. J., Gardeström, P. 2001. The role of photorespiration in redox and energy balance of photosynthetic plant cells: a study with a barley mutant deficient in glycine decarboxylase. Physiol Plant 111:427–438.

Igarashi, D., Miwa, T., Seki, M., Kobayashi, M., Kato, T., Tabata, S., Shinozaki, K., Ohsumi, C. 2003. Identification of photorespiratory *glutamate:glyoxylate aminotransferase* (*GGAT*) in *Arabidopsis*. Plant J 33:975–987.

Ishizaki, K., Larson, T. R., Schauer, N., Fernie, A. R., Graham, I. A., Leaver, C. J. 2005. The critical role of *Arabidopsis* electron-transfer flavoprotein:ubiquinone oxidoreductase during dark-induced starvation. Plant Cell 17:2587–2600.

Ishizaki, K., Schauer, N., Larson, T. R., Graham, I. A., Fernie, A. R., Leaver, C. J. 2006. The mitochondrial electron transfer flavoprotein complex is essential for survival of *Arabidopsis* in extended darkness. Plant J 47:751–760.

Islam, M. S., Niwa, Y., Takagi, S. 2009. Light-dependent intracellular positioning of mitochondria in *Arabidopsis thaliana* mesophyll cells. Plant Cell Physiol 50:1032–1040.

Jamai, A., Salomé, P. A., Schilling, S. H., Weber, A. P. M., McClung, C. R. 2009. *Arabidopsis* photorespiratory serine hydroxymethyltransferase activity requires the mitochondrial accumulation of ferredoxin-dependent glutamate synthase. Plant Cell 21:595–606.

Jans, F., Mignolet, E., Houyoux, P.-A., Cardol, P., Ghysels, B., Cuiné, S., Cournac, L., Peltier, G., Remacle, C., Franck, F. 2008. A type II NAD(P)H dehydrogenase mediates light-independent plastoquinone reduction in the chloroplast of *Chlamydomonas*. Proc Natl Acad Sci USA 105:20546–20551.

Juszczuk, I. M., Flexas, J., Szal, B., D browska, Z., Ribas-Carbo, M., Rychter, A. M. 2007. Effect of mitochondrial genome rearrangement on respiratory activity, photosynthesis, photorespiration and energy status of MSC16 cucumber (*Cucumis sativus*) mutant. Physiol Plant 131:527–541.

Keech, O., Pesquet, E., Ahad, A., Askne, A., Nordvall, D., Vodnala, S. M., Tuominen, H., Hurry, V., Dizengremel, P., Gardeström, P. 2007. The different fates of mitochondria and chloroplasts during dark-induced senescence in *Arabidopsis* leaves. Plant Cell Environ 30:1523–1534.

Keys, A. J., Leegood, R. C. 2002. Photorespiratory carbon and nitrogen cycling: evidence from studies of mutant and transgenic plants. In Photosynthetic Nitrogen Assimilation and Associated Carbon and Respiratory Metabolism, eds. C. H. Foyer, G. Noctor, pp. 115–134. Dordrecht: Kluwer.

Kozaki, A., Takeba, G. 1996. Photorespiration protects C3 plants from photooxidation. Nature 384:557–560.

Kristensen, B. K., Askerlund, P., Bykova, N. V., Egsgaard, H., Møller, I. M. 2004. Identification of oxidized proteins in the matrix of rice leaf mitochondria by immunoprecipitation and two-dimensional liquid chromatography-tandem mass spectrometry. Phytochemistry 65:1839–1851.

Krömer, S. 1995. Respiration during photosynthesis. Annu Rev Plant Physiol Plant Mol Biol 46:45–70.

Krömer, S., Heldt, H. W. 1991. On the role of mitochondrial oxidative phosphorylation in photosynthesis metabolism as studied by the effect of oligomycin on photosynthesis in protoplasts and leaves of barley. Plant Physiol 95:1270–1276.

Krömer, S., Malmberg, G., Gardeström, P. 1993. Mitochondrial contribution to photosynthetic metabolism. Plant Physiol 102:947–955.

Laloi, C., Rayapuram, N., Chartier, Y., Grienenberger, J.-M., Bonnard, G., Meyer, Y. 2001. Identification and characterization of a mitochondrial thioredoxin system in plants. Proc Natl Acad Sci USA 98:14144–14149.

Lemaitre, T., Urbanczyk-Wochniak, E., Flesch, V., Bismuth, E., Fernie, A. R., Hodges, M. 2007. NAD-dependent isocitrate dehydrogenase mutants of *Arabidopsis* suggest the enzyme is not limiting for nitrogen assimilation. Plant Physiol 144:1546–1558.

Linka, M., Weber, A. P. M. 2005. Shuffling ammonia between mitochondria and plastids during photorespiration. Trends Plant Sci 10:461–465.

Liu, Y.-J., Norberg, F. E. B., Szilágyi A., De Paepe, R., Akerlund, H.-E., Rasmusson, A. G. 2008. The mitochondrial external NADPH dehydrogenase modulates the leaf NADPH/$NADP^+$ ratio in transgenic *Nicotiana sylvestris*. Plant Cell Physiol 49:251–263.

Martí, M. C., Olmos, E., Calvete, J. J., Díaz, I., Barranco-Medina, S., Whelan, J., Lázaro, J. J., Sevilla, F., Jiménez, A. 2009. Mitochondrial and nuclear localization of a novel pea thioredoxin: identification of its mitochondrial target proteins. Plant Physiol 150:646–657.

Matsuo, M., Obokata, J. 2006. Remote control of photosynthetic genes by the mitochondrial respiratory chain. Plant J 47:873–882.

Maxwell, D. P., Wang, Y., McIntosh, L. 1999. The alternative oxidase lowers mitochondrial reactive oxygen production in plant cells. Proc Natl Acad Sci USA 96:8271–8276.

McDonald, A. E. 2008. Alternative oxidase: an inter-kingdom perspective on the function and regulation of this broadly distributed 'cyanide-resistant' terminal oxidase. Funct Plant Biol 35:535–552.

Michalecka, A. M., Svensson, A. S., Johansson, F. I., Agius, S. C., Johansson, U., Brennicke, A., Binder, S., Rasmusson, A. G. 2003. *Arabidopsis* genes encoding mitochondrial type II NAD(P)H dehydrogenases have different evolutionary origin and show distinct responses to light. Plant Physiol 133:642–652.

Millar, A. H., and Leaver, C. J. 2000. The cytotoxic lipid peroxidation product, 4-hydroxy-2-nonenal, specifically inhibits decarboxylating dehydrogenases in the matrix of plant mitochondria. FEBS Lett 481:117–121.

Molen, T. A., Rosso, D., Piercy, S., Maxwell, D. P. 2006. Characterization of the alternative oxidase of *Chlamydomonas reinhardtii* in response to oxidative stress and a shift in nitrogen source. Physiol Plant 127:74–86.

Møller, I. M., Kristensen, B. K. 2006. Protein oxidation in plant mitochondria detected as oxidized tryptophan. Free Radic Biol Med 40:430–435.

Niessen, M., Thiruveedhi, K., Rosenkranz, R., Kebeish, R., Hirsch, H.-J., Kreuzaler, F., Peterhänsel, C. 2007. Mitochondrial glycolate oxidation contributes to photorespiration in higher plants. J Exp Bot 58:2709–2715.

Noctor, G., De Paepe, R., Foyer, C. H. 2007. Mitochondrial redox biology and homeostasis in plants. Trends Plant Sci 12:125–134.

Noguchi, K., Taylor, N. L., Millar, A.H., Lambers, H., Day, D. A. 2005. Response of mitochondria to light intensity in the leaves of sun and shade species. Plant Cell Environ 28:760–771.

Noguchi, K., Terashima, I. 2006. Responses of spinach leaf mitochondria to low N availability. Plant Cell Environ 29:710–719.

Noguchi, K., Yoshida, K. 2008. Interaction between photosynthesis and respiration in illuminated leaves. Mitochondrion 8:87–99.

Nunes-Nesi, A., Carrari, F., Gibon, Y., Sulpice, R., Lytovchenko, A., Fisahn, J., Graham, J., Ratcliffe, R. G., Sweetlove, L. J., Fernie, A. R. 2007a. Deficiency of mitochondrial fumarase activity in tomato plants impairs photosynthesis via an effect on stomatal function. Plant J 50:1093–1106.

Nunes-Nesi, A., Carrari, F., Lytovchenco, A., Smith, A. M. O., Loureiro, M. E., Ratcliffe, R. G., Sweetlove, L. J., Fernie, A. R. 2005. Enhanced photosynthetic performance and growth as a consequence of decreasing mitochondrial malate dehydrogenase activity in transgenic tomato plants. Plant Physiol 137:611–622.

Nunes-Nesi, A., Sweetlove, L., Fernie, A. R. 2007b. Operation and function of the tricarboxylic acid cycle in the illuminated leaf. Physiol Plant 129:45–56.

Nunes-Nesi, A., Sulpice, R., Gibon, Y., Fernie, A. R. 2008. The enigmatic contribution of mitochondrial function in photosynthesis. J Exp Bot 59:1675–1684.

Oliver, D. J. 1994. The glycine decarboxylase complex from plant mitochondria. Annu Rev Plant Physiol Plant Mol Biol 45:323–337.

Oliver, D. J., Gardeström, P. 2004. Photorespiration: photosynthesis in the mitochondria. In Plant Mitochondria: From Genome to Function, eds. D. A. Day, A. H. Millar, J. Whelan, pp. 293–306. London: Kluwer.

Padmasree, K., Raghavendra, A. S. 1999a. Importance of oxidative electron transport over oxidative phosphorylation in optimizing photosynthesis in mesophyll protoplasts of pea (*Pisum sativum* L.). Physiol Plant 105:546–553.

Padmasree, K., and Raghavendra, A. S. 1999b. Response of photosynthetic carbon assimilation in mesophyll protoplasts to restriction on mitochondrial oxidative metabolism: metabolites related to the redox status and sucrose biosynthesis. Photosynth Res 62:231–239.

Pärnik, T., Keerberg, O. 2007. Advanced radiogasometric method for the determination of the rates of photorespiratory and respiratory decarboxylations of primary and stored photosynthates under steady-state photosynthesis. Physiol Plant 129:34–44.

Peltier, G., Cournac, L. 2002. Chlororespiration. Annu Rev Plant Biol 53:523–550.

Priault, P., Fresneau, C., Noctor, G., De Paepe, R., Cornic, G., Streb, P. 2006a. The mitochondrial CMSII mutation of *Nicotiana sylvestris* impairs adjustment of photosynthetic carbon assimilation to higher growth irradiance. J Exp Bot 57:2075–2085.

Priault, P., Tcherkez, G., Cornic, G., De Paepe, R., Naik, R., Ghashghaie, J., Streb, P. 2006b. The lack of mitochondrial complex I in a CMSII mutant of *Nicotiana sylvestris* increases photorespiration through an increased internal resistance to CO_2 diffusion. J Exp Bot 57:3195–3207.

Price, J., Laxmi, A., Martin, S. K. S., Jang, J.-C. 2004. Global transcription profiling reveals multiple sugar signal transduction mechanisms in *Arabidopsis*. Plant Cell 16:2128–2150.

Rachmilevitch, S., Cousins, A. B., Bloom, A. J. 2004. Nitrate assimilation in plant shoots depends on photorespiration. Proc Natl Acad Sci USA 101: 11506–11510.

Raghavendra, A. S., Padmasree, K. 2003. Beneficial interactions of mitochondrial metabolism with photosynthetic carbon assimilation. Trends Plant Sci 8:546–553.

Raghavendra, A. S., Reumann, S., Heldt, H. W. 1998. Participation of mitochondrial metabolism in photorespiration: reconstituted system of peroxisomes and mitochondria from spinach leaves. Plant Physiol 116:1333–1337.

Rasmusson, A. G., Escobar, M. A. 2007. Light and diurnal regulation of plant respiratory gene expression. Physiol Plant 129:57–67.

Rasmusson, A. G., Fernie, A. R., van Dongen, J. T. 2009. Alternative oxidase: a defence against metabolic fluctuations? Physiol Plant 137:371–382.

Rasmusson, A. G., Geisler, D. A., Møller, I. M. 2008. The multiplicity of dehydrogenases in the electron transport chain of plant mitochondria. Mitochondrion 8:47–60.

Reichheld, J.-P., Meyer, E., Khafif, M., Bonnard, G., Meyer, Y. 2005. AtNTRB is the major mitochondrial thioredoxin reductase in *Arabidopsis thaliana*. FEBS Lett 579:337–342.

Rhoads, D. M., Subbaiah, C. C. 2007. Mitochondrial retrograde regulation in plants. Mitochondrion 7:177–194.

Rhoads, D. M., Umbach, A. L., Subbaiah, C. C., Siedow, J. N. 2006. Mitochondrial reactive oxygen species. Contribution to oxidative stress and interorganellar signaling. Plant Physiol 141:357–366.

Ribas-Carbo, M., Lennon, A. M., Robinson, S. A., Giles, L., Berry, J. A., Siedow, J. N. 1997. The regulation of electron partitioning between the cytochrome and alternative pathways in soybean cotyledon and root mitochondria. Plant Physiol 113:903–911.

Ribas-Carbo, M., Robinson, S. A., Gonzàlez-Meler, M. A., Lennon, A. M., Giles, L., Siedow, J. N., Berry, J. A. 2000. Effects of light on respiration and oxygen isotope fractionation in soybean cotyledons. Plant Cell Environ 23: 983–989.

Robinson, S. A., Ribas-Carbo, M., Yakir, D., Giles, L., Reuveni, Y., Berry, J. A. 1995. Beyond SHAM and cyanide: opportunities for studying the alternative oxidase in plant respiration using oxygen isotope discrimination. Aust J Plant Physiol 22:487–496.

Rochaix, J.-D. 2007. Role of thylakoid protein kinases in photosynthetic acclimation. FEBS Lett 581:2768–2775.

Sabar, M., De Paepe, R., de Kouchkovsky, Y. 2000. Complex I impairment, respiratory compensations, and photosynthetic decrease in nuclear and mitochondrial male sterile mutants of *Nicotiana sylvestris*. Plant Physiol 124:1239–1249.

Sakamoto, W. 2006. Protein degradation machineries in plastid. Annu Rev Plant Biol 57:599–621.

Scheibe, R. 2004. Malate valves to balance cellular energy supply. Physiol Plant 120:21–26.

Schneidereit, J., Häusler, R. E., Fiene, G., Kaiser, W. M., Weber, A. P. M. 2006. Antisense repression reveals a crucial role of the plastidic 2-oxoglutarate/malate translocator DiT1 at the interface between carbon and nitrogen metabolism. Plant J 45:206–224.

Schönfeld, C., Wobbe, L., Borgstädt, R., Kienast, A., Nixon, P. J., Kruse, O. 2004. The nucleus-encoded protein MOC1 is essential for mitochondrial light acclimation in *Chlamydomonas reinhardtii*. J Biol Chem 279:50366–50374.

Shen, W., Wei, Y., Dauk, M., Tan, Y., Taylor, D. C., Selvaraj, G., Zou, J. 2006. Involvement of a glycerol-3-phosphate dehydrogenase in modulating the NADH/NAD^+ ratio provides evidence of a mitochondrial glycerol-3-phosphate shuttle in *Arabidopsis*. Plant Cell 18:422–441.

Shikanai, T. 2007. Cyclic electron transport around photosystem I: genetic approaches. Annu Rev Plant Biol 58:199–217.

Siedow, J. N., Umbach, A. L. 2000. The mitochondrial cyanide-resistant oxidase: structural conservation amid regulatory diversity. Biochim Biophys Acta 1459:432–439.

Sienkiewicz-Porzucek, A., Nunes-Nesi, A., Sulpice, R., Lisec, J., Centeno, D. C., Carillo, P., Leisse, A., Urbanczyk-Wochniak, E., Fernie, A. R. 2008. Mild reduction in mitochondrial citrate synthase activity result in a compromised nitrate assimilation and reduced leaf pigmentation but have no effect on photosynthetic performance or growth. Plant Physiol 147:115–127.

Smith, A. M. O., Ratcliffe, R. G., Sweetlove, L. J. 2004. Activation and function of mitochondrial uncoupling protein in plants. J Biol Chem 279:51944–51952.

Strodtkötter, I., Padmasree, K., Dinakar, C., Speth, B., Niazi, P. S., Wojtera, J., Voss, I., Do, P. T., Nunes-Nesi, A., Fernie, A. R., Linke, V., Raghavendra, A. S., Scheibe, R. 2009. Induction of the AOX1D isoform of alternative oxidase in *A. thaliana* T-DNA insertion lines lacking isoform AOX1A is insufficient to optimize photosynthesis when treated with antimycin A. Mol Plant 2:284–297.

Studart-Guimarães, C., Fait, A., Nunes-Nesi, A., Carrari, F., Usadel, B., Fernie, A. R. 2007. Reduced expression of succinyl-coenzyme A ligase can be compensated for by up-regulation of the *r*-aminobutyrate shunt in illuminated tomato leaves. Plant Physiol 145:626–639.

Svensson, A. S., Rasmusson, A. G. 2001. Light-dependent gene expression for proteins in the respiratory chain of potato leaves. Plant J 28:73–82.

Sweetlove, L. J., Foyer, C. H. 2004. Roles for reactive oxygen species and antioxidants in plant mitochondria. In Plant Mitochondria: From Genome to Function, eds. D. A. Day, A. H. Millar, J. Whelan, pp. 307–320. London: Kluwer.

Sweetlove, L. J., Heazlewood, J. L., Herald, V., Holtzapffel, R., Day, D. A., Leaver, C. J., Millar, A. H. 2002. The impact of oxidative stress on *Arabidopsis* mitochondria. Plant J 32:891–904.

Sweetlove, L. J., Lytovchenko, A., Morgan, M., Nunes-Nesi, A., Taylor, N. L., Baxter, C. J., Eickmeier, I., Fernie, A. R. 2006. Mitochondrial uncoupling protein is required for efficient photosynthesis. Proc Natl Acad Sci USA 103:19587–19592.

Taira, M., Valtersson, U., Burkhardt, B., Ludwig, R. A. 2004. *Arabidopsis thaliana GLN2*-encoded glutamine synthetase is dual targeted to leaf mitochondria and chloroplasts. Plant Cell 16:2048–2058.

Taniguchi, M., Taniguchi, Y., Kawasaki, M., Takeda, S., Kato, T., Sato, S., Tabata, S., Miyake, H., Sugiyama, T. 2002. Identifying and characterizing plastidic 2-oxoglutarate/malate and dicarboxylate transporters in *Arabidopsis thaliana*. Plant Cell Physiol 43:706–717.

Taylor, N. L., Day, D. A., Millar, A. H. 2002. Environmental stress causes oxidative damage to plant mitochondria leading to inhibition of glycine decarboxylase. J Biol Chem 277:42663–42668.

Taylor, N. L., Heazlewood, J. L., Day, D. A., Millar, A. H. 2005. Differential impact of environmental stresses on the pea mitochondrial proteome. Mol Cell Proteomics 8:1122–1133.

Tcherkez, G., Bligny. R., Gout, E., Mahé, A., Hodges, M., Cornic, G. 2008. Respiratory metabolism of illuminated leaves depends on CO_2 and O_2 conditions. Proc Natl Acad Sci USA 105: 797–802.

Tcherkez, G., Cornic, G., Bligny, R., Gout, E., Ghashghaie, J. 2005. In vivo respiratory metabolism of illuminated leaves. Plant Physiol 138:1596–1606.

Timm, S., Nunes-Nesi, A., Pärnik, T., Morgenthal, K., Wienkoop, S., Keerberg, O., Weckwerth, W., Kleczkowski, L. A., Fernie, A. R., Bauwe, H. 2008. A cytosolic pathway for the conversion of hydroxypyruvate to glycerate during photorespiration in *Arabidopsis*. Plant Cell 20:2848–2859.

Tovar-Mendéz, A., Miernyk, J. A., Randall, D. D. 2003. Regulation of pyruvate dehydrogenase complex activity in plant cells. Eur J Biochem 270:1043–1049.

Turko, I. V., Li, L., Aulak, K. S., Stuehr, D. J., Chang, J. V., Murad, F. 2003. Protein tyrosine nitration in the mitochondria from diabetic mouse heart. Implications to dysfunctional mitochondria in diabetes. J Biol Chem 278:33972–33977.

Umbach, A. L., Fiorani, F., Siedow, J. N. 2005. Characterization of transformed *Arabidopsis* with altered alternative oxidase levels and analysis of effects on reactive oxygen species in tissue. Plant Physiol 139:1806–1820.

Vanlerberghe, G. C., Cvetkovska, M., Wang, J. 2009. Is the maintenance of homeostatic mitochondrial signaling during stress a physiological role for alternative oxidase? Physiol Plant 137:392–406.

Vanlerberghe, G. C., Ordog, S. H. 2002 Alternative oxidase: integrating carbon metabolism and electron transport in plant respiration. In Photosynthetic Nitrogen Assimilation and Associated Carbon and Respiratory Metabolism, eds. C. H. Foyer, G. Noctor, pp. 173–191. Dordrecht: Kluwer.

Vercesi, A. E., Borecký, J., Maia, I. G., Arruda, P., Cuccovia, I. M., Chaimovich, H. 2006. Plant uncoupling mitochondrial proteins. Annu Rev Plant Biol 57:383–404.

Verniquet, F., Gaillard, J., Neuburger, M., Douce, R. 1991. Rapid inactivation of plant aconitase by hydrogen peroxide. Biochem J 276:643–648.

Watanabe, C. K., Hachiya, T., Terashima, I., Noguchi, K. 2008. The lack of alternative oxidase at low temperature leads to a disruption of the balance in carbon and nitrogen metabolism, and to an up-regulation of anti-oxidant defense systems in *Arabidopsis thaliana* leaves. Plant Cell Environ 31:1190–1202.

Winger, A. M., Millar, H., Day, D. A. 2005. Sensitivity of plant mitochondrial terminal oxidases to the lipid peroxidation product 4-hydroxy-2-nonenal (HNE). Biochem J 387:865–870.

Winger, A. M., Taylor, N. L., Heazlewood, J. H., Day, D. A., Millar, A. H. 2007a. Identification of intra- and intermolecular disulphide bonding in the plant mitochondrial proteome by diagonal gel electrophoresis. Proteomics 7:4158–4170.

Winger, A. M., Taylor, N. L., Heazlewood, J. H., Day, D. A., Millar, A. H. 2007b. The cytotoxic lipid peroxidation product 4-hydroxy-2-nomenal covalently modifies a selective range of proteins linked to respiratory function in plant mitochondria. J Biol Chem 282:37436–37447.

Woodson, J.D., Chory, J. 2008. Coordination of gene expression between organellar and nuclear genomes. Nat Rev Genet 9: 383–395.

Xue, X., Gauthier, D. A., Turpin, D. H., Weger, H. G. 1996. Interactions between photosynthesis and respiration in the green alga *Chlamydomonas reinhardtii*: characterization of light-enhanced dark respiration. Plant Physiol 112:1005–1014.

Yoshida, K., Noguchi, K. 2009. Differential gene expression profiles of the mitochondrial respiratory components in illuminated *Arabidopsis* leaves. Plant Cell Physiol 50:1449–1462.

Yoshida, K., Shibata, M., Terashima, I., Noguchi, K. 2010. Simultaneous determination of in vivo plastoquinone and ubiquinone redox states by HPLC-based analysis. Plant Cell Physiol 51:836–841.

Yoshida, K., Terashima, I., Noguchi, K. 2006. Distinct roles of the cytochrome pathway and alternative oxidase in leaf photosynthesis. Plant Cell Physiol 47:22–31.

Yoshida, K., Terashima, I., Noguchi, K. 2007. Up-regulation of mitochondrial alternative oxidase concomitant with chloroplast over-reduction by excess light. Plant Cell Physiol 48:606–614.

Yoshida, K., Watanabe, C., Kato, Y., Sakamoto, W., Noguchi, K. 2008. Influence of chloroplastic photo-oxidative stress on mitochondrial alternative oxidase capacity and respiratory properties: a case study with *Arabidopsis yellow variegated 2*. Plant Cell Physiol 49:592–603.

Chapter 16
Plant Mitochondrial Retrograde Regulation

David M. Rhoads

Abstract Changes in plant cellular metabolism require altered gene expression. Cellular adjustments and altered gene expression also occur in response to environmental stresses. Both of these situations involve changes in mitochondrial activities. Changes in the status of the mitochondria and resulting inputs for altered nuclear gene expression are communicated to the nucleus by interorganellar signaling called mitochondrial retrograde regulation (MRR). The study of plant MRR is a young field and the mechanisms and components are just beginning to be discovered. Evidence suggests that reactive oxygen species (ROS) can be involved in plant MRR, but this does not mean that they are involved in all cases. Calcium level changes, redox changes, and changes in metabolite levels are leading candidates for nonprotein signaling components. Protein signaling components like kinases, phosphatases, and transcription factors are likely to be involved, but only the recent discovery of the involvement of transcription factor Abscisic Acid Insensitive 4 in Arabidopsis provides a specific protein example. Growing evidence indicates overlaps of MRR with other signaling pathways, including those from chloroplasts. MRR could be linked to metabolic signaling and/or ROS production, but these would still originate in mitochondria and be components of MRR. The emerging view is that plant mitochondria are stress sensors that contribute to decisions regarding cell fate during stresses and that this is conveyed to the nucleus by MRR.

Keywords Inter-organellar communication • Mitochondria • Mitochondria-to-nucleus signaling • Retrograde regulation • Retrograde signaling

Abbreviations

AOX Mitochondrial alternative oxidase
CMS Cytoplasmic male sterility

D.M. Rhoads (✉)
School of Plant Sciences, The University of Arizona, Forbes Building, Room 303, Tucson, AZ 85721-0036, USA
e-mail: drhoads@cals.arizona.edu

F. Kempken (ed.), *Plant Mitochondria*, Advances in Plant Biology 1,
DOI 10.1007/978-0-387-89781-3_16,

COX	Cytochrome *c* oxidase
CRR	Chloroplastic retrograde regulation
ETC	Mitochondrial electron transport chain
HR	Hypersensitive response
HSP	Heat shock/stress protein
MRR	Mitochondrial retrograde regulation
mt	Mitochondrial
NCS	Non-chromosomal stripe
OGE	Organellar gene expression
PCD	Programmed cell death
Redox	Reduction/oxidation
ROS	Reactive oxygen species
sHSP	Small heat shock/stress protein
TCA	Tricarboxylic acid

16.1 Introduction

The pattern of nuclear gene expression (from transcription through protein targeting) is a major factor in determining organellar function. Communication from organelles back to the nucleus to influence the pattern of nuclear gene expression is referred to as retrograde regulation (Fig. 16.1). Changes in the status of the mitochondria and resulting inputs for altered nuclear gene expression are communicated to the nucleus by mitochondrial retrograde regulation (MRR). The study of MRR is a young field and the mechanisms and components of this communication are just beginning to be defined. Growing evidence illustrates overlaps with other signaling pathways, including those from chloroplasts. As has been observed in the more extensively studied MRR in yeast and animal systems (Butow and Avadhani 2004; Box 16.1), plant MRR could be linked to metabolic signaling and/or reactive oxygen species (ROS) production, but these signals would still originate in mitochondria and would be considered part of MRR. Further, the view that plant mitochondria are stress sensors that contribute to decisions regarding cell fate during stresses in a way that differs from animal systems is emerging. Here, current concepts in the rapidly evolving plant MRR field are presented.

16.2 MRR Overlaps with Chloroplast Retrograde Regulation

Chloroplasts and mitochondria are intimately connected through, for example, metabolism, energy status, and reduction/oxidation (redox) status (Raghavendra and Padmasree 2003; van Lis and Atteia 2004). Therefore, it is logical that overlapping/interacting mechanisms of retrograde regulation exist and these are now beginning to be identified (see Fig. 16.1). Of course, this does not preclude separate retrograde

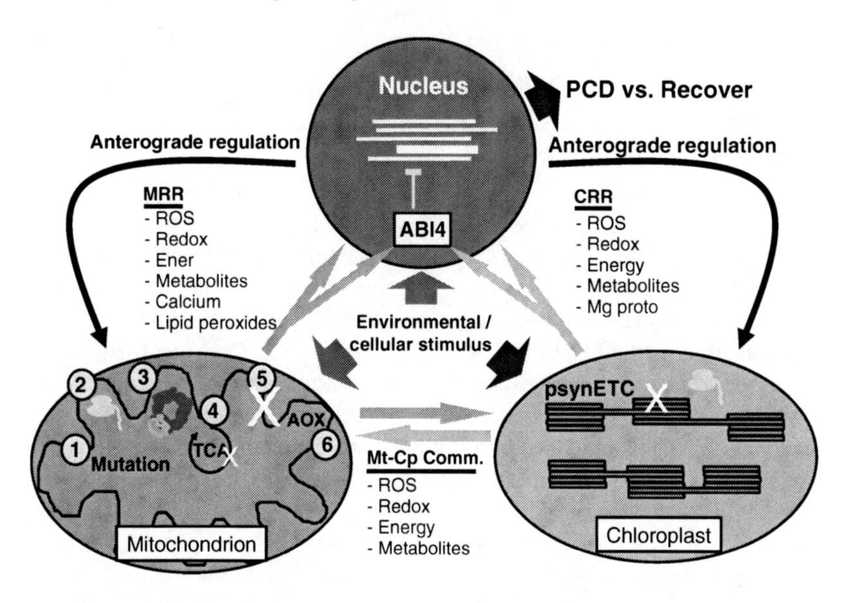

Fig. 16.1 Schematic representation of inter-organellar communication in plants. Environmental and/or cellular changes cause altered functions in all organelles. Changes in nuclear gene expression can result in alteration of function of organelles; anterograde regulation (*black arrows*). Changes in chloroplastic (*cp*) and mitochondrial (*mt*) function signal to the nucleus to alter gene expression and this is chloroplastic retrograde regulation (*CRR*) and mitochondrial retrograde regulation (*MRR*). Changes in mitochondrial function known to alter nuclear gene expression include: (1) mitochondrial mutations; (2) altered organellar gene expression (OGE; only in concert with cpOGE); (3) presence of high levels of mt small heat shock protein (only during heat stress); (4) altered TCA cycle function, including inhibition; (5) altered mitochondrial electron transport chain (*mtETC*) function, including inhibition and mutations; (6) altered levels of alternative oxidase (AOX). Signaling mechanisms are largely unknown, but CRR and MRR may overlap in ROS and reduction-oxidation (*redox*) changes, energy, and metabolites, which also may be sources of direct communication (*Comm.*). Ultimately, inputs from the environmental (including oxygen deprivation and aluminum) and cellular (including nuclear mutations, developmental and hormonal changes) stimuli result in changes in nuclear gene expression directed at recovery or programmed cell death (*PCD*). *Mg proto* Mg protoporphyrins, *psyn* photosynthetic

signaling from these organelles as well. Chloroplastic retrograde regulation (CRR) pathways in plants have been relatively well studied due to their significance and readily observable phenotypes of CRR mutants affected in the greening process. There seem to be several potential, but likely overlapping, CRR pathways, including those involved in photomorphogenesis, altered function during chloroplast biogenesis, and redox poise (Rodermel 2001; Woodson and Chory 2008). The importance of the interactions between mitochondria and chloroplasts is apparent when considering that, in several well documented cases; mitochondrial dysfunction due to mitochondrial mutations can lead to failed chloroplast biogenesis and chlorosis (Newton et al. 2004). Interactions are illustrated further by a very interesting molecular phenotype of down-regulation of nuclear genes encoding photosynthetic proteins reported in double mutant Arabidopsis plants that lack expression of both

Box 16.1 So far, MRR targets and mechanisms appear to differ between plants and yeast

MRR has been most extensively studied in the yeast *Saccharomyces cerevisiae*. It is now clear that there are several MRR pathways in yeast and that they are involved in important processes. In the best studied example, MRR contributes to the response of yeast to impaired mitochondrial respiration by directing an alteration in nuclear gene expression leading to altered expression of aerobic genes and nitrogen and carbon metabolism, including transport of pathway intermediates (Butow and Avadhani 2004). One result of this is increased expression of genes that ultimately result in trying to restore levels of the TCA cycle intermediates oxaloacetate and acetyl-CoA (Butow and Avadhani 2004). This is accomplished by proliferation of peroxisomes and increased expression of genes encoding enzymes for fatty acid oxidation. There is also increased expression of genes encoding enzymes for the glyoxylate cycle, such as citrate synthase (including the *CIT2* gene, which was used to uncover signaling mechanisms) and a directed shift toward maintaining levels of glutamate, for which oxaloacetate is an upstream precursor. Analogous situations for altered gene expression leading to these metabolic shifts following inhibition of respiration in plants has not been documented and results to date suggest novel strategies in plant cells. Analyses of yeast mutants in subunits of TCA cycle enzymes suggests MRR directed at shifting metabolism away from oxidative metabolism and suggests that TCA cycle intermediates are signaling molecules – or at least that their levels are monitored in some way that can affect nuclear gene expression (McCammon et al. 2003; Butow and Avadhani 2004).

Activation of genes directed at restoring TCA cycle intermediates in yeast generally involves regulation by RTG transcription factors (the so-called RTG-dependent pathway; Poyton and McEwen 1996; Sekito et al. 2000; Butow and Avadhani 2004), which were the first signaling components involved in retrograde regulation to be identified in any organism. There do not appear to be homologs in plants. Several other MRR pathways have been identified in yeast. Mitochondrial dysfunction induces PDR5, a nuclear gene encoding an ABC membrane transporter involved in multiple or pleitropic-drug resistance, through transcription factors Pdr1p and Pdr3p (Devaux et al. 2002). Mitochondrial dysfunction can also cause increased expression of ATO3, which encodes a plasma membrane ammonium transporter. Induction is via transcription factor Gcn4p, which is involved in amino acid control (Guaragnella and Butow 2003). A signaling pathway in which lack of mtDNA alters nuclear gene expression also appears to be independent of the RTG pathway (Woo et al. 2009). Further, there is a separate hypoxia-response signaling pathway between mitochondria and the nucleus (Kwast et al. 1998; Castello et al. 2006).

(continued)

Box 16.1 (continued)

MRR is also thought to be involved in developmental changes such as the transition to filamentous growth state in yeast. This transition occurs when nutrients are limited. Finally, an RTG-dependent signaling pathway initiated by mitochondrial dysfunction in cells lacking mtDNA was proposed to be involved in increased longevity observed in yeast (Kirchman et al. 1999). However, it has become clear that this is not necessarily the case and that it is possible that the lack of mtDNA specifically is responsible for the altered nuclear gene expression that results in increased longevity (Woo and Poyton 2009). In summary, it is clear that altered nuclear gene expression is important for the response of yeast cells to various environmental and developmental cues and that several distinct MRR pathways are involved in directing this altered expression. It will be vary interesting to determine if there are any overlaps in plants.

a chloroplast form of prolyl-tRNA synthetase and a mitochondrial form, affecting gene expression at the translational level in each of these organelles (Pesaresi et al. 2006). Double mutant plants show the molecular phenotype while each mutation alone does not result in the molecular phenotype. This shows that it is the tandem interaction of the effects on organellar gene expression (OGE) that is necessary. Finally, an overlap in MRR and CRR signal transmission has recently been uncovered. Transcription factor Abscisic Acid Insensitive 4 (ABI4), which is a suggested common component of multiple CRR pathways (Woodson and Chory 2008) as well as abscisic acid, sugar, and developmental signaling (Rook et al. 2006), also appears to be involved in MRR as a promoter-binding, negative regulator of gene expression that allows derepression during MRR (Giraud et al. 2009).

16.3 Mitochondrial Dysfunction and MRR

Studies of MRR that communicates mitochondrial dysfunction to the nucleus generally fall into one of five categories: (1) MRR due to mutations that cause mitochondrial dysfunction; (2) chemically induced mitochondrial dysfunction – usually through use of inhibitors; (3) altered mitochondrial function caused by expression of (a) transgene(s); (4) mitochondrial dysfunction caused by biotic stress; and (5) mitochondrial dysfunction caused by abiotic stress. Each of these approaches has advantages and disadvantages. Mutants have attained a "new homeostasis" (relative to wild-type plants) in which whole cell metabolism has adjusted from the time of fertilization to the new status imposed by the mitochondrial dysfunction. Therefore, any nuclear gene expression changes may be due to this life-long, whole cell adjustment rather than to specific MRR. On the other hand, the changes may be driven primarily by the mitochondrial dysfunction and be consistent and robust enough to be used to identify mechanisms or targets of MRR. Inhibitors may act at sites (perhaps even undefined ones) other than the intended mitochondrial targets, though there is no clear evidence to support that this is actually occurring when the commonly used inhibitors are applied under the proper conditions. Proteins expressed from transgenes may create unusual characteristics in the mitochondria, but this, like mutations, may reveal signaling activities that would otherwise be difficult to observe – especially if they are inducible. Abiotic and biotic stresses may affect mitochondrial function and this could lead to MRR, but the difficulty is to distinguish nuclear gene expression changes caused by effects on mitochondria versus those caused by other effects on cells from the stress. Nevertheless, evidence suggests that MRR contributes to plant responses to biotic and abiotic stresses and that the altered gene expression contributes to the determination of cell and plant fate in response to the stress.

Although many important functions occur in mitochondria and any of these could affect or be affected by MRR, several mutations or stresses affect the mitochondrial electron transport chain (mtETC) and this requires some discussion because some aspects of the plant mtETC differ significantly from most other eukaryotes. Among the most unique features is the presence of several "extra" components. The plant mtETC includes the "standard" Complexes I-IV (though the protein compositions may differ; see Chap. 13) as well as cytochrome *c* and the ubiquinone/ubiquinol pool, comprising the minimal cytochrome respiratory pathway (Siedow and Umbach 1995; Vanlerberghe and McIntosh 1997). But, in all plants studied to date, the mtETC also contains a second terminal oxidase, called the alternative oxidase (AOX) (see Chap. 14), which can oxidize ubiquinol and function along with Complex IV (cytochrome *c* oxidase, COX) to reduce oxygen to water (Finnegan et al. 2004). The combination of Complex I, the ubiquinone/ubiquinol pool, and the AOX comprise a much-simplified, alternative respiratory pathway. Another group of common additional components are the inner-membrane-associated NAD(P)H dehydrogenases. These enzymes are present on both the internal and external surfaces of the inner mitochondrial membrane and function alongside Complex I, the "usual" NADH dehydrogenase that extracts electrons from NADH produced by metabolic pathways, such as the tricarboxylic acid (TCA) cycle, and uses them to reduce ubiquinone to ubiquinol (Finnegan et al. 2004; Rasmusson et al. 2004). The various isoforms of AOX and the additional NAD(P)H dehydrogenases are nucleus-encoded and none of these mitochondrial inner membrane-associated enzymes is proton pumping. Functioning along with the ubiquinone/ubiquinol pool, these enzymes can: (1) form a complete respiratory pathway that consists of only two distinct protein components that are nucleus-encoded and that can transfer electrons from NAD(P)H to oxygen, and (2) help remove electrons from the mtETC when the oxidation rate is excessive or when the ATP/ADP ratio is high and the normal mtETC becomes adenylate restricted. These functions can allow continued carbon metabolism during adenylate restriction, can help prevent ROS formation from release of electrons from various points along the mtETC when function is impaired (Møller 2001), and can be critical when the cytochrome respiratory pathway is affected by metabolic changes or by biotic or abiotic stresses.

16.3.1 Mutations Can Cause Mitochondrial Dysfunction Resulting in MRR

Plant mitochondrial and nuclear mutants that affect mitochondrial function frequently exhibit both cytoplasmic male sterility (CMS) (see Chap. 18) and altered nuclear gene expression. Several such mutants of various plant species have been discovered. One such "family" of mutants is the maize nonchromosomal stripe (NCS) mutants (Newton et al. 2004). The following categories of NCS mutants have been isolated: (1) NCS2 mutant, which has decreased expression of Complex I of the mtETC due to a mitochondrial DNA (mtDNA) rearrangement; (2) NCS5 and NCS6 mutants, which express low levels of the COX subunit 2 due to partially deleted versions of the mitochondrial gene; and (3) NCS3 and NCS4 mutants, which lack expression of a ribosomal protein due to mitochondrial gene deletion, so that this is another example of a general organellar gene expression mutant (Newton et al. 2004). In each of the NCS mutants, leaves of plants are variegated because they contain dysfunctional mitochondria with all or mostly mutated mtDNA in the affected pale sectors and mitochondria that function normally in the normal-looking green sectors. Cells in the green sectors have greater amounts of normal mtDNA and lack severe effects on chloroplast development. In the affected sectors, the mitochondrial mutations result in abnormal chloroplast development (Newton et al. 2004). In each of the two types of mtETC mutants of NCS maize there is a unique pattern of induction of alternative oxidase genes (Karpova et al. 2002). Likewise, the inhibitors of mtETC complexes that correspond to each mutation cause a similar pattern of induction of AOX genes as the mutations. So, in both the affected sectors of the NCS2 mutant (deficient in Complex I) and in leaves treated with rotenone, an inhibitor of Complex I, the *aox2* gene is induced and in NCS6 (deficient in cytochrome *c* oxidase) and from cyanide, an inhibitor of cytochrome *c* oxidase, *aox3* is induced. Because of the similarities between the patterns of gene expression alteration between the NCS mutants and the inhibitors, the long-term alteration of AOX gene expression established in the mutants is likely a result of MRR from the specific mitochondrial dysfunction rather than a general shifting of cellular homeostasis. Likewise, this supports the proposal that the inhibitors are acting specifically on the mitochondrial activities and that this is the source of MRR that alters the AOX gene expression.

Expression of nucleus-encoded heat shock proteins (HSPs) commonly follows perturbation of plant mitochondrial function. This is true of the NCS mutants (Kuzmin et al. 2004) as well as following mitochondrial dysfunction caused by harpin, a bacterial elicitor (see following text). A heat-stress-associated MRR in Arabidopsis also alters expression of HSPs, though it is unclear if this is due to mitochondrial dysfunction or an alternative mechanism originating in the mitochondria (see further). Because these examples of altered expression of nuclear genes are initiated by mitochondrial perturbations (though initiated from seemingly distinct stimuli), the responses are considered to be caused by MRR.

When an unedited (compare Chap. 7), and therefore mutant, form of ATP synthase subunit 9 (ATP9) is expressed in transgenic Arabidopsis (for which flower-specific promoters were used) or tobacco (for which a constitutive promoter was used), these plants become male sterile (Hernould et al. 1998; Gomez-Casati et al. 2002). These mutant plants exhibit (1) reduced mitochondrial respiration, (2) altered mitochondrial morphology, and (3) increased expression of nuclear genes encoding Complex I subunits. Clearly the status of the mitochondria is conveyed to the nucleus in some manner in these plants, whether through direct MRR or through secondary means such as shifted metabolism or altered redox status established by the mitochondrial dysfunction.

The CMSII mutant of tobacco, which lacks functional Complex I, is also male sterile and exhibits altered nuclear gene expression (Gutierres et al. 1997; Dutilleul et al. 2003). The mutation results in an adjustment of the cellular redox state and, therefore, it is not surprising that mutant plants have increased expression at the level of transcripts for redox enzymes such as ascorbate peroxidase, superoxide dismutase and catalase in the light. They also show increased expression of AOX and increased alternative NAD(P)H dehydrogenase activity (Gutierres et al. 1997; Dutilleul et al. 2003). Although NAD, NADH, and NADPH levels are elevated and glutathione level is slightly higher in these mutants, there are no significant changes in the ratios of reduced to oxidized glutathione or ascorbate and they display significantly lower hydrogen peroxide levels (Dutilleul et al. 2003). This suggests that these plants have adjusted their redox levels to a new homeostasis in response to impaired mitochondrial function. Interestingly, this new homeostasis results in increased resistance to ozone stress and pathogen attack (Dutilleul et al. 2003). A new pattern of nuclear gene expression is established in these plants based on the dysfunctional mitochondria, whether this is direct or indirect.

A mitochondrial ABC transporter, encoded by the nuclear STARIK gene (*STA1*), is believed to be involved in maturation of Fe/S proteins in Arabidopsis mitochondria. *Starik* mutant plants have an elevated level of free (nonheme, nonprotein) iron in the mitochondria, indicating that cellular iron homeostasis is dependent upon STA1 function (Kushnir et al. 2001). The mutant plants display increased expression of a nuclear gene encoding a putative mitochondrial L-cysteine desulfurase, which is thought to be involved in mitochondrial Fe/S cluster biosynthesis.

Therefore, it seems that a change in iron homeostasis can cause altered nuclear gene expression that can result in compensation for the mitochondrial perturbation, which suggests some specificity in directing nuclear gene expression.

16.3.2 Mitochondrial Dysfunction from Enzyme Inhibitors Results in MRR

As discussed above, when the mtETC cannot function properly, electrons flow is disrupted and electrons can leak out of this system and end up in oxygen, creating ROS. In order to help alleviate this potentially damaging situation, plants induce expression of AOX genes and, in the extensively studied Arabidopsis system, genes for an alternative NADH dehydrogenase (NDB2) that is associated with the outer surface of the inner mitochondrial membrane. The induction of alternative oxidase genes following mtETC restriction has been extensively studied (Mackenzie and McIntosh 1999; Rhoads and Vanlerberghe 2004; Rhoads and Subbaiah 2007). AOX gene induction following inhibition of Complex III (cytochrome bc_1 complex) by the inhibitor antimycin A has been demonstrated in tobacco (the *NtAox1* gene), soybean (the *GmAox1* gene), maize (the *aox1*, *aox2* and *aox3* genes), and Arabidopsis (the *AtAOX1a* gene) (Rhoads and Subbaiah 2007). AOX gene induction following inhibition of the TCA cycle enzyme aconitase by monofluoroacetate has been demonstrated in Arabidopsis and tobacco (Vanlerberghe and McIntosh 1996; Dojcinovic et al. 2005; Zarkovic et al. 2005).

Through the use of transgenic tobacco lines with increased expression of AOX or that lack expression of the primary form of alternative oxidase, the importance of mtROS production in antimycin A-directed MRR was established (see below; Maxwell et al. 1999, 2002). In tobacco cells, monofluoroacetate also causes increased cellular ROS levels, but this has not yet been determined to be due to increased ROS produced in the mitochondria (Gray et al. 2004). However, there is no detectable increase in cellular ROS levels following TCA cycle inhibition in Arabidopsis cells (Dojcinovic 2003) or Arabidopsis leaves (Umbach et al. unpublished). Increased mitochondrial citrate could be the actual underlying cause of altered nuclear gene expression from inhibition of the TCA cycle because (1) inhibition of aconitase causes an increase in citrate levels (Vanlerberghe and McIntosh 1996) and (2) addition of exogenous citrate alone induces expression of an AOX gene in suspension-cultured tobacco or soybean cells (Vanlerberghe and McIntosh 1996; Djajanegara et al. 2002; Gray et al. 2004). This may be through a distinct MRR pathway compared to antimycin A-induced MRR (see further). The antimycin A- and monofluoroacetate-induced expression of *AtAOX1a* both appear to be directed by the *AtAOX1a* promoter and the promoter regions responding appear to overlap and be complex (Dojcinovic et al. 2005; Zarkovic et al. 2005). The promoter regions of *AtAOX1a* and *NDB2* have been analyzed fairly extensively for *cis*-acting regulatory elements and shown to be complex in potential responses to MRR (Dojcinovic et al. 2005; Ho et al. 2008). A genetic screen allowed identification of plant nuclear mutants lacking MRR signaling pathway(s) from antimycin-A inhibition of the mtETC. Some of the MRR deficient (*mrrd*) mutants display lack of response of *AtAOX1a* to inhibition of the mtETC but can respond to inhibition of the TCA cycle, indicating that MRR pathways from these inhibitions only partially overlap (Zarkovic et al. 2005).

16.4 MRR and Abiotic Stresses

16.4.1 Oxygen Deprivation and Sensing by Mitochondria May Result in MRR

Flooding can result in hypoxia and anoxia and this can cause a rise in cytoplasmic calcium concentration and trigger altered nuclear gene expression (Bailey-Serres and Chang 2005 and references therein; Kreuzwieser et al. 2009). It has been known for some time that metabolism is quickly shifted to glycolysis and fermentation and away from the oxidative TCA cycle during anoxia/hypoxia (Geigenberger 2003). Commonly known changes accompanying this shift include increased expression of alcohol dehydrogenase and sucrose synthase. The targeted changes in gene expression can also result in lowered levels of TCA cycle enzymes and mtETC components and this repression can be reversed when oxygen levels begin to return to normal (Tsuji et al. 2000; Millar et al. 2004; Branco-Price et al. 2005; Gonzali et al. 2005; Huang et al. 2005). Because mitochondria are the primary sites of oxygen consumption and targets of metabolic changes, it would be logical for oxygen sensing to occur in these organelles. Studies that show that increased cytoplasmic calcium concentration during oxygen deprivation is due to release of calcium from mitochondria suggest the importance of MRR in response to this abiotic stress (Subbaiah et al. 1994, 1998; Bailey-Serres and Chang 2005). In addition, ROS production may influence the response to oxygen deprivation and this may include mtROS production (Bailey-Serres and Chang 2005; Rhoads and Subbaiah 2007). The mechanisms for how the low oxygen concentration is detected and calcium is released and/or ROS are produced are unknown. The observation that inhibition of the mtETC leads to induction of hypoxic genes (Nie and Hill 1997) suggests that the mtETC could be inhibited by the low oxygen concentration (i.e., act as a sensor of low oxygen concentration) and initiate signaling. However, genes that respond to hypoxia are induced at higher oxygen concentrations than the Km for oxygen of cytochrome *c* oxidase (Millar et al. 1994). This suggests that Complex IV would not be inhibited at these oxygen concentrations and that there is a lower-affinity sensor. Thus, although it is commonly thought that mitochondrial components could act as oxygen sensors, the mechanism(s) of oxygen sensing, which could include combinations of sensors and altered cell homeostasis (as in respiratory mutants), and the role of mitochondria and MRR pathways are only beginning to be unveiled in plants (Bailey-Serres and Chang 2005).

16.4.2 MRR and Heat Stress

A heat stress-associated MRR was discovered in transgenic Arabidopsis plants with constitutively high expression of a maize mitochondrial small heat shock protein (mt sHSP). These plants exhibit super-induction of several HSP genes (compared to the normal induction of these genes that follows heat stress), including for the gene (*AtHSP23.6*) encoding an endogenous mt sHSP, and shifts in induction kinetics – but *only* during heat stress (Rhoads et al. 2005). Because increased thermotolerance was also observed, these results suggest that plants with altered MRR can exhibit enhanced stress tolerance. However, it is not clear whether the increased expression of the mt sHSP (i.e., a direct result of a transgene expression) or the overall increased expression of HSPs, seemingly due to MRR, is the source of the increased thermotolerance. Because the profile of altered gene expression in this system is distinct from the profile of gene expression change from MRR resulting from other mitochondrial perturbations, this is likely a distinct form of MRR. On the other hand, this heat stress-associated MRR resembles a communication pathway between the endoplasmic reticulum and nucleus that exists in both animals and plants and that is referred to as the unfolded protein response (Patil and Walter 2001; Martinez and Chrispeels 2003). A similar signaling pathway between mitochondria and the nucleus in animals that appears to be initiated by unfolded proteins in mitochondria has also been identified (Zhao et al. 2002). The underlying mechanisms of this newly discovered mt sHSP and heat-stress-associated MRR in Arabidopsis have yet to be determined.

16.4.3 MRR and Other Abiotic Stresses

Oxidative stress is a common component of abiotic stresses and can result in mitochondrial damage (Møller 2001). Mitochondrial components that are sensitive to oxidative damage have been identified (see further). Specifically for abiotic stresses, oxidative damage induced by chilling or drought damages glycine decarboxylase, an enzyme in the photorespiratory pathway (Taylor et al. 2002). As discussed above, the mtETC can be the source of ROS production. Both the mtROS and ROS produced outside the mitochondria can damage mitochondrial enzymes. On the other hand, these ROS can be signaling components.

When suspension-cultured tobacco cells or pea roots are treated with aluminum, a disruption of mitochondrial membrane potential is followed by ROS production and ATP depletion, which negatively affect growth (Yamamoto et al. 2002). Aluminum treatment is also known to increase expression of anti-oxidant enzymes, so it is not surprising that less lipid peroxidation and less of a negative effect on root growth were observed in transgenic plants with constitutively high expression of a mitochondrial superoxide dismutase compared to controls following aluminum treatment (Basu et al. 2001). Taken altogether, these observations suggest that aluminum stress causes mitochondrial dysfunction and production of mtROS leading to MRR to increase expression of antioxidant enzymes to alleviate the resulting oxidative stress.

16.5 MRR and Biotic Stresses

The relationship between mitochondrial function and biotic stress has been difficult to clearly establish, in part, due to the fact that events during pathogen attack could be occurring on a more-or-less individual cell scale. One cell may be undergoing cell death to prevent spread of the infection while the cell next to it may be mounting a preventative response aimed at survival. However, there is strong evidence that mitochondria play important roles during pathogen attack, including the likelihood for the involvement of MRR. While an increase in respiration may occur during pathogen attack (Baker et al. 2000, 2001), this may simply be due to increased demand for energy to, for example, produce secondary metabolites known to be produced during pathogen attack. On the other hand, increased electron flow through the mtETC could result in increased production of mtROS that could be used to initiate MRR.

16.5.1 The Hypersensitive Response and Plant Mitochondria

Ever since the early observations that salicylic acid induces AOX gene expression and is involved in the hypersensitive response (HR; a form of programmed cell death that occurs at the site of infection to limit pathogen spread) (Zaninotto et al. 2006), it was postulated that mitochondria, and specifically the alternative respiratory pathway, play an important role in HR. While strong evidence that the alternative pathway is critical for the HR is lacking, there is evidence to support the hypothesis that it contributes to the HR. For example, infection by tobacco mosaic virus causes a HR and over-expression of AOX in tobacco plants results in smaller lesions (Ordog et al. 2002). In addition, reduced lesion size and number following tobacco mosaic virus inoculation compared to control plants was observed in mitochondrial mutants with altered redox status and increased expression of alternative oxidase and other antioxidant enzymes (Dutilleul et al. 2003). One aspect that could connect salicylic acid and AOX in the HR in tobacco is that salicylic acid can inhibit the mtETC, acting as an uncoupler at low concentrations and an inhibitor of electron transfer at higher concentrations (Xie and Chen 1999; Norman et al. 2004). Exogenously added salicylic acid induces expression of AOX in tobacco (Rhoads and McIntosh 1993) and this could be via MRR initiated by mtETC inhibition. If salicylic acid acts through inhibition of the mtETC and subsequent MRR affecting nuclear genes, then the profiles of the genes induced should be similar. This is true for some target genes, but not others. Therefore, the overlap in the mechanisms is not clear.

Polyamines (putrescine, spermidine, and spermine) are thought to have roles in the HR (Takahashi et al. 2004a, b). Spermine affects mitochondrial function through an unknown mechanism that is proposed to involve ROS and calcium and act through mitogen-activated protein kinases and this results in altered nuclear gene expression (Takahashi et al. 2003, 2004b). Because a mitochondrial uncoupler, HR and spermidine alter expression of an overlapping set of genes, a role for MRR in the HR response in tobacco can be proposed (Takahashi et al. 2003, 2004b).

16.5.2 Bacterial and Fungal Elicitors Cause Mitochondrial Dysfunction, Resulting in MRR

The usefulness of MRR as a response mechanism to pathogen attack is obvious when it is realized that several compounds produced by plant pathogens (called elicitors because they elicit a response by the plants to the pathogens) disrupt mitochondrial function and induce nuclear gene expression. One of the best known examples of an elicitor is harpin, which is produced by bacterial pathogens that induce a HR, including *Pseudomonas syringae*. Exogenously applied harpin disrupts mitochondrial ATP synthesis in both Arabidopsis and tobacco cells, both of which are hosts for bacterial pathogens that produce harpin (Xie and Chen 2000; Krause and Durner 2004). Harpin induces expression of genes for AOX, genes encoding sHSPs (including mt sHSPs), genes encoding plant defense proteins, and increases activities of mitogen-activated protein kinases that are dependent on increased ROS level (Desikan et al. 1998, 2001; Krause and Durner 2004; Samuel et al. 2005). Some induced gene expression from harpin is through ROS-dependent mechanisms. Thus, one intriguing possibility is that the mtROS produced due to mitochondrial dysfunction is involved in harpin-initiated MRR during pathogen attack.

The pathogenic fungus *Cochliobolus victoriae* is the causal agent of victoria blight of oat (Curtis and Wolpert 2002) and it produces victorin, another well studied elicitor. Treatment of oat cells with victorin results in disruption of the inner mitochondrial membrane potential, production of mtROS, and programmed cell death (PCD; Curtis and Wolpert 2002; Yao et al. 2002). Clearly this is a system in which MRR could assist in the response to pathogen attack, but the validity of this has not been established.

In a similar example, the pathogenic fungus *Phytophthora cryptogea* produces the protein elicitor cryptogein, which causes reduced glucose uptake, decreased respiration and mitochondrial membrane depolarization in tobacco cells (Bourque et al. 2002). It is thought that cryptogein induces hypersensitive cell death through these effects (Bourque et al. 2002). Thus, disruption of mitochondrial function and the resulting MRR could be involved in the plant response to *P. cryptogea* infection.

An additional connection between pathogenesis and mitochondria is that secondary metabolites of plant pathogens can inhibit the mtETC (Kim et al. 2004). Descriptions of gene expression changes that occur in these last two examples have not been detailed, but the mitochondrial dysfunction associated with each provides the possibility that mitochondria-directed nuclear gene expression changes occur in these systems.

Mitochondrial dysfunction seems to be a common feature of pathogen attack. A MRR response to infection is suggested by the observations that (1) altered nuclear gene expression follows elicitor treatment and mitochondrial dysfunction and (2) inhibition of the mtETC by antimycin A (itself a fungal toxin) causes a transcriptome response by plant cells similar to the transcriptome response to pathogen attack (Yu et al. 2001; Maxwell et al. 2002).

16.6 Components of Plant MRR

16.6.1 Reactive Oxygen Species

ROS are both damaging molecules and important signaling components for several forms of abiotic and biotic stress (Dat et al. 2000; Desikan et al. 2005). Production of reactive oxygen species at a very high level will likely overwhelm the cell and cause damage, but a specific type of ROS produced at low levels and at specific sites could result in a specific and helpful response to a particular stress. It is clear, in a general sense, that production of cellular ROS results in altered nuclear gene expression (Inzé and Van Montagu 1995; Kubo et al. 1999; Jones 2000; Kuzniak and Urbanek 2000; Taylor et al. 2002; Vranová et al. 2002). The fact that ROS are produced at a high level during various abiotic and biotic stresses suggests the importance of ROS, but adds to the difficulty to study the specific effects. The involvement of ROS in signaling for any stress response can be difficult to decipher because: (1) individual ROS can be difficult to quantify precisely; (2) some ROS are transient; (3) ROS are metabolized by various enzymes; (4) ROS interact with various cellular components; and (5) the same ROS may be performing multiple functions simultaneously. Indeed, ROS can alter nuclear gene expression, but various treatments that produce ROS result in uniquely altered transcriptomes (Kubo et al. 1999; Gadjev et al. 2006). It has been proposed that this is due to production of a unique type or a unique combination of ROS for each stress (i.e., a unique ROS profile or "signature") (Mahalingam and Fedoroff 2003; Gadjev et al. 2006). Further, a unique ROS could be produced at a unique cellular location in each individual stress case. In addition, there could be a specific (perhaps even specifically induced) ROS detection component in each case. To complicate the situation even more, other signals (e.g., hormones and calcium) contribute to the response along with the ROS change profile to cause gene expression changes (Desikan et al. 2005). Several lines of evidence suggest that production of mtROS can contribute to the ROS signature produced by stresses (Rhoads et al. 2006), but the exact contribution in each case will be difficult to determine precisely. However, determining the contribution and purpose of mtROS (and ROS produced at other specific sites such as chloroplasts and NADPH oxidase) for each stress response is an important goal. Here, some observations directed at understanding the roles of mtROS in plant MRR are presented.

Inhibition of the plant cytochrome respiratory pathway by inhibitors such as antimycin A cause increased cellular ROS and, in each case so far examined, increased alternative oxidase expression (Maxwell et al. 1999; Djajanegara et al. 2002; Gray et al. 2004; Rhoads and Vanlerberghe 2004). For tobacco cells, the source of the cellular ROS level increase is increased production of mtROS (Maxwell et al. 1999). The importance of the ROS in transmitting the signal of mitochondrial distress to the nucleus was demonstrated by blocking antimycin A induction of AOX with antioxidants (Minagawa et al. 1992; Maxwell et al. 1999; Vanlerberghe et al. 2002). Two lines of evidence suggest an important role of AOX for helping to control the level of ROS produced following mtETC inhibition: (1) the presence of AOX decreases mtROS production that follows antimycin A or cyanide inhibition of the cytochrome respiratory pathway (Maxwell et al. 1999; Umbach et al. 2005); and (2) inhibition or lack of AOX activity causes increased ROS accumulation, especially following perturbations (Popov et al. 1997; Maxwell et al. 1999). Thus, cytochrome pathway inhibition causes an increase in mtROS, which likely triggers MRR signaling and increased AOX gene expression, resulting in more AOX activity in the mitochondria and decreased mtROS production. However, mtROS is likely not required for all plant MRR resulting from mitochondrial dysfunction. For example, inhibition of the TCA cycle strongly induces AOX gene expression (Dojcinovic 2003; Zarkovic et al. 2005), but does not cause a dramatic increase in cellular ROS during the same time frame (Dojcinovic 2003; Rhoads and Vanlerberghe 2004).

Oxidative stress can cause mitochondrial damage and dysfunction such as inhibition of cytochrome respiratory pathway components, aconitase in the TCA cycle and ATP synthase (Møller 2001; Sweetlove et al. 2002; Taylor et al. 2002). These perturbations likely result in MRR, which would be predicted to cause a transcriptome response from cytochrome respiratory pathway inhibition that would overlap to some degree with transcriptome responses to oxidative stress treatments. Although aconitase is inhibited by oxidative stresses such as hydrogen peroxide application, it is not inhibited by inhibition of the cytochrome respiratory pathway by antimycin A, which clearly causes an increase in cellular hydrogen peroxide (Vanlerberghe and McIntosh 1996). This illustrates the complexity of reactive oxygen species effects and suggests that the source of the ROS is critical in determining the outcome. Although a great deal of evidence suggests an important role for mtROS in MRR signaling, no downstream ROS detection mechanisms or other signaling components have been identified for any plant MRR signaling pathway.

Downstream primary or secondary signaling components may include compounds that react with ROS, such as lipids, forming lipid peroxides that are known to be signaling components in animals (Kumagai et al. 2000; Tang et al. 2002) or ROS scavenging systems, such as the glutathione-ascorbate cycle (see further). The observation that hydroxyalkenal levels increase following antimycin A inhibition of the cytochrome respiratory pathway, implicates lipid peroxides as signaling components in plant MRR (Winger et al. 2005). The possibility that lipid peroxides are involved in plant MRR signaling requires further investigation.

16.6.2 Redox Signals in Plant MRR

Mitochondria, like chloroplasts, can contribute to the establishment of the cellular redox state, which is important for stress responses (Foyer and Noctor 2003; Foyer and Noctor 2005).

Mitochondria contribute to the cellular redox status through the TCA cycle, which reduces NAD to NADH; the production of ROS from the mtETC; and their glutathione-ascorbate cycle (Jimenez et al. 1997; Mittova et al. 2000). Inhibition/restriction of the mtETC can cause cell oxidative stress through the production of mtROS. This could result in the typical shifts in cellular redox components, such as an increase in the NADP/NADPH ratio because of the requirement for NADPH as the ultimate reductant donor for H_2O_2 removal (Sweetlove and Foyer 2004; Foyer and Noctor 2005).

Glutathione and ascorbate themselves are likely redox signaling components in plants. They are thought to act as signaling molecules in responses to plant pathogens

(Kocsy et al. 2001; Dutilleul et al. 2003; Foyer and Noctor 2003, 2005), in signaling during PCD (de Pinto et al. 2002), and in cold stress signaling (Kocsy et al. 2001). Redox signals such as glutathione and ascorbate could act downstream of any mtROS increase from mitochondrial dysfunctions. Redox changes could also occur through distinct mechanisms, such as inhibition/restriction of the TCA cycle, resulting in partially overlapping signaling between MRR induced by inhibition/restriction of the mtETC (which could have an additional signaling component from the mtROS produced) and MRR induced by inhibition/restriction of the TCA cycle. Evidence that redox status is an important signaling component of plant MRR is currently lacking.

16.6.3 Calcium and Plant MRR

One aspect of MRR that appears to be especially common is the involvement of calcium as a signaling component (Butow and Avadhani 2004). Therefore, it is not surprising that several lines of evidence implicate calcium in plant MRR signaling, particularly in association with oxygen deprivation. For example, in suspension-cultured maize cells, anoxia causes a rise in cytoplasmic calcium and the calcium appears to originate from mitochondria (Subbaiah et al. 1994, 1998). Further, addition of ruthenium red, a mitochondrial calcium channel blocker, suppresses the induction of anoxia-responsive genes (Subbaiah et al. 1994, 1998). Similarly, anoxia-induced expression of nuclear genes of barley is inhibited by ruthenium red, again suggesting a role for mitochondria-derived calcium (Nie and Hill 1997; Nie et al. 2006). Calcium involvement in oxygen-deprivation responses is also observed in Arabidopsis and rice (Sedbrook et al. 1996; Chung and Ferl 1999; Tsuji et al. 2000; Nie et al. 2006).

Increased cytoplasmic calcium levels could result from release of cellular stores (including release from organelles or from sequestration in the cytoplasm) or lack of uptake by dysfunctional mitochondria as has been seen in animal cells (Butow and Avadhani 2004). Calcium also likely acts as a signaling component in plant stresses other than oxygen deprivation. For example, oxidative stress causes increased cytoplasmic calcium levels, but the mechanism for this increase is currently unknown (Price et al. 1994; Rentel and Knight 2004; Evans et al. 2005). Increased cytoplasmic calcium levels and increased mitochondrial calcium levels are both observed in Arabidopsis leaf cells during cold stress and osmotic stress (Logan and Knight 2003). These observations indicate that calcium level changes in various cellular compartments may be stress-specific and demonstrate the complexity of the involvement of calcium as a signaling component in plant stresses. Another connection between calcium levels and mitochondria was observed in *Fucus* (seaweed) rhizoid cells. In this system, a "calcium wave" causes mtROS production in response to hyperosmotic stress (Coelho et al. 2002).

Further evidence for the involvement of calcium in MRR following mtETC inhibition comes from the observations that ruthenium red partially inhibits the induction of AOX capacity by inhibition of respiration in tobacco cells (Vanlerberghe et al. 2002). Taken altogether, these data suggest mitochondria can affect cellular calcium fluxes, which are important for plant stress responses.

How any mitochondria-directed change in cellular calcium profile is transmitted to the nucleus to affect nuclear gene expression is currently unknown. The most likely scenario is that commonly known calcium concentration detectors are involved. In addition to the involvement of calcium binding proteins such as calcium-dependent protein kinases and calmodulin, this could include transcription factors and regulatory proteins such as 14-3-3 family members (Lu et al. 1994; Trewavas 1999; Sanders et al. 2002). Genes encoding such proteins are observed to increase following mitochondrial perturbations. Another possibility is that the altered cellular calcium concentration is detected in the nucleus. Increased cellular calcium concentration can result in increased nuclear calcium concentration during anoxia (Subbaiah et al. 1998) and altered nuclear calcium concentration can affect nuclear gene expression (Hardingham et al. 1997; Pusl et al. 2002). Whether this is a MRR signaling mechanism in plants remains to be determined.

16.6.4 Potential Protein Components of Plant MRR

The involvement of ABI4 in repression of AOX gene expression (discussed previously) is the strongest evidence of a specific protein that is involved in plant MRR (Giraud et al. 2009). There are several other candidates for protein signaling components in plants, including several signal transduction proteins, such as protein kinases and transcription factors, whose genes are induced following inhibition of the cytochrome respiratory pathway and/or the TCA cycle in Arabidopsis plants (Yu et al. 2001; Umbach et al. unpublished).

Numerous reports detail the involvement of protein kinases in many plant abiotic and biotic stress responses (Kovtun et al. 2000; Romeis et al. 2001; Yuasa et al. 2001; Zhang and Klessig 2001; Cheng et al. 2002; Koiwa et al. 2002; Zhu 2002; Ahlfors et al. 2004; Miles et al. 2005) and protein kinases can be modulated by ROS (Pitzschke and Hirt 2006). In terms of direct indications of involvement in MRR, activities of mitogen-activated protein kinases are induced by mitochondrial dysfunction in tobacco cells caused by spermine (Takahashi et al. 2003, 2004b). The likely involvement of calcium in at least some MRR suggests a role for calcium-dependent protein kinases, which are involved in many plant abiotic and biotic stress responses (Romeis et al. 2001; Cheng et al. 2002).

The comparatively high number of genes encoding transcription factors in plants (Riechmann et al. 2000; Chen et al. 2002) suggests a high degree of regulation of gene responses to stresses at the transcriptional level in plants. Transcription factors are likely involved in MRR, and, as discussed above, a recent report provides direct evidence that this is the case (Giraud et al. 2009). Because they are involved in biotic and abiotic stress responses, including oxidative stress, WRKY, bZIP, and Dof transcription factors are additional candidates to be involved in plant MRR (Zhang and

Singh 1994; Chen and Singh 1999; Chen et al. 2002). Because there are putative binding sites in promoter regions for AOX genes for bZIP, Dof, and WRKY transcription factors (Dojcinovic et al. 2005) in addition to other abscisic acid-responsive transcription factors (Ho et al. 2008), these transcription factors are good candidates for signaling components in the response of AOX genes to mitochondrial dysfunction in Arabidopsis. Several other *cis*-acting regulatory elements that appear to have roles in MRR in response to inhibition of respiration have also been identified in the AtAOX1a and/or NDB2 promoters (Ho et al. 2008). Thus, the transcription factors that interact with these elements may be involved in MRR in Arabidopsis.

16.7 Mitochondria, MRR and Programmed Cell Death

Mitochondrial status seems to influence programmed cell death in plant cells (see also Chap. 17) and there is evidence that mtROS and cell redox state (see Chap. 14) are involved (Jones 2000; Sweetlove and Foyer 2004). PCD in plant cells shares several characteristics with PCD in animal cells, including the apparent formation of a permeability transition pore (Xie and Chen 2000; Arpagaus et al. 2002; Tiwari et al. 2002; Krause and Durner 2004), though the exact components of the plant permeability transition pore have not been identified. Cytochrome *c* release from mitochondria also appears to be a mechanism in PCD shared by both plant and animals (Balk et al. 1999; Sun et al. 1999; Balk and Leaver 2001; Robson and Vanlerberghe 2002; Tiwari et al. 2002; Vacca et al. 2006). Finally, there appear to be proteases that function as caspases in plants (Woltering et al. 2002; Vacca et al. 2006), although direct homologs to animal caspase genes have not been identified in plants (Woltering et al. 2002). There are several examples of how mitochondria can affect PCD in plant cells (Sun et al. 1999; Xie and Chen 2000; Arpagaus et al. 2002; Robson and Vanlerberghe 2002; Tiwari et al. 2002; Krause and Durner 2004; Vacca et al. 2006), but the specific involvement of MRR in these cases is unclear. If PCD causes cellular changes outside the mitochondria that affect mitochondrial function and this directs changes in nuclear gene expression, then MRR is involved in PCD.

16.7.1 MRR Initiated by Plant Pathogens May Cause PCD/HR in Some Cases

As discussed above, several elicitors of pathogens cause mitochondrial dysfunction, including dissipation of the mitochondrial membrane potential, reduced respiration, reduced ATP production, and release of mtROS. The elicitors initiate PCD/HR and cause altered nuclear gene expression, presumably due, at least in part, to MRR (Curtis and Wolpert 2002; Yao et al. 2002). Therefore, it is reasonable to suggest that, in these cases of pathogen attack, the MRR is a way to signal to the nucleus to initiate programmed cell death and/or a means to regulate it.

16.7.2 MRR and the Decision Between Recovery and Cell Death

ROS produced outside the mitochondria and mtROS can influence plant programmed cell death. For example, inhibition of the cytochrome respiratory pathway triggers ROS production that appears to direct an effort to prevent cell death by inducing expression of AOX and, at least in the case of antimycin A, alternative NAD(P)H dehydrogenases. Further, cells that cannot induce alternative oxidase expression instead initiate a cell death in response to inhibition of the cytochrome respiratory pathway (Vanlerberghe et al. 1995, 2002; Robson and Vanlerberghe 2002; Zarkovic et al. 2005). In this case, accumulation of reactive oxygen species likely shifts cells to PCD rather than recovery because the MRR-directed attempt to recover is overridden.

So, plant cells respond to a stress by either initiating cell death or by attempting to recover – or, perhaps, initiating both if the "best" outcome is unclear initially. As has been pointed out previously (Jones 2000; Rhoads and Subbaiah 2007; Van Aken et al. 2009; Vanlerberghe et al. 2009), evidence suggests that inputs from mitochondria, in concert with other inputs (or, perhaps at times super ceding other inputs), contribute to the decision during each stress whether to initiate PCD or attempt to recover. There is increasing evidence that AOX is a component of the decision-making process (Arnholdt-Schmitt et al. 2006; Van Aken et al. 2009; Vanlerberghe et al. 2009).

16.8 Multiple MRR Pathways in Plants

Several results indicate that there are multiple MRR signaling pathways in plants (Gray and McIntosh 1998; Djajanegara et al. 2002; Karpova et al. 2002; Kuzmin et al. 2004; Zarkovic et al. 2005). First, as discussed previously, the maize NCS mutants have mutations in separate mitochondrial genes encoding components of the separate respiratory complexes and this, along with inhibitors of the complexes affected by the mutations, induces distinct AOX genes (Karpova et al. 2002). Second, a mutant isolated from a genetic screen for lack of induction of an Arabidopsis AOX gene by cytochrome respiratory pathway inhibition exhibits induction following inhibition of the TCA cycle (Zarkovic et al. 2005). Third, the same Arabidopsis AOX gene induced by cytochrome pathway inhibition is also induced by inhibitors of mitochondrial ATP synthase, which causes increased respiration (Saisho et al. 2001). Fourth, citrate addition, which does not cause detectable ROS increase and is assumed to affect mitochondrial metabolism, induces AOX gene expression in tobacco and soybean cells (Djajanegara et al. 2002; Gray et al. 2004). Also, in soybean cells, AOX gene induction by citrate is blocked by a protein kinase inhibitor, but induction by inhibition of respiration is not (Djajanegara et al. 2002). Further, the mitochondrial contribution to the overall retrograde regulation observed in the double mutant of prolyl-tRNA synthase discussed above

(Pesaresi et al. 2006) is expected to be independent of other forms of MRR from other mitochondrial perturbations. Finally, the mt sHSP and heat-associated MRR discussed above strongly induces expression of genes encoding HSPs, whereas cytochrome respiratory pathway and TCA cycle inhibition do not (Yu et al. 2001; Rhoads et al. 2005; Umbach et al. unpublished). Thus, evidence favors the likelihood of multiple MRR pathways in plants, but pathways have yet to be defined and delineated.

16.9 Summary

Cellular metabolism shifts during plant growth and development, and this requires altered gene expression. In addition, plants experience environmental stresses that require cellular adjustments and altered gene expression. Both these situations involve changes in mitochondrial activities. Evidence increasingly indicates that mitochondria can sense metabolic changes and abiotic and biotic environmental stresses and can then influence nuclear gene expression through MRR, which could act as an input to the overall transcriptome response. Some signaling components that function in at least some MRR pathways have been identified, but much more about MRR mechanisms and components is yet to be discovered. It seems certain that ROS can be involved in plant MRR, but this does not mean that they are involved in all cases. Calcium level changes, redox changes, and changes in metabolite levels are leading candidates for nonprotein signaling components. Usual protein signaling components like kinases, phosphatases, and transcription factors are likely to be involved, but only the recent discovery of the involvement of ABI4 provides a specific example of a protein component of plant MRR. Future research will focus on identification of mechanisms and components of plant MRR as well as the potential functions of MRR in determining the fate of cells during development and environmental stresses.

Acknowledgments I thank Ann Umbach for many helpful discussions leading to the preparation of this chapter. This work was supported by grants from the National Science Foundation (IOS-0822521) and the US Department of Agriculture (2009-35100-05007).

Glossary

Adenylate restriction: When the cellular level of ADP becomes so low that the formation of ATP becomes restricted, which has effects on many cellular processes including mitochondrial respiration.

Hypersensitive response: A plant defense response to infection that directs death of cells adjacent to the infected cells so as to prevent spreading of the infection by depriving the pathogen of resources; a type of programmed cell death in plants.

Programmed cell death: The well-orchestrated cellular death program that involves, in most cases, defined morphological changes and biochemical shifts aimed at controlled organism recovery of cellular components.

Retrograde regulation: Communication pathways initiated by organelles that result in altered nuclear gene expression.

References

Ahlfors, R., Macioszek, V., Rudd, J., Brosche, M., Schlichting, R., Scheel, D., Kangasjarvi, J. 2004. Stress hormone-independent activation and nuclear translocation of mitogen-activated protein kinases in *Arabidopsis thaliana* during ozone exposure. Plant J. 40:512–522.

Arnholdt-Schmitt, B., Costa, J. H., de Melo, D. 2006. AOX – a functional marker for efficient cell reprogramming under stress? Trends Plant Sci. 11:281–287.

Arpagaus, S., Rawyler, A., Braendle, R. 2002. Occurrence and characteristics of the mitochondrial permeability transition in plants. J. Biol. Chem. 277:1780–1787.

Bailey-Serres, J., Chang, R. 2005. Sensing and signalling in response to oxygen deprivation in plants and other organisms. Ann. Bot. 96:507–518.

Baker, C. J., Orlandi, E. W., Deahl, K. L. 2000. Oxygen metabolism in plant/bacteria interactions: characterization of the oxygen uptake response of plant suspension cells. Physiol. Mol. Plant Pathol. 57:159–167.

Balk, J., Leaver, C. J. 2001. The PET1-CMS mitochondrial mutation in sunflower is associated with premature programmed cell death and cytochrome c release. Plant Cell 13:1803–1818.

Balk, J., Leaver, C. J., McCabe, P. F. 1999. Translocation of cytochrome c from the mitochondria to the cytosol occurs during heat-induced programmed cell death in cucumber plants. FEBS Lett. 463:151–154.

Basu, U., Good, A. G., Taylor, G. J. 2001. Transgenic *Brassica napus* plants overexpressing aluminum-induced mitochondrial manganese superoxide dismutase cDNA are resistant to aluminum. Plant Cell Environ. 24:1269–1278.

Bourque, S., Lemoine, R., Sequeira-Legrand, A., Fayolle, L., Delrot, S., Pugin, A. 2002. The elicitor cryptogein blocks glucose transport in tobacco cells. Plant Physiol.130:2177–2187.

Branco-Price, C., Kawaguchi, R., Ferreira, R. B., Bailey-Serres, J. 2005. Genome-wide analysis of transcript abundance and translation in Arabidopsis seedlings subjected to oxygen deprivation. Ann. Bot. 96:647–660.

Butow, R. A., Avadhani, N. G. 2004. Mitochondrial signaling, the retrograde response. Mol. Cell 14:1–15.

Castello, P. R., David, P. S., McClure, T., Crook, Z., Poyton, R. O. 2006. Mitochondrial cytochrome oxidase produces nitric oxide under hypoxic conditions: implications for oxygen sensing and hypoxic signaling in eukaryotes. Cell Metab. 3:277–287.

Chen, W., Singh, K. B. 1999. The auxin, hydrogen peroxide and salicylic acid induced expression of the Arabidopsis GST6 promoter is mediated in part by an ocs element. Plant J. 19:667–678.

Chen, W., Provart, N. J., Glazebrook, J., Katagiri, F., Chang, H. S., Eulgem, T., Mauch, F., Luan, S., Zou, G., Whitham, S. A., Budworth, P. R., Tao, Y., Xie, Z., Chen, X., Lam, S., Kreps, J. A., Harper, J. F., Si-Ammour, A., Mauch-Mani, B., Heinlein, M., Kobayashi, K., Hohn, T., Dangl, J. L., Wang, X., Zhu, T. 2002. Expression profile matrix of Arabidopsis transcription factor genes suggests their putative functions in response to environmental stresses. Plant Cell 14:559–574.

Cheng, S. H., Willmann, M. R., Chen, H. C., Sheen, J. 2002. Calcium signaling through protein kinases. The Arabidopsis calcium-dependent protein kinase gene family. Plant Physiol. 129:469–485.

Chung, H. J., Ferl, R. J. 1999. Arabidopsis alcohol dehydrogenase expression in both shoots and roots is conditioned by root growth environment. Plant Physiol. 121:429–436.
Coelho, S. M., Taylor, A. R., Ryan, K. P., Sousa-Pinto, I., Brown, M. T., Brownlee, C. 2002. Spatiotemporal patterning of reactive oxygen production and Ca^{2+} wave propagation in *Fucus* rhizoid cells. Plant Cell 14:2369–2381.
Curtis, M. J., Wolpert, T. J. 2002. The oat mitochondrial permeability transition and its implication in victorin binding and induced cell death. Plant J. 29:295–312.
Dat, J., Vandenabeele, S., Vranova, E., Van Montagu, M., Inzé, D., Van Breusegem, F. 2000. Dual action of the active oxygen species during plant stress responses. Cell Mol. Life Sci. 57:779–795.
de Pinto, M. C., Tommasi, F., De Gara, L. 2002. Changes in the antioxidant systems as part of the signaling pathway responsible for the programmed cell death activated by nitric oxide and reactive oxygen species in tobacco Bright-Yellow 2 cells. Plant Physiol. 130:698–708.
Desikan, R., Hancock, J., Neill, S. 2005. Reactive oxygen species as signalling molecules. In Antioxidants and reactive oxygen species in plants, ed. N. Smirnoff. pp. 169–196. Oxford: Blackwell.
Desikan, R., Reynolds, A., Hancock, J., Neill, S. J. 1998. Harpin and hydrogen peroxide both initiate programmed cell death but have differential effects on defence gene expression in Arabidopsis suspension cultures. Biochem. J. 330:115–120.
Devaux, F., Carvajal, E., Moye-Rowley, S., Jacq, C. 2002. Genome wide studies on the nuclear PDR3 controlled response to mitochondrial dysfunction in yeast. FEBS Lett. 515:25–28.
Djajanegara, I., Finnegan, P. M., Mathieu, C., McCabe, T., Whelan, J., Day, D.A. 2002. Regulation of alternative oxidase gene expression in soybean. Plant Mol. Biol. 50:735–742.
Dojcinovic, D. 2003. Analysis of cis-regions and trans-factors regulating the expression of the *AtAOX1a* gene. Master of Science thesis, Arizona State University.
Dojcinovic, D., Krosting, J., Harris, A. J., Wagner, D. J., Rhoads, D. M. 2005. Identification of regions of the *Arabidopsis AtAOX1a* promoter important for developmental and mitochondrial retrograde regulation of expression. Plant Mol. Biol. 58:159–175.
Dutilleul, C., Garmier, M., Noctor, G., Mathieu, C., Chetrit, P., Foyer, C. H., De Paepe, R. 2003. Leaf mitochondria modulate whole cell redox homeostasis, set antioxidant capacity, and determine stress resistance through altered signaling and diurnal regulation. Plant Cell 15:1212–1226.
Evans, N. H., McAinsh, M. R., Hetherington, A. M., Knight, M. R. 2005. ROS perception in Arabidopsis thaliana, the ozone-induced calcium response. Plant J. 41:615–626.
Finnegan, P. M., Soole, K. L., Umbach, A. L. 2004. Alternative mitochondrial electron transport proteins in higher plants. In Advances in Photosynthesis and Respiration, Vol. 17, eds. D. A. Day, A. H. Millar, J. Whelan. pp. 163–230. Dordrecht: Kluwer.
Foyer, C. H., Noctor, G. 2003. Redox sensing and signaling associated with reactive oxygen in chloroplasts, peroxisomes and mitochondria. Physiol. Plant. 119:355–364.
Foyer, C. H., Noctor, G. 2005. Redox homeostasis and antioxidant signaling: a metabolic interface between stress perception and physiological responses. Plant Cell 17:1866–1875.
Gadjev, I., Vanderauwera, S., Gechev, T. S., Laloi, C., Minkov, I. N., Shulaev, V., Apel, K., Inzé, D., Mittler, R., Van Breusegem, F. 2006. Transcriptomic footprints disclose specificity of reactive oxygen species signaling in Arabidopsis. Plant Physiol.141:436–445.
Geigenberger, P. 2003. Response of plant metabolism to too little oxygen. Curr. Op. Plant Biol. 6:247–256.
Giraud, E., Van Aken, O., Ho L. H. M., Whelan, J. 2009. The transcription factor ABI4 is a regulator of mitochondrial retrograde expression of ALTERNATIVE OXIDASE1a. Plant Physiol. 150:1286–1296.
Gómez-Casati, D. F., Busi, M. V., Gonzalez Schain, N., Mouras, A., Zabaleta, E. J., Araya, A. 2002. A mitochondrial dysfunction induces the expression of nuclear encoded Complex I genes in engineered male sterile *Arabidopsis thaliana*. FEBS Lett. 532:70–74.
Gonzali, S., Loreti, E., Novi, G., Poggi, A., Alpi, A., Perata, P. 2005. The use of microarrays to study the anaerobic response in Arabidopsis. Ann. Bot. 96:661–668.

Gray, G. R., Maxwell, D. P., Villarimo, A. R., McIntosh, L. 2004. Mitochondria/nuclear signaling of alternative oxidase gene expression occurs through distinct pathways involving organic acids and reactive oxygen species. Plant Cell Rep. 23:497–503.

Gray, G. R., McIntosh, L. 1998. Respiratory carbon flux, NAD(P)+-dependent isocitrate dehydrogenases. In Plant Mitochondria, From Gene to Function, eds. I. M. Møller, P. Gardeström, K. Glimelius, E. Glaser, pp. 353–357. Leiden: Backhuys.

Guaragnella, N., Butow, R. A. 2003. ATO3 encoding a putative outward ammonium transporter is an RTG independent retrograde responsive gene regulated by GCN4 and the Ssy1 Ptr3 Ssy5 amino acid sensor system. J. Biol. Chem. 278:45882–45887.

Gutierres, S., Sabar, M., Lelandais, C., Chetrit, P., Diolez, P., Degand, H., Boutry, M., Vedel, F., de Kouchkovsky, Y., De Paepe, R. 1997. Lack of mitochondrial and nuclear-encoded subunits of Complex I and alteration of the respiratory chain in *Nicotiana sylvestris* mitochondrial deletion mutants. Proc. Natl. Acad. Sci. USA 94:3436–3441.

Hardingham, G. E., Chawla, S., Johnson, C. M., Bading, H. 1997. Distinct functions of nuclear and cytoplasmic calcium in the control of gene expression. Nature 385:260–265.

Hernould, M., Suharsono, Zabaleta, E., Carde, J. P., Litvak, S., Araya, A., Mouras, A. 1998. Impairment of tapetum and mitochondria in engineered male-sterile tobacco plants. Plant Mol. Biol. 36:499–508.

Ho, L. H. M., Giraud, E., Uggalla, V., Lister, R., Clifton, R., Glen, A., Thirkettle-Watts, D., Van Aken, O., Whelan, J. 2008. Identification of regulatory pathways controlling gene expression of stress- responsive mitochondrial proteins in Arabidopsis. Plant Physiol. 147:1858–1873.

Huang, S., Greenway, H., Colmer, T. D., Millar, A. H. 2005. Protein synthesis by rice coleoptiles during prolonged anoxia: implications for glycolysis, growth and energy utilization. Ann. Bot. 96:703–715.

Inzé, D., Van Montagu, M. 1995. Oxidative stress in plants. Curr. Op. Biotech. 6:153–158.

Jimenez, A., Hernandez, J. A., delRio, L. A., Sevilla, F. 1997. Evidence for the presence of the ascorbate-glutathione cycle in mitochondria and peroxisomes of pea leaves. Plant Physiol. 114:275–284.

Jones, A., 2000. Does the plant mitochondrion integrate cellular stress and regulate programmed cell death? Trends Plant Sci. 5:225–230.

Karpova, O. V., Kuzmin, E. V., Elthon, T. E., Newton, K. J. 2002. Differential expression of alternative oxidase genes in maize mitochondrial mutants. Plant Cell 14:3271–3284.

Kim, J. H., Mahoney, N., Chan, K. L., Molyneux, R. J., Campbell, B. C. 2004. Secondary metabolites of the grapevine pathogen *Eutypa lata* inhibit mitochondrial respiration, based on a model bioassay using the yeast *Saccharomyces cerevisiae*. Curr. Microbiol. 49:282–287.

Kirchman, P. A., Kim, S., Lai, C. Y., Jazwinski, S. M. 1999. Interorganelle signaling is a determinant of longevity in *Saccharomyces cerevisiae*. Genetics 152:179–190.

Kocsy, G., Galiba, G., Brunold, C. 2001. Role of glutathione in adaptation and signalling during chilling and cold acclimation in plants. Physiol. Plant. 113:158–164.

Koiwa, H., Barb, A. W., Xiong, L. M., Li, F., McCully, M. G., Lee, B. H., Sokolchik, I., Zhu, J. H., Gong, Z. Z., Reddy, M., Sharkhuu, A., Manabe, Y., Yokoi, S., Zhu, J. K., Bressan R. A., Hasegawa, P. M. 2002. C-terminal domain phosphatase-like family members (AtCPLs) differentially regulate *Arabidopsis thaliana* abiotic stress signaling, growth, and development. Proc. Natl. Acad. Sci. USA 99:10893–10898.

Kovtun, Y., Chiu, W.-L., Guillaume, T., Sheen, J. 2000. Functional analysis of oxidative stress-activated mitogen-activated protein kinase cascade in plants. Proc. Natl. Acad. Sci. USA 97:2940–2945.

Krause, M., Durner J. 2004. Harpin inactivates mitochondria in Arabidopsis suspension cells. Mol. Plant Microbe Interact. 17:131–139.

Kreuzwieser, J., Hauberg, J., Howell, K. A., Carroll, A., Rennenberg, H., Millar, A. H., Whelan, J. 2009. Differential response of gray poplar leaves and roots underpins stress adaptation during hypoxia. Plant Physiol. 149:461–473.

Kubo, A., Aono, M., Nakajima, N., Saji, H., Tanaka, K., Kondo, N. 1999. Differential responses in activity of antioxidant enzymes to different environmental stresses in *Arabidopsis thaliana*. J. Plant Res. 112:279–290.

Kumagai, T., Kawamoto, Y., Nakamura, Y., Hatayama, I., Satoh, K., Osawa, T., Uchida, K. 2000. 4-hydroxy-2-nonenal, the end product of lipid peroxidation, is a specific inducer of cyclooxygenase-2 gene expression. Biochem. Biophys. Res. Commun. 273:437–441.

Kushnir, S., Babiychuk, E., Storozhenko, S., Davey, M. W., Papenbrock, J., De Rycke, R., Engler, G., Stephan, U. W., Lange, H., Kispal, G., Lill, R., Van Montagu, M. 2001. A mutation of the mitochondrial ABC transporter Sta1 leads to dwarfism and chlorosis in the Arabidopsis mutant *starik*. Plant Cell 13:89–100.

Kuzmin, E. V., Karpova, O. V., Elthon, T. E., Newton, K. J. 2004. Mitochondrial respiratory deficiencies signal up-regulation of genes for heat shock proteins. J. Biol. Chem. 279:20672–20677.

Kuzniak, E., Urbanek, H. 2000. The involvement of hydrogen peroxide in plant responses to stresses. ACTA Physiol. Plant 22:195–203.

Kwast, K. E., Burke, P. V., Poyton, R. O. 1998. Oxygen sensing and the transcriptional regulation of oxygen responsive genes in yeast. J. Exp. Biol. 201:1177–1195.

Logan, D. C., Knight, M. R. 2003. Mitochondrial and cytosolic calcium dynamics are differentially regulated in plants. Plant Physiol. 133:21–24.

Lu, G., Sehnke, P. C., Ferl, R. J. 1994. Phosphorylation and calcium binding properties of an Arabidopsis GF14 brain protein homolog. Plant Cell 6:501–510.

Mackenzie, S., McIntosh, L. 1999. Higher plant mitochondria. Plant Cell 11:571–586.

Mahalingam, R., Fedoroff, N. 2003. Stress response, cell death and signaling: the many faces of reactive oxygen species. Physiol. Plant. 119:56–68.

Martinez, I. M., Chrispeels, M. J. 2003. Genomic analysis of the unfolded protein response in Arabidopsis shown its connection to important cellular processes. Plant Cell 15:561–576.

Maxwell, D. P., Nickels, R., McIntosh, L. 2002. Evidence of mitochondrial involvement in the transduction of signals required for the induction of genes associated with pathogen attack and senescence. Plant J. 29:269–279.

Maxwell, D. P., Wang, Y., McIntosh, L.1999. The alternative oxidase lowers mitochondrial reactive oxygen production in plant cells. Proc. Natl. Acad. Sci. USA 96:8271–8276.

McCammon, M. T., Epstein, C. B., Przybyla-Zawislak, B., McAlister-Henn, L., Butow, R. A. 2003. Global transcription analysis of Krebs tricarboxylic acid cycle mutants reveals an alternating pattern of gene expression and effects on hypoxic and oxidative genes. Mol. Biol. Cell 14:958–972.

Miles, G. P., Samuel, M. A., Zhang, Y., Ellis, B. E. 2005. RNA interference-based (RNAi) suppression of AtMPK6, an Arabidopsis mitogen-activated protein kinase, results in hypersensitivity to ozone and misregulation of AtMPK3. Environ. Pollut. 138:230–237.

Millar, A. H., Bergersen F. J., Day D. A. 1994. Oxygen affinity of terminal oxidases in soybean mitochondria. Plant Physiol. Biochem. 32:847–852.

Millar, A. H., Trend, A. E., Heazlewood, J. L. 2004. Changes in the mitochondrial proteome during the anoxia to air transition in rice focus around cytochrome-containing respiratory complexes. J. Biol. Chem. 279:39471–39478.

Minagawa, N., Koga, S., Nakano, M., Sakajo, S., Yoshimoto, A. 1992. Possible involvement of superoxide anion in the induction of cyanide-resistant respiration in *Hansenula anomala*. FEBS Lett. 302:217–219.

Mittova, V., Volokita, M., Guy, M., Tal, M. 2000. Activities of SOD and the ascorbate-glutathione cycle enzymes in subcellular compartments in leaves and roots of the cultivated tomato and its wild salt-tolerant relative *Lycopersicon pennellii*. Physiol. Plant 110:42–51.

Møller, I. M. 2001. Plant mitochondria and oxidative stress, Electron transport, NADPH turnover, and metabolism of reactive oxygen species. Annu. Rev. Plant Physiol. Plant Mol. Biol. 52:561–591.

Newton, K. J., Gabay-Laughnan, S., DePaepe, R. 2004. Mitochondrial mutations in plants. In Advances in Photosynthesis and Respiration, Vol. 17, eds. D. A. Day, A. H. Millar, J. Whelan, pp. 121–142. Dordrecht: Kluwer.

Nie, X., Durnin, D. C., Igamberdiev, A. U., Hill, R. D. 2006. Cytosolic calcium is involved in the regulation of barley hemoglobin gene expression. Planta 223:542–549.

Nie, X., Hill, R. D. 1997. Mitochondrial respiration and hemoglobin gene expression in barley aleurone tissue. Plant Physiol. 114:835–840.

Norman, C., Howell, K. A., Millar, A. H., Whelan, J., Day, D.A. 2004. Salicylic acid is an uncoupler and inhibitor of mitochondrial electron transport. Plant Physiol. 134:492–501.

Ordog, S. H., Higgins, V. J., Vanlerberghe, G. C. 2002. Mitochondrial alternative oxidase is not a critical component of plant viral resistance but may play a role in the hypersensitive response. Plant Physiol. 129:1858–1865.

Patil, C., Walter, P. 2001. Intracellular signaling from the endoplasmic reticulum to the nucleus, the unfolded protein response in yeast and mammals. Curr. Opin. Cell Biol. 13:349–355.

Pesaresi, P., Masiero, S., Eubel, H., Braun, H.P., Bhushan, S., Glaser, E., Salamini, F., Leister, D. 2006. Nuclear photosynthetic gene expression is synergistically modulated by rates of protein synthesis in chloroplasts and mitochondria. Plant Cell 18:970–991.

Pitzschke, A., Hirt, H. 2006. Mitogen-activated protein kinases and reactive oxygen species signaling in plants. Plant Physiol. 141:351–356.

Popov, V. N., Simonian, R. A., Skulachev, V. P., Starkov, A. A. 1997. Inhibition of the alternative oxidase stimulates H_2O_2 production in plant mitochondria. FEBS Lett. 415:87–90.

Poyton, R. O., McEwen, J. E. 1996. Crosstalk between nuclear and mitochondrial genomes. Ann. Rev. Biochem. 65:563–607.

Price, A. H., Taylor, A., Ripley, S. J., Griffiths, A., Trewavas, A. J., Knight, M. R. 1994. Oxidative signals in tobacco increase cytosolic calcium. Plant Cell 6:1301–1310.

Pusl, T., Wu, J. J., Zimmerman, T. L., Zhang, L., Ehrlich, B. E., Berchtold, M. W., Hoek, J. B., Karpen, S. J., Nathanson, M. H., Bennett, A. M. 2002. Epidermal growth factor-mediated activation of the ETS-domain transcription factor Elk-1 requires nuclear calcium. J. Biol. Chem. 277:27517–27527.

Raghavendra, A. S., Padmasree, K. 2003. Beneficial interactions of mitochondrial metabolism with photosynthetic carbon assimilation. Trends Plant Sci. 8:546–553.

Rasmusson, A. G., Soole, K. L., Elthon, T. E. 2004. Alternative NAD(P)H dehydrogenases of plant mitochondria. Annu. Rev. Plant Biol. 55:23–39.

Rentel, M. C., Knight, M. R. 2004. Oxidative stress-induced calcium signaling in Arabidopsis. Plant Physiol. 135:1471–1479.

Rhoads, D.M., McIntosh, L. 1993. Cytochrome and alternative pathway respiration in tobacco: effects of salicylic acid. Plant Physiol. 103:877–883.

Rhoads, D. M., Subbaiah, C. C. 2007. Mitochondrial retrograde regulation in plants. Mitochondrion. 7:177–194.

Rhoads, D. M., Umbach, A. L., Subbaiah, C. C., Siedow, J. N. 2006. Mitochondrial ROS, contribution to oxidative stress and inter-organellar signaling. Plant Physiol. 141:357–366.

Rhoads, D. M., Vanlerberghe, G. C. 2004. Mitochondria-nucleus interactions, Evidence for mitochondrial retrograde communication in plant cells. In Advances in Photosynthesis and Respiration, Vol. 17, eds. D. A. Day, A. H. Millar, J. Whelan, pp. 83–106. Dordrecht: Kluwer.

Rhoads, D. M., White, S. J., Zhou, Y., Muralidharan, M., Elthon, T. E. 2005. Altered gene expression in plants with constitutive expression of a mitochondrial small heat shock protein suggests the involvement of retrograde regulation in the heat stress response. Physiol. Plant. 123:435–444.

Riechmann, J. L., Heard, J., Martin, G., Reuber, L., Jiang, C.-Z., Keddie, J., Adam, L., Pineda, O., Ratcliffe, O. J., Samaha, R. R., Creelman, R., Pilgrim, M., Broun, P., Zhang, J. Z., Ghandehari, D., Sherman, B. K., Yu, G.-L. 2000. Arabidopsis transcription factors: genome-wide comparative analysis among eukaryotes. Science 290:2105–2110.

Robson, C. A., Vanlerberghe, G. C. 2002. Transgenic plant cells lacking mitochondrial alternative oxidase have increased susceptibility to mitochondria-dependent and -independent pathways of programmed cell death. Plant Physiol. 129:1908–1920.

Rodermel, S. 2001. Pathways of plastid-to-nucleus signaling. Trends Plant Sci. 6:471–478.

Romeis, T., Ludwig, A. A., Martin, R., Jones, J. D. G. 2001. Calcium-dependent protein kinases play an essential role in a plant defence response. EMBO J. 20:5556–5567.

Rook, F., Hadingham, S. A., Li, Y., Bevan, M. W. 2006. Sugar and ABA response pathways and the control of gene expression. Plant Cell Environ. 29:426–434.

Saisho, D., Nakazono, M., Tsutsumi, N., Hirai, A. 2001. ATP synthesis inhibitors as well as respiratory inhibitors increase steady-state level of alternative oxidase mRNA in *Arabidopsis thaliana*. J. Plant Physiol. 158:241–245.

Samuel, M. A., Hall, H., Krzymowska, M., Drzewiecka, K., Hennig, J., Ellis, B. E. 2005. SIPK signaling controls multiple components of harpin-induced cell death in tobacco. Plant J. 42:406–416.

Sanders, D., Pelloux, J., Brownlee, C., Harper, J. F. 2002. Calcium at the crossroads of signaling. Plant Cell. 14(Suppl):S401–S417.

Sedbrook, J. C., Kronebusch, P. J., Borisy, G. G., Trewavas, A. J., Masson, P. H. 1996. Transgenic *AEQUORIN* reveals organ-specific cytosolic Ca^{2+} responses to anoxia in Arabidopsis thaliana seedlings. Plant Physiol. 111:243–257.

Sekito, T., Thornton, J., Butow, R. A. (2000) Mitochondria to nuclear signaling is regulated by the subcellular localization of the transcription factors Rtg1p and Rtg3p. Mol Biol Cell. 11: 2103–2115.

Siedow, J. N., Umbach, A. L. 1995. Plant mitochondrial electron transfer and molecular biology. Plant Cell. 7:821–831.

Subbaiah, C. C., Bush, D. S., Sachs, M. M. 1994. Elevation of cytosolic calcium precedes anoxic gene expression in maize suspension-cultured cells. Plant Cell. 6:1747–1762.

Subbaiah, C. C., Bush, D. S., Sachs, M. M. 1998. Mitochondrial contribution to the anoxic Ca^{2+} signal in maize suspension-cultured cells. Plant Physiol. 118:759–771.

Sun, Y. L., Zhao, Y., Hong, X., Zhai, Z. H. 1999. Cytochrome *c* release and caspase activation during menadione-induced apoptosis in plants. FEBS Lett. 462:317–321.

Sweetlove, L. J., Foyer, C. H. 2004. Roles for reactive oxygen species and antioxidants in plant mitochondria. In Advances in Photosynthesis and Respiration, Vol. 17, eds. D. A. Day, A. H. Millar, J. Whelan, pp. 307–320. Dordrecht: Kluwer.

Sweetlove, L. J., Heazlewood, J. L., Herald, V., Holtzapffel, R., Day, D. A., Leaver, C. J., Millar, A. H. 2002. The impact of oxidative stress on Arabidopsis mitochondria. Plant J. 32:891–904.

Takahashi, Y., Berberich, T., Miyazaki, A., Seo, S., Ohashi, Y., Kusano, T. 2003. Spermine signaling in tobacco: activation of mitogen-activated protein kinases by spermine is mediated through mitochondrial dysfunction. Plant J. 36:820–829.

Takahashi, Y., Berberich, T., Yamashita, K., Uehara, Y., Miyazaki, A., Kusano, T. 2004. Identification of tobacco HIN1 and two closely related genes as spermine-responsive genes and their differential expression during the Tobacco mosaic virus-induced hypersensitive response and during leaf- and flower-senescence. Plant Mol. Biol. 54:613–622.

Takahashi, Y., Uehara, Y., Berberich, T., Ito, A., Saitoh, H., Miyazaki, A., Terauchi, R., Kusano, T. 2004a. A subset of hypersensitive response marker genes, including HSR203J, is the downstream target of a spermine signal transduction pathway in tobacco. Plant J. 40:586–595.

Tang, D. G., La, E., Kern, J., Kehrer, J. P. 2002. Fatty acid oxidation and signaling in apoptosis. Biol. Chem. 383:425–442.

Taylor, N. L., Day, D. A., Millar, A. H. 2002. Environmental stress causes oxidative damage to plant mitochondria leading to inhibition of glycine decarboxylase. J. Biol. Chem. 277:42663–42668.

Tiwari, S. B., Belenghi, B., Levine, A. 2002. Oxidative stress increased respiration and generation of reactive oxygen species, resulting in ATP depletion, opening of mitochondrial permeability transition, and programmed cell death. Plant Physiol. 128:1271–1281.

Trewavas, A., 1999. Le calcium, c'est la vie, calcium makes waves. Plant Physiol. 120:1–6.

Tsuji, H., Nakazono, M., Saisho, D., Tsutsumi, N., Hirai, A., 2000. Transcript levels of the nuclear-encoded respiratory genes in rice decrease by oxygen deprivation: evidence for involvement of calcium in expression of the alternative oxidase 1a gene. FEBS Lett. 471:201–204.

Umbach, A. L., Fiorani, F., Siedow, J. N. 2005. Characterization of transformed Arabidopsis with altered alternative oxidase levels and analysis of effects on reactive oxygen species in tissue. Plant Physiol. 139:1806–1820.

Vacca, R. A., Valenti, D., Bobba, A., Merafina, R. S., Passarella, S., Marra, E. 2006. Cytochrome *c* is released in a reactive oxygen species-dependent manner and is degraded via caspase-like proteases in tobacco Bright-Yellow 2 cells en route to heat shock-induced cell death. Plant Physiol. 141:208–219.

Van Aken, O., Giraud, E., Clifton, R., Whelan, J. 2009. Alternative oxidase: a target and regulator of stress responses. Physiol. Plant. 137:354–361.

Vanlerberghe, G. C., Cvetkovska, M., Wang, J. 2009. Is the maintenance of homeostatic mitochondrial signaling during stress a physiological role for alternative oxidase? Physiol. Plant 137:392–406.
Vanlerberghe, G. C., Day, D. A., Wiskich, J. T., Vanlerberghe, A. E., McIntosh, L. 1995. Alternative oxidase activity in tobacco leaf mitochondria. Dependence on tricarboxylic acid cycle-mediated redox regulation and pyruvate activation. Plant Physiol. 109:353–361.
Vanlerberghe, G. C., McIntosh, L. 1996. Signals regulating the expression of the nuclear gene encoding alternative oxidase of plant mitochondria. Plant Physiol. 111:589–595.
Vanlerberghe, G. C., McIntosh, L. 1997. Alternative oxidase, from gene to function. Annu. Rev. Plant Physiol. Plant Mol. Biol. 48:703–734.
Vanlerberghe, G. C., Robson, C. A., Yip, J. Y. 2002. Induction of mitochondrial alternative oxidase in response to a cell signal pathway down-regulating the cytochrome pathway prevents programmed cell death. Plant Physiol. 129:1829–1842.
van Lis, R., Atteia, A. 2004. Control of mitochondrial function via photosynthetic redox signals. Photosynth. Res. 79:133–148.
Vranová, E., Inzé, D., Van Breusegem, F. 2002. Signal transduction during oxidative stress. J. Exp. Bot. 53:1227–1236.
Winger, A. M., Millar, A. H., Day, D. A. 2005. Sensitivity of plant mitochondrial terminal oxidases to the lipid peroxidation product 4-hydroxy-2-nonenal (HNE). Biochem. J. 387:865–870.
Woltering, E. J., van der Bent, A., Hoeberichts, F. A. 2002. Do plant caspases exist? Plant Physiol. 130:1764–1769.
Woo, D. K., Phang, T. L., Trawick, J. D., Poyton, R. O. 2009. Multiple pathways of mitochondrial-nuclear communication in yeast: intergenomic signaling involves ABF1 and affects a different set of genes than retrograde regulation. Biochim. Biophys. Acta Gene Reg. Mech. 1789:135–145.
Woo, D. K., Poyton, R. O. 2009. The absence of a mitochondrial genome in rho0 yeast cells extends lifespan independently of retrograde regulation. Exp. Geront. 44:390–397.
Woodson, J. D., Chory J. 2008. Coordination of gene expression between organellar and nuclear genomes. Nat. Rev. Genet. 9:383–395.
Xie, Z., Chen, Z. 1999. Salicylic acid induces rapid inhibition of mitochondrial electron transport and oxidative phosphorylation in tobacco cells. Plant Physiol. 120:217–225.
Xie, Z., Chen, Z. 2000. Harpin-induced hypersensitive cell death is associated with altered mitochondrial functions in tobacco cells. Mol. Plant Microbe Interact. 13:183–190.
Yamamoto, Y., Kobayashi, Y., Devi, S. R., Rikiishi, S., Matsumoto, H. 2002. Aluminum toxicity is associated with mitochondrial dysfunction and the production of reactive oxygen species in plant cells. Plant Physiol. 128:63–72.
Yao, N., Tada, Y., Sakamoto, M., Nakayashiki, H., Park, P., Tosa, Y., Mayama, S. 2002. Mitochondrial oxidative burst involved in apoptotic response in oats. Plant J. 30:567–579.
Yu, J., Nickels, R., McIntosh, L. 2001. A genome approach to mitochondrial-nuclear communication in Arabidopsis. Plant Physiol. Biochem. 39:345–353.
Yuasa, T., Ichimura, K., Mizoguchi, T., Shinozaki, K. 2001. Oxidative stress activates ATMPK6, an Arabidopsis homologue of MAP kinase. Plant Cell Physiol. 42:1012–1016.
Zaninotto, F., La Camera, S., Polverari, A., Delledonne, M. 2006. Cross talk between reactive nitrogen and oxygen species during the hypersensitive disease resistance response. Plant Physiol. 141:379–383.
Zarkovic, J., Anderson, S. L., Rhoads, D. M. 2005. A reporter gene system used to study developmental expression of alternative oxidase and isolate mitochondrial retrograde regulation mutants in Arabidopsis. Plant Mol. Biol. 57:871–888.
Zhang, S. Q., Klessig, D. F. 2001. MAPK cascades in plant defense signaling. Trends Plant Sci. 6:520–527.
Zhang, B., Singh, K. 1994. Ocs element promoter sequences are activated by auxin and salicylic acid in Arabidopsis. Proc. Natl. Acad. Sci. USA 91:2507–2511.
Zhao, Q., Wang, J., Levichkin, I. V., Stasinopoulos, S., Ryan, M. T., Hoogenraad, N. J. 2002. A mitochondrial specific stress response in mammalian cells. EMBO J. 21:4411–4419.
Zhu, J. K. 2002. Salt and drought stress signal transduction in plants. Annu. Rev. Plant Physiol. Plant Mol. Biol. 53:247–273.

Chapter 17
Mitochondrial Regulation of Plant Programmed Cell Death

Mark Diamond and Paul F. McCabe

Abstract Programmed cell death is a cellular process that is an essential component of a plants normal growth and development, its defence mechanism to counter pathogen attack and its response to stress conditions. In both plant and animal cells there are various recognized types of programmed cell death; apoptosis or apoptotic-like, autophagy, or necrosis, and there is a certain amount of overlap between these different cell death processes. This overlap is unsurprising when one considers that a recurrent common feature between several types of programmed cell death is the role of the mitochondria in orchestrating the cascade of events that lead to the death of the cell. It has been shown that the mitochondria can coordinate death signals that lead to the initiation of cell death and subsequently release molecules that drive the destruction of the cell. For example, intermembrane space molecules such as cytochrome *c* are released from the mitochondria, into the cytoplasm, at an extremely early stage of the death process in both plant and animal cells. This release appears to be controlled in part by selective opening of the permeability transition pore in the mitochondria, and the magnitude of the opening of the permeability transition pore may be a determining factor in the type of programmed cell death a cell undergoes. Similarly, reactive oxygen species (ROS) production by mitochondria (possibly as a result of cytochrome *c* release) has been identified as a driver of cell death in plant cells and again the type of death program that is activated may be due to the levels of ROS produced in the dying cell as apoptosis, autophagy, or necrosis can be initiated by oxidative stress.

Keywords Mitochondria • Programmed cell death • Autophagy • Apoptosis • Necrosis • Cytochrome *c* • ROS • Alternative oxidase

---

P.F. McCabe (✉)
School of Biology and Environmental Science, University College Dublin, Dublin, 4, Ireland
e-mail: paul.mccabe@ucd.ie

F. Kempken (ed.), *Plant Mitochondria*, Advances in Plant Biology 1,
DOI 10.1007/978-0-387-89781-3_17,

Abbreviations

AIF	Apoptosis inducing factor
ANT	Adenine nucleotide transporter
AOX	Alternative oxidase
CsA	Cyclosporine A
CypD	Cyclophilin D
HR	Hypersensitive response
IMS	Intermembrane space
MPT	Mitochondrial permeability transition
OMM	Outer mitochondrial membrane
PCD	Programmed cell death
PTP	Permeability transition pore
ROS	Reactive oxygen species
TE	Tracheary element
TMV	Tobacco mosaic virus
VDAC	Voltage-dependent anion channel
VIGS	Virus induced gene silencing

17.1 Introduction

17.1.1 Programmed Cell Death

17.1.1.1 History

Research into cell death took a giant leap forward following the publication of Kerr and coworkers seminal work identifying apoptosis; an "active and inherently programmed" form of cell death (Kerr et al. 1972). This work led to a shift in the way cell death was viewed and resulted in the classification of cell death into a regulated form (apoptosis) and an unregulated form (necrosis). The authors defined the morphologic changes occurring in the cell during apoptosis as occurring in two main stages. First, the nucleus and cytoplasm condense and eventually break up into many membrane-bound fragments, which they termed apoptotic bodies. Next, these apoptotic bodies are ingested by other cells and are rapidly degraded. Since the publication of this study the majority of research into the mechanisms of cell death has been carried out in animal systems, and in a relatively short time some of the genetic machinery of apoptosis was revealed in the nematode *Caenorhabditis elegans*. In 1986, Ellis and Horvitz characterized *ced-3* and *ced-9* as two genes whose wild-type functions were required for PCD during *C. elegans* development (Ellis and Horvitz 1986). Following this work, studies were carried out in mammalian systems in which functional homologues of *ced-3*, *ced-9* and other apoptosis regulators were identified. These included the B-cell lymphoma 2 (Bcl-2) protein family, the apoptotic protease-activating factor (Apaf-1), and the cysteine-dependent aspartate-specific protease (caspase) family. These and other proteins; orchestrated by the mitochondrion, regulate the initiation and execution of apoptosis.

Since the initial identification of apoptosis, several types of programmed cell death (PCD) have been identified. The most commonly used PCD classification system was proposed by Clarke and characterizes animal PCD according to lysosomal involvement, thereby dividing animal PCD into three categories (Clarke 1990; Bras et al. 2005). Apoptosis (Type I PCD) is usually mediated by the caspase family of cysteine proteases as well as certain proteins from the mitochondria, and leads to cell shrinkage, oligonucleosomal DNA fragmentation, chromatin condensation (leading to the appearance of pyknotic nuclei), and regulated disintegration of the cell via the formation of apoptotic bodies. This form of PCD involves heterophagocytosis, and the apostrophe in cell's own lysosome plays no apparent role. Autophagy (Type II PCD) – which literally means "self-eating," is mediated through the formation of autophagic vesicles that sequester cytoplasmic components and ultimately fuse with the cells own lysosomes, where the cytoplasmic components are degraded prior to heterophagocytosis of cellular remains. This form of cell death is also characterized by the dilation of the endoplasmic reticulum (ER) and mitochondria, and slight enlargement of the Golgi. Finally, necrosis-like PCD (Type III PCD) is characterized by the apparent absence of lysosomal involvement and involves the swelling of organelles, rupture of the plasma membrane, and breakdown of the cytoplasm to varying degrees.

17.1.1.2 Importance

Research in animal systems has demonstrated the importance of different forms of cell death at various stages of the life cycle of the organism. Apoptosis is by far the most extensively studied form of cell death. It is involved in the maintenance of cellular homeostasis, and has been linked to foetal developmental events such as the formation of independent digits (Thompson 1995; Zuzarte-Luis and Hurle 2002). In addition, many important diseases have been linked to either an increase (e.g., Parkinson's disease or Alzheimer's disease), or a decrease (e.g., some types of cancer, systemic lupus erythematosus, herpes), in rates of apoptosis (Thompson 1995). Autophagic cell death is less well studied than apoptosis and seems to occur less often during development. However, it is known to contribute to the development of insect salivary glands, and has been linked to heart failure in humans (Knaapen et al. 2001; Berry and Baehrecke 2007). Necrosis-like neuronal cell death has been reported in dying neurons of *C. elegans* (Hall et al. 1997) and also occurs in mammalian cells when caspase activation is blocked or apoptosis is aborted following death program activation (Leist and Jaattela 2001).

In plants, PCD is also vital to normal growth and development. For example, tracheary element (TE) cells undergo PCD at maturity in order to fully differentiate into the water-conducting vessels of the xylem. After secondary wall synthesis these cells initiate a PCD that is characterized by complete autolysis, leaving a hollow cell corpse (Fukuda 2000; Yu et al. 2002). In angiosperms, two sister cells arise from the asymmetric division of the zygote during embryogenesis, and go on to form either the embryo or vacuolated suspensor cells that signal to the early embryo (McCabe et al. 1997a). Using a carrot somatic embryogenesis system, it was shown that the suspensor-type cell undergoes a PCD with some of the features of apoptosis, namely condensation and shrinkage of the cytoplasm and nucleus, and fragmentation of DNA (McCabe et al. 1997b). During cytoplasmic male sterile (CMS)-associated death in sunflower, cellular condensation and nuclear DNA cleavage were also observed (Balk and Leaver 2001) (see also Chap. 18). In common with animal mitochondria after apoptosis induction, the mitochondria in these sunflower cells were found to release cytochrome *c* and to persist until very late stages in the death process. This indicated a role for the mitochondrion in regulated CMS-associated PCD. In *Papaver rhoeas*, the prevention of inbreeding (so-called "self-incompatibility") is maintained through the recognition and subsequent growth inhibition of the plants own pollen. This is achieved through the interaction of two multiallelic gene products PrsS (stigma S locus determinant) and PrpS (pollen S locus determinant). Upon interaction of the pollen PrpS extracellular loop with the PrsS in the stigma of self plants, a Ca^{2+}-dependent signaling network is triggered, which results in growth inhibition and PCD in the pollen tube (Thomas and Franklin-Tong 2004; Wheeler et al. 2009).

As well as plant development, PCD plays a significant role in plant stress responses. Apoptotic-like PCD morphology (cellular condensation, which in plant cells is observed as a retraction of the protoplast away from the cell wall) has been shown to occur in cells subjected to various abiotic stress-inducing treatments, such as treatment of carrot or *Arabidopsis thaliana* cells with H_2O_2 (Kawai-Yamada et al. 2004), tunicamycin or brefaldin A treatment of sycamore cells (Crosti et al. 2001; Malerba et al. 2004), and treatment of *A. thaliana* cell cultures with ceramides (Townley et al. 2005). In addition, heat shock is a potent inducer of apoptotic-like PCD morphology in carrot, *A. thaliana*, and tobacco (McCabe et al. 1997b; McCabe and Leaver 2000; Vacca et al. 2004). Indeed, McCabe et al. (1997b) showed that the level of apoptotic-like PCD morphology observed in heat-shocked carrot cells correlated with the temperature. The authors found that at lower temperatures (25–35°C) cells remained alive. When cells were treated with increasingly higher temperatures (35–55°C), there was a proportional increase in cell death (up to 100% at 55°C) that resulted in a shrunken apoptotic-like cell corpse. While all cells died with temperature treatments above 55°C (65–85°C), the incidence of this corpse morphology declined rapidly and was completely absent after temperature treatments above 75°C, indicating that these cells died *via* necrosis. This suggested that cell shrinkage is an active process resulting from PCD rather than an uncontrolled collapsing of the cell.

PCD is also a crucial component of responses associated with the biotic stresses caused by pathogen attack. Host cell death occurs in response to many, but not all, plant–pathogen interactions. This cell death can lead to plant resistance or susceptibility to the invading pathogen, largely depending on the food preference of the pathogen (Greenberg and Yao 2004). The hypersensitive response (HR) to avirulent pathogens often terminates in the rapid death of infected or challenged cells, which can result in arrest of pathogen growth (Heath 2000). Elicitors inducing the HR, such as cryptogein, induce an apoptotic-like PCD morphology in challenged soybean and tobacco cells in the HR lesion zone (Levine et al. 1996). Many virulent pathogens also induce programmed cell death. Yao and coworkers, studying the effects of many different pathogens (fungi, bacteria, and viruses) in oat, found apoptotic-like PCD morphologic features occurred in infected cells, and sometimes neighboring cells, at various time points depending on the type of infectious agent (Yao et al. 2002).

17.2 The Mitochondrion and PCD in Animal Systems

Due to the association between PCD and cancer as well as other important diseases, the vast majority of research into the mechanisms of PCD has been carried out in animal systems. Therefore, for the purposes of this review we will also evaluate the role of the mitochondrion in different forms of animal PCD. We approach this by separating PCD into the three types suggested by Clarke (1990). This is purely for the purposes of convenience, as these classification systems do not fully reflect the sophisticated interplay that can occur between these pathways (Bras et al. 2005). For example, both apoptosis and autophagic cell death are induced simultaneously in response to proteotoxic stress inducing agents (Gozuacik et al. 2008), and cell death that usually occurs with an apoptotic morphology can be shifted toward a more necrotic morphology when caspase activation is inhibited (Kroemer and Martin 2005). This lends itself to the argument put forward by Bras and coworkers (2005) that there is a certain amount of interdependence between the PCD modes, and that these pathways may be initiated simultaneously. The cell's destiny would depend therefore on the nature and degree of the death stimulus, the condition of the cell, and/or the presence or absence of inhibitors of the death machinery. Whatever the case may be, there is evidence for mitochondrial involvement in the three death modes, which we will evaluate here.

17.2.1 Apoptosis

Many studies examining the involvement of mitochondria in plant PCD have focused on possible similarities with mammalian apoptosis. The mitochondrion is a central component during this apoptosis, coordinating death signals during the induction phase and initiating the death program through the release of proapoptotic molecules during the effector phase. During the apoptotic signaling pathway the mitochondria are believed to be a central node that can integrate the plethora of death signals (Bras et al. 2005). Classical apoptosis, mediated by caspases, can be triggered by release of cytochrome *c* from the intermembrane space of the mitochondrion (Leist and Jaattela 2001). Following its release, cytochrome *c* forms a complex with pro-caspase 9, dATP, and the apoptosis protease-activating factor (Apaf-1), which leads to the creation of the apoptosome (see Fig. 17.1). A caspase activation cascade is then initiated by the apoptosome and this leads to the apoptotic

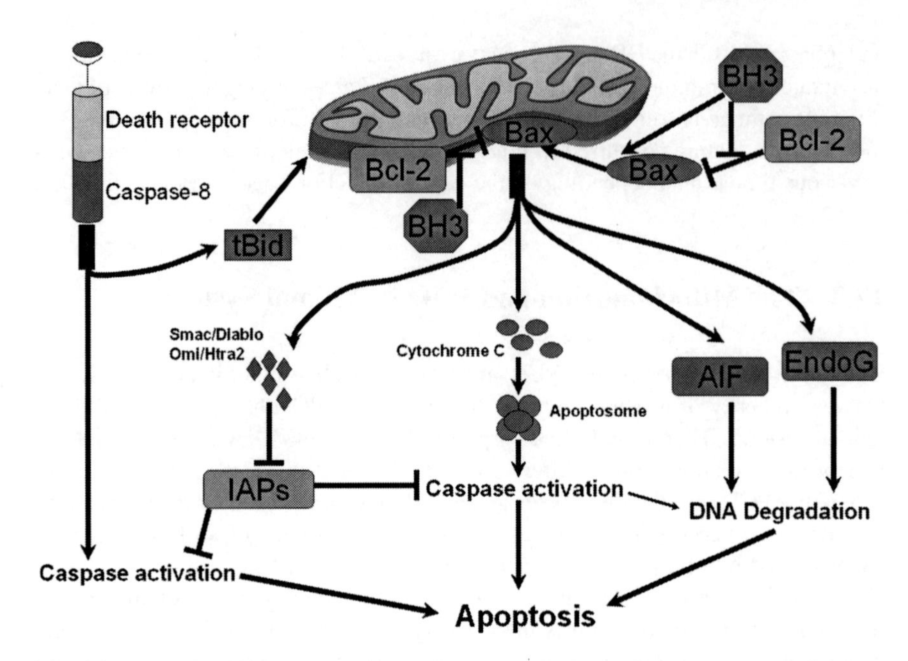

Fig. 17.1 Activation of apoptotic cell death in animals. In the intrinsic pathway the mitochondrion perceives a death signal and can release apoptogenic factors. This occurs either through the permeabilization/rupture of the outer mitochondrial membrane (*OMM*) or through the action of pro-apoptotic Bcl-2 family members, such as Bax, under the regulation of BH1234 (like Bcl-2) and BH3-only proteins. Cytochrome *c* is released and it complexes with Apaf-1, pro-caspase 9 and dATP to form the apoptosome. The apoptosome then instigates a caspase-activation cascade, which results in the apoptotic dismantling of the cell. IAP's can inhibit caspase activation, but these themselves can be inhibited through the action of Smac/Diablo and Omi/Htra2, which are also released from mitochondria. Other molecules released from the mitochondria include AIF and EndoG, which can instigate DNA degradation and apoptosis in a caspase-independent fashion. In the extrinsic pathway, death receptors can activate apoptosis directly through the action of caspase-8 and caspase-3, or interact with the mitochondrion via tBid, which can lead to apoptogenic protein release

degradation of key cellular components and to the death of the cell. Multiple checks and balances exist both to prevent the unnecessary activation of apoptosis and equally to prevent the inhibition of apoptosis when cell suicide is favorable. The activation of caspases can be negatively regulated by inhibitor of apoptosis proteins (IAPs). These IAPs bind and inhibit Caspase 3, 7, and/or 9; however, upon apoptosis induction these IAPs can themselves be inhibited through the antagonizing efforts of two mitochondrial proteins SMAC/DIABLO and Omi/HtrA2 (Schimmer 2004).

Further checks and balances are in place at the level of proapoptotic protein release from the mitochondrion and are mediated by the B cell lymphoma 2 (Bcl-2) family of proteins, which contain both pro- and anti-apoptotic members (Kroemer et al. 2007). The antiapoptotic members of the family (of which Bcl-2 and Bcl-X_L are the prototypes) contain four Bcl-2 homology (BH) regions (BH1234). The proapoptotic members consist of multidomain proteins containing three BH (BH123) domains (such as Bax and Bak), or single domain BH3-only proteins (such as Bid and Bad). Under normal physiologic conditions the proapoptotic BH123 members are localized either in the outer mitochondrial membrane (OMM) in the case of Bak, or in the cytosol (Bax). However, upon apoptosis induction Bax inserts into the OMM. There it can form supramolecular openings in the mitochondrion either alone or in combination with other proapoptotic members such as Bak. Such openings can facilitate the release of apoptogenic proteins from the intermembrane space (IMS). The anti-apoptotic members of the Bcl-2 family are mainly localized in the OMM, presumably so they can bind and neutralize pro-apoptotic Bcl-2 family proteins to prevent mitochondrial membrane permeabilization and apoptogenic protein release. BH3-only proteins can carry out their pro-apoptotic function in two distinct ways. They can either directly activate the BH123 proteins to induce permeabilization (for example by stimulating the relocation of Bax to the OMM), or they can indirectly facilitate apoptosis by binding to the BH1234 proteins keeping them separated from other proapoptotic Bcl-2 family proteins (Letai et al. 2002).

The extrinsic signaling pathways are primarily mediated by so-called "death receptors" at the cell surface, which become activated after association with a particular ligand. Upon activation, adaptor molecules can bind to their cytosolic ends to form a death-inducing signaling complex (DISC). Procaspase-8; which strongly localizes to the region around the receptor-adaptor complex, can then undergo autocatalytic activation in a process that is facilitated by the DISC. In this way the death receptors can directly activate the caspases and bring about apoptosis without the need for mitochondrial involvement, as active caspase-8 can then activate the downstream effector caspases (Scaffidi et al. 1998). However, if the death receptors do not generate a strong enough signal to induce death on their own, mitochondrion-dependent apoptotic pathways can amplify the signal. In this scenario, caspase-8 can cleave the BH3-only protein Bid and this truncated Bid can then initiate the complete release of cytochrome *c* from the mitochondrial IMS, which leads to the initiation of apoptosis in a manner that does not seem to cause flagrant mitochondrial swelling (Luo et al. 1998).

Two additional mitochondrial apoptotic signaling pathways involve the apoptosis inducing factor (AIF) and endonuclease G (EndoG). Under normal conditions AIF is localized in the IMS either in the form of a soluble monomer or attached to the inner membrane. In the IMS it is required for the assembly and maintenance of respiratory chain complex I, and for optimal detoxification of ROS by-products (Vahsen et al. 2004; Kroemer et al. 2007). Following apoptosis induction, AIF is released from the mitochondria and migrates to the nucleus, where it initiates chromatin condensation and cleavage of DNA into fragments of approximately 50 kb in size (Balk et al. 2003). AIF is thought to act independently of the caspases, as inhibition of caspase activity or activation does not abrogate AIF's cell death-promoting activity (Cande et al. 2004). Endo G is a 30-kDa mitochondrion-specific nuclease that, similarly to AIF, migrates to the nucleus after death signal perception (Li et al. 2001). From there it brings about internucleosomal DNA fragmentation, and apparently works together with AIF and the caspase activated DNase (CAD) to mediate chromatin condensation and nuclear degradation (Bras et al. 2005).

17.2.2 Autophagic Cell Death

Autophagy, which translates literally as "self-eating," is actually recognized above all as a pro-survival pathway in most eukaryotic cells. Autophagy mediates the lysosome or vacuole-dependent processing and recycling of macromolecules, or even whole organelles, and is important for the metabolism of lipids under nutrient-deficient conditions (Thompson and Vierstra 2005; Geng and Klionsky 2008; Singh et al. 2009). It is also responsible for the removal of toxic cellular components such as damaged organelles (including mitochondria) or abnormal proteins (Bassham 2007; Zhang et al. 2007). The most studied form of autophagy is macroautophagy, which is distinguished from other forms of autophagy (microautophagy and chaperone-mediated autophagy) as it occurs through *de novo* double membrane vesicle formation that does not take place at the vacuolar membrane (Geng and Klionsky 2008). The formation of these vesicles (called autophagosomes) is governed by the protein products of the Atg (autophagy) genes, which were originally identified through mutant screening in *Saccharomyces cerevisiae* (Klionsky et al. 2003). While the cellular origin of the autophagosome is unknown, some researchers believe that the mitochondrion might contribute to their formation (Luo et al. 2009). During their formation these autophagosomes sequester their cytoplasmic cargo in cage-like structures that eventually fuse together, forming the double membrane (Thompson and Vierstra 2005). The autophagosome then travels to the vacuole where its outer membrane fuses with the tonoplast and the inner vesicle (autophagic body) is released, along with its sequestered cytoplasmic cargo, into the vacuole where it is degraded. Interestingly, one of the key proteins in regulating autophagosome formation; Beclin 1 (ATG6), was identified in mammals originally as a Bcl-2-interacting protein (Liang et al. 1998). Since that time, it has been shown that Bcl-2 binding of Beclin 1 causes an inhibition of Beclin 1-dependent autophagy. It has been suggested that the two proteins function in a rheostat manner ensuring autophagy normally operates at homeostatic levels, enabling the cells to cope with low nutrient conditions but not enough to cause cell death (see Fig. 17.2) (Pattingre et al. 2005; Kroemer et al. 2007). Therefore when autophagy is hyper-activated or Bcl-2 activity is inhibited (or its levels reduced), Bcl-2 may no longer be able to prevent high levels of autophagy, which can lead to cell death.

As previously mentioned, autophagy is primarily regarded as a prosurvival process, and it has been proposed that autophagic cell death is a misnomer (Kroemer and Levine 2008). However, there are instances of autophagic cell death during development (Berry and Baehrecke 2007), or in response to toxic agents (Chen et al. 2008). It is perhaps the case that the intensity of autophagic activation is important in determining whether or not a cell undergoes autophagic cell death (Levine and Yuan 2005; Kang et al. 2007; Kang and Avery 2008). Autophagy may indeed be primarily activated to protect the cell but when overstimulated, it instead consumes it. In any case, autophagy leading to cell death does occur in animals, and there is evidence for mitochondrial involvement in this process. The release of high levels of ROS from mitochondria leads to oxidative stress and oxidative stress is

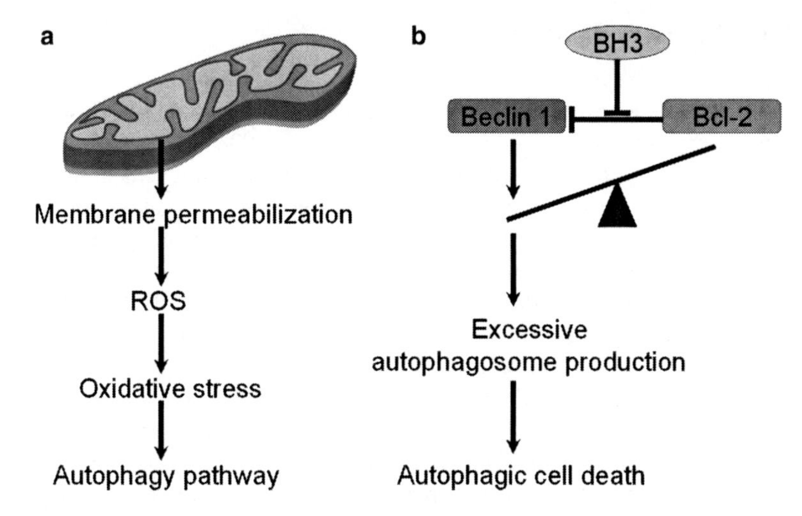

Fig. 17.2 A model for mitochondrial induction of autophagic cell death. (**a**) The mitochondrion is one of the main sources of reactive oxygen species (*ROS*) in the cell. Upon membrane permeabilization, MPT occurs and this can lead to the excessive production and release of ROS. The resulting oxidative stress is a known activator of autophagy. (**b**) In this rheostat model, homeostatic levels of autophagy may be maintained in mammalian cells through the inhibition of the key autophagic component Beclin 1 by Bcl-2. However, this inhibition could be prevented by BH3-only proteins (like BNIP3), which can compete with Beclin 1 for interaction with Bcl-2. When there is more active Beclin 1 available, autophagy could become superinduced, which would cause excessive autophagosomal-mediated cellular digestion and autophagic cell death

known to induce autophagic cell death (Chen et al. 2008). Additionally, loss of mitochondrial permeability transition seems to induce autophagy (Elmore et al. 2001). There is also evidence that Bcl-2 family members play a role in autophagic cell death. The BH3-only protein BNIP3 is known to form stable homodimerization complexes that localize to the OMM following stress (Chinnadurai et al. 2008). Studies have shown that over-expression of BNIP3 caused a delayed cell death, which was originally labeled as apoptotic (Chen et al. 1997; Yasuda et al. 1998; Imazu et al. 1999). However, it has been suggested that these studies employed methods that were not capable of distinguishing between death modes (Chinnadurai et al. 2008). Subsequent ultrastructural studies have revealed that BNIP3 over-expression resulted in a cell death characterized by excessive numbers of cytoplasmic vacuoles that resembled autophagosomes (Vande Velde et al. 2000). More recently it has been reported that chronic exposure to hypoxia in several apoptosis-competent cancer lines results in BNIP3-dependent autophagy and cell death (Azad et al. 2008). In these instances a depletion of BNIP3 resulted in reduced autophagy activation and a reduced level of cell death. It is now thought that BNIP3 may facilitate the activation of autophagic cell death by inhibiting the interaction of Bcl-2 and Beclin 1, as pharmacologic BH3 mimetics competitively inhibited this interaction and BNIP3 can itself compete with Beclin 1 for interaction with Bcl-2 (reviewed by Chinnadurai et al. 2008).

17.2.3 Necrosis-Like Cell Death

Necrosis has been described as an unregulated cell death caused by an overwhelming stress or severe injury to the cell (Reape and McCabe 2008). It is characterized by cellular swelling and rupture of the plasma membrane, which results in the release of cellular contents into the extracellular space, frequently damaging neighboring cells. Recently the classification of necrosis as only accidental and stimulus driven has come into question, as several studies have shown that the susceptibility to undergo necrosis is partially determined by the cell (Kroemer et al. 2007). In agreement with this concept, the overexpression of Bcl-2, which stabilizes mitochondrial membranes and is a negative regulator of apoptosis, leads to a reduction or complete inhibition of necrotic cell death in response to cyanide (Kroemer et al. 2007). Furthermore, mitochondrial membrane permeabilization can determine whether or not a cell undergoes necrosis-like cell death because knocking out cyclophilin D (CypD), a mitochondrial component frequently associated with this permeabilization, prevents H_2O_2 or A23187 (Ca^{2+} ionophore) induced necrotic death of liver cells (Nakagawa et al. 2005).

17.3 Release Mechanisms for Mitochondrial Factors

The mitochondrion releases factors into the cytosol during PCD in both plants and animals. As reviewed previously (Diamond and McCabe 2007), there are two main ways in which pro-death factors can be released from mitochondria in animal cells. Mitochondrial proteins can be released if the permeability transition pore (PTP) opens in response to cellular stress or death signals, causing the mitochondria to swell and resulting in OMM rupture. The PTP is made up of a polyprotein complex that includes (but is not limited to) CypD from the matrix, inner membrane proteins such as the adenine nucleotide transporter (ANT) and the outer membrane voltage-dependent anion channel (VDAC) protein (Green and Reed 1998; Saviani et al. 2002). As well as the release of pro-death factors, PTP opening likely results in mitochondrial permeability transition (MPT), the collapse of mitochondrial transmembrane potential ($\Delta\psi_m$), production of ROS, termination of ATP synthesis, and the release of Ca^{2+} and other matrix solutes (Curtis and Wolpert 2004). A second way mitochondrial proteins can be released is if the BH123 protein Bax interacts with VDAC and ANT to induce membrane permeabilization. This activity is positively regulated by another BH123 protein Bak and a truncated version of the BH3-only protein Bid (truncated = tBid) and negatively regulated by the BH1234 proteins Bcl-2 and Bcl-X_L (Bras et al. 2005). However, according to some reports Bax can cause mitochondrial membrane permeabilization and cytochrome *c* release independently of any of the critical components of the PTP complex (Kuwana et al. 2002; Kroemer et al. 2007).

In plants, no homologues of the Bcl-2 family of proteins have been identified to date; however, there is evidence that functional homologues may exist, as both Bcl-X_L and Bax can fulfill apoptotic roles when engineered into plant cells (Jones 2000). For example, mammalian Bax was expressed in tobacco under the control of a tobacco mosaic virus vector and triggered a form of PCD with the characteristic features of the HR (Lacomme and Cruz 1999). Another research group expressed the antiapoptotic BH1234 protein Bcl-X_L in tobacco plants under the control of a strong promoter (Mitsuhara et al. 1999). They found that overexpression of Bcl-X_L prevented cell death induced by UV-B irradiation, methyl viologen, or HR, in response to tobacco mosaic virus infection in a fashion that depended on the amount of Bcl-X_L protein expressed. Lacomme and Cruz (1999) demonstrated that Bax localized to the mitochondria, and it was later established that Bcl-2 and Bcl-X_L can localize to mitochondria, chloroplasts, and nuclei in plants (Chen and Dickman 2004). However, to our knowledge it has not been established in any studies thus far, if Bax, Bcl-2, or Bcl-X_L actually interact with mitochondria or regulate the release of mitochondrial proteins in plants. It remains a possibility that while Bcl-2 family proteins can function in plants in a pro-, or anti-survival fashion, functional homologues do not exist in plants that can interact with the mitochondrion.

However, homologues do exist in plants for the major constituents of the PTP complex, VDAC, ANT, and CypD (Yao et al. 2004). It is therefore possible that OMM permeabilization is mediated by the PTP in plants. In animals, when the mitochondrial PTP opens, solutes of <1.5 kDa can enter and go through the inner mitochondrial membrane in what is known as MPT. When this happens, $\Delta\psi_m$ collapses, oxidative phosphorylation and electron transport are dissociated, ions and solutes are released, and a high amplitude swelling can occur, which results in OMM rupture and the release of IMS-localized compounds (Arpagaus et al. 2002). PTP opening can be induced by a number of factors including, ROS, the accumulation of Ca^{2+} in the mitochondrial matrix, $\Delta\psi_m$, phosphate (Pi) and ATP levels, fatty acids, and pH. This opening can be inhibited by the pharmacologic application of cyclosporine A (CsA), an agent that is now known to displace the binding of CypD to ANT (Crompton 1999). In plants, much of the research undertaken into the role of the PTP in PCD has used CsA as a diagnostic tool.

Arpagaus and coworkers (2002) investigated the effects of Ca^{2+} and other factors on mitochondrial morphology in potato tubers and the role of the PTP in these effects. The authors were able to demonstrate that, in the presence of Pi, the addition of 5 mM $CaCl_2$ caused an initial fast shrinkage, then a lag phase and finally a pronounced swelling of mitochondria and the release of a significant amount of proteins, including cytochrome *c*. Neither swelling nor release of proteins was observed when 5 mM $MgCl_2$ was used in place of $CaCl_2$ and all of the $CaCl_2$ mediated effects were effectively counteracted through the addition of CsA in combination with dithioerythritol. These results provided the initial evidence that MPT can occur in plants and that the PTP is probably involved. Evidence supporting a role for MPT and the PTP in plant PCD came from an investigation into betulinic acid-induced PCD in TE cells of *Zinnia elegans* (Yu et al. 2002). In this study the authors showed that inhibiting PTP opening with CsA prevented PCD even though cytochrome *c* was released. Betulinic acid induces PTP opening and CsA counteracts this so perhaps the effect of betulinic acid was strong enough to cause cytochrome *c* release but not strong enough; in the presence of CsA, to induce PCD. Additionally, the authors stated that CsA could inhibit TE formation, which strongly suggests a role for the PTP and mitochondria in the PCD occurring at TE maturation. Further studies demonstrated the ability of CsA to inhibit cell death induced by nitric oxide (NO), ROS and the phytotoxin fusicoccin (Saviani et al. 2002; Tiwari et al. 2002; Contran et al. 2007). The fact that these studies were carried out on cell cultures of *A. thaliana*, sweet orange, and sycamore (in addition to the previously mentioned studies carried out in *Z. elegans* and potato) indicates the importance of the PTP (and MPT) in PCD occurring during development, or in response to diverse toxic agents.

Investigations into the role of the PTP in apoptosis has progressed through the use of genetic screening, by knocking out individual components of the PTP such as CypD and ANT. One study used mice in which the two isoforms of ANT were eliminated in liver cells (Kokoszka et al. 2004; Forte and Bernardi 2005). ANT deletion led to the selective loss of PTP regulation by ANT ligands such as atractyloside (which can promote pore opening) and ADP (which can promote pore closure). Furthermore, Ca^{2+} concentrations of threefold higher than normal were required to open the PTP in ANT-deficient mitochondria. Taken together these results suggested that ANT is required for PTP opening by regulating the sensitivity of the PTP to Ca^{2+} levels and ANT ligands. CypD elimination also resulted in a desensitization of the PTP to Ca^{2+} (Basso et al. 2005). PTP in CypD-deficient mitochondria required a twofold Ca^{2+} concentration for opening to occur; however, responses to pH, adenine nucleotides, oxidative stress, and depolarization remained unaffected. This demonstrated that CypD participates in regulating the PTP in response to Ca^{2+}, but not in response to other important mediators of PTP opening. However, it was later determined that CypD is required for MPT in the mediation of some forms of necrotic cell death but not apoptotic cell death (Baines et al. 2005; Nakagawa et al. 2005). This would seem to suggest that the Bcl-2 family mediates apoptosis and that necrosis-like PCD is mediated by the PTP, but these results do not exclude the possibility that the proteins forming the PTP complex mediate some forms of apoptosis.

Similar investigations into the roles of PTP complex members in plants should prove beneficial in elucidating the role of the PTP in plant PCD. Knockouts (T-DNA insertion lines) of PTP complex homologues are available, and virus induced gene silencing (VIGS) treatments have been carried out successfully on VDAC (Palmieri et al. 2008; Kusano et al. 2009). One study demonstrated that tobacco VDACs can regulate cell death in response to human Bax expression (Tateda et al. 2009). The authors showed that VDAC-silenced plants displayed delayed cell death in response to Bax expression in comparison to controls. Additionally, they showed that silencing of tobacco VDACs compromised the plants defence against *Pseudomonas cichorii*, in particular the production of H_2O_2.

However, further investigations testing VDAC-silenced plants with more natural plant stresses are needed to determine the relevancy of VDAC for plant PCD.

In animal systems it is known that binding of mitochondrial-associated hexokinases to VDAC hinders the interaction of Bax with VDAC and can inhibit cytochrome *c* release and apoptosis (Hofius et al. 2007). It has recently been demonstrated that plant hexokinases can play a similarly important role in PCD. In a study utilizing VIGS of a hexokinase in tobacco, the authors showed that hexokinase 1-silenced plants spontaneously formed necrotic lesions in their leaves (Kim et al. 2006). This hexokinase was associated with mitochondria and its expression could be upregulated by various death-inducing stimuli. Furthermore, overexpression of mitochondria-associated hexokinases in *A. thaliana* led to an enhanced resistance to oxidative-stress induced death, and the addition of purified hexokinase to isolated mitochondria prevented the release of cytochrome *c* and the loss of membrane potential. These data suggest a direct link between plant hexokinases and the PCD process. Because hexokinases occupy a key position at the entry point of carbon into respiratory pathways and are a component of the sugar-sensing pathway, some researchers suggest that hexokinase is ideally placed to monitor cellular carbon metabolism and link this information to cell death pathways (Sweetlove et al. 2007).

17.4 Structural and Physiological Changes to Mitochondria During Plant PCD

Alterations in the physiology and structure of mitochondria frequently occur during PCD and necrosis. Physiologic changes include a reduced $\Delta\psi_m$, a reduction in ATP production (usually during necrosis), and enhanced ROS production and accumulation (Diamond and McCabe 2007). There does not, however, appear to be a uniform type of mitochondrial structural change that occurs during PCD. Here we will give a few examples of the structural and physiologic changes occurring in plant mitochondria during developmental and stress-induced PCD.

In some plants inbreeding is prevented through the induction of PCD in pollen, which is recognized as "self," in what is known as self-incompatibility (SI) (Thomas and Franklin-Tong 2004). In one study, during the SI response of *Papaver rhoeas*, mitochondrial ultrastructural changes were observed that included swelling, loss of electron density, loss of cristae and blebbing in all mitochondria within 1 h of SI induction (Geitmann et al. 2004). At 2 h after SI induction all mitochondria were either extremely swollen with a reduced cristae number, or displayed a blebbing and local ballooning of the cristae and/or OMM. After 4 h the mitochondria were unrecognizable. In another study the mitochondrial changes during H_2O_2-induced cell death in soybean cells was examined (Casolo et al. 2005). Using two different concentrations of H_2O_2 (5 or 20 mM) the authors were able to induce cell death with an apoptotic-like or a necrotic morphology. The authors then measured changes in mitochondrial integrity and found that 5 mM H_2O_2 caused a slight rupture of the OMM and a loss of mitochondrial functionality, while oxygen consumption was only slightly disrupted and ATP levels remained unaffected. By contrast, cells treated with 20 mM H_2O_2 displayed a pronounced rupture of the OMM while oxygen consumption and ATP levels were greatly reduced.

Physiologic changes to mitochondria were also noted during salt stress-induced ovule abortion in *A. thaliana* (Box 17.1) (Hauser et al. 2006).

Box 17.1 Does the ER interact with the mitochondrion to promote plant PCD?

In a study examining salt stress-induced ovule abortion in *A. thaliana*, Hauser and coworkers (2006) observed membrane depolarization in gametophyte mitochondria within 24 h of salt stress. ROS accumulation in almost all ovules followed. In this report structural changes occurred in the mitochondria; however, they were nonuniform and therefore deemed nonimportant. More interestingly, the authors noted concentric rings of endoplasmic reticulum (ER) encircling many organelles, including mitochondria. Concentric rings of ER surrounding mitochondria and other organelles were also noted in a previous study during SI-induced PCD in *P. rhoeas* (Geitmann et al. 2004). In animal systems it is known that mitochondria and ER are often closely associated where they maintain a local Ca^{2+} communication conduit, allowing highly efficient and effective transfer of Ca^{2+} between the two organelles (Walter and Hajnoczky 2005). Furthermore, it is known that ER stress can activate a Ca^{2+} efflux from the ER and a redistribution to the mitochondria, leading to PCD, and that this PCD can be inhibited by agents affecting Ca^{2+}-uptake to the mitochondria. It is therefore possible that the close proximity of the ER to mitochondria in pollen grains undergoing SI-induced PCD (or indeed in ovules undergoing salt stress-induced PCD) could result in the opening of the PTP and PCD via the translocation of high levels of Ca^{2+} to mitochondria. One possible regulator of this interaction may be the evolutionarily conserved protein Bax inhibitor 1 (BI1), which is known to reside in the ER (Kawai-Yamada et al. 2001). Although the precise mode of action of this protein in plants is unknown at this time, BI1 overexpression reduces death in response to many stimuli, and this cytoprotective effect has been correlated with a reduced level of releasable luminal Ca^{2+} (reviewed in Cacas 2010). BI1 could theoretically act to determine cell fate by controlling the release of Ca^{2+} from the ER, thereby limiting mitochondrial uptake and downstream PCD.

17.5 Plant PCD and Apoptogenic Proteins

In stress induced mammalian apoptosis, the mitochondria releases numerous molecules that can activate death cascades. These apoptosis-triggering factors include cytochrome *c*, AIF, and EndoG, which can activate caspases or initiate caspase-independent cell death (Reape and McCabe 2008). Here we will examine the evidence for a similar role for mitochondrial factors in plant PCD.

17.5.1 Cytochrome c

Cytochrome *c* is an ancient and highly conserved protein, normally found associated with the inner mitochondrial membrane (IMM), where it shuttles electrons between the IMM and OMM at the final step in aerobic energy production (Goodsell 2004). In animal cells, release of cytochrome *c* from the mitochondria results in it binding to Apaf-1 and this complex then interacts with, and promotes, the activation of procaspase-9. The activation of caspase-9 leads to a caspase-activation cascade and apoptosis. In plants there is significant data demonstrating either the release of cytochrome *c*, or activation of caspase-like activity, after PCD induction (Diamond and McCabe 2007; Reape and McCabe 2008).

Cytochrome *c* translocation has been shown to occur following a 55°C heat shock treatment of cucumber cotyledons (Balk et al. 1999), D-mannose treatment of maize suspension cultures (Stein and Hansen 1999), menadione treatment of tobacco protoplasts (Sun et al. 1999), ceramide-treated *A. thaliana* protoplasts (Yao et al. 2004), and in heat-shocked *A. thaliana* and tobacco cell cultures (Balk et al. 2003; Vacca et al. 2006). This release of cytochrome *c* following PCD induction appeared to be selective, as the matrix protein fumarase and the OMM protein porin remained associated with mitochondria during heat shock-induced PCD (Balk et al. 1999, 2003). However, other investigations have demonstrated that cytochrome *c* release does not occur during all forms of PCD (Xu and Hanson 2000; Duval et al. 2005), or that cytochrome *c* release does not necessarily lead to PCD (Yu et al. 2002). Further evidence in support of a divergent role for cytochrome *c* during plant PCD came from studies using an *A. thaliana* cell free system containing purified nuclei incubated with mitochondria and/or cytosolic extract (Balk et al. 2003). Addition of broken mitochondria to the system resulted in DNA degradation by two different mechanisms. One was dependent on the presence of the cytosol and resulted in DNA fragmentation after 12 h. The other did not require the presence of cytosol and resulted in cleavage of DNA into large fragments (≈30 kb) after 3 h and chromatin condensation after 6 h. Both activities were caspase-independent as they were insensitive to caspase inhibitors. The addition of purified *A. thaliana* cytochrome *c* did not result in any nuclear degradation, however, suggesting that cytochrome *c* release does not directly activate cytoplasmic executioner proteases in plants. On the other hand, due to its normal physiologic role as an electron shuttle carrier in the mitochondrial electron transport chain, its release would cause disruption to the electron transport chain and increased production of ROS. This could in turn amplify death signals via a feedback loop mechanism (Jabs 1999).

17.5.2 AIF and EndoG

Although cytochrome *c* does not seem to directly activate death-inducing proteases in plants, there is a possibility that functional homologues of either AIF or EndoG are a component of plant PCD. Balk and coworkers (2003) identified five homologues of AIF in the *A. thaliana* genome and demonstrated that one of these homologues was localized in the mitochondrial matrix. After purifying a recombinant version of this protein (minus the mitochondrial targeting sequence) the authors added it to purified *A. thaliana* nuclei and determined that it did not cause the DNA cleavage they had previously observed after the addition of broken mitochondria. At this time, it has not been determined whether an AIF-like molecule is involved in plant PCD. Balk and coworkers (2003) carried out further analysis examining submitochondrial fractions for DNase activities capable of cleaving a DNA substrate. The authors found that the IMS contained a strong Mg^{2+}-dependent nuclease activity, which seemed to directly initiate cleavage of DNA into large fragments and play a role in nuclear condensation. The localization and Mg^{2+}-dependency are reminiscent of EndoG, and the authors suggested that during eukaryotic evolution an endonuclease derived from the α-protobacterial ancestor of mitochondria was co-opted to participate in the regulation of plant PCD.

17.6 Reactive Oxygen Species and Alternative Oxidase

ROS are often produced as a result of perturbations to the finely tuned mitochondrial electron transport chain (see also Chap. 14). PCD-inducing agents can lead to such perturbations. One way this can occur is via the opening of the PTP, which allows the diffusion of water and low molecular weight compounds into the mitochondrion, leading to matrix swelling and OMM rupture. This causes a severe disruption to the electron transport chain and a high accumulation and release of ROS. The ROS can then facilitate the opening of PTP in other mitochondria, leading to the production of more ROS and the release of apoptogenic proteins. As we have already discussed, PCD-inducing agents can also induce the release of cytochrome *c* in plants (Balk et al. 2003), and while this does not seem to activate caspase-like death proteases it could lead to the production of increased amounts of ROS. Under normal circumstances the cell can eliminate superoxide, H_2O_2 and other damaging radicals by producing ROS scavengers such as catalase. However when the cell is under stress the levels of ROS produced can exceed the cell's ROS scavenging capacity. Furthermore, cytochrome *c* (and other metal containing proteins) can interact with the relatively mild superoxide and H_2O_2 radicals to form hydroxyl radicals, which are amongst the most reactive and mutagenic molecules known (Blackstone and Kirkwood 2003). Therefore cytochrome *c* release does not only lead to an increased generation of ROS, it can also increase the damage caused by ROS by converting them to hydroxyl radicals. Thus relatively minor perturbations to mitochondrial respiration can be augmented and result in an accelerated triggering of PCD (Bras et al. 2005).

In plants, PCD can be induced by the direct application of ROS (Solomon et al. 1999) and ROS scavengers can inhibit PCD (Vacca et al. 2004). Furthermore, plants lacking the ability to produce ROS scavengers (cytosolic ascorbate peroxidase or catalase) are more susceptible to abiotic and biotic stresses and develop necrotic lesions (reviewed by Jabs 1999). Other studies in plants have demonstrated that ROS can activate signaling pathways, which are important for the tolerance of biotic and abiotic stresses. In one such study, treatment of *A. thaliana* with H_2O_2 activated ANP1 (a specific mitogen-activated protein kinase (MAPK) kinase and a homologue of tobacco NPK1 (*Nicotiana* protein kinase 1)), which in turn initiated a phosphorylation cascade involving two stress MAPKs: AtMPK3 and AtMPK6 (Kovtun et al. 2000). Additionally, the authors demonstrated that the overexpression of NPK1 led to an enhanced tolerance to abiotic stresses such as freezing, heat shock and salt stress. Another study demonstrated the induction of a serine/threonine kinase (OXI1–oxidative signal-inducible 1) in response to a variety of H_2O_2-generating stimuli in *A. thaliana* (Rentel et al. 2004). These OXI1-inducing stimuli included H_2O_2 itself, wounding, cellulase treatment, as well as *Peronospora parasitica* infection and normal root growth. Mutants lacking OXI1 were much more susceptible to *P. parasitica* and also had impaired root growth. Furthermore, OXI1 was shown to be required for the normal activation of the aforementioned MAPKs: AtMPK3 and AtMPK6. These two proteins are involved in a MAPK cascade that is activated in response to the perception of bacterial flagellin by the receptor-like kinase FLS2, and which initiates the induction of defence responses that are effective against bacterial and fungal pathogens (Asai et al. 2002). More recently, OXI1 has also been linked to protection against *Pseudomonas syringae* infection (Petersen et al. 2009).

All of this data indicates that ROS signaling forms the basis of plant responses to a variety of stresses capable of inducing PCD. These include heat, salt stress, freezing, wounding and bacterial as well as fungal pathogen attack. It seems plausible then that ROS may function as signallers in the effector phase of plant PCD. Whatever the case may be, it is unlikely that ROS can execute PCD directly, as *de novo* transcription and translation are required for PCD in which ROS are induced (Solomon et al. 1999; Vacca et al. 2004).

On the other side of the coin is the IMM-localized enzyme alternative oxidase (AOX), which is present in plant, but not animal mitochondria (see also Chap. 14). AOX catalyses the flow of electrons from ubiquinol directly to oxygen, thereby generating a shunt that circumvents complexes III and IV of the electron transport chain resulting in a cyanide insensitive electron transfer pathway (Lam et al. 2001). Many studies implicate AOX involvement in plant PCD. Treatment of tobacco cells with cysteine triggered a signaling pathway, culminating in a significant loss in the capacity of the mitochondrial cytochrome pathway but also a significant up-regulation of AOX levels and activity (Vanlerberghe et al. 2002). Transgenic AS8 cells, which lacked the ability to induce AOX, lost all respiratory capacity and underwent PCD following cysteine treatment. In a separate study, AS8 tobacco cells showed increased susceptibility to salicylic acid, H_2O_2, and the protein phosphatase inhibitor cantharidin, and underwent PCD, which was characterized by DNA cleavage (Robson and Vanlerberghe 2002). The authors suggested that the ability of AOX to inhibit PCD may relate to its capacity to sustain mitochondrial respiratory function after treatment with the toxic agent; however, it may also relate to its ability to prevent ROS build-up within the mitochondria. Subsequently it was shown that AOX can influence the response of potato cells to β-glucan defence elicitors (Mizuno et al. 2005). The authors demonstrated that H_2O_2 accumulated in response to elicitor treatment, but that simultaneous inhibition of catalase and AOX activities resulted in a further increase in H_2O_2 accumulation, reduced $\Delta\psi_m$, and PCD. All of these data support the previously held view that AOX can act as a safety valve (Lam et al. 2001), regulating ROS generation in the plant mitochondrion during PCD. However, there are conflicting data; for example, overexpression of AOX in tobacco plants increases the plant's sensitivity to ozone (Pasqualini et al. 2007). It now appears that the role of AOX is more complex than first envisioned, as the absence of AOX can lead to changes in the plant transcriptome even under normal conditions (Giraud et al. 2008). These changes included increases in transcript abundance of genes involved in antioxidant defence. Therefore it seems that artificially altering AOX expression levels can have far-reaching and unexpected effects on the plants response to different stresses.

17.7 Autophagy

Autophagy is primarily regarded as a pro-survival process in plants (Cacas and Diamond 2009). Using plants with deletions in the orthologues of *S. cerevisiae* ATG genes it has been demonstrated that the absence of the autophagy pathway leaves plants more sensitive to low nutrient conditions (Doelling et al. 2002; Hanaoka et al. 2002; Surpin et al. 2003; Yoshimoto et al. 2003; Xiong et al. 2005; Phillips et al. 2008). Mutant plants commonly experience an accelerated senescence and appear more sensitive to chronic oxidative stress (Doelling et al. 2002; Xiong et al. 2007b). Autophagy-deficiency also results in constitutive oxidative stress (Xiong et al. 2007a). It is known in animals that autophagy is required to remove oxidized proteins and damaged mitochondria (Zhang et al. 2007). If this holds true in plants, it would suggest that the reason autophagy-deficient plants undergo premature senescence is because of an inability to clean up toxic cellular components, such as damaged mitochondria.

Autophagy is also an important component of normal defence responses against invading pathogens (Liu et al. 2005; Patel et al. 2006; Patel and Dinesh-Kumar 2008;

Hofius et al. 2009; Yoshimoto et al. 2009). However, there are conflicting reports as to the actual role it plays in plant defence. For example, Patel and Dinesh-Kumar (2008) reported that autophagy-deficient (antisense At-*Atg6*) *A. thaliana* plants underwent an uncontrolled HR response after infection with avirulent *P. syringae* pv. *tomato* (*Pto*) DC3000 (*AvrRpm1*), where the lesion that normally remains localized to the point of infection, instead spread throughout the leaf and killed it. Similar results were previously observed in tobacco plants deficient in Nt-atg6/Beclin1 or Nt-atg7 (after VIGS) following infection with tobacco mosaic virus (Liu et al. 2005). This would suggest that autophagy is regulating the PCD that occurs during HR by limiting its spread. However, more recently Hofius and coworkers have demonstrated that autophagy can also be a positive regulator of PCD during the HR (Hofius et al. 2009). Confusingly, infection of autophagy-deficient *A. thaliana* knockout plants with avirulent *Pto* DC3000 (*AvrRpm1*) did not lead to an uncontrolled HR lesion spread. In fact in this instance, cell death in autophagy-deficient mutants was actually reduced. Yoshimoto and coworkers suggest that the differences between these studies may relate to differences in pathogen-treated plant or leaf age (Yoshimoto et al. 2009). Whatever the case may be, it seems that in plants, as in animals, autophagy can act in a protective capacity, but can also function as a form of PCD.

Oxidative stress is also known to activate autophagy (Xiong et al. 2007b). Until recently it was thought that autophagy functioned purely in a pro-survival capacity. However, it has also been demonstrated that the direct application of H_2O_2 can induce autophagic cell death in plants (Diamond and Lawton, unpublished data). As mitochondria are one of the main locations in plant cells for the production of ROS, it is possible that plant mitochondria could directly initiate autophagic cell death. In the future it will be interesting to test the effects of factors that inhibit MPT in plants (such as CsA) on the induction of autophagic cell death. This will help to determine what role, if any, the mitochondrion plays in plant autophagic cell death.

17.8 Concluding Remarks

It seems probable that the mitochondrion is a crucial mediator of PCD in most, or possibly all, eukaryotic cells. However, it is also clear that the pathways by which it orchestrates PCD have diverged in plants and animals. While research has uncovered similarities between plant and animal PCD systems, the data suggests there are also key differences. Although there are proteases with caspase-like activities in plants, their method of activation is unknown, and none of the proteases identified to date have yet been shown to have a widespread role in plant PCD. Furthermore, to date there have been no members of the Bcl-2 family, or even functional homologues, identified in plants. Although apoptotic-like PCD characteristics are evident in many forms of plant PCD it would seem that the major plant PCD pathway(s) do differ from apoptosis. However, there does seem to be commonality in regulation of cell death by the mitochondria in plants and this seems to involve mitochondrial

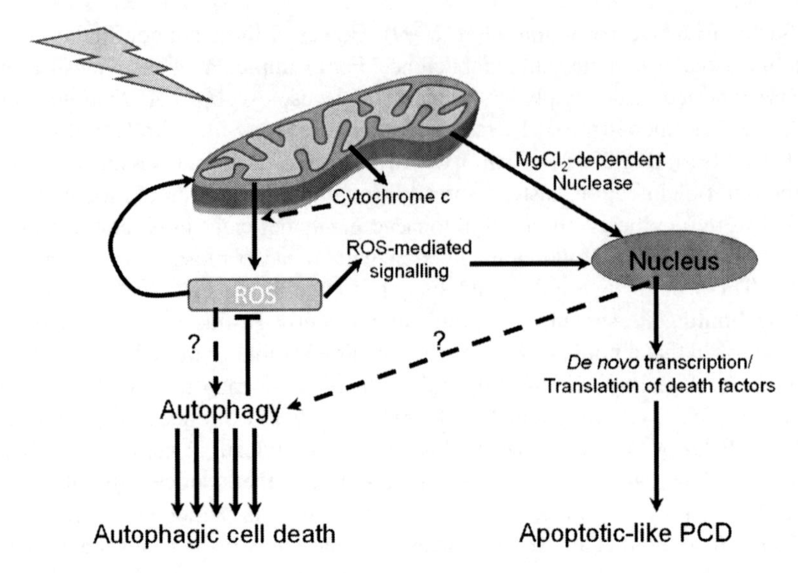

Fig. 17.3 A model for mitochondrial induction of plant programmed cell death (*PCD*). Death signal perception can lead to the release of factors from the mitochondria. In plants these factors include cytochrome *c*, reactive oxygen species (*ROS*) and an $MgCl_2$-dependent nuclease. The nuclease can cleave DNA into large (≈30 kb) fragments possibly leading to apoptotic-like PCD. Cytochrome *c* release causes a disturbance in mitochondrial respiration and leads to the production of more ROS. Mitochondrial ROS release may lead to ROS-mediated PCD signaling. However, ROS cannot induce apoptotic-like PCD directly and requires *de novo* transcription and translation of death factors. ROS release also results in oxidative damage to the cell, which is known to induce autophagy. If an excess of autophagy is induced, then autophagic cell death can occur

production of ROS (Fig. 17.3), which may either: induce autophagic cell death or signal to the nucleus and induce apoptotic-like PCD with a dependence on *de novo* transcription or translation of death factors.

Glossary

Apoptosis: A caspase-dependent form of programmed cell death in animal cells.

Apoptotic-like programmed cell death: A regulated form of cell death in plants that results in a characteristic cell morphology caused by cell shrinkage.

Autophagy: "Self-eating" – lysosome or vacuole-dependent processing and recycling of macromolecules.

Intermembrane space (IMS): The intermembrane space (space between outer and inner mitochondrial membranes) contains a number of proteins that can be released to activate cell death cascades.

Mitochondrial permeability transition (MPT): An increase in the permeability of the mitochondrial membranes that can lead to collapse of mitochondrial membrane potential, production of ROS, termination of ATP synthesis, and release of matrix solutes. MPT results from opening of mitochondrial permeability transition pores.

Necrosis: An unregulated form of cell death – often as a result of severe stress.

Outer mitochondrial membrane (OMM): Localization site for a number of PCD controlling proteins.

Permeability transition pore (PTP): A polyprotein complex formed at contact sites between the mitochondrial inner and outer membranes.

Programmed cell death (PCD): The process of organized destruction of the cell.

Reactive oxygen species (ROS): Any of a number of highly reactive forms of oxygen. At low levels, ROS may function in cell signaling processes. At higher levels, ROS will damage cellular macromolecules.

Tracheary element (TE): Xylem cells, which die at maturity, leaving their lignified remains, which conduct water and minerals throughout plant.

References

Arpagaus, S., Rawyler, A., Braendle, R. 2002. Occurrence and characteristics of the mitochondrial permeability transition in plants. J. Biol. Chem. 277:1780–1787.

Asai, T., Tena, G., Plotnikova, J., Willmann, M. R., Chiu, W. L., Gomez-Gomez, L., Boller, T., Ausubel, F. M., Sheen, J. 2002. MAP kinase signalling cascade in *Arabidopsis* innate immunity. Nature 415:977–983.

Azad, M. B., Chen, Y, Henson, E. S, Cizeau, J., McMillan-Ward, E., Israels, S. J., Gibson, SB. 2008. Hypoxia induces autophagic cell death in apoptosis-competent cells through a mechanism involving BNIP3. Autophagy 4:195–204.

Baines, C. P., Kaiser, R. A., Purcell, N. H., Blair, N. S., Osinska, H., Hambleton, M. A., Brunskill, E. W., Sayen, M. R., Gottlieb, R. A., Dorn, G. W., Robbins, J., Molkentin, J. D. 2005. Loss of cyclophilin D reveals a critical role for mitochondrial permeability transition in cell death. Nature 434:658–662.

Balk, J., Leaver, C. J. 2001. The PET1-CMS mitochondrial mutation in sunflower is associated with premature programmed cell death and cytochrome *c* release. Plant Cell 13:1803–1818.

Balk, J., Leaver, C. J., McCabe, P. F. 1999. Translocation of cytochrome *c* from the mitochondria to the cytosol occurs during heat-induced programmed cell death in cucumber plants. FEBS Lett. 463:151–154.

Balk, J., Chew, S. K., Leaver, C. J., McCabe, P. F. 2003. The intermembrane space of plant mitochondria contains a DNase activity that may be involved in programmed cell death. Plant J. 34:573–583.

Bassham, D. C. 2007. Plant autophagy-more than a starvation response. Curr. Opin. Plant Biol. 10:87–593.

Basso, E., Fante, L., Fowlkes, J., Petronilli, V., Forte, M. A., Bernardi, P. 2005. Properties of the permeability transition pore in mitochondria devoid of cyclophilin D. J. Biol. Chem. 280:18558–18561.

Berry, D. L., Baehrecke, E. H. 2007. Growth arrest and autophagy are required for salivary gland cell degradation in Drosophila. Cell 131:1137–1148.

Blackstone, N. W., Kirkwood, T. B. L. 2003. Mitochondria and programmed cell death: "slave revolt" or community homeostasis? In Genetic and Cultural Evolution of Cooperation, ed. P. Hammerstein. pp. 309–325. Cambridge: The MIT Press.

Bras, M., Queenan, B., Susin, S. A. 2005. Programmed cell death via mitochondria: different modes of dying. Biochem.-Moscow 70:231–239.

Cacas, J.-L. 2010. Devil Inside: does plant programmed cell death involve the endomembrane system? Plant Cell Environ. 33:1453–1473.

Cacas, J.-L., Diamond, M. 2009. Is the autophagy machinery an executioner of programmed cell death in plants? Trends Plant Sci. 14:299–300.

Cande, C., Vahsen, N., Garrido, C., Kroemer, G. 2004. Apoptosis-inducing factor (AIF): caspase-independent after all. Cell Death Differ. 11:591–595.

Casolo, V., Petrussa, E., Krajnakova, J., Macri, F., Vianello, A. 2005. Involvement of the mitochondrial K-ATP(+) channel in H_2O_2- or NO-induced programmed death of soybean suspension cell cultures. J. Exp. Bot. 56:997–1006.

Chen, S. R., Dickman, M. B. 2004. Bcl-2 family members localize to tobacco chloroplasts and inhibit programmed cell death induced by chloroplast-targeted herbicides. J. Exp. Bot. 55:2617–2623.

Chen, G., Ray, R., Dubik, D., Shi, L. F., Cizeau, J., Bleackley, R. C., Saxena, S., Gietz, R. D., Greenberg, A. H. 1997. The E1B 19 K Bcl-2-binding protein Nip3 is a dimeric mitochondrial protein that activates apoptosis. J. Exp. Med. 186:1975–1983.

Chen, Y., McMillan-Ward, E., Kong, J., Israels, S. J., Gibson, S. B. 2008. Oxidative stress induces autophagic cell death independent of apoptosis in transformed and cancer cells. Cell Death Differ. 15:171–182.

Chinnadurai G., Vijayalingam S., Gibson S. B. 2008. BNIP3 subfamily BH3-only proteins: mitochondrial stress sensors in normal and pathologic functions. Oncogene 27:S114–S127.

Clarke, P. G. H. 1990. Developmental cell death – morphological diversity and multiple mechanisms. Anat. Embryol. 181:195–213.

Contran, N., Cerana, R., Crosti, P., Malerba, M. 2007. Cyclosporin A inhibits programmed cell death and cytochrome *c* release induced by fusicoccin in sycamore cells. Protoplasma 231:193–199.

Crompton, M. 1999. The mitochondrial permeability transition pore and its role in cell death. Biochem. J. 341:233–249.

Crosti, P., Malerba, M., Bianchetti, R. 2001. Tunicamycin and Brefeldin A induce in plant cells a programmed cell death showing apoptotic features. Protoplasma 216:31–38.

Curtis, M. J., Wolpert, T. J. 2004. The victorin-induced mitochondrial permeability transition precedes cell shrinkage and biochemical markers of cell death, and shrinkage occurs without loss of membrane integrity. Plant J. 38:244–259.

Diamond, M., McCabe, P. F. 2007. The mitochondrion and plant programmed cell death. In Annual Plant Reviews: Plant Mitochondria, ed. D. C. Logan. pp. 308–334. Oxford: Blackwell.

Doelling, J. H., Walker, J. M., Friedman, E. M., Thompson, A. R., Vierstra, R. D. 2002. The APG8/12-activating enzyme APG7 is required for proper nutrient recycling and senescence in *Arabidopsis thaliana*. J. Biol. Chem. 277:33105–33114.

Duval, I., Brochu, V., Simard, M., Beaulieu, C., Beaudoin, N. 2005. Thaxtomin A induces programmed cell death in *Arabidopsis thaliana* suspension-cultured cells. Planta 222:820–831.

Ellis, H. M., Horvitz, H. R. 1986. Genetic-control of programmed cell-death in the nematode *C. elegans*. Cell 44:817–829.

Elmore, S. P., Qian, T., Grissom, S. F., Lemasters, J. J. 2001. The mitochondrial permeability transition initiates autophagy in rat hepatocytes. FASEB J. 15:2286–2287.

Forte, M., Bernardi, P. 2005. Genetic dissection of the permeability transition pore. J. Bioenerg. Biomembr. 37:121–128.

Fukuda, H. 2000. Programmed cell death of tracheary elements as a paradigm in plants. Plant Mol. Biol. 44:245–253.

Geitmann, A., Franklin-Tong, V. E., Emons, A. C. 2004. The self-incompatibility response in *Papaver rhoeas* pollen causes early and striking alterations to organelles. Cell Death Differ. 11:812–822.

Geng, J. F., Klionsky, D. J. 2008. The Atg8 and Atg12 ubiquitin-like conjugation systems in macroautophagy. EMBO Rep. 9:859–864.

Giraud, E., Ho, L. H. M., Clifton, R., Carroll, A., Estavillo, G., Tan, Y. F., Howell, K. A., Ivanova, A., Pogson, B. J., Millar, A. H., Whelan, J. 2008. The absence of ALTERNATIVE OXIDASE1a in Arabidopsis results in acute sensitivity to combined light and drought stress. Plant Physiol. 147:595–610.

Goodsell, D. S. 2004. The molecular perspective: Cytochrome *c* and apoptosis. Oncologist 9:226–227.

Gozuacik, D., Bialik, S., Raveh, T., Mitou, G., Shohat, G., Sabanay, H., Mizushima, N., Yoshimori, T., Kimchi, A. 2008. DAP-kinase is a mediator of endoplasmic reticulum stress-induced caspase activation and autophagic cell death. Cell Death Differ. 15:1875–1886.

Green, D. R., Reed, J. C. 1998. Mitochondria and apoptosis. Science 281:1309–1312.

Greenberg, J. T., Yao, N. 2004. The role and regulation of programmed cell death in plant-pathogen interactions. Cell. Microbiol. 6:201–211.

Hall, D. H., Gu, G., Garcia-Anoveros, J., Gong, L., Chalfie, M., Driscoll, M. 1997. Neuropathology of degenerative cell death in *Caenorhabditis elegans*. J. Neurosci. 17:1033–1045.

Hanaoka, H., Noda, T., Shirano, Y., Kato, T., Hayashi, H., Shibata, D., Tabata, S., Ohsumi, Y. 2002. Leaf senescence and starvation-induced chlorosis are accelerated by the disruption of an Arabidopsis autophagy gene. Plant Physiol. 129:1181–1193.

Hauser, B. A., Sun, K., Oppenheimer, D. G., Sage, T. L. 2006. Changes in mitochondrial membrane potential and accumulation of reactive oxygen species precede ultrastructural changes during ovule abortion. Planta 223:492–499.

Heath, M. C. 2000. Hypersensitive response-related death. Plant Mol. Biol. 44:321–334.

Hofius, D., Tsitsigiannis, D. I., Jones, J. D. G., Mundy, J. 2007. Inducible cell death in plant immunity. Semin. Cancer Biol. 17:166–187.

Hofius, D., Schultz-Larsen, T., Joensen, J., Tsitsigiannis, D. I., Petersen, N. H. T., Mattsson, O., Jorgensen, L. B., Jones, J. D. G., Mundy, J., Petersen, M. 2009. Autophagic components contribute to hypersensitive cell death in Arabidopsis. Cell 137:773–783.

Imazu, T., Shimizu, S., Tagami, S., Matsushima, M., Nakamura, Y., Miki, T., Okuyama, A., Tsujimoto, Y. 1999. Bcl-2/E1B 19 kDa-interacting protein 3-like protein (Bnip3L) interacts with Bcl-2/Bcl-x(L) and induces apoptosis by altering mitochondrial membrane permeability. Oncogene 18(32):4523–4529.

Jabs, T. 1999. Reactive oxygen intermediates as mediators of programmed cell death in plants and animals. Biochem. Pharmacol. 57:231–245.

Jones, A. 2000. Does the plant mitochondrion integrate cellular stress and regulate programmed cell death? Trends Plant Sci. 5:225–230.

Kang, C. H., Avery, L. 2008. To be or not to be, the level of autophagy is the question. Autophagy 4:82–84.

Kang, C., You, Y. J., Avery, L. 2007. Dual roles of autophagy in the survival of *Caenorhabditis elegans* during starvation. Genes Dev. 21:2161–2171.

Kawai-Yamada, M., Jin, L., Yoshinaga, K., Hirata, A., Uchimiya, H. 2001. Mammalian Bax-induced plant cell death can be downregulated by overexpression of *Arabidopsis* Bax inhibitor-1 (AtBI-1). Proc. Natl. Acad. Sci. USA 98:12295–12300.

Kawai-Yamada, M., Ohori, Y., Uchimiya, H. 2004. Dissection of Arabidopsis Bax inhibitor-1 suppressing Bax-, hydrogen peroxide-, and salicylic acid-induced cell death. Plant Cell 16:21–32.

Kerr, J. F. R., Wyllie, A. H., Currie, A. R. 1972. Apoptosis: a basic biological phenomenon with wide-ranging implications in tissue kinetics. Br. J. Cancer 26:239–257.

Kim, M., Lim, J. H., Ahn, C. S., Park, K., Kim, G. T., Kim, W. T., Pai, H. S. 2006. Mitochondria-associated hexokinases play a role in the control of programmed cell death in *Nicotiana benthamiana*. Plant Cell 18:2341–2355.

Klionsky, D. J., Cregg, J. M., Dunn, W. A., Emr, S. D., Sakai, Y., Sandoval, I. V., Sibirny, A., Subramani, S., Thumm, M., Veenhuis, M., Ohsumi, Y. 2003. A unified nomenclature for yeast autophagy-related genes. Dev. Cell 5:539–545.

Knaapen, M. W. M., Davies, M..J., De Bie, M., Haven, A. J., Martinet, W., Kockx, M. M. 2001. Apoptotic versus autophagic cell death in heart failure. Cardiovasc. Res. 51:304–312.

Kokoszka, J. E., Waymire, K. G., Levy, S. E., Sligh, J. E., Cal, J. Y., Jones, D. P., MacGregor, G. R., Wallace, D. C. 2004. The ADP/ATP translocator is not essential for the mitochondrial permeability transition pore. Nature 427:461–465.

Kovtun, Y., Chiu, W. L., Tena, G., Sheen, J. 2000. Functional analysis of oxidative stress-activated mitogen-activated protein kinase cascade in plants. Proc. Natl. Acad. Sci. USA 97:2940–2945.

Kroemer, G., Levine, B. 2008. Autophagic cell death: the story of a misnomer. Nat. Rev. Mol. Cell Biol. 9:1004–1010.

Kroemer, G., Martin, S. J. 2005. Caspase-independent cell death. Nat. Med. 11:725–730.

Kroemer, G., Galluzzi, L., Brenner, C. 2007. Mitochondrial membrane permeabilization in cell death. Physiol. Rev. 87:99–163.

Kusano, T., Tateda, C., Berberich, T., Takahashi, Y. 2009. Voltage-dependent anion channels: their roles in plant defense and cell death. Plant Cell Rep. 28:1301–1308.

Kuwana, T., Mackey, M. R., Perkins, G., Ellisman, M. H., Latterich, M., Schneiter, R., Green, D. R., Newmeyer, D.D. 2002. Bid, Bax, and lipids cooperate to form supramolecular openings in the outer mitochondrial membrane. Cell 111:331–342.

Lacomme, C., Cruz, S. S. 1999. Bax-induced cell death in tobacco is similar to the hypersensitive response. Proc. Natl. Acad. Sci. USA 96:7956–7961.

Lam, E., Kato, N., Lawton, M. 2001. Programmed cell death, mitochondria and the plant hypersensitive response. Nature 411:848–853.

Leist, M., Jaattela, M. 2001. Four deaths and a funeral: from caspases to alternative mechanisms. Nat. Rev. Mol. Cell Biol. 2:589–598.

Letai, A., Bassik, M. C., Walensky, L. D., Sorcinelli, M. D., Weiler, S., Korsmeyer, S. J. 2002. Distinct BH3 domains either sensitize or activate mitochondrial apoptosis, serving as prototype cancer therapeutics. Cancer Cell 2:183–192.

Levine, B., Yuan, J. Y. 2005. Autophagy in cell death: an innocent convict? J. Clin. Invest. 115:2679–2688.

Levine, A., Pennell, R. I., Alvarez, M. E., Palmer, R., Lamb, C. 1996. Calcium-mediated apoptosis in a plant hypersensitive disease resistance response. Curr. Biol. 6:427–437.

Li, L.Y., Luo, L., Wang, X. D. 2001. Endonuclease G is an apoptotic DNase when released from mitochondria. Nature 412:95–99.

Liang, X. H., Kleeman, L. K., Jiang, H. H., Gordon, G., Goldman, J. E., Berry, G., Herman, B., Levine, B. 1998. Protection against fatal Sindbis virus encephalitis by Beclin, a novel Bcl-2-interacting protein. J. Virol. 72:8586–8596.

Liu, Y., Schiff, M., Czymmek, K., Talloczy, Z., Levine, B., Dinesh-Kumar, S. P. 2005. Autophagy regulates programmed cell death during the plant innate immune response. Cell 121:567–577.

Luo, X., Budihardjo, I., Zou, H., Slaughter, C., Wang, X. D. 1998. Bid, a Bcl2 interacting protein, mediates cytochrome *c* release from mitochondria in response to activation of cell surface death receptors. Cell 94:481–490.

Luo S. M., Chen, Q., Cebollero, E., Xing, D. 2009. Mitochondria: one of the origins for autophagosomal membranes? Mitochondrion 9:227–231.

Malerba, M., Cerana, R., Crosti, P. 2004. Comparison between the effects of fusicoccin, Tunicamycin, and Brefeldin A on programmed cell death of cultured sycamore (*Acer pseudoplatanus* L.) cells. Protoplasma 224:61–70.

McCabe, P. F., Leaver, C. J. 2000. Programmed cell death in cell cultures. Plant Mol. Biol. 44:359–368.

McCabe, P. F., Valentine, T. A., Forsberg, L. S., Pennell, R. I. 1997a. Soluble signals from cells identified at the cell wall establish a developmental pathway in carrot. Plant Cell 9:2225–2241.

McCabe, P. F., Levine, A., Meijer, P. J., Tapon, N. A., Pennell, R. I. 1997b. A programmed cell death pathway activated in carrot cells cultured at low cell density. Plant J. 12:267–280.

Mitsuhara, I., Malik, K. A., Miura, M., Ohashi, Y. 1999. Animal cell-death suppressors Bcl-x(L) and Ced-9 inhibit cell death in tobacco plants. Curr. Biol. 9:775–778.

Mizuno, M., Tada, Y., Uchii, K., Kawakami, S., Mayama, S. 2005. Catalase and alternative oxidase cooperatively regulate programmed cell death induced by beta-glucan elicitor in potato suspension cultures. Planta 220:849–853.

Nakagawa, T., Shimizu, S., Watanabe, T., Yamaguchi, O., Otsu, K., Yamagata, H., Inohara, H., Kubo, T., Tsujimoto, Y. 2005. Cyclophilin D-dependent mitochondrial permeability transition regulates some necrotic but not apoptotic cell death. Nature 434:652–658.

Palmieri, L., Santoro, A., Carrari, F., Blanco, E., Nunes-Nesi, A., Arrigoni, R., Genchi, F., Fernie, A. R., Palmieri, F. 2008. Identification and characterization of adnt1, a novel mitochondrial adenine nucleotide transporter from Arabidopsis. Plant Physiol. 148:1797–1808.

Pasqualini, S., Paolocci, F., Borgogni, A., Morettini, R., Ederli, L. 2007. The overexpression of an alternative oxidase gene triggers ozone sensitivity in tobacco plants. Plant Cell Environ. 30:1545–1556.

Patel, S., Dinesh-Kumar, S. P. 2008. Arabidopsis ATG6 is required to limit the pathogen-associated cell death response. Autophagy 4:20–27.

Patel, S., Caplan, J., Dinesh-Kumar, S. P. 2006. Autophagy in the control of programmed cell death. Curr. Opin. Plant Biol. 9:391–396.

Pattingre, S., Tassa, A., Qu, X. P., Garuti, R., Liang, X. H., Mizushima, N., Packer, M., Schneider, M. D., Levine, B. 2005. Bcl-2 antiapoptotic proteins inhibit Beclin 1-dependent autophagy. Cell 122:927–939.

Petersen, L. N., Ingle, R. A., Knight, M. R., Denby, K. J. 2009. OXI1 protein kinase is required for plant immunity against *Pseudomonas syringae* in Arabidopsis. J. Exp. Bot. 60:3727–3735.

Phillips, A. R., Suttangkakul, A., Vierstra, R. D. 2008. The ATG12-conjugating enzyme ATG10 is essential for autophagic vesicle formation in *Arabidopsis thaliana*. Genetics 178:1339–1353.

Reape, T. J., McCabe, P. F. 2008. Apoptotic-like programmed cell death in plants. New Phytol. 180:13–26.

Rentel, M. C., Lecourieux, D., Ouaked, F., Usher, S. L., Petersen, L., Okamoto, H., Knight, H., Peck, S. C., Grierson, C. S., Hirt, H., Knight, M. R. 2004. OXI1 kinase is necessary for oxidative burst-mediated signalling in Arabidopsis. Nature 427:858–861.

Robson, C. A., Vanlerberghe, G. C. 2002. Transgenic plant cells lacking mitochondrial alternative oxidase have increased susceptibility to mitochondria-dependent and -independent pathways of programmed cell death. Plant Physiol. 129:1908–1920.

Saviani, E. E., Orsi, C. H., Oliveira, J. F. P., Pinto-Maglio, C. A. F., Salgado, I. 2002. Participation of the mitochondrial permeability transition pore in nitric oxide-induced plant cell death. FEBS Lett. 510:136–140.

Scaffidi, C., Fulda, S., Srinivasan, A., Friesen, C., Li, F., Tomaselli, K. J., Debatin, K. M., Krammer, P. H., Peter, M. E. 1998. Two CD95 (APO-1/Fas) signaling pathways. EMBO J. 17:1675–1687.

Schimmer, A. D. 2004. Inhibitor of apoptosis proteins: translating basic knowledge into clinical practice. Cancer Res. 64:7183–7190.

Singh, R., Kaushik, S., Wang, Y., Xiang, Y., Novak, I., Komatsu, M., Tanaka, K., Cuervo, A. M., Czaja, M. J. 2009. Autophagy regulates lipid metabolism. Nature 458:1131–1135.

Solomon, M., Belenghi, B., Delledonne, M., Menachem, E., Levine, A. 1999. The involvement of cysteine proteases and protease inhibitor genes in the regulation of programmed cell death in plants. Plant Cell 11:431–443.

Stein, J. C., Hansen, G. 1999. Mannose induces an endonuclease responsible for DNA laddering in plant cells. Plant Physiol. 121:71–79.

Sun, Y. L., Zhao, Y., Hong, X., Zhai, Z. H. 1999. Cytochrome *c* release and caspase activation during menadione-induced apoptosis in plants. FEBS Lett. 462:317–321.

Surpin, M., Zheng, H. J., Morita, M. T., Saito, C., Avila, E., Blakeslee, J. J., Bandyopadhyay, A., Kovaleva, V., Carter, D., Murphy, A., Tasaka, M., Raikhel, N. 2003. The VTI family of SNARE proteins is necessary for plant viability and mediates different protein transport pathways. Plant Cell 15:2885–2899.

Sweetlove, L. J., Fait, A., Nunes-Nesi, A., Williams, T., Fernie, A. R. 2007. The mitochondrion: an integration point of cellular metabolism and signalling. Crit. Rev. Plant Sci. 26:17–43.

Tateda, C., Yamashita, K., Takahashi, F., Kusano, T., Takahashi, Y. 2009. Plant voltage-dependent anion channels are involved in host defense against *Pseudomonas cichorii* and in Bax-induced cell death. Plant Cell Rep. 28:41–51.

Thomas, S. G., Franklin-Tong, V. E. 2004. Self-incompatibility triggers programmed cell death in Papaver pollen. Nature 429:305–309.

Thompson, C. B. 1995. Apoptosis in the pathogenesis and treatment of disease. Science 267: 1456–1462.

Thompson, A. R., Vierstra, R. D. 2005. Autophagic recycling: lessons from yeast help define the process in plants. Curr. Opin. Plant Biol. 8:165–173.

Tiwari, B. S., Belenghi, B., Levine, A. 2002. Oxidative stress increased respiration and generation of reactive oxygen species, resulting in ATP depletion, opening of mitochondrial permeability transition, and programmed cell death. Plant Physiol. 128:1271–1281.

Townley, H. E., McDonald, K., Jenkins, G. I., Knight, M. R., Leaver, C.J. 2005. Ceramides induce programmed cell death in Arabidopsis cells in a calcium-dependent manner. Biol. Chem. 386:161–166.

Vacca, R. A., de Pinto, M. C., Valenti, D., Passarella, S., Marra, E., De Gara, L. 2004. Production of reactive oxygen species, alteration of cytosolic ascorbate peroxidase, and impairment of mitochondrial metabolism are early events in heat shock-induced programmed cell death in tobacco bright-yellow 2 cells. Plant Physiol. 134:1100–1112.

Vacca, R. A., Valenti, D., Bobba, A., Merafina, R. S., Passarella, S., Marra, E. 2006. Cytochrome *c* is released in a reactive oxygen species-dependent manner and is degraded via caspase-like proteases in tobacco bright-yellow 2 cells en route to heat shock-induced cell death. Plant Physiol. 141:208–219.

Vahsen, N., Cande, C., Briere, J., Benit, P., Rustin, P., Kroemer, G. 2004. The absence of apoptosis-inducing factor AIF induces complex I deficiency. Biochim. Biophys. Acta-Bioenerg. 1657:83–83.

Vande Velde, C., Cizeau, J., Dubik, D., Alimonti, J., Brown, T., Israels, S., Hakem, R., Greenberg, A. H. 2000. BNIP3 and genetic control of necrosis-like cell death through the mitochondrial permeability transition pore. Mol. Cell. Biol. 20:5454–5468.

Vanlerberghe, G. C., Robson, C. A., Yip, J. Y. H. 2002. Induction of mitochondrial alternative oxidase in response to a cell signal pathway down-regulating the cytochrome pathway prevents programmed cell death. Plant Physiol. 129:1829–1842.

Walter, L., Hajnoczky, G. 2005. Mitochondria and endoplasmic reticulum: the lethal interorganelle cross-talk. J. Bioenerg. Biomembr. 37:191–206.

Wheeler, M. J., de Graaf, B. H. J., Hadjiosif, N., Perry, R. M., Poulter, N. S., Osman, K., Vatovec, S., Harper, A., Franklin, F. C. H., Franklin-Tong, V. E. 2009. Identification of the pollen self-incompatibility determinant in *Papaver rhoeas*. Nature 459:992–995.

Xiong, Y., Contento, A. L., Bassham, D. C. 2005. AtATG18a is required for the formation of autophagosomes during nutrient stress and senescence in *Arabidopsis thaliana*. Plant J. 42: 535–546.

Xiong, Y., Contento, A. L., Bassham, D. 2007a. Disruption of autophagy results in constitutive oxidative stress in Arabidopsis. Autophagy 3:257–258.

Xiong, Y., Contento, A. L., Nguyen, P. Q., Bassham, D. C. 2007b. Degradation of oxidized proteins by autophagy during oxidative stress in Arabidopsis. Plant Physiol. 143:291–299.

Xu, Y., Hanson, M. R. 2000. Programmed cell death during pollination-induced petal senescence in petunia. Plant Physiol. 122:1323–1333.

Yao, N., Imai, S., Tada, Y., Nakayashiki, H., Tosa, Y., Park, P., Mayama, S. 2002. Apoptotic cell death is a common response to pathogen attack in oats. Mol. Plant-Microbe Interact. 15:1000–1007.

Yao, N., Eisfelder, B. J., Marvin, J., Greenberg, J. T. 2004. The mitochondrion – an organelle commonly involved in programmed cell death in *Arabidopsis thaliana*. Plant J. 40:596–610.

Yasuda, M., D'Sa-Eipper, C., Gong, X. L., Chinnadurai, G. 1998. Regulation of apoptosis by a *Caenorhabditis elegans* BNIP3 homolog. Oncogene 17:2525–2530.

Yoshimoto, K., Hanaoka, H., Noda, T., Sato, S., Kato, T., Tabata, S., Ohsumi, Y. 2003. Analysis of AtAPG8 families in plant autophagy. Plant Cell Physiol. 44:S202.

Yoshimoto, K., Jikumaru, Y., Kamiya, Y., Kusano, M., Consonni, C., Panstruga, R., Ohsumi, Y., Shirasu, K. 2009. Autophagy negatively regulates cell death by controlling npr1-dependent salicylic acid signaling during senescence and the innate immune response in *Arabidopsis*. Plant Cell 21:2914–2927.

Yu, X. H., Perdue, T. D., Heimer, Y. M., Jones, A. M. 2002. Mitochondrial involvement in tracheary element programmed cell death. Cell Death Differ. 9:189–198.

Zhang, Y., Qi, H. Y., Taylor, R., Xu, W. H., Liu, L. F., Jin, S. K. 2007. The role of autophagy in mitochondria maintenance: characterization of mitochondrial functions in autophagy-deficient *S. cerevisiae* strains. Autophagy 3:337–346.

Zuzarte-Luis, V., Hurle, J. M. 2002. Programmed cell death in the developing limb. Int. J. Dev. Biol. 46:871–876.

Part V
Mitochondrial Dysfunction & Repair

Chapter 18
Cytoplasmic Male-Sterility and Nuclear Encoded Fertility Restoration

Jenny Carlsson and Kristina Glimelius

Abstract Cytoplasmic male-sterility (CMS) is caused by mutations, rearrangements, or recombinations in the mitochondrial genome resulting in a CMS-inducing gene (or genes) leading to malfunctioning, but not mutated, nuclear genes causing abnormal or inhibited development of stamens or pollen. This trait is inherited via the mitochondria, i.e., CMS is by large maternally inherited. CMS has been observed in at least 150 different species, here exemplified by, e.g., the *Zea mays* CMS-T, *Brassica napus* (Ogu-INRA), and *Nicotiana tabacum* CMS systems. The CMS-associated genes studied so far have been chimeric and affect the nuclear genome through retrograde signaling, resulting in a male-sterile phenotype ranging from pollen abortion to homeotically converted stamens into carpeloid or petaloid structures or the lack of stamens. Male-fertility can be restored by nuclear encoded *Restorer-of-fertility* genes.

Keywords Cytoplasmic male sterility • Anther development • Maternal inheritance • Chimeric genes • Restorer genes

18.1 Introduction

Cytoplasmic male-sterility (CMS) is a category of male-sterility that is caused by the mitochondria and can be obtained by combining different nuclear and mitochondrial (mt) backgrounds. It is an interaction between the mitochondria and the nucleus that entails an erroneous development of stamens and flowers resulting in inhibited pollen production. Besides affecting development of the stamens, the mitochondrial signaling can also cause aberrations in other floral organs, reduced yield, modified starch content, or give rise to chlorotic leaves. Male-fertility can be restored by introducing nuclear encoded *Restorer-of-fertility* (*Rf*) genes into the CMS-line (Fig. 18.1).

K. Glimelius (✉)
Department of Plant Biology and Forest Genetics,
Swedish University of Agricultural Sciences, Uppsala BioCenter,
Box 7080, SE 750 07 Uppsala, Sweden
e-mail: Kristina.Glimelius@vbsg.slu.se

F. Kempken (ed.), *Plant Mitochondria*, Advances in Plant Biology 1,
DOI 10.1007/978-0-387-89781-3_18,

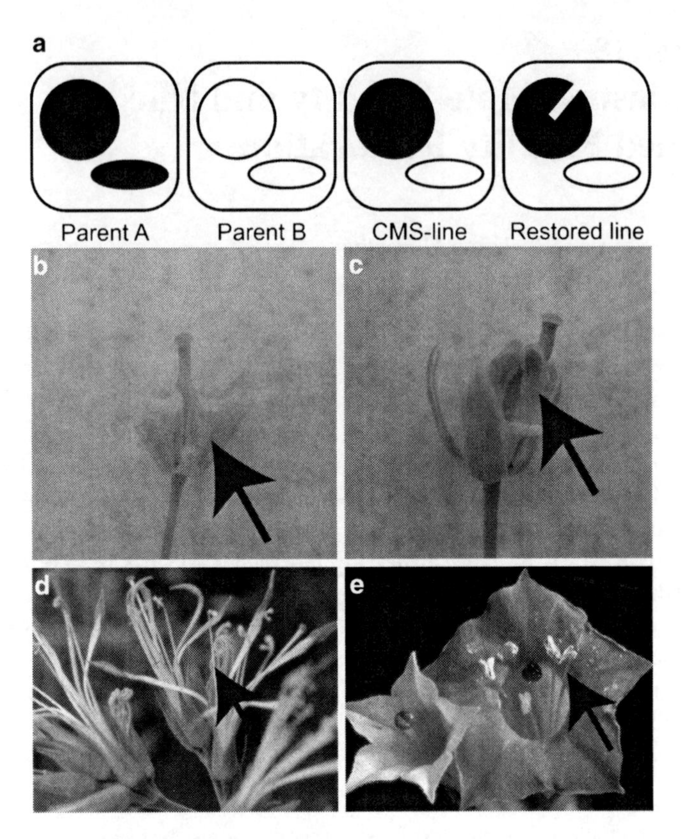

Fig. 18.1 (**a**) A model describing a complete CMS-system with parents, a CMS-line, and a restored line. The two parents A and B have distinct (*black or white*) nuclear (*circles*) and mitochondrial (*ellipses*) genomes. The two parents can originate from the same species or two different species. Parent A (sometimes called maintainer line) has donated the nuclear genome to the CMS line. The CMS line is usually backcrossed to this line. Parent B has contributed with the male-sterility inducing mitochondrial genome. The CMS line has been obtained by sexual crosses (alloplasmic hybrid) or protoplast fusions (somatic hybrid) and reoccurring backcrosses to parent A (to obtain a pure nuclear background). The restored line is isogeneic to the CMS line except for the addition of an *Rf* gene (*white line in the black nucleus*). The *Rf* gene usually originates from parent B and restores male-fertility. (**b–e**) Examples of CMS-flower phenotypes from *B. napus* (B, C) and *N. tabacum* (D, E). (**b**) An almost normal *B. napus* flower phenotype with shrunken stamens lacking pollen (*arrow*) is observed in Ogu-INRA, while *B. napus* (*A. thaliana*) line 4:19 (**c**) have carpeloid whorl three organs instead of stamens (*arrow*). The *N. tabacum* CMS lines display, e.g., filamentous stamens lacking proper anthers (*arrow*) and a split corolla (**d**) or short or missing stamens with white petals (**e**, *left flower*). The pink flower (**e**, *right flower*) is a wild-type *N. tabacum* flower with normal pollen-producing anthers (*arrow*)

CMS might also be one of the genetic mechanisms that could explain the phenomenon of natural occurring plant populations with co-existing females and hermaphrodites, so called gynodioecious populations (Budar and Berthomé 2007; Dufaÿ et al. 2007; McCauley and Olson 2008). To keep a gynodioecious population there must be a female advantage (a mechanism driving the development towards male-sterile individuals, e.g., CMS individuals). CMS-inducing genes could be part of such a regulation. In addition, a negative effect of male-sterility, e.g., CMS, on female fitness must exist (a mechanism keeping hermaphroditic individuals; (Dufaÿ et al. 2007). Similarly the *Rf*-genes could be involved in that regulation. To keep male-fertile and normal cytoplasms as well as male-sterile cytoplasms in the gynodioecious population, a cost of restoration should exist (Dufaÿ et al. 2007). If CMS gives a high female fitness (e.g., increased seed production compared to the hermaphroditic counterpart) the CMS-inducing gene will quickly be fixed in the population (McCauley and Olson 2008), and in the end the population will disappear due to the lack of pollen-parents. Based on the observation of paternal leakage of organelles (Azhagiri 2007; Elansary et al. 2010) it may be assumed that if a small percentage of such leakage is allowed, this will result in a polymorphic population. The conclusion when using different mathematical models was that paternal leakage of mitochondria could contribute to the evolutionary stability of gynodioecious systems (McCauley and Olson 2008).

CMS has mainly been studied and developed in crop species of large economical importance, such as *Zea mays*, *Beta vulgaris*, *Brassica napus*, *Sorghum bicolor*, *Triticum aestivum*, and *Helianthus annuus*. The reason behind this is that CMS is of clear importance for hybrid seed production (Havey 2004), since the utilization of CMS systems facilitates robust breeding systems for the production of hybrids for the agriculture. The relative lack of genomic and molecular information regarding crop species in comparison to model systems has, however, made it difficult to experimentally verify the genetic mechanisms underpinning CMS. Here we present what is known about the genetics and the mechanisms behind CMS and the effects of the mitochondrial influence on stamen development, and how they were obtained. However, due to the limited space several CMS-systems will be set aside.

18.2 Definition of Cytoplasmic Male-Sterility

CMS is caused by mutations, rearrangements, or recombinations in the mt-genome resulting in a CMS-inducing gene (or genes) leading to malfunctioning, but not mutated, nuclear genes (e.g., reduced expression of the *APETALA3* and *PISTILATA* homologues) causing abnormal or inhibited development of stamens or pollen (see Fig. 18.1). The CMS-associated genes studied so far have been chimeric. Mitochondria are in most cases maternally inherited; thus, CMS is also maternally inherited (Hanson and Bentolila 2004). Nevertheless, evidence for paternal leakage of mitochondria and heteroplasmic mitochondria has been found that complicates the idea that CMS is strictly maternally inherited (McCauley and Olson 2008). For example, CMS-inducing mitochondria could be introduced via pollen into a nonrestoring nuclear background, giving rise to male-sterility. Perhaps mitochondria that do not induce CMS that have been transferred via pollen can restore the fertility of the stamens. This leakage could explain how CMS can occur in gynodioecious populations. The CMS-inducing gene affects the nuclear genome through retrograde signaling (i.e., the signaling from the mitochondria to the nucleus), resulting in a male-sterile phenotype ranging from pollen abortion to homeotically converted stamens into carpeloid or petaloid structures or the lack of stamens (Carlsson et al. 2008). Male-fertility can be restored by introducing nuclear encoded *Rf*-genes into the CMS-line (Hanson and Bentolila 2004).

18.3 CMS-Associated Genes

The CMS-inducing or CMS-associated genes appear to arise from recombination events creating novel open reading frames (ORF) that often are recombinants between a mt-gene and an unknown *orf* or coding region. Often the CMS-associated genes or loci are located close to an *atp* gene or contain parts of a gene encoding an ATPase subunit and are cotranscribed with flanking mt-genes (Hanson and Bentolila 2004). Many of the CMS-associated genes have a similar expression profile as the standard mt-gene that they are cotranscribed with, and they are furthermore expressed throughout the plant (Budar and Berthomé 2007). An intriguing observation is that despite the similarities in phenotype, i.e., male-sterility and effects on the stamens of different CMS-systems, the CMS-associated genes show little or no structural relationship. The question then is whether the common features of the CMS-associated genes are important for the establishment of CMS-inducing genes or not. Must a CMS-inducing gene be close to, or contain a part of an ATPase encoding gene, or could it be any mitochondrial encoded gene? It has been suggested that an altered ATP-content could cause CMS (Bergman et al. 2000; Sabar et al. 2003; Teixeira et al. 2005b), based on the assumption that a reduced energy production would impair stamen development and pollen production. With that in mind it could be reasonable to assume that CMS-inducing genes containing parts of ATPase genes lead to a malfunctioning ATPase and disturbed ATP production. Or could it be other malfunctioning enzymes that cause the male-sterility? It has been shown in *Nicotiana tabacum* that the silencing of the mitochondrial pyruvate dehydrogenase leads to male-sterility (Yui et al. 2003). The authors speculate that this could be due to alterations in the activity of the citric acid cycle, leading to altered metabolite- and energy content in the tapetal cells. It has also been shown in CMS-lines of *Nicotiana sylvestris* that their mt-genome lacks the *nad7* subunit of complex I (Gutierres et al. 1997; Sabar et al. 2000). These lines were suggested to have a complex I dysfunction that contributes to the male-sterility phenotype. In addition, there is a mechanism or pathway that can override this dysfunction to provide survival of the male-sterile plants. A more simple explanation is that the CMS-inducing gene needs a promoter and/or other regulatory elements to become active. However, any recombined mt-gene is not necessarily a CMS-inducing gene. There are examples of recombined genes or genes with altered expression levels or patterns that do not have a direct link to CMS (Leino et al. 2005; Carlsson et al. 2007b) and within the sequenced plant mt-genomes many chimeric genes are found that have no apparent correlation with CMS (Budar and Berthomé 2007). For most CMS-inducing or CMS-associated genes described today, little or nothing is known of their function, mechanism, and signaling pathways, unfortunately.

18.4 Retrograde Signaling

The mitochondria have the ability, through the CMS-inducing genes, to influence nuclear gene expression via retrograde signaling (also called retrograde communication), retrograde stress signaling or retrograde regulation (see Chap. 16; Liao and Butow 1993; Patil and Walter 2001; Rodermel 2001; Butow and Avadhani 2004; Rhoads and Vanlerberghe 2004). Molecular studies of, e.g., alterations in the mt-genomes, inhibition of the electron transport chain, and different stresses have shown that the nuclear gene expression is affected in such conditions. CMS is one of the most well-known examples of retrograde signaling in plants, but alterations in the mt-genome can also lead to, e.g., chlorotic leaves or embryo lethality (Newton et al. 2004; Rhoads and Subbaiah 2007). Chlorotic leaves have been found in cybrids of *N. tabacum* (*Petunia hybrida*) (Bonnett et al. 1993), *T. aestivum* (*Triticum aegilops*) (Mukai and Tsunewaki 1976) and *Solanum lycopersicum* (*Lycopersicon pennellii*) (Bonnema et al. 1995), but also in *Arabidopsis thaliana* (Sakamoto et al. 1996). In *Z. mays* the so called nonchromosomal stripe (NCS) mutants have been characterized, and shown to carry specific mt DNA deletions (Newton and Coe 1986). Altered mitochondrial backgrounds have also been demonstrated to affect the starch content in several *Solanum tuberosum* hybrids (Lössl et al. 1994) and in *B. napus* cybrids (Teixeira et al. 2005b). Still, no retrograde pathway specific for CMS has been described.

18.4.1 Restorer-of-Fertility *Genes and Restoration Mechanisms*

Nuclear encoded genes called *Restorers-of-fertility* (*Rf* genes), often found in the nuclear genome of the cytoplasmic donor species, can suppress the CMS flower phenotype and restore pollen production (Hanson and Bentolila 2004). Restored lines have been produced through introgression of parts of nuclear DNA from the cytoplasmic donor species to the CMS line in, e.g., *N. tabacum* (Burns et al. 1978; Gerstel et al. 1978), *T. aestivum* (*Triticum thimophevii*) (Livers 1964), and numerous *Brassica juncea* (Prakash et al. 2001; Banga et al. 2003; Pathania et al. 2003) CMS lines. Several *Rf* genes have been shown to encode proteins from the pentatricopeptide repeat (PPR) family (Bentolila et al. 2002; Brown et al. 2003; Desloire et al. 2003; Koizuka et al. 2003; Klein et al. 2005; Wang et al. 2006), but there are exceptions, e.g., the *Rf2* gene that restores fertility in CMS-T *Z. mays*, which is a putative aldehyde dehydrogenase (Cui et al. 1996; Liu et al. 2001).

Interestingly, the majority of PPR proteins in *A. thaliana* are predicted or have been shown to be localized to the mitochondria or the plastids (Lurin et al. 2004). Several have been shown to bind to RNA and mediate RNA stability, splicing, cleavage, degradation, and editing and to have a role in both translation and transcription (Kotera et al. 2005; Schmitz-Linneweber et al. 2005; Okuda et al. 2006; Schmitz-Linneweber et al. 2006; Wang et al. 2006; Hattori et al. 2007). These properties of the PPR proteins constitute most probably the expected characteristics of the *Rf* genes, and it has been shown that the CMS-associated transcripts have been altered in restored lines (Tang et al. 1996; Dill et al. 1997; Li et al. 1998; Wen and Chase 1999). The PPR-B protein restores fertility in *B. napus* (Ogu-INRA), and has been shown to be associated with RNA from the CMS-associated gene *orf138*. It is not clear how PPR-B interacts with *orf138* RNA, but the PPR-B protein is correlated with the removal of ORF138 protein and a restored male-fertility (Uyttewaal et al. 2008). Perhaps PPR-B regulates translation of the *orf138* RNA into a protein.

18.5 CMS System

The CMS systems described in the literature broadly fall into two categories. One category is due to rearranged or recombined mt-genomes, e.g., the *B. napus* Ogu-INRA (Pelletier et al. 1983), the *B. napus* (*A. thaliana*) (Leino et al. 2003), and the *N. tabacum* (*Nicotiana repanda*) (Bonnett et al. 1991; Kofer et al. 1991) CMS systems. Usually these are obtained via production of somatic hybrids or cybrids. Another category has been found in alloplasmatic hybrids combining the nuclear genome of one species with the mt-genome of another species or sub-species. e.g., *Daucus carota* (Linke et al. 1999; Nothnagel et al. 2000; Linke et al. 2003) and *N. tabacum* (Burns et al. 1978; Gerstel et al. 1978).

18.6 Somatic Hybrids

Male-sterility is often observed in somatic hybrids (Glimelius 1999), i.e., hybrids generated from protoplast fusions between two different lines from in general two different species. Somatic hybrids have been generated to facilitate transfer of genetic material between sexually incompatible species. Somatic hybrids can be symmetric, i.e., containing all the nuclear genetic information from both parents, or asymmetric, i.e., containing the complete nuclear genome from one of the parents but only parts of the genome from the other parent (Sundberg et al. 1987; Fahleson et al. 1988; Forsberg et al. 1994). In general, a sorting out of organelles occurs, resulting in lines with chloroplasts originating from one of the parents (Sundberg et al. 1987; Forsberg et al. 1994). The mitochondria can also be derived from one of the parents, but in contrast to the chloroplast genome the mt-genomes are usually recombined or rearranged with DNA from both of the parents (Forsberg et al. 1994). Thus, somatic hybridization facilitates the recombination of mt DNA and gives new opportunities to combine a specific nuclear genome with a new cytoplasmic genome, with the potential to give rise to novel CMS-lines.

There are several hypotheses regarding on how CMS-inducing genes can be manifested in somatic hybrids. One is the transfer of known CMS-inducing mitochondria (with known or unknown CMS-associated genes) into a new nuclear background. Further, it could be due to species incompatibility, i.e., simply an incompatibility between mitochondria from one species and the nucleus from another species leading to male-sterility. A third hypothesis could be that there actually is an active CMS-inducing gene in one of the species that is silenced by an *Rf* gene. If these CMS mitochondria are transferred to another species that lacks the *Rf* gene, CMS will occur. Finally, it is possible that the rearrangements and recombination events that occur in the mitochondria during protoplast fusion could create novel CMS-inducing genes or simply activate silenced CMS-inducing genes.

18.6.1 The Ogura CMS Systems

A well-known and exploited CMS-inducing cytoplasm is the Ogura cytoplasm, initially discovered in *Raphanus sativus*. The Ogura cytoplasm was observed and described in Japan in the 1960s (Ogura 1968). This cytoplasm has been used to induce CMS in several species (Bannerot et al. 1974; Heath et al. 1994) and is today a carefully studied as well as highly utilized CMS-system. The Ogura type of male-sterility was successfully introduced into several different *Brassica* species. However, other negative characteristics were introduced as well, such as yellowing of leaves, low chlorophyll content, and poorly developed nectaries with low nectar production. These traits appeared to be caused by the Ogura *R. sativus* chloroplasts and were removed by protoplast fusions between *Brassica* lines carrying the Ogura cytoplasm and the line of interest (Menczel et al. 1987; Pelletier et al. 1987; Jarl et al. 1989; Gourret et al. 1992). Through the fusion event, a sorting out of chloroplasts from *R. sativus* and replacement of *B. napus* chloroplasts were obtained.

The Ogura-cytoplasm has been transferred to *B. oleracea* and *B. napus* (Bannerot et al. 1974) as well as to *B. juncea* (Kirti et al. 1995). Besides inhibited pollen production (see Fig. 18.1b) this transfer modified flower development. In the *B. oleracea* (Ogura) CMS lines, petaloid and carpeloid stamens were occasionally observed (McCollum 1979, 1981). In *B. juncea* (Ogura) CMS lines, the Ogura cytoplasm induced floral abnormalities like development of petaloid anthers and stamens with aborted microspores (Kirti et al. 1995; Meur et al. 2006). In lines of *B. napus* cv. Westar with Ogura cytoplasm the flowers developed different types of stamens depending on the temperature. Under low temperature the stamens developed into carpeloid structures bearing stigmatoid surfaces and external ovules. Normal anthers were produced at high temperatures even though microspore development was affected (Polowick and Sawhney 1987).

The CMS-system *B. napus* (Ogu-INRA) is today effectively used in hybrid seed production, and was originally produced through protoplast fusions between a *B. napus* cultivar with *B. napus* nuclear, chloroplast, and mt-genomes and a *B. napus* CMS-line with a *B. napus* nuclear genome in a CMS-inducing Ogura *R. sativus* cytoplasm (Pelletier et al. 1983). This was done to correct for flower abnormalities and to remove the varying phenotypes that were obtained in the original protoplast fusions (Pelletier et al. 1983; Vedel et al. 1986; Gourret et al. 1992).

The CMS-inducing mitochondria of Ogura-type carries *orf138* that causes the CMS-phenotype (Bonhomme et al. 1991, 1992; Grelon et al. 1994). The stamens of *R. sativus* carrying the Ogura cytoplasm display abnormal division and vacuolization of tapetal cells and degeneration of microspore cells (Ogura 1968; Yang et al. 2008). It has been shown that the expression of *chalcone synthase* (*CHS*), encoding an enzyme in the flavonoid biosynthetic pathway, is lower in the male-sterile *R. sativus* line MS-G (Yang et al. 2008). The authors speculate that the decreased expression of *CHS* could lead to a lower flavonoid content, which in turn could negatively affect pollen development. The CMS-associated gene *orf138* encodes a 20 kDa protein that oligomerizes and is part of a 750 kDa complex localized to the inner mt membrane (Bonhomme et al. 1991, 1992; Grelon et al. 1994; Duroc et al. 2005). The ORF138 protein has a toxic effect on *E. coli*, but not by affecting the respiration.

An *Rf*-gene named *Rfo* or *orf687* has been characterized (Delourme et al. 1998; Brown et al. 2003; Desloire et al. 2003; Koizuka et al. 2003) and introduced into several *B. napus* lines (Pellan-Delourme and Renard 1988; Sakai et al. 1996; Primard-Brisset et al. 2005). *Rfo* is a *PPR*-gene named *PPRB* found in the *Rf* locus of Ogu-INRA CMS lines regulating a protein that is detected in high levels in anthers of young buds (Uyttewaal et al. 2008). The presence of the PPR-B protein in the anthers is correlated with the removal of the CMS-associated protein ORF138 (Bellaoui et al. 1999; Uyttewaal et al. 2008). The elimination of ORF138 proteins from the tapetal layers is correlated with restoration of male-fertility, even though mRNA of *orf138* still is found in the tapetal cells in restored lines, suggesting that the restorer protein acts on the translational level.

Another *R. sativus* CMS-system is the Kosena system, which genetically is the same as Ogura (Iwabuchi et al. 1999; Koizuka et al. 2000, 2003). The putative CMS-associated gene, *orf125*, has a 39bp deletion compared to the Ogura homologue, *orf138*, and the putative protein carries a two amino acid substitution. Interestingly, the *Rf* gene is highly similar or identical to the corresponding gene *PPR-B* in Ogu-INRA (Koizuka et al. 2003).

18.6.2 The CMS System **B. napus (A. thaliana)**

Several somatic hybrids were produced between *B. napus* and *A. thaliana* in the 1990s (Forsberg et al. 1994, 1998a, b). CMS was observed in a number of lines. A subset of the somatic hybrid lines was divided into three groups (male-sterile, semi-sterile, and fertile) and studied in more detail regarding the mt-genome genome and its expression, nuclear gene expression, as well as phenotypic, physiologic, and metabolic features (Leino et al. 2003; Carlsson et al. 2007b).

Each line carried the expected number or close to the expected number of floral organs (Leino et al. 2003; Teixeira et al. 2005a; Carlsson et al. 2007b). The morphology and color of the sepals were similar in all lines studied and the size varied only slightly (see Fig. 18.1c). The same observation was made for the pistil,

Box 18.1 Flower development and the ABC Model

The basic model of an angiosperm flower is four whorls with distinct organs (Fig. 18.2). The two outermost (first and second) whorls consist of sepals and petals respectively, i.e., the sterile perianth. The third whorl contains the male reproductive organs, the stamens; and the forth and innermost whorl comprises the female reproductive organs, the carpels. The genetic regulation of flower development is to a large extent similar for all angiosperms (Soltis et al. 2002; Jack 2004; Kramer and Hall 2005; Krizek and Fletcher 2005).

Through studies of the two plant species *Arabidopsis thaliana* and *Antirrhinum majus*, the genetic regulation of flower development has been elucidated and presented as the so called ABC model (Schwarz-Sommer et al. 1990; Coen and Meyerowitz 1991; Theißen 2001). The floral organs are under

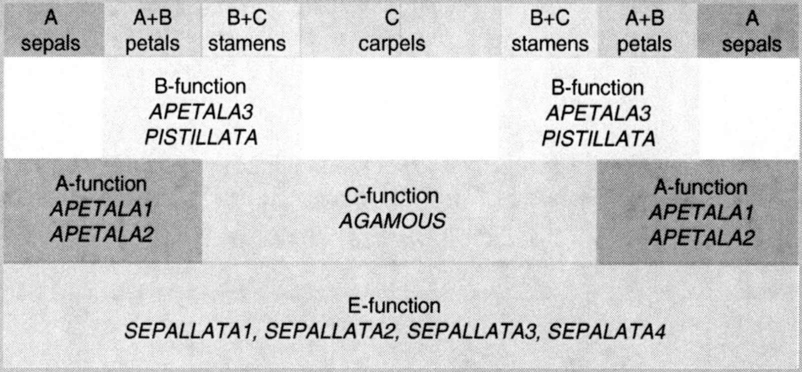

A sepals	A+B petals	B+C stamens	C carpels	B+C stamens	A+B petals	A sepals
	B-function *APETALA3* *PISTILLATA*			B-function *APETALA3* *PISTILLATA*		
A-function *APETALA1* *APETALA2*			C-function *AGAMOUS*		A-function *APETALA1* *APETALA2*	
			E-function *SEPALLATA1, SEPALLATA2, SEPALLATA3, SEPALATA4*			

Fig. 18.2 An overview of the extended ABCmodel where each gene function is exemplified by the *A. thaliana* genes. The corresponding *A. majus* genes are: A-function, *LIPLESS1,2* (*LIP1,2*); B-function, *DEFICIENS* (*DEF*) and *GLOBOSA* (*GLO*); C-function, *PLENA* (*PLE*) and *FARINELLI* (*FAR*; Krizek and Fletcher 2005)

control of three gene functions (A, B, and C) in this model. The A function (e.g., *APETALA1*, *AP1* (Irish and Sussex 1990; Gustavson-Brown et al. 1994; Mandel et al. 1992) and *APETALA2*, *AP2*, (Bowman et al. 1989, 1991; Kunst et al. 1989)) alone specifies the sepals, while the A and B (e.g., *APETALA3*, *AP3*, and *PISTILLATA*, *PI* (Bowman et al. 1989, 1991; Jack et al. 1992; Goto and Meyerowitz 1994)) functions together specify the petals. B together with C (e.g., *AGAMOUS*, *AG* (Bowman et al. 1989, 1991; Yanofsky et al. 1990; Mizukami and Ma 1992)) defines stamens and the C function alone specifies the carpels. The C function also prevents an indeterminate floral meristem growth. Furthermore, the A and C functions are antagonistic to each other. The ABC model has been described in a variety of species, e.g., *Petunia hybrida* (Angenent et al. 1992), *Oryza sativa* (Nagasawa et al. 2003) and *Zea mays* (Whipple et al. 2004). The ABC model has been extended with an E-function

(continued)

Box 18.1 (continued)

(*SEPALLATA1,2,3,4*, *SEP1*, *SEP2*, *SEP3*, *SEP4*, Ma et al. 1991; Purugganan et al. 1995; Pelaz et al. 2000; Malcomber and Kellogg 2005), and the regulation of the ovule development is sometimes referred to as the D-function (*AG*, *SEEDSTICK*, *STK* (Rounsley et al. 1995) and *SHATTERPROOF1,2*, *SHP1*, *SHP2* (Flanagan et al. 1996; Ma et al. 1991; Savidge et al. 1995)).

If the A function is removed, the theoretical flower phenotype will develop carpels in whorls one and four, and stamens in whorls two and three. The *ap2* mutant in *A. thaliana* has this phenotype, while the *ap1* mutant has additional roles in floral meristem identity and often displays an extra set of miniature flowers within the original flower. A deletion of the B-function will give flowers with sepals in the two outermost whorls and carpels in the remaining two whorls. This phenotype is found in *ap3* and *pi* mutants. The removal of the C-function will give flowers with only sepals and petals, and this phenotype is found in the *ag* mutant. Finally, when the E-function is completely deleted, i.e., in the quadruple *sep* mutant, the flower only produces leaf-like structures. Intriguingly, CMS lines sometimes obtain homeotic conversions of the stamens that are transformed into carpeloid or petaloid structures (Kofer et al. 1991; Zubko et al. 1996; Ogihara et al. 1997; Linke et al. 1999, 2003; Murai et al. 2002; Leino et al. 2003) that partly resemble the B- and C-mutants found in *A. thaliana*.

although the pistil in some male-sterile flowers appeared to have a crocked style. The male-sterile lines lacked true stamens; these organs were homeotically converted into carpeloid organs carrying stigmatoid and ovule-like structures. The carpeloid organs were shorter than the true pollen-producing stamens found in the fully fertile lines, but of equal size of the short pollen-producing stamens found in the semi-sterile lines. The size and morphology of the petals and stamens were correlated with the fertility of the flowers, i.e., fully fertile flowers were large, "normal" and producing pollen, while the male-sterile flowers were small carrying carpeloid structures instead of pollen-producing stamens. The phenotype of the male-sterile lines resembled that of the *apetala3* (*ap3*) and *pistillata* (*pi*) mutants found in *A. thaliana* (Box 18.1), and indeed the two B-class genes *AP3* and *PI* had a reduced expression in the male-sterile lines (Teixeira et al. 2005a; Carlsson et al. 2007a, b) directly or indirectly caused by the CMS-inducing mitochondria. Taking the results from the studies together, it is clear that the carpeloid phenotype of the whorl three organs of the male-sterile lines was caused by the reduced *AP3*- and *PI*-expression. Three mitochondrial *orfs* (*orf139*, *orf240a*, and *orf294*) and *cox3* have been shown to be expressed in all or most semi- and male-sterile lines (Leino et al. 2005; Carlsson 2007). One or several of these three orfs could be the CMS-associated gene(s).

The original population from which one of the CMS-lines was isolated segregated into male-sterile plants (i.e., CMS-line 4:19) and male-fertile plants (Leino et al. 2004). The male-sterile trait was stably inherited, while the male-fertile plants segregated into male-fertile and male-sterile plants respectively, generation after generation. The male-fertile plants all contained the markers from *A. thaliana* chromosome three but no markers from the four other chromosomes. It was observed that the restored line shared all *A. thaliana* mtDNA markers with the CMS line 4:19 but not always the gene expression (Leino et al. 2005; Carlsson et al. 2007b). The flower phenotype of the restored line was clearly restored as compared to the CMS-line 4:19. Not only the flower phenotype but also the nuclear gene expression of, e.g., *AP3* and *PI* was restored in the restored line. Furthermore, the three mitochondrial *orfs* had a reduced expression (Carlsson et al. 2007a, b).

18.7 Alloplasmic, Sexual Hybrids

It is also possible to obtain CMS-lines by sexual crossings giving alloplasmic lines. Compared to somatic hybrids, the mitochondrial genome is identical to the mt-genome of the maternal parent and has most likely not recombined with the mt-genome of the paternal parent. Of course there is a possibility that some rearrangements could occur due to the paternal leakage of mitochondria described above (McCauley and Olsen 2008). However, through the sexual crosses a new combination of a nuclear genome from one species with the mt-genome from another is obtained. The first cross is followed by several back-crosses with the nuclear donor to obtain a pure nuclear genome. The crosses can be made between or within species. Within the species *D. carota* several crosses have been performed between different subspecies leading to CMS (Linke et al. 1999, 2003; Nothnagel et al. 2000). Crosses have also been performed between different *Daucus* species. The different *Daucus* CMS lines display a wide variety of stamen phenotypes: shrunken anthers with disturbed pollen production, rudimentary or absent stamens with no pollen, petaloid whorl three organs with no pollen. It is not only within the genus *Daucus* that crosses have been formed but also within *Nicotiana* (Burns et al. 1978; Gerstel et al. 1978). Male-sterility has also been obtained through crosses between different *Triticum* cultivars and different *Triticum* and *Aegilops* species. The resulting alloplasmic hybrids were generated after several back-crosses (Mukai and Tsunewaki 1976).

18.7.1 Zea mays *CMS-T*

There are at least three CMS-inducing cytoplasms in *Z. mays* (Kaul 1988): Texas sterile cytoplasm (T-cytoplasm, CMS-T); USDA sterile cytoplasm (S-cytoplasm, CMS-S); and C-cytoplasm (CMS-C). These three CMS systems are distinct and have different *Rf* genes and unique mt-genomes (Allen et al. 2007). In addition at least two distinct male-fertile mt-genomes, NA and NB, have been reported (Clifton et al. 2004; Allen et al. 2007). The five mt-genomes clearly differ in size and the amount of repeated sequences, with the smallest genome found in CMS-T (535 825 bp) and the largest found in CMS-C (739 719 bp). Duplications are responsible for most of the size differences among the five genomes. The five mt genomes contain more or less the same set of genes, except for the CMS-associated genes; *T-urf13* (only in CMS-T) and *orf355/orf77* (only in CMS-S). In addition, 8–10% of each mt-genome contains *orfs*. It is tempting to speculate that the chimeric *orfs*, e.g., the CMS-inducing *orfs*, might have occurred due to rearrangements, genome duplications, or loss of sequences or that the CMS traits were obtained due to ancient crosses between *Z. mays* and related species resulting in an alloplasmy.

In CMS-T lines the tapetum layer starts to degenerate at a stage when the young microspores are released into the locule cavity (Warmke and Lee 1977; Lee and Warmke 1979). The premature degradation of the tapetum layer is accompanied with vacuolization of the tapetal cells and a more irregular shape of these cells. The microspores of the male-sterile plants also become vacuolated; in addition they are irregular and have thinner walls compared to the fertile microspores. When the pollen of the male-fertile plants reach maturity, the CMS-T plants have desiccated, empty, and collapsed anthers. By the tetrad stage and later stages, the mitochondria in the degrading tissues are disorganized and have no cristae. The mitochondria appear to have a diffuse matrix or to be empty. The mitochondria are also swollen or enlarged (Warmke and Lee 1977; Lee and Warmke 1979). In *Z. mays* besides correlating mitochondrial mutations with CMS, correlations have also been made with abnormal growth, defective kernels, and striped leaves (Shumway and Bauman 1967; Newton and Coe 1986).

The CMS-associated gene in CMS-T, *T-urf13*, encodes a pore-forming protein located in the inner mt membrane that not only induces male-sterility but also makes the plant more sensitive to a toxin produced by *Cochliobolus heterostophus* (asexual stage of *Bipolaris maydis*), race T (T-toxin), and the carbamate insecticide methomyl (Budar and Berthomé 2007). The T-URF13 protein accumulates in all tissues studied, but in the absences of T-toxin and methomyl a modified phenotype is only observed in the stamens. CMS-T has at least four unlinked *Rf* genes, *Rf1*, *Rf2*, *Rf8* and *Rf**, that act in a combinatorial manner (Dill et al. 1997). The genes have similar but distinct effects on the *T-urf13* expression. The combined action of *Rf1* and *Rf2* genes restore male-fertility in CMS-T lines. *Rf1*, *Rf8*, and *Rf** modify the transcripts or transcription of *T-urf13* at independent sites (Dill et al. 1997). *Rf8* and *Rf** can partially substitute *Rf1* when restoring the male-fertility in CMS-T (Dill et al. 1997; Wise et al. 1999).

The *Rf2* gene encodes an aldehyde dehydrogenase, the RF2 protein or mtALDH, with a mitochondrial targeting signal, which is cleaved off (Cui et al. 1996; Liu et al. 2001). This enzyme accumulates in most organs and is necessary for a normal anther development in both CMS-T and normal fertile *Z. mays* lines. RF2 is found in the mitochondrial matrix. These results indicate that RF2 might not be a true Rf-protein, but is still crucial for normal stamen development.

18.7.2 CMS in Nicotiana

Alloplasmic lines as well as protoplast fusions have been produced between various species within the genus *Nicotiana* and between *Nicotiana* and other genera within the *Solanaceae* family, both for breeding purposes and for genetic studies (Burk 1967; Burns et al. 1978; Gerstel et al. 1978; Kofer et al. 1991; Zubko et al. 1996). The flower phenotypes of the male-sterile lines vary widely depending on which mt-genome is combined with the *N. tabacum* nuclear genome (see Fig. 18.1d–e); no stamens, stamens without anthers but with filaments, petaloid stamens/whorl three organs, carpeloid or stigmatoid stamens/whorl three organs, flowers lacking true corolla and stamens, fasciated pistils, separated pistils, and normal flowers not producing fertile pollen (Bonnett et al. 1991; Kofer et al. 1991; Zubko et al. 1996). The different phenotypes resemble those found in plants with mutations in the nuclear encoded genes responsible for the B-function (see Box 18.1) and mutant phenotypes described for the *A. thaliana SUPERMAN* (*SUP*) gene. Indeed, it was shown that *N. tabacum GLOBOSA* (*NtGLO*) (Zubko et al. 2001) and *NtSUP* (Bereterbide et al. 2001, 2002; Hernould et al. 2002) had an altered expression. Furthermore, an abnormal accumulation of cotranscribed *atp1* and *orf274* has been correlated to male-sterility in *N. tabacum* (*Nicotiana repanda*) CMS lines (Bergman et al. 2000; Edqvist and Bergman 2002). For some of the *N. tabacum* CMS lines, altered DNA, RNA, and protein profiles have been observed (Håkansson et al. 1988, 1991).

18.7.3 **Triticum aestivum** *CMS*

T. aestivum has been combined with different cytoplasms from a variety of related species, resulting in a range of flower phenotypes and male-sterility (Livers 1964; Endo 1980; Kaul 1988). Observed phenotypes include photoperiod-sensitive CMS (Murai and Tsunewaki 1993) and homeotically converted stamens into pistil-like structures (Murai and Tsunewaki 1993; Murai et al. 2002). The pistil-like phenotype of the stamens can be explained by alterations in the expression of the *T. aestivum* homologues ABC-genes (see Box 18.1). For example, *T. aestivum AP3* (*WAP3*) and *T. aestivum PI* (*WPI1*) in the male-sterile line were expressed only in the lodicules but not in the primordia of the carpeloid stamens, while *WAP3* and *WPI1* in fully fertile *T. aestivum* were expressed in the lodicules and in the stamen primordia (Murai et al. 2002; Hama et al. 2004). Other ABC genes and flower development genes also show an altered expression, such as the D-class gene *T. aestivum SEEDSTICK* (*WSTK*; Mizumoto et al. 2009; Yamada et al. 2009).

Alterations in the mt-genome and the gene expression of the male-sterile *T. aestivum* lines with the *Aegilops crassa* cytoplasm have been observed (Ogihara et al. 1997) indicating that *orf25* could be the CMS-associated gene (Ogihara et al. 1999) or even more likely *orf260*^{cra} (Zhu et al. 2008). In the *T. aestivum* CMS line with the *Ae. crassa* cytoplasm, the *Rf* gene, *Rfd1*, is located on the long arm of chromosome 7B (Murai and Tsunewaki 1994; Murai et al. 2002; Meguro et al. 2003). The expression of *Rdf1* is correlated with a male-fertile phenotype and the decrease of *orf260*^{cra} (Zhu et al. 2008). Zhu et al. (2008) found *orfs* in other *Triticum* and *Aegilops* species that shared high sequence homology with *orf260*^{cra}. Probably the sequence differences between the *orfs*, or maybe the differences in expression, account for the possibilities to induce male-sterility, which is supported by the observation of identical copies of *orf260*^{cra} in *Aegilops juvenalis* and *Aegilops vavilovii*, two cytoplasms known to induce male-sterile and pistilody whorl three organs (Zhu et al. 2008). In addition, other CMS-associated genes are found in, e.g., *T. timopheevi* and *Aegilops speltoides*. These two species carry the CMS-associated gene *orf256* that induce male-sterility in *T. aestivum* (Hedgcoth et al. 2002).

18.8 Does Cell Death Have a Role in CMS?

Empty, enlarged mitochondria without or with reduced amount of cristae have been observed in cells in male-sterile organs or tissues in several male-sterile lines (Warmke and Lee 1977; Lee and Warmke 1979; Hernould et al. 1998; Farbos et al. 2001; Teixeira et al. 2005a; González-Melendi et al. 2008). The first sign of disturbed development in Ogu-INRA is observed in the tapetum layer at the tetrad stage (González-Melendi et al. 2008). The microspore cells become prematurely vacuolated eventually leading to the abortion of pollen. These cells not only contain large vacuoles, but also diffuse mitochondria. The morphology of the mitochondria becomes more and more abnormal during the development of the cells supposed to form pollen. The mitochondria observed in the tapetal and pollen producing cells, i.e., the cells responsible for the male-sterility, are lacking the characteristic morphology of mitochondria found in fertile flowers. Swollen mitochondria lacking cristae and with a clear matrix have been found in male-sterile tissues in a CMS-line of *N. tabacum* (*N. repanda*) (Farbos et al. 2001) and a male-sterile *N. tabacum* line carrying a construct consisting of the unedited *N. tabacum atp9* mt-gene and the *coxIV* mitochondrial presequence from *Saccharomyces cerevisiae* (Hernould et al. 1998). Similar cell and mitochondrial morphologies have been observed in *Z. mays* CMS-T (Warmke and Lee 1977; Lee and Warmke 1979) and in two male-sterile lines from the *B. napus* (*A. thaliana*) CMS-system (Teixeira et al. 2005a). Taking these observations together it appears as if programmed cell death (PCD; see Chap. 17) or necrosis in the cells giving rise to stamens, anthers, or pollen is part of the mechanism leading to CMS. This has also has been discussed by Budar and Berthomé (2007) and Chase (2007). In addition, it has been shown that PCD occurs in the *H. annuus* CMS-system *PET1* (Balk et al. 2001; Nizampatnam et al. 2009).

Even though much is known about CMS, such as the structure and function of several CMS-associated genes and *Rf* genes, stamen morphologies of the male-sterile lines, how to transfer the male-sterility, and how to restore the male-fertility, several questions remain to be answered. For example, why is mainly stamens affected and how can the male-sterility have so many different phenotypes? Which retrograde signal, mechanism, or pathway transfers the male-sterility signal from the mitochondria to the nucleus? By utilizing a well characterized CMS-system with CMS-inducing genes and *Rf* genes, it would be possible to answer the questions and learn more about the involvement of the mitochondria in plant and flower development and as a result further develop the use of CMS in plant breeding for establishing hybrids.

Acknowledgments We thank Dr. Jens F. Sundström for exciting discussions and critically reading the manuscript. This work was supported by the NL-faculty at the Swedish University of Agricultural Sciences, SLU, Sweden.

Glossary

Alloplasmic: The nucleus from one line combined with the cytoplasm from another line. Often used for describing the genomic content of the offspring from sexual crosses between two different species.

CMS: Short for cytoplasmic male-sterility. A type of male-sterility observed in plants caused by the influence of the mitochondria on nuclear encoded genes.

Gynodioecy: The co-occurrence of females and hermaphrodites in a plant population.

Retrograde signaling: The signaling processes from the mitochondria to the nucleus.

Rf: Short for *restorer-of-fertility*. An *Rf* gene is a nuclear encoded gene (or set of genes) that can restore male-fertility in CMS-plants.

Ogu-INRA: A well-known French CMS-system in *Brassica napus*, based on the Ogura-system.

Ogura: A Japanese CMS system in *Raphanus sativus*. The Ogura CMS was first presented by H. Ogura in 1968, hence the name.

PPR: Short for pentatricopeptide repeat. The PPR-family is a large gene family where the members have RNA-binding and modifying properties. Many *PPR*-genes are targeted to the chloroplasts or the mitochondria. Some *PPR* genes are *Rf* genes.

Somatic hybrid: A hybrid generated from a protoplast fusion between two different lines, in general two different but related plant species.

References

Allen, J. O., Fauron, C. M., Minx, P., Roark, L., Oddiraju, S., Lin, G. N., Meyer, L., Sun, H., Kim, K., Wang, C., Du, F., Xu, D., Gibson, M., Cifrese, J., Clifton, S. W., Newton, K. J. 2007. Comparisons among two fertile and three male-sterile mitochondrial genomes of maize. Genetics 177:1173–1192.

Angenent, G. C., Busscher, M., Franken, J., Mol, J. N. M., Tunen, A. J. V. 1992. Differential expression of two MADS box genes in wild-type and mutant Petunia flowers. Plant Cell 4:983–993.

Azhagiri, A. K., Maliga, P. 2007. Exceptional paternal inheritance of plastids in Arabidopsis suggests that low-frequency leakage of plastids via pollen may be universal in plants. Plant J 52:817–823.

Balk, J., Leaver, C. J. 2001. The PET1-CMS mitochondrial mutation in sunflower is associated with premature programmed cell death and cytochrome c release. Plant Cell 13:1803–1818.

Banga, S. S., Deol, J. S., Banga, S. 2003. Alloplasmic male-sterile *Brassica juncea* with *Enarthrocarpus lyratus* cytoplasm and the introgression of gene(s) for fertility restoration from cytoplasm donor species. Theor Appl Genet 106:1390–1395.

Bannerot, H., Loulidard, L., Cauderon, Y., Tempe, J. 1974. Transfer of cytoplasmic male sterility from *Raphanus sativus* to *Brassica oleracea*. Cruciferae 1974 Eucarpia Meeting, Dundee, Scotland, pp. 52–54.

Bellaoui, M., Grelon, M., Pelletier, G., Budar, F. 1999. The restorer *Rfo* gene acts post-translationally on the stability of the ORF138 Ogura CMS-associated protein in reproductive tissues of rapeseed cybrids. Plant Mol Biol 40:893–902.

Bentolila, S., Alfonso, A. A., Hanson, M. R. 2002. A pentatricopeptide repeat-containing gene restores fertility to cytoplasmic male-sterile plants. Proc Natl Acad Sci USA 99:10887–10892.

Bereterbide, A., Hernould, M., Castera, S., Mouras, A. 2001. Inhibition of cell proliferation, cell expansion and differentiation by the *Arabidopsis SUPERMAN* gene in transgenic tobacco plants. Planta 214:22–29.

Bereterbide, A., Hernould, M., Farbos, I., Glimelius, K., Mouras, A. 2002. Restoration of stamen development and production of functional pollen in an alloplasmic CMS tobacco line by ectopic expression of the *Arabidopsis thaliana SUPERMAN* gene. Plant J 29:607–615.

Bergman, P., Edqvist, J., Farbos, I., Glimelius, K. 2000. Male-sterile tobacco displays abnormal mitochondrial *atp1* transcript accumulation and reduced floral ATP/ADP ratio. Plant Mol Biol 42:531–544.

Bonhomme, S., Budar, F., Férault, M., Pelletier, G. 1991. A 2.5 kb *Nco*I fragment of Ogura radish mitochondrial DNA is correlated with cytoplasmic male-sterility in *Brassica* cybrids. Curr Genet 19:121–127.

Bonhomme, S., Budar, F., Lancelin, D., Small, I., Defrance, M.-C., Pelletier, G. 1992. Sequence and transcription analysis of the *Nco2.5* Ogura-specific fragment correlated with cytoplasmic male sterility in *Brassica* hybrids. Mol Gen Genet 235:340–348.

Bonnema, A. B., Castillo, C., Reiter, N., Cunningham, M., Adams, H. P., O'Connell, M. 1995. Molecular and ultrastructural analysis of a nonchromosomal variegated mutant. Plant Physiol 109:385–392.

Bonnett, H. T., Djurberg, I., Fajardo, M., Glimelius, K. 1993. A mutation causing variegation and abnormal development in tobacco is associated with an altered mitochondrial DNA. Plant J 3:519–525.

Bonnett, H. T., Kofer, W., Håkansson, G., Glimelius, K. 1991. Mitochondrial involvement in petal and stamen development studied by sexual and somatic hybridization of *Nicotiana* species. Plant Sci 80:119–130.

Bowman, J. L., Smyth, D. R., Meyerowitz, E. M. 1989. Genes directing flower development in *Arabidopsis*. Plant Cell 1:37–52.

Bowman, J. L., Smyth, D. R., Meyerowitz, E. M. 1991. Genetic interactions among floral homeotic genes of *Arabidopsis*. Development 112:1–20.

Brown, G. G., Formanová, N., Jin, H., Wargachuk, R., Dendy, C., Patil, P., Laforest, M., Zhang, J., Cheung, W. Y., Landry, B. S. 2003. The radish *Rfo* restorer gene of Ogura cytoplasmic male sterility encodes a protein with multiple pentatricopeptide repeats. Plant J 35:262–272.

Budar, F., Berthomé, R. 2007. Cytoplasmic male sterilities and mitochondrial gene mutations in plants. Logan, D. C., ed. In Plant mitochondria. Blackwell, Oxford, pp. 278–307.

Burk, L. G. 1967. An interspecific bridge-cross *Nicotiana repanda* through *N. sylvestris* to *N. tabacum*. J Hered 58:215–218.

Burns, J. A., Gerstel, D. U., Sand, S. A. 1978. Cytoplasmic male sterility in *Nicotiana*, restoration of fertility, and the nucleolus. II. *N. debneyi* cytoplasm. Genetics 90:151–159.

Butow, R. A., Avadhani, N. G. 2004. Mitochondrial signaling: the retrograde response. Mol Cell 14:1–15.

Carlsson, J. 2007 The mitochondrial influence on nuclear gene expression in cytoplasmic male-sterile *Brassica napus*. Swedish University of Agricultural Sciences, Uppsala.

Carlsson, J., Lagercrantz, U., Sundström, J., Teixeira, R., Wellmer, F., Meyerowitz, E. M., Glimelius, K. 2007a. Microarray analysis reveals altered expression of a large number of nuclear genes in developing cytoplasmic male-sterile *Brassica napus* flowers. Plant J 49:452–462.

Carlsson, J., Leino, M., Glimelius, K. 2007b. Mitochondrial genotypes with variable parts of *Arabidopsis thaliana* DNA affect development in *Brassica napus* lines. Theor Appl Genet 115:627–641.

Carlsson, J., Leino, M., Sohlberg, J., Sundström, J. F., Glimelius, K. 2008. Mitochondrial regulation of flower development. Mitochondrion 8:74–86.

Chase, C. D. 2007. Cytoplasmic male sterility: a window to the world of plant mitochondrial–nuclear interactions. TRENDS in Genetics 23:81–90.

Clifton, S. W., Minx, P., Fauron, C. M.-R., Gibson, M., Allen, J. O., Sun, H., Thompson, M., Barbazuk, W. B., Kanuganti, S., Tayloe, C., Meyer, L., Wilson, R. K., Newton, K. J. 2004. Sequence and comparative analysis of the maize NB mitochondrial genome. Plant Physiol 136:3486–3503.

Coen, E., Meyerowitz, E. M. 1991. The war of the whorls: genetic interactions controlling flower development. Nature 353:31–37.

Cui, X., Wise, R. P., Schnable, P. S. 1996. The *rf2* nuclear restorer gene of male-sterile T-cytoplasm maize. Science 272:1334–1336.

Delourme, R., Foisset, N., Horvais, R., Barret, P., Champagne, G., Cheung, W. Y., Landry, B. S., Renard, M. 1998. Characterisation of the radish introgression carrying the *Rfo* restorer gene for the I-INRA cytoplasmic male sterility in rapeseed (*Brassica napus* L.). Theor Appl Genet 97:129–134.

Desloire, S., Gherbi, H., Laloui, W., Marhadour, S., Clouet, V., Cattolico, L., Falentin, C., Giancola, S., Renard, M., Budar, F., Small, I., Caboche, M., Delourme, R., Bendahmane, A. 2003. Identification of the fertility restoration locus, *Rfo*, in radish, as a member of the pentatricopeptide repeat protein family. EMBO Rep 4:588–594.

Dill, C. L., Wise, R. P., Schnable, P. S. 1997. *Rf8* and *Rf** mediate unique T-*urf13*-transcript accumulation, revealing a conserved motif associated with RNA processing and restoration of pollen fertility in T-cytoplasm maize. Genetics 147:1367–1379.

Dufaÿ, M., Touzet, P., Maurice, S., Cuguen, J. 2007. Modelling the maintenance of male-fertile cytoplasm in a gynodioecious population. Heredity 99:349–356.

Duroc, Y., Gaillard, C., Hiard, S., Defrance, M.-C., Pelletier, G., Budar, F. 2005. Biochemical and functional characterization of ORF138, a mitochondrial protein responsible for Ogura cytoplasmic male sterility in Brassiceae. Biochimie 87:1089–1100.

Edqvist, J., Bergman, P. 2002. Nuclear identity specifies transcriptional initiation in plant mitochondria. Plant Mol Biol 49:59–68.

Elansary, H. O., Müller, K., Olson, M. S., Storchová, H. 2010. Transcription profiles of mitochondrial genes correlate with mitochondrial DNA haplotypes in a natural population of Silene vulgaris. BMC Plant Biology 10:11.

Endo, T. R. 1980. Genetic constancy of the cytoplasm. Tsunewaki, K., ed. In Genetic diversity of the cytoplasm in Triticum and Aegilops. Japan Society for the Promotion of Science, Tokyo, pp. 13–48.

Fahleson, J., Dixelius, J., Sundberg, E., Glimelius, K. 1988. Correlation between flow cytometric determination of nuclear DNA content and chromosome number in somatic hybrids within *Brassicaceae*. Plant Cell Rep 7:74–77.

Farbos, I., Mouras, A., Bereterbide, A., Glimelius, K. 2001. Defective cell proliferation in the floral meristem of alloplasmic plants of *Nicotiana tabacum* leads to abnormal floral organ development and male sterility. Plant J 26:131–142.

Flanagan, C. A., Hu, Y., Ma, H. 1996. Specific expression of the *AGL1* MADS-box gene suggests regulatory functions in *Arabidopsis* gynoecium and ovule development. Plant J 10:343–353.

Forsberg, J., Dixelius, C., Lagercrantz, U., Glimelius, K. 1998a. UV dose-dependent DNA elimination in asymmetric somatic hybrids between *Brassica napus* and *Arabidopsis thaliana*. Plant Sci 131:65–76.

Forsberg, J., Lagercrantz, U., Glimelius, K. 1998b. Comparison of UV light, X-ray and restriction enzyme treatment as tools in production of asymmetric somatic hybrids between *Brassica napus* and *Arabidopsis thaliana*. Theor Appl Genet 96:1178–1185.

Forsberg, J., Landgren, M., Glimelius, K. 1994. Fertile somatic hybrids between *Brassica napus* and *Arabidopsis thaliana*. Plant Sci 95:213–223.

Gerstel, D. U., Burns, J. A., Burk, L. G. 1978. Cytoplasmic male sterility in *Nicotiana*, restoration of fertility, and the nucleolus. Genetics 89:157–169.

Glimelius, K. 1999. Somatic hybridization. Gómez-Campo, C., ed. In Biology of Brassica coenospecies. Elsevier Science B.V., Amsterdam, pp. 107–148.

González-Melendi, P., Uyttewaal, M., Morcillo, C. N., Mora, J. R. H., Fajardo, S., Budar, F., Lucas, M. M. 2008. A light and electron microscopy analysis of the events leading to male sterility in Ogu-INRA CMS of rapeseed (*Brassica napus*). J Exp Bot 59:827–838.

Goto, K., Meyerowitz, E. M. 1994. Function and regulation of the *Arabidopsis* floral homoetic gene *PISTILLATA*. Genes Dev 8:1548–1560.

Gourret, J.-P., Delourme, R., Renard, M. 1992. Expression of *ogu* cytoplasmic male sterility in cybrids of *Brassica napus*. Theor Appl Genet 83:549–556.

Grelon, M., Budar, F., Bonhomme, S., Pelletier, G. 1994. Ogura cytoplasmic male-sterility (CMS)-associated *orf138* is translated into a mitochondrial membrane polypeptide in male-sterile *Brassica* cybirds. Mol Gen Genet 243:540–547.

Gustavson-Brown, C., Savidge, B., Yanofsky, M. F. 1994. Regulation of the *Arabidopsis* floral homeotic gene *APETALA1*. Cell 76:131–143.

Gutierres, S., Sabar, M., Lelandais, C., Chetrit, P., Diolez, P., Degand, H., Botry, M., Vedel, F., de Kouchkovsky, Y., De Paepe, R. 1997. Lack of mitochondrial- and nuclear-encoded subunits of complex I and alteration of the respiratory chain in Nicotiana sylvestris mitochondrial deletion mutants. Proc Natl Acad Sci U S A 94:3436–3441.

Hama, E., Takumi, S., Ogihara, Y., Murai, K. 2004. Pistillody is caused by alterations to the class-B MADS-box gene expression pattern in alloplasmic wheats. Planta 218:712–720.

Hanson, M. R., Bentolila, S. 2004. Interactions of mitochondrial and nuclear genes that affect male gametophyte development. Plant Cell 16:S154–S169.

Hattori, M., Miyake, H., Sugita, M. 2007. A pentatricopeptide repeat protein is required for RNA processing of *clpP* pre-mRNA in moss chloroplasts. J Biol Chem 282:10773–10782.

Havey, M. J. 2004. The use of cytoplasmic male sterility for hybrid seed production. Daniell, H., Chase, C., eds. In Molecular biology and biotechnology of plant organelles. Springer, Dordrecht, pp. 623–634.

Håkansson, G., Glimelius, K. 1991. Extensive nuclear influence on mitochondrial transcription and genome structure in male-fertile and male-sterile alloplasmic Nicotiana materials. Mol Gen Genet 229:380–388.

Håkansson, G., van der Mark, F., Bonnett, H. T., Glimelius K. 1988. Variant mitochondrial protein and DNA patterns associated with cytoplasmic male-sterile lines of Nicotiana. Theor Appl Genet 76:431–437.

Heath, D. W., Earle, E. D., Dickson, M. H. 1994. Introgressing cold-tolerant Ogura cytoplasm from rapeseed into pak choi and Chinese cabbage. Hort Sci 29:202–203.

Hedgcoth, C., El-Shehawi, A. M., Wei, P., Clarkson, M., Tamalis, D. 2002. A chimeric open reading frame associated with cytoplasmic male sterility in alloplasmic wheat with Triticum timopheevi mitochondria are present in several Triticum and Aegilops species, barley, and rye. Curr Genet 41:357–365.

Hernould, M., Bereterbide, A., Farbos, I., Glimelius, K., Mouras, A. 2002. Cytoplasmic male sterility in tobacco: a plant genetic disease curable by gene therapy. Flowering Newsl 34: 1–10.

Hernould, M., Suharsono, Zabaleta, E., Carde, J. P., Litvak, S., Araya, A., Mouras, A. 1998. Impairment of tapetum and mitochondria in engineered male-sterile tobacco plants. Plant Mol Biol 36:499–508.

Irish, V. F., Sussex, I. M. 1990. Function of the *apetala-1* gene during *Arabidopsis* floral development. Plant Cell 2:741–753.

Iwabuchi, M., Koizuka, N., Fujimoto, H., Sakai, T., Imamura, J. 1999. Identification and expression of kosena radish (*Raphanus sativus* cv. Kosena) homologue of the ogura radish CMS-associated gene, *orf138*. Plant Mol Biol 39:183–188.

Jack, T. 2004. Molecular and genetic mechanisms of floral control. Plant Cell 16:S1–S17.

Jack, T., Brockman, L. L., Meyerowitz, E. M. 1992. The homeotic gene *APETALA3* of *Arabidopsis thaliana* encodes a MADS box and is expressed in petals and stamens. Cell 68:683–697.

Jarl, C. I., Grinsven, M. Q. J. M. v., Mark, F. v. d. 1989. Correction of chlorophyll-defective male-sterile winter oilseed rape (*Brassica napus*) through organelle exchange: molecular analysis of the cytoplasm of parental fines and corrected progeny. Theor Appl Genet 77:135–141.

Kaul, M. L. H. 1988. Male-sterility in higher plants. Springer-Verlag, Berlin.

Kirti, P. B., Banga, S. S., Prakash, S., Chopra, V. L. 1995. Transfer of *Ogu* cytoplasmic male-sterility to *Brassica juncea* and improvement of the male-sterile line through somatic cell fusion. Theor Appl Genet 91:517–521.

Klein, R. R., Klein, P. E., Mullet, J. E., Minx, P., Rooney, W. L., Schertz, K. F. 2005. Fertility restorer locus *Rf1*of sorghum (*Sorghum bicolor* L.) encodes a pentatricopeptide repeat protein not present in the colinear region of rice chromosome 12. Theor Appl Genet 111:994–1012.

Kofer, W., Glimelius, K., Bonnett, H. T. 1991. Modifications of mitochondrial DNA cause changes in floral development in homeotic-like mutants of tobacco. Plant Cell 3:759–769.

Koizuka, N., Imai, R., Fujimoto, H., Hayakawa, T., Kimura, Y., Kohno-Murase, J., Sakai, T., Kawasaki, S., Imamura, J. 2003. Genetic characterization of pentatricopeptide repeat protein gene, *orf687*, that restores fertility in the cytoplasmic male-sterile Kosena radish. Plant J 34:407–415.

Koizuka, N., Imai, R., Iwabuchi, M., Sakai, T., Imamura, J. 2000. Genetic analysis of fertility restoration and accumulation of ORF125 mitochondrial protein in the kosena radish (*Raphanus sativus* cv. Kosena) and a *Brassica napus* restorer line. Theor Appl Genet 100:949–955.

Kotera, E., Tasaka, M., Shikanai, T. 2005. A pentatricopeptide repeat protein is essential for RNA editing in chloroplasts. Nature 433:326–330.

Kramer, E. M., Hall, J. C. 2005. Evolutionary dynamics of genes controlling floral development. Curr Opin Plant Biol 8:13–18.

Krizek, B. A., Fletcher, J. C. 2005. Molecular mechanisms of flower development: an armchair guide. Nat Rev Genet 6:688–698.

Kunst, L., Klenz, J. E., Martinez-Zapater, J., Haughn, G. W. 1989. *AP2* gene determines the identity of perianth organs in flowers of *Arabidopsis thaliana*. Plant Cell 1:1195–1208.

Lee, S.-L. J., Warmke, H. E. 1979. Organelle size and number in fertile and T-cytoplasmic male-sterile corn. Am J Bot 66:141–148.

Leino, M., Landgren, M., Glimelius, K. 2005. Alloplasmic effects on mitochondrial transcriptional activity and RNA turnover result in accumulated transcripts of *Arabidopsis orfs* in cytoplasmic male-sterile *Brassica napus*. Plant J 42:469–480.

Leino, M., Teixeira, R., Landgren, M., Glimelius, K. 2003. *Brassica napus* lines with rearranged *Arabidopsis* mitochondria display CMS and a range of developmental aberrations. Theor Appl Genet 106:1156–1163.

Leino, M., Thyselius, S., Landgren, M., Glimelius, K. 2004. *Arabidopsis thaliana* chromosome III restores fertility in a cytoplasmic male-sterile *Brassica napus* line with *A. thaliana* mitochondrial DNA. Theor Appl Genet 109:272–279.

Li, X.-Q., Jean, M., Landry, B. S., Brown, G. G. 1998. Restorer genes for different forms of *Brassica* cytoplasmic male-sterility map to a single nuclear locus that modifies transcripts of several mitochondrial genes. Proc Natl Acad Sci USA 95:10032–10037.

Liao, X., Butow, R. A. 1993. *RTG1* and *RTG2*: two yeast genes required for a novel path of communication from the mitochondria to the nucleus. Cell 72:61–71.

Linke, B., Nothnagel, T., Börner, T. 1999. Morphological characterization of modified flower morphology of three novel alloplasmic male-sterile carrot sources. Plant Breed 118:543–548.

Linke, B., Nothnagel, T., Börner, T. 2003. Flower development in carrot CMS plants: mitochondria affect the expression of MADS box genes homologous to GLOBOSA and DEFICIENS. Plant J 34:27–37.

Liu, F., Cui, X., Horner, H. T., Weiner, H., Schnable, P. S. 2001. Mitochondrial aldehyde dehydrogenase activity is required for male fertility in maize. Plant Cell 13:1063–1078.

Livers, R. W. 1964. Fertility restoration and its inheritance in cytoplasmic male-sterile wheat. Science 144:420.

Lössl, A., Frei, U., Wenzel, G. 1994. Interaction between cytoplasmic composition and yield parameters in somatic hybrids of *S. tuberosum* L. Theor Appl Genet 89:873–878.

Lurin, C., Andrés, C., Aubourg, S., Bellaoui, M., Bitton, F., Bruyère, C., Caboche, M., Debast, C., Gualberto, J., Hoffman, B., Lecharny, A., Ret, M. L., Martin-Magniette, M.-L., Mireau, H., Peeters, N., Renou, J.-P., Szurek, B., Taconnat, L., Small, I. 2004. Genome-wide analysis of *Arabidopsis* pentatricopeptide repeat proteins reveals their essential role in organelle biogenesis. Plant Cell 16:2089–2103.

Ma, H., Yanofsky, M. F., Meyerowitz, E. M. 1991. *AGL1-AGL6*, an *Arabidopsis* gene family with similarity to floral homeotic and transcription factor genes. Genes Dev 5:484–495.

Malcomber, S. T., Kellogg, E. A. 2005. *SEPALLATA* gene diversification: brave new whorls. Trends Plant Sci 10:427–435.

Mandel, M. A., Gustavson-Brown, C., Savidge, B., Yanofsky, M. F. 1992. Molecular characterization of the *Arabidopsis* floral homeotic gene *APETALA1*. Nature 360:273–277.

McCauley, D. E., Olson, M. S. 2008. Do recent findings in plant mitochondrial molecular and population genetics have implications for the study of gynodioecy and cytonuclear conflict? Evolution 62:1013–1025.

McCollum, G. D. 1979. Sterility in successive backcrosses of *Raphanobrassica* (2n = 4x = 36) with recurrent *Brassica oleracea* (2n = 2x = 18). Can J Genet Cytol 21:479–485.

McCollum, G. D. 1981. Induction of an alloplasmic male-sterile *Brassica oleracea* by substituting cytoplasm from 'Early Scarlet Globe' radish (*Raphanus sativus*). Euphytica 30:855–859.

Meguro, A., Takumi, S., Ogihara, Y., Murai, K. 2003. *WAG*, a wheat *AGAMOUS* homolog, is associated with development of pistil-like stamens in alloplasmic wheats. Sex Plant Reprod 15:221–230.

Menczel, L., Morgan, A., Brown, S., Maliga, P. 1987. Fusion-mediated combination of Ogura-type cytoplasmic male-sterility with *Brassica napus* plastids using X-irradiated CMS protoplasts. Plant Cell Rep 6:98–101.

Meur, G., Gaikwad, K., Bhat, S. R., Prakash, S., Kirti, P. B. 2006. Homeotic-like modifications of stamens to petals is associated with aberrant mitochondrial gene expression in cytoplasmic male-sterile Ogura *Brassica juncea*. J Genet 85:133–139.

Mizukami, Y., Ma, H. 1992. Ectopic expression of the floral homeotic gene *AGAMOUS* in transgenic *Arabidopsis* plants alters floral organ identity. Cell 71:119–131.

Mizumoto, K., Hatano, H., Hirabayashi, C., Murai, K., Takumi, S. 2009. Altered expression of wheat *AINTEGUMENTA* homolog, *WANT-1*, in pistil and pistil-like transformed stamen of an alloplasmic line with *Aegilops crassa* cytoplasm. Dev Genes Evol 219:175–187.

Mukai, Y., Tsunewaki, K. 1976. Genetic diversity of the cytoplasm in *Triticum* and *Aegilops* IV. Distribution of the cytoplasm inducing variegation in common wheat. Theor Appl Genet 48:9–16.

Murai, K., Takumi, S., Koga, H., Ogihara, Y. 2002. Pistillody, homeotic transformation of stamens into pistil-like structures, caused by nuclear-cytoplasm interaction in wheat. Plant J 29:169–181.

Murai, K., Tsunewaki, K. 1993. Photoperiod-sensitive cytoplasmic male-sterility in wheat with *Aegilops crassa* cytoplasm. Euphytica 67:41–48.

Murai, K., Tsunewaki, K. 1994. Genetic analysis on the fertility restoration by *Triticum aestivum* cv. Chinese Spring against photoperiod-sensitive cytoplasmic male-sterilty. Jpn J Genet 69:195–202.

Nagasawa, N., Miyoshi, M., Sano, Y., Satoh, H., Hirano, H., Sakai, H., Nagato, Y. 2003. *SUPERWOMAN1* and *DROOPING LEAF* genes control floral organ identity in rice. Development 130:705–718.

Newton, K. J., Coe, E. H. 1986. Mitochondrial DNA changes in abnormal growth (nonchromosomal stripe) mutants of maize. Proc Natl Acad Sci USA 83:7363–7366.

Newton, K. J., Gabay-Laughnan, S., De Paepe, R. 2004. Mitochondrial mutations in plants. Day, D. A., Millar, A. H., Whelan, J., eds. In Plant mitochondria: from genome to function. Kluwer, Dordrecht, pp. 121–141.

Nizampatnam, N. R., Doodhi, H., Narasimhan, Y. K., Mulpuri, S., Viswanathaswamy, D. K. 2009. Expression of sunflower cytoplasmic male-sterility-associated open reading frame, orfH522 induces male-sterility in transgenic tobacco plants. Planta 229:987–1001.

Nothnagel, T., Straka, P., Linke, B. 2000. Male-sterility in populations of *Daucus* and the development of alloplasmic male-sterile lines of carrot. Plant Breed 119:145–152.

Ogihara, Y., Futami, K., Tshji, K., Murai, K. 1997. Alloplasmic wheats with *Aegilops crassa* cytoplasm which express photoperiod-sensitive homeotic transformations of anthers, show alterations in mitochondrial DNA structure and transcription. Mol Genet Genomics 255:45–53.

Ogihara, Y., Kurihara, Y., Futami, K., Tsuji, K., Murai, K. 1999. Photoperiod-sensitive cytoplasmic male-sterility in wheat: nuclear-mitochondrial incompatibility results in differential processing of the mitochondrial *orf25* gene. Curr Genet 36:354–362.

Ogura, H. 1968. Studies on the new male-sterility in Japanese radish, with special reference to the utilization of this sterility towards the practical raising of hybrid seeds. Mem Fac Agric Kagoshima Univ 6:39–78.

Okuda, K., Nakamura, T., Sugita, M., Shimizu, T., Shikanai, T. 2006. A pentatricopeptide repeat protein is a site recognition factor in chloroplast RNA editing. J Biol Chem 281:37661–37667.

Pathania, A., Bhat, S. R., Dinesh Kumar, V., Ashutosh, Kirti, P. B., Prakash, S., Chopra, V. L. 2003. Cytoplasmic male-sterility in alloplasmic *Brassica juncea* carrying *Diplotaxis catholica* cytoplasm: molecular characterization and genetics of fertility restoration. Theor Appl Genet 107:455–461.

Patil, C., Walter, P. 2001. Intracellular signaling from the endoplasmic reticulum to the nucleus: the unfolded protein response in yeast and mammals. Curr Opin Cell Biol 13:349–356.

Pelaz, S., Ditta, G. S., Baumann, E., Wisman, E., Yanofsky, M. F. 2000. B and C floral organ identity functions require *SEPALLATA* MADS-box genes. Nature 405:200–203.

Pellan-Delourme, R., Renard, M. 1988. Cytoplasmic male-sterility in rapeseed (*Brassica napus* L.): female fertility of restored rapeseed with "Ogura" and cybrid cytoplasms. Genome 30:234–238.

Pelletier, G., Primard, C., Vedel, F., Chetrit, P., Remy, R., Rousselle, Renard, M. 1983. Intergeneric cytoplasmic hybridization in cruciferae by protoplast fusion. Mol Gen Genet 191:244–250.

Pelletier, G., Primard, C., Vedel, F., Chetrit, P., Renard, M., Pellan-Delourme, R., Mesquida, J. 1987. Molecular, phenotypic and genetic characterization of mitochondrial recombinants in rapeseed. Paper presented at: Proceedings of the 7th International Rapeseed Conference, Poznan, Poland.

Polowick, P. L., Sawhney, V. K. 1987. A scanning electron microscopic study on the influence of temperature on the expression of cytoplasmic male-sterility in *Brassica napus*. Can J Bot 65:807–814.

Prakash, S., Ahuja, I., Upreti, H. C., Dinesh Kumar, V., Bhat, S. R., Kirti, P. B., Chopra, V. L. 2001. Expression of male-sterility in alloplasmic *Brassica juncea* with *Erucastrum canariense* cytoplasm and the development of a fertility restoration system. Plant Breed 120:479–482.

Primard-Brisset, C., Poupard, J. P., Horvais, R., Eber, F., Pelletier, G., Renard, M., Delourme, R. 2005. A new recombined double low restorer line for the *Ogu*-INRA CMS in rapeseed (*Brassica napus* L.). Theor Appl Genet 111:736–746.

Purugganan, M. D., Rounsley, S. D., Schmidt, R. J., Yanofsky, M. F. 1995. Molecular evolution of flower development: diversification of the plant MADS-box regulatory gene family. Genetics 140:345–356.

Rhoads, D. M., Subbaiah, C. C. 2007. Mitochondrial retrograde regulation in plants. Mitochondrion 7:177–194.

Rhoads, D. M., Vanlerberghe, G. C. 2004. Mitochondria-nucleus interactions: evidence for mitochondrial retrograde communication in plant cells. Day, D. A., Millar, A. H., Whelan, J., eds. In Plant mitochondria: from genome to function. Kluwer, Dordrecht, pp. 83–106.

Rodermel, S. 2001. Pathways of plastid-to-nucleus signalling. Trends Plant Sci 6:471–478.

Rounsley, S. D., Ditta, G. S., Yanofsky, M. F. 1995. Diverse roles for MADS box genes in *Arabidopsis* development. Plant Cell 7:1259–1269.

Sabar, M., De Paepe, R., de Kouchkowsky, Y. 2000. Complex I impairment, respiratory compensations, and mitochondrial male-sterile mutants of Nicotiana sylvestris. Plant Physiol 124:1239–1249.

Sabar, M., Gagliardi, D., Balk, J., Leaver, C. J. 2003. ORFB is a subunit of F_1F_0-ATP synthase: insight into the basis of cytoplasmic male-sterility in sunflower. EMBO Rep 4:381–386.

Sakai, T., Liu, H. J., Iwabuchi, M., Kohno-Murase, J., Imamura, J. 1996. Introduction of a gene from fertility restored radish (*Raphanus sativus*) into *Brassica napus* by fusion of X-irradiated protoplasts from a radish restorer line and iodacetoamide-treated protoplasts from a cytoplasmic male-sterile cybrid of *B. napus*. Theor Appl Genet 93:373–379.

Sakamoto, W., Kondo, H., Murata, M., Motoyoshi, F. 1996. Altered mitochondrial gene expression in a maternal distorted leaf mutant of *Arabidopsis* induced by *chloroplast mutator*. Plant Cell 8:1377–1390.

Savidge, B., Rounsley, S. D., Yanofsky, M. F. 1995. Temporal relationship between the transcription of two *Arabidopsis* MADS box genes and the floral organ identity genes. Plant Cell 7:721–733.

Schmitz-Linneweber, C., Williams-Carrier, R., Barkan, A. 2005. RNA immunoprecipitation and microarray analysis show a chloroplast pentatricopeptide repeat protein to be associated with the 59 region of mRNAs whose translation it activates. Plant Cell 17:2791–2804.

Schmitz-Linneweber, C., Williams-Carrier, R. E., Williams-Voelker, P. M., Kroeger, T. S., Vichas, A., Barkan, A. 2006. A pentatricopeptide repeat protein facilitates the *trans*-splicing of the maize chloroplast *rps12* pre-mRNA. Plant Cell 18:2650–2663.

Schwarz-Sommer, Z., Huijser, P., Nacken, W., Saedler, H., Sommer, H. 1990. Genetic control of flower development by homeotic genes in *Antirrhinum majus*. Science 250:931–936.

Shumway, L. K., Bauman, L. F. 1967. Nonchromosomal stripe of maize. Genetics 55:33–38.

Soltis, D. E., Soltis, P. S., Albert, V. A., Oppenheimer, D. G., dePamphilis, C. W., Ma, H., Frohlich, M. W., Theißen, G. 2002. Missing links: the genetic architecture of flower and floral diversification. Trends Plant Sci 7:22–31.

Sundberg, E., Landgren, M., Glimelius, K. 1987. Fertility and chromosome stability in *Brassica napus* resynthesised by protoplast fusions. Theor Appl Genet 75:96–104.

Tang, H. V., Pring, D. R., Shaw, L. C., Salazar, R. A., Muza, F. R., Yan, B., Schertz, K. F. 1996. Transcript processing internal to a mitochondrial open reading frame is correlated with fertility restoration in male-sterile sorghum. Plant J 10:123–133.

Teixeira, R. T., Farbos, I., Glimelius, K. 2005a. Expression levels of meristem identity and homeotic genes are modified by nuclear-mitochondrial interactions in alloplasmic male-sterile lines of *Brassica*. Plant J 42:731–742.

Teixeira, R. T., Knorpp, C., Glimelius, K. 2005b. Modified sucrose, starch, and ATP levels in two alloplasmic male-sterile lines of *B. napus*. J Exp Bot 56:1245–1253.

Theißen, G. 2001. Development of floral organ identity: stories from the MADS house. Curr Opin Plant Biol 4:75–85.

Uyttewaal, M., Arnal, N., Quadrado, M., Martin-Canadell, A., Vrielynck, A., Hiard, S., Gherbi, H., Bendahmane, A., Budar, F., Mireau, H. 2008. Characterization of *Raphanus sativus* pentatricopeptide repeat proteins encoded by the fertility restorer locus for Ogura cytoplasmic male-sterility. Plant Cell 20:3331–3345.

Vedel, F., Chétrit, P., Mathieu, C., Pelletier, G., Primard, C. 1986. Several different mitochondrial DNA regions are involved in intergenomic recombination in *Brassica napus* cybrid plants. Curr Genet 11:17–24.

Wang, Z., Zou, Y., Li, X., Zhang, Q., Chen, L., Wu, H., Su, D., Chen, Y., Guo, J., Luo, D., Long, Y., Zhong, Y., Liua, Y.-G. 2006. Cytoplasmic male-sterility of rice with Boro II cytoplasm is caused by a cytotoxic peptide and is restored by two related PPR motif genes via distinct modes of mRNA silencing. Plant Cell 18:676–687.

Warmke, H. E., Lee, S.-L. J. 1977. Mitochondrial degeneration in Texas cytoplasmic male-sterile corn anthers. J Hered 68:213–222.

Wen, L., Chase, C. D. 1999. Pleiotropic effects of a nuclear restorer-of-fertility locus on mitochondrial transcripts in male-fertile and S male-sterile maize. Curr Genet 35:521–526.

Whipple, C. J., Ciceri, P., Padilla, C. M., Ambrose, B. A., Bandong, S. L., Schmidt, R. J. 2004. Conservation of B-class floral homeotic gene function between maize and *Arabidopsis*. Development 131:6083–6091.

Wise, R. P., Gobelman-Werner, K., Pei, D., Dill, C. L., Schnable, P. S. 1999. Mitochondrial transcript processing and restoration of male-fertility in T-cytoplasm maize. J Hered 90:380–385.

Yamada, K., Saraike, T., Shitsukawa, N., Hirabayashi, C., Takumi, S., Murai, K. 2009. Class D and B_{sister} MADS-box genes are associated with ectopic ovule formation in the pistil-like stamens of alloplasmic wheat (*Triticum aestivum* L.). Plant Mol Biol 71:1–14.

Yang, S., Terachi, T., Yamagishi, H. 2008. Inhibition of *chalcone synthase* expression in anthers of *Raphanus sativus* with Ogura male-sterile cytoplasm. Ann Bot 102:483–489.

Yanofsky, M. F., Ma, H., Bowman, J. L., Drews, G. N., Feldmann, K. A., Meyerowitz, E. M. 1990. The protein encoded by the *Arabidopsis* homeotic gene *agamous* resembles transcription factors. Nature 346:35–39.

Yui, R., Iketani, S., Mikami, T., Kubo, T. 2003. Antisense inhibition of mitochondrial pyruvate dehydrogenase E1α subunit in anther tapetum cause male-sterility. Plant J 34:57–66.

Zhu, Y., Saraike, T., Yamamoto, Y., Hagita, H., Takumi, S., Murai, K. 2008. $orf260^{cra}$, a novel mitochondrial gene, is associated with the homeotic transformation of stamens into pistil-like structures (pistillody) in alloplasmic wheat. Plant Cell Physiol 49:1723–1733.

Zubko, M. K., Zubko, E. I., Patskovsky, Y. V., Khvedynich, O. A., Fisahn, J., Gleba, Y. Y., Schieder, O. 1996. Novel "homeotic" CMS patterns generated in *Nicotiana* via cybridization with *Hyoscyamus* and *Scopolia*. J Exp Bot 47:1101–1110.

Zubko, M. K., Zubko, E. I., Ruban, A. V., Adler, K., Mock, H.-P., Misera, S., Gleba, Y. Y., Grimm, B. 2001. Extensive developmental and metabolic alterations in cybrids *Nicotiana tabacum* (+ *Hyoscyamus niger*) are caused by complex nucleo-cytoplasmic incompatibility. Plant J 25:627–693.

Chapter 19
Human Mitochondrial Mutations and Repair

Nadiya M. Druzhyna, Glenn L. Wilson, and Susan P. LeDoux

Abstract One of the unique features of mitochondria is that they have their own genome. Mitochondrial DNA, just like its counterpart in the nucleus, is constantly exposed to damaging agents such as ionizing radiation, environmental toxins, as well as many therapeutic drugs. Mitochondrial DNA is also the main target of endogenous ROS. Oxidative damage and mutations are common in mtDNA. A wide spectrum of pathogenic mutations of mtDNA has been demonstrated and associated with common diseases such as diabetes, neurodegeneration, cancer, and aging. Although some of these mutations are inherited, more and more attention is being focused on the accumulation of mitochondrial DNA mutations in somatic cells, particularly terminally differentiated cells. Mutations can be the results of unrepaired damage to mtDNA. Evidence now clearly shows that mitochondria contain the machinery to repair the damage to their genomes caused by certain endogenous or exogenous damaging agents. In this review we provide general information about the human mitochondrial genome and susceptibility of mtDNA to damage, show the association of human mtDNA mutations with aging and diseases, describe the pathways and the proteins involved in mammalian mtDNA repair in normal and pathologic states, and discuss how modulation of mtDNA repair could be a powerful tool to better understanding of the biologic significance of mtDNA repair mechanisms for cellular defenses.

Keywords Mitochondrial DNA (mtDNA) • mtDNA mutations • Base excision repair (BER) • DNA repair enzymes • Mitochdrial theory of aging

Abbreviations

4NQO 4-Nitroquinoline-1 Oxide
8-oxoG 8-Oxoguanine

N.M. Druzhyna (✉)
Department of Cell Biology and Neuroscience, College of Medicine, University of South Alabama, Mobile, AL 36688, USA
e-mail: ndruzhyna@jaguar1.usouthal.edu

F. Kempken (ed.), *Plant Mitochondria*, Advances in Plant Biology 1,
DOI 10.1007/978-0-387-89781-3_19,

AD	Alzheimer's Disease
ALS	Amyotrophic Lateral Sclerosis
APE	Apurinic/Apyrimidinic Endonuclease
ATP	Adenosine Triphosphate
BER	Base Excision Repair
bp	Base Pair
COX	Cytochrome C Oxidase
DR	Direct Repair/Reversal
dsb	Double Strand Break
ETC	Electron Transport Chain
HD	Huntington's Disease
HR	Homologous Recombination
LPBER	Long-Patch Base Excision Repair
MELAS	Mitochondrial Encephalopathy Lactic Acidosis and Stroke
MERRF	Myoclonic Epilepsy with Ragged Red Fibers
MGMT	O^6-Methy Guanine DNA Methyltransferase
MMR	Mismatch Repair
MnSOD	Manganese Superoxide Dismutase
MPG	N-Methylpurine DNA Glycosylase
mtDNA	Mitochondrial DNA
mtSSB	Mitochondrial Single Stranded Binding Protein
MYH	MutY Homolog DNA Glycosylase
nDNA	Nuclear DNA
NER	Nucleotide Excision Repair
NHEJ	Nonhomologous End Joining
OGG1	8-Oxo-Guanine DNA Glycosylase
PD	Parkinson's Disease
pol γ	DNA Polymerase Gamma
ROS	Reactive Oxygen Species
SOD	Superoxide Dismutase
TFAM	Transcriptional Factor A
tg	Thymine Glycol
tRNA	Transfer RNA
UNG	Uracil DNA Glycosylase

19.1 Introduction

Human cells rely on adenosine triphosphate (ATP), which is a source of chemical energy, for growth, differentiation, and response to physiologic stimuli and environmental challenge. Mitochondria, membrane-enclosed organelles that are found in eukaryotic cells, generate most of the cell's supply of ATP. The synthesis of ATP occurs through the respiratory or electron transport chain (ETC), which is located at the inner mitochondrial membrane and consists of five protein complexes (complex I to

V; see also Chapters 13 and 14). Mitochondrial respiration accounts for approximately 90% of cellular oxygen consumption. Most of the oxygen that is consumed is reduced to water through four consecutive one-electron reductions. During this process, a small proportion of the oxygen molecules (1–2%) are converted to superoxide anion radicals. Initially, it was believed that there were two main sites of superoxide production in the respiratory chain, Complex I and Complex IV (Turrens 1997). However, recent studies have identified at least nine submitochondrial reactive oxygen species (ROS) generating sites (Andreyev et al. 2005; see also Chapter 14). The superoxide that is formed then dismutates either spontaneously or enzymatically through the action of superoxide dismutase to form hydrogen peroxide (Cadenas and Davies 2000). Hydrogen peroxide then can diffuse throughout the cell and decompose to form noxious hydroxyl radicals, which can injure the cell through interactions with macromolecules. Because of these processes, mitochondria are a major source for the endogenous production of ROS. Although much attention has been focused on the harmful effects of ROS, it now has become apparent that mitochondrially-generated ROS also are involved in the regulation of intracellular signal transduction pathways leading to cellular activities such as proliferation (Hsu et al. 2000). In addition to supplying ATP, mitochondria are required for numerous other cellular functions, including biosynthesis of heme, cholesterol, and phospholipids (Bauer et al. 1999). Also, mitochondria play a major role in initiating the process of apoptosis (Gupta 2001) (see Chapters 17 and 18 for plant related processes).

One of the unique features of mitochondria is that they have their own genome. The human mitochondrial DNA (mtDNA) encodes for 13 proteins exclusively involved in ETC and genes for tRNA and rRNA for the mitochondrion's translation machinery. Mitochondria replicate and transcribe their DNA semiautonomously. Mitochondrial DNA, just like its counterpart in the nucleus, is constantly exposed to damaging agents such as ionizing radiation, environmental toxins, as well as many therapeutic drugs. Mitochondrial DNA is also the main target of endogenous ROS. Oxidative damage and mutations are common in mtDNA. Until recently, interests in alterations of mtDNA were confined to relatively rare inborn mitochondrial neuromyopathies. However, a wide spectrum of pathogenic mutations (see Box 19.1) of mtDNA has been demonstrated and associated with common diseases such as diabetes, neurodegeneration, cancer, and aging. Although some of these mutations are inherited, more and more attention is being focused on the accumulation of mitochondrial DNA mutations in somatic cells, particularly terminally differentiated cells. Mutations can be the result of unrepaired damage to mtDNA. The prevailing concept for many years was that mtDNA molecules suffering an excess of damage would simply be degraded to be replaced by newly generated successors copied from undamaged genomes. However, evidence now clearly shows that mitochondria contain the machinery to repair the damage to their genomes caused by certain endogenous or exogenous damaging agents. In this review we provide general information about the human mitochondrial genome and susceptibility of mtDNA to damage, show the association of human mtDNA mutations with aging and diseases, describe the pathways and the proteins involved in mammalian mtDNA repair in normal and pathologic states, and discuss how modulation of mtDNA repair could be a powerful tool to better understanding of the biologic significance of mtDNA repair mechanisms for cellular defenses.

Box 19.1 DNA Damage

Injuries in DNA introduce deviations from its normal double-helical conformation. These changes include structural distortions that interfere with replication and transcription, as well as point mutations that disrupt base pairs and exert damaging effects. Broadly speaking, there are two large classes of damage, endogenous DNA damage and exogenous DNA damage.

Endogenous DNA damage – originates within the cell; also referred to as Spontaneous:

1. Mismatches that arise during DNA replication
2. Damage resulting from the inherent instability of DNA (incorporation of uracil, deamination of bases, depurination and depyrimidination)
3. Oxidative damage

Exogenous or Environmental DNA damage – originates from the environment.

1. Ionizing radiation produces base damage and strand breaks
2. UV radiation produces cyclobutane pyrimidine dimers and other photoproducts
3. Alkylating agents alkylate bases

19.2 Mitochondrial Genome

Mammalian cells contain one to several thousand copies of mtDNA per cell, of the genome with 1–11 copies per organelle (Cavelier et al. 2000). Mitochondrial DNA comprises 0.1–2% of the total DNA in the most mammalian cells. The inheritance of mtDNA is almost exclusively maternal, although some important exceptions have been reported (Gyllensten et al. 1991; Schwartz and Vissing 2003). Human mtDNA is a circular, negatively supercoiled, double-stranded molecule that is 16,569 bp long. It encodes 2 rRNAs, 22 tRNAs, and 13 polypeptides, of which 7 are components of Complex I (NADH dehydrogenase), 3 are components of Complex IV (cytochrome c oxidase), and 2 are subunits of Complex V (ATP synthase) and cytochrome b (a subunit of Complex III) (Attardi and Schatz 1988). The rest of the 87 polypeptides of the ETC complexes, including all the subunits of Complex II (succinate dehydrogenase), as well as 1,500 other proteins that function in mitochondria are encoded by nuclear genes, synthesized in the cytosol, and imported into mitochondria through various protein import systems. The two mtDNA strands can be separated by denaturing cesium chloride gradient centrifugation (Kasamatsu and Vinograd 1974). Most of the information is encoded in the heavy (purine-rich) strand. The light (pyrimidine-rich) strand contains genetic information for only one polypeptide and 8 tRNAs. Mitochondrial genes have no introns and intergenic sequences are absent or limited to a few bases. Some genes overlap and in some instances, termination codons are not encoded (Ojala et al. 1981). mtDNA is totally dependent upon nuclear-encoded proteins for its maintenance and transcription. Mitochondrial DNA replicates throughout the lifespan of an organism in both proliferating and postmitotic cells in order to maintain a constant supply of genetic material so that the organelles can undergo continuous turnover. The mean lifetime of mtDNA molecules has been estimated to be 2 and 4 weeks in rat liver and brain cells, respectively (Gross et al. 1969). The replication of mtDNA is conducted by DNA polymerase gamma (DNA pol γ). Among the 16 known eukaryotic DNA polymerases, DNA pol γ is the only one to have been detected in mammalian mitochondria (Graziewicz et al. 2006). According to the strand-asymmetric model, mtDNA replication occurs bidirectionally, initiated at two spatially and temporally distinct origins of replications, O_H and O_L, for the heavy and light strands, respectively (Taanman 1999). However, this paradigm recently has been challenged, and evidence has been offered suggesting the presence of conventional duplex mtDNA replication intermediates, indicative of coupled leading and lagging-strand DNA synthesis (Yang et al. 2002). Mitochondria do not have histones. However, mammalian mtDNA is organized in nucleoids, which can be seen under the microscope as punctuate structures containing mtDNA and proteins, which localize to the matrix surface of the mitochondrial inner membrane. mtDNA binding proteins such as transcriptional factor A (TFAM), mtSSB, and Twinkle have been shown to colocalize with mtDNA in intramitochondrial foci in living cells (Garrido et al. 2003). The molecular structure of the nucleoid is, however, little elucidated even for yeast mtDNA, much less for a mammalian one.

Mitochondria are dynamic structures subject to fusion and fission and the rates of these processes can differ between cell types. Thus, a given mtDNA copy does not "belong" to a given mitochondrion, but rather is part of the fluid mitochondrial landscape of a cell (Legros et al. 2004) (see also Chapter 2).

19.3 Susceptibility of Mitochondrial DNA to Damage

Like nuclear DNA, mitochondrial DNA (mtDNA) is constantly exposed to DNA damaging agents. It is now recognized that the mitochondria produce nearly 90% of the free radicals generated in the living cells and are also the prime targets of these reactive oxygen species. Mitochondrial DNA is vulnerable to oxidative damage from both exogenous and endogenous ROS due to its proximity to the ETC, and to the lack of protective histones (Ames et al. 1993). Reactive oxygen radicals can cause a wide variety of lesions in DNA. Currently, more than 50 oxidative DNA base lesions have been characterized (reviewed in Cooke et al. 2003; Bjelland and Seeberg 2003), and many of these were identified in vitro or in vivo following exposure of mammalian cells to ROS. Other lesions include DNA-protein cross-links, sugar modifications, and single and double-strand breaks (Cooke et al. 2003). One of the most studied lesions is 8-oxoguanine (8-oxoG), which is formed by a variety of oxidative treatments. 8-oxoG has been found to be a mutagenic lesion.

Mispairing of 8-oxoG with adenine results in a G-C to T-A transversion during the subsequent round of replication. Early studies, using HPLC-electrochemical detection of the 8-oxoG, described 16-fold higher levels of 8-oxoG in mtDNA than in nDNA (Richter et al. 1988). Further studies revealed that 8-oxoG accumulates with age in nDNA and to a greater extent in mtDNA (Ames et al. 1993; Mecocci et al. 1993). However, others reported that the damage in mitochondria was overestimated, and its level is comparable to that of nuclear DNA (Anson et al. 2000; Lim et al. 2005). In more definitive studies, Yakes and Van Houten (1997), employing the method of quantitative extended-length PCR (QXL-PCR), showed that mtDNA damage is more extensive and persists longer than nDNA damage in human cells following oxidative stress.

Mitochondrial DNA also suffers damage from toxic chemicals to a greater extent than does nuclear DNA (Bandy and Davidson 1990). Because the mitochondrial membrane potential generates a negative charge on the matrix-side of the inner membrane, lipophilic cations tend to accumulate in mitochondria, specifically in mitochondrial membranes. Mitochondria import lipophilic cations from the cytosol and concentrate them up to 1,000-fold (Singer and Ramsay 1990). Many drugs and biologically toxic chemicals are lipophilic and have positive charges, and are thus concentrated in mitochondria and are a threat to their components. For example, mtDNA is modified by alkylating agents at least tenfold more than is nuclear DNA (nDNA) (Bandy and Davidson 1990).

Damage to mtDNA has been suggested to have a greater impact on cellular function than damage to nuclear DNA, despite the presence of multiple copies of mtDNA per mitochondrion and per cell (Bohr 2002a). The greater damage naturally would be expected to cause a higher mutation rate in mtDNA. Consistent with this expectation, genes encoded by mtDNA evolve tenfold faster than nuclear genes (Brown et al. 1979). It has been shown, using denaturing gradient gel electrophoresis, that the mutation rate in human mtDNA is several hundredfold higher than nuclear gene mutation rates (Khrapko et al. 1997). Furthermore, Taylor et al. (2003) estimated that the mutation rate of mtDNA would be about 100-fold higher than that of nuclear DNA by calculating the incidence of cytochrome c oxidase activity-negative cryptic cells of colon epithelia. Mitochondrial DNA encodes polypeptides of ETC or tRNA and rRNA required for their synthesis. Therefore, any coding mutations in mtDNA will affect the ETC as a whole, and the defects in the ETC can have pleiotropic effects because they affect the ability of the cell to produce energy.

19.4 Human Mitochondria Mutation in Aging and Disease

It has long been considered that mitochondrial DNA disease is a relatively rare genetic disorder causing myopathies. In the late 1980s it was discovered that mutations in mtDNA could be pathogenic. Since that time a number of human diseases have been linked to mutations in the mitochondrial genome. Among them are common pathologic states, including premature aging, cancer, diabetes mellitus, and neurodegenerative disorders. The recent release of the Mitomap (Human Mitochondrial Genome Database) lists almost 200 pathogenic point mutations, single nucleotide deletions, and insertions (http://mitomap.org). Although some of these mutations are inherited, more and more attention is being focused on the accumulation of mitochondrial DNA mutations in somatic cells. In contrast to diseases caused by mutations in nuclear DNA, mutations in mtDNA might not be fully expressed. This is because each mammalian cell contains 100 to 1,000 mitochondria, each of which carries 2 to 10 copies of mtDNA. If all of the mtDNA in a cell has a single sequence (whether mutated or normal) the cell is homoplasmic. If some of the mtDNA copies have a normal base sequence and others have a mutated sequence the cell is heteroplasmic. Somewhat paradoxically, it is common for somatic mitochondrial mutations (including non-detrimental silent mutations) to clonally expand within cells, (the progeny of a single mutant molecule is favored over the non-mutant and eventually comprises a large portion or even the entire population of mtDNA molecules). Clonal mtDNA mutation expansions can extend beyond the single cell level. The mechanisms of expansion remain a matter of debate. With mitochondrial diseases the concept of threshold burden has been proposed. This means that when the percentage of mutated genomes rises above a certain level, noticeable symptoms will be observed. When the percentage of mutated DNA far exceeds this threshold then an acute (and sometimes lethal) condition will be observed (Berdanier and Everts 2001). Many mitochondrial diseases have a delayed onset and progressive course and they result in many of the same clinical manifestations that are observed in age-related diseases.

19.4.1 Aging

There has been a longstanding debate as to whether aging is a process of damage accumulation or an ordered, programmed process that is determined by changes in gene regulation. Accumulation of oxidative damage is the basis of Harman's free-radical theory of aging (Harman 1956, 2001). The main source for ROS in the cell is oxidative phosphorylation in mitochondria, so that the free-radical theory of aging is essentially a mitochondrial theory of aging. Mutations of mtDNA are thought to arise due to its close proximity to the oxidative phosphorylation system, which is located on the inner mitochondrial membrane, thus making it vulnerable to damage through leakage of ETC ROS. mtDNA mutations accumulate progressively during life and are directly responsible for a measurable deficiency in cellular oxidative phosphorylation activity, leading to enhanced ROS production. In turn, increased ROS production results in progressive mtDNA damage and mutagenesis, thus causing a "vicious cycle" of exponentially increasing oxidative damage and cellular dysfunction, which ultimately culminates in death. An alternative theory is that mtDNA mutations occur throughout life either due to oxidative damage or errors of the mtDNA polymerase, and it is these mutations that clonally expand to cause cellular dysfunction rather than accumulation of new mutations as predicted by the vicious cycle.

Numerous studies have shown that mtDNA accumulates oxidative damage in an age-dependent manner in skeletal muscle, cardiac muscle, brain, and liver (Sastre et al. 2003; Alexeyev et al. 2004). Specifically, the increase in 8-oxoG levels in mtDNA with age appears to be a general phenomenon and has been reported by de Souza-Pinto et al. (2001a) and Hudson et al. (1998). The steady-state concentration of 8-oxoG in mtDNA, but not in nuclear, was shown to be inversely correlated with maximum lifespan in the heart and brain of mammals. Slowly aging mammals show lower 8-oxoG levels in mtDNA than rapidly aging ones (Barja and Herrero 2000). Persistence of this damage would be expected to cause a high mutation rate in mtDNA. Thus, it is not surprising that mutations, including deletions, duplications, and point mutations, have been found to accumulate in mtDNA in a variety of tissues during aging in humans, monkeys, and rodents (Wei and Lee 2002; Khaidakov et al. 2003) and cause a mosaic pattern of respiratory chain deficiency in pre- and postmitotic tissues. The most frequent and best-characterized age-associated mtDNA mutation is a 4977-bp deletion also called "common deletion." The 4977-bp deletion is located in the major arc between the two origins of replication (O_H and O_L), and removes seven genes for polypeptides and five tRNA genes. The common deletion was shown to be present in a wide variety of aged tissues, and was not detected in the same tissues from young individuals (Greaves and Turnbull 2009). Several mechanisms have been proposed to explain the formation of 4977-bp deleted mtDNA, including replication error and inefficient repair of damaged mtDNA (Krishnan et al. 2008). Recently, two independent studies showed that with age mtDNA deletions accumulate to high levels in neurons from the substantia nigra, and these high levels of mtDNA deletions were associated with a respiratory chain defect (Kraytsberg et al. 2006; Bender et al. 2006). Point mutations in the control/D-loop region of human mtDNA accumulate in age-dependent manner, and age-related large rearrangements of mtDNA also have been reported (Wei and Lee 2002). Some of these aging-associated point mutations were originally observed in the affected tissues of patients with mitochondrial diseases. Among them were the A3243G mutation associated with mitochondrial myopathy, encephalopathy, lactic acidosis, and stroke (MELAS) and A8344G transition commonly associated with Myoclonic Epilepsy with Ragged Red Fibers (MERRF). Analysis of these transitions showed them to be undetectable in infant tissue but present in adult tissue (Greaves and Turnbull 2009). Michikawa et al. (1999) reported that a T414G transversion in the D-loop of mtDNA is accumulated in 25–50% of mtDNA in the skin fibroblasts from elderly subjects. This mutation is located at the control region of mtDNA and may impair the replication and transcription of mtDNA in cells of the elderly human subjects. However, subsequent studies have failed to show the presence of this mutation accumulating with age in the brain (Murdock et al. 2000; Simon et al. 2001). This suggests that the mutation may be tissue specific. It has been established that many of these mtDNA mutations start to occur after the mid-thirties and they accumulate with age. The proportions of mutated mtDNA in aging human tissues rarely exceed 1%. To explain how these mutations can be functionally relevant in aging there are debates in the literature about mosaic patterns of mtDNA segregation, a clonal expansion of many individual mtDNA point mutations, as well as methodologic pitfalls, that may lead to misinterpretation of some of the results (Wei and Lee 2002; Szibor and Holtz 2003; Kang and Hamasaki 2005; Singh 2006; Berneburg et al. 2006; Alberio et al. 2007). A definitive explanation remains elusive.

To establish a casual relationship between mtDNA mutations and aging, Trifunovic et al. (2004), recently developed knock-in mice that express a proofreading deficient version of DNA pol γ and studied these mice for mtDNA mutations and mitochondrial age-related changes. They observed a three- to fivefold increase in the level of point mutations and deletions in mtDNA. Clinically, these mutant mice developed progeroid symptoms such as reduced lifespan, weight loss, reduced subcuntaneous fat, osteoporosis, alopecia, kyphosis, and reduced fertility (Trifunovic et al. 2004). Another group (Kujoth et al. 2005) confirmed these observations but could not detect elevated levels of ROS in their knock-out mice. However, they did find increased apoptosis in tissues with rapid cellular turnover. They suggested that the aging phenotype in these mice could be due to apoptosis, which in this case could come from decreased maintenance of mtDNA through restricted repair of replicating mtDNA (Kujoth et al. 2005). These findings suggest a causative link between mtDNA mutations and aging phenotypic changes in mammals.

19.4.2 Diabetes Mellitus

It is well understood that normal mitochondrial function is essential for insulin secretion in pancreatic beta cells. When insulin-secreting cells are depleted of mtDNA, they lose the ability to secrete insulin upon glucose stimulation (Hayakawa et al. 1998). Diabetes mellitus often accompanies congenital encephalomyopathies with mtDNA defects (DiMauro and Moraes 1993), suggesting that pancreatic beta cells are also prone to accumulate mtDNA molecules with mutations. Of the 1% of diabetes that is caused by a mitochondrial DNA defect, A to G substitution at position 3243 is the most common subtype, causing over 85% of such forms of diabetes (Murphy et al. 2008). Other, rare mitochondrial point mutations associated with diabetes also occur within tRNA genes for either leucine: at positions 3256C>T, 3264T>C, 3254C>G, and 3271T>C; lysine: 8269A>G, 8344A>G, and 8356T>C; arginine: 10438A>G; serine 12258C>A; or glutamine 1409T>C. Other point mutations are found in the ND1 subunit of the respiratory chain complex 1, such as 3394T>C and 3398T>C, or in the D loop, at 14577T>C. Most of these include prominent other neurologic manifestations (Murphy et al. 2008). To date 42 mutations in the mtDNA have been found that phenotype as diabetes in humans. The population with mitochondrial diabetes is not large. Estimate of the percentage of the diabetic population with mtDNA mutations range from 0.1 to 10% (Berdanier and Everts 2001). Because hundreds to thousands of copies of mtDNA molecules are present in one cell, wild-type and mutant mtDNA normally coexist. The degree of heteroplasmy varies among cells and tissues. Defects in mitochondrial respiration do not become apparent until the mutant mtDNA molecules occupy a large part of the total mtDNA population. Additionally, it was shown that diabetic hyperglycemia, which is a major factor involved in progression of pancreatic beta cell dysfunction, induced mitochondrial superoxide anion overproduction and impaired mitochondrial respiration in beta-cell (Nishikawa et al. 2000). Beta cells are vulnerable to damage caused by ROS. In animal models of type 2 diabetes, 8-oxoG (Ihara et al. 1999) and 4-hydroxynonen (HNE)-modified proteins (Gorogawa et al. 2002) increase. (HNE is a degradation product of lipid peroxides and is frequently used as a marker of lipid peroxidation). Alloxan, which is used in drug induced diabetes animal models, causes oxidative DNA damage in a mouse beta cell line (Driggers et al. 1993). It can be suggested, that mitochondrial oxidative stress and related mtDNA damage play a critical role in the onset and progression of diabetes mellitus.

19.4.3 Neurodegeneration

Mitochondrial oxidative stress and accumulation of the mtDNA mutations are believed to be particularly devastating to postmitotic, terminally differentiated cells such as neurons. Mitochondria are central components of synapses, where they provide the energy required for synaptic activities (Ly and Verstreken 2006). Damage to mtDNA could potentially result in bioenergetic dysfunction and consequently aberrant neuronal function. Neurodegenerative diseases are associated with a progressive loss of neurons through apoptosis and/or necrosis.

There have been a number of reports of mtDNA mutations associated with neurodegenerative disease. Several studies have demonstrated that mitochondrial function, especially the activity of cytochrome c oxidase (COX), is decreased in brain tissues of patients with Alzheimer's disease (AD) (reviewed in Atamna and Frey 2007). In a recent study, 20 point mutations were detected in the mitochondrial-encoded COX subunits I, II, and III genes in AD patients (Hamblet et al. 2006). Other works showed two missense mutations in the mtDNA of COX an high aggregate burden of somatic mtDNA point mutations in the patients with AD (Qiu et al. 2001; Lin et al. 2002); however, direct sequencing of the complete mtDNA coding regions has not identified disease-specific mutations (Elson et al. 2006). The content of 4977-bp deleted mtDNA in brain tissues samples taken from eldery AD patients was about 6.5-fold higher than that in normal age-related subjects (Hamblet and Castora 1997). Although DNA damage is elevated in both nuclear and mitochondrial DNA in AD brains, mtDNA in AD brains was shown to contain between three to tenfold higher levels of oxidized bases than nDNA (Wang et al. 2005). It has been suggested that mitochondrial respiratory dysfunction that results from mtDNA mutations may play a role in the progression and pathophysiology of AD. There is some evidence that mtDNA mutations may also be involved in amyotrophic lateral sclerosis (ALS), and Huntington's disease (HD). Ro et al. (2003) analyzed the 4977-bp deleted mtDNA in muscle specimens from ALS patients and matched controls, and found that the frequency of occurrence and average level of the "common deletion" in the ALS patients were significantly higher than those in the controls. Elevated levels of 8-oxoG have been found in the cortex of ALS patients (Bogdanov et al. 2000). Also, it was shown that HD patients have higher frequencies of mtDNA deletions in lymphocytes and leukocytes in comparison to the controls, and the number of mutations is directly related to the size of CAG repeats expansion (Banoei et al. 2007; Chen et al. 2007).

On the other hand, Parkinson's disease (PD) has been reported to be associated with deficiency of the respiratory enzyme complex I. To correlate with this, an accumulation of point mutations in complex I genes in PD has been reported (Vives-Bauza et al. 2002; Simon et al. 2004; Smigrodzki et al. 2004). Similarly, reduced complex I activity and increased susceptibility to mitochondrial processing peptidase were also observed in cybrids containing mtDNA from PD patients, suggesting mitochondrial DNA encoded defects in PD (Swerdlow et al. 1996; Gu et al. 1998). More mtDNA molecules with the common deletion are found in PD than in age-matched controls (Ikebe et al. 1990), suggesting that mtDNA is also more strongly damaged in this disease. An increased level of 8-oxoG (Shimura-Miura et al. 1999), as well as increase in COX-deficient neurons (Bender et al. 2006) was also seen in the substantia nigra from patients with PD compared with control subjects. The recent studies explored the characteristics of the mtDNA deletions detected in single substantia nigra neurons in patients with PD, age-matched controls, and a patient with a multiple mtDNA deletions (Reeve et al. 2008). There was no difference between the size of mtDNA deletion, length, and nature of the repeat sequence at the deletion breakpoint. This suggests that mechanism of mtDNA deletion formation is similar under a variety of different conditions and emphasizes the role of mtDNA mutations in aging and neurodegeneration.

19.4.4 Cancer

Mitochondrial defects have long been suspected to play an important role in the development and progression of cancer. One of the first indications that mtDNA suffer modifications in cancer was provided by Clayton and Vinograd (1967). They reported that 26% of the mtDNA molecules in human leukemia leukocytes were circular dimers. Horton et al. (1996) reported a renal adenocarcinoma in which 50% of the mtDNAs contained a 264 bp in-frame deletion in the *ND1* gene, which was preferentially transcribed, showing for the first time, that an mtDNA mutation could be functionally significant. Polyak et al. (1998) detected somatic mutations in 70% of investigated colorectal tumors. Over 90% of these mutations were T-C and G-A transitions. Zhu et al. (2005) found mutated mtDNA in 93% of breast cancer tissues, with the frequency of mutations being higher in the coding regions and D-loop regions than in the other tested loci. During the last decade mitochondrial DNA aberrations, which include point mutations, instability of mono- or dinucleotide repeats, insertions, and deletions or quantitative alterations, have been found in solid tumors such as breast, colon, stomach, liver, kidney, bladder, prostate, skin, and lung cancer (reviewed in Chatterjee et al. 2006a; Brandon et al. 2006; Santos et al. 2008). They have also been found in hematologic malignancies, such as leukemia and lymphoma (Fontenay et al. 2006). Mitochondrial DNA mutations often are found in primary tumors but not in surrounding tissues. According to Carew and Huang (2002), the main features of mtDNA mutations common to all tumor types are: (1) the majority of the mutations are base substitutions; (2) mutations occur in all protein-coding mitochondrial genes; (3) the D-loop region is the most frequent site of somatic mutations across most tumor types; (4) the mutations are homoplasmic in nature. The homoplasmic nature of mutated mtDNA raises the possibility that some mutations are involved in tumorigenesis itself by affecting energy metabolism and/or ROS production. Mitochondrial DNA mutations in tumors also could be due to the effect of aging (Trifunovic et al. 2004: Kujoth et al. 2005). Shidara et al. (2005) showed that specific point mutations in mtDNA accelerate growth and reduce apoptosis in a variety of tumors, supporting the notion that some mtDNA mutations in tumors have functional advantages that promote tumor growth. Additionally, mitochondria are key players in tumor control by apoptosis. A demonstration of the physiologic significance of mtDNA mutations found in cancer was provided recently by studies with transmitochondrial cybrids ("chimeric" cells in which their own mtDNA is destroyed and replaced by mtDNA from other cells). A known pathogenic mtDNA mutation in ATP6 was introduced into PC3 cells, a prostate cancer cell line via cybrid transfer. Injection of the resulting ATP6 mutant cybrids into nude mice resulted in the generation of tumors that were seven times larger than those generated from wild type cybrids (Petros et al. 2005). In addition, the ATP6 mutant cybrid tumors generated significantly more ROS than their wild type counterparts. These findings are a part of a growing body of evidence that indicates that cancer cells are under increased intrinsic oxidative stress (Pelicano et al. 2004). Since mitochondria are the major source for ROS production in cells, the vulnerability of mtDNA to ROS-mediated damage appears to be a mechanism for amplifying ROS stress in cancer cells. Also, mitochondrial ROS may play a role in carcinogenesis. This notion is supported by the finding that mice that are heterozygous for mitochondrial MnSOD ($MnSOD^{+/-}$) have a 100% increase in tumor incidence, accompanied by an increase in oxidative damage in nuclear and mtDNA (van Remmen et al. 2003). The mutated mtDNA found in cancer is 100-fold more abundant than mutated nuclear p53 DNA (Fliss et al. 2000), indicating that the mitochondrial mutations could be feasible as a molecular marker for detection of cancer.

19.5 Mitochondrial DNA Repair

It is well established that to maintain normal cell function, the preservation of DNA requires efficient repair systems. Characterization of DNA repair mechanisms has generally focused on these processes in nuclear DNA. Several major repair pathways processing a wide variety of DNA lesions are operative in the nucleus (Larsen et al. 2005). They include nucleotide excision repair (NER), base excision repair (BER), direct repair/reversal (DR), mismatch repair (MMR), homologous recombination (HR), nonhomologous end joining (NHEJ), and translesion synthesis. These incredibly complex systems, involving dozens of proteins and multiprotein complexes, continue to be actively investigated.

Initially, it was thought that DNA repair mechanisms were either nonexistent or very inefficient in mitochondria and that damaged DNA molecules were simply degraded, and undamaged copies served as templates for new mtDNA synthesis. Such views were based on early experiments showing that UV-induced pyrimidine dimers were not repaired in mtDNA (Clayton et al. 1974). In later studies, it was observed that mitochondria also did not repair complex alkyl damage caused by nitrogen mustard and intrastrand cross-links induced by cisplatin (LeDoux et al. 1992). Psoralen-induced cross-links also appeared to persist in mitochondria (Magana-Schwencke et al. 1982). The first report of mtDNA repair in mammalian cells came from studies by Pettepher et al. (1991), where they showed the removal of alkali-labile sites from mtDNA caused by streptozocin in an insulinoma cell line. Evidence indicating that mitochondria do possess some repair systems started accumulating at this time. Subsequent studies have demonstrated that certain types of damage to DNA bases resulting from deamination, simple alkylation, and oxidation can be efficiently repaired in mitochondria (reviewed in Croteau et al. 1999; LeDoux et al. 1999; Sawyer and Van Houten 1999). Thus, the major DNA repair mechanism acting in mitochondria is BER. In addition to BER, there is some evidence that mammalian mitochondria possess other DNA repair pathways. Myers et al. (1988) showed that mitochondrial extracts from liver cells effectively remove O^6-methyl-guanine from mtDNA, suggesting that direct repair mechanism is present in mammalian mitochondria. Repair of double strand breaks (DSB) in mtDNA may be conducted by recombination or nonhomologous end joining, as it occurs in nucleus. Based on studies involving in vivo induction of multiple DSBs in mtDNA, it was suggested that infrequent intermolecular recombination is a potential consequence of DSBs in mtDNA (Bacman et al. 2009). Also, it was shown that mammalian mitochondria can rejoin blunt-ended and cohesive linearized plasmid DNA at a low level (Lakshmipathy and Campbell 1999a). Mismatch repair (MMR) is a postreplicative DNA repair system, which corrects base mismatches and small loops. A very recent study (Souza-Pinto et al. 2009) showed that human mitochondrial extracts do support mismatch repair and the pathway is independent of nuclear MMR factors but involves the activity of polymerase γ and the multifunctional YB-1 protein. However, the nature of these mtDNA repair mechanisms is still a fertile area for investigation.

Base excision repair is relatively well described in mitochondria. In vitro repair studies have shown that mitochondria not only contain all the enzymes required for BER, but also this process is very similar to the short-patch pathway of BER in eukaryotic nucleus (Sawyer and Van Houten 1999). Complete BER has been reconstituted using mitochondrial extracts (Pinz and Bogenhagen 1998). In general, BER starts with recognition and removal of a damaged or inappropriate base by a DNA glycosylase that cleaves the N-glycosylic bond between the base and the sugar. Many glycosylases have an associated lyase activity that breaks the phosphodiester backbone 3′ to the abasic site. Apurinic/apyrimidinic (AP) endonuclease then cleaves the 5′ phosphodiester bond leaving a one-nucleotide gap. For glycosylases like those that remove uracil without an associated lyase activity, the action of AP endonuclease generates a deoxyribose-phosphate moiety that must be removed by the lyase activity in DNA polymerase γ (Longley et al. 1998). In both cases, the resulting 3′-hydroxyl moiety can be extended by DNA polymerase γ and DNA ligase activity completes the repair event by joining the free DNA ends. The sequence of events described above illustrates the short-patch (SP) BER, which involves the removal of the DNA lesion and incorporation of a single nucleotide. Nuclear BER can also proceed via long-patch (LP) BER, which involves the incorporation of two to twelve nucleotides, followed by cleavage of the resulting 5′ flap and ligation. Until recently, it was believed that only SP-BER occurred in mitochondria. However, intense studies of the last years suggest that long-patch BER is also involved in repairing mtDNA lesions in mammals (Akbari et al. 2008; Liu et al. 2008; Szczesny et al. 2008; Kalifa et al. 2009). Over the past several years, an active investigation has started to clarify the nature and mechanisms of action of the particular DNA repair enzymes involved in mammalian mtDNA BER. Recently, it was shown that mitochondrial BER proteins are not freely soluble, but strongly associated with an inner membrane-containing particulate fraction (Stuart et al. 2005).

19.5.1 DNA-Glycosylases in Mitochondria

The first step of the BER is catalyzed by DNA glycosylases. Up to eleven different DNA glycosylases have been characterized in mammals (Robertson et al. 2009), but only a few of them have been detected in mitochondria. The first DNA glycosylase activity identified in mitochondria of human cells was uracil DNA glycosylase (UNG; UDG) (Anderson an Friedberg 1980). This enzyme initiates the removal of uracil from DNA resulting from deamination of cytosine or misincorporation of dUMP. Mitochondrial UDG activity represents about 3–6% of the total cellular activity. The mitochondrial isoform of UNG (UNG 1) is transcribed from a separate promoter in the *UNG* gene and is further processed by alternative splicing (Krokan et al. 2001). UNG1 is the only mtDNA glycosylase that can act on both single-strand and double-strand DNA, whereas all others require double-strand DNA for activity. Moreover, UNG1 is the only uracil-DNA-glycosylase known to be present in mammalian mitochondria (Akbari et al. 2008). In the knockout mice model, it has been shown that *Ung−/−* mice develop normally and have no obvious phenotype after 18 months (Nilsen et al. 2000). These mice are deficient in UNG activity, but display low uracil-glycosylase activities in extracts from different organs, suggesting the existence of an unrecognized uracil-DNA-glycosylase (Krokan et al. 2001). Another group showed that *UDG−/−* mice are susceptible to neuronal injury following ischemia-reperfusion (Endres et al. 2004). The authors suggest that these phenotypes are due to mtDNA damage on the basis of apparent mitochondrial dysfunction and the observation that mitochondrial (but not nuclear) UDG activity is dramatically upregulated following exposure. It may be suggested that uracil-DNA-glycosylase plays an important role in mitochondrial function and in enhancing cell survival under extreme oxidative stress.

One of the most studied DNA glycosylase in mammalian cells is 8-oxo-guanine DNA glycosylase (OGG1). OGG1 is the principle enzyme identified as initiating the repair of 8-oxo-G. It is a bi-functional DNA glycosylase with an associated AP lyase activity. Two isoforms, αOgg1 and βOGG1 of OGG1 exist in human, which both arise from alternative splicing of the nuclear encoded transcript (Nishioka et al. 1999). Both isoforms contain a mitochondrial targeting sequence (MTS); however, hOGG1 contains a nuclear localization signal (NLS). While hOGG1 has been found exclusively in mitochondria (Nishioka et al. 1999). It has been described that the hOGG1 isoform is present to a lesser extent in the mitochondria as well (Hashiguchi et al. 2004). The importance of OGG1 in repair of mtDNA has been emphasized by several studies. One of these showed that mitochondria isolated from OGG1-deficient mice were defective in incision of 8-oxoG lesions, while the nDNA pool of 8-oxoG remained unaffected (Klungland et al. 1999). Additionally, it has been estimated that mtDNA extracted from *Ogg−/−* mice contained a 20-fold increase in 8-oxoG compared to wt mice, while nuclear DNA from Ogg null mice showed only twofold increase. This indicates that OGG1 is much more important for removal of 8-oxoG from mtDNA than nDNA (de Souza-Pinto et al. 2001b). The rate of repair of 8-oxoG DNA lesions in the actively transcribed strand of mtDNA is approximately twofold faster than from nDNA (Bohr 2002b). The study conducted in our laboratory revealed that overexpression of OGG1 in mitochondria of HeLa cells, which show reduced mtDNA repair, improves repair and enhances cell survival following oxidative stress (Dobson et al. 2000). The reported increase in 8-oxoG levels in mammalian mitochondria in association with aging and neurodegenerative diseases (Kang and Hamasaki 2005) can be attributed to inadequate or dysfunctional OGG1 activity in mitochondria. This hypothesis is strengthened by studies on overexpression of mitochondrially targeted OGG1 in the cells of central nervous system. Oligodendrocytes expressing OGG1 showed a significant enhancement in repair of oxidative lesions in mtDNA and were better protected against caspase 9-dependent apoptosis after menadione-induced oxidative stress (Druzhyna et al. 2003) and cytokines-mediated damage (Druzhyna et al. 2005). Several studies have indicated that OGG1 activity in liver and heart mitochondrial extracts from old rats is higher compared with extracts from young animals de Souza-Pinto et al. 2001a; (Stevnser et al. 2002). Other studies have proposed that there is an age-dependent deficiency in the mitochondrial import of h OGG1 (Szczesny et al. 2004).

Another DNA glycosylase that has been found in mitochondria of mammalian cells is a monofunctional MutY homolog DNA glycosylase (MYH) (Parker et al. 2000). In *E. coli,* the MutY glycosylase removes adenine from 8-oxoG/A mispair. Human MYH (hMYH), in contrast to *E. coli,* also can remove 2-OH-A opposite any normal base in the template (Ohtsubo et al. 2000). A single human *MYH* gene produces multiple transcripts, which results in the generation of a N-terminal MTS in a number of products (Nakabeppu 2001). However, the actual size of the active mitochondrial form of hMYH requires further clarification. Mice deficient in MYH activity are viable but show an increased occurrence of cancers in various tissues 1.5 years after birth (Nakabeppu et al. 2004). In humans, germline defects in MYH cause multiple colorectal adenoma and carcinoma phenotypes (Sieber et al. 2003). Increased ROS production is accompanied by an increase in MYH levels in mitochondria (Wang et al. 2000; Englander et al. 2002), suggesting that the level of MYH in mitochondria is specifically regulated in response to mtDNA damage.

Oxidative damage to mtDNA can lead to several forms of oxidized pyrimidine like thymine glycol (Tg), which can block both DNA and RNA polymerases. NTH1 glycosylase is the main homologue of *Escherichia coli.* Endonuclease III has been cloned and characterized in humans (Ikeda et al. 1998). It is a bifunctional glycosylase/AP lyase that recognizes and removes a number of oxidative pyrimidine adducts such as formamidopyrimidine, dihydroxyuracil, dihydroxycytosine, and urea (Luna et al. 2000). Several studies have reported that human NTH1 can localize both to the nucleus and mitochondria (Ikeda et al. 1998; Takao et al. 1998). The presence of NTH1 in mitochondria also is supported by a study that showed that incision activity against Tg was absent in liver mitochondrial extracts from mice deficient in NTH1 (Karahalil et al. 2003). The importance of NTH1 activity in the protection of mitochondria against oxidative stress was observed in a study where HeLa cells expressing Endonuclease III from *E. coli* showed enhanced repair of mtDNA as well as increased cell survival after treatment with the oxidizing agent menadione (Rachek et al. 2004). Homozygous *nth−/−* mutant mice showed no phenotypical abnormalities (Ocampo et al. 2002). Tissue extracts from these mice contained enzymatic activities against Tg, although biochemically different from NTH1, which suggests that there are back-up mechanisms in mammalian cells that compensate for the loss of NTH1.

A new group of DNA glycosylases, named NEIL, have recently been identified in mammalian tissues (Hazra et al. 2002). They are homologous of the *E. coli* DNA glycosylase Fpg/Nei, and primarily excise oxidized purines. NEIL1, NEIL2, and NEIL3 have been identified in mammalian genomes. Although NEILs were initially localized in the nucleus, Hu et al. (2005) have reported the presence of NEIL1 in mouse liver mitochondria, and NEIL activity has recently been observed in mouse brain mitochondria as well (Gredilla et al. 2008).

19.5.2 AP Endonuclease (APE)

Recognition and removal of a DNA lesion by glycosylase results in an abasic site, which is subsequently cleaved by AP endonuclease (next step in BER pathway) creating a 3′-OH and 5′-deoxyribose phosphate group on either side of the lesion. The major hydrolytic AP endonuclease in mammalian cells is APE1. A second AP endonuclease protein (APE2) has been described in human cells (Tsuchimoto et al. 2001), however, with no detectable APE activity (Ribar et al. 2004). APE1 also possess a 3′-diesterase activity specific for the 3′-dRP products generated by β-elimination by bifunctional glycosylases (Demple and Harrison 1994). In addition, APE1 has redox activity (Tell et al. 2005) that is involved in signal transduction. Unlike knock-out mice for most of the individual DNA glycosylases, which do not exhibit a significant phenotype, APE1 knock-out mice show embryonic lethality. Tomkinson et al. (1988) first identified APE activity in pure mitochondrial preparations from mouse cells. They found that 1–3% of the total cellular APE activity is associated with the mitochondria. The study by Tell et al. (2001) demonstrated that APE/Ref-1 was present in both the nucleus and mitochondria in rat thyroid gland cells despite the lack of an MTS. Tsuchimoto et al. (2001) showed that a putative NH_2-terminal MTS sequence is present in the human *APE2* gene and, in HeLa cells, GFP-tagged APE2 is located in mitochondria. The level of mitochondrial APE/Ref-1 in human Raji cells is increased after treatment with H_2O_2 (Frossi et al. 2002). Several studies regarding BER activity in neurodegenerative disorders showed increased expression of APE1 in cortex extracts of AD patients (Davydov et al. 2003) and in motor cortex of individuals with ALS (Shaijh and Martin 2002).

19.5.3 Mitochondrial Polymerase γ

Of the 16 DNA polymerases in the eukaryotic cells, only polymerase γ (pol γ) is known to function in the mitochondria. It is involved in replication of mtDNA as well as its repair. In higher eukaryotes, pol γ contains a 120–140 kDa catalytic subunit related to the family-A DNA polymerases, such as *E. coli* DNA polI, and a homodimeric form of accessory subunit of 35–50 kDa each that serves as a processivity factor (Graziewicz et al. 2006). The mammalian pol γ possesses AP-lyase activity, which allows it to remove 5′-terminal dRP sugar moieties created by APE or bifunctional glycosylases (Pinz and Bogenhagen 2000). This allows pol γ to function in the place of pol γ in the mitochondrial BER pathway. However, the catalytic rate of dRP lyase by pol γ is slower than that of pol γ, suggesting that only a low frequency of abasic sites can be processed in mtDNA compared to nDNA (Pinz and Bogenhagen 2000). Similar to APE1 knock-out mice, pol γ knock-out mice show embryonic lethality (Hance et al. 2005). In 2001, the first disease mutations were identified in *POLG* gene, encoding the catalytical subunit of pol γ (Van Goethem et al. 2001). These mutations were associated with progressive external ophthalmoplegia. To date, approximately 150 disease mutations in *POLG* have been identified, which places *POLG* as a major locus for mitochondrial disease (reviewed in Chan and Copeland 2009). As mentioned earlier, mice carrying a homozygous loss of function mutation in the exonuclease domain of pol γ display a higher rate of mtDNA base substitutions and deletions, which leads to the early onset of a wide range of age-related disorders (Trifunovic et al. 2004). DNA polymerase γ itself is a target of oxidative damage and has been shown to be one of the major oxidized mitochondrial matrix proteins (Graziewicz et al. 2006). This damage can lead to a detectable decline in polymerase activity, a condition that can cause a reduction in mtDNA replication and repair capacities.

19.5.4 *Mitochondrial DNA Ligase*

The final step in the BER pathway is sealing of the nick left behind by pol γ, which requires a mtDNA ligase activity. Mitochondrial DNA ligase was first identified by Levin and Zimmerman (1976), and was shown to be an ATP-dependent enzyme, similar to the nuclear DNA ligase. Subsequent studies revealed that the only DNA ligase in mitochondria is the product of DNA ligase III gene. The mitochondrial localization signal is generated on the N-terminus of mtDNA ligase during translation through an alternative upstream translation initiation site (Lakshmipathy and Campbell 1999b). The participation of DNA ligase III in mtDNA replication and repair was confirmed using an antisense strategy to decrease levels of ligase III. DNA ligase III antisense mRNA-expressing cells had reduced mtDNA content, accumulated single-strand nicks and did not repair mtDNA damage by gamma-irradiation (Lakshmipathy and Campbell 2001). Unlike nuclear DNA ligase III, which requires an interaction with accessory nuclear protein XRCC1 for stability, mtDNA ligase has been shown to act independently (Lakshmipathy and Campbell 2000). Further research is needed to see if the level or activity of DNA ligase III changes during aging and other pathologic conditions.

19.6 Modulation of mtDNA Repair

The important question remains to be answered, "What is the biologic significance of mtDNA repair for cellular defenses in normal and pathologic state?" Considering the differential sensitivities to DNA damaging agents, which exist among cells and tissues, it is intriguing whether there are cell- and tissue- specific differences in mtDNA repair. For example, in neurodegeneration, studies of whole brain regions do not differentiate between neurons and glial cells. However, evidence from studies using cells in culture showed that the repair efficiency of mtDNA was significantly decreased in oligodendrocytes and microglia compared with astrocytes after treatment with methylnitrosourea, an alkylating agent (LeDoux et al. 1998). Moreover, the induction of apoptosis correlated with this decrease. These studies were the first to demonstrate a cell-specific difference in repair of mtDNA damage in cells from the central nervous system (CNS), and indicated that this difference correlated with the induction of programmed cell death (LeDoux et al. 1998). In a similar study, Hollensworth et al. (2000) showed that after exposure to oxidative DNA damage, oligodendrocytes and microglia accumulated more mtDNA damage, and they repaired the damage less efficiently than astrocytes. The differential susceptibility of glial cell types to oxidative damage and apoptosis did not appear related to cellular antioxidant capacity, because astrocytes had lower total glutathione content and superoxide dismutase (SOD) activity than did oligodendrocytes and microglia (Hollensworth et al. 2000). In a subsequent study, primary cerebellar granule cells were used to determine if mitochondrial DNA repair efficiencies correlated with oxidative stress-induced apoptosis in neuronal cells (Harrison et al. 2005). Primary cerebellar granule cells had increased basal levels of glutathione and APE1 and were more sensitive to oxidative stress, resulting in less efficient repair of oxidative mtDNA lesions when compared with astrocytes. Of interest, however, is that the glycosylase and APE1 activities in the neurons were significantly higher with a reduction in polymerase γ activity, suggesting that the granule cells have an imbalance in the mitochondrial BER pathway. It is this imbalance that leads to the observed increase in sensitivity to oxidative stress (Harrison et al. 2005). Thus, while antioxidant and repair capacity are both involved in protecting cells from oxidant insults, it appears that cells with an efficient DNA repair may be spared, even in the presence of very low antioxidant levels, and that cells with less efficient repair are susceptible to death even in the presence of higher levels of antioxidants. So, it can be concluded that mtDNA repair plays a pivotal role in cellular defense mechanisms. This leads to the notion that modulation of mtDNA repair could have a profound effect on cell survival under conditions of genotoxic stress. Indeed, in the study, mentioned previously, targeting hOGG1 glycosylase to mitochondria in HeLa cells protects them against oxidative damage (Dobson et al. 2000). A similar protection is afforded by targeting hOGG1 into mitochondria of rat pulmonary artery endothelial cells (Dobson et al. 2002), and primary culture of rat oligodendrocytes (Druzhyna et al. 2003). A recent study from another group showed that mitochondrial over-expression of hOGG1 protects HeLa cells from H_2O_2-induced oxidative stress, but fails to defend cells from 4NQO damage, a UV-mimetic agent that induces a wide range of DNA lesions (Chatterjee et al. 2006b). These lesions are mainly repaired through nucleotide excision repair, a mechanism apparently absent in mitochondria. It was reported that targeting O^6-MeG DNA methyltransferase (MGMT) to mitochondria of breast non-tumorigenic epithelial cells protected these cells against alkylation induced cell death (Rasmussen and Rasmussen 2005). On the other hand, targeting of specific repair enzymes into mitochondria also can lead to an imbalance in BER, thus making otherwise resistant cells (for example in cancer) more vulnerable to the effects of cytotoxic drugs. It was shown that the expression of Exonuclease III from *E. coli* in mitochondria of breast cancer cells diminishes mtDNA repair capacity and cellular viability following oxidative stress (Shokolenko et al. 2003). Also targeting and over-expressing of N-methylpurine DNA glycosylase (MPG) in mitochondria of the breast cancer cell line resulted in increased mtDNA damage following treatment with the alkylating agent methyl methansulfonate, as well as increased apoptosis levels in these cells (Fishel et al. 2003). New methodologic approaches have been developed in a cell delivery system through protein transduction (Wadia and Dowdy 2005). It was successfully used for the purpose of mitochondrial targeting of repair enzymes (Shokolenko et al. 2005). In the latest study, performed in our lab, mitochondrial targeted hOGG1-TAT transduction fusion protein was used to protect HeLa cells from Chk2-associated cell cycle arrest initiated by oxidative mtDNA damage (Koczor et al. 2009). The results of this research show the new role of mtDNA damage in cell cycle progression. Chk2-associated cell cycle control by mtDNA integrity links to the relationship between mitochondria and development of cancer. Rapidly developing methods of modulation of mtDNA repair efficiency can be a powerful tool in future research of investigating the role of mtDNA repair in cellular processes in vitro and in vivo, as well, as therapeutic approaches.

19.7 Conclusion

Interest in DNA damage, mutations, and repair in mitochondria reflects an increasing awareness of the central role of that these organelles play in cellular physiology in both normal and disease states. The human mitochondrial genome has been completely sequenced and mapped. Research in the past several years has shown that mtDNA is more susceptible to various carcinogens and ROS than nDNA. Damage to mtDNA and mtDNA mutations, whether inherited or somatic, are implicated in a variety of common pathologic states, including metabolic and neurodegenerative diseases, and cancer. Additionally, the accumulation of the damage in mitochondrial genomes plays a central role in the mitochondrial theory of aging. The major protective mechanism that cells use to deal with this damage is DNA repair. Mammalian mitochondria clearly posses the ability to repair endogenous damages such as abasic sites and oxidized bases through BER mechanisms. Studies into the mechanisms involved in mtDNA repair have only just begun and already it has become apparent how efficient and well-coordinated these processes can be. Our knowledge is still limited in many aspects of organization, physiologic regulation, and biologic significance of mitochondrial BER. Undoubtedly, future research will address such topics as: a more precise definition of the components involved in mtDNA repair, a better comprehension of how they are regulated, and a more thorough measure the contribution that mtDNA repair pathway makes to maintaining mitochondrial genomic stability in health and disease.

Glossary

Base excision repair (BER): Is a cellular mechanism that repairs damaged DNA. It is primarily responsible for removing small, non-helix distorting base lesions from genome. BER is important for removing damaged bases that could otherwise cause mutations by mispairing or lead to breaks in DNA during replication.

Mitochondrial DNA: Mitochondria have a small genome that, in mammals generally takes the form of covalently closed circles approximately 16 kb in circumference. The mitochondrial genome encodes 13 proteins that are essential for oxidative phosphorylation, 22 tRNAs, and 2 rRNAs.

Mutations: Are permanent changes in the DNA sequence of a cell's genome. Mutations in a gene's DNA sequence either have no effect, alter the amino acid sequence of the protein encoded by the gene, or prevent the gene from functioning.

Reactive oxygen species (ROS): Molecules or ions formed by the incomplete one-electron reduction of oxygen. These reactive oxygen intermediates include singlet oxygen; superoxides; peroxides; hydroxyl radical; and hypochlorous acid. They contribute to the microbicidal activity of phagocytes, regulation of signal transduction and gene expression, and the oxidative damage to nucleic acids, proteins, and lipids.

References

Akbari, M., Visnes, T., Krokan, H. E., Otterlei, M. 2008. Mitochondrial base excision repair of uracil and AP sites takes place by single-nucleotide insertion and long-patch DNA synthesis. DNA Repair (Amst.) 7(4):605–616.

Alberio, S., Mineri, R., Tiranti, V., Zeviani, M. 2007. Depletion of mtDNA: syndromes and genes. Mitochondrion 7:6–12.

Alexeyev, M. F., Ledoux, S. P., Wilson, G. L. 2004. Mitochondrial DNA and aging. Clin. Sci. (Lond.) 107(4):355–264.

Ames, B. N., Shigenaga, M. K., Hagen, T. M. 1993. Oxidants, antioxidants, and the degenerative diseases of aging. Proc. Natl Acad. Sci. USA 90:7915–7922.

Anderson, C. T., Friedberg, E. C. 1980. The presence of nuclear and mitochondrial uracil-DNA glycosylase in extracts of human KB cells. Nucleic Acids Res. 8(4):875–888.

Andreyev, A. Yu., Kushnareva, Yu. E., Starkov, A. A. 2005. Mitochondrial metabolism of reactive oxygen species. Biochemistry (Moscow) 70(2):200–214.

Anson, R. M., Hudson, E., Bohr, V. A. 2000. Mitochondrial endogenous oxidative damage has been overestimated. FASEB J. 14:355–360.

Atamna, H., Frey, W. H. 2nd. 2007. Mechanisms of mitochondrial dysfunction and energy deficiency in Alzheimer's disease. Mitochondrion 7(5):297–310.

Attardi, G., Schatz, G. 1988. Biogenesis of mitochondria. Annu. Rev. Cell Biol. 4:289–333.

Bacman, S. R., Williams, S. L., Moraes, C. T. 2009. Intra- and inter-molecular recombination of mitochondrial DNA after in vivo induction of multiple double-strand breaks. Nucleic Acids Res. 37(13):4218–4226.

Bandy, B., Davidson, A. J. 1990. Mitochondrial mutations may increase oxidative stress: implication for carcinogenesis and aging? Free Radic. Biol. Med. 8:523–539.

Banoei, M. M., Houshmand, M., Panahi, M. S., Shariati, P., Rostami, M., Manshadi, M. D., Majidizadeh, T. 2007. Huntington's disease and mitochondrial DNA deletions: event or regular mechanism for mutant huntingtin protein and CAG repeats expansion?! Cell. Mol. Neurobiol. 27(7):867–875.

Barja, G., Herrero, A. 2000. Oxidative damage to mitochondrial DNA is inversely related to maximum life span in the heart and brain of mammals. FASEB J. 14:312–318.

Bauer, M. F., Gempel, K., Hofman, S., Jaksch, M., Philbrook, C., Gerbitz, K. D. 1999. Mitochondrial disorders. A diagnostic challenge in clinical chemistry. Clin. Chem. Lab. Med. 37:855–876.

Bender, A., Krishnan, K. J., Morris, C. M., Taylor, G. A., Reeve, A. K., Perry, R. H., Jaros, E., Hersheson, J. S., Betts, J., Klopstock, T., Taylor, R. W., Turnbull, D. M. 2006. High levels of

mitochondrial DNA deletions in substantia nigra neurons in aging and Parkinson disease. Nat. Genet. 38(5):515–517.
Berdanier, C. D., Everts, H. B. 2001. Mitochondrial DNA in aging and degenerative disease. Mutat. Res. 475:169–184.
Berneburg, M., Kamenisch, Y., Krutmann, J., Rocken, M. 2006. "To repair or not to repair – no longer a question": repair of mitochondrial DNA shielding against age and cancer. Exp. Dermatol. 15:1005–1015.
Bjelland, S., Seeberg, E. 2003. Mutagenicity, toxicity and repair of DNA base damage induced by oxidation. Mutat. Res. 531:37–80.
Bogdanov, M., Brown, R. H., Matson, W., Smart. R., Hayden, D., O'Donnell, H., Flint Beal M., Cudkowicz, M. 2000. Increased oxidative damage to DNA in ALS patients. Free Radic. Biol. Med. 29:652–658.
Bohr, V. A. 2002a. DNA damage and its processing. Relation to human disease. J. Inherit. Metab. Dis. 25:215–222.
Bohr, V. A. 2002b. Repair of oxidative DNA damage in nuclear and mitochondrial DNA, and some changes with aging in mammalian cells. Free Radic. Biol. Med. 32(9):804–812.
Brandon, M., Baldi, P., Wallace, D. C. 2006. Mitochondrial mutations in cancer. Oncogene 25(34):4647–4662.
Brown, W. M., George, M. Jr., Wilson, A. C. 1979. Rapid evolution of animal mitochondrial DNA. Proc. Natl Acad. Sci. USA 76:1967–1971.
Cadenas, E., Davies, K. J. A. 2000. Mitochondrial free radical generation, oxidative stress, and aging. Free Radic. Biol. Med. 29:222–230.
Carew, J. S., Huang, P. 2002. Mitochondrial defects in cancer. Mol. Cancer 1:9.
Cavelier, L., Johannisson, A., Gyllensten, U. 2000. Analysis of mtDNA copy number and composition of single mitochondrial particles using flow cytometry and PCR. Exp. Cell. Res. 259:79–85.
Chan, S. S., Copeland, W. C. 2009. DNA polymerase gamma and mitochondrial disease: understanding the consequence of POLG mutations. Biochim. Biophys. Acta. 1787(5):312–319.
Chatterjee, A., Mambo, E., Sidransky, D. 2006a. Mitochondrial DNA mutations in human cancer. Oncogene 25(34):4663–4674.
Chatterjee, A., Mambo, E., Zhang, Y., DeWeese, T., Sidransky, D. 2006b. Targeting of mutant hoggl in mammalian mitochondria and nucleus: effect on cellular survival upon oxidative stress. BMC Cancer 6:235.
Chen, C. M., Wu, Y. R., Cheng, M. L., Liu, J. L., Lee, Y. M., Lee, P. W., Soong, B. W., Chiu, D. T. 2007. Increased oxidative damage and mitochondrial abnormalities in the peripheral blood of Huntington's disease patients. Biochem. Biophys. Res. Commun. 359(2):335–340.
Clayton, D. A., Vinograd, J. 1967. Circular dimer and catenate forms of mitochondrial DNA in human leukaemic leucocytes. Nature 216(5116):652–657.
Clayton, D. A., Doda, J. N., Friedberg, E. C. 1974. The absence of a pyrimidine dimmer repair mechanism in mammalian mitochondria. Proc. Natl Acad. Sci. USA 71:2777–2778.
Cooke, M. S., Evans, M. D., Dizdaroglu, M., Lunec, J. 2003. Oxidative DNA damage: mechanisms, mutation, and disease. FASEB J. 17:1195–1214.
Croteau, D. L., Stierum, R. H., Bohr, V. A. 1999. Mitochondrial DNA repair pathway. Mutat. Res. 434:137–148.
Davydov, V., Hansen, L. A., Shackelford, D. A. 2003. Is DNA repair compromised in Alzheimer's disease. Neurobiol. Aging 24:953–968.
de Souza-Pinto, N. C., Hogue, B. A., Bohr, V. A. 2001a. DNA repair and aging in mouse liver: 8-oxoG glycosylase activity increase in mitochondrial but not in nuclear extracts. Free Radic. Biol. Med. 30:916–923.
de Souza-Pinto, N. C., Eide, L., Hogue, B. A., Thybo, T., Stevnsner, T., Seeberg, E., Klungland, A., Bohr, V. A. 2001b. Repair of 8-oxodeoxyguanosine lesions in mitochondrial DNA depends on the oxoguanine DNA glycosylase (OGG1) gene and 8-oxoguanine accumulates in the mitochondrial DNA of OGG1-defective mice. Cancer Res. 61(14):5378–5381.

de Souza-Pinto, N. C., Mason, P. A., Hashiguchi, K., Weissman, L., Tian, J., Guay, D., Lebel, M., Stevnsner, T. V., Rasmussen, L. J., Bohr,. V. A. 2009. Novel DNA mismatch-repair activity involving YB-1 in human mitochondria. DNA Repair (Amst.) 8(6):704–719.

Demple, B., Harrison, L. 1994. Repair of oxidative damage to DNA: enzymology and biology. Annu. Rev. Biochem. 63:915–48.

DiMauro, S., Moraes, C. T. 1993. Mitochondrial encephalomyopathies. Arch. Neurol. 50(11):1197–208.

Dobson, A. W., Xu, Y., Kelley, M. R., LeDoux, S. P., Wilson, G. L. 2000. Enhanced mtDNA repair and cellular survival following oxidative stress by targeting the hOGG repair enzyme to mitochondria. J. Biol. Chem. 275:37518–37523.

Dobson, A. W., Grishko, V., LeDoux, S. P., Kelley, M. R., Wilson, G. L., Gillespie, M. N. 2002. Enhanced mtDNA repair capacity protects pulmonary artery endothelial cells from oxidant-mediated death. Am. J. Physiol. Lung Cell. Mol. Physiol. 283(1):L205–210.

Driggers, W. J., LeDoux, S. P., Wilson, G. L. 1993. Repair of oxidative damage within the mitochondrial DNA of RINr 38 cells. J. Biol. Chem. 268:22042–22045.

Druzhyna, N. M., Hollensworth, S. B., Kelley, M. R., Wilson, G. L., LeDoux, S. P. 2003. Targeting human 8-Oxoguanine glycosylase to mitochondria of oligodendrocytes protects against menadione-induced oxidative stress. Glia 42:370–378.

Druzhyna, N. M., Musiyenko, S. I., Wilson, G. L., LeDoux, S. P. 2005. Cytokines induce NO-mediated mtDNA damage and apoptosis in oligodendrocytes. Protective role of targeting 8-oxoguanine glycosylase to mitochondria. J. Biol. Chem. 280:21673–21679.

Elson, J. L., Herrnstadt, C., Preston, G., Thal, L., Morris, C. M., Edwardson, J. A., Beal, M. F., Turnbull, D. M., Howell, N. 2006. Does the mitochondrial genome play a role in the etiology of Alzheimer's disease? Hum. Genet. 119(3):241–254.

Endres, M., Biniszkiewicz, D., Sobol, R. W., Harms, C., Ahmadi, M., Lipski, A., Katchanov, J., Mergenthaler, P., Dirnagl, U., Wilson, S. H., Meisel, A., Jaenisch, R. 2004. Increased postischemic brain injury in mice deficient in uracil-DNA glycosylase. J. Clin. Invest. 113(12):1711–1721.

Englander, E. W., Hu, Z., Sharma, A., Lee, H. M., Wu, Z. H., Greeley, G. H. 2002. Rat MYH, a glycosylase for repair of oxidatively damaged DNA, has brain-specific isoforms that localize to neuronal mitochondria. J. Neurochem. 83(6):1471–1480.

Fishel, M. L., Seo, Y. R., Smith, M. L., Kelley, M. R. 2003. Imbalancing the DNA base excision repair pathway in the mitochondria; targeting and overexpressing N-Methylpurine DNA glycosylase in mitochondria leads to enhanced cell killing. Cancer Res. 63:608–615.

Fliss, M. S., Usadel, H., Caballero, O. L., Wu, L., Buta, M. R., Eleff, S. M., Jen, J., Sidransky, D. 2000. Facile detection of mitochondrial DNA mutations in tumors and bodily fluids. Science 287(5460):2017–2019.

Fontenay, M., Cathelin, S., Amiot, M., Gyan, E., Solary, E. 2006. Mitochondria in hematopoiesis and hematological diseases. Oncogene 25(34):4757–4767.

Frossi, B., Tell, G., Spessotto, P., Colombatti, A., Vitale, G., Pucillo, C. 2002. H(2)O(2) induces translocation of APE/Ref-1 to mitochondria in the Raji B-cell line. J. Cell. Physiol. 193(2):180–186.

Garrido, N., Griparic, L., Jokitalo, E., Wartiovaara, J., van der Bliek, A. M., Spelbrink, J. N., 2003. Composition and dynamics of human mitochondrial nucleoids. Mol. Biol. Cell 14:1583–1596.

Gorogawa, S., Kajimoto, Y., Umayahara, Y., Kaneto, H., Watada, H., Kuroda, A., Kawamori, D., Yasuda, T., Matsuhisa, M., Yamasaki, Y., Hori, M. 2002. Probucol preserves pancreatic beta-cell function through reduction of oxidative stress in type 2 diabetes. Diabetes Res. Clin. Pract. 57(1):1–10.

Graziewicz, M. A., Longley, M. J., Copeland, W. C. 2006. DNA Polymerase gamma in mitochondrial DNA replication and repair. Chem. Rev. 106:383–405.

Greaves, L. C., Turnbull, D. M. 2009. Mitochondrial DNA mutations and ageing. Biochim. Biophys. Acta. 1790(10):1015–1020.

Gredilla, R., Garm, C., Holm, R., Bohr, V. A., Stevnsner, T. 2008. Differential age-related changes in mitochondrial DNA repair activities in mouse brain regions. doi:10.1016/j. neurobiolaging2008.07004.

Gross, N. J., Getz, G. S., Rabinowitz, M. 1969. Apparent turnover of mitochondrial deoxyribonucleic acid and mitochondrial phospholipids in the tissues of the rat. J. Biol. Chem. 244:1552–1562.

Gu, M., Cooper, J. M., Taanman, J. W., Schapira, A. H. 1998. Mitochondrial DNA transmission of the mitochondrial defect in Parkinson's disease. Ann. Neurol. 44(2):177–186.

Gupta, S., 2001. Molecular steps of death receptor and mitochondrial pathways of apoptosis. Life Sci. 69(25–26):2957–2964.

Gyllensten, U., Wharton, D., Josefsson, A., Wilson, A. C. 1991. Parental inheritance of mitochondrial DNA in mice. Nature 352:255–257.

Hamblet, N. S., Castora, F. J. 1997. Elevated levels of the Kearns-Sayre syndrome mitochondrial DNA deletion in temporal cortex of Alzheimer's patients. Mutat. Res. 379(2):253–262.

Hamblet, N. S., Ragland, B., Ali, M., Conyers, B., Castora, F. J. 2006. Mutations in mitochondrial-encoded cytochrome c oxidase subunits I, II, and III genes detected in Alzheimer's disease using single-strand conformation polymorphism. Electrophoresis 27(2):398–408.

Hance, N., Ekstrand, M. I., Trifunovic, A. 2005 Mitochondrial DNA polymerase gamma is essential for mammalian embryogenesis. Hum. Mol. Genet. 14(13):1775–1883.

Harman, D. 1956. Aging: a theory based on free radical and radiation chemistry. J. Gerontol. 11:298–300.

Harman, D. 2001. Aging: overview. Ann. NY Acad. Sci. 928:1–21.

Harrison, J. F., Hollensworth, S. B., Spitz, D. R., Copeland, W. C., Wilson, G. L., LeDoux, S. P. 2005. Oxidative stress-induced apoptosis in neurons correlates with mitochondrial DNA base excision repair pathway imbalance. Nucleic Acids Res. 22(14):4660–4671.

Hashiguchi, K., Stuart, J. A., de Souza-Pinto, N. C., Bohr, V. A. 2004. The C-terminal alphaO helix of human Ogg1 is essential for 8-oxoguanine DNA glycosylase activity: the mitochondrial beta-Ogg1 lacks this domain and does not have glycosylase activity. Nucleic Acids Res. 32(18):5596–5608.

Hayakawa, T., Noda, M., Yasuda, K., Yorifuji, H., Taniguchi, S., Miwa, I., Sakura, H., Terauchi, Y., Hayashi, J., Sharp, G. W., Kanazawa, Y., Akanuma, Y., Yazaki, Y., Kadowaki, T. 1998. Ethidium bromide-induced inhibition of mitochondrial gene transcription suppresses glucose-stimulated insulin release in the mouse pancreatic beta-cell line betaHC9. J. Biol. Chem. 273(32):20300–20307.

Hazra, T. K., Izumi, T., Boldogh, I., Imhoff, B., Kow, Y. W., Jaruga, P., Dizdaroglu, M., Mitra, S. 2002. Identification and characterization of a human DNA glycosylase for repair of modified bases in oxidatively damaged DNA. Proc. Natl Acad. Sci. USA 99(6):3523–3528.

Hollensworth, B. S., Shen, C., Sim, J. E., Spitz, D. R., Wilson, G. L., LeDoux, S. P. 2000. Glial cell type-specific responses to menadione-induced oxidative stress. Free Radic. Biol. Med. 28:1161–1174.

Horton, T. M., Petros, J. A., Heddi, A., Shoffner, J., Kaufman, A. E., Graham, S. D. Jr., Gramlich, T., Wallace, D. C. 1996. Novel mitochondrial DNA deletion found in a renal cell carcinoma. Genes Chromosom. Cancer 15(2):95–101.

Hsu. T. C., Young, M. R., Cmarik, J., Colburn, N. H. 2000. Activator protein 1 (AP-1)- and nuclear factor kappaB (NF-kappaB)-dependent transcriptional events in carcinogenesis. Free Radic. Biol. Med. 28:1338–1348.

Hu, J., de Souza-Pinto, N. C., Haraguchi, K., Hogue, B. A., Jaruga, P., Greenberg, M. M., Dizdaroglu, M., Bohr, V. A. 2005. Repair of formamidopyrimidines in DNA involves different glycosylases: role of the OGG1, NTH1, and NEIL1 enzymes. J. Biol. Chem. 280(49):40544–40551.

Hudson, E. K., Hogue, B. A., Souza-Pinto, N. C. et al. 1998. Age-associated change in mitochondrial DNA damage. Free Radic. Res. 29:573–579.

Ihara, Y., Toyokuni, S., Uchida, K., Odaka, H., Tanaka, T., Ikeda, H., Hiai, H., Seino, Y., Yamada, Y. 1999. Hyperglycemia causes oxidative stress in pancreatic beta-cells of GK rats, a model of type 2 diabetes. Diabetes 48(4):927–932.

Ikebe, S., Tanaka, M., Ohno, K., Sato, W., Hattori, K., Kondo, T., Mizuno, Y., Ozawa, T. 1990. Increase of deleted mitochondrial DNA in the striatum in Parkinson's disease and senescence. Biochem. Biophys. Res. Commun. 170(3):1044–1048.

Ikeda, S., Biswas, T., Roy, R., Izumi, T., Boldogh, I., Kurosky, A., Sarker, A. H., Seki, S., Mitra, S. 1998. Purification and characterization of human NTH1, a homolog of *Escherichia coli* endonuclease III. Direct identification of Lys-212 as the active nucleophilic residue. J. Biol. Chem. 273(34):21585–21593.

Kalifa, L., Beutner, G., Phadnis, N., Sheu, S. S., Sia, E. A. 2009. Evidence for a role of FEN1 in maintaining mitochondrial DNA integrity. DNA Repair (Amst.) 8(10):1242–1249.

Kang, D., Hamasaki, N. 2005. Alterations of mitochondrial DNA in common diseases and disease states: aging, neurodegeneration, heart failure, diabetes and cancer. Curr. Med. Chem. 12:429–441.

Karahalil, B., de Souza-Pinto, N. C., Parsons, J. L., Elder, R. H., Bohr, V. A. 2003. Compromised incision of oxidized pyrimidines in liver mitochondria of mice deficient in NTH1 and OGG1 glycosylases. J. Biol. Chem. 278(36):33701–33707.

Kasamatsu, H., Vinograd, J. 1974. Replication of circular DNA in eukaryotic cells. Annu. Rev. Biochem. 43:695–719.

Khaidakov, M., Heflich, R. H., Manjanatha, M. G., Myers, M. B., Aidoo, A. 2003. Accumulation of point mutations in mitochondrial DNA of aging mice. Mutat. Res. 526(1–2):1–7.

Khrapko, K., Coller, H. A., André, P. C., Li, X. C., Hanekamp, J. S., Thilly, W. G. 1997. Mitochondrial mutational spectra in human cells and tissues. Proc. Natl Acad. Sci. USA 94(25):13798–13803.

Klungland, A., Rosewell, I., Hollenbach, S., Larsen, E., Daly, G., Epe, B., Seeberg, E., Lindahl, T., Barnes, D. E. 1999. Accumulation of premutagenic DNA lesions in mice defective in removal of oxidative base damage. Proc. Natl Acad. Sci. USA 96(23):13300–13305.

Koczor, C. A., Shokolenko, I. N., Boyd, A. K., Balk, S. P., Wilson, G. L., Ledoux, S. P. 2009. Mitochondrial DNA damage initiates a cell cycle arrest by a Chk2-associated mechanism in mammalian cells. J. Biol. Chem. 284(52):36191–36201.

Kraytsberg, Y., Kudryavtseva, E., McKee, A. C., Geula, C., Kowall, N. W., Khrapko, K. 2006. Mitochondrial DNA deletions are abundant and cause functional impairment in aged human substantia nigra neurons. Nat. Genet. 38(5):518–520.

Krishnan, K. J., Reeve, A. K., Samuels, D. C., et al. 2008. What causes mitochondrial DNA deletions in human cells? Nat. Genet. 40:275–279.

Krokan, H. E., Otterlei, M., Nilsen, H., Kavli, B., Skorpen, F., Andersen, S., Skjelbred, C., Akbari, M., Aas, P. A., Slupphaug, G. 2001. Properties and functions of human uracil-DNA glycosylase from the *UNG* gene. Prog. Nucleic Acid Res. Mol. Biol. 68:365–386.

Kujoth, G. C., Hiona, A., Pugh, T. D., Someya, S., Panzer, K., Wohlgemuth, S. E., Hofe, T., Seo, A. Y., Sullivan, R., Jobling, W. A., Morrow, J. D., Van Remmen, H., Sedivy, J. M., Yamasoba, T., Tanokura, M., Weindruch, R., Leeuwenburgh, C., Prolla, T. A. 2005. Mitochondrial DNA mutations, oxidative stress, and apoptosis in mammalian aging. Science 309(5733):481–484.

Lakshmipathy, U., Campbell, C. 1999a. Double strand break rejoining by mammalian mitochondrial extracts. Nucleic Acids Res. 27(4):1198–1204.

Lakshmipathy, U., Campbell, C. 1999b. The human DNA ligase III gene encodes nuclear and mitochondrial proteins. Mol. Cell. Biol. 19:3869–3876.

Lakshmipathy, U., Campbell, C. 2000. Mitochondrial DNA ligase III function is independent of Xrcc1. Nucleic Acids Res. 28(20):3880–3886.

Lakshmipathy, U., Campbell, C. 2001. Antisense-mediated decrease in DNA ligase III expression results in reduced mitochondrial DNA integrity. Nucleic Acids Res. 29(3):668–676.

Larsen, N. B., Rasmussen, M., Rasmussen, L. J. 2005. Nuclear and mitochondrial DNA repair: similar pathways? Mitochondrion 5:89–108.

LeDoux, S. P., Wilson, G. L., Beecham, E. J., Stevnsner, T., Wassermann, K., Bohr, V. A. 1992. Repair of mitochondrial DNA after various types of DNA damage in Chinese hamster ovary cells. Carcinogenesis 13:1967–1973.

LeDoux, S. P., Shen, C., Grishko, V. I., Fields, P. A., Gard, A. L., Wilson, G. L. 1998. Glial cell-specific differences in response to alkylation damage. Glia 24:304–312.

LeDoux, S. P., Driggers, W. J., Hollensworth, B. S., Wilson, G. L. 1999. Repair of alkylation and oxidative damage in mitochondrial DNA. Mutat. Res. 434:149–159.

Legros, F., Malka, F., Frachon, P., Lombès, A., Rojo, M. 2004. Organization and dynamics of human mitochondrial DNA. J. Cell Sci. 117:2653–2662.

Levin, C. J., Zimmerman, S. B A DNA ligase from mitochondria of rat liver. 1976. Biochem. Biophys. Res. Commun. 69(2):514–520.

Lim, K. S., Jeyaseelan, K., Whiteman, M., Jenneer, A., Halliwell, B. 2005. Oxidative damage in mitochondrial DNA is not extensive. Ann. NY Acad. Sci. 1042:210–220.

Lin, M. T., Simon, D. K., Ahn, C. H., Kim, L. M., Beal, M. F. 2002. High aggregate burden of somatic mtDNA point mutations in aging and Alzheimer's disease brain. Hum. Mol. Genet. 11(2):133–145.

Liu, P., Qian, L., Sung, J. S., de Souza-Pinto, N. C., Zheng, L., Bogenhagen, D. F., Bohr, V. A., Wilson, D. M. 3rd, Shen, B., Demple, B. 2008. Removal of oxidative DNA damage via FEN1-dependent long-patch base excision repair in human cell mitochondria. Mol. Cell. Biol. 28(16):4975–4987.

Longley, M. J., Prasad, R., Srivastava, D. K., Wilson, S. H., Copeland, W. C. 1998. Identification of 5′-deoxyribose phosphate lyase activity in human DNA polymerase g and its role in mitochondrial base excision repair. Proc. Natl Acad. Sci. USA 95:12244–12248.

Luna, L., Bjørås, M., Hoff, E., Rognes, T., Seeberg, E. 2000. Cell-cycle regulation, intracellular sorting and induced overexpression of the human NTH1 DNA glycosylase involved in removal of formamidopyrimidine residues from DNA. Mutat. Res. 460(2):95–104.

Ly, C. V., Verstreken, P. 2006. Mitochondria at the synapse. Neuroscientist 12:291–299.

Magaña-Schwencke, N., Henriques, J. A., Chanet, R., Moustacchi, E. 1982. The fate of 8-methoxypsoralen photoinduced crosslinks in nuclear and mitochondrial yeast DNA: comparison of wild-type and repair-deficient strains. Proc. Natl Acad. Sci. USA 79(6):1722–1726.

Mecocci, P., MacGarvey, U., Kaufman, A. E., Koontz, D., Schoffner, J. M., Wallace, D. C., Beal, M. F. 1993. Oxidative damage to mitochondrial DNA shows marked age-dependent increases in human brain. Ann. Neurol. 34:609–616.

Michikawa, Y., Mazzucchelli, F., Bresolin, N., Scarlato, G., Attardi, G. 1999. Aging-dependent large accumulation of point mutations in the human mtDNA control region for replication. Science 286(5440):774–779.

Murdock, D. G., Christacos, N. C., Wallace, D. C. 2000. The age-related accumulation of a mitochondrial DNA control region mutation in muscle, but not brain, detected by a sensitive PNA-directed PCR clamping based method. Nucleic Acids Res. 28(21):4350–4355.

Murphy, R., Turnbull, D. M., Walker, M., Hattersley, A. T. 2008. Clinical features, diagnosis and management of maternally inherited diabetes and deafness (MIDD) associated with the 3243A>G mitochondrial point mutation. Diabet. Med. 25(4):383–399.

Myers, K. A., Saffhill, R., O'Connor, P. J. 1988. Repair of alkylated purines in the hepatic DNA of mitochondria and nuclei in the rat. Carcinogenesis 9(2):285–292.

Nakabeppu, Y. 2001. Regulation of intracellular localization of human MTH1, OGG1, and MYH proteins for repair of oxidative DNA damage. Prog. Nucleic Acid Res. Mol. Biol. 68:75–94.

Nakabeppu, Y., Tsuchimoto, D., Ichinoe, A., Ohno, M., Ide, Y., Hirano, S., Yoshimura, D., Tominaga, Y., Furuichi, M., Sakumi, K. 2004. Biological significance of the defense mechanisms against oxidative damage in nucleic acids caused by reactive oxygen species: from mitochondria to nuclei. Ann. NY Acad. Sci. 1011:101–111.

Nilsen, H., Rosewell, I., Robins, P., Skjelbred, C. F., Andersen, S., Slupphaug, G., Daly, G., Krokan, H. E., Lindahl, T., Barnes, D. E. 2000. Uracil-DNA glycosylase (UNG)-deficient mice reveal a primary role of the enzyme during DNA replication. Mol. Cell. 5(6):1059–1065.

Nishikawa, T., Edelstein, D., Du, X. L., Yamagishi, S., Matsumura, T., Kaneda, Y., Yorek, M. A., Beebe, D., Oates, P. J., Hammes, H. P., Giardino, I., Brownlee, M. 2000. Normalizing mitochondrial superoxide production blocks three pathways of hyperglycaemic damage. Nature 404(6779):787–790.

Nishioka, K., Ohtsubo, T., Oda, H., Fujiwara, T., Kang, D., Sugimachi, K., Nakabeppu, Y. 1999. Expression and differential intracellular localization of two major forms of human 8-oxoguanine

DNA glycosylase encoded by alternatively spliced OGG1 mRNAs. Mol. Biol. Cell. 10(5):1637–1652.
Ocampo, M. T., Chaung, W., Marenstein, D. R., Chan, M. K., Altamirano, A., Basu, A. K., Boorstein, R. J., Cunningham, R. P., Teebor, G. W. 2002. Targeted deletion of mNth1 reveals a novel DNA repair enzyme activity. Mol. Cell. Biol. 22(17):6111–6121.
Ohtsubo, T., Nishioka, K., Imaiso, Y., Iwai, S., Shimokawa, H., Oda, H., Fujiwara, T., Nakabeppu, Y. 2000. Identification of human MutY homolog (hMYH) as a repair enzyme for 2-hydroxyadenine in DNA and detection of multiple forms of hMYH located in nuclei and mitochondria. Nucleic Acids Res. 28(6):1355–1364.
Ojala, D., Montoya, J., Attardi, G. 1981. tRNA punctuation model of RNA processing in human mitochondria. Nature 290:465–470.
Parker, A., Gu, Y., Lu, A. L. 2000. Purification and characterization of a mammalian homolog of *Escherichia coli* MutY mismatch repair protein from calf liver mitochondria. Nucleic Acids Res. 28(17):3206–3215.
Pelicano, H., Carney, D., Huang, P. 2004. ROS stress in cancer cells and therapeutic implications. Drug Resist. Updat. 7:97–110.
Petros, J. A., Baumann, A. K., Ruiz-Pesini, E., et al. 2005. mtDNA mutations increase tumorigenicity in prostate cancer. Proc. Natl Acad. Sci. USA 102:719–724.
Pettepher, C. C., LeDoux, S. P., Bohr, V. A., Wilson, G. L. 1991. Repair of alkali-labile sites within the mitochondrial DNA of RINr 38 cells after exposure to the nitrosourea streptozotocin. J. Biol. Chem. 266(5):3113–3117.
Pinz, K. G., Bogenhagen, D. F. 1998. Efficient repair of abasic sites in DNA by mitochondrial enzymes. Mol. Cell. Biol. 18:1257.
Pinz, K. G., Bogenhagen, D. F. 2000. Characterization of a catalytically slow AP lyase activity in DNA polymerase gamma and other family A DNA polymerases. J. Biol. Chem. 275(17):12509–12514.
Polyak, K., Li, Y., Zhu, H., Lengauer, C., Willson, J. K., Markowitz, S. D., Trush, M. A., Kinzler, K. W., Vogelstein, B. 1998. Somatic mutations of the mitochondrial genome in human colorectal tumours. Nat. Genet. 20(3):291–293.
Qiu, X., Chen Y., Zhou, M. 2001. Two point mutations in mitochondrial DNA of cytochrome c oxidase coexist with normal mtDNA in a patient with Alzheimer's disease. Brain Res. 893:261–263.
Rachek, L. I., Grishko, V. I., Alexeyev, M. F., Pastukh, V. V., LeDoux, S. P., Wilson, G. L. 2004. Endonuclease III and endonuclease VIII conditionally targeted into mitochondria enhance mitochondrial DNA repair and cell survival following oxidative stress. Nucleic Acids Res. 32(10):3240–3247.
Rasmussen, A. K., Rasmussen, L. J. 2005. Targeting of O6-MeG DNA methyltransferase (MGMT) to mitochondria protects against alkylation induced cell death. Mitochondrion 5:411–417.
Reeve, A. K., Krishnan, K. J., Elson, J. L., Morris, C. M., Bender, A., Lightowlers, R. N., Turnbull, D. M. 2008. Nature of mitochondrial DNA deletions in substantia nigra neurons. Am. J. Hum. Genet. 82(1):228–235.
Ribar, B., Izumi, T., Mitra, S. 2004. The major role of human AP-endonuclease homolog Apn2 in repair of abasic sites in Schizosaccharomyces pombe. Nucleic Acids Res. 32(1):115–126.
Richter, C., Park, J. W., Ames, B. N. 1988. Normal oxidative damage to the mitochondrial and nuclear DNA is extensive. Proc. Natl Acad. Sci. USA 85:6465–6467.
Ro, L. S., Lai, S. L., Chen, C. M., Chen, S. T. 2003. Deleted 4977-bp mitochondrial DNA mutation is associated with sporadic amyotrophic lateral sclerosis: a hospital-based case-control study. Muscle Nerve 28(6):737–743.
Robertson, A. B., Klungland, A., Rognes, T., Leiros, I. 2009. DNA repair in mammalian cells: base excision repair: the long and short of it. Cell. Mol. Life Sci. 66(6):981–993.
Santos, C., Martínez, M., Lima, M., Hao, Y. J., Simões, N., Montiel, R. 2008. Mitochondrial DNA mutations in cancer: a review. Curr. Top. Med. Chem. 8(15):1351–1366.

Sastre, J., Pallardó, F. V., Viña, J. 2003. The role of mitochondrial oxidative stress in aging. Free Radic. Biol. Med. 35(1):1–8.
Sawyer, D. E., Van Houten, B. 1999. Repair of DNA damage in mitochondria. Mutat. Res. 434:161–176.
Schwartz, M. Vissing, J. 2003. New patterns of inheritance in mitochondrial disease. Biochem. Biophys. Res. Commun. 310:247–251.
Shaijh, A. Y., Martin, L. J. 2002. DNA base-excision repair enzyme apurinic/apyrimidinic endonuclease/redox factor-1 is increased and competent in the brain and spinal cord of individuals with amyotrophic lateral sclerosis. NeuroMol. Med. 2:47–60.
Shidara, Y., Yamagata, K., Kanamori, T., et al. 2005. Positive contribution of pathogenic mutations in the mitochondrial genome to the promotion of cancer by prevention from apoptosis. Cancer Res. 65:1655–1663.
Shimura-Miura, H., Hattori, N., Kang, D., Miyako, K., Nakabeppu, Y., Mizuno, Y. 1999. Increased 8-oxo-dGTPase in the mitochondria of substantia nigral neurons in Parkinson's disease. Ann. Neurol. 46:920–924.
Shokolenko, I. N., Alexeyev, M. F., Robertson, F. M., LeDoux, S. P., Wilson, G. L. 2003. The expression of Exonuclease III from *E. coli* in mitochondria of breast cancer cells diminishes mitochondrial DNA repair capacity and cell survival after oxidative stress. DNA Repair (Amst.) 2:471–482
Shokolenko, I. N., Alexeyev, M. F., LeDoux, S. P., Wilson, G. L. 2005. TAT-mediated protein transduction and targeted delivery of fusion proteins into mitochondria of breast cancer cells. DNA Repair (Amst.) 4:511–518.
Sieber, O. M., Lipton, L., Crabtree, M., Heinimann, K., Fidalgo, P., Phillips, R. K., Bisgaard, M. L., Orntoft, T. F., Aaltonen, L. A., Hodgson, S. V., Thomas, H. J., Tomlinson, I. P. 2003. Multiple colorectal adenomas, classic adenomatous polyposis, and germ-line mutations in MYH. New Engl. J. Med. 348(9):791–799.
Simon, D. K., Lin, M. T., Ahn, C. H., Liu, G. J., Gibson, G. E., Beal, M. F., Johns, D. R. 2001. Low mutational burden of individual acquired mitochondrial DNA mutations in brain. Genomics 73(1):113–116.
Simon, D. K., Lin, M. T., Zheng, L., Liu, G. J., Ahn, C. H., Kim, L. M., Mauck, W. M., Twu, F., Beal, M. F., Johns, D. R. 2004. Somatic mitochondrial DNA mutations in cortex and substantia nigra in aging and Parkinson's disease. Neurobiol. Aging 25(1):71–81.
Singer, T. P., Ramsay, R. R. 1990. Mechanism of the neurotoxicity of MPTP. An update. FEBS Lett. 274:1–8.
Singh, K. 2006. Mitochondria damage checkpoint, aging, and cancer. Ann. NY Acad. Sci. 1067:182–190.
Smigrodzki, R., Parks, J., Parker, W. D. 2004. High frequency of mitochondrial complex I mutations in Parkinson's disease and aging. Neurobiol. Aging 25(10):1273–1281.
Stevnser, T., Thorslund, T., de Souza-Pinto, N. C., Bohr. V. A. 2002. Mitochondrial repair of 8-oxoguanine and changes with aging. Exp. Gerontol. 37:1189–1196.
Stuart, J. A., Mayard, S., Hashiguchi, K., Souza-Pinto, N. C., Bohr, V. A. 2005. Localization of mitochondrial DNA base excision repair to an inner membrane-associated particulate fraction. Nucleic Acids Res. 33(12):3722–3732.
Swerdlow, R. H., Parks, J. K., Miller, S. W., Tuttle, J. B., Trimmer, P. A., Sheehan, J. P., Bennett, J. P. Jr., Davis, R. E., Parker, W. D. Jr. 1996. Origin and functional consequences of the complex I defect in Parkinson's disease. Ann. Neurol. 40(4):663–671.
Szczesny, B., Bhakat, K. K., Mitra, S., Boldogh, I. 2004. Age-dependent modulation of DNA repair enzymes by covalent modification and subcellular distribution. Mech. Ageing Dev. 125:755–765.
Szczesny, B., Tann, A. W., Longley, M. J., Copeland, W. C., Mitra, S. 2008. Long patch base excision repair in mammalian mitochondrial genomes. J Biol. Chem. 283(39):26349–25356.
Szibor, M., Holtz, J. 2003. Mitochondrial ageing. Basic Res. Cardiol. 98(4):210–218.
Taanman, J. W. 1999. The mitochondrial genome: structure, transcription, translation and replication. Biochim. Biophys. Acta. 1410:103–123.
Takao, M., Aburatani, H., Kobayashi, K., Yasui, A. 1998. Mitochondrial targeting of human DNA glycosylases for repair of oxidative DNA damage. Nucleic Acids Res. 26(12):2917–2922.

Taylor, R. W., Barron, M. J., Borthwick, G. M., Gospel, A., Chinnery, P. F., Samuels, D. C., Taylor, G. A., Plusa, S. M., Needham, S. J., Greaves, L. C., Kirkwood, T. B., Turnbull, D. M. 2003. Mitochondrial DNA mutations in human colonic crypt stem cells. J. Clin. Invest. 112(9):1351–1360.

Tell, G., Crivellato, E., Pines, A., Paron, I., Pucillo, C., Manzini, G., Bandiera, A., Kelley, M. R., Di Loreto, C., Damante, G. 2001. Mitochondrial localization of APE/Ref-1 in thyroid cells. Mutat. Res. 485(2):143–52.

Tell, G., Damante, G., Caldwell, D., Kelley, M. R. 2005. The intracellular localization of APE1/Ref-1: more than a passive phenomenon? Antioxid. Redox Signal. 7(3–4):367–384.

Tomkinson, A. E., Bonk, R. T., Linn, S. 1988. Mitochondrial endonuclease activities specific for apurinic/apyrimidinic sites in DNA from mouse cells. J. Biol. Chem. 263(25):12532–12537.

Trifunovic, A., Wredenberg, A., Falkenberg, M., Spelbrink, J. N., Rovio, A. T., Bruder, C. E., Bohlooly, Y. M., Gidlöf, S., Oldfors, A., Wibom, R., Törnell, J., Jacobs, H. T., Larsson, N. G. 2004. Premature ageing in mice expressing defective mitochondrial DNA polymerase. Nature 429(6990):417–423.

Tsuchimoto, D., Sakai, Y., Sakumi, K., Nishioka, K., Sasaki, M., Fujiwara, T., Nakabeppu, Y. 2001. Human APE2 protein is mostly localized in the nuclei and to some extent in the mitochondria, while nuclear APE2 is partly associated with proliferating cell nuclear antigen. Nucleic Acids Res. 29(11):2349–2360.

Turrens, J. F. 1997. Superoxide production by the mitochondrial respiratory chain. Biosci. Rep. 17:3–8.

Van Goethem, G., Dermaut, B., Löfgren, A., Martin, J. J., Van Broeckhoven, C. 2001. Mutation of POLG is associated with progressive external ophthalmoplegia characterized by mtDNA deletions. Nat. Genet. 28(3):211–212.

Van Remmen, H., Ikeno, Y., Hamilton, M., Pahlavani, M., Wolf, N., et al. 2003. Life-long reduction in MnSOD activity results in increased DNA damage and higher incidence of cancer but does not accelerate aging. Physiol. Genomics 16:29–37.

Vives-Bauza, C., Andreu, A. L., Manfredi, G., Beal, M. F., Janetzky, B., Gruenewald, T. H., Lin, M. T. 2002. Sequence analysis of the entire mitochondrial genome in Parkinson's disease. Biochem. Biophys. Res. Commun. 290(5):1593–1601.

Wadia, J. S., Dowdy, S. F. 2005. Transmembrane delivery of protein and peptide drugs by TAT-mediated transduction in the treatment of cancer. Adv. Drug Deliv. Rev. 57(4):579–596.

Wang, G., Hazra, T. K, Mitra, S., Lee, H. M, Englander, E. W. 2000. Mitochondrial DNA damage and a hypoxic response are induced by CoCl(2) in rat neuronal PC12 cells. Nucleic Acids Res. 28(10):2135–2140.

Wang, H., Xiong, S., Xie, C., Markesbery, W. R., Lovell, M. A. 2005. Increased oxidative damage in nuclear and mitochondrial DNA in Alzheimer's disease. J. Neurochem. 93:953–962.

Wei, Y. H., Lee, H. C. 2002. Oxidative stress, mitochondrial DNA mutation, and impairment of antioxidant enzymes in aging. Exp. Biol. Med. 227:671–682.

Yakes, M. F. and Van Houten, B. 1997. Mitochondrial DNA damage is more extensive and persists longer than nuclear DNA damage in human cells following oxidative stress. Proc. Natl Acad. Sci. USA 94:514–519.

Yang, M. Y., Bowmaker, M., Reyes, A., Vergani, L., Angeli, P., Gringeri, E., Jacobs, H. T., Holt, I. J., 2002. Biased incorporation of ribonucleotides on the mitochondrial L-strand accounts for apparent strand-asymetric DNA replication. Cell 111:495–505.

Zhu, W., Qin, W., Bradley, P., Wessel, A., Puckett, C. L., Sauter, E. R. 2005. Mitochondrial DNA mutations in breast cancer tissue and in matched nipple aspirate fluid. Carcinogenesis 26(1):145–152.

Index

F. Kempken (ed.), *Plant Mitochondria*, Advances in Plant Biology 1,
DOI 10.1007/978-0-387-89781-3, © Springer Science+Business Media, LLC 2011

N

Q

R

Lightning Source UK Ltd.
Milton Keynes UK
UKOW032312090413

208963UK00003B/81/P